# AFRICA
# ENVIRONMENT
# OUTLOOK

*Past, present and future perspectives*

UNEP

# AEO

## COLLABORATING CENTRES

### AMCEN/UNEP IN COLLABORATION WITH:

Association pour le Développement de l'Information
Environnementale (ADIE)
B.P. 4080 Libreville, Gabon
Tel: +241 763040/763019/763032   Fax: +241 774261
E-mail: jpvandeweghe@adie-plgie.org

National Environment Managemental Authority (NEMA)
6th Floor Communications House, 1 Colville Street
P.O. Box 22255, Kampala, Uganda
Tel: +256 41 251064/251065   Fax: +256 41 232680/257521
E-mail: bgowa@nemaug.org
www.nemaug.org

Centre for Environment and Development for the
Arab Region & Europe (CEDARE)
21/23 Giza Street, Nile Tower Building 13th Floor, P.O.
Box 52, Orman, Giza, Egypt
Tel: +202 570 1859/570 0979   Fax: +202 570 3242
www.cedare.org.eg

Network for Environment and
Sustainable Development (NESDA)
CBP 95 Guichet Annexe BAD, Abidjan, Côte d'Ivoire
Tel: +225 20 54 19   Fax: +225 20 59 22
www.rri.org/nesda

Indian Ocean Commission
Regional Environment Programme (IOC- REP)
Q4, Avenue Sir Guy Forget, Quatre Bornes,
Republic of Mauritius
Tel: +230 425 9564   Fax: +230 425 2709

Southern African Research and
Documentation Centre (SARDC)
Musokotwane Environment Resource Centre for
Southern Africa (IMERCSA)
15 Downie Avenue, Belgravia, P. O. Box 5690
Harare, Zimbabwe
Tel: +263 4 738 894/5   Fax: +263 4 738 693

# ACKNOWLEDGEMENTS

UNEP acknowledges the contributions made by the many individuals and institutions that have contributed to *Africa Environment Outlook*. A full list of names is included in the latter pages of this report. Special thanks are extended to:

● **AEO Collaborating Centres**

Association pour le Développement de l'Information Environnementale— Programme Régional de Gestion de l'Information Environnementale (ADIE-PRGIE), Cameroon

Centre for Environment and Development for the Arab Region & Europe (CEDARE), Egypt

Indian Ocean Commission (IOC), Mauritius

National Environment Management Authority (NEMA), Uganda

Network for Environment and Sustainable Development in Africa (NESDA), Côte d'Ivoire

Southern African Research and Documentation Centre (SARDC-IMERCSA), Zimbabwe

● **Funding support**

The Governments of Denmark, Belgium, Luxembourg and the UNEP Environment Fund have provided funding support for the first phase of the Africa Environment Outlook Process.

## Africa Environment Outlook Production Team

**AEO Coordinating Team with support from Collaborating Centre representatives**
Kagumaho Kakuyo
Thomas Tata
Jacquie Chenje

**Collaborating Centre Representatives**
Ahmed Abdel-Rehim (CEDARE)
Abou Bamba (NESDA)
Clever Mafuta (SARDC-IMERCSA)
Rajendranath Mohabeer (IOC)
Leonard Ntonga Mvondo (ADIE-PRGIE)
Charles Sebukeera (NEMA)

**AEO Authors**
Bola Ayeni (Chapter 4)
Anna Ballance (Chapter 2)
Munyaradzi Chenje (Chapters 1 and 3)
Tabeth Matiza Chiuta (Chapter 1)
Debbie Manzolillo Nightingale (Chapter 2)
Elton Laisi (Chapter 1)
Yakobo Moyini (Chapter 5)
Thomas Tata (Chapter 3)
Charl de Villiers (Chapter 3)

**Editors**
Geoffrey Bird
Sarah Medina

**AEO Support Team**
Marion Cheatle
Munyaradzi Chenje
Volodymyr Demkine
Salif Diop
Sheila Edwards
Tessa Goverse
Strike Mkandla
Naomi Poulton
Megumi Seki
David Smith
Anna Stabrawa
Laura Williamson
Sekou Toure
Rungano Karimanzira

**Specialist Support**
John Roberts
Richard Fuggle
Conmany Wesseh
Osama Salem

**Graphics, cover and page design**
Words and Publications, UK
(www.words.co.uk)

# CONTENTS

**CHAPTER 2** *(continued)*

# LIST OF ILLUSTRATIONS

# LIST OF BOXES

# List of Tables

# FOREWORD

About two decades age, African environment ministers met in the Egyptian capital, Cairo, to lay the foundation of the African Ministerial Conference on the Environment (AMCEN), the supreme continental forum responsible for articulating authoritative perspectives on Africa's environment and its place in the global arena. AMCEN was established against a backdrop of deteriorating state of environment as well as increasing social and economic inequality and their impacts on the region's environment. However, and from the outset, the Organisation of African Unity (OAU) and the United Nations Economic Commission for Africa (ECA) have been particularly supportive of the work of AMCEN as the environmental voice of conscience of the region, thereby highlighting the interdependence of environment, social and economic issues and the political commitment to work in concert in the interest of the well-being of the peoples of Africa.

Despite the achievements registered since its establishment, AMCEN still faces daunting challenges, including amongst others: harmonization of sub-regional and regional environmental issues in order that they receive equal attention at these levels; translation of global environmental concerns into practical, feasible and achievable programmes of action at national, sub-regional and regional levels; the positioning of AMCEN within the framework of new and emerging issues at regional and global level (the African Union, NEPAD, etc.); enhancing AMCEN's advocacy role in the new global economic order and, in particular, voicing Africa's concerns in the intergovernmental organizations (e.g. WTO, IMF, WB) that are beginning to place environmental considerations very high in their decision-making processes; and, promoting and according environmental concerns priority status within national development processes.

The efforts of AMCEN to address these challenges are the subject of the first ever regional comprehensive report on the state of Africa's environment—*Africa Environment Outlook* (AEO). The report, which was specifically requested by AMCEN, traces environment and development trends since the 1972 United Nations Conference on the Human Environment; provides a comprehensive analysis of the status and trends of the environment in Africa, integrated with the impacts of policies, laws and regional agreements; and proposes alternative policy options for the future as well as recommends concrete policy actions for follow-up at national and sub-regional levels.

It is our hope that this report will provide a valuable opportunity for AMCEN to take stock of its policy performance and effectiveness since its inception in 1985; to look into the future and access the various policy options for consideration at national, sub-regional and regional levels; and to serve as a basis for AMCEN to reorient its focus and programmes in light of the findings and recommendations contained in the report.

The AEO report also highlights some of the major issues to be addressed at the World Summit on Sustainable Development (WSSD) to be held in Johannesburg, South Africa in late August 2002. It is the basis upon which Africa and its cooperating partners, both bilateral and multilateral can engage each other to determine how best to tackle some of the pressing challenges facing Africa. It is also our hope that emerging initiatives under the African Union and the New Partnership for African Development (NEPAD) and its environmental component will be able to take advantage of the information contained in this report, in particular the policy recommendations, in order to advance their programmes of work for the future. These new initiatives provide a challenge for the future AEO process as the African development paradigm shifts beyond WSSD.

The success of the AEO process requires special mention in view of the unique approach adopted in the production of the report. It has engendered participation from a wide range of stakeholders, built consensus on several issues and findings, cultivated a sense of ownership and clearly demonstrated the need to build from the bottom up. It is through this unique approach that the AEO report has established a strong foundation for the harmonization of integrated environmental assessment and reporting processes in the Africa region. It is hoped that in the short term the assessment and reporting methodologies will be adopted and become fully incorporated in national level environmental management practices.

The AEO report and process are clear testimony of Africa's capacity to undertake specialized scientific work for itself and not rely on northern-based institutions to analyse, articulate and make recommendations on Africa's own issues. AEO is the basis for the African renaissance in environmental terms.

I trust the report will be useful to all who subscribe to the attainment of sustainable development in Africa for the benefit of present and future generations.

*Muhammad Kabir Sai'd*
**President,
African Ministerial Conference on the
Environment (AMCEN)**

# PREFACE

The very first *Africa Environment Outlook* report is a significant milestone in the collaboration between the United Nations Environment Programme (UNEP) and the African Ministerial Conference on the Environment (AMCEN). It should serve as a roadmap, supporting the environmental component, on Africa's journey to sustainable development. It is clear that poverty is inextricably linked to the environment, and action to protect and care for the environment must be taken. Otherwise the poison of poverty will continue to plague the continent.

Africa faces a number of critical challenges. The environment continues to deteriorate; social and economic inequality is increasing; and globalization is sweeping across the world, largely leaving Africa behind. Rapid changes in the global economy, in consumption patterns and in population and demographics are having a negative impact on the environment.

Without sustainable development we cannot solve the problems. It is not enough to simply say that we have a conservation plan for nature and natural resources. We have also to look at the impact of environmental change on people. The people who are struggling to survive are most vulnerable to environmental change, and suffer most from its effects. We must give these people a chance to live a better life.

Extensive consultations with African experts and collaborating institutions in the AEO process have concluded that the challenges Africa faces require new thinking and vision. There is an urgent need for all stakeholders within Africa, and globally, to act coherently in order to meet the challenges.

The AEO is a unique tool, providing an analysis of the state of the environment over the past 30 years, the driving forces behind environmental change, and the consequences for social and economic development. These consequences are presented both in terms of impacts on ecosystems, and vulnerability of human populations to floods, droughts, earthquakes, pests and diseases. The links between environmental change and poverty are explored, and appropriate intervention points identified. The AEO assessment methodology is modelled on UNEP's Global Environment Outlook process. It builds on sound data, information and science, and input from all stakeholders to identify priority issues.

Africa is a continent of great natural riches, biodiversity and vast unspoiled landscapes. These assets must be valued and preserved, in order to bring benefits to all. The decisions taken today and tomorrow will define the kind of environment this and future generations will enjoy. Building on the analysis of the past, AEO outlines a series of policy approaches for the future, leading to different outcomes over the next 30 years. *Africa Environment Outlook* concludes with recommendations on the road that Africa should take.

The AEO report is, therefore, a substantive tool for African policy makers to use in their assessment of the pressing environmental issues facing the region. The information in the report can serve as a firm foundation for discussions at the World Summit on Sustainable Development. I trust that many of you will find it a useful aid in preparing for the summit and beyond. I hope that it will inspire you, and all its readers to increase your commitment to care for the environment, especially in Africa.

*Klaus Töpfer*

**United Nations Under-Secretary General and**
**Executive Director, United Nations Environment Programme**

 # THE AEO PROJECT

## ORIGINS

The Eighth Session of the African Ministerial Conference on the Environment (AMCEN) which was held in Abuja, Nigeria in April 2000 approved AMCEN's medium-term programme, a key element of which was the production of the *Africa Environment Outlook* (AEO) report. This decision was affirmed at the AMCEN Inter-sessional Committee, which met in Malmo, Sweden in May 2000. In response to this, the AMCEN Secretariat—the United Nations Environment Programme Regional Office for Africa (ROA)—in collaboration with the Division of Early Warning and Assessment (DEWA), embarked on a process to produce the *Africa Environment Outlook* report.

## THE AEO PROCESS
### Partnership

The AEO report process has been based on wide consultation and participation between UNEP and various partners in the Africa region. It therefore reflects a variety of sub-regional perspectives and priorities. The AEO process involves partnership with six collaborating centres (see page v) responsible for producing sub-regional state of the environment and policy retrospective reports for Central Africa, Eastern Africa, Northern Africa, Southern Africa, Western Africa and the Western Indian Ocean Islands. These centres engaged individual and institutional experts at the national and sub-regional level to provide inputs into the process.

Experts from specialized organizations were also involved in providing inputs for sections of the report and in its review to ensure sub-regional balance, scientific credibility, and comprehensiveness. They include, among others, The UN Economic Commission for Africa (ECA), African Development Bank (ADB), the Organisation for African Unity (OAU), Southern African Development Community (SADC), Intergovernmental Authority on Development (IGAD), Economic Community of West African States (ECOWAS), Permanent Interstate Committee for Drought Control in the Sahel (CILSS), AMU and the Indian Ocean Commission (IOC).

### Sources of information

In compiling the sub-regional inputs, national level information and data sources were used. These data sources were then compared and harmonized with data available from regional sources such as the UN Food and Agriculture Organization (FAO), UN Development Programme (UNDP), Africa Development Bank (ADB), the World Bank and the World Resources Institute, and others. A meeting of experts from the Collaborating Centres took place to agree on harmonization of information and standardization of data sources, to ensure consistency in the report.

### Capacity building

The AEO report process has also successfully built capacity in state of the environment reporting, policy analysis, scenario development and integrated reporting, at national, sub-regional and regional levels in Africa. Capacity-building workshops were organized at sub-regional level for national experts and NGOs on the methodologies of state of the environment/policy retrospective reporting using the Pressure, State, Impacts and Responses (PSIR) framework, including methods of data management. A scenario development workshop was also held.

## THE AEO REPORT

*Africa Environment Outlook* is the first comprehensive integrated report on the African environment. The AEO assessment methodology is derived from UNEP's cutting-edge Global Environment Outlook (GEO) Process. The AEO process was initiated incorporating key attributes of the GEO process, such as the:

- use of sound data, information and science;
- incorporation of regional and sub-regional perspectives;
- inclusion of multi-stakeholder perspectives;
- identification of priority and emerging issues such as human vulnerability to environmental change;
- provision of early warning of impending threats; and
- orientation toward sustainable development.

## Why integrated environmental assessment reporting?

- It is a process of identifying environmental trends and conditions integrated with the assessment of key driving forces, while identifying leverage points to decision makers.
- It goes beyond the scope of traditional state of the environment reporting (SOE) which falls short of integrating the assessment of key driving forces and policies that cause or influence environmental trends.
- It answers four consecutive questions that are key to effective decision making. They are:
  1. What is happening to the environment?
  2. Why is it happening?
  3. What can we do, and what are we doing about it?
  4. What will happen if we do not act now?
- It brings together information and insight that is usually dispersed across disciplines and institutions.
- It is a tool to aid communication between science and policy.

*Africa Environment Outlook* aims to provide comprehensive, credible environmental information in a way that is relevant to policy making. The structure, which combines comprehensive environmental information with policy analysis, within an overall context of socio-economic conditions and development imperatives, is thus ideally suited to this purpose.

It provides recommendations for international cooperation and action, and can therefore be used by sub-regional organizations and national environment departments in developing national policies and international agreements.

The AEO report responds directly to Agenda 21, Chapter 40, which states:

*'While considerable data already exist, as the various sectoral chapters of Agenda 21 indicate, more and different types of data need to be collected, at the local, provincial, national and international levels, indicating the status and trends of the planet's ecosystem, natural resource, pollution and socio-economic variables. The gap in the availability, quality, coherence, standardization and accessibility of data between the developed and the developing world has been increasing, seriously impairing the capacities of countries to make informed decisions concerning environment and development.*

*'There is a general lack of capacity, particularly in developing countries, and in many areas at the international level, for the collection and assessment of data, for their transformation into useful information, and their dissemination. There is also need for improved coordination among environmental, demographic, social and developmental data and information activities.'*

# SYNTHESIS

The *Africa Environment Outlook* (AEO) report provides a comprehensive and integrated analysis of Africa's environment. AEO contains a detailed assessment of the current state of the environment in the region, indicates discernible environmental trends and examines the complex interplay between natural events and the impacts of human actions on the environment. Against this background, the report analyses the effects of environmental change in terms of human vulnerability and security, presents a set of scenarios for Africa's future and gives recommendations for concrete policy actions to steer the region, ultimately, towards the most favourable of those scenarios.

## ENVIRONMENT AND DEVELOPMENT 1972–2002

The historical focus of AEO is the 30-year period since the United Nations Conference on the Human Environment, held in Stockholm, Sweden in 1972. However, much of the degradation of Africa's environment today is part of a legacy from less favourable times, including the periods of the slave trade and colonialism. The historical scope of AEO therefore widens to discuss that legacy, and to show how the march of history has often overshadowed traditional African ways of life and knowledge that were inherently more respectful of the environment than some modern forms of development.

The 'winds of change' that began to blow across Africa in the early 1960s and the gathering momentum of African countries' struggle for independence are also described. An understanding of this process is vital to appreciate the emergence of a common African will to address the problems of environmental change and sustainable development.

In the 1970s, it was largely as a result of the 1972 Stockholm Conference that environmental concerns moved centre stage in the social and political debate in most parts of the world, and its conclusions helped to set the modern environmental agenda in Africa as elsewhere. But the conclusions of the Stockholm Conference have, perhaps, special significance for Africa. First, because they state firmly that a healthy environment is not only a fundamental right but that it is one that cannot be attained while apartheid, racial segregation or colonial domination persist, and second, because they call for the Earth's resources to be protected for the benefit of present and future generations. Such a call had immediate relevance for a region that was throwing off colonial ties and where many people are poor and therefore rely directly on natural resources for their livelihood. AEO traces the efforts made by African organizations, governments and institutions throughout the focus period to meet the challenge of that call, to translate a common will into careful, planned and appropriate management of Africa's huge wealth of natural resources, and to set the region on a course for sustainable development.

## STATE OF THE ENVIRONMENT 1972–2002

The causes of environmental change up to 2002 are examined, including those relating to policy and governance. The impacts of environmental change on the functioning of ecosystems, and on social and economic development, are also considered, in seven major areas.

● **Atmosphere:** Africa is extremely vulnerable to climate variability and climate change. Variations in rainfall patterns have led to incidences of drought and flooding, often with disastrous consequences for populations and for the environment. The predicted consequences of global climate change—worsening impacts of drought, desertification, flooding and sea level rise—may well worsen the situation of Africa's people, even though the region's greenhouse gas emissions are, on the whole, negligible. Analysis of the consequences of activities such as deforestation, inappropriate coastal development and poor land management shows that these can exacerbate the effects of climate variability and climate change. Air quality is an emerging issue of concern in many parts of Africa, especially in expanding urban areas where concentrations of population, industry and vehicles are increasing air pollution.

● **Biodiversity:** Africa's biological resources are declining rapidly as a result of habitat loss, overharvesting of selected resources, and illegal activities. Formal protection has been strengthened at the national and international level over the past 30 years. However, additional measures are required including: additional research and documentation, particularly of indigenous knowledge; implementation of strategies for sustainable harvesting and trade; wider involvement of stakeholders; and more equitable sharing of benefits.

● **Coastal and marine habitats:** Coastal and marine habitats and resources in Africa are under threat from pollution, overharvesting of resources, inappropriate development in the coastal zone, and poor inland land-management. Oil pollution is a major threat to resources, habitats and economies along the African coastline. Policies and regulations for sustainable coastal development and use of marine resources are in place but require sustained resources such as trained personnel, equipment, financial resources, and more effective policing, monitoring, administration and enforcement.

● **Forests:** Africa has the fastest rate of deforestation anywhere in the world. In addition to its ecological impacts, deforestation also means definitive loss of vital resources, causing communities to lose their livelihoods and vital energy sources. Political commitment to protection of indigenous forests, sustainable harvesting practices and community ownership require strengthening. Development of alternative energy sources is also a priority.

● **Freshwater:** Lack of availability and low quality of freshwater are the two most limiting factors for development in Africa, constraining food production and industrial activities, and contributing significantly to the burden of disease.

● **Land:** Degradation of soil and of vegetation resources is largely a result of increasing population pressures, inequitable land access and tenure policies, poor land management and widespread poverty. The results are: declining agricultural yields, affecting economies and food security; desertification of arid areas, raising competition for remaining resources; and increased potential for conflict. Land tenure reform, international cooperation, and integration of land resource management with development goals are required.

● **Urbanization:** Although most Africans currently live in rural areas, the region's rates of urbanization are among the highest in the world. Poor economic growth and low investment in infrastructure have left provision of housing and basic services in urban areas lagging far behind rates of inward migration, resulting in a proliferation of informal settlements in urban Africa.

## ENVIRONMENTAL CHANGE AND HUMAN VULNERABILITY

The poverty of most Africans, and their consequent direct dependence on natural resources for their livelihoods, increases their vulnerability to environmental change. Over the past 30 years, poverty has continued to worsen in Africa, and the region's environment has continued to deteriorate, making Africans even more vulnerable to environmental change.

In sub-Saharan Africa, 61 per cent of the population lives in ecologically vulnerable areas characterized by a high degree of sensitivity and low degree of resilience (IDS 1991). This is not necessarily by choice, but by force of circumstance, because other options are either unavailable or have been exhausted.

Rapid population growth and overexploitation of natural resources, deepening poverty and increasing food insecurity in sub-Saharan Africa have brought about human-induced environmental change. Human mismanagement of environmental resources and processes significantly exacerbates the impacts of disasters and their effects on natural resources.

●

*In sub-Saharan Africa, 61 per cent of the population lives in ecologically vulnerable areas characterized by a high degree of sensitivity and low degree of resilience (IDS 1991). This is not necessarily by choice, but by force of circumstance, because other options are either unavailable or have been exhausted.*

●

Other factors, such as poor economic performance, and weak institutional and legal frameworks, have left most Africans with limited choices and low coping capacity. Interventions addressing human vulnerability to environmental change must therefore be translated into integrated responses that reflect the multi-dimensional nature of the causes and states of vulnerability.

## Outlook: 2002–2032

Focusing on the next three decades, AEO considers a number of policy options that are likely to have the most significant impact on environment and socio-economic development. Four scenarios are presented, based on different environmental and social situations that are likely to result from alternative policy interventions. These scenarios are not predictions of the future, but aim to illustrate the range of possible outcomes based upon four policy choices and their interface with environment and developmental conditions, driving forces and management interventions. The identified driving forces most likely to shape the future are: demographics, economics (including poverty), social, culture, environment, technology and governance.

The four scenarios are:
- Market Forces
- Policy Reform
- Fortress World
- Great Transitions

The **Market Forces** scenario presents market-driven global development, leading to a dominant western-style economy. The environmental impact of this style of development in Africa will be a series of gains tempered by further environmental and social problems, and continued low economic growth.

Potential outcomes of this scenario include: increased incidence of drought and floods; reduction in agricultural production; increased health problems due to continued depletion of the ozone layer; intensification of migration to urban areas; spread of invasive species; increased deforestation; further decrease in the availability of freshwater; increased water-borne diseases; and intensified degradation of coastal and marine resources.

The **Policy Reform** scenario sees policy adjustments steer conventional development towards poverty-reduction goals. While more significant progress is made in terms of social and economic development, it is largely at the cost of further exploitation of natural resources and environmental degradation.

Potential outcomes of this scenario include: gradual decline in atmospheric pollution; expansion of the tourism industry in Africa; more energy choices and hence less dependence on biomass for fuel; meeting of water resource needs; reduced rural urban migration; and reduction of conversion of fragile ecosystems into agricultural land.

The **Fortress World** scenario is a future where socio-economic and environmental stresses mount, the world descends towards fragmentation, extreme inequality in power and socio-economic status, resulting in widespread conflicts, both within Africa and between Africa and other regions.

Potential outcomes of this scenario include: increased vulnerability to climate change; decline in urban air quality and higher incidence of respiratory diseases; declining productivity of grazing and agricultural land and overexploitation of water, land, forest and pasture resources; poor water quality and poor health; depletion of groundwater resources; increased vulnerability of coral reefs and mangrove forests; deterioration of the urban economy; and higher incidence of crime.

The **Great Transitions** scenario describes new development paradigms emerging in response to the challenges of sustainability, new values, pluralism and planetary solidarity. As this new ethical code is translated into policies that are, in turn, implemented in an integrated fashion, social and political stability permeate throughout Africa. Renewed ecosystem health and vitality ensure abundant resources and services, sustaining the lives and livelihoods of new generations.

Potential outcomes of this scenario include: improved urban air quality and energy use efficiency; increased equitable access to land; rehabilitation of marginal and degraded lands; decline in biodiversity losses and strengthening of ecosystem integrity; empowerment of communities to manage resources; less pressure on coastal zones; marked increase in access to water and sanitation; and a marked decrease in people living in urban slums and upland settlements.

## CALL TO ACTION

Most African countries face many challenges, such as the need to reduce poverty and improve the quality of life of their people and to improve the state of the environment. These need to be addressed from a policy perspective in order for the region to move closer to sustainable development.

A key aim of AEO is to recommend 'achievable action items' to AMCEN, as Africa's environmental body, and to other relevant policy officials. Urgent actions are required to reverse the current trend in environmental degradation in Africa. The key issues for action are poverty reduction and reversal of the direct causes of environmental degradation, by addressing environment and development together. There are also a number of cross-cutting issues that affect Africa's quest for sustainable development.

Specific actions are summarized below in the following categories:
- eradicating poverty;
- halting and reversing environmental degradation; and
- promoting actions on cross-cutting issues.

### Eradicating poverty

Poverty is a complex issue requiring a multi-dimensional approach and there is no uniform solution to its eradication. In Africa, poverty is considered to be both an agent and a consequence of environmental degradation. Because poverty reduction is pivotal to sustainable development, African countries are called upon to direct attention and resources to poverty challenges in the region through:
- endorsement and promotion of principles of sustainable development;
- acceleration of industrial development;
- securing food-self sufficiency and food security; and
- reversing the health crisis, including overcoming the HIV/AIDS pandemic.

### Halting and reversing environmental degradation

The problems of socio-economic development in Africa are inextricably linked to people, resources and the environment. Environmental conservation thus relates directly to the structure and functioning of the economy given that the majority of African people derive their livelihood directly from natural resources. The region is, however, losing its resources at relatively rapid rates, thereby leaving millions of people vulnerable to adverse environmental change. Future strategies and actions aimed at halting and reversing environmental degradation must include and prioritize:
- halting and reversing desertification and land degradation;
- conservation and management of biodiversity and forest resources, including wetland and cross-border ecosystems;
- climate change mitigation and improvement of air quality;
- improvement of access to and quality of freshwater resources;
- conservation of coastal and marine ecosystems and resources; and
- promotion of environmentally-sound management of toxic wastes.

### Promoting action on cross-cutting issues

A number of cross-cutting issues require urgent attention in Africa in order to halt and reverse environmental degradation, and to reduce vulnerability. These are wide-ranging and must include such important areas as:
- mobilization of domestic and international financial resources for sustainable development;
- promoting trade;
- promoting peace building, good governance and human rights;
- enhancement of scientific and technological base;
- accelerating regional cooperation and integration;
- promoting the role of civil society; and
- promoting the development of human resources.

# AEO SUB-REGIONS

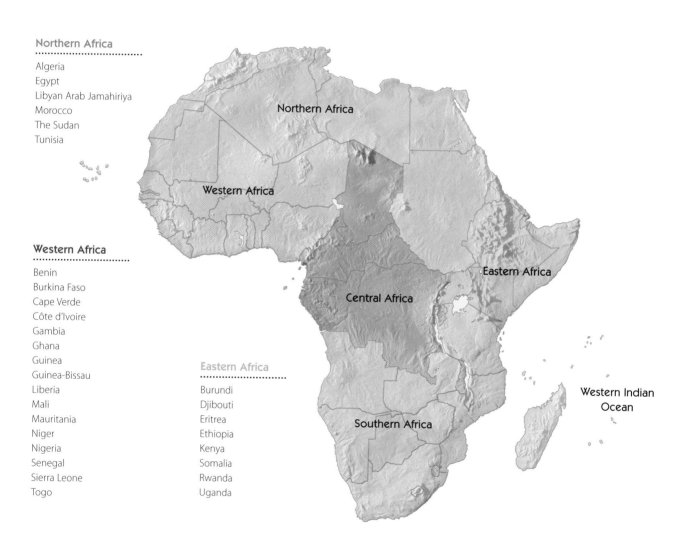

### Northern Africa

Algeria
Egypt
Libyan Arab Jamahiriya
Morocco
The Sudan
Tunisia

### Western Africa

Benin
Burkina Faso
Cape Verde
Côte d'Ivoire
Gambia
Ghana
Guinea
Guinea-Bissau
Liberia
Mali
Mauritania
Niger
Nigeria
Senegal
Sierra Leone
Togo

### Eastern Africa

Burundi
Djibouti
Eritrea
Ethiopia
Kenya
Somalia
Rwanda
Uganda

### Central Africa

Cameroon
Central African Republic
Chad
Congo
Democratic Republic of Congo
Equatorial Guinea
Gabon
Sao Tomé and Principe

### Southern Africa

Angola
Botswana
Lesotho
Malawi
Mozambique
Namibia
South Africa
Swaziland
United Republic of Tanzania
Zambia
Zimbabwe

### Western Indian Ocean Islands

Comoros
Madagascar
Mauritius
Réunion (France)
Seychelles

# CHAPTER 1

## ENVIRONMENT AND DEVELOPMENT IN AFRICA: 1972–2002

# CHAPTER 1

# ENVIRONMENT AND DEVELOPMENT IN AFRICA: 1972–2002

## INTRODUCTION

Chapter 1 of the *Africa Environment Outlook* (AEO) provides an overview of developments in Africa, particularly over the three decades up to 2002. It highlights social and economic policies and programmes which have impacted on the environment, and which have influenced various policy responses.

While the chapter focuses mainly on the 30 years since the 1972 Stockholm Conference on the Human Environment—which laid the foundation for international action on the environment—it also reviews some of the developments before 1972 which had a major bearing on Africa's political economy.

**Figure 1.1 The political regions and boundaries of Africa**

Such developments include: the slave trade; colonization; decolonization; and the struggle for independence. This historical background information serves to place the evolution of environmental management in Africa in its proper perspective.

Social, economic and environmental issues in Africa are discussed in three major sections, each covering approximately one decade, namely: the 1970s; the 1980s; and the 1990s and beyond.

The section on the 1970s also deals with some developments before 1972, including: colonization; decolonization; and the establishment of the Organization of African Unity (OAU) in 1963. The legal and institutional framework for environmental management is also covered.

The section on the 1980s highlights the economic decline of Africa, including: the debt problem; famine in the mid-1980s; and African attempts at revival, with the adoption of the Lagos Plan of Action at the beginning of the decade. The creation in 1985 of the African Ministerial Conference on the Environment (AMCEN) and the work of the World Commission on Environment and Development (WCED) are also highlighted. The WCED popularized the concept of sustainable development. By embracing both sustainable development and the World Conservation Strategy, which spawned many national conservation strategies, Africa has claimed its position as a major global player at the international level in terms of environmental management.

The section on the 1990s and beyond lowers the curtains on the decolonization process, with the abolition of apartheid in South Africa and the creation of a new political dispensation. The section also covers the end of the Cold War, which had made the region a theatre for both Western and Soviet geopolitical games. The 1990s also saw the evolution of new political systems in many African countries. In terms of the environment, many countries adopted new constitutions, which enshrined environmental rights. The section looks at the impact of the 1992 United Nations Conference on Environment and Development (UNCED), and at preparations for the World Summit on Sustainable Development to be held in South Africa in 2002.

## BACKGROUND

Since time immemorial, the environment has been woven into the lives of African people. Traditional and cultural values among varied and disparate communities across the region have governed the way in which people interact with the environment, and the way in which natural resources are used and managed. In many sub-regions, the people's relationship with natural resources is strong, and there have been traditional regulatory mechanisms covering natural resource management. Box 1.1 highlights how such traditional rules have facilitated resource use and conservation in parts of Africa.

Africa is also rich in indigenous knowledge systems, some of which have survived two of the major events that have contributed to defining modern Africa: slavery and colonization.

The slave trade facilitated the shipping of millions of Africans across the Atlantic Ocean to work in plantations in North and South America and the Caribbean, and as domestic servants in Europe. Arabs and Europeans, on both the east and the west coast of Africa, bartered simple commodities—such as cloth, gunpowder, salt, beads and others—for slaves.

The abolition of the slave trade beckoned in a new form of commerce. The discovery of large tracts of 'empty lands', and the discovery of rich reserves of gold and other minerals further inland, enticed Europeans to explore Africa. Africa's vast natural resources promised cheap raw material for European industries. Charters were established with local chiefs, and Europeans were offered great tracts of land. In return, the chiefs and their people were promised protection from invading armies as well as European commodities. A new era of subjugation—colonialism—took root in Africa. The plunder of Africa's natural resources and the environment had begun, leading to the scramble for colonies by European countries.

Colonial policies led to heightened conflicts between users, and to assaults on the environment, through the destruction of natural forests for timber, cropland, fuelwood, pasture and urbanization. For

## Box 1.1 Guardians of tradition

Traditional rules, once established, controlled the access of African people to natural resources. Rules prohibited, for example: cutting particular trees; some methods of gathering certain fruits and other tree by-products; and access to sacred groves and mountains.

Cutting fruit trees, in particular, was prohibited. In Zimbabwe, it was almost inconceivable for anyone under traditional tenure to cut *Uacapa kirkiana* without the express permission of the guardians of the land. Other trees, such as *Sclerocarya birrea* and *Parinari curatellifolia*, were directly linked to ancestral spirits and rituals, and were protected by a standing penalty system, which was enforced by a chief and his lineage.

Traditional rules regarding gathering fruit facilitated the conservation of fruit trees. Most fruits were supposed to be harvested for use in the home, and not for sale. Rules governing fruit gathering included the following:

- Never pick up a [*Uacapa kirkiana*] fruit with two hands.
- Shake the tree, using a stone or another instrument, as a way to dislodge the fruit.
- Do not curse or express delight about the quality or quantity of fruit.

Other rules limited the quantity of unripe fruits leaving the forest, so that fruit picking did not damage the trees. It was generally understood that if any of the offences were committed, the person who committed them would disappear in the forest.

In terms of woodland management, the traditional rules went even further: tree cutting was banned in designated places. The declaration of such places, and their subsequent protection, lay in the land-guardian relationship.

*Source: SADC/IUCN/SARDC 2000*

example, colonial forestry policies tended to focus on plantations, in order to meet the growing and specialized demands of industry and commerce in Europe. This practice led to loss of species diversity as large areas were cleared of indigenous trees and substituted with exotic ones. Under colonialism, African people had little say regarding how their resources were exploited, and they benefited little from the region's natural assets.

The situation has changed, particularly over the past 30 years, with African countries attaining national independence, and adopting their own social, political, economic and environmental policies and programmes. Progress has been achieved on many fronts, but many challenges remain. Some of the developments that have shaped Africa's socio-economic and environment agendas are explored in the following sections.

## THE 1970s—WINDS OF CHANGE

### DECOLONIZATION

The decolonization of Africa was described in the 1960s by the former British Prime Minister Harold Macmillan as the 'winds of change'. Decolonization of the region began in 1957, with the independence of Ghana. It gathered momentum through the 1960s and beyond, to 1994 and the eradication of apartheid in South Africa.

The momentum which fanned African nationalism strengthened relations between former colonies, leading to stronger voices in favour of pan-Africanism. In 1963, the founding fathers of African independence established the OAU, whose main objectives were to:

● promote the unity and solidarity of African states;
● coordinate and intensify cooperation and efforts among African states, in order to achieve a better life for the people; and
● defend the sovereignty of African states, their territorial integrity and their independence.

In many instances, independence did not mean political stability. Many countries which gained independence in the late 1950s and the 1960s—such as Chad, Congo, Ghana, Sierra Leone, Nigeria, Rwanda, Uganda, Somalia and Sudan—underwent phases of political instability which, in some cases, have

continued up to the present. Major developments which took place in the 1970s, and which have influenced policies in Africa, are listed in Table 1.1.

The Cold War is arguably one of the major events that had the greatest impact on Africa, in terms of its socio-economic alliances and development. Environmental management was generally not considered paramount during that period as it is today, even though Africa has a long track record in terms of the sustainable use of natural resources. The two dominant development paradigms during the Cold War were capitalism and socialism. This sometimes led to tensions between African countries, often resulting in armed conflict between them and civil war in others. Armed conflict led to a refugee problem, which saw the number of refugees grow from a low figure of 23 500 people at the end of the 1950s to a high of about 50 million refugees at the end of the 1990s (UNHCR 2000). Political unrest and the resultant refugee situation in many parts of Africa during the past decades have led to many problems, including:

● deforestation, resulting from massive land clearance for agriculture and fuelwood;
● rapid urbanization, particularly in coastal areas;
● widespread poverty;
● poor economic performance;
● trade policies which are not conducive towards peace and development;
● inadequate technology base to satisfy existing demand;
● increased civil strife; and
● growth in illegal trade in minerals and other natural resources.

### ENVIRONMENTAL AGENDA

Five years after the OAU was established, African countries adopted the African Convention on the Conservation of Nature and Natural Resources, in Algiers in September 1968. The main objective of the Algiers Convention was to encourage individual and joint action for the conservation, utilization and development of soil, water, flora and fauna, for the present and future welfare of humankind.

The main principle of the Algiers Convention states: 'The contracting states shall undertake to adopt the measures necessary to ensure conservation, utilization

| Table 1.1  Major developments which shaped policies in Africa in the 1970s | |
|---|---|
| **Year** | **Developments** |
| 1971 | ● The Convention on Wetlands of International Importance especially as Waterfowl Habitat is adopted in Ramsar, Iran |
| 1972 | ● The UN Conference on the Human Environment is held in Stockholm, Sweden<br><br>● The United Nations Environment Programme (UNEP) is established, with its headquarters in Nairobi, Kenya<br><br>● The Convention Concerning the Protection of the World Cultural and Natural Heritage is adopted in Paris, France<br><br>● The Convention on the Prevention of Marine Pollution by Dumping of Wastes and other Substances is adopted in London, United Kingdom, and Mexico City, Mexico |
| 1973 | ● The 'oil weapon' is first used on the world oil market by the Arab oil exporting countries. This has a devastating impact, especially on the economies of developing countries, including those of Africa<br><br>● The Convention on the International Trade in Endangered Species of Wild Fauna and Flora (CITES) is adopted in Washington, D.C., USA<br><br>● The International Convention for the Prevention of Pollution from Ships is adopted in London, United Kingdom |
| 1975 | ● Mozambique becomes an independent state on 25 June, followed by Seychelles on 29 June and by Angola on 11 November<br><br>● The Convention on Wetlands of International Importance Especially as Waterfowl Habitat enters into force<br><br>● The Convention Concerning the Protection of the World Cultural and Natural Heritage enters into force<br><br>● The Convention on the Prevention of Marine Pollution by Dumping of Wastes and Other Substances enters into force |
| 1976 | ● The Convention on the Prohibition of Military or Any Other Hostile Use of Environmental Modification Techniques |
| 1977 | ● The Convention on the Prohibition of Military or Any Other Hostile Use of Environmental Modification Techniques is opened for signature in Geneva, Switzerland in May |
| 1978 | ● The Protocol relating to the International Convention for the Prevention of Pollution from Ships modifying provisions, adopted in 1973, is adopted in London, United Kingdom<br><br>● The Convention on the Prohibition of Military or Any Other Hostile Use of Environmental Modification Techniques enters into force |
| 1979 | ● Protracted negotiations are held at Lancaster House between the British government and the Patriotic Front. The talks led to the independence of Zimbabwe in the following year<br><br>● The Convention on the Conservation of Migratory Species of Wild Animals is adopted in Bonn, Germany in June |

*Sources: SADC/IUCN/SARDC (1998) and UNEP/Sida (undated)*

and development of soil, water, floral and faunal resources in accordance with scientific principles and with due regard to the best interests of the people.'

The Algiers Convention also demands that parties undertake to:

- adopt effective measures to conserve and improve the soil; and to control erosion and land use;
- establish policies to conserve, utilize and develop water resources; to prevent pollution; and to control water use;
- protect flora and ensure its best utilization; ensure good management of forests; and control burning, land clearance and overgrazing;
- conserve fauna resources and use them wisely; manage populations and habitats; control hunting, capture and fishing; and prohibit the use of poisons, explosives and automatic weapons in hunting;
- tightly control traffic in trophies, in order to prevent trade in illegally killed and illegally obtained trophies; and
- reconcile customary rights with the convention.

Well after the Algiers Convention, the modern environmental agenda—which was first set at the 1972 Stockholm Conference on the Human Environment—also shaped environmental policies and programmes in the region. For example, African governments have responded positively through policy implementation to global, regional and sub-regional environmental problems and challenges, although the success of policy implementation has varied from one sub-region to another. At the national level, the Stockholm Conference influenced the establishment of the first environment ministry in 1975 in what was then Zaire (now the Democratic Republic of Congo). More environment ministries have been established in other African countries over the past three decades. At the global level, the Stockholm Conference led to the establishment of the United Nations Environment Programme (UNEP), with its headquarters in Nairobi.

The global environmental, political, economic and social issues of the 1960s and early 1970s influenced the preparations for, and the final decisions of, the 1972 Stockholm Conference. For Africa, the Stockholm Declaration on the Human Environment stands out as the defining document in terms of 'soft law' on environment and development issues. The Stockholm Declaration laid the foundation in terms of:

- Environmental rights.
- Environmental education.
- The sovereign rights of states to 'exploit their own resources', in terms of their own environmental policies and their responsibility to ensure that activities in their territory do not harm the environment of other states.
- Calling for the 'elimination and complete destruction' of nuclear weapons and 'all other means of mass destruction'.
- Speaking strongly against 'apartheid, racial segregation, discrimination, colonial and other forms of oppression and foreign domination'.
- Highlighting nature conservation, including wildlife, as important in planning for economic development.
- The sustainable utilization of non-renewable resources, to ensure that they benefit all humankind.
- Recognizing the importance of 'substantial quantities of financial and technological assistance' to developing countries, in order to tackle environmental deficiencies caused by underdevelopment and natural disasters.
- The need for environmental policies of all countries to enhance, and not to adversely affect, the present or future development potential of developing countries.
- Rational planning to reconcile any conflict between the needs of development and the need to protect and improve the environment.
- Appropriate demographic policies 'which are without prejudice to basic human rights'.
- The application of science and technology to identify, avoid and control environmental risks.

The 1972 Stockholm Conference rekindled the African spirit of living in harmony with each other and with the environment, as was stated by Professor Mostafa K. Tolba (who later became the second UNEP executive director) at that conference (see Box 1.2).

In addition to the Algiers Convention, African countries are party to some of the following international agreements, which were adopted in the 1970s:

- The 1971 Convention on Wetlands of International Importance Especially as Waterfowl Habitat (Ramsar).

---

**Box 1.2  Opening a new window in global environmental management**

'One of our prominent responsibilities in this conference is to issue an international declaration on the human environment; a document with no binding legislative imperatives, but—we hope—with moral authority, that inspire in the hearts of men, the desire to live in harmony with each other, and with their environment.'

Professor Mostafa K. Tolba, President of the Academy of Scientific Research and Technology, and head of the Egyptian delegation at the 1972 Stockholm Conference on the Human Environment

---

- The 1972 Convention Concerning the Protection of the World Cultural and Natural Heritage (World Heritage).
- The 1973 Convention on International Trade in Endangered Species of Wild Fauna and Flora (CITES).
- The 1979 Convention on the Conservation of Migratory Species of Wild Animals (CMS).

## THE 1980s—STAGNATION

By the close of the 1970s, some countries in Africa were still under colonial rule. South Africa was still struggling to eliminate apartheid and, in what is now Zimbabwe, a liberation war was raging against the minority government, which had pronounced a Unilateral Declaration of Independence (UDI) from Britain in 1965. South West Africa, now Namibia, was also yet to achieve independence. Elsewhere in the region, the territories of Western Sahara and Eritrea were fighting for self-determination.

Civil and political strife in Africa were taking a large toll on human life and on natural resources, which were being plundered to finance wars. In Mozambique, for example, the civil war intensified in the 1980s, forcing millions to become refugees in the neighbouring countries of Malawi, South Africa, Swaziland, Tanzania, Zambia and Zimbabwe. At the height of the war, Malawi was host to more than 1 million Mozambicans—about 10 per cent of the country's population.

## SOCIO-ECONOMIC ISSUES

Since independence, many African countries have persistently faced social and economic challenges. Economic growth for most African countries has been sluggish or negative, impacting heavily on the welfare of the people, especially the rural population. In the 1980s, Africa underwent many economic experiments, such as economic Structural Adjustment Programmes (SAPs), which have been blamed in some countries for exacerbating poverty. The region's continued dependence on external aid, and increasing external debt, illustrate the complete failure of some of its social and economic policies, a number of which were prescribed by the World Bank and the International Monetary Fund (IMF). SAPs in the region led to, among other things, the removal by governments of subsidies on essential services, such as education, health and transport; and a severe reduction of jobs in the public service sector. These policies have resulted in: a reduction in real income and purchasing power; an increase in the importance of the informal economy and family labour; an increase in the relative price of many basic goods and services; and a reduction in the quality of public services.

The negative impacts of SAPs have been heaviest on: the urban poor, who rely most heavily on employment, consumer subsidies and public services; and rural smallholder farmers, who relied on subsidies for their farm inputs. In urban areas, wages and job opportunities declined considerably following the introduction of SAPs.

The external debt problem in Africa heightened during the 1980s. The decade between 1985–87 and 1995–97 saw 41 sub-Saharan countries sinking deeper into debt (see Table 1.2). In some cases, the debt rose by more than 150 per cent, as in the case of Angola, Chad and Lesotho. Debt-related issues in the region are covered in more detail in Chapter 3 of this report.

## DISASTERS

The major environmental disasters in Africa are recurrent droughts and floods. Their socio-economic and ecological impacts are devastating to African countries, because most of the countries do not have real-time forecasting technology, or resources for post-disaster rehabilitation. The impacts of disasters include: massive displacement of people, as happened in

## Table 1.2 Percentage change in indebtedness by African countries

| Country | Total external debt US$ (million) 1985–87 | Total external debt US$ (million) 1995–97 | Percentage change |
|---|---|---|---|
| Angola | 4 035 | 10 739 | 166 |
| Benin | 1 012 | 1 611 | 59.2 |
| Botswana | 438 | 626 | 42.9 |
| Burkina Faso | 659 | 1 286 | 95.1 |
| Burundi | 598 | 1 117 | 86.8 |
| Cameroon | 4 003 | 9 394 | 135 |
| Central African Republic | 474 | 921 | 94.3 |
| Chad | 275 | 975 | 255 |
| Congo | 3 625 | 5 439 | 50.0 |
| Congo, D. R. | 7 373 | 12 799 | 73.6 |
| Côte d'Ivoire | 11 562 | 18 010 | 55.8 |
| Equatorial Guinea | 162 | 286 | 76.5 |
| Eritrea | - | 52 | - |
| Ethiopia | 6 234 | 10 155 | 62.9 |
| Gabon | 1 923 | 4 318 | 125 |
| Gambia | 281 | 437 | 55.5 |
| Ghana | 2 779 | 5 992 | 116 |
| Guinea | 1 767 | 3 334 | 88.7 |
| Guinea Bissau | 390 | 918 | 135 |
| Kenya | 4 841 | 6 922 | 43.0 |
| Lesotho | 211 | 669 | 217 |
| Liberia | 1 461 | 2 091 | 43.1 |
| Madagascar | 3 073 | 4 191 | 36.4 |
| Malawi | 1 182 | 2 253 | 90.6 |
| Mali | 1 749 | 2 970 | 69.8 |
| Mauritania | 1 740 | 2 405 | 38.2 |
| Mozambique | 3 496 | 5 833 | 66.8 |
| Namibia | - | - | - |
| Niger | 1 411 | 1 567 | 11.0 |
| Nigeria | 23 392 | 31 318 | 33.9 |
| Rwanda | 474 | 1 061 | 124 |
| Senegal | 3 275 | 3 725 | 13.7 |
| Sierra Leone | 870 | 1 169 | 34.4 |
| Somalia | 1 816 | 2 628 | 44.7 |
| South Africa | - | 25 543 | - |
| Sudan | 9 945 | 16 967 | 70.6 |
| Tanzania | 6 506 | 7 345 | 12.9 |
| Togo | 1 078 | 1 427 | 32.4 |
| Uganda | 1 522 | 3 652 | 140 |
| Zambia | 5 655 | 6 933 | 22.7 |
| Zimbabwe | 2 631 | 5 006 | 90.3 |

Ethiopia in the mid-1980s; increased erosion and sedimentation of reservoirs; degradation of coastal zones; and general changes in habitats. These impacts negatively affect both people and wildlife.

In addition to drought and floods, tropical cyclones cause havoc, especially in the Western Indian Ocean Islands. Islands states, such as Comoros, Madagascar, Mauritius, Seychelles, Reunion and others, and coastal states, such as Mozambique, are also vulnerable.

Poor land management practices, which lead to land degradation and deforestation, contribute to increased flood disasters in some risk areas. The effects of droughts and floods are exacerbated by ineffective policies. For instance, where governments are aware that a large percentage of their people rely heavily on wood for energy, and yet do not provide adequate energy resources, people are forced to cut trees for charcoal, which is sold primarily in urban areas. This contributes to deforestation in Africa. Unless alternative energy sources are made available, the deforestation trend is likely to continue, exposing more and more people to risk from disasters related to environmental change. Human vulnerability to environmental change is discussed in more detail in Chapter 3 of this report.

Major developments which took place in the 1980s, and which have influenced policies in Africa, are listed in Table 1.3.

## ENVIRONMENTAL POLICY

Although the 1980s have been referred to as the 'lost decade' for Africa, it was also the decade in which governments in the region consolidated efforts to set their countries on a path of sustainable development. Various environmental initiatives were undertaken during this period, at both regional and global levels, and these greatly influenced environmental policy in Africa.

### Emergence of African common resolve

Meetings under the auspices of the OAU, such as the 1980 Extraordinary Summit of Heads of State and Government, which led to the adoption of the Lagos Plan of Action, helped to highlight the challenges facing the region. Under the Lagos Plan of Action, African leaders emphasized that 'Africa's huge resources must be applied principally to meet the needs and purposes of its people.' They also emphasized the need for Africa's

## Table 1.3  Major events which shaped policies in Africa in the 1980s

| Year | Developments |
|---|---|
| 1980 | • Zimbabwe attains independence from Britain<br><br>• The Organization of African Unity (OAU) adopts the Lagos Plan of Action<br><br>• Nine southern African countries – Angola, Botswana, Lesotho, Malawi, Mozambique, Swaziland, Tanzania, Zambia and Zimbabwe –establish a political and economic bloc called the Southern Africa Development Coordination Conference (SADCC), now the Southern African Development Community (SADC)<br><br>• The World Conservation Strategy is published by the World Conservation Union (IUCN), introducing the concept of sustainable development, and becomes a blueprint for national conservation strategies (NCS) |
| 1982 | • The Convention on Wetlands of International Importance especially as Waterfowl Habitat is amended in Paris, France<br><br>• The United Nations Convention on the Law of the Sea is adopted |
| 1983 | • The United Nations establishes the World Commission on Environment and Development (WCED)<br><br>• The Protocol relating to the International Convention for the Prevention of Pollution from Ships enters into force<br><br>• The Convention on the Conservation of Migratory Species of Wild Animals enters into force<br><br>• The International Tropical Timber Agreement is adopted in Geneva, Switzerland in November. This agreement was later succeeded by the International Tropical Timber Agreement (1994)<br><br>• The first incidence of HIV/AIDS is recorded in Africa |
| 1985 | • The Vienna Convention for the Protection of the Ozone Layer is adopted in Vienna, Austria<br><br>• The Convention for the Protection, Management and Development of the Marine and Coastal Environment of the Eastern African Region is adopted in Nairobi, Kenya<br><br>• The Protocol Concerning Protected Areas and Wild Fauna and Flora in the Eastern African Region is adopted in Nairobi, Kenya |
| 1986 | • The Convention on Wetlands of International Importance Especially as Waterfowl Habitat enters into force |
| 1987 | • The Brundtland Commission publishes *Our Common Future*, which advocates sustainable development<br><br>• The Montreal Protocol on Substances that Deplete the Ozone Layer is adopted in Montreal, Canada<br><br>• The Agreement on the Action Plan for the Environmentally Sound Management of the Common Zambezi River System is adopted in Harare, Zimbabwe, and enters into force |
| 1988 | • The Vienna Convention for the Protection of the Ozone Layer enters into force |
| 1989 | • Parties to CITES ban international trade in ivory and other elephant products. Some Southern African countries put up a strong opposition<br><br>• The Montreal Protocol on Substances that deplete the Ozone Layer enters into force, in January<br><br>• The Basel Convention on the Control of Transboundary Movement of Hazardous Wastes and their Disposal is adopted in Basle, Switzerland |

Sources: SADC/IUCN/SARDC (1998) and UNEP/Sida (undated)

apparent 'total reliance on the export of raw materials' to change, and the need to mobilize its entire human and material resources for the development of the region (OAU 1980). The Lagos Plan of Action (see Table 1.4) is one of many measures adopted by the region which set either qualitative or quantitative targets. Unfortunately, many of these targets remain unmet.

## African Ministerial Conference on the Environment

The first meeting of AMCEN, organized by UNEP in close collaboration with the UN Economic Commission for Africa (UNECA) and the OAU, was held in Cairo, Egypt in December 1985. In addition to being Africa's direct response to the 1972 Stockholm Conference, the establishment of AMCEN was also part of UNEP's response to Africa's environmental crisis. The objective of the AMCEN programme, which was adopted in Cairo, is to mobilize national, sub-regional and regional cooperation in four priority areas:

● halting environmental degradation;
● enhancing Africa's food producing capacity;
● achieving self-sufficiency in energy; and
● correcting the imbalance between population and resources.

As part of its programme, AMCEN focuses on environmental, social and economic inequality, and their impact on the environment. It also focuses on the pace of economic globalization and its environmental impact on Africa. The AMCEN meeting in Abuja in 2000 marked a turning point for AMCEN. At this meeting, African governments committed themselves to:

● keeping a constant review of policy actions that would enable Africa to address environmental challenges, especially new and emerging issues;
● building capacity to deal with major concerns;
● forging strategic partnerships with the public and private sectors, with civil society, with non-governmental organizations (NGOs) and with the international community in preparing and implementing AMCEN policies and programmes;
● coordinating the implementation of environmental treaties, in accordance with environmental and development priorities; and
● cooperating with relevant regional and sub-regional bodies in preparing a common position for the World Summit on Sustainable Development, to be held in Johannesburg in 2002.

Through its partnership with UNEP, AMCEN has committed itself to keeping under review the state of the environment, and emerging environmental issues and trends, in Africa. It also aims to provide early warning signals, and to promote government and public access to environmental information, as a basis for policy development, programme responses and action to achieve environmental security.

For almost 20 years, AMCEN has facilitated the broadening of the political and public policy legitimacy of environmental concerns, through the growth of civil society organizations, and their active participation in international and national environmental activities. Some of the milestones that AMCEN has achieved include the following:

● the adoption in January 1991 of the Bamako Convention on Hazardous Wastes;
● the adoption in Abidjan in November 1992 of the African Common Position, which was subsequently submitted to the UNCED Secretariat;
● the establishment and promotion of eight networks, in the areas of: environmental monitoring; climatology; soils and fertilizers; energy; water resources; genetic resources; environmental education and training; and science and technology;
● the establishment of four committees related to the development and improvement of the environment of the five African ecosystems, namely: deserts and arid lands; rivers and lake basins; forests and woodlands; regional seas; and island ecosystems;
● the harmonization of Africa's position on global environmental issues, through the Convention on Biological Diversity (CBD), the UN Framework Convention on Climate Change (UNFCCC) and the UN Convention to Combat Desertification in Countries Experiencing Serious Drought and/or Desertification, Particularly in Africa (UNCCD); and
● the strengthening of cooperation between African member states.

## World Conservation Strategy

The 1980 World Conservation Strategy (WCS), which was developed by the World Conservation Union (IUCN), introduced the concept of sustainable development.

## Table 1.4  Goals of the Lagos Plan of Action, 1980–2000

| *Issue* | *Action* |
|---|---|
| Environment | • Adopt a plan of action, which should incorporate the development of policies, strategies, institutions and programmes, for the protection of the environment. |
| | • Utilize urban wastes to produce biogas, in order to save energy; and convert rubbish into manure; combat water-borne diseases; control water pollution from agricultural and industrial effluents. |
| | • Introduce measures to control marine pollution from land-based industrial wastes and oil from shipping. |
| | • Implement stricter control of fish exploitation in economic exclusion zones by foreign transnationals. |
| | • Establish programmes to rehabilitate mined-out sites, by removing earth tailings; filling up ponds to eradicate water-borne diseases; and controlling toxic heavy metal poisoning. |
| | • Establish stations to monitor air pollutants from factories, cars, and electrical generators using coal. |
| | • Control the importation of pollutive industries (cement, oil refineries, tanneries and so on). |
| | • Create national programmes in environmental education. |
| | • Improve legislation and law enforcement, in order to protect the environment. |
| | • Plan and manage the rational use of land, water and forest resources as part of the campaign against desertification. |
| | • Develop innovate approaches in drought management and desertification control. |
| | • Collect and disseminate environmental data, in order to monitor the state of the environment. |
| | • Facilitate the establishment of techniques for the proper exploitation of natural resources, in order to prevent water and air pollution. |
| | • Facilitate the establishment of techniques to manage and use forests and grasslands, in order to prevent the exposure of the land to soil and wind erosion. |
| Food and agriculture | • Achieve a 50 per cent reduction in post-harvest food losses. |
| | • Attain food self-sufficiency in the next decades. |
| | • Set up national strategic food reserves, at 10 per cent of total food production. |
| | • Increase production from African waters by 1 million tonnes by 1985. |
| | • Develop a national food policy in each country. |
| | • Establish an inventory of forest resources. |
| | • Promote indigenous research, and the study of indigenous species in particular ecological areas. |
| | • Expand areas under forestry regeneration by 10 per cent annually up to 1985. |
| | • Expand forest reserves by 10 per cent by 1985. |
| Water resources | • Establish an inventory of surface and groundwater sources. |
| | • Develop special techniques for managing water resources, that is, collect data on water availability and quality; forecast demand in various rural sectors; and develop and use technologies for recovery and recycling. |
| | • Develop technologies for collecting water in rural areas, for distribution, irrigation, treating polluted water and disposal of waste water. |
| | • Establish river basin organizations. |
| | • Strengthen existing sub-regional organizations, such as river and lake basin commissions. |

*Source: Field-Juma (1996), OAU (1995, 2001)*

The WCS influenced African governments to undertake their own national conservation strategies, satisfying one of the objectives of the 1972 Stockholm Conference, that is, to incorporate the environment in development planning. While such policy documents became common, particularly in the 1980s, the environment did not immediately become part of mainstream activity, as indicated by the small annual budget allocations for environmental management.

## World Commission on Environment and Development

The WCED was established in 1983 in response to the United Nations General Assembly (UN GA) Resolution 38/161, which mandated the WCED to:

- examine the critical environment and development issues, and formulate realistic proposals for dealing with them;
- propose new forms of international cooperation on issues that would influence policies and events in the direction of necessary change; and
- raise levels of understanding and commitment to the action of individuals, voluntary organizations, business institutions and governments.

The UN GA asked the WCED to formulate 'A Global Agenda for Change' on the environment and development. The WCED's Environmental Perspectives examine issues in their relationship to the challenges of social and economic development; set out goals for environmentally sound and sustainable development; and call upon governments, international organizations, industry, financial institutions and NGOs to take specific actions to achieve those goals (UNEP/OAU 1991). The WCED, or the Brundtland Commission, popularized sustainable development in its 1987 report, *Our Common Future*. The WCED's definition of sustainable development—development that meets the needs of the present without compromising the ability of future generations to meet their own needs—is now part of the environment lexicon. The WCED process also popularized public participation in environmental issues, because it convened many public meetings in Africa, and in other developed and developing regions.

Some of the actions recommended by the WCED for African countries are shown in Box 1.3.

### Box 1.3 Key issues faced by Africa

The WCED has defined sustainable development as 'a process in which the exploitation of resources, the direction of investments, the orientation of technological development and institutional change are made consistent with future as well as present needs'. For Africa, this entails massive house-cleaning exercises, and international negotiations yet unmatched among national governments, including:

- providing more resources to meet the priority needs of the people, rather than satisfying the needs of international creditors;
- utilizing the ability and the aspirations of the people in development plans, so that poverty alleviation becomes a core element as they move towards sustainable development;
- setting up democratic domestic mechanisms to harmonize the activities of NGOs operating in Africa with development policies defined by governments;
- negotiating for commodity prices which reflect the real cost of production for Africa; and
- carrying out intensive intra-Africa trade.

These are the issues that put the destiny of Africa at stake. They can only be ignored at great cost to an environmentally sound future for the region.

Source: UNEP/OAU 1991

### First African Regional Conference on Environment and Development

In response to the UN GA resolution adopted in 1987, and by the recommendation of the WCED, the First African Regional Conference on Environment and Development was convened in Kampala, Uganda in June 1989. Ministers responsible for economic planning, education and environment, as well as NGOs, youth and women, attended the conference.

The Kampala Conference undertook to integrate environmental concerns into all existing and future economic and sectoral policies, in order to ensure that they protect and improve the environment and the natural resource base on which the health and welfare of the African people depend. The conference also adopted the Kampala Agenda for Action on Sustainable Development in Africa. The Kampala Conference was a synthesis of the programmes and plans of action taken on the environment since the Lagos Plan of Action was adopted in 1980 (see Figure 1.2). The priority issues adopted by the Kampala Conference were:

● managing demographic change and pressures;
● achieving food self-sufficiency and food security;
● ensuring efficient and equitable use of water resources;
● securing greater energy self-sufficiency;
● optimizing industrial production;
● maintaining species and ecosystems; and
● preventing and reversing desertification.

The Kampala Conference was further endorsed by the OAU Pan-African Conference on Environment and Development, held in Bamako in January 1991.

## THE 1990s TO 2002— TOWARDS REVITALIZATION

Following global trends, significant and positive achievements, including political liberalization, spread across Africa during the 1990s. Pluralism and accountability were more evident than ever before. One-party dictatorships and military regimes were swept out of power, as Africans exercised their right to elect their governments. Leaders who accepted the will of the people at the ballot box began to emerge. In most countries in the region, civil society grew in strength, with significant movements towards decentralization, and with popular participation in the development process.

### POLITICAL DEVELOPMENTS

The 1990s saw a further shift in the development paradigm for Africa. The 'real issue' for the 1990s centred on good governance. One of the major political events of the last decade of the 20th century was the abolition of apartheid in South Africa. The photograph of the release of Nelson Mandela from prison in February 1990 perhaps best represents the most lasting icon not only of the decade, but also of decolonization. The first democratic elections in the country in 1994, which elected Mandela into power as the first black South African president, essentially marked the end of the decolonization process for Africa, even though many hotspots continue to exist in the region.

During the period 1992–2002, the OAU also recognized the importance of cooperation in

**Figure 1.2 Some international conventions and the number of countries participating in each since the 1970 Lagos Plan of Action**

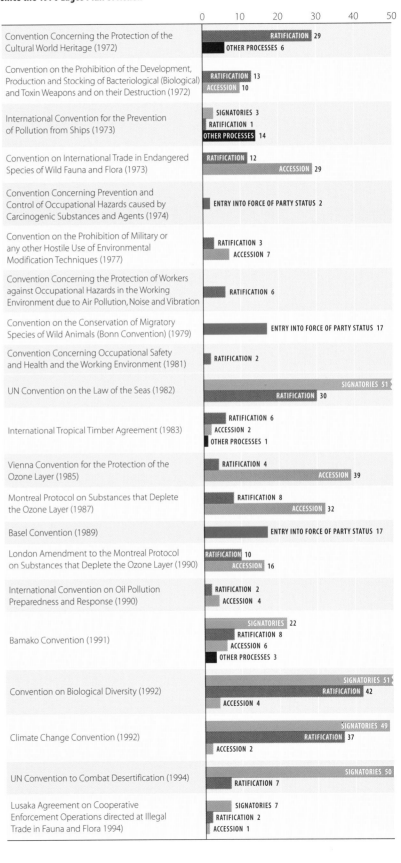

environmental management. It established broad-based agreements, such as Articles 56–59 of the Treaty Establishing the African Economic Community, which relate to: natural resources; energy; the environment; and the control of hazardous wastes (UNEP 1999). The OAU and many governments in Africa have adopted instruments or national constitutions which recognize the environment as a fundamental right. This is, perhaps, a direct achievement of the 1972 Stockholm Conference, which articulated, in Principle 1, the right of people to live 'in an environment of a quality that permits a life of dignity and well-being'.

## REGIONAL INSTITUTIONS

A number of regional and sub-regional institutions have been established in Africa during the 1990s, in order to introduce and to strengthen sustainable development programmes. Some of the institutions and initiatives are highlighted in the following paragraphs.

### African Economic Community

Efforts to strengthen regional cooperation in the sustainable use and management of natural resources and the environment have never been more inclusive and holistic than is provided for under the Treaty Establishing the African Economic Community, adopted by OAU member states in Abuja in June 1991. The Treaty aims at ensuring the harmonization and coordination of environmental protection policies among member states. The objectives of the African Economic Community (AEC), which was launched in Harare in 1997, are to promote economic, social and cultural development, and the integration of African economies, in order to increase economic self-reliance and to promote self-sustained development. Specifically, the Abuja Treaty obliges parties to:

- coordinate and harmonize their policies and programmes in the field of energy and natural resources, and to promote new and renewable forms of energy;
- promote a healthy environment; adopt national, sub-regional and regional policies, strategies and programmes; and establish appropriate industries for environmental development and protection;
- take appropriate measures to ban the importation and dumping of hazardous wastes in AEC

territories; and cooperate in the transboundary movement, management and processing of such wastes, where these emanate from a member state;

- cooperate in the development of river and lake basins; in the development and protection of marine and fishery resources; and in plant and animal protection;
- ensure the development within the borders of member states of certain basic industries (for example, forestry and energy) which are conducive to collective self-reliance and to modernization; and
- ensure the proper application of science and technology to a number of sectors, including energy and environmental conservation.

### Intergovernmental Authority on Development

In 1986, six drought-stricken countries—Djibouti, Ethiopia, Kenya, Somalia, Sudan and Uganda—created the Inter-governmental Authority on Drought and Development (IGADD), in order to coordinate development in the Horn of Africa. IGADD was later renamed the Intergovernmental Authority on Development (IGAD). Eritrea became the seventh member of IGAD in September 1993. In April 1996, the IGAD Council of Ministers identified three priority areas of cooperation:

- conflict prevention, management and resolution, and humanitarian affairs;
- infrastructure development in the areas of transport and communications; and
- food security and environmental protection.

### Common Market for East and Southern Africa

In December 1994, the Common Market for East and Southern Africa (COMESA) succeeded the Preferential Trade Area (PTA), which had been established in 1981. COMESA's main focus has been on the formation of a large economic and trading unit which is capable of overcoming some of the barriers that are faced by individual states. COMESA's strategy is 'economic prosperity through regional integration'.

### African Union and the New African Initiative

Some 38 years after the establishment of the OAU, African heads of state meeting in Sirte, Libya in March 2001 declared the establishment of the African Union

(AU). The main thrust of the AU is to build capacities, in order to enhance the economic, political and social integration and development of the African people.

Crowning the birth of the AU is the New African Initiative, unanimously adopted by the Lusaka Summit on 11 July 2001. The New African Initiative represents a merger between the Millennium Partnership for the African Recovery Programme (MAP) and the OMEGA Plan. It is a pledge by African leaders based on a common vision to eradicate poverty. The New African Initiative is also a conviction by African leaders to place their countries, both individually and collectively, on a path of sustainable growth and development, and to participate actively in the world economy and body politic. It is a call for a new relationship—one of partnership—between Africa and the international community, in order to overcome the development chasm.

## CONFLICT PREVENTION AND PEACE-BUILDING

Despite making some noticeable efforts towards progress in the region, setbacks have also been encountered during the past decade. Wars in countries such as Angola, Liberia, Sierra Leone, Ethiopia-Eritrea and the Democratic Republic of Congo have not only led to a resurgence of the serious problems of refugees, but also to the plunder of natural resources. The United Nations (UN) has produced a number of reports on this issue, and some countries have been sanctioned over the trade in illegal diamonds and other minerals and natural resources.

African countries have taken bold steps in solving conflicts themselves at a regional level. For instance, the sub-regional defence bloc for West Africa, the Economic Community of West African States (ECOWAS) Monitoring Group (ECOMOG), has been instrumental in bringing peace to Liberia and Sierra Leone. Elsewhere in the region, there are similar bodies that are entrusted with sub-regional security, such as the Southern African Development Community (SADC) Organ on Politics, Defence and Security.

## SOCIO-ECONOMIC ISSUES

The 1990s were also characterized by state involvement in the shift towards a market economy in Africa. This is the decade when many African countries

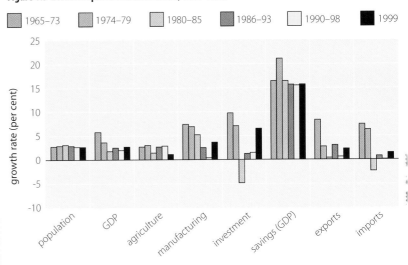

**Figure 1.3 Economic performance in Africa, 1965–2000**

liberalized their economies. Privatization of state-owned infrastructure was introduced in many countries. There were mixed results but, overall, job losses were evident in those countries. Figure 1.3 illustrates Africa's economic performance between 1965 and 2000.

### Structural Adjustment Programmes (SAPs)

During the past decade, many African countries continued with economic reform through SAPs. While economic liberalization may have triggered economic recovery, there have been indications that economic growth will worsen, as opposed to improving, environmental conditions (UNDP/UNEP/World Bank/ WRI 1996). The general trend between 1995 and 1998 shows a declining economic situation, with GNP per capita falling. Indications, however, reveal that, in some countries of the region, the economy may have started to pick up again. According to the World Bank, only nine out of 48 countries have annual per capita income of more than US$1 000, and only five countries in the region—Botswana, Gabon, Mauritius, Seychelles and South Africa—have annual per capita levels of more than US$2 500 (Kappel 2001).

### Value of Africa's natural resources

One of the challenges facing Africa is the failure of economic markets to capture and promote the real value of Africa's natural resources. Many African countries are in the same position today as they were at independence, that is, dependent on capital from natural resources for economic development and growth. Commodity prices on exports from developing countries are determined by the World Trade Organization (WTO) through a quota system. This puts African countries at a great disadvantage. In most countries, national policies and market activities fail to reflect the full economic value and potential of their natural resources. This has led to the degradation and overexploitation of natural resources, as industry and commerce have generally focused on maximizing profits at the expense of sound environmental management and protection. The greatest problem lies in the imbalance in the use of the natural resources, which results from a combination of factors, such as: lack of investment capital; inappropriate technologies; and poor management.

### Debt

External debt continues to be a major impediment to the achievement of accelerated economic development and sustainable environmental management in the region. Africa's external debt has been growing since the 1980s (Expanded Joint Secretariat 2001). The total debt stock stood at US$313 000 million in 1994, equivalent to 234 per cent of income from exports, and 83 per cent of gross domestic product (GDP) (UNEP 1999). Between 1980 and 1995, 22 African countries renegotiated their commercial bank debts 58 times. During the same period, 35 African countries renegotiated their external debt with their creditors in the Paris Club a total of 151 times (Expanded Joint Secretariat 2001). Some 33 of the 41 most heavily indebted countries in the world (in relative terms) are in Africa. The debt issue is covered in more detail in Chapter 3 of this report.

In many countries of the region, natural resource utilization is driven by the demand on governments to earn foreign exchange from exports of primary commodities. Trade liberalization and the pressures to service foreign debts may exacerbate environmental degradation, if appropriate regulatory policies and laws are not instituted. In the face of declining export earnings and debt burdens, many governments have tried to boost the exploitation of natural resources and cash crop production. This has led to widespread environmental damage, as rural communities are forced to cultivate fragile and marginal areas.

### Globalization

Globalization has resulted in the removal of barriers to trade, capital mobility and technological advances,

mostly in the developed world. While globalization may drive future economic prosperity, including poverty reduction, it has, however, advanced the interests of developed countries to the detriment of developing countries, particularly in the areas of trade, finance and technology.

The Millennium Report of the United Nations Secretary-General, published in September 2000, states that although globalization is transforming the world today, there are dangers associated with it, including crime, narcotics, terrorism, diseases and weapons. The benefits and the opportunities of globalization are concentrated in a small number of countries. An imbalance has arisen between the creation and enforcement of rules which have facilitated the expansion of global markets, while environmental activities or social programmes—such as labour standards, human rights or poverty reduction—have received no support (Expanded Joint Secretariat 2001).

### Health and HIV/AIDS

The United Nations Development Programme (UNDP) has established that most African countries in sub-Saharan Africa experienced declines in per capita incomes during the past decade as a result of HIV/AIDS. In many African countries, HIV/AIDS has already had a devastating impact on many development sectors, such as agriculture, health and education. Of the world's 36 million people living with HIV/AIDS, more than 23 million, or 64 per cent of them, are in sub-Saharan Africa (UN 2000). It has been estimated that, by 2010, there could be 40 million orphans in sub-Saharan Africa as a result of HIV/AIDS.

### Poverty

Poverty encompasses a range of deprivations, including: lack of access to natural resources, health care and education; inability to access the political process; vulnerability to catastrophes; and the denial of opportunities and choices that are basic to human development. An estimated 40 per cent of people in sub-Saharan Africa live below the poverty line, and both income and human poverty are increasing (UNDP 1997). Poverty is a major factor in accelerating environmental degradation in the region. This is because the majority of the poor are heavily dependent on land and its resources for livelihood. The poor are

## Table 1.5  HDI ranking of African countries in 2000

| Sub-regional grouping | African countries by HDI levels | | |
|---|---|---|---|
| | *Low HDI* | *Medium HDI* | *High HDI* |
| Northern Africa | Sudan (143) | Libya (72)<br>Tunisia (101)<br>Algeria (107)<br>Egypt (119)<br>Morocco (124) | none |
| Western Africa | Togo (145)<br>Mauritania (147)<br>Nigeria (151)<br>Côte d'Ivoire (154)<br>Senegal (155)<br>Benin (157)<br>Gambia (161)<br>Guinea (162)<br>Mali (165)<br>Guinea Bissau (169)<br>Burkina Faso (172)<br>Niger (173)<br>Sierra Leone (174) | Cape Verde (105)<br>Ghana (129) | none |
| Central Africa | Democratic Republic of<br>Congo (152)<br>Central African Republic<br>(166)<br>Chad (167) | Gabon (123)<br>Equatorial Guinea (131)<br>Sao Tome & Principe (132)<br>Cameroon (134)<br>Congo (139) | none |
| Southern Africa | Zambia (153)<br>Tanzania (156)<br>Angola (160)<br>Malawi (163)<br>Mozambique (168) | South Africa (103)<br>Swaziland (112)<br>Namibia (115)<br>Botswana (122)<br>Lesotho (127)<br>Zimbabwe (130) | none |
| Eastern Africa | Djibouti (149)<br>Uganda (158)<br>Eritrea (159)<br>Rwanda (164)<br>Burundi (170)<br>Ethiopia (171) | Kenya (138) | none |
| Western Indian Ocean States | Madagascar (140) | Seychelles (53)<br>Mauritius (71)<br>Comoros (137) | none |

Source: UNDP 2000 (Figures in brackets refer to world HDI rankings, from the highest ranking of 1, for Canada, to the lowest of 174, for Sierra Leone)

forced to overexploit resources, such as fisheries, forests and water, in a desperate struggle to survive. Environmental degradation contributes markedly to many health threats, including: air and water pollution; poor sanitation; and diseases, such as malaria.

The Human Development Index (HDI) is a way of measuring quality of life, defined by the UNDP. It is clear from Table 1.5 that, in 2000, there were no African countries in the high (HDI) group. A number of countries were in the medium HDI group, while the majority were ranked in the low HDI group.

African governments now generally acknowledge that the overriding objective of development hinges on poverty reduction in the short term, and on its complete eradication in the long term. In reducing poverty, there is a need for strong political commitment combined with specific policy instruments targeting the poorest segments of society. To achieve average growth rates of 7 per cent per year—which estimates suggest would resuscitate Africa's economic performance and put the region on a path of sustainable development—an additional investment of 33 per cent of GDP is required.

## ENVIRONMENTAL AGENDA

### United Nations Conference on Environment and Development

The major environmental highlight of the past decade was the 1992 UNCED, or Earth Summit, held in Rio de Janeiro, Brazil. Africa played a major role both during the process leading up to UNCED and at the conference itself.

---

**Box 1.4 A turning point in global environment and development**

'It is indeed an historic conference. Possibly, future generations will call it a turning point, a moment in history when a major correction was introduced in the process of the industrial revolution which started, less than 200 years ago, to transform so profoundly conditions on our planet … While the environment is an emerging new and very serious problem, we must not forget that development is still the highest priority and an unreached objective.'

UN Secretary-General, Butros Butros Ghali, opening the 1992 Earth Summit

---

**Box 1.5 Abuja Treaty—Article 58 (Environment)**

'Member states undertake to promote a healthy environment. To this end, they shall adopt national, regional and continental policies, strategies and programmes and establish appropriate institutions for the protection and enhancement of the environment. For purposes of paragraph 1 of this article, member states shall take the necessary measures to accelerate the reform and innovation process leading to ecologically rational, economically sound and socially acceptable development policies and programmes.'

Source: OAU 1991

---

The region, through the OAU, presented the African Common Position on Environment and Development, which highlighted the region's environment and development priorities. The environmental challenges facing Africa and the rest of the world were articulated by the then UN Secretary-General, Butros Butros Ghali, at the opening of UNCED (see Box 1.4).

Perhaps the most defining decision of the 1992 Earth Summit was the granting of equal footing, in the Rio Declaration, to both the environment and to development. This was a significant departure from the 1972 Stockholm Conference, which gave prominence to the environment, despite its groundbreaking decisions on political, social and economic issues. While the Stockholm Conference defined an environmental right, the Earth Summit not only reaffirmed this right, but also balanced it with 'the right to development', which it said must be fulfilled 'to equitably meet developmental and environmental needs of present and future generations'. This reaffirmation echoes the 1991 Abuja Treaty Establishing the African Economic Community (see Box 1.5), which sets out Africa's obligations towards natural resources and development.

At the Earth Summit, the eradication of poverty was identified as an indispensable requirement for sustainable development, in order to decrease the disparities in standards of living and to better meet the needs of most people in the world. The Earth Summit also called for:
- elimination of unsustainable patterns of production and consumption;
- enhancement of the development, adaptation, diffusion and transfer of technologies, including new and innovative technologies;

- recognition that environmental issues are best handled with the participation of all concerned citizens at the relevant level;
- enactment of effective environmental legislation;
- implementation of environmental impact assessments (EIAs) before projects are undertaken;
- recognition of the vital role women play in environmental management and development, and the need to ensure their full participation; and
- recognition of the vital role of indigenous people and their communities, and of other local communities, in environmental management.

Major developments which took place in the 1990s, and which have influenced policies in Africa, are listed in Table 1.6.

## Agenda 21 and Multilateral Environmental Agreements

The recommendations set out in the blueprint for environment and development—Agenda 21—adopted at the Rio Earth Summit included:

- the integration of environment and development in policies, plans and management;
- the provision of an effective legal and regulatory framework;
- making effective use of economic instruments and other incentives; and
- the establishment of systems for integrated environment and economic accounting.

The Multilateral Environmental Agreements (MEAs) of the 1990s, to which African countries are party, include the following:

- the 1992 UN Framework Convention on Climate Change (UNFCCC);
- the 1992 Convention on Biological Diversity (CBD);
- the 1994 UN Convention to Combat Desertification in Countries Experiencing Serious Drought and/or Desertification, Particularly in Africa (UNCCD); and
- the 1997 Kyoto Protocol (not yet in force).

## Sub-regional policy responses and actions

Many of Africa's policy responses to the environmental issues and challenges of the 1972 Stockholm Conference and the 1992 Earth Summit are found in the various sub-regional frameworks and agreements that have been developed since 1972. These responses are based on sub-regional political and economic groupings and priorities.

### Central Africa

The Central Africa sub-region has a number of economic units, namely: the Economic and Monetary Community of Central Africa (CEMAC); the Economic Community of Central African States (ECCAS); the Lake Chad Basin Commission (LCBC); and the African Timber Organization (ATO). The primary aim of these organizations is to promote economic cooperation and sound environmental management in the sub-region. In the past three decades, Central Africa has also seen the emergence of intergovernmental organizations responsible for the development and management of shared rivers in the sub-region. However, the performance of these institutions remains far below their potential. Part of the underperformance of the sub-region's institutions relate to governance. Furthermore, much of Central Africa is, or has been, involved in civil war at one time or another, and this has affected the progress of policy responses. Poverty and lack of cooperation have also combined to affect the progress and success of sub-regional efforts.

### Eastern Africa

In Eastern Africa, policy responses have largely been based on ecosystems, rather than political and economic groupings. A number of sub-regional initiatives have been developed and implemented, including: the Eastern Africa Biodiversity Support Programme; the Nile Basin Initiative (NBI); the Eastern Africa Wetlands Programme; Integrated Coastal Zone Management (ICZM); and the Lake Victoria Global Environment Facility (GEF) Project. The Eastern African Convention on the Protection of Coastal and Marine Environment has been very important in bringing together the coastal countries of the sub-region to discuss and address issues of common interest. The Nile Basin Initiative is one of the most successful regional initiatives. It has 21 projects, whose main focus is: integrated water resources planning and management; capacity building; training; harmonization of legislation; and environmental protection. Famine and civil strife have slowed down the progress of Eastern Africa's environmental policy responses. Despite the presence of the East Africa Economic

## Table 1.6  Major events which shaped policies in Africa in the 1990s

| Year | Developments |
|---|---|
| 1990 | • After 27 years of political detention by apartheid in South Africa, Nelson Mandela is finally released from prison, and preparations for a new political dispensation begin<br>• The Montreal Protocol on Substances that deplete the Ozone Layer is amended in London, United Kingdom<br>• The International Convention on Oil Pollution Preparedness, Response and Cooperation is signed |
| 1991 | • Leaders of the Organization of African Unity (OAU) sign a Treaty establishing the African Economic Community (AEC)<br>• The Protocol to the Antarctic Treaty on the environment, reaffirming the status of the Antarctic as a special conservation area, is adopted in Madrid, Spain<br>• The Bamako Convention on the Ban of the Import into Africa and the Control of the Transboundary Movement and Management of Hazardous Waste within Africa is adopted |
| 1992 | • The Conference on the Environment is held in Dublin, Ireland; many African countries are present, and demand that water should be recognized both as a social and an economic good<br>• The United Nations Conference on the Environment and Development (UNCED), also known as the Earth Summit, is held in Rio de Janeiro, Brazil<br>• Agenda 21 is adopted by the international community<br>• The Montreal Protocol on Substances that Deplete the Ozone Layeras amended in London, enters into force<br>• The United Nations Framework Convention on Climate Change is adopted in New York, USA<br>• The Montreal Protocol on Substances that Deplete the Ozone Layer further amended in Copenhagen, Denmark<br>• The Basel Convention on the Control of Transboundary Movement of Hazardous Wastes and their Disposal enters into force<br>• The Convention on Biological Diversity is adopted |
| 1993 | • The Convention on Biological Diversity enters into force, in December<br>• The Convention on the Prohibition of the Development, Production, Stockpiling and Use of Chemical Weapons and on their Destruction |
| 1994 | • The United Nations Convention to Combat Desertification in Those Countries Experiencing Serious Drought and/or Desertification, Particularly in Africa<br>• The Convention on Nuclear Safety<br>• South Africa abolishes its apartheid laws, general elections are held and Nelson Mandela becomes the first black president of a multiracial society<br>• The Montreal Protocol on Substances that Deplete the Ozone Layer as amended in Copenhagen, enters into force<br>• The Lusaka Agreement on Cooperative Enforcement Operations Directed at Illegal Trade in Wild Fauna and Flora |
| 1995 | • The SADC Protocol on Shared Watercourse Systems is adopted and signed by member states<br>• The Protocol Concerning Protected Areas and Wild Fauna and Flora in the Eastern African Region enters into force |
| 1996 | • The Convention for the Protection, Management and Development of the Marine and Coastal Environment of the Eastern African Region enters into force, 11 years after its adoption |
| 1997 | • The United Nations Convention on the Law of the Non-Navigational Uses of International Watercourses<br>• Parties to CITES gather in Harare, Zimbabwe, where they agree to relax the ban in international trade in ivory and other elephant products<br>• The Rio +5 Summit is held in New York to review progress made since the Rio Earth Summit in 1992 |
| 1998 | • The Bamako Convention comes into effect<br>• The Protocol on Energy in the Southern African Development Community (SADC) region enters into force in April |

Sources: SADC/IUCN/SARDC (1998) and UNEP/Sida (undated)

Community (EAEC), regional integration and cooperation in Eastern Africa has been weak. The frameworks for cooperation are generally limited in the sub-region.

### Northern Africa

In Agenda 21, UNCED adopted the river basin as the unit of analysis for integrated water resources management (IWRM). Since then, Northern African has started to create an enabling environment for IWRM, including formulation of the legal framework governing the development and preservation of freshwater resources and of the institutional framework for conducting this approach.

At the multinational level, the NBI is an excellent example of cooperation within the framework of IWRM between and among the ten Nile riparian states. The Joint Authority for the Nubian Sandstone Aquifer is another example of cooperation, between Chad, Egypt, Libya and Sudan (CEDARE 2000).

### Southern Africa

The SADC formulated a Regional Policy and Strategy for Food, Agriculture and Natural Resources in the 1990s. The main aim of this regional policy and strategy is to ensure the efficient and sustainable use of natural resources, and their effective management and conservation. It incorporates environmental considerations in all policies and programmes, and integrates the sustainable use of natural resources with development needs. Other environmental policies that have been put in place in the sub-region include: the SADC Wildlife Policy; the SADC Wildlife Protocol; the Forestry Sector Policy and Development Strategy; the Protocol on Shared Watercourse Systems; the Southern African Power Pool; and the Southern Africa Trade Protocol. Running parallel to these initiatives are other important sub-regional responses, including: the SADC Environmental Information Systems Programme; the SADC Wetlands Conservation Programme; the SADC Environmental Education Programme; and the Southern Africa Biodiversity Support Programme. Although regional frameworks for coordination and cooperation exist on paper, their implementation on the ground has been weak, as a result of lack of funding and institutional problems.

### Western Africa

ECOWAS brings together Western African countries as a single grouping and sub-region. Most of the policy responses to environmental issues in Western Africa have been at the national level. However, a few initiatives have been taken, based on river basins and ecosystems. These include: the creation of river basin authorities in the Senegal, Gambia and Niger basins; and the West Africa Wetlands Programme, supported by the IUCN and Wetlands International. One of the oldest African intergovernmental organizations, the Niger River Basin Authority, is in Western Africa. The main aim of this authority, and of numerous others in the sub-region, is to promote cooperation among member countries and to ensure integrated development in the river basins.

### Western Indian Ocean Islands

Policy responses are few in the Western Indian Ocean Islands sub-region, with the exception of an environmental education programme. An ICZM programme, a five-year project supported by the European Union, is being implemented. The goal of the programme is to enable the sustainable development of coastal zones. In addition to this programme, other sub-regional projects, funded by GEF, include: the Western Indian Ocean Marine Biodiversity Conservation programme; the transboundary Diagnostic Analysis and Strategic Action Plan for Marine and Coastal Environments; and the Western Indian Ocean Oil Spill Contingency Planning Project. Due to the isolated nature of the countries in this sub-region, environmental responses at the sub-regional level are few. Apart from the common problem of the seas, there are few common agendas. Funding is also a problem within the sub-region.

### National responses and actions

Africa's policy responses to environmental concerns are most pronounced at the national level. The policy responses vary from one country to another, depending on the priority environmental issues. In most countries, the responses take the form of: policy frameworks; resource use planning regulations; public awareness programmes; and the promotion of private sector involvement in natural resources management issues.

National Environment Action Plans (NEAPs), National Conservation Strategies (NCSs), National Plans of Action to Combat Desertification (NPACDs), National Tropical Forestry Action Plans (NTFPAs), Country Environmental Strategy Papers (CESPs), National Energy Assessments and Country Programmes for the Phasing out of Ozone-Depleting Substances under the Montreal Protocol are playing significant roles in integrating environment and development in many African countries. At present, about 80 per cent of the countries in sub-Saharan Africa are involved in the NEAP process, and other countries are preparing or implementing similar kinds of environmental strategies (World Bank 1995).

Success stories continue to be seen in Africa, especially with regard to moving towards strengthening the institutional frameworks that deal with the environment. High-level coordination for environmental management has been created in some countries, such as: Benin, Botswana, Ethiopia, Ghana, Lesotho, Madagascar, Malawi, Mozambique, Namibia, Nigeria, South Africa, the Gambia, Uganda, Zambia and Zimbabwe. EIAs have also been implemented in many countries, and EIA guidelines and procedures are being prepared as a follow-up to appropriate legislation.

## 2002 WORLD SUMMIT ON SUSTAINABLE DEVELOPMENT

In August–September 2002, Africa will host the World Summit on Sustainable Development (WSSD) in Johannesburg, South Africa—a major milestone for a region which faces many environmental challenges. The main objective of the WSSD is to review progress made on sustainable development since the 1992 Rio Earth Summit. The main focus points for Africa are:

- Eradicating poverty in African countries, through the formulation and implementation of policies that are conducive to the enhancement of domestic savings, as well as allowing external resource inflows, such as foreign direct investment.
- Promoting education, which will require African countries to establish new institutions, or to strengthen existing institutions, in order to enhance their ability to respond to new and longer-term challenges, instead of concentrating on immediate problems.
- Providing new, and improving on existing, healthcare institutions, in order to reduce the

incidence of disease in Africa. This will require a broad development approach, with health sector reforms going hand-in-hand with poverty reduction, conflict prevention and community participation. Among the most urgent areas for action in African countries is the HIV/AIDS pandemic, which is a serious impediment to sustainable development and growth.

- Taking new and strong measures in the management and use of biodiversity, including forest and marine ecosystems and resources. Strategies to implement these programmes could include:
  - developing national forest programmes, in accordance with each country's national conditions, objectives and priorities;
  - strengthening political commitment in the management, conservation and sustainable development of forests;
  - undertaking concrete actions to share equitably the benefits arising from the use of genetic resources;
  - recognizing the role of women in the conservation of biological diversity and the sustainable use of biological resources;
  - providing necessary support to integrate the conservation of biological diversity and the sustainable use of biological resources into national development plans; and
  - promoting the involvement of communities, the private sector, NGOs and other stakeholders in the management of forests, with a view to ensuring the equitable sharing of the benefits accruing from forest resources.
- Ratifying the 1997 UNFCCC Kyoto Protocol to control greenhouse gas emissions, which are believed to be a major factor in climate change, which threatens African countries.
- Switching to higher value-added resource industries, where African countries will have a comparative advantage, and promoting the diversification of industrial production.
- Taking all necessary measures to guarantee a pollution-free environment, especially with respect to toxic wastes, including:
  - ratification and implementation of all relevant conventions;

- development of skilled personnel and testing equipment required for the effective detection and monitoring of the movements of hazardous wastes; and
- strengthening the respective institutions and enacting legislation, in order to facilitate the smooth implementation of the conventions.
- Reviewing national development planning options.
- Promoting communication, and removing the digital gap that currently exists between Africa and the rich nations.
- Promoting trade with targeted strategies, including:
  - developing higher value-added resource-based industries;
  - broadening the production base;
  - allowing the establishment of, and the strengthening of, regional trade;
  - forging ahead with regional integration, in order to increase Africa's share of global trade; and
  - integrating environmental and resource management policies which take into account the effects on sustainable development of trade liberalization programmes.
- Promoting the role of civil society.
- Establishing a single centralized political council which meets regularly, and which is responsible for environmental policy and governance.
- Promoting peace, democracy and human rights, and moving away from the position in which Africa is characterized by conflicts, political strife and human suffering (Expanded Joint Secretariat 2001).

## CONCLUSION

This chapter has highlighted some of the major policy issues which have influenced development in Africa, particularly over the past three decades, and which have contributed to the region's responses to growing environmental challenges. Many of these issues are discussed in greater detail in the following chapters.

- Chapter 2, 'The State of Africa's Environment and Policy Analysis', provides more in-depth analysis of the environmental issues facing Africa, and looks at how countries in the region have tried to address them.

- Chapter 3, 'Human Vulnerability to Environmental Change', explains how African people are particularly vulnerable to changes in the environment.
- Chapter 4, 'Outlook 2002–2032', uses four scenarios to explore possible alternative futures in the region, depending on the policy decisions taken to address particular problems.
- Chapter 5, 'Policy Responses, Analysis and Action', presents some of the policy responses needed to resolve some of the environmental and developmental challenges facing the region.

## REFERENCES

CEDARE (2000). *GEO-3 Draft Report*. CEDARE, Cairo

Expanded Joint Secretariat (2001). *Assessment of Progress on Sustainable Development in Africa since Rio (1992)*. African Preparatory Conference for the World Summit on Sustainable Development, 15–18 October 2001, Nairobi

Field-Juma, A. (1996). 'Governance, Private Property and Environment.' In Juma, C. and Ojwang, J. B. (eds.) (1996). *In Land We Trust: Environment, Private Property and Constitutional Change*. Initiatives Publishers/SED Books, Nairobi/London

Kappel, Robert (2001). The End of the Great Illusion: Most African Countries Face Uncertain Future. In *Development & Co-operation*, No. 2/2001, March/April 2001, DSE

OAU (1980). *Lagos Plan of Action*. OAU, Addis Ababa

OAU (1991). Treaty Establishing the African Economic Community. OAU. Abuja

OAU (1995) *Re-launching Africa's Economic and Social Development: The Cairo Agenda for Action*. http://www.oau-oua.org/document/Treaties 27 June 2001

OAU (2001), *The African Economic Community, Addis Ababa, Ethiopia*, http://www.oau-oua.org/document/documents/AEC.htm 27 June 2001

SADC/IUCN/SARDC (1998). *Reporting the Southern African Environment—A Media Handbook*. SADC/IUCN/SARDC, Harare

SADC/IUCN/SARDC (2000). *Biodiversity of Indigenous Forests and Woodlands in Southern Africa*. SADC/IUCN/SARDC, Harare/Maseru, Zimbabwe/Lesotho

UN (2000). *We the Peoples—The Role of the United Nations in the 21st Century*. United Nations, New York

UNDP (1997). *Human Development Report 1997*. Oxford University Press, Oxford

UNDP/UNEP/World Bank/WRI (1996). *World Resources 1996–97: The Urban Environment*. Oxford University Press. New York/Oxford

UNDP/UNEP/World Bank/WRI (2000). *World Resources 2000*. Oxford University Press, Oxford

UNDP/UNEP/World Bank/WRI (2000). *World Resources 2000–2001: People and Ecosystems – The Fraying Web of Life*. World Resources Institute. Washington D.C.

UNEP (1999). *Global Environment Outlook 2000*. Earthscan Publications Limited, London

UNEP/OAU (1991). *Regaining the Lost Decade: A Guide to Sustainable Development in Africa*. UNEP/OAU, Nairobi

UNEP/Sida (undated). *Multilateral Environmental Agreements: Relevance, Implications and Benefits to African States*. UNEP/Sida, Nairobi

UNHCR (2000). *The State of the World Refugees: Fifty Years of Humanitarian Action*. UNHCR, Geneva

World Bank (1995). *Towards Environmental Sustainable Development in Sub-Saharan Africa: A World Bank Agenda*. World Bank, Washington, D.C.

# CHAPTER 2

## THE STATE OF AFRICA'S ENVIRONMENT AND POLICY ANALYSIS

# CHAPTER 2

## THE STATE OF AFRICA'S ENVIRONMENT AND POLICY ANALYSIS

### INTRODUCTION

The material presented here constitutes the first comprehensive, integrated report on the state of the environment in Africa. Following on from the review and assessment of development policies and progress described in Chapter 1, attention is now turned to the environmental context that underlies policy and forms the background for progress.

Improved understanding of the causes, patterns and consequences of environmental change can contribute to more effective design and implementation of mechanisms to tackle the negative impacts of such change. This report helps to improve understanding by presenting detailed retrospective analyses of Africa's environment, and by describing discernible environmental trends against the backdrop of human activities and management practices of the past 30 years. It thus provides a basis for learning from past experience and lays the groundwork for more effective implementation of Agenda 21 and for sustainable development of Africa's environmental, social and economic resources.

The framework used to assess the state of Africa's environment is called a 'Pressure-State-Impact-Response' (PSIR) framework:

● 'Pressures' are the root causes of environmental change (natural or resulting from human activities).
● 'State' reflects the current situation (and qualitative or quantitative trends over the past 30 years).
● 'Impacts' are the consequences of environmental change on human and ecological systems, and on social and economic development potential.
● 'Responses' include regional agreements and

strategies for cooperation, national policies, awareness and education programmes, and community-level projects, aimed at addressing both the causes and impacts of environmental change.

The PSIR framework allows analysis of policies and activities relating to specific environmental issues; reveals positive and negative impacts of economic and development policies on the environment; and shows how consideration of the environment can drive policy. Examples and case studies are used to highlight particular issues of concern and instances of good practice, and to illustrate the links between environmental components and issues. Pressures, states, impacts and responses are discussed in an integrated manner for each issue.

It is not possible to cover all of the environmental issues that have arisen over the past 30 years. Rather, this chapter aims to draw attention to priority issues for Africa, as identified through regional and sub-regional consultations. Hotspots of environmental degradation, and potential and emerging issues, are highlighted where relevant. The 'bottom-up' means of assessment—in which information comes from national or sub-national activities and is synthesized in sub-regional or regional analyses—and the extensive consultation and review processes ensure that a wide range of perspectives and studies are incorporated, and that the review of the state of the environment is as balanced and objective as possible.

The information is presented under seven environmental themes, namely:

● Atmosphere
● Biodiversity
● Coastal and marine environments
● Forests

- Freshwater
- Land
- Urban areas.

Key issues identified for each theme are introduced and regional perspectives presented. Further details are then given for each sub-region. The cross-cutting nature of environmental issues is emphasized at every opportunity, with links between issues, themes, sub-regions and causes or impacts of change being highlighted. After this sub-regional analysis, the chapter ends with a *Concluding Summary* which ties together the different themes at regional level and examines both present and future priorities for action.

The breakdown of Africa into sub-regions, presented in Chapter 1, is along political and economic lines. Its borders do not, therefore, always coincide with those of an ecological breakdown. There is, therefore, some overlap in the sub-regional analyses presented below.

## PART A: ATMOSPHERE

UNEP

## REGIONAL OVERVIEW

Africa is faced with three major issues where the atmosphere is concerned, namely:
- climate variability;
- climate change; and
- air quality.

### CLIMATE VARIABILITY

Climate variability means the seasonal and annual variations in temperature and rainfall patterns within and between regions or countries. For Africa, it is determined by prevailing patterns of sea surface temperature, atmospheric winds, regional climate fluctuations in the India and Atlantic Oceans, and by the El Niño Southern Oscillation (ENSO) phenomenon—the natural shift in ocean currents and winds off the coast of South America which occurs every two to seven years. ENSO events bring above average rainfall to some regions and reduced rainfall to others.

Africa is characterized by considerable climatic variations, both spatial and temporal, and extreme events such as flooding and drought have been recorded for thousands of years (Verschuren, Laird and Cumming 2000). The equatorial belt generally has high rainfall, whereas northern and southern African countries, and those in the Horn of Africa, are typically arid or semi-arid (see Figure 2a.1). All parts of Africa, even those that usually have high rainfall, experience climatic variability and extreme events such as floods or droughts. Most of Eastern, Central and Southern Africa, as well as the Western Indian Ocean Islands, are affected by the ENSO phenomenon. The 1997–98 ENSO triggered very high sea surface temperatures in the south-western Indian Ocean, causing high rainfall,

**Figure 2a.1 Map of rainfall variability in Africa**

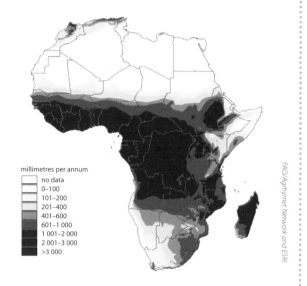

FAO/Agrhymet Network and ESRI

millimetres per annum

no data
0–100
101–200
201–400
401–600
601–1 000
1 001–2 000
2 001–3 000
>3 000

•

*Rainfall records from the early 1900s to mid-1980s show that Africa's average annual rainfall has decreased since 1968, and has been fluctuating around a notably lower mean level, as shown in Figure 2a.2 (UNEP 1985). There is also some evidence that natural disasters have increased in frequency and severity over the past 30 years, particularly drought in the Sahel*

•

cyclones, flooding and landslides across most of Eastern Africa, whereas south-western Africa experienced drier conditions. The higher sea temperatures also caused extensive bleaching of corals on the Eastern African coast and in the Western Indian Ocean Islands (Obura, Suleiman, Motta and Schleyer 2000, PRE/COI 1998).

Rainfall records from the early 1900s to mid-1980s show that Africa's average annual rainfall has decreased

since 1968, and has been fluctuating around a notably lower mean level, as shown in Figure 2a.2 (UNEP 1985). There is also some evidence that natural disasters have increased in frequency and severity over the past 30 years, particularly drought in the Sahel (OFDA 2000). The most prolonged and widespread droughts occurred in 1973 and 1984 (when almost all African countries were affected), and in 1992, although in this case drought was mainly restricted to Southern Africa. The impacts of the 1984 and 1992 droughts were alleviated to some extent by increased preparedness of some countries, even though the droughts themselves were more severe (Gommes and Petrassi 1996). Countries most regularly affected by drought include Botswana, Burkina Faso, Chad, Ethiopia, Kenya, Mauritania and Mozambique (FAO 2001a).

Human activities, such as deforestation and inappropriate management of land and water resources, can contribute to the frequency and impacts of natural climatic events. For example, clearing of tropical forests in Central and Western Africa alters local climate and rainfall patterns and increases the risk of drought. Clearing of vegetation increases run-off and soil erosion, and damming of rivers and draining of wetlands reduces the environment's natural ability to absorb excess water, increasing the impacts of floods.

Africa's people and economies are heavily dependent on rain-fed agriculture (for commercial export and subsistence) and are therefore vulnerable to rainfall fluctuations. It is usually the poor who suffer most from flood- or drought-induced crop failure, as they are forced to cultivate marginally productive land and cannot accumulate reserves for times of hardship. Malnutrition and famine have resulted from both droughts and floods in Africa, and associated food imports and dependency on food aid have contributed to limited economic growth of the countries affected. Additional impacts include loss of infrastructure and disruption of economic activities, outbreaks of disease and sometimes population displacements, both internal and international. Over the past 30 years, millions of Africans have sought refuge from natural disasters, often settling in fragile ecosystems and/or experiencing social tensions with neighbouring communities. Ecological impacts of drought and flooding include: land degradation and desertification; loss of natural habitat or changes in distribution of

**Figure 2a.2 Rainfall fluctuations in Africa 1900–2000**

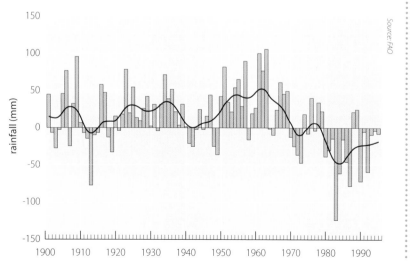

Source: FAO

rainfall (mm)

biodiversity; increased soil erosion; and silting of rivers, dams and coastal ecosystems.

## Strategies for coping with climate variability in Africa

Fifty-two African countries are parties to the United Nations Convention to Combat Desertification in Countries Experiencing Serious Drought and/or Desertification, Particularly in Africa (UNCCD). The Convention, signed in 1992, calls for international cooperation and a partnership approach, and focuses on improving land productivity, rehabilitation of land, conservation, and sustainable management of land and water resources. By 2001, sixteen African countries had produced National Action Plans (NAPs) in accordance with the UNCCD, and action plans had been developed for all sub-regions. Food reserve programmes have also been successfully established to provide additional resources for emergencies.

Other African initiatives to predict and cope with rainfall fluctuations include the establishment of early warning systems (EWS). These comprise climate-monitoring centres which can assess the likelihood of drought or flood and can alert the relevant departments or agencies as to requirements for food imports, requests to cull livestock, or orders to evacuate vulnerable areas. However, despite significant progress in recent decades in predicting seasonal fluctuations in rainfall (by monitoring interactions between the oceans and the atmosphere), much further investigation will be needed before fluctuations can be predicted accurately or their effects on food production or other human systems fully appreciated.

Longer-term responses include crop research to develop more resistant strains of staple crops, improved housing design and construction, and better urban planning to reduce the vulnerability of human populations.

## CLIMATE CHANGE

Climate change is now recognized as a pressing global environmental issue. It is the result of higher mean temperatures caused by increased amounts of greenhouse gases (GHG) in the Earth's atmosphere, of which the most important is carbon dioxide ($CO_2$) released during burning of fossil fuels. Since the start of

Snow cover on Mt Kilimanjaro, Tanzania, April 1993 (top) and 2001 (bottom)

*D. Manzolillo Nightingale*

the industrial revolution (in about 1750) some 270 000 million tonnes of carbon have been released globally from the consumption of fossil fuels and cement production. Half of these emissions have occurred since the mid-1970s, although there was a slight decline of 0.3 per cent between 1997 and 1998 (Marland, Boden and Andres 2000).

It is predicted that, globally, increased temperatures will lead to rising sea levels, accompanied by displacement of people living in low-lying areas, and loss of some island states; shifts and reductions in agricultural production; the possibility of more frequent and more severe climatic events such as droughts and flooding; and possible shifts in health problems with vector-borne diseases being re-introduced or introduced into different areas.

According to the Intergovernmental Panel on Climate Change (IPCC), global average temperatures have risen by 0.6 °C over the past century, and the 1990–99 period was probably the warmest decade since the 1860s (IPCC 2001a). In addition, records indicate that snow and ice cover has decreased, and sea levels have risen by 10–20 cm over the past century.

*It is predicted that, globally, increased temperatures will lead to rising sea levels, accompanied by displacement of people living in low-lying areas, and loss of some island states; shifts and reductions in agricultural production; the possibility of more frequent and more severe climatic events such as droughts and flooding ...*

**Figure 2a.3 Africa's contribution to global carbon dioxide emissions, 1998**

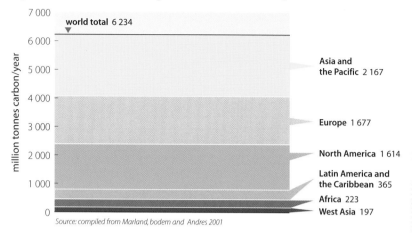

Source: compiled from Marland, bodem and Andres 2001

For example, the glaciers of Mount Kilimanjaro have shrunk by over 70 per cent over the last century (WorldWatch Institute 2000).

Africa's $CO_2$ emissions from the use of fossil fuels are low in relation to other regions, in both absolute and per capita terms, as shown in Figure 2a.3. Despite the region's total emissions having risen to 223 million metric tonnes of carbon in 1998 (eight times the level in 1950), this is still less than the emissions for the United States, mainland China, Russia, Japan, India or Germany. Per capita emissions also increased three-fold over the same period, reaching 0.3 metric tonnes of carbon, only 5.7 per cent of the comparable value for North America.

A handful of African nations account for the bulk of the region's emissions from fossil fuels: South Africa

accounts for 42 per cent, with another 35.5 per cent coming from Egypt, Nigeria and Algeria combined. This is illustrated in Figure 2a.4 (Marland and others 2000).

Although Africa contributes very little to global GHG emissions the region is highly susceptible to the impacts of climate change because of its dependency on agriculture and limited financial resources for development of mitigation strategies. The IPCC predicts that the greater variability and unpredictability of temperature and rainfall cycles in Africa resulting from climate change would alter the area of suitable land for agricultural or livestock production, and increase the frequency of flooding and drought. Grain yields are expected to decline due to increased rainfall variability, especially in the Horn and Southern Africa, and desertification rates may increase (IPCC 2001b). In Central Africa and parts of Eastern Africa, increased rainfall and reduced frosts are expected, resulting in an increase in the area suitable for cultivation, possibly at the expense of natural habitat.

Climate change may also have a devastating impact on human settlements and infrastructure development in Africa. Low-lying coastal areas are at particular risk from sea level rise, and many coastal urban developments are of unsuitable design or are inadequately equipped to cope with storms and flooding. In particular, the Gulf of Guinea, Senegal, Egypt, Gambia, the eastern African coast and the Western Indian Ocean Islands are at risk from sea level rise (IPCC 2001b).

Africa's natural environment could also be severely affected by climate change with impacts including changes to forest cover and grassland distribution that would result from temperature rises of 1 °C or more. These, in turn, may result in significant changes in the abundance and diversity of species. In particular, species living in arid zones will be less able to adapt because they already exist at the very limits of their environmental tolerance (IPCC 1998). Significant extinction of plants and animals is anticipated, impacting on rural livelihoods and tourism (IPCC 2001b). For example, hartebeest, wildebeest and zebra in the Kruger National Park (South Africa), the Okavango Delta (Botswana) and Hwange National Park (Zimbabwe) could be severely threatened by the anticipated 5 per cent drop in rainfall that would affect grazing distribution (WWF 2000). Marine environments

**Figure 2a.4 Sub-regional comparison of carbon dioxide emissions 1972–98**

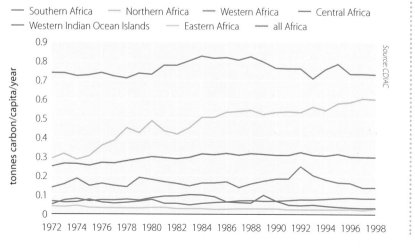

Source: CDIAC

could also be affected severely—a sea temperature rise of 1–2 °C could cause extensive coral bleaching in the Western Indian Ocean, affecting the economies of the coastal countries and islands.

Climate change also poses a threat to human health in Africa, through reduced nutrition and possible expansion or creation of new habitats for disease-carrying organisms such as mosquitoes (IPCC 1998). Warmer temperatures and altered rainfall patterns could open up new areas to diseases such as malaria, yellow fever, dengue fever and trypanosomiasis (IPCC 1998).

## Climate change mitigation and adaptation strategies

African countries are faced with the dual challenge of meeting economic development needs without increasing dependence on fossil fuels or inefficient technologies while simultaneously mitigating the diverse and complex impacts of climate change. All African countries, with the exceptions of Angola, Liberia and Somalia, have ratified the United Nations Framework Convention on Climate Change (UNFCCC) and its proposed mechanism for implementation, the Kyoto Protocol, agreed in Bonn, in 2001, by 180 countries from all over the world.

African countries stand to benefit from the Kyoto Protocol and from the funding streams it proposes, namely the Special Climate Change Fund and the fund for Least Developed Countries. Under the Protocol's mechanisms, developed countries will be able to offset some of their emissions by paying for carbon-saving projects such as tree planting and forest conservation schemes in developing countries. Funds will also be available to help developing countries to convert to cleaner technologies such as solar and wind power or fuel-cell-operated vehicles, currently too expensive for many African nations. Additional funds will be available to assist developing countries in adapting and mitigating the impacts of climate change by means, for example, of flood defence systems and appropriate infrastructure design. Many countries (including Algeria, Botswana, Cape Verde, Côte D'Ivoire, Egypt, Ghana, Lesotho, Mali, Mauritius, Niger, Senegal, Seychelles, South Africa and Zimbabwe) have embarked on National Communications Strategies to provide detailed inventories of emissions and $CO_2$ 'sinks' and programmes to mitigate the impacts of climate change.

In both Northern and Southern Africa, options for further exploitation of alternative sources of energy (for example, solar, wind, micro-hydro, geothermal and biomass) are being explored as additional means to combat climate change.

## AIR QUALITY

Air quality in Africa is an issue that has emerged over the past few decades, particularly in large urban centres. It has been identified as a priority issue for action because rates of urbanization in Africa are the highest in the world, and there are enormous economic pressures for continued industrial growth.

The air in Africa's urban centres is polluted by emissions from industry, households and vehicles. Major air pollutants from these sources include sulphur dioxide, carbon monoxide, carbon dioxide, particulates, lead and organic compounds. In most countries, economic pressures to increase industrial output have contributed to rising levels of pollution, and this trend is likely to continue if current development patterns persist (SEI 1998). Policy measures, such as high taxes on fuel and on importation of new vehicles, have also contributed to emissions, by encouraging the use of dirty fuels and a prevalence of old and more polluting

Air quality is an issue in large urban centres.

*UNEP*

vehicles. Studies in the United States have shown that a car manufactured today produces 70 to 90 per cent less pollution over its lifetime than a car made in the 1970s (CARB 2001a). Conversion to cleaner vehicles is therefore a priority for improvement of urban air quality. Diesel is a commonly used fuel in Africa, especially in commercial trucks and public vehicles, because it is cheaper than petrol. However, the particulate matter emitted from diesel exhausts is small enough to reach deep into the lungs and contains toxic substances such as arsenic, benzene, formaldehyde, nickel and polycyclic aromatic hydrocarbons (PAHs) that have been shown to cause cancer in humans and animals (CARB 2001b). The use of leaded fuel in Africa is also of concern, as very few countries have promoted conversion to unleaded fuel.

In both urban and rural areas, air pollution from domestic combustion of wood, coal, paraffin, crop residues and refuse is a major health issue. Use of these traditional energy sources is driven by lack of investment in rural electrification and high costs of electricity and electrical appliances. These traditional fuels emit toxic pollutants including sulphur dioxide, carbon dioxide, nitrogen oxides, aldehydes, dioxin, PAHs and respirable particulate matter. Exposure to such emissions is associated with acute respiratory infections, chronic obstructive lung diseases (such as asthma and chronic bronchitis), lung cancer and pregnancy-related problems. Women are particularly susceptible because of their traditional roles as cook, which means that they spend more time indoors and close to the pollution sources. Children are also at risk if they spend long periods indoors with their mothers. Other environmental impacts of air pollution include accelerated corrosion rates of buildings, and toxicity to water, soil and vegetation.

### Towards improving air quality in Africa

Recent responses to try to improve air quality in Africa include the establishment of air quality standards and guidelines and monitoring of ambient air quality. In addition to promulgation of laws and standards, the 'polluter pays principle', which advocates fines for companies exceeding certain emission levels, has been adopted, in theory, by many African governments. However, most countries lack the capacity and resources to enforce regulations or to enter into lengthy and costly legal proceedings with offenders.

In 1998, the Air Pollution Impact Network for Africa (APINA) convened a meeting to discuss the prevention and control of transboundary air pollution. APINA is a network of scientists, policy makers and non-governmental organizations (NGOs) established to provide information on air pollution, methodologies and databases, and to bridge the gap between information and policy making. It also provides training courses and workshops with specific groups such as the mining sector. Both APINA and NGOs have been instrumental in promoting the use of more efficient stoves, although with mixed results.

The city of Cairo, where only unleaded fuel has been sold since 1997, is a notable exception to the general use of leaded fuel in Africa. Other urban centres in Egypt are scheduled to follow suit by 2002. Some countries are upgrading public transport systems and have imposed age limits on private and commercial vehicles or are subsidizing the conversion to unleaded fuels. The World Bank's Clean Air Initiative in sub-Saharan African Cities is one such programme, involving phasing out of leaded petrol, and the revision of transport policies. Responses to address traffic congestion have included upgrading roads and incentives for car-share schemes.

Adoption of cleaner technologies to reduce industrial emissions is planned in most parts of Africa, but has, until now, been too expensive to adopt. With funding under the Kyoto Protocol mechanisms, cleaner technologies should become more widely accessible to African countries.

## NORTHERN AFRICA

Northern Africa is one of the most arid areas of the world, with rainfalls that vary widely in terms of both temporal and geographical distribution. Northern African countries are also the most urbanized in Africa and are heavily dependent on fossil fuels for energy. The major issues of concern in the sub-region are, therefore, climate variability, climate change, and air quality in urban centres.

### CLIMATE VARIABILITY IN NORTHERN AFRICA

Northern Africa experiences highly variable rainfall and recurrent droughts. The sub-region receives only 7 per cent of Africa's total precipitation and this is not evenly distributed. Egypt, for example, receives just 18 mm/yr of rain (FAO 1995); Algeria, Morocco and Tunisia experienced 6–7 years of drought between 1980 and 1993; and Morocco has experienced a drought in one year out of every three over the past century (Swearingen and Bencherifa 1996).

Flash floods—short-lived but very rapid rises in the level of water courses or filling of dry beds—are a typical feature of several African countries. Often following a brief downpour, flash floods rapidly erode soils, particularly where natural vegetation cover has been cleared from slopes (Swearingen and Bencherifa 1996). In Egypt, flash floods are often accompanied by mudflows which can be more disastrous than the floods themselves (Nemec 1991). Prior to the building of the Aswan High Dam, Egypt was also subject to frequent floods in the Ethiopian plateau during the rainy season (August through October), and to water shortages in years with below normal rainfall.

Swearingen and Bencherifa (1996) have suggested that the hazard of drought in Northern Africa has increased mainly as a result of expansion of cereal cultivation to drought-prone rangeland and reduction of fallow systems. This process was fostered during the colonial period by large-scale land expropriation and by displacement of peasants to marginal lands. It was also influenced by incentive-raising policies for cereal production, mechanization of agriculture and by increased demand for food associated with rapid population growth.

Drought has major socio-economic significance in Northern Africa because rain-fed cereal cultivation is predominant. In 1997, for example, Algeria's cereal harvest decreased sharply as a result of severe drought. In Morocco, agricultural output recorded losses in 1992, 1995 and 1997. Drought also aggravates the effects of overgrazing, increasing degradation of natural vegetation and soils. The Nile River and Delta were much affected by the drought in the 1980s which resulted in loss of output from agriculture and fisheries and a drop in water level in Lake Nasser which exacerbated the country's existing irrigation problems (Abdel-Rahman, Gad and Younes 1994). In the Sudano-Sahelian area, hundreds of thousands of people were affected by famine, infectious diseases and displacement (OFDA 1987).

**Figure 2a.5  Predicted inundation of the Nile Delta**

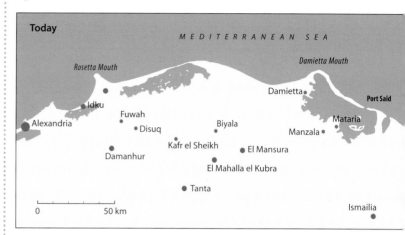

## Strategies for coping with climate variability in Northern Africa

All countries of the Northern Africa sub-region have ratified the UNCCD. Tunisia has produced a National Action Programme, and the Union du Maghreb Arabe has produced a sub-regional Action Plan for the Arab countries of Northern Africa (UNCCD 2001). Additional government policies to avoid or mitigate drought include changing agricultural practices and technology, and rearranging land-use patterns through reduction of cultivation in marginal areas and reduction of fallow periods.

Long-term solutions include an initiative by the Moroccan government to finance job-generating projects in rural areas to keep farmers from abandoning their lands. Officials are also working on expanding irrigation systems to cultivated areas and the use of supplementary irrigation in rain-fed areas (PEACENET 2000). However, policies are needed to improve the ability of societies to understand and detect drought, as well as to improve their ability to cope with its effects.

Improved monitoring and forecasting systems, such as modern remote sensing technology, are also required. For example, the impacts of the 1980s drought could have been alleviated if more effective detection and mitigation methods had been available (White 1992).

Improved integrated water resource management can assist in mitigating the intensity and impacts of flooding. Maintenance of wetlands, vegetation cover (especially riparian vegetation) and flood plains improves the ability to soak up excess water and thus reduce the intensity of flooding. Mitigating measures of the research and planning type, such as mapping of hydrological, hydrogeological and land-use patterns, could help to avoid urbanization in areas where there is a risk of flash floods.

## CLIMATE CHANGE IN NORTHERN AFRICA

Northern African countries account for a large proportion of all of Africa's GHG emissions, although emissions are lower than those of most European or North American countries. In 1996, Northern African countries emitted a total of 280 million metric tonnes of $CO_2$, 37 per cent of Africa's total (African Development Bank 2001). The highest total emissions were from Algeria and Egypt, although Libya had the highest per capita emissions of any country in Africa (8 tonnes). Northern Africa is also responsible for about 20 per cent of Africa's anthropogenic methane emissions (African Development Bank 2001).

Predicted impacts of climate change for Northern Africa include: decreased run-off water; increased desertification; and increased frequency of flooding and drought (IPCC 1998). In the drylands, predominant throughout the sub-region, population growth will push people into marginal lands which are highly vulnerable to desertification.

Management of changes in water resources resulting from climate change will be very difficult, because the sub-region is already suffering from problems of water availability and distribution. However, the most dramatic impacts will be seen in the coastal zone as a result of sea level rise. For example, studies have shown that a sizeable portion of the Nile

Delta will be lost through inundation and erosion, with consequent loss of agricultural production, damage to infrastructure and displacement of people. A one-metre rise in sea level would inundate large areas of land in the Nile Delta area, and the Egyptian city of Alexandria would also be severely affected (see 'Coastal and marine environment').

### Climate change mitigation and adaptation strategies in Northern Africa

All countries in the Northern Africa sub-region are parties to the UNFCCC and, by October 2001, Egypt had signed the Kyoto Protocol. Northern African countries are investigating the application of stricter efficiency measures, improved quality of fuels, and advances in the use of higher efficiency engines and industrial plants to reduce emissions, including GHGs (Unified Arab Economic Report 1999). Improved energy efficiency, brought about through energy-pricing reforms and technological adaptations, could also make a substantial contribution to curbing $CO_2$ emissions. Alternative sources of energy—such as solar energy which is abundant in the region—are being explored, especially in the rural areas that are home to about 50 per cent of the region's population. With international assistance, solar power could increasingly become a commercially viable option for many applications.

Northern African countries are also benefiting from Global Environment Facility (GEF) funding for climate change mitigation. For example, Sudan has projects for rangeland rehabilitation to enhance carbon sequestration and to reduce GHG emissions, and Tunisia has received funds for renewable energy development (GEF 1999).

## AIR QUALITY IN NORTHERN AFRICA

Declining air quality, especially in urban centres, continues to be one of the most serious local environmental problems in Northern Africa and a continuing threat to human health. The three principal anthropogenic causes of declining air quality are energy generation, emissions from vehicles and industrial production, all of which have increased in the past 30 years.

Much of Northern Africa's industrial base was developed in the 1960s, and the capital stock of most industries is, therefore, old and highly polluting. The situation is further complicated by protective trade regimes, foreign exchange constraints and by dominance of the public sector in industry, all of which provide little incentive for adopting more efficient and cleaner industrial technologies. Few enterprises have air emission controls, and a lack of maintenance and spare parts impairs the performance of existing systems (World Bank 1995).

In many cities in Northern Africa—especially those close to refineries and oil-fired power plants using high-sulphur fuel—sulphur dioxide levels reach more than 100 micrograms/m$^3$, double the World Health Organization (WHO) standard (World Bank 1999). Industry and the power sector are the sub-region's major sources of sulphur dioxide and total suspended particulates, and are large contributors to nitrogen oxide emissions (90, 80 and 60 per cent respectively). Sulphur dioxide, either dry or dissolved in rain water, causes erosion of infrastructure and monuments as well as soil and water acidification. For example, studies in 1985 and 1995 concluded that sulphur dioxide is a major factor contributing to corrosion of metallic structures in central Cairo and to deterioration of the Sphinx (Hewehy 1993).

Atmospheric concentrations of lead and particulates in Northern Africa often exceed WHO guidelines by a multiple of two to five in large cities. The cement and steel industries produce 50 per cent of the total particulate emissions (World Bank 1995). In Egypt, the Cairo Air Improvement Project (CAIP) reported greatly increased concentrations of lead in industrial districts during 1998, because of excessive lead emissions from smelters. It has been estimated that, prior to 1997, lead emissions from motor vehicles in Cairo were about 700 to 1000 tonnes per annum.

The number of motor vehicles in most countries in the sub-region has nearly doubled in the past 10 to 15 years and many vehicles are of older types. Older vehicles emit 20 times more hydrocarbons and carbon monoxide and four times more nitrogen oxides than new vehicles. Particulate emissions from poorly maintained diesel-fueled buses and trucks are five to seven times higher than those from similar but well-maintained vehicles (World Bank 1995, Larsen 1995).

City dwellers in Northern Africa, particularly those in congested metropolitan centres, are exposed to a variety of toxic and carcinogenic compounds including heavy metals and PAHs, which contribute to the incidence of respiratory diseases such as asthma, bronchitis and emphysema. If combined with other pollutants such as sulphur dioxide, suspended particulates can cause bronchitis and other lung diseases. For example, statistics from Egypt's Ministry of Health indicate that, in the areas of Ma'adi and Helwan, chest diseases are the second cause of death after communicable diseases.

## Towards improving air quality in Northern Africa

In 1991, having identified the major sources and established levels of emissions of the major air contaminants, the Council of Arab Ministers Responsible for the Environment (CAMRE) adopted the concept of sustainable development as the basis for development in the 21st century. Control of air pollution, especially in urban centres, was identified as one of the main objectives in Algeria, Libya and Morocco. More liberal trade policies and the increased production of more affordable vehicles will lead to the gradual replacement of highly polluting, older vehicles. Another target that is expected to be of higher priority is the reduction of the sulphur content of fuels. Algeria, Morocco and Tunisia have included electrified railways in their transportation infrastructures, and Egypt has built an underground metro system that has made a considerable contribution to reducing surface mass public transit, thus reducing emissions from vehicles.

Industrial pollution abatement initiatives have been introduced in Northern Africa and have begun to reduce $CO_2$ emissions. For example, conversion of industrial enterprises to natural gas has been financed. In Tunisia, a solar water-heating project will promote commercialization of solar water heating technology in the residential sector. A repowering project in Morocco has also been approved, but is awaiting construction of a natural gas pipeline from the Algeria-Portugal pipeline. Other projects at the preparation stage include solar, wind and waste-to-energy projects in Algeria, Egypt and Morocco.

In Egypt, the Cairo Air Improvement Project (CAIP) aims to initiate and implement measures to reduce air pollutants that have the most serious impacts on human health in Greater Cairo, especially suspended particulates and lead (see Box 2a.1). The CAIP will also

*Atmospheric concentrations of lead and particulates in Northern Africa often exceed WHO guidelines by a multiple of two to five in large cities*

### Box 2a.1 The Cairo Air Improvement Project

Recognising the seriousness of air pollution, the Egyptian government has embarked on a comprehensive programme to improve air quality throughout Egypt. In 1993, the country's Ministry of Petroleum introduced unleaded fuel as a part of this programme and within two years 85 per cent of the country's fuel supply was converted. The Cairo Air Improvement Project (CAIP), set up in 1997 and tasked with monitoring ambient air quality, reported that levels of airborne lead in Cairo had decreased by up to 88 per cent in just one year. Other initiatives under the project include testing of compressed natural gas as an alternative to diesel for public buses, and running a public awareness campaign.

*Source: USAID 2001a*

monitor the effectiveness of pollution abatement schemes implemented by the Egyptian Environment Affairs Agency, CAIP and other organizations.

## EASTERN AFRICA

Large parts of Eastern Africa are arid and semi-arid, and annual rainfalls below 500 mm are common. Amounts of rain and its distribution are highly unpredictable, both from year to year and in terms of distribution within a given year (FAOSTAT 2000). These conditions make the sub-region particularly vulnerable to the impacts of climate change on food production and security of livelihoods. These are, accordingly, priority issues.

Air pollution is not currently a major problem in the sub-region, as rates of urbanization and industrial production are relatively low at present. However, these rates are rising rapidly in comparison with other parts of Africa, and effective long term development plans are needed to prepare for potential increases in emissions.

### CLIMATE VARIABILITY IN EASTERN AFRICA

Eastern Africa has experienced at least one major drought in each decade over the past 30 years. There were serious droughts in 1973–74, 1984–85, 1987, 1992–94, and in 1999–2000, and there is some evidence of increasing climatic instability in the sub-region, and increasing frequency and intensity of

drought (FAOSTAT 2000). For example, records of dry and wet years for Uganda between 1943 and 1999 show a marked increase in the frequency of very dry years over the past 30 years (Department of Meteorology 2000). Rainfall records also indicate that, in some parts of the sub-region, the drought in 2000 was worse than that experienced in 1984 (DMC 2000).

Persistent deficits in rainfall in Eastern Africa have had serious impacts, including total crop failure, which has led to increasing food prices and dependency on food relief in Burundi, Ethiopia, Kenya and Uganda (DMC 2000). In Ethiopia, the 1984 drought caused the deaths of about 1 million people, 1.5 million head of livestock perished, and 8.7 million people were affected in all. In 1987, more than 5.2 million people in Ethiopia, 1 million in Eritrea and 200 000 in Somalia were severely affected (DMC 2000). Severe water shortages and rationing, continued reductions in water quantity and quality, increased conflicts over water resources, and the drying up of some rivers and small reservoirs have contributed to death of livestock from hunger, thirst and disease, and to increased conflicts over grazing belts.

Additional impacts in the sub-region include persistently low water levels in rivers, underground aquifers and reservoirs, impacting on hydrology, biodiversity, and the use of water for domestic, industrial and irrigation purposes. Low reservoir levels have also reduced the potential for hydropower generation, leading to the introduction of power rationing in the domestic and commercial sectors, which has caused

Drought impacts on soil

*UNEP*

interruptions of economic activities and declines in manufacturing output. This was the case in Kenya where low rainfall between 1998 and 2000 led to reductions in hydropower generation and the need for drastic rationing schedules. The Kenya Power and Lighting Company was estimated to have lost US$20 million (IRI Climate Digest 2000), and there were additional losses to the economy resulting from the enforced closure of industrial facilities.

By contrast, some areas have experienced above-average rainfall, triggered by the ENSO phenomenon. The very warm ENSO event during the rainy season of 1997 resulted in record rainfall in some areas (averaging 5 to 10 times more than normal in many areas) and disastrous flooding. Thousands of people were displaced and extensive damage to property was caused. In Uganda, about 525 people died and another 11 000 were hospitalized and treated for cholera which broke out after flooding and landslides. About 1 000 more people were reported to have died in flood-related accidents and 150 000 were displaced from their homes (NEMA 1999). About 40 per cent of Uganda's nationwide 9 600-km feeder road network was destroyed and the country experienced widespread crop failure resulting in dependency on food imports and aid.

### Strategies for coping with climate variability in Eastern Africa

All Eastern African governments except Somalia have signed and ratified the UNCCD. Djibouti, Ethiopia and Uganda have produced National Action Plans, and the Intergovernmental Authority on Development (IGAD) (see Chapter 1) has produced a sub-regional action plan for the countries in the Horn of Africa (UNCCD 2001). All Eastern African countries (except Rwanda and Burundi) belong to IGAD. Monitoring and early warning systems have been put in place, through IGAD, to improve the ability to cope with climate variability. ENSO-related events can also now be detected, as a result of research conducted under the World Meteorological Organization's (WMO) Tropical Ocean and Global Atmosphere programme. The WMO issues monthly statements (*El Niño Update*) to provide effective, accurate, and timely information to all concerned, to allow them to take mitigatory action. However, most of the national institutions in the sub-

---

**Box 2a.2  Traditional strategies for coping with drought**

Drought is extremely difficult to predict and the variable duration and extent of the phenomenon make its effects difficult to manage. For pastoralists, following the rains and pasture is a natural part of their system, and setting aside of areas for grazing reserves and splitting of herds to minimize risk are part of their coping mechanisms. However, exclusion from some traditional grazing areas has compromised their ability to cope during dry periods and drought.

A project conducted by the African Centre for Technology Studies (Kenya) aims to identify traditional means of reducing vulnerability to environmental change in dryland Africa, and incorporate them into commercial food production systems. Field studies were conducted to gather information on ways in which rural households use indigenous plants in responding to drought, and how national environmental policies affect their practices.

*Source: ACTS 2001*

---

region are under-resourced making adequate early warning dependent on donor support.

In April 2000, an Inter-Agency Task Force on the UN Response to Long Term Food Security, Agricultural Development and Related Aspects in the Horn of Africa was launched. The Task Force has produced a strategy for the Elimination of Hunger in the Horn of Africa aimed at broadening opportunities for sustainable livelihoods, and formulating and implementing country food security programmes. In Kenya, research into traditional methods of coping with climate variability is also underway, with the aim of applying traditional knowledge to commercial enterprises (see Box 2a.2).

### CLIMATE CHANGE IN EASTERN AFRICA

Low levels of industrialization and urbanization in Eastern Africa mean that the sub-region's contribution to global $CO_2$ emissions is negligible—less than 2 per cent of Africa's total emissions in 1996 (African Development Bank 2001). However, the impacts of climate change, particularly on food security, are a priority concern for this sub-region.

Climate change impacts include reduced rainfall (a reduction of 10 per cent by 2050 is anticipated for the

Horn of Africa), increased temperatures, and increased evaporation (IPCC 1998, IPCC 2001b). Resulting shifts in vegetation zones will be felt particularly in the areas of agriculture, tourism, energy, industry and commerce (Ottichilo and others 1991).

ENSO events may also be influenced by climate change, but it is not known at this stage whether flood frequency and intensity in Eastern Africa will increase. Increased rainfall intensity, together with degraded vegetation cover, would make the sub-region more prone to landslides associated with flooding.

Rises in sea levels and sea temperatures could have devastating consequences on the Eastern African coast, with risk of inundation of many important commercial centres, loss of infrastructure and population displacement. For example, port activities at Dar es Salaam and Mombasa may be interrupted, and tourism activities and potential activities foreclosed. The coral reefs in the Western Indian Ocean are particularly at risk from sea temperature rise—almost 90 per cent of the sub-region's corals suffered bleaching during the 1997–8 warm ENSO event (Obura and others 2000).

### Climate change mitigation and adaptation strategies in Eastern Africa

In response to growing concern over climate change in the sub-region, almost all governments have ratified the UNFCCC, and Burundi has ratified the Kyoto Protocol.

The Kenyan government has established a National Climate Change Activities Coordinating Committee, with members drawn from the ministries of agriculture and forestry, energy, planning, finance and industry, and from research institutes, municipal councils, public universities, the private sector and NGOs. The Council will coordinate and facilitate research, response strategies, policy options, public information and awareness, and liaison with the IPCC.

Several countries have developed national strategies for disaster prevention, preparedness, and management. In Ethiopia, this was followed by the establishment of a five-year plan for the Federal Disaster Prevention and Preparedness Commission, in 1998. Successes of the plan's implementation include a ten-fold increase in food reserves (WFP 2000).

### AIR QUALITY IN EASTERN AFRICA

Emissions from vehicles, manufacturing, mining and industrial activities (including diesel-powered generators, copper smelters, ferro-alloy works, steel works, foundries, and cement and fertilizer plants) contain carbon, sulphur and nitrogen oxides, as well as hydrocarbons and particulates, causing localized smog.

Domestic combustion of 'biofuels' poses a risk for human health. Figures since 1980 show that the traditional use of biomass as a source of energy still accounts for more than 70 per cent of total energy consumption in Eastern Africa (UNDP 2000), and consumption of biomass is predicted to increase over the next 20 years (FAO 2001b).

Demand for vehicular transport is on the increase in Eastern Africa, and many of the vehicles presently on the road are old and inefficient. For example, as shown in Figure 2a.6, Uganda had about 44 500 vehicles on the road in 1971; by 1998 this number had climbed to more than 182 400, an approximately four-fold increase in less than 30 years (MoWTC 2000). In Ethiopia, the capital, Addis Ababa, accounts for 41 per cent of all of the country's petrol consumption, indicating the concentration of vehicles and their emissions in Addis. Many of these vehicles are old and do not, therefore, have filtering systems (NESDA 2000).

Additional sources of air pollutants include both legal and illegal waste dumps. For example, in 1998, methane emissions from the municipal dump in Addis

**Figure 2a.6  Number of vehicles registered in Uganda in 1971–99**

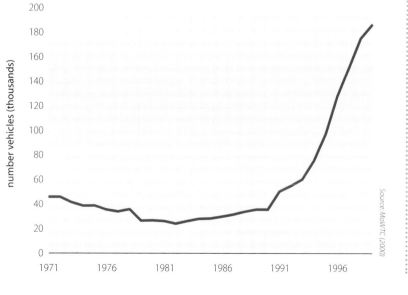

number vehicles (thousands)

*Source: MoWTC (2000)*

Ababa were estimated to be more than 9 Gg (1 Gg = 1 x 10$^9$ grams) (NESDA 2000).

### Towards improving air quality in Eastern Africa

Air quality standards for all major pollutants have now been established for most countries in Eastern Africa, but lack of resources renders enforcement less than optimum.

Addis Ababa has recently acquired a new pollution-measuring laboratory, costing US$286 000, which will help to identify the type and amount of pollutants emitted by factories in the city, assess the impacts on soil and water, and recommend measures to prevent further environmental pollution (PanAfrican News Agency 2001).

Tropical cyclone *Ando* in the Indian Ocean

*EUMETSAT*

## WESTERN INDIAN OCEAN ISLANDS

The Western Indian Ocean Islands lie between the tropics, with the exception of a small part of Madagascar which falls south of the tropic of Capricorn. They are subject to about ten tropical storms or cyclones each year in the period between November and May (four per year in Madagascar). The sub-region also experiences inter-annual variations in rainfall as well as periodic flooding and droughts.

Although early warning systems are well developed in the Western Indian Ocean Islands, the threat of increased climate variability and of sea level rise, arising from climate change, are issues of priority and concern.

Air pollution in urban areas is emerging as a problem for human health and for the sub-region's ecology. Preventive action is required.

### CLIMATE VARIABILITY IN THE WESTERN INDIAN OCEAN ISLANDS

Cyclones—with high winds gusting at over 200 km/hour—demolish lightweight buildings, damage overhead cables, uproot trees, and are a threat to life and property. Cyclones also cause heavy swells in the Western Indian Ocean which, in turn, cause significant rises in sea levels that affect coastal infrastructures such as roads and settlements; undermine beach stability; and cause vertical scouring of up to two metres (Ragoonaden 1997). The heavy rains resulting from cyclones cause destruction of crops and vegetation,

flooding, soil erosion and contamination of freshwater supplies, with risk to humans and to animal life. At the height of a cyclone, most outdoor human activity comes to a halt, schools and workplaces close down, emergency shelter has to be found for those whose homes are destroyed or damaged, and calls are issued for community aid programmes.

In the aftermath of a cyclone, communities may be temporarily prevented from returning to normal activity because people, domestic animals, crops, services and buildings have been lost (UNEP 1999). In some cases, the damage may be so severe that countries are obliged to seek international relief aid. Madagascar is an example of this (FAO 1984, UNDHA 1994).

The ENSO phenomenon is also a major factor in climate variability in the sub-region, causing floods and droughts. Mauritius, for example, is prone to drought, especially in the dry season, while Madagascar is most affected by desertification, with sandstorms that cause sand dunes to invade the interior along the coast, covering houses and crops (UNEP 1999). In these countries and elsewhere in the sub-region, the pressure of an increasing population is resulting in the use of marginal land close to rivers, of sand dunes and of land reclaimed from the sea for residential and industrial purposes. Conditions in these marginal or reclaimed lands are more precarious than in other areas, and such encroachment puts more people and jobs at risk from the effects of climate change and natural disasters.

Coral reefs in the sub-region are also at risk. In 1997 and 1998, the ENSO phenomenon caused abnormal increases in sea and air temperatures that led to bleaching and death of coral reefs. In Seychelles, more than 80 per cent of coral reefs were lost and, in the same period, a prolonged drought caused temporary closure of the Seychelles Breweries and the Indian Ocean Tuna Company (UNEP 1999).

### Strategies for coping with climate variability in the Western Indian Ocean Islands

Cyclones cannot be controlled, but their impacts on lives, livelihoods, crops and infrastructure can be minimized through adequate preparation and efficient, accurate warnings by meteorological services. For example, Mauritius has a cyclone warning system that was established in the 1950s, at a time when the island's economy was dominated by sugar production (a crop highly vulnerable to the wind and rain damage associated with cyclones). The subsequent rapid growth in the country's population, and its economic and agricultural development, mean that more people and infrastructure are now at risk from cyclone impacts, and the system is used to give a series of warnings, allowing for preparations or evacuation (see Box 2a.3). The regional Tropical Cyclone Warning System in the southwest Indian Ocean is also being upgraded and observation and telecommunications systems are being enhanced. Meteorologists and hydrologists are being given advanced training on early warning systems.

*Cyclones cannot be controlled, but their impacts on lives, livelihoods, crops and infrastructure can be minimized through adequate preparation and efficient, accurate warnings by meteorological services*

Cyclone-proof buildings are becoming a common feature in the sub-region (incidentally creating a demand for building sand which, because it is often dug from beaches, exacerbates erosion and damage to the delicate reef ecology). Wind resistant crops are also being developed on the islands.

Regional programmes can contribute to better protection from and response to crises, but effective intra-regional collaboration is vital to ensure sharing of technical skills, training, information, research, and collaboration in response.

### CLIMATE CHANGE IN THE WESTERN INDIAN OCEAN ISLANDS

The contribution of the Western Indian Ocean Islands to global climate change is small (less than half of one per cent of Africa's total $CO_2$ emissions in 1996), but vulnerability to the impacts of change is high, as is the case for other Small Island Developing States (African Development Bank 2001).

Seychelles has the highest per capita emission rate for $CO_2$ emissions in the sub-region (2.2 tonnes of $CO_2$ per inhabitant per year in 1996), but the country's natural capacity for removal of the gas from the atmosphere is estimated to be four times higher than the levels it emits (African Development Bank 2001, UNDP 2000).

Energy production and industry are important sources of GHGs, and commercial energy consumption is increasing in Madagascar by 1.6 per cent per year and in Mauritius by 2.6 per cent per year. Per capita consumption of electricity has more than doubled in Mauritius and Seychelles since 1984, while in Comoros and Madagascar it has remained constant or fallen. Between 1991 and 1995, total $CO_2$ emissions from industrial processes increased by 5 per cent in Madagascar and by 23 per cent in Mauritius (UNEP 1999).

The sub-region is particularly vulnerable to the impacts of global climate change because of its low lying flat lands, narrow coastal strips, atolls, reefs and sandy beaches, and the concentration of human populations, tourism, infrastructure, transport and industrial activities in coastal zones (Leatherman 1997). It is estimated that a one-metre rise in sea level would submerge most of the Seychelles islands with a loss of

---

### Box 2a.3 Cyclone warnings and preparation in Mauritius

Mauritius' Meteorological Office rates cyclones on a four-point scale of likelihood that the cyclone will hit the island. Warnings are issued to indicate that the cyclone is heading towards the island between 6 and 12 hours before it hits, and once the windspeed reaches 120 km/hr. There are fully equipped shelters and emergency procedures that are tested regularly. Plentiful advice on stocking up with provisions and water is also given.

*Source: US Embassy in Mauritius 2001*

Coral sand beaches with unique granite formations are unique to the Seychelle islands.

*W. Norbert/Still Pictures*

70 per cent of their land area (Shah 1995, Shah 1996). The same rise would result in a loss of 5 km² of land in Mauritius, 0.3 per cent of its surface area (IPCC 1995). For the sub-region as a whole, it has been estimated that, in total, 22 per cent of the population of the islands would be at risk from ocean rises and that rises would affect fishing and tourism, and jeopardize economic viability.

### Climate change mitigation and adaptation strategies in the Western Indian Ocean Islands

Responses to climate variability and climate change include ratification of the UNFCCC and, in the case of Seychelles, the Kyoto Protocol. This has been supported at the national level by the development of National Communications Strategies, as well as National Environment Action Plans. Mauritius has also undertaken studies of GHG sources and sinks.

The Western Indian Ocean Marine Applications Programme is a regional branch of the Global Ocean Observing System (GOOS), a permanent global system for observation, modelling and analysis of marine and ocean variables to support operational ocean services worldwide. The GOOS provides information on the current state of the oceans and forecasts on future conditions as the basis for climate change modelling.

In May 1991, the Mauritius Meteorological Services established a National Climate Committee (NCC) involving all parties with an interest in climate change. The Committee was tasked with monitoring progress on the scientific understanding of climate change, and evaluating the possible economic impacts on agriculture, coastal zones, energy and water resources, and human health and welfare.

### AIR QUALITY IN THE WESTERN INDIAN OCEAN ISLANDS

The Western Indian Ocean Islands are well ventilated and have yet to experience levels of air pollution with lasting effects on ecology and human health. However, concerns are growing about air pollution from industry and transport, especially in Seychelles and Mauritius. Traffic density is increasing alarmingly and the transport sector, the largest energy user in the sub-region as a whole, currently accounts for 43 per cent of total energy consumption. Vehicles also tend to be older—in Madagascar, the average age of buses is 11 years, that of cars 7–8 years (ONE 1997). In Mauritius, urban pollution, caused largely by vehicle exhaust emission, often exceeds WHO guidelines (UNEP 1999). Government subsidies for public transport and import duties on vehicles have failed to stem the growth in numbers of private vehicles and have resulted in a fleet that is increasing both in age and inefficiency. Lack of policies and of incentives to use cleaner fuels has also added to the burden of pollution.

Increasing disposable income in the sub-region will continue to drive up both energy use and numbers of vehicles. In Mauritius, for example, accelerated industrial development has resulted in increased industrial emissions and commercial road traffic, even though the capital, Port Louis, escapes concentrations of urban pollution because of its coastal position. In Madagascar, more than 70 per cent of industrial enterprises—principally agro-processing, textiles, clothing, leather, wood, paper, chemicals, mineral and metal products and industrial boilers—are located around the capital, Antananarivo, causing pollution problems. Air quality and pollution in Antananarivo are also influenced by strong sunshine and an absence of air movement, putting additional pressure on human respiratory health (ONE 1997).

The domestic use of wood and charcoal for cooking is causing concern over air pollution in Comoros and

*It is estimated that a one-metre rise in sea level would submerge most of the Seychelles islands with a loss of 70 per cent of their land area*

Madagascar (UNEP 1999). In Mauritius, where this problem also exists, its impacts are compounded by cigarette consumption which is at significant levels among males (UNDP 2000).

The practice of burning sugar cane before harvesting, in Mauritius and elsewhere in the sub-region, causes severe, though often localized, air pollution with release of fly ash, ozone, $CO_2$, carbon monoxide, methane and volatile organic compounds. Pollutants such as these can have an impact on respiratory health and can acidify ecosystems, thereby affecting the quality of soil, land and water.

The sub-region has no heavy, power-intensive industrial sector, but odours from the fish processing industry are a source of complaint.

Environmental quality is the key factor on which the sub-region's tourism and other inward investment depend. The main priority is therefore to avoid further deterioration of air quality and to reduce current sources of pollution.

### Towards improving air quality in the Western Indian Ocean Islands

Even though natural ventilation in these islands disperses much of the pollution, there is still a need for additional control measures. Some steps have been taken, including:

● Monitoring of pollution along main roads and the introduction of public awareness and school programmes highlighting the risks of air pollution to the environment and to human health.

● Use of some renewable energy sources including hydro-electricity in Madagascar and Mauritius and solar power in most countries of the sub-region. Bagasse from sugar cane is used to produce electricity in Mauritius (UNEP 1999).

● Environmental protection legislation is in place in each country, but its application is at the 'educational' rather than the enforcement stage.

In general, establishing the nature of the problem in the sub-region and quantifying it are difficult because monitoring tools are inadequate. In addition, control of sources of pollution is weak and there is a general laxness in regulation of the industrial sector. Social costs and impacts on health have not been studied, public sensitivity to environmental issues has yet to be stimulated and the sub-region is characterized by a general lack of awareness of the threats to the ecology and to human health.

## SOUTHERN AFRICA

The main issues of concern regarding the atmosphere in Southern Africa are the occurrence of flooding and droughts arising from climate variability, impacts of climate change on vegetation systems, biodiversity, freshwater availability, food production and localized air quality problems associated with emissions from industry, vehicles and use of domestic fuels.

### CLIMATE VARIABILITY IN SOUTHERN AFRICA

Rainfall in Southern Africa is strongly influenced by the Inter-Tropical Convergence Zone (ITCZ), a zone close to the equator where massive rain-bearing clouds form when the South East Trade Wind (from the south-east of the region) meets the North East Monsoon Winds. The ITCZ changes position during the year, oscillating between the Equator and the Tropic of Capricorn, and its southward movement usually marks the beginning of a rainy season. The further south the zone moves, the more promising this is

Maize rotting in the fields after the February–March 2000 floods in Mozambique

considered to be for the rainy season. In a normal season, the ITCZ can exert an influence between mid-Tanzania and southern Zimbabwe and is associated with favourable rainfall. Another system, the Botswana High, often tends to push the ITCZ away, resulting in periods of drought.

The ENSO also influences Southern Africa's climate, tending to bring either heavy rains often accompanied by severe floods (as in 1999–2000 when Mozambique was exceptionally hard hit), or drought (as in 1982–83 when much of Southern Africa was severely affected) (National Drought Mitigation Center 2000).

In the wet season, normal rainfall in Southern Africa ranges from 50 mm to more than 1000 mm. Recent weather patterns have, however, been erratic with severe droughts recorded in 1967–73, 1981–83, 1986–87, 1991–92 and 1993–94. Floods have also been observed, most notably across most of Southern Africa in 1999–2000 (Chenje and Johnson 1994, WMO 2000).

The drought of 1991–92 was the severest on record, causing a 54 per cent reduction in cereal harvest and exposing more than 17 million people to risk of starvation (Calliham, Eriksen and Herrick 1994). Zimbabwe alone imported an additional 800 000 tonnes of maize, 250 000 tonnes of wheat, and 200 000 tonnes of sugar (Makarau 1992). Water and electricity shortages resulted in a 9 per cent reduction in manufacturing output and a 6 per cent reduction in foreign exchange (Benson and Clay 1994).

Cyclone Eline, which hit south-eastern Africa in 1999–2000, affected 150 000 families and wreaked havoc in Mozambique where it caused US$273 million worth of physical damage, and cost US$295 million in lost production and US$31 million in food imports (Mozambique National News Agency 2000).

A combination of dry spells, severe floods, and disruption of farming activities between 1999 and 2001 has left Southern Africa with meagre food reserves. Several of the sub-region's countries have faced food shortages (FAO 2001a).

### Strategies for coping with climate variability in Southern Africa

The Southern African Development Community (SADC) has developed a Sub-Regional Action Programme to Combat Desertification in southern Africa, in line with the UNCCD. All of the countries in the Southern African

### Box 2a.4 Early warning in Southern Africa

The SADC Regional Early Warning Unit, the Regional Remote Sensing Project, the Drought Monitoring Centre and the Famine Early Warning System Project all advise governments on drought preparedness. During the 1991–1992 drought, the unit altered national agencies and donors in time to increase food imports and request food aid.

Source: Masundire 1993 Photo: UNEP

sub-region are party to the UNCCD, and Lesotho, Malawi, Swaziland, Tanzania, and Zimbabwe have also produced National Action Plans (UNCCD 2001).

Early warning and response strategies for mitigating the impacts of climate variability are relatively well developed in the sub-region (see Box 2a.4) and a drought fund is in place to mitigate the effects of poor rainfall (Chimhete 1997). However, monitoring, research, and preparedness strategies need further strengthening. This is evidenced by the response to cyclone Eline—according to some sources, the early warning systems did not supply sufficient information about the extent of the impacts of the cyclone, resulting in the loss of many lives that could have been saved (CNN 2000).

### CLIMATE CHANGE IN SOUTHERN AFRICA

Emissions of GHGs by Southern Africa (in particular South Africa) are higher than those for other sub-regions of Africa, although they represent only 2 per cent of the world total. The sub-region's emissions are, however, projected to rise as countries develop, including a threefold increase in Zimbabwe over the next 50 years (Southern Centre 1996). South Africa already has a net positive GHG emission level—it accounts for the majority of emissions from the sub-region and 42 per cent of all emissions from Africa (World Bank 2000a, Marland and others 2000).

The majority of Southern Africa's primary energy comes from fossil fuels, in the form of coal and petroleum, and its level of industrialization is high compared to other parts of Africa. It is therefore imperative that the sub-region adopt a two-pronged approach, introducing cleaner technologies and renewable sources of energy production while simultaneously improving coping capacities to deal with the anticipated impacts of an increasingly variable and unpredictable climate.

Southern Africa's background of climate variability, food insecurity and water stress makes it one of the most vulnerable areas to climate change (IPCC 1998, IPCC 2001b). The sub-region is expected to experience a mean temperature rise of 1.5 °C, and increased rainfall variability and insecurity (Hulme 1995). The expected impacts of these changes include reductions in the extent of grasslands and expansion of thorn savannas and dry forest, together with a general increase in the extent of desertification across the sub-region. This will, in turn, affect the distribution of wildlife and some of the major national parks could suffer economic losses through reduced tourism potential (IPCC 2001b). Crop yields are also expected to vary, dropping by as much as 10–20 per cent in some parts of the sub-region (GLOBE 2001). It is also predicted that the malaria-carrying *Anopheles* female mosquito will spread to parts of Namibia and South Africa, where it has not been found before (IPCC 1998).

According to the IPCC, global sea levels are expected to rise by between 10 cm and 100 cm by 2100, due to snow and ice melt. This could inundate up to 2 117 km$^2$ and erode up to 9 km$^2$ of the coast of Tanzania, with a cost of more than US$50 million (IPCC 1998). The economies of most Southern African countries are highly vulnerable to a changing climate because of their dependence on commercial agriculture. In addition, more than 50 per cent of the sub-region's population is based in rural areas, and is directly dependent on small-scale cultivation and rearing of livestock.

### Climate change mitigation and adaptation strategies in Southern Africa

To date, all of the 11 countries of the Southern African sub-region have signed and ratified the UNFCCC, signalling the importance with which they regard climate change. Several countries have embarked on the production of National Communications Strategies to document GHG emissions and sinks, and have developed strategies for mitigating climate change.

A sub-regionwide initiative is also underway to develop a strategy for GHG reduction in Southern Africa. Tanzania's Centre for Energy, Environment,

Science and Technology, in collaboration with Zimbabwe's Southern Centre for Energy and Environment and Zambia's Centre for Energy, Environment and Engineering, are coordinating studies for this initiative.

South Africa, the sub-region's largest contributor to GHG emissions, has established a National Committee for Climate Change to advise its Minister of Environmental Affairs on the implications of climate change and policy options for reducing emissions. In 1996, the government of Zimbabwe reviewed the country's environmental legislation and incorporated climate change issues. A climate change office was set up within the then Ministry of Mines, Environment and Tourism to coordinate activities on the subject (Government of Zimbabwe 1999).

In Mozambique, plans are underway for a tree-planting scheme which should qualify for funds from the Cleaner Development Mechanism of the Kyoto Protocol. This project aims to restore native Miombo woodlands in Gorongosa National Park and to provide economic opportunities to local communities through sales of timber and development of eco-tourism.

### AIR QUALITY IN SOUTHERN AFRICA

Although GHG emissions for Southern Africa are generally low, emissions of toxic air pollutants, especially in urban

Smog over Cape Town, South Africa

*Jonathan Kaplan/Still Pictures*

centres, are cause for concern. Examples include the heavy metals mercury and lead, synthetic chemicals such as polychlorinated biphenyls (PCB), polycyclic organic matter (POM), dioxins and benzene. These pollutants arise from industrial and domestic sources and from use of vehicles, and because Southern Africa lacks affordable cleaner technologies and fuel sources. Weather conditions in cities in the sub-region also aggravate the levels of pollution observed so that air pollution levels observed in certain cities in winter exceed WHO levels (Chenje and Johnson 1994).

Traditional fuels also cause pollution problems. For example, a study in Tanzania found that children under the age of five who died of acute respiratory infections were almost three times more likely to have been exposed to burning of traditional fuels in the home than healthy children in the same age group (World Bank 2000b).

In South Africa, sulphur dioxide emissions from power stations travel significant distances, and acidification of water and forests has been recorded in the north-eastern part of the country.

## Towards improving air quality in Southern Africa

Actions to improve air quality in Southern Africa include air quality monitoring and establishment of air quality guidelines, as well as requirements for environmental impact assessments to be made for any new development with potentially damaging effects on air quality. Botswana, South Africa and Zimbabwe have air pollution legislation, although it is enforced to varying degrees, being limited by lack of resources. Tanzania's National Environment Management Council commissioned a study of air quality in Dar es Salaam, to assess emissions from traffic as well as noise pollution, sulphur dioxide levels and particulate concentrations, measured against national and international standards. The purpose of the study was to find ways to reduce emissions and to improve transport systems. In August 2000, the United Nations Motor Vehicle Emissions Agreement came into force, to develop globally uniform regulations, promoting energy efficiency and vehicle safety. South Africa was the first country to sign this agreement; other Southern African countries are encouraged to follow suit.

## CENTRAL AFRICA

Central Africa is faced with similar challenges to other sub-regions in terms of climate variability: periodic extreme weather events, impacts of climate change on food production, sea level rise and localized air quality problems in urban areas. These challenges are partly natural and require effective impact mitigation strategies but they are, in some cases, also exacerbated by human activities and thus require an integrated environmental management approach.

### CLIMATE VARIABILITY IN CENTRAL AFRICA

Rainfall and temperature patterns in Central Africa vary considerably, with unpredictable seasonal variations. Rainfall is relatively high and reliable over the central and coastal parts of the sub-region but tends to diminish and become more variable towards the north. For example, Douala, in coastal Cameroon, has an average rainfall of 3 850 mm/yr while Djamena, in Chad, receives only 500 mm/yr, and suffers periodic drought. Temperatures in the low-lying coastal forests vary little because persistent cloud keeps mean annual temperatures between 26 °C and 28 °C. In the high-relief mountainous areas, mean annual temperatures are low, between 19 °C and 24 °C. In the semi-arid zone of Cameroon and Chad, clear skies lead to strong insolation during the day and massive heat losses by emission of longer wavelength radiation at night.

Droughts in the Central African Sahelian zone have become more frequent since the late 1960s, and food security is declining, particularly among the poor who are forced to cultivate marginal lands and are unable to accumulate food reserves (IPCC 1998). Flooding is common in the more humid areas of Central Africa, especially where forests and natural vegetation have been cleared for cultivation or human settlements.

In the past 30 years, development policies and activities, such as commercial logging, commercial or subsistence agriculture and collection of firewood, have led to extensive clearing of forests in Central Africa. These changes have disturbed the sub-region's microclimate, increasing vulnerability to rainfall fluctuations. Furthermore, the dense humid tropical forests of the sub-region are important sinks for atmospheric $CO_2$ (more so than forests of equal area in temperate zones). Reduction in their area therefore limits carbon sequestration and thus contributes to global climate change. Reduction of vegetation cover also exposes the soil, and worsens the impacts of drought and flooding.

### Strategies for coping with climate variability in Central Africa

Central African countries have ratified the UNCCD, and Chad has also produced a National Action Plan. The Permanent Inter-State Committee for Drought Control in the Sahel (CILSS) has developed a sub-regional action plan to combat desertification (UNCCD 2001). CILSS and the Club du Sahel have also developed a new vision for food security in the Sahel, and have established a Food Crisis Prevention network to improve coordination between countries.

## CLIMATE CHANGE IN CENTRAL AFRICA

GHG emissions are minimal in Central Africa, contributing just 2 per cent of Africa's total emissions in 1996. They form a negligible part of global emissions (African Development Bank 2001). Emissions of gaseous pollutants such as carbon monoxide, carbon dioxide and methane result from a variety of sources, including dumping of gas-producing garbage, use of

traditional fuels in domestic energy production, and from slash-and-burn agriculture. The occasional eruption of Mount Cameroon also contributes to gaseous emissions.

However, global atmospheric and climatic changes will impact on Central African countries, increasing rainfall and temperature fluctuations, and thus affecting security of food and water resources. Most of Central Africa will experience increased precipitation, soil moisture and run-off. This could result in a net increase in forest cover, although increased suitability of land for agriculture may lead to accelerated rates of forest clearance. Shifting distribution of natural habitat could also have important consequences for the unique biological resources of the Central African forests, such as the endangered Mountain Gorilla. Other perturbations of hydrological systems may change flooding patterns, increase the risk of contamination of freshwater supplies and encourage outbreaks of water-borne diseases. Malaria and trypanosomiasis may spread to new areas, particularly to areas at higher altitudes where their presence was previously limited by low temperatures, and to drier areas where increased rainfall is predicted (IPCC 1998).

Freshwater availability may decrease in the arid and semi-arid parts of Cameroon, Central African Republic and Chad because of decreased rainfall and increased evaporation. Over the past 30 years the level of water in Lake Chad has dropped greatly under the combined pressures of rainfall fluctuations and continued withdrawals, and the water level today is around only one-twentieth of what it was 30 years ago (NASA GSFC 2001). The millions of people currently dependent on the lake's resources could suffer enormous economic and food security losses if its level decreases further.

Sea-level rise and increased vulnerability to inundation and storm surges will render some of the coastal areas of Central Africa uninhabitable, displace millions of people and threaten low-lying urban areas, such as Douala in Cameroon (IPCC 1998, IPCC 2001b).

### Climate change mitigation and adaptation strategies in Central Africa

In response to the challenges of adverse climatic variations, all countries in Central Africa have joined the international community in signing and ratifying the UNFCCC. At the national level, capacity building amongst stakeholder groups and revision of policies

Large-scale logging of hardwood in Cameroon

## Box 2a.5 The role of Central African forests in mitigation of climate change

Central Africa's extensive tropical forests are important carbon sinks and can assist in mitigating GHG emissions. In the Republic of Congo a research study has been launched to evaluate the carbon sink potential of the tropical forests and develop strategies for forest conservation. Additional studies are underway as part of the Central African Regional Programme for the Environment (CARPE) to develop and implement means of reducing deforestation and biodiversity loss in the Congo Basin.

The Conference on Ecosystems of Dense Humid Forests in Central Africa (CEFDHAC) is another initiative linked to CARPE, established to improve sub-regional cooperation on issues of forest management in the Congo Basin.

and legislation for enhanced environmental protection is underway, including protection of the sub-region's important forest reserves (see Box 2a.5). Although this action demonstrates political commitment across the region to addressing the problem, Central African countries emit negligible amounts of GHG and—because they are classed as developing countries under the Kyoto Protocol—they are not yet required to reduce emissions. Climate variability and the impacts of climate change are also being tackled through the establishment of climate modelling programmes and early warning of rainfall variations. These are, however, in the early stages of development and little information on their effectiveness is available.

### Air quality in Central Africa

There have been few studies of the ecological impacts of air pollution in Central Africa, as emission levels are generally low. However, generic studies of urban air quality have linked pollutants arising from domestic combustion of traditional fuels to increased rates of respiratory diseases, particularly among children. Industrial emissions, although currently below international averages, must be considered a potential threat as the needs of economic development continue to exert a pressure for greater industrial output. Of particular concern are the

extensive oil and gas reserves in the sub-region, which could be exploited for domestic energy production and export to other regions. Exploitation would involve clearing of forests and disturbance of marine ecosystems, as well as considerable increases in the sub-region's GHG emissions.

### Towards improving air quality in Central Africa

It is clear that cleaner technologies are required urgently if Central Africa is to meet its energy requirements and satisfy its economic development needs without increasing pressure on the environment and human health. The sub-region has abundant renewable energy resources, including solar, hydro and wind, which have not yet been fully exploited. The transport sector consumes the greatest proportion of energy and strategies for improving fuel efficiency of vehicles and public transport systems are therefore priorities for action. The World Bank's Sub-Saharan Africa Transport Policy Programme, launched in 1998, provides a framework for phasing out leaded fuel and improving transport infrastructure. It could be instrumental in improving air quality in Central African countries.

## WESTERN AFRICA

Western Africa faces challenges arising from climate variability—especially in the arid Sahelian zone where drought is recurrent—and from the predicted impacts of climate change on food production, freshwater availability and desertification. Localized problems of air quality have also been identified as priorities for action.

### Climate variability in Western Africa

The climate of the Western Africa sub-region varies greatly from north to south and is mainly governed by the seasonal movements of the ITCZ. Desert and semi-desert regimes, characterized by annual precipitation of 100–300 mm, prevail along the border of the Sahel covering Mauritania, northern Senegal, Mali and Niger. There is substantial inter-annual variability in precipitation with no perennial run-off and flash floods occurring in small basins during rainy periods. In addition, evaporation rates are very high (more than 4 m/year). Further south, temperatures and evaporation

Desert landscape, Mali

*Romano Cagnoni/Still Pictures*

African States (ECOWAS), together with the CILSS, has produced a Sub-regional Action Plan (UNCCD 2001).

Other recent projects in the sub-region include an assessment of the vulnerability of production systems in Burkina Faso, Mali, Niger and Senegal. A Sahel Sahara Observatory has been set up, within the framework of the UNCCD, to act as a sub-regional coordinating body and to provide improved access to and sharing of information, and to implement scientific and technical capacity-building projects in land and water resource management. Areas of specific interest for this project include development of data banks, preparation of manuals and handbooks, facilitating of north-south exchanges of experience and knowledge, and provision for the transfer of know-how between the various areas of Africa.

rates are lower, although widespread flooding occurs because of the marked hydrological degradation, relatively high run-off and large areas of flat land. The coastal zone experiences warmer, wetter conditions, with annual rainfall in excess of 1 000 mm/yr, and greater reliability of rainfall, although floods are not uncommon.

Drought is a recurrent problem in the Sahelian zone of Western Africa, although the equatorial zone is rarely affected. The last Sahelian drought, one of the worst on record, persisted for a whole decade from 1972–84, and reduced precipitation was even noted in the equatorial zone. During this period, more than 100 000 people died, and more than 750 000 people in Mali, Niger and Mauritania were totally dependent on food aid in 1974 (Wijkman and Timberlake 1984). The drought also resulted in power shortages in Benin, Chad, Mali and Nigeria because of hydropower failures at the Kainji Dam on the River Niger (IPCC 1998). Desertification is also a problem in the region, especially in the arid and semi-arid zones, but it also affects the sub-humid zones. For example, as in Central Africa, the falling level of Lake Chad could have major economic implications for the millions of people in Western Africa who depend on its resources.

### Strategies for coping with climate variability in Western Africa

All the countries of Western Africa have ratified the UNCCD. Benin, Burkina Faso, Cape Verde, Gambia, Mali, Niger and Senegal have produced National Action Plans, and The Economic Community of Western

## CLIMATE CHANGE IN WESTERN AFRICA

The pressures resulting in climate variability and climate change originate outside of Western Africa, given that the sub-region contributes little to global emissions of $CO_2$ and other GHGs. Nigeria is the largest contributor in the sub-region, accounting for 11 per cent of Africa's total emissions in 1996 (African Development Bank 2001). Gas flaring in Nigeria consumes large amounts of natural gas every day during oil extraction. This not only contributes to atmospheric pollution by releasing considerable quantities of $CO_2$, but also wastes important resources. The Nigerian gas reserves are estimated to be sufficient to provide power to the whole of Western Africa (United Nations 1999). Overall, however, it is the transport sector which contributes the most to carbon emissions in the sub-region, followed by the industrial sector. Widespread use of biomass for energy results in fewer emissions from electricity generation than in Northern and Southern African countries.

Equatorial zones are expected to experience a 5 per cent increase in annual rainfall and temperature rise of 1.4 °C as a result of climate change. Although parts of the Sahel could receive increased rainfall, the Sahel in general will be more prone to desertification because of increased evaporation and run-off (IPCC 1998, IPCC 2001b). In addition, increased intensity of rainfall may increase soil erosion, nutrient leaching and crop damage. Loss of rainfall in marginal and vulnerable areas would exacerbate drought and desiccation problems and increase the risk of bushfires, in turn threatening forest

●

*The last Sahelian drought, one of the worst on record, persisted for a whole decade from 1972–84; during this period, more than 100 000 people died, and more than 750 000 people in Mali, Niger and Mauritania were totally dependent on food aid in 1974*

●

margins. Changes in timing and length of growing seasons may lead to planning problems for agriculture. The recent history of droughts in this zone has reduced national food reserves, and people in the sub-region are now more than ever vulnerable to food shortages. Changing patterns of distribution of forests and other natural habitat may place more animal and plant species under threat of extinction, and inland fisheries are at risk from changing rainfall and flood regimes. In addition to increasing vulnerability in the Sahelian zone, there is increased vulnerability of the equatorial zone to sea level rises as a consequence of climate change. Côte d'Ivoire, The Gambia, Nigeria and Senegal are among the most vulnerable countries in Western Africa (IPCC 1998).

### Climate change mitigation and adaptation strategies in Western Africa

In response to climate variability and climate change, all of the countries of Western Africa (with the exception of Liberia) have ratified the UNFCCC. Ratification of the Kyoto Protocol—under which countries stand to benefit from funds for forest conservation and adoption of cleaner technologies—is expected shortly.

The need for coping strategies to mitigate the impacts of climate variability and drought is equally important, and significant progress has been made in this area in the past 30 years. The AGRHYMET Regional Centre was established in 1974 by the Permanent Inter-State Committee for Drought Control in the Sahel (CILSS) following the Sahelian drought of the 1970s. Its primary function is to act as regional producer of raw data and analysed information, and to provide training in agrometeorology, hydrology, monitoring and evaluation, and plant protection. The nine member states of CILSS participating in the AGHRYMET activities are Burkina Faso, Cape Verde, Chad, Gambia, Guinea-Bissau, Mali, Mauritania, Niger and Senegal. AGRHYMET's Major Programme on Information aims to produce information relating specifically to food security, the fight against desertification, and livestock management.

### AIR QUALITY IN WESTERN AFRICA

Rapid urbanization and concentration of economic activities in Western Africa's urban centres is leading to air pollution from industry, vehicle emissions and quarrying activities. Combustion of traditional fuels for

Smoke from fires can cause or enhance respiratory diseases

*Ron Giling/Still Pictures*

domestic energy needs is another major source of air pollution in both urban and rural areas. For example, children in the Gambia exposed to smoky stoves were six times more likely to develop acute respiratory infections than unexposed children (World Bank 2000b). Inadequate urban planning is a significant driving force behind rising emission levels, because residential and commercial centres are often far apart, forcing mass movement of workers on a daily basis. Poor economic development has also contributed to air pollution by creating dependence on old vehicles and dirty fuels.

Pollutants such as sulphur oxides, nitrogen oxides, hydrocarbons and heavy metals, together with particulate matter, form dense concentrations of smog in urban centres, causing respiratory diseases, contamination of vegetation and water resources, and corrosion of buildings.

### Towards improving air quality in Western Africa

Most countries in Western Africa have introduced standards and regulations to control the atmospheric pollution in cities, but lack of resources makes enforcement of these emission standards and regulations weak.

In Accra, Ghana a project is under way to analyse and monitor sources of air pollution and to compare ambient pollution levels in commercial and residential areas, affluent suburbs and slums (Accra Mail 2001). A recent study of transport in Dakar, Senegal, estimated

## Box 2a.6  Tackling vehicle emissions in Senegal

Senegal has become a major importer of used cars in recent years and these now represent 84 per cent of all vehicles in the Dakar region where they constitute a major source of air pollution. The average age of vehicles in Dakar is approximately 15 years for cars and 20 years for buses. More than 40 per cent of the cars have diesel engines, which have particularly toxic emissions. The high level of diesel use results from a combination of many diesel engine cars being imported and many owners replacing petrol engines with diesel engines after importation, because diesel is cheaper than petrol.

The Senegal Ministry of Environment has proposed a new law to provide new clean air standards to limit emissions from vehicles. The law also re-introduces a requirement for customers to lodge request for cars before they are imported, a measure which was scrapped in 1996. It also rules that imported cars must be under five years old.

Air quality monitoring stations are to be set up around Dakar, to record ambient pollution levels on a daily basis. The ministry also intends to establish an environmental police force, backed up by the national police force, to impose compliance with the regulations and track down polluting vehicles.

*Source: Abdourahmane Ndiaye, Advisor to the Ministry of the Environment (quoted in Africa Online News, 21 January 2002)*

---

the costs associated with death or injury, hours wasted through congestion, and health costs relating to air pollution to be equivalent to 5 per cent of GDP (World Bank 2001). This study recommended that planning should be improved to ease the traffic flow and that mass transport should be re-organized. The small French locomotives known as 'Petit Train Bleu' have recently been rehabilitated and this has helped to alleviate urban congestion and pollution as well as providing reliable, secure, affordable mass transit and creating jobs (UNCHS 2001). Other measures to improve air quality include tighter controls on importation of second hand cars from Europe (see Box 2a.6).

### CONCLUSION

Climate variability is common throughout Africa and it is increasingly limiting development. Extremes of rainfall are the most damaging aspect of climate variability, and

*Africa's limited economic and infrastructure resources for mitigating or coping with shifting patterns of food production, increased frequency and severity of natural disasters and rising seal levels make the region one of the most vulnerable to climate change*

Africa frequently suffers the devastating impacts of floods and drought. Lives, livelihoods, crops, livestock and infrastructure are lost during these events, and the financial cost is well beyond the means of African countries, meaning that they are neither prepared for such events, nor able to afford to repair the damage caused. It is the poor who are the most vulnerable because they have no alternative source of income to their direct, subsistence level dependence on natural resources (either through cultivation, livestock rearing, or harvesting resources from natural habitats). Most countries have developed and implemented strategies for coping with climate variability, including ratification of the UNCCD, and development of National Action Plans. There are also effective climate and hazard monitoring programmes and early warning systems in many sub-regions. However, given the additional impacts of climate change, these systems may need additional trained staff, financial resources and equipment. The main priorities for action are, therefore, to strengthen coping strategies for effective management of the impacts of extreme events, increase food security and maintain healthy ecosystems.

Africa's limited economic and infrastructure resources for mitigating or coping with shifting patterns of food production, increased frequency and severity of natural disasters and rising seal levels make the region one of the most vulnerable to climate change. Although many African countries have ratified the UNFCCC and the Kyoto Protocol, most countries (with the exception of those of Northern Africa and South Africa) have negligible GHG emissions. Developing countries such as those in Africa stand to gain from the proposed mechanisms for emissions trading, reforestation schemes, and cleaner development. Emphasis in the global arena should therefore be placed on their implementation as soon as possible, to assist in meeting development objectives through sound environmental management.

Above all, it is critical that action be taken immediately. Further delays in curbing the trend of increasing atmospheric pollution will only add to the uncertainties and insecurities that natural climate variability causes. Developing countries also need to invest in disaster preparedness strategies in the short term, and to diversify their economies away from the heavy dependence on agriculture in the long term.

Ambient air pollution, particularly in urban centres, is emerging as an issue of concern for human health in many African countries. Increasing levels of toxic pollutants, such as sulphur dioxide, nitrogen oxides, lead and particulates, are a result of industrial emissions and vehicle exhausts (particularly from older vehicles), as well as burning of coal, wood or other fuels to meet domestic energy requirements. Population growth over the past 30 years has increased the demand for energy and industrialization, thus raising emissions of pollutants. Population growth and the pattern of human settlements have also put pressure on transport systems, increasing vehicle exhaust emissions. Radical changes in technology are required if economic development is to proceed without adding to the existing environmental challenges, and if stringent emission controls are to be avoided. Removal of subsidies, expansion of electrification programmes, promotion of unleaded fuels and conversion to cleaner fuels are measures that have been implemented successfully in parts of Africa recently. In particular, the energy industry, manufacturing industries and transport systems will have to undergo fundamental changes in order to supply sustainable energy, material and mobility to future generations.

# References

Abdel-Rahman, S.I, Gad, A., & Younes, H.A. (1994). Monitoring Of Drought on Lake Nasser Region Using Remote Sensing

Accra Mail (2001). Research into Air Pollution Underway. Accra Mail 1 June 2001

ACTS (2001). African Centre for Technology Studies. http://www.acts.or.ke/

Africa Online News (2002). Used Cars Pollute Dakar's Air. Article 21st January 2002

African Development Bank (2001). African Indicators: Gender, Poverty, and Environmental Indicators for African Countries 2000–2001

Benson, C., & Clay, E. (1994). The Impact of Drought on Sub-Saharan African Economies. IDS Bulletin 25(4); 24–32

Calliham, D.M., Eriksen, J.H., & Herrick, A.B. (1994). *Famine Averted: The United States Government Response to the 1991/92 Southern Africa Drought: Evaluation Synthesis Report*. Management Systems International, Washington

CARB (2001)b. Toxic Air Contaminant Identification Process: Toxic Air Contaminant Emissions from Diesel-Fueled Engines. California Air Resources Board 2001. Published on www.arb.ca.gov/html/fslist.htm

CARB (2001a). Buyers Guide to Cleaner Cars. California Air Resources Board 2001. Published on http://www.arb.ca.gov/html/fslist.htm

Chenje, M., & Johnson, P. (eds., 1994). *State of the Environment in Southern Africa*. SADC/IUCN/SARDC, Maseru and Harare

Chimhete, C. (1997). Southern Africa Prepares For El Nino-Induced Drought. Southern African Research & Documentation Centre, Harare

CNN (2000). Archives on the Mozambique Flood. http://www.cnn.com

Department of Meteorology (2000). Databank 2000. Ministry of Water, Lands and Environment. Kampala

DMC (2000). *DEKAD 19 Report (1-10 July, 2000) Ten Day Bulletin. No. DMCN/01/337/19/07/2000*. Drought Monitoring Centre, Nairobi

FAO (1995). Irrigation in Africa in Figures. Water Report 7, FAO, Rome

FAO (1984). *Madagascar Post Cyclone Evaluation In Agriculture*. FAO, Rome

FAO (1997). *The State of the World's Forests*. FAO, Rome

FAO (2001a). 17 Countries Are Facing Exceptional Food Emergencies in Sub-Saharan Africa- FAO Concerned About Deteriorating Food Situation in Sudan, Somalia and Zimbabwe. Press Release 01/48

FAO (2001b). Woodfuel Consumption and Projections 1970–2030. FAO, Rome

FAOSTAT ( 2000). FAO STATISTICS Database. United Nations Food and Agriculture Organization, Rome

GEF (1999). Conclusions of the GEF Heads of Agencies Meeting. 11 March 1999. Global Environment Facility, Washington D.C.

GLOBE (2001). Global Legislators Organisation for a Balanced Environment. Southern Africa Newsletter, Issue 2; March-April, 2001

Gommes, R. and Petrassi, F. (1996). Rainfall Variability and Drought in Sub-Saharan Africa Since 1960. *FAO Agrometeorology Series Working Paper No 9*. FAO, Rome

GOZ (1999). Government of Zimbabwe Factbook 1999. Government Printers, Harare

Hewehy, M.A.(1993). Impacts Of Air Pollution Upon Cultural Resources In Cairo, 13th Annual Meeting and International Conference of IAIA, Shanghai

Hulme, M.(1995). *Climate Change and Southern Africa: An Exploration of Some Potential Impacts and Implications in the SADC Region*. WWF, Harare

IPCC (1995). *Climate Change 1995. Impacts, Adaptations and Mitigation of Climate Change: Scientific and Technical Analysis,* Working Group II to the second Assessment Report of the Intergovernmental Panel on Climate Change, Edited by Robert T Watson, Marufu C Zinyowera and Richard H Moss. WMO and UNEP Cambridge University Press, Cambridge

IPCC (1998). *The Regional Impacts of Climate Change: An Assessment of Vulnerability*. IPCC, Geneva

IPCC (2001a). *Climate Change 2001: The Scientific Basis*. IPCC Working Group I, Geneva

IPCC (2001b). *Climate Change 2001: Impacts, Vulnerability, and Adaptation*. IPCC Working Group II, Geneva

IRI Climate Digest (2000). Climate Impacts October 2000. Published by the International Research Institute for Climate Prediction. Available on http://iri.columbia.edu/climate/cid/Oct2000/impacts.html

Larsen, B. (1995). Natural Resource Extraction, Pollution, Intensive Spending and Inequities in the Middle East and North Africa. Working Paper Series. World Bank

Leatherman, S.P. (1997). Sea Level Rise and Small Island States. *Journal of Coastal Research* Special Issue 4

Makarau, A. (1992). National Drought and Desertification Policies: The Zimbabwe Situation. SADC Regional Workshop on Climate Change, 1992. Windhoek

Marland, G., T.A. Boden, & R.J. Andres (2001). Global, Regional, and National CO2 Emissions. *Trends: A Compendium of Data on Global Change.* Carbon Dioxide Information Analysis Center, Oak Ridge National Laboratory, U.S. Department of Energy, Oak Ridge

Masundire, R.T. (1993). The Drought in Southern Africa and How REWS Enabled Timely Management

SADC Regional Early Warning System. Available on http://www/fao.org

MoWTC (2000). *Vehicle Database.* Ministry of Works, Transport and Communications. Kampala

Mozambique National News Agency (2000). *AIM Reports,* Issue No. 194, 6 November 2000

NASA Global Earth Observing System (2001). A Shadow of a Lake: Africa's Disappearing Lake Chad. GSFC on-line news. http://www.gsfc.nasa.gov/gsfc/earth/environ/lakechad/chad.htm

National Drought Mitigation Center (2000). Understanding ENSO and Forecasting Drought. Available on http://drought.unl.edu/ndmc/enigma/elnino.htm

NEMA (1999). *State of Environment Report for Uganda 1998.* Kampala

Nemec, J. (1991). Mitigation Of Hazards, Planning Of Preparedness And Management. Selected Papers Of The UNDRO/UNDP/Government of Egypt. Training Seminars, Cairo (2-7 March, 1991), pp 61-66

NESDA (2000). Ethiopia: The Extent And Impact Of Air Pollution In Addis Ababa. *Ecoflash 10*; July-August 2000

Obura, D., Suleiman, M., Motta, H., and Schleyer M. (2000). Status Of Coral Reefs In East Africa: Kenya, Mozambique, South Africa And Tanzania Pp. 65–76 In: Wilkinson, C. *Status of Coral Reefs of the World: 2000.* Australian Institute of Marine Science and Global Coral Reef Monitoring Network, Townsville

OFDA (1987). Disaster History; Major Disasters World-Wide. United States Office for Disaster Assistance

OFDA (2000). Statistics Database. United States Office for Disaster Assistance

ONE (1997). Rapport Sur L'Environment Urbain. Cas de la Zone d'Antananarivo. Edition 1997. Banque Mondiale IDA 2125 MAG. Office National Pour L'Environnement, Antananarivo

Ottichilo, W.K., and others (1991). *Weathering the Storm—Climate Change and Investment in Kenya.* African Centre for Technology Studies. Nairobi

PanAfrican News Agency (2001). *Addis Ababa Gets Laboratory To Detect Pollution.* PanAfrican News Agency, 14 May, 2001

PEACENET (2000). *Morocco: Drought Threatens Economy.* PEACENET Headlines 11-14 April. Available on: http://www.icg.org/icg/pn/hl/1000411275/hl8.html

PRE/COI (1998). Rapport Régional sur les Récifs. Programme Régional Environnement, Commission de l'Ocean Indien, Quatre Bornes, Mauritius. Avril 1998

Ragoonaden, S. (1997). Impact of sea level rise in Mauritius. In Island at Risk; Global Climate Change, Development and Population. Leatherman, S.P., (Ed) *Journal of Coastal Research*, Special Issue 4

SEI (1998). Regional Air Pollution in Africa. http://www.york.ac.uk/inst/sei/africa/afpol4.html

Shah, N.J. (1995). Managing Coastal Areas in the Seychelles. *Nature and Resources* 31 (4) 16-33. UNESCO, Paris

Shah, N.J. (1996). Climate Change Poses Grave Danger to African Island States. *Splash* 12 (1) 14-15. SADC

Southern Centre (1996). *Climate Change Mitigation in Southern Africa, Methodological Development, Regional Implementation Aspects, National Mitigation Analysis and Institutional Capacity Building in Botswana, Tanzania, Zambia and Zimbabwe.* Southern Centre, Harare

Swearingen, W.D., & Bencherif, A. (eds.) T*he North African Environment at Risk.* Westview Press

UNCCD (2001). Action Programme to Combat Desertification: Africa. Updated list November 2001, available on http://www.unccd.int/actionprogrammes/africa/africa.php

UNCHS (2001). *The State of the World's Cities 2001.* United Nations Centre for Human Settlements (HABITAT), Nairobi

UNDHA (1994). *Madagascar Cyclone January, February and March 1994.* UNDHA Reports 1–67 United Nations Division for Humanitarian Affairs. http://wwwnotes.reliefweb.int

UNDP (2000). *Human Development Report 2000.* United Nations Development Programme. Oxford University Press, New York

UNEP (1985). First African Ministerial Conference on the Environment, Cairo, 16–18 December 1985

UNEP (1999). *Western Indian Ocean Environment Outlook.* United Nations Environment Programme, Nairobi

Unified Arab Economic Report (1999)

United Nations (1999). Harnessing Abundant Gas Reserves. *Africa Recovery* Vol 13(1); June 1999

US Embassy in Mauritius (2001). Cyclone Preparedness in the Indian Ocean. http://usembassymauritius.mu/Consular/cyclone.htm

USAID (2001a). Protecting Egypt's Environment. Perspectives From the Field. Available on http://www.usaid.gov/regions/ane/newpages/perspectives/egypt/egenv.htm

USAID (2001b). Global Climate Change and Africa Webpage. Available on: http://africagcc.gecp.virginia.edu/USAID/Fut_dir.htm

Verschuren, D., Laird, K.R., & Cumming, B.F. (2000). Rainfall and Drought in Equatorial East Africa During the Past 1,100 years. *Nature* 403: 410-414

WFP (2000). *Drought in the Horn of Africa: Getting Food to the Hungry.* World Food Programme ReliefWeb, 13 June 2000

White, G.F. (1992). Natural Hazards Research. *Natural Hazards Observer.* Vol. XVI. No3 1-2 January, 1992

Wijkman, A., & Timberlake, L. (1984). *Natural Disasters: Acts of God or Acts of Man?* Earthscan, Nottingham

WMO (2000). WMO Statement on the Status of the Global Climate in 2000. WMO, Geneva, 19 December 2000

World Bank (1995). Middle East and North Africa. Environmental Strategy. Report No. 13601- MNA

World Bank (1999). *World Development Indicators.* The World Bank, Washington D.C.

World Bank (2000a). World Development Report 2000/2001. World Bank, Washington D.C.

World Bank (2000b). *Indoor Air Pollution: Energy and Health for the Poor.* Issue No. 1; September 2000

World Bank (2001). Urban Transport Dysfunction and Air Pollution in Dakar. *Findings* No 184, June 2001

WorldWatch Institute (2000). The Melting of the World's Ice. http://www.worldwatch.org/mag/2000/136c.pdf

WWF (2000). Climate Change and Southern Africa. Available on: http://www.livingplanet.org/resources/publications/climate/Africa_issue/page1.htm

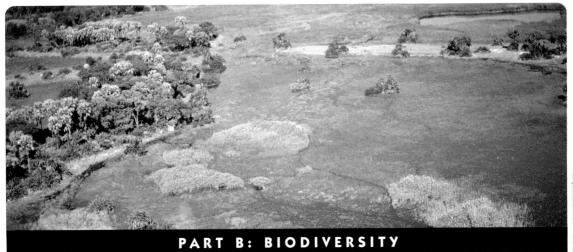

*François Suchel/Still Pictures*

# PART B: BIODIVERSITY

## REGIONAL OVERVIEW

Biological diversity, or 'biodiversity', means the variety of plant and animal life at the ecosystem, community or species level, and even at the genetic level. Biodiversity is most commonly measured and reported at species level with characteristics such as species richness (number of species), species diversity (types of species) and endemism (uniqueness of species to a certain area) being the most useful elements for comparison.

Only a fraction of the species inhabiting the earth have been identified and studied to date, and the roles they play in influencing the environment are still often poorly understood. Studies have tended to concentrate on higher plants and mammals, and such species have also been the focus of most conservation efforts. This can give a misleading impression of the importance of 'lower organisms', such as bacteria, insects and fungi, which play vital ecological roles—for example, in nutrient cycling, and regulation of water, soil and air quality. Lack

**Figure 2b.1 Map of Africa showing biodiversity hotspots and potential hotspots**

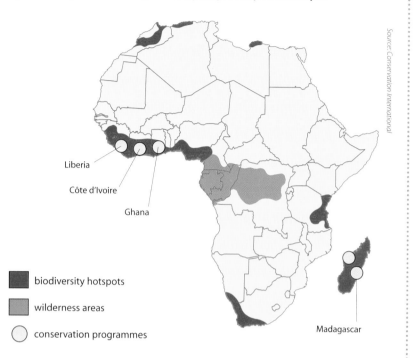

*Source: Conservation International*

Liberia

Côte d'Ivoire

Ghana

Madagascar

■ biodiversity hotspots

▨ wilderness areas

○ conservation programmes

atmosphere into green plants—greater than 800 g carbon/m$^2$/yr). They also support an estimated 1.5 million species. By contrast, Africa's arid areas are among the harshest environments in the world. The Sahara and Namib deserts and the Sahel, for example, have NPP of just 100 g carbon/m$^2$/yr (WCMC 2000). But even under these conditions, many plant and animal species manage to thrive.

The designation of some areas as 'biodiversity hotspots' is a useful concept developed in recent years as a means of prioritizing habitats for conservation (Myers 1990). Hotspots are areas where species diversity and endemism are particularly high, and where there is an extraordinary threat of loss of species or habitat. There are 25 internationally recognized hotspots, six of them are in Africa (Mittermeier, Myers, Gil and Mittermeier 2000). These are shown in Figure 2b.1 and are described below:

- **The Mediterranean Basin Forests** constitute just 1.5 per cent of the world's forests, yet are home to 25 000 plant species and 14 endemic genera (Quézel, Médail, Loisel and Barbero 1999).
- **The Western Indian Ocean Islands** have extremely high levels of endemism due to their isolation. This is especially true for Madagascar which has the highest number of endemic species in Africa (including 700 endemic vertebrate species), and ranks 6th in the world (UNEP 1999).
- **The Cape Floristic Region** in South Africa is the smallest and richest of the world's floral kingdoms, with 68 per cent of the 8 700 plant species endemic to the region (Low and Rebelo 1996).
- **The Succulent Karoo**, shared between South Africa and Namibia, is the richest desert in the world—40 per cent of its 4 849 species are endemic (Low and Rebelo 1996).
- **The Guinean Forest** hotspot is a strip of fragmented forest running parallel to the coast of Western Africa through 11 countries from Guinea to Cameroon. It has the highest mammalian diversity of all of the world's 25 hotspots (551 species out of the 1 150 mammalian species in the African region) and contains 2 250 plant species, 90 bird species, 45 mammal species and 46 reptile species found nowhere else (Conservation International 2002, Mittermeier and others 2000).

of understanding of the role of such organisms can lead to their being 'sidelined' when conservation efforts or commercial utilization are being considered. It is worth noting that about 1 million of the 1.75 million species described so far are insects and myriapods, and that the total number of such species is estimated to be about 8 million. In other words, only one-eighth of insect and myriapod species have been identified and recorded to date. It is also estimated that there are 1.5 million species of fungi, of which 72 000 have been described, and 1 million species of bacteria, of which a mere 4 000 have been described (WCMC 2000).

Africa has rich and varied biological resources, forming the region's natural wealth on which its social and economic systems are based. These resources also have global importance, for the world's climate and for the development of agriculture or industrial activities such as pharmaceutics, tourism or construction, to name but a few of the most important areas.

Africa is also a continent of extremes, both in terms of physical features and climatic conditions and, therefore, in terms of the life it supports. The humid tropical forests of equatorial Africa are among the most productive ecosystems in the world, with Net Primary Productivity (NPP)—the net flux of carbon from the

- **The Eastern Arc Mountain Forests** of Eastern Africa are 30 million years old and are thought to have evolved in isolation for at least 10 million years. As a result, more than 25 per cent of the plant species are endemic (Lovett 1998).

Africa also has several areas where both species richness and degree of threat are high, but endemism is lower. These 'potential hotspots' include the highlands of Ethiopia; the forests of the Albertine Rift in eastern Congo, Rwanda, Burundi and adjacent parts of Uganda and Kenya; the western escarpment of Angola; and the Miombo Woodlands of interior Southern Africa (Mittermeier and others 2000).

Africa also has a range of aquatic habitats with very high levels of biodiversity. Marine ecosystems tend to be more diverse than terrestrial ones and reach even higher levels of diversity in warmer tropical waters than in cooler seas. The coasts of many African countries have rich ecosystems, such as coral reefs, seagrass meadows, mangrove forests, estuaries and floodplain swamps (Martens 1995). Rivers, lakes (freshwater and soda) and riverine edge swamps, valley swamps, seasonal floodplains, ponds and high altitude peat-forming wetlands all contribute to a wide variety of aquatic ecosystems in Africa that support an extensive range of resident as well as migratory species (Harper and Mavuti 1996).

The national and sub-regional boundaries that characterize present day Africa are the result of geographical and human activities often determined by political or economic factors, they therefore seldom reflect the boundaries of ecological systems. This difference between political and ecological units is significant—when the boundaries of ecosystems extend beyond territorial boundaries, the protection of the natural resources within those ecosystems requires management strategies that are coordinated jointly between nations or sub-regions (Westing 1993). The disparities between political and ecological boundaries also imply some overlap in discussion of biological systems in the sub-regions referred to in this report. For example, Tanzania forms part of the Southern African sub-region, but shares major ecological systems with Kenya and Uganda. It is therefore discussed in both the Southern Africa and Eastern Africa analyses.

## ECOLOGICAL, ECONOMIC AND SOCIAL VALUES OF BIOLOGICAL RESOURCES

Biological resources are the backbone of the African economy as well as the life-support system for most of Africa's people. A variety of resources, both plant and animal, are used for food, construction of houses, carts and boats, household utensils, clothing and as raw materials for manufactured goods. Many resources, such as timber and agricultural produce, are traded commercially, and others are used in traditional crafts such as basket weaving and carving. In addition, many species with medicinal properties are harvested by local communities and pharmaceutical multinationals alike. These include the African potato (*Hypoxis rooperi*) in Southern Africa (Natures Truth 2001), the rosy periwinkle (*Catharanthus roseus*) from Madagascar and Mozambique, and prunus (*Prunus africana*) from Cameroon, Democratic Republic of Congo (DRC), Kenya and Madagascar (Sheldon, Balick and Laird 1997). Other species provide the genetic resources for improved agricultural products such as disease- or drought-

Madagascar: picking rosy periwinkle, *Catharantus roseus*, used to make drugs to treat cancer

*D. Halleux/Still Pictures*

resistant crops. For example, an African species of rice has been used in the development of a high-productivity, drought-resistant variety (Science in Africa 2001) and the native Mauritian caffeine-free *Coffea* species could be used to develop coffee cultivars with low caffeine content (GOM/ERM 1998). The richness and diversity of ecosystems in Africa also provide opportunities for tourism which many African countries have successfully exploited. The coral reefs of the Red Sea, Eastern African coast and Western Indian Ocean Islands, for example, are among the most famous in the world, and the savannas of Eastern and Southern Africa are popular destinations for safari-goers.

Africa's biodiversity is under threat from four main sources, described below. These are:

● natural habitat loss;

● loss of species or subspecies;

● invasion by alien (non-native) species; and

● lack of recognition of indigenous knowledge and property rights.

These issues, individually and in combination, constitute forces that are restricting full realization of the value of Africa's biodiversity and use of the resources it provides for Africa's own development.

## NATURAL HABITAT LOSS

Natural habitats in Africa are being degraded or lost owing to a number of 'proximate' and 'ultimate' (or root) causes. Proximate causes include clearing for alternative land uses (mainly agriculture and human settlements) and overharvesting of resources (most notably timber in the forests of Central and Western Africa). More than 211 million hectares of African forest have been lost since 1970, amounting to almost 30 per cent of the original extent. In the same period, the land area under cultivation has increased by 36 million hectares, or 21 per cent (FAOSTAT 2000). Other threats to terrestrial habitats include bush fires which are commonly used in agriculture to prepare the soil, but which can get out of control and destroy large areas of forest or woodland. On the other hand, fire (along with grazing) is also considered to be one of the most important factors determining the structure of savanna ecosystems (Gichohi, Gakahu and Mwangi 1996).

Coastal habitats are being reduced by proximate causes, such as: overharvesting of resources; physical alterations, and urban and industrial developments; siltation; pollution; introduction of alien species; and global climate change (Martens 1995). Inland wetland areas have suffered in the same way, with additional problems stemming from drainage for conversion to agriculture and salinization of soils due to irrigation, as well as overextraction of water from lakes or the rivers feeding them (Harper and Mavuti 1996).

The ultimate causes of habitat loss in Africa are human population growth and the resulting demand for space, food and other resources; widespread poverty; a dependence on natural resources; and economic pressures to increase exports, particularly of agricultural produce, timber and mineral products. A lack of awareness of the value of wild biological resources, lack of knowledge of biodiversity and of how to apply that knowledge, and failure to enforce conservation policies have also contributed to a decline in area of natural habitat. For example, development and expansion of settlements, infrastructure and logging activities have not, until recently, been coordinated with respect to natural habitat or sensitive ecosystems and were not subject to environmental impact assessment requirements. In some cases, protected areas have been established without consultation of local people, without their consent and, frequently, without being based on rigorous biological inventory studies. This has led to protected areas which are not only less effective in their contribution to conservation but which have also failed to earn the respect of local communities. In some cases, communities have encroached on the protected areas; in others, they have deliberately raided resources (FAO 1995, Fay, Palmer, and Timmermans 2000). Weak legal and institutional structures, corruption, conflict and civil strife, as well as market factors, can also contribute indirectly (or even directly) to habitat degradation and loss (Martens 1995, Rodgers, Salehe and Olsen 2000).

Loss of habitat threatens biodiversity at all levels, from the 'ecosystem' to the 'genetic'. Three main processes are at work, namely:

● reduction in the total size of habitat;

● fragmentation of habitat; and

● change in the structure or characteristics of the habitat.

Plant and animal communities need habitats that are of sufficient size to provide food and water and to

*In some cases, protected areas have been established without consultation of local people, without their consent and, frequently, without being based on rigorous biological inventory studies. This has led to protected areas which are not only less effective in their contribution to conservation but which have also failed to earn the respect of local communities*

## Table 2b.1 Protected areas in Africa, 1999

| | Nationally Protected Areas | | | | Internationally Protected Areas | | | | | |
| | Terrestrial | | | Marine | Biosphere Reserves* | | World Heritage Sites | | Ramsar Sites | |
| | No. | Area (000 ha) | % Land area | No. | No. | Area (000 ha) | No. | Area (000 ha) | No. | Area (000 ha) |
|---|---|---|---|---|---|---|---|---|---|---|
| Central | 69 | 31 161 | 33.1 | 10 | 11 | 3 034 | 7 | 9 121 | 8 | 4 228 |
| Eastern | 119 | 11 981 | N/A | 16 | 7 | 1 126 | 5 | 454 | 5 | 105 |
| Northern | 72 | 15 862 | 7.8 | 50 | 13 | N/A | 2 | >13 | 22 | >2 000 |
| Southern | 578 | 65 014 | N/A | 44 | 8 | N/A | 10 | 7 850 | 27 | 12 026 |
| Western | 123 | 28 724 | 68.2 | 25 | 15 | 31 112 | 10 | 1 2003 | 37 | 3 674 |
| WIOI | 89 | N/A | N/A | 3 | 3 | N/A | 3 | N/A | 4 | 53 |
| Total | 1050 | N/A | N/A | 148 | 57 | N/A | 37 | >29 441 | 103 | >22 086 |

Source: Ramsar 2002, UNDP and others 2000, UNEP 1999, UNESCO 2002 Data not available for Burundi, Cape Verde, Comoros, Djibouti, Sao Tome & Principe, Seychelles & Swaziland
*Some Biosphere Reserves are also World Heritage Sites or Ramsar sites

find a mate or nesting site. If the total size of a habitat is reduced, the sizes of the populations of many species, particularly large mammals and top predators, are forced to decline. In some cases, population sizes have reached the minimum viable population, that is to say the lower limits beyond which the species will not be able to breed successfully (Gilpin and Soule 1986).

Fragmentation of habitats is a particular problem for large animals and top predators which require extensive ranges. Even if the total area of a natural habitat is large, some species may still be threatened if it is divided into fragments that are too small or are lacking in some vital components (Harris 1984, Gilpin and Soule 1986, Skole and Tucker 1993). This factor has not always been taken into account when creating protected areas, and migration routes of large herbivores have been cut off, preventing them from finding adequate food or water resources in times of stress.

Fragmentation and selective harvesting can also change the nature of a habitat, making it unsuitable for certain species. Forests, for example, have distinct 'edge species' and 'interior species'. Edge species are those that are suited to conditions at the edge of the forest (where there is typically more exposure to light, wind, and predation); interior species are more suited to conditions inside the forest. Reduction of forest area or size of forest patches increases the ratio of edge to interior, and thus changes the species composition (Lovejoy, Bierregaard, Rylands, Malcolm, Quintela, Harper, Brown, Powell, Powell, Schubart and Hays 1986). Too much edge threatens interior species and natural forest dynamics can collapse.

### Mitigation of natural habitat loss

The typical response to warnings of loss of natural habitat in the past 30 years has been to increase the number and extent of protected areas. Although initially the establishment of protected areas was not always based on biodiversity assessments or threats to certain habitats, in recent years, tools such as Geographic Information Systems (GIS) have been used to identify areas of particularly high conservation priority, such as unique habitats or especially species-rich habitats (Burgess, de Klerk, Fjeldså, Crowe and Rahbek 2000). Protected areas in Africa are shown in Table 2b.1

Only six African countries (Botswana, Burkina Faso, Namibia, Rwanda, Senegal and Tanzania) have more than the international target of 10 per cent of their land area under protection (World Bank 2001). Absolute percentage area is not, however, the only important consideration in conservation efforts. International efforts and regional partnerships have contributed to biological conservation both within and outside of protected areas. Wetlands, for example, have received much attention through the signing of the Convention on Wetlands of International Importance, Especially as Waterfowl Habitat, signed in Ramsar, Iran in 1971 (and known as the 'Ramsar Convention'). The Convention provides the framework for national action and international cooperation for the conservation and wise use of wetlands and their resources. Thirty-three African countries are party to the Ramsar Convention (February 2002), and there are 103 'Ramsar sites' in Africa with a combined area of more than 20 million hectares.

In recent years, the concepts of World Heritage Sites, Biosphere Reserves, and Transborder Parks have also been influential in establishing conservation priorities. World Heritage Sites are sites considered to be of global ecological and cultural significance. There are 35 of these in Africa, totalling 37 million hectares (UNDP, UNEP, World Bank and WRI 2000).

The concept of Biosphere Reserves—developed in 1971 by the United Nations Educational, Scientific and Cultural Organization (UNESCO), together with Conservation International—established biosphere reserves to protect whole ecosystems rather than selected species. Biosphere reserves include areas in which various types of human activity are allowed. There are currently 50 such reserves in Africa covering a total of 52 million hectares (UNDP and others 2000).

Transborder parks are protected areas that over-run national boundaries and in which the relevant countries share the conservation activities, as well as the benefits. The first of these, the Kgalagadi Transfrontier Park, was established in 1998, between Botswana and South Africa, allowing free migration of species within the Kalahari Desert. Table 2b.1 shows protected areas in Africa.

Protected areas do not, however, meet all of Africa's needs in terms of conservation of natural habitat. In some countries war, poaching and encroachment by refugees or local communities claiming traditional ownership of the land have contributed to loss and degradation of vegetation, water and species composition. Lack of resources to enforce protection of protected areas also constitutes a major barrier to their effectiveness. In addition, concern has been expressed at the potential loss of species (or local extinctions) within such areas if they become too

insularized. One study has shown that six species of large diurnal mammals have become locally extinct in Tanzanian parks in the last 80 years. This problem could be alleviated by the establishment of corridors, or through protection outside of parks, to facilitate recolonization (Newmark 1999). Furthermore, the purpose of conserving habitats is to allow present and future generations to benefit from the resources and services they provide. Programmes of sustainable use of natural resources should therefore also be considered in addition to exclusion of human activities from some areas. To this end, Community Based Natural Resource Management Programmes (CBNRM) have been implemented in parts of Africa such as Zimbabwe, Kenya and Uganda, with varying degrees of success in terms of socio-economic development of surrounding communities and protection of threatened habitat (Hulme and Murphree 2001).

Private reserves have also been created as a means of promoting habitat protection. Although there are few assessments of the effectiveness of this approach, surveys carried out in 1989 and 1993 in Latin America and Africa concluded that private reserves were generating substantial employment and that their profitability was increasing. They also noted that African reserves were larger than those in Latin America (with an average size of 11 436 ha). A follow-up study concluded that the effectiveness and profitability of private reserves were sufficient for them to warrant greater support as agents of sustainable development and conservation (Alderman 1991, Langholz 1996).

Fifty-two African countries are party to the Convention on Biological Diversity and most have shown their commitment at the national level through the development of National Environmental Action Plans (NEAPS) and National Conservation Strategies. Financial assistance through the World Conservation Union (IUCN), the World Bank, UNEP and UNDP's Global Environment Facility (GEF) offer opportunities to overcome some of these barriers and to promote sub-regional cooperation in conservation.

## SPECIES LOSS

Individual species are under threat from a variety of pressures in addition to loss of their natural habitat. Recent estimates show that a total of 126 recorded

Bushmeat trader, Abidjan

*Anna Giovenito*

animal species have become extinct (or extinct from the wild) in Africa, and that there are 2 018 threatened animal species across the region. Some 123 plants are recorded as extinct and 1 771 are threatened (IUCN 1997).

The reasons for such high rates of species loss or endangerment include:

- habitat loss;
- illegal hunting for food;
- medicinal, or commercial use; and
- national and international trade.

A recent study found that the bushmeat trade in Central and Western Africa is contributing significantly to the decline in populations of gorillas, chimpanzees, elephants, bush pigs and forest antelopes. Bushmeat is a traditional supplement to the diets of many African communities, but the increasing human populations and commercial trade are pressurizing these species to the extent of a million tonnes of bushmeat a year (Greenwire 2001). Bushmeat is now a major source of animal protein in many towns and cities in tropical Africa (Fa, Garcia Yuste and Castello, 2000). In addition, activities such as logging and mining are improving access to previously remote areas, making collection from the wild more profitable. A recent study showed that road density is linked to habitat fragmentation, deforestation and intensified bushmeat hunting (Wilkie, Shaw, Rotberg, Morelli and Auzel 2000).

Selective harvesting of medicinal plants is also

taking its toll on species diversity and abundance. The World Health Organization (WHO) has estimated that 80 per cent of people in the developing world are reliant on traditional medicines. Eighty-five per cent of these medicines use plant extracts, so it is estimated that around 3 000 million people around the world rely on plants for traditional health treatments. The number is even larger if plant-derived commercial drugs are included (Sheldon and others 1997). In Africa, 80 per cent of both rural and urban populations depend on medicinal plants for their health needs (and those of their livestock) either because they prefer them for cultural or traditional reasons, because such remedies are effective in treating certain diseases, or because there is a lack of affordable alternatives (Baquar 1995, Ole Lengisugi and Mziray 1996). Overharvesting of these plants from the wild and loss of their habitat are threatening many species (as well as human and livestock health) and alternative strategies, such as cultivation of medicinal species in nurseries, are being considered (Dery and Otsyina 2000). The *Acacia senegal* tree is another highly sought-after species because of the high economic value of the gum arabic which it produces. The species has been commercially exploited, particularly in Sudan (Abu-Zeid 1995).

The exotic pet trade is another powerful international driving force for species reduction, as is demand for animal products such as ivory, rhino horn, skins, furs and other trophies. Many species are protected by restrictions on international trade, but a thriving black market creates a demand, and extreme levels of poverty tempt local people to meet it (CITES 2002).

Loss of species means loss of economic opportunities, both now and in the future. For example, the possibility of developing new strains of drought- or disease-resistant crops, crops with higher protein content or synthetic manufacture of pharmaceutical products of plant origin may be lost. Different breeds of livestock are also important components of global biodiversity, because they have genes that are or may be useful for agricultural production. Locally adapted strains of livestock often have greater potential for increased productivity, due to their ability to thrive under specific conditions. For example, Botswana and Namibia have the greatest proportional diversity of breeds, but these indigenous breeds could be

•

*Recent estimates show that a total of 126 recorded animal species have become extinct (or extinct from the wild) in Africa, and that there are 2 018 threatened animal species across the region. Some 123 plants are recorded as extinct and 1 771 are threatened*

•

African elephant *(Loxodonta africana)* in Amboseli National Park, Kenya

*M. and C. Denis-Huot/Still Pictures*

threatened if the livestock husbandry systems under which they have been developed are replaced or changed, or if exotic breeds or their semen are imported under aid programmes or livestock development policies. Loss of traditional livestock management regimes and the genetic diversity that they provide could result in increases in the cost of food production in those areas (Hall and Ruane 1993, Hall and Bradley 1995). On the other hand, properly planned and managed *in-situ* conservation of livestock breeds could satisfy demands for intensified production without the need for massive external inputs and foreign aid (Hall and Bradley 1995).

Species loss or extinction can also affect local and regional communities, either because opportunities for tourism are lost or because of security problems arising from heavy poaching. Local communities also lose valuable sources of food and other raw materials, as well as losing irreplaceable cultural and spiritual assets.

Ecological communities are impacted by species loss, because this affects predator/prey relationships, and removes agents of pollination, seed dispersal or germination such as insects, elephants, primates or other animals. Species loss reduces the integrity of ecosystems in ways that are still far from being fully understood and reduces the environment's capacity to regulate climate, water and soil quality.

## Combating species loss

Forty-eight African countries are party to the Convention on International Trade in Endangered Species of Flora and Fauna (CITES) which regulates trade in endangered species or products through trade prohibitions or suspensions. This has had varying degrees of success in Africa. For example, there has been significant growth in national herds of elephant in Botswana and Zimbabwe in recent years after a ban was introduced on trading in ivory, although it is argued in some quarters that this has more to do with good conservation measures than trade restrictions on ivory. Southern African countries now want to resume some level of trade in order to capitalize on economic opportunities from successful conservation. In Kenya, however, the decline in poaching of elephants (from 3 000 deaths a year to less than 50) which followed a dramatic drop in prices in ivory (from US$60 to US$5 per kilo in Somalia) indicated direct links between the ivory trade, prices and criminal activities (Western 1997). East African countries feel that any relaxation of restrictions will further endanger their elephant populations which have not recovered so quickly. The black rhino, another endangered species, is still threatened throughout Africa by illegal hunting and populations have not recovered to pre-1960s levels.

Species reintroduction and *ex-situ* plant propagation in nurseries are some of the additional efforts underway to counteract the recent rapid loss of species in Africa. In the Western Indian Ocean Islands, successful conservation measures resulted in growth of populations of the Mauritian Kestrel and the Pink Pigeon.

## ALIEN INVASIVE SPECIES

A further threat to biodiversity comes from invasion by non-native, or alien, species of plants and animals. These are species that have been introduced both accidentally and intentionally, and that are free from their natural predators or other natural limitations to their population growth. They are thus able to dominate plant and animal communities, either by out-competing native species for space, light or nutrients, or through predation.

Southern and Eastern Africa and island nations have been particularly affected by the introduction of alien species. The Nile perch (*Lates nilotica*), introduced into Lake Victoria 30 years ago to stimulate the fisheries of Uganda, Kenya and Tanzania, is a striking example. In the 1960s, the Nile perch accounted for about 1 per cent of fish catch; it is now dominant in the lake, representing close to 80 per cent of annual fish harvests, and its introduction is believed to have caused the loss of more than 200 endemic species (MoFPED 2000). Lack of control on introduction of species has also led to the rapid spread and domination of the water hyacinth (*Eichornia crassipes*) in freshwater bodies across Africa, including Lake Victoria and Lake Kariba. This weed blocks water channels, alters hydrological regimes, and renders surrounding areas more vulnerable to flooding.

Islands are particularly vulnerable to invasions by alien species, particularly predators, because many island species have evolved in isolation from predators such as cats and rodents. The Western Indian Ocean Islands have experienced dramatic changes to their ecology through the introduction of a range of species, including monkeys, pigs, rats, mice, rabbits, privet, Chinese guava and wild pepper (UNEP 1999).

Lake Victoria: dense mats of water hyacinths accidentally introduced from Latin America

*Hartmut Schwarzbach /Still Pictures*

Invasion by alien species reduces biodiversity, either through predation, competition or smothering. In some cases, alien plants form such dense infestations and produce so many seeds that are dispersed so widely that it is virtually impossible to control them. They also change the dynamics of the natural system and may produce toxic chemicals, inhibiting the growth of native species. In other cases, they threaten native species and functioning of ecosystems through an excessive consumption of resources such as water. In Southern Africa, pines, eucalyptus and acacias have been introduced for commercial forestry, but have invaded natural habitats where they threaten ecological integrity by using many times more water than native species (Working for Water 2000).

**Prevention and control of invasion by alien species**

Options to control the introduction and spread of alien species include tightening controls on importation of products of animal or plant origin. However, lack of resources to police borders and entry points, and to enforce fines for breach of regulations, results in continued threats to biodiversity.

Infestations of alien plants can be physically removed by hand, by mechanical or chemical means, or by a combination of these. Such approaches have been adopted in Lake Victoria and in parts of Southern Africa. Biological control is gaining in popularity as a means of containing populations of alien plants and animals, as it is less harmful to other species and to the surrounding environment. However, it is a more lengthy process, because release of control organisms into an environment has to be rigorously tested to ensure that release has no adverse effects.

## INDIGENOUS KNOWLEDGE AND PROPERTY RIGHTS

Efforts to protect and conserve Africa's biodiversity are constrained by a lack of research into and documentation of the continent's biodiversity. This is especially true regarding indigenous knowledge of, for example, the properties of selected species and their traditional uses, and of natural resource management and conservation practices. The relatively low level of knowledge in the science of biodiversity has resulted from a lack of investment in research and development, compounded in some cases by a failure to assign monetary values to the ecological functions of plant and animal species, thus fostering the belief that natural resources are free and in unlimited supply.

Lack of knowledge has two main results:

● conservation efforts are constrained by lack of understanding of species and ecological systems; and

● opportunities for commercial exploitation and economic gains are missed or are taken up by international companies without the benefits returning to the original holders of the knowledge.

Some efforts have been made to address this situation. In recent years, influential publications have increasingly helped to improve understanding by emphasizing the links between cultural diversity and biological diversity, by describing how indigenous peoples use their knowledge to manage their natural resources, including plants and animals (Warren 1995). Studies on the value that indigenous people place on forest and wildlife resources indicate that they ascribe a high value to their resources and understand that—in addition to being sources of food, fuel, medicine and building materials—these resources can also provide ecosystem services (Ntiamoa-Baidu 1995, Olsen, Rodgers and Salehe 1999).

Increasing efforts are, therefore, being made to understand indigenous knowledge systems and to promote their continued application, thereby mainstreaming such knowledge and knowledge-holders and bringing them into new development projects. This new form of partnership includes the application of indigenous knowledge to environmental assessment and to project implementation (Croal 2000, Emery 2000).

The second result mentioned above—missing of commercial opportunities by communities—relates to the right of control over traditional lands and resources, intellectual property, and accommodation of customary laws and practices. Article 14 of the International Labour Organization's Convention 169 on Indigenous and Tribal Peoples (1989) states that *'the rights of ownership and possession of the peoples concerned over the lands which they traditionally occupy shall be recognized. In addition, measures shall be taken in appropriate cases to safeguard the right of the peoples concerned to use lands not exclusively occupied by them, but to which they have traditionally had access*

*for the subsistence and traditional activities. Particular attention shall be paid to the situation of nomadic peoples and shifting cultivators in this respect.'*

The Convention on Biological Diversity also recognizes the value of traditional knowledge. Its Article 8(j) states that parties should, as far as possible, *'...respect, preserve and maintain knowledge, innovations and practices of indigenous and local communities embodying traditional lifestyles relevant for the conservation and sustainable use of biological diversity and promote their wider application with the approval and involvement of the holders of such knowledge, innovations and practices and encourage the equitable sharing of the benefits arising from the utilization of such knowledge innovations and practices.'*

Many governments are now in the process of implementing this principle through their national biodiversity action plans, strategies and programmes. The means employed include adoption of relevant legislation, policies, and administrative arrangements for protecting traditional knowledge through prior informed consent. Although such means are usually applied to medicinal plants, cultivars, and some traditional practices, it has been pointed out that protection of traditional knowledge should also include traditional stock raiser rights and the protection of animal genetic resources (Kohler-Rollefson and McCorkle 2000). It has been suggested that livestock projects should be required to present 'genetic impact statements' to anticipate the effect on the local gene pool of the introduction of exotic livestock breeds. If a negative impact is anticipated, the projects could perhaps include a provision for maintaining a purebred population of the indigenous livestock breed (Hall and Bradley 1995).

The World Bank's programme on Indigenous Knowledge in Africa has established 15 resource centres across Africa, focusing on identification and dissemination of indigenous or traditional knowledge and practices. Policies have also been developed under this programme to protect indigenous knowledge, and to apply it in agricultural, medicinal and conservation practices.

Given the many and diverse pressures on Africa's biological resources, and the enormous untapped potential for economic and social development, implementation of strategies for effective conservation and sustainable use of these resources is of paramount and immediate importance. Such strategies must be activated on a number of fronts, and existing measures at national, sub-regional and regional levels will have to be reinforced, upgraded and opened up to further innovative measures to deal with changes in the status of habitats or species. Compliance with regional and international agreements must be strengthened through enforcement of legislation, policies and plans, and through institutional reform. Further research and application of research findings is required to strengthen *in-situ* and *ex-situ* conservation systems, particularly those involving local and indigenous innovations and interventions. Systematic valuation of resources and incorporation of those values into national accounting systems and development policies are also required. Capacity building and awareness programmes are critical components for each activity, and at each level (Mugabe 1998).

A view over Haute Atlas, Morocco

*Michael Gunther/Still Pictures*

## NORTHERN AFRICA

Northern Africa's extreme climatic conditions and physical features greatly influence the sub-region's biological elements. Most of Northern Africa falls within arid and semi-arid environments, but there is also a range of geomorphological features and sub-climatic zones which have created diverse ecosystems and extremely rich communities of flora and fauna. For example, long shores with vast areas of coastal land, oases in the Sahara and varying landforms create diversity of habitat and remarkable biological diversity, including a large number of varieties and strains of field crops. Morocco, for example, has 3 675 recorded species of higher plants, and Sudan has 267 recorded species of mammals and 938 recorded species of birds. The sub-region's reptiles and amphibians are still under assessment in most countries, but Egypt has already recorded 83 species of reptile and 6 species of amphibians. The sub-region has 1 129 endemic species, 22 endemic species of mammals, one endemic bird species, 20 endemic species of reptiles and 4 endemic species of amphibians, with Morocco having the greatest level of endemism (WCMC 1992). Northern Africa also encompasses the biota of the semi-closed Mediterranean and Red Seas. The

Mediterranean Basin is one of the 25 internationally recognized biodiversity hotspots, with extraordinary plant diversity and endemism.

There are also five regionally recognized hotspots in Northern Africa. These are:

- The Imatong Mountains and surrounding lowlands on the border between Sudan and Uganda. This area has nearly half of the total flora of Sudan and 12 endemic species of plants.
- The isolated Jebel Marra volcanic massifs near the Sudan border with Chad, with about 950 plant species.
- Jebel Elba, a mountainous ecosystem bordering the Red Sea between Egypt and Sudan. This is a transitional area between Afrotropical and palaearctic biogeographic realms and has an estimated three to four times as many plant species as desert areas further north.
- Tassili d'Ajjer, a highland area in Algeria, where several plant species are recorded as near endemic and one is strictly endemic.
- The High Atlas Mountains which extend along the northern part of Morocco and into Tunisia, home to more than one-third of all endemic species in Northern Africa.

## ECOLOGICAL, ECONOMIC AND SOCIAL VALUES OF BIOLOGICAL RESOURCES IN NORTHERN AFRICA

The richness and diversity of species in Northern Africa constitutes a wealth of biological resources (Hegazy 2000a, 2000b). Plant biodiversity in the region has supported the grazing herds of camels, sheep and goats led by nomadic pastoralists for millennia, and agricultural advances have promoted the use of many high-yielding cultivars adapted to the arid environment that predominates in the sub-region. Some species offer opportunities for biotechnological modification to improve agricultural, medicinal and industrial applications. About 70 per cent of wild plants in the region are known to be of potential value, more than 10 per cent have the potential for commercial exploitation, and 35 per cent of useful plants are either under-utilized or can be used for more than one purpose. These under-utilized and multipurpose species have potential value as sources of food, forage for livestock, medicine and pharmaceuticals and for agro-forestry (Ucko and Dimbleby 1969, WWF and IUCN 1994, UNESCO/UCO 1998).

## THREATS TO BIODIVERSITY IN NORTHERN AFRICA

Threats to Northern Africa's natural habitats include rapid population growth with a consequent demand for space and resources, agricultural and urban expansion, poverty and unsustainable use of biota. Depletion of groundwater resources is also a problem in many countries and has led to the deterioration and loss of unique wetland habitats with their associated biota.

Coral reef life in Egypt's Red Sea

*Rafel Al Ma'ary / Still Pictures*

Natural, macro-scale stresses such as drought also have the potential to change ecosystem dynamics and species composition over time.

Some specific threats have been recorded in the sub-region. For example, the Imatong Mountains have been threatened by the civil war in Sudan, bush fires, fuelwood collection and by conversion of land to agricultural plantations. There are also threats to individual species from overharvesting, as evidenced by the deforestation of the *Acacia senegal* tree in Sudan. The *Acacia senegal* is the source of gum arabic, and Sudan is the world's main producer. In the 1970s, the Sudanese government set up a company to control prices and exports of the gum. However, flawed pricing policies led to low producer prices and farmers cut down their trees for sale as firewood. In an attempt to slow the rate of deforestation, the government responded by allowing the price to rise by 300 per cent over the next two years. Producers, realizing that they could now make large economic gains rapidly, increased their production to such an extent that 80 per cent of the remaining trees were overtapped and died (Larson and Bramley 1991). Other threats to species in Northern Africa include pollution from industrial emissions and agricultural chemicals, and pressure from hunting. The cheetah (*Acinonyx jubatus*) is endangered in Northern Africa because of hunting and of reductions in populations of its prey caused by recurrent drought.

Marine habitats are also threatened—by overfishing, intensive tourism and invasion by alien species. Species common to the Red Sea have recently been found in the Mediterranean, where it is feared that their introduction (probably through discharges of ballast water by ships) could disturb the ecological balance. Exotic algae species such as *Caulerpa taxifolia* have also been found in the Mediterranean and Red Seas where they have formed toxic algal blooms.

A further, emerging threat to biodiversity is the introduction of genetically modified species, which may result in lowered genetic diversity through hybridization, competition, or predation (Hegazy, Diekman and Ayad 1999).

As a result of the pressures outlined above, a total of 139 species of mammals, birds, reptiles, invertebrates and plants are currently threatened with extinction in Northern African, and each country in the sub-region has witnessed the extinction of at least one

### Table 2b.2  Threatened species in Northern Africa, 2000

| Country | Mammals | Birds | Reptiles | Amphibians | Fishes | Inverts | Plants | Total |
|---------|---------|-------|----------|------------|--------|---------|--------|-------|
| Algeria | 13 | 6 | 2 | 0 | 0 | 11 | 2 | 34 |
| Egypt | 12 | 7 | 6 | 0 | 0 | 1 | 2 | 28 |
| Libya | 9 | 1 | 3 | 0 | 0 | 0 | 1 | 14 |
| Morocco | 16 | 9 | 2 | 0 | 0 | 7 | 2 | 36 |
| Sudan | 24 | 6 | 2 | 0 | 0 | 1 | 17 | 50 |
| Tunisia | 11 | 5 | 3 | 0 | 0 | 5 | 0 | 24 |

Source: IUCN 2000a

animal species (IUCN 2000a). This situation is summarized in Table 2b.2.

The numbers of extinctions and of threatened species are set to rise over the next 30 years. Up to 5 per cent of plant species will disappear from Algeria and Morocco, 16 per cent of mammals are expected to disappear from Libya, and 13 per cent of mammals from Tunisia. About 12 per cent of bird species in Egypt and Libya, and 8 per cent in Morocco and Tunisia, are threatened with extinction. It is also expected that Egypt will lose 2 per cent of its reptile species (WCMC 1992, WWF and IUCN 1994, World Bank 1996).

## TOWARDS SUSTAINABLE MANAGEMENT AND CONSERVATION OF BIODIVERSITY IN NORTHERN AFRICA

Arab cultures have traditionally practiced biodiversity conservation, as evidenced by 'Hema', the traditional Bedouin practice of rangeland conservation and management of grazing areas. Return to traditional control of rangelands has proven successful as a conservation and rehabilitation strategy, for example in Syria, where a programme of cooperatives was implemented over several years. Applications by tribal units for control over their former traditional grazing lands were granted by the government, and now approximately two thirds of Syria's Bedouin population are member of *hema* cooperatives and associated schemes. The members benefit from greater security and incentives for conservative practices, and the natural resource base benefits from reduced pressure (Chatty 1998). Other traditional conservation measures include the forest reserves, known as 'Harags', dating from Mediaeval Egypt, and protection of oases in Morocco and Andalusia (Draz 1969, Kassas 1972, Ghabbour 1975, UNESCO 1996). In Islam, hunting is prohibited during certain months of the year, 'Al-Ash-hur Al-Hurum'.

In more recent times, schemes have been introduced to establish protected areas and biosphere reserves such as those set up under the Arab Man and Biosphere (ArabMAB) Network. ArabMAB reserves are areas of terrestrial and coastal ecosystem in which solutions are promoted that reconcile the conservation of biodiversity with its sustainable use. There are 12 such reserves in Northern Africa, covering an area of about 13 million hectares.

At present, there are 72 terrestrial protected areas in Northern Africa, with a combined area of more than

### Table 2b.3  Nationally protected areas in Northern Africa

| Country | Terrestrial | | | Marine |
|---------|-------------|--------------|--------------|--------|
| | Number | Area (000 ha) | % land area | Number |
| Algeria | 18 | 5 891 | 2.5 | 8 |
| Egypt | 16 | 794 | 0.8 | 18 |
| Libya | 8 | 173 | 0.1 | 5 |
| Morocco | 12 | 317 | 0.7 | 10 |
| Sudan | 11 | 8 642 | 3.4 | 2 |
| Tunisia | 7 | 45 | 0.3 | 7 |
| **Total** | **72** | **15 862** | **7.8** | **50** |

**Table 2b.4  Internationally protected areas in Northern Africa**

| Country | Biosphere Reserves* | | World Heritage Sites | | Ramsar Sites | |
| --- | --- | --- | --- | --- | --- | --- |
| | Number | Area (000 ha) | Number | Area (000 ha) | Number | Area (000 ha) |
| Algeria | 3 | N/A | 1 | N/A | 13 | 1 866 |
| Egypt | 2 | 2 577 | 0 | 0 | 2 | 106 |
| Libya | 0 | | 0 | 0 | 2 | N/A |
| Morocco | 2 | N/A | 0 | 0 | 4 | 14 |
| Sudan | 2 | 1 901 | 0 | 0 | 0 | |
| Tunisia | 4 | 32 | 1 | 13 | 1 | 13 |
| **Total** | **13** | | **2** | **>13** | **22** | **>1 999** |

Source: Ramsar 2002, UNDP and others 2000, UNESCO 2002.

*Some Biosphere Reserves are also World Heritage Sites or Ramsar sites

15 million hectares and 50 marine protected areas (World Bank 2001a). Details of nationally protected areas are given in Table 2b.3. Internationally protected areas are shown in Table 2b.4. Many more sites are proposed for protection (Hegazy, Fahmy and Mohamed 2001). However, in spite of such efforts, the total area officially declared as protected in Northern Africa remains less than the international target of 10 per cent, although some countries are aiming to increase their protected areas to more than 15 per cent within the next three decades.

Between 1993 and 1999, more than 30 regional meetings were convened to promote inter-Arab cooperation on biodiversity conservation, with regular participation by most of the countries involved. In 1996, the IUCN sponsored a regional programme for Northern Africa and, the Arab League produced a comprehensive policy programme for the Council of Arab Ministers of the Environment meeting, in November 1997. Trans-border conservation is an issue that has received recent attention, and plans for protection are underway between Egypt and Sudan, and between Morocco and Algeria.

Conservation measures through sustainable use of resources include four pilot projects by Algeria's National Agency for the Conservation of Nature. One project aims to protect, document, and establish nurseries for medicinal plants, another aims to conserve and manage cheetah populations, and two are designed to raise awareness among local farming communities in and around protected areas. Working with communities has resulted in more widespread use and cultivation of hardy species, and less intensive harvesting of endangered species from the wild.

In Egypt, researchers and conservationists have been working with Bedouins to document and conserve medicinal plants. So far, they have published a book on the wild Medicinal Plants of Egypt and have established nurseries with the Bedouins to provide a source of income from sustainable use of these resources (IUCN 2000b).

The Moroccan Association for the Protection of the Environment has established a project with women in rural areas, to relieve the pressure on the environment from fuelwood collection and to lessen the burden on women of searching and collecting firewood. As an interim measure, all women in a village have been provided with cookers, and regular workshops are held to showcase their traditional knowledge and promote the apprenticeship of their skills (IUCN 2000b).

A recent GEF-funded project in Dinder National Park, Sudan aims to preserve biodiversity by encouraging species conservation and the sustainable use of resources through the integration of local communities in the utilization and management of natural resources. Dinder National Park lies along Sudan's border with Ethiopia and serves as a vital habitat for terrestrial migratory species which spend the dry season in the park. The park's extensive wetlands also provide refuge for a large number of migratory birds. The project will develop and implement an integrated management plan, in partnership with the impoverished surrounding communities and with equitable sharing of conservation benefits (IUCN 2000b). The fauna and flora of the park will receive protection and there are plans to reintroduce certain species which have been exterminated, such as the Nile Crocodile.

It is imperative that biodiversity conservation efforts in Northern Africa incorporate modern knowledge as well as traditional protection systems if they are to be acceptable and successful. The pressures of urbanization, industrialization, growing population, abuse of agrochemicals, and uncontrolled fishing and hunting are expected to increase in Northern Africa over the next decade. Protection of critical sites and creation of national parks are, therefore, needed urgently, together with more sustainable agricultural, forestry and fisheries practices.

## EASTERN AFRICA

Eastern Africa is well known for its rich and diverse biological resources, and for its variety of habitats, which range from high montane forests and afroalpine ecosystems to dense tropical lowland forests, plains and savannas, freshwater and soda lakes, coastal forests and mangroves. Eastern Africa is also home to

Mountain gorilla

*Phil Ward/FLPA*

**Figure 2b.2: economic benefits of biological resources, Uganda**

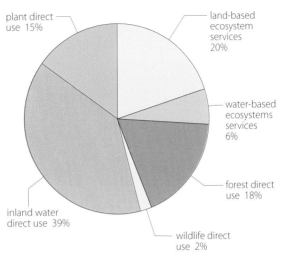

*Source: Emerton & Muramira 1999*

the world's population of 320 Mountain Gorillas (*Gorilla beringei beringei*) and to other critically endangered primates in Kenya, Uganda and Rwanda (Moyini and Uwimbabazi, 2000, Mbora and Weizckowski 2001, Butynski 2001). The sub-region's grassland savanna parks have large populations of antelope, buffalo and other ungulates, as well as elephant, rhino, hippos, crocodiles and large cats. The coral reefs along Eastern Africa's coasts are among the most spectacular in the world, and the sub-region's freshwater lakes have remarkable levels of species richness and endemism.

### Ecological, economic and social values of biological resources in Eastern Africa

Many of Eastern Africa's biological resources are used for agricultural, pharmaceutics, construction, clothing and ornamental products, and have high local, national and global economic value. For example, the economic benefits of biological resources in Uganda have been estimated at about US$741 million annually (Emerton and Muramira 1999). The breakdown of economic benefits to Uganda is illustrated by Figure 2b.2.

The agricultural biodiversity of the sub-region is also rich as evidenced by Ethiopia, one of the world's 12 centres of genetic diversity known as 'Vavilov Centres'. Ethiopia is the sole or the most important centre of genetic diversity for arabica coffee, tef, enset (*Ensete ventricosum*) and anchote (*Coccinia abyssinica*), and for sorghum, finger millet, field pea, chickpea, cow pea, perennial cotton, safflower, castor bean and sesame. Genetic erosion in other parts of the world has led to Ethiopia now also being the most important centre of genetic diversity for durum wheat, barley and linseed. The Plant Genetic Resources Centre of Ethiopia has been entrusted with safeguarding this wealth of genetic resources and, by 1994, had a collection of 53 625 specimens of 100 crop types in its gene bank. The Centre also keeps substantial *ex-situ* collections of arabica coffee and is involved with a number of farming communities in promoting the *in-situ* conservation of crops (EPA/MEDC 1997).

Eastern Africa's biological resources also make the sub-region a desirable destination for tourists, and makes a significant contribution to economic development. For example, Kenya's tourism industry is the country's second largest earner of foreign exchange, contributing 19 per cent to the country's GDP (World Bank 2000).

## THREATS TO BIODIVERSITY IN EASTERN AFRICA

Natural habitats in Eastern Africa are under threat from a rapidly increasing human population and from the accompanying demands for space, agricultural produce, and economic gains from commercial and industrial exploitation. With the sub-region's population growing at about 3 per cent per annum (World Bank 2001a), the pressures on biological resources are likely to increase in the near future.

Destruction of natural habitat in Eastern Africa is a threat to wildlife and to the biological resources that are both the basis of survival for local communities and the mainstay of the economy for many countries. Clearing of natural habitat forces wildlife to invade human settlements; the invading species become crop pests, predators on livestock, and a danger to humans and, in turn, become threatened with trapping, shooting and poisoning. Wild animals may also cross-breed with domesticated species, which can alter their genetic make-up and thus affect their status as a species. Ethiopia, for example, is witnessing the hybridization of the Ethiopian Wolf (*Canis simiensis*) with domestic dogs (EPA/MEDC 1997). The Ethiopian Wolf is the most endangered canid in the world and, in addition to the problem of hybridization, is threatened by exposure to canine pathogens prevalent among domestic dogs (Laurenson, Sillero-Zubiri, Thomson, Shiferaw, Thirgood

and Malcolm 1998, Vigne 1999). Conversely, wildlife-related diseases can be transmitted to domestic animals. A regional research effort covering Kenya, Tanzania and Uganda is underway to model wildlife, livestock and human interactions.

Loss of natural habitat and species could have a negative impact on tourism and on the foreign exchange earnings this generates. However, in the short run, tourism is more likely to be affected by issues such as bad publicity, lack of security and poor infrastructure. Such issues can lower the amount of income earned from the sub-region's substantial natural assets, meaning that less investment goes back into the areas supporting concentrations of biodiversity on which the tourism industry depends.

The lack of an adequate legal framework for protection has also contributed to the problem of biodiversity loss in Eastern Africa. For example, of the 38 wildlife conservation areas in Ethiopia, only two are 'gazetted', meaning they have legal protection (EPA/MEDC 1997). Human settlements are encroaching on protected areas such as national parks and forest reserves, as a result of weak law enforcement and low monitoring capacity resulting from inadequate funding. In some cases, critical ecosystems have been damaged

Ethiopian Wolf—a highly endangered species

*Michel Gunther/Still Pictures*

*Clearing of natural habitat forces wildlife to invade human settlements; the invading species become crop pests, predators on livestock, and a danger to humans and, in turn, become threatened with trapping, shooting and poisoning. Wild animals may also cross-breed with domesticated species, which can alter their genetic make-up and thus affect their status as a species*

## Table 2b.5 Threatened species in Eastern Africa 2000

| Country | Mammals | Birds | Reptiles | Amphibians | Fishes | Inverts | Plants | Total |
|---------|---------|-------|----------|------------|--------|---------|--------|-------|
| Burundi | 5 | 7 | 0 | 0 | 0 | 3 | 2 | 17 |
| Djibouti | 4 | 5 | 0 | 0 | 0 | 0 | 2 | 11 |
| Eritrea | 12 | 7 | 6 | 0 | 0 | 0 | 3 | 28 |
| Ethiopia | 34 | 16 | 1 | 0 | 0 | 4 | 22 | 77 |
| Kenya | 51 | 24 | 5 | 0 | 18 | 15 | 98 | 211 |
| Rwanda | 8 | 9 | 0 | 0 | 0 | 2 | 3 | 22 |
| Somalia | 19 | 10 | 2 | 0 | 3 | 1 | 17 | 52 |
| Uganda | 19 | 13 | 0 | 0 | 27 | 10 | 33 | 102 |

*Source: IUCN 2000aa*

beyond repair. For example, most parts of the Gambella National Park have been converted to irrigated agricultural land and another part has been settled by refugees from Sudan (EPA/MEDC 1997).

Although the countries of Eastern Africa are signatories to the Convention on Biological Diversity and have ratified it, their individual efforts at meeting the provisions of the Convention are clearly inadequate, as is their strategic planning for protected area management. There are exceptions to this situation such as Uganda's Protected Area System Plan, described below. However, even where plans exist, their implementation is often hampered by lack of funds, and the high revenues from tourism—even though this is based on conservation—tend to be absorbed by other government activities rather than being invested in further conservation. Funding requirements for effective conservation include well-trained, well-remunerated and equipped staff, security equipment, monitoring and assessment equipment, and maintenance of infrastructure.

Threatened species in Eastern Africa include the African Wild Dog, Grevy's Zebra, Lion, Dugong, the Black Rhinoceros, the Imperial Eagle, the Greater Spotted Eagle, the African Green Broadbill, the Turkana Mud Turtle, the West African Dwarf Crocodile and the Kyoga Flameback (IUCN 1997). Critically endangered species include the Ethiopian Wolf (*Canis simiensis*) and several primates: the Mountain Gorilla (*Gorilla beringei beringei*) in the Virungas, the Tana River Red Colobus (*Procolobus rufomitratus*) and the Tana River Crested Mangabey

(*Cercocebus galeritus*) in Kenya (Butynski 2001, Mbora and Weiczkowski 2001). In Uganda, the Northern White Rhino and the Black Rhino have been poached to extinction (NEIC 1994), and populations of large mammals were reported to have decreased from 141 300 in the 1960s to about 41 000 by 1995 (MUIENR 2000). This situation is summarized in Table 2b.5.

Loss of biodiversity in Eastern Africa is being further exacerbated by changes in institutional mandates, and by political instability. In Uganda and Ethiopia, protracted civil wars destroyed a lot of infrastructure necessary for management of protected areas. For example, four Ethiopian national parks lost all their facilities, including ranger camps and equipment. In Uganda, two parks are currently closed to both management operations and tourism.

Many non-native animal and plant species that have been introduced to Eastern Africa and have become invasive or problematic. These include *Tonna ciliate*, *Cassia spectabilis* and *Cedrella mexicana*, *Broussonetia papyrifera*, and various eucalyptus species. As already mentioned, introduction of the Nile perch into Lake Victoria is believed to have led to the disappearance of more than 200 endemic species of fish. Some of the ameliorative measures that have been suggested include the reduction of eutrophication in the Lake and the establishment of 'fish parks'. Water hyacinth, *Eichornia crassipes*, is another introduced species causing havoc on Lake Victoria. It forms dense mats on the surface of the lake, creating a hazard to boats and impeding flow of water, reducing sunlight and

## Table 2b.6 Nationally protected areas in Eastern Africa

| Country | Terrestrial Number | Terrestrial Area (000 ha) | Terrestrial % land area | Marine Number |
|---------|--------|-----------------|--------------|--------|
| Eritrea | 3 | 501 | 4.3 | |
| Ethiopia | 21 | 5 518 | 5.0 | |
| Kenya | 50 | 3 507 | 6.0 | 14 |
| Rwanda | 6 | 362 | 13.8 | |
| Somalia | 2 | 180 | 0.3 | 2 |
| Uganda | 37 | 1 913 | 7.9 | |
| **Total** | **119** | **11 981** | | **16** |

*Data not available for Burundi and Djibouti.*

nutrient availability to species below the surface, and when it dies, releasing compounds into the water that are toxic to other species. Water hyacinth infestations also threaten to block the turbines of the Owen Falls Hydroelectric Facility in Uganda, to interrupt shipping and commerce, and disrupt artisanal fisheries (Olal, Muchilwa and Woomer 2001). The Lake Victoria Environment Management Programme (LVEMP), a regional conservation management programme funded by the GEF, is contributing to development of income from profitable fisheries and to control of the water hyacinth through manual, chemical, and biological control methods.

## TOWARDS SUSTAINABLE MANAGEMENT AND CONSERVATION OF BIODIVERSITY IN EASTERN AFRICA

In 1994, the combined size of the 95 protected areas in East Africa was about 12 million hectares, bigger than the combined areas of Djibouti, Rwanda, Burundi and Eritrea (UNDP 2000), although this calculation included the sizeable reserves of Tanzania. In 1999 another assessment was undertaken, which excluded Burundi and Djibouti (due to lack of information), and Tanzania (now placed in Southern Africa). This survey found 119 nationally protected terrestrial areas and 16 marine protected areas (World Bank 2001a). The slight change in overall size results from the exclusion of Tanzania from this calculation and the increase in number of terrestrial protected areas in the other countries. There are now also 17 internationally protected areas, although Burundi, Eritrea and Somalia have yet to designate any internationally important sites. National protected sites in Eastern Africa are shown in Table 2b.6; internationally protected ones are shown in Table 2b.7.

While policies, laws and regulations are putting greater emphasis on community participation in biodiversity conservation in Eastern Africa, the results have been less than satisfactory in some areas. For example, in Ethiopia, weak law enforcement has resulted in the encroachment of protected areas by neighbouring communities and refugees. Areas that are suffering include the Abijatta-Shalla Lakes National

## Table 2b.7 Internationally protected areas in Eastern Africa

| Country | Biosphere Reserves* Number | Biosphere Reserves* Area (000 ha) | World Heritage Sites Number | World Heritage Sites Area (000 ha) | Ramsar Sites Number | Ramsar Sites Area (000 ha) |
|---------|--------|--------------|--------|--------------|--------|--------------|
| Burundi | 0 | | 0 | 0 | 0 | |
| Eritrea | 0 | | 0 | 0 | 0 | |
| Ethiopia | 0 | | 1 | 22 | 0 | |
| Kenya | 5 | 891 | 2 | 300 | 4 | 90 |
| Rwanda | 1 | 15 | 0 | 0 | 0 | |
| Somalia | 0 | | 0 | 0 | 0 | |
| Uganda | 1 | 220 | 2 | 132 | 1 | 15 |
| **Total** | **7** | **1 126** | **5** | **454** | **5** | **105** |

*Data not available for Djibouti. *Some Biosphere Reserves are also World Heritage Sites or Ramsar sites*

Park, one of the most heavily settled protected areas in Africa now completely overrun by people who are now permanently settled. Awash National Park is being severely degraded as a result of illegal occupancy, especially in the northern part, where nomadic pastoralists contribute to overgrazing (WCMC 1991).

A successful example of protection is the Mgahinga Bwindi Impenetrable Forest Conservation Trust (MBIFCT), a GEF-funded endowment, which supports communities around the two parks of Uganda, the Mgahinga Gorilla National Park and the Bwindi Impenetrable National Park. The Trust is intended to convey management responsibility and long-term proprietorship to the Government of Uganda and local communities, and provides a chance to pilot test conservation and development partnerships between multiple stakeholders. The Trust funds community development projects, research projects and park management activities, and these have been successful so far (MBIFCT 1994).

Uganda has also recently completed a Protected Area System Plan (UWA 2000). This was developed in recognition of the fact that protected areas were neglected during the wars and conflicts of the 1970s and 1980s, and that wildlife populations were reduced or almost wiped out in some areas and large areas settled by displaced communities. The plan was developed with the active participation of these communities, and various types of management were agreed for different areas, including Community Wildlife Management Areas, forestry reserves managed by the Forestry Department, and animal sanctuaries managed by the Uganda Wildlife Authority. The plan is currently awaiting approval by Uganda's parliament.

In 1999, Kenya enacted a very comprehensive Environmental Management and Coordination Act. Under this Act, the new guidelines for environmental impact assessment (EIA), will contain a provision for the inclusion of traditional holders of knowledge in a technical advisory committee for the EIA process. The EIA will take into account the potential impacts of projects on local cultures and the views of local communities in the assessment process as well as in the review process (Berger and Mugo 2001). Uganda has also formulated a national policy on Indigenous Knowledge, under the World Bank's Programme on Indigenous Knowledge in Africa.

Little Bee Eaters, Masai Mara, Kenya

*Gunter Ziesler /Still Pictures*

Kenya has established a wide range of institutions for the managemnt of biodiversity. These research institutions specialize in areas such as agriculture and livestock, as well as monitoring and research in forestry, fisheries, wildlife and rangelands. The National Museums of Kenya (NMK) houses a Centre for Biodiversity, as well as the East Africa Natural History Society, and the Kenya Research Centre for Indigenous Knowledge. NMK is also affiliated with the Institute for Primate Research. The Kenya Wildlife Service is the major body whose madate is the conservation and management of wildlife and Kenya's National Park system.

Because most of Kenya's wildlife exists outside of its National Parks, Kenyans have been experimenting with a range of community and private initiatives to set up tourism or ecotourism enterprises on their land. These ventures may involve the setting aside of game viewing areas on large private ranches, or on group ranches, and sometimes collaboration between groups and individual ranchers to increase the total area of land available for conservation and ecotourism activities. Il Ngwesi, a 16 500 acre group ranch in the northern area of Kenya, opened a small lodge in 1996, grossed US$40,000 in 2000 and is expecting more than double that in 2001 (Johnstone, 2001). The Mara Conservancy, in the Trans Mara area of Kenya, is a new initiative aimed at ensuring that the local community actually receives the proceeds from tourism activities carried out in their area. Several group ranches have subcontracted a private company to manage their game viewing area, collect entry fees, reinvest a proportion into the infrastructure of the reserve and pay the rest into the community (Daily Nation, 13 December 2001).

Ecotourism and conservation enterprises can be part of an effort by pastoral communities to stop the sub-division, sale and subsequent encroachment of settlement onto their common grazing land, which interferes with their traditional livelihood activities. Some of the Maasai residents of a former conservation area located next to Nairobi National Park (and part of the wildlife dispersal area) are attempting to re-integrate use of the riverine forest and grazing land to attract tourists and dissuade members of the scheme from selling off small plots of land for settlement (Ole Kaasha 2001). On a different tack, the Loita Maasai of Kenya responded firmly to attempts to include the sacred Loita Forest as part of the Maasai Mara game reserve, to be developed for tourism. As part of a series of efforts to prevent the alienation of this forest and the loss of its biodiversity, the Loita Maasai went to court to seek legal entitlement, invoking Article 8(j) of the Convention on Biological Diversity (Stephenson 1999).

Other conservation measures include the reintroduction of species to areas from which they had formerly been eradicated, and management of wildlife-livestock interactions. For example, giraffes were successfully reintroduced from Kenya into Kidepo Valley National Park, Uganda, and there are plans underway to reintroduce black rhino to Uganda. In July 2001, the Kenya Wildlife Service translocated 56 elephants from the Sweetwaters Rhino Sanctuary in Laikipia district, to Meru National Park. Meru's elephant herds were decimated during the 1960s and 1970s. This is the sixth major operation of this kind in Eastern Africa since 1996, and is one way that the KWS is handling wildlife-human conflicts (Situma 2001). The Ethiopian Wolf Conservation Programme, the IUCN/SSC Canid Specialist Group and the Ethiopian Wildlife

Conservation Organization have started a programme to vaccinate dogs against rabies, canine distemper and canine parvovirus in and around the Bale Mountains National Park. The programme will not only help curb the spread of diseases that are a threat to this critically endangered species, but it will also benefit local communities, by reducing the number of deaths of humans and their livestock from rabies. Another initiative is to encourage responsible dog ownership, and the neutering of both male and female dogs (Laurenson and others 1998, Vigne 1999).

## WESTERN INDIAN OCEAN ISLANDS

The importance of conservation within oceanic islands is perhaps best expressed by the fact that they are home to around one-sixth of all plant species, and that one in three of all known threatened plant species are island endemics. The Western Indian Ocean Islands, uninhabited until the 16th and 17th centuries, are a typical example where rich land-based flora and fauna have evolved in isolation from human intervention and from the intrusion of other alien species that the human presence so often brings

Madagascar has the highest number of endemic species of any country in the Africa region, and ranks

Ring-tailed lemurs, Madagascar

sixth in the world. Up to 8 000 of the 9 500 species of higher plants and more than 50 per cent of all vertebrate species found in the island are known or are thought to be endemic (UNEP 1999). In Mauritius, about 50 per cent of all higher plants, mammals, birds, reptiles and amphibians are endemic to the island, and Seychelles has the highest level of amphibian endemism of any island in the world (11 of the 12 species are found nowhere else) (WCMC 1992).

## ECOLOGICAL, ECONOMIC AND SOCIAL VALUES OF BIOLOGICAL RESOURCES IN THE WESTERN INDIAN OCEAN ISLANDS

This extraordinary biodiversity of the Western Indian Ocean Islands not only contributes to their unique ecological conditions, but it also provides valuable raw materials for local and commercial use. In addition to use for food, construction, clothing and shelter, many plant species are used medicinally, and several species are being researched for commercial agricultural or pharmaceutical use. For example, the native Coffea sp. is being investigated for commercial production of naturally caffeine-free coffee (GOM/ERM 1998).

The Islands' internationally renowned coral reefs support a booming tourism industry and subsistence and commercial fisheries. There are also ten species of mangroves, providing stabilization of the coastal zone, creating spawning and nursery grounds for many fish species, mitigating storm impacts, and providing resources for construction, weaving, and food.

## THREATS TO BIODIVERSITY IN THE WESTERN INDIAN OCEAN ISLANDS

Human population growth, the selective and accidental introduction of alien species, urbanization, rapid change in use of land for cultivation, hunting, pollution and degradation of soil have all taken their toll on endemic species of plants, mammals, birds, reptiles and amphibians, and the Islands' biodiversity is now seriously under threat. All of the Western Indian Ocean Islands are important endemic bird areas, and four islands in Seychelles are bird sanctuaries. The greatest threat to their biodiversity is habitat destruction resulting from expansion of agriculture and of human settlements, with additional threats from natural processes such as coastal erosion, bush fires and seawater intrusion. The continued extension of agriculture into the natural forest and woodland areas exerts a continuous pressure on already endangered species of flora and fauna in the islands. Grazing by introduced species of animals such as deer, goats and cattle has diminished plants not adapted to grazing and has led to a predominance of exotic grasses. In addition, sugar cane, coffee and tea estates now use large areas of land for these imported species, and the high dependency on wood and charcoal for household use has further depleted forest and woodland areas (UNEP 1999).

In Madagascar, grassland fires are commonly used to clear land at the end of the dry season, and this is contributing to the destruction of woodlands and threatening the habitat of birds, insects and mammals which depend on the forest cover for survival. In 1995, more than 1.2 million hectares of Madagascar's forest were destroyed by grassland fires (4 900 ha of natural forest and 10 287 of plantation forest) (Republic of Madagascar 1997). In Mauritius, burning of sugar cane is a common cause of habitat destruction for insects, birds and reptiles, as well as a source of air pollution.

Modification of freshwater habitats through pollution is another cause of reductions in biodiversity on the islands, as are selective harvesting and overharvesting. For example, collection of fuelwood is a form of selective harvesting that is contributing to loss of quality of habitat and to reduction of species diversity. Overharvesting of land tortoises and marine turtles over the past three decades has contributed to their decline (RFIC 1998).

Western Indian Ocean Island coastal and marine habitats and species are under threat from destructive fishing practices, overharvesting and intense tourism. Mangrove clearance means loss of buffering against ocean swells and is accelerating coastal erosion and saltwater intrusion. Nursery grounds for shrimp, crab and other species are also being lost as mangroves are reduced. Mining of coral and sand for use in construction is also damaging habitats, and threatening the biodiversity they support. Intensive tourism is thought to be damaging the coral reef habitats by pollution from boats, hotels and other infrastructure and facilities, and by excessive walking on coral or its removal for souvenirs.

Island species are particularly vulnerable to competition and predation by invasive alien species. Animal species introduced to the Western Indian Ocean Islands include rats and mice, rabbits, pigs and the long-tailed macaque (*Macaca fascicularis*). This latter species is a problem for farmers because it regularly damages crops. It is also contributing to the extinction of many wild bird species through predation on their nests. Rodents are also problem predators and have contributed to the demise of birds and reptiles. The two most prolific alien plant species in Mauritius are the Chinese guava (*Psidium cattleianum*), from South America, and the privet (*Ligustrum robustrum*) from Asia. Neither species is kept in check by natural consumers or competition and, therefore, they form dense thickets in the upland forests, preventing other species from regenerating. Lowland forests are being invaded by liane cerf, wild pepper and aloe.

As a result of these pressures, a significant number of plant and animal species in the Western Indian Ocean Islands are threatened with extinction or have become extinct—the most notorious case being that of the Dodo of which the demise is attributed to overhunting and the introduction of alien species. Throughout the Western Indian Ocean Islands, populations of endangered marine species, such as the Green Turtle, the Hawksbill Turtle, the Coelacanth and the Dugong, have declined in recent years. In Mauritius, 62 animal species—mainly birds, reptiles and a large number of molluscs—have become extinct, and several species now only survive in small populations under protection schemes. Mauritius and Seychelles are ranked second and third in the world in terms of the percentage of native plants threatened (UNEP 1999). The numbers of threatened species in the sub-region are given in Table 2b.8.

•

*Mauritius and Seychelles are ranked second and third in the world in terms of the percentage of native plants threatened*

•

In response to the threats to natural habitats in the Western Indian Ocean Islands, protected areas have been established inland, and in the coastal and marine zones. In 1999, there were: 1 reserve in Comoros; 44 protected areas covering 2.9 per cent of the land area in Madagascar; 18 reserves covering 3.7 per cent of the land area in Mauritius; and 26 sites covering 47 per cent of the land area in Seychelles (UNEP 1999). The Andringitra Reserve, newly created in Madagascar with the assistance of WWF, is thought to be one of the world's most biologically rich areas, and to be most representative of the island. The Andohahela National Park, Bexa Mahafaly Special Reserve and Lac Tsimanampetsotsa are areas designated as reserves with the specific aim of protecting the dry forest and spiny thicket habitats which are unique to Madagascar.

The sub-region also has two Biosphere Reserves in Madagascar and one in Mauritius, three World Heritage sites (two in Seychelles and one in Madagascar), and four Ramsar sites (one in Comoros, two in Madagascar, and one in Mauritius) (Ramsar 2002, UNESCO 2002, UNEP 1999). Mauritius has also signed the Ramsar Convention, but has not yet designated a site. All of the Western Indian Ocean Islands are parties to the Convention on Biological Diversity, and efforts are underway to develop a conceptual framework for coral reef conservation. The Indian Ocean Commission has initiated a regional project with a view to achieving sustainable management of natural resources. This aims to protect resources and integrate management in the coastal zone, and to protect and conserve endangered endemic flora.

More than 20 endangered species in the Western Indian Ocean Islands, are protected by official programmes within national protected areas. Madagascar has 10 programmes for conservation of species in protected areas and there are 8 in Mauritius and 3 in Seychelles. Particular success has been achieved in the Seychelles with the protection of the Aldabra Tortoise through an Australasian Species Management Programme—there are now 155 000 specimens in the wild. In Mauritius, the Pink Pigeon has been successfully conserved through a Jersey Wildlife Preservation Trust project and some 300 are now surviving in the wild. In Seychelles, populations of the Brush Warbler, a species that was critically threatened,

**Table 2b.8 Threatened species in the Western Indian Ocean Islands (percentage of known species)**

| Country | Higher plants | Mammals | Birds | Reptiles | Amphibians |
|---|---|---|---|---|---|
| Comoros | 1 | 18 | 6 | 8 | 0 |
| Madagascar | 5 | 28 | 10 | 5 | 1 |
| Mauritius | 71 | 100 | 37 | 55 | 27 |
| Seychelles | 8 | 8 | 7 | 27 | 33 |

have grown to 250 and, in Mauritius, the Rodrigues Flying Fox has successfully recovered under the North American Species Survival Plan. There are now 350 flying foxes surviving in the wild (UNEP 1999).

Eradication of introduced problem predators such as rats, mice and macaques has been achieved in some of the smaller islands, by means including poison pellets, wax blocks and trapping. Infestations of alien plants are being controlled within conservation management areas by manual weeding, and by erecting barriers to pigs and other animals that may disperse the seeds. Airports and seaports are also carefully monitored and incoming traffic is sprayed with insecticides and herbicides to reduce the risk of accidental introductions. There are also strict regulations on animal and plant products entering the country. Reducing populations to manageable levels will, however, take a long time and will require considerable resources.

All of the Western Indian Ocean Islands have ratified CITES, and Madagascar, Mauritius and Seychelles have established Management and Scientific Authorities to regulate the granting of import and export permits. National programmes have also been established to encourage sustainable use and trade in certain wildlife products (including shells, turtle products, seabirds and their eggs, and certain plants), but these frequently suffer from insufficient resources to properly implement the restrictions. A major sub-regionwide policy, giving highest priority to conservation of endangered species, is an immediate requirement. Alternative strategies to individual species conservation programmes and creation of protected areas are also required. Local communities and national economies need alternative resources or means of support, and a culture of sustainable harvesting needs to be implemented.

## SOUTHERN AFRICA

Southern Africa has rich biological resources in a variety of ecosystems, which range from moist tropical forests in Angola and Zambia to savannas, coastal forests and mangroves, deserts and semi-deserts, and to the extraordinary diversity of plants of the Cape Floral Region, in South Africa. The sub-region boasts an average of 57 mammalian species and 136 breeding bird species per 10 000 km$^2$ (UNDP and others 2000). South Africa ranks as the third most biologically diverse country in the world, mainly because of the richness of its plant life—more than 18 000 species of vascular plants of which more than 80 per cent are endemic. In terms of numbers of endemic species of mammals, birds, reptiles and amphibians, South Africa is the fifth richest country in Africa and the 24th richest in the world (DEA&T 1997).

### ECOLOGICAL, ECONOMIC AND SOCIAL VALUES OF BIOLOGICAL RESOURCES IN SOUTHERN AFRICA

Southern Africa's rich biological resources play an important role in ensuring long-term food security. Access to genetic resources for crop and animal breeding purposes is also seen as a critical factor. Many species of plants and animals have medicinal properties and most of these are used in traditional healing. Some are being investigated for commercial production. Approximately 10 per cent of Southern African plants (roughly 3 000 species) are used medicinally, and 10 per cent of these (about 350 species) are commonly and widely used (van Wyk, Van Oudtshoorn and Gericke 1997). They include *Warburgia salutaris*, a plant of which the root and bark are used to treat coughs, headaches and stomach problems, and which is fast disappearing in Southern Africa (Cunningham 1993). The locally known 'African Potato' (*Hypoxis sp*) is being researched for the extraction of hypoxicide, a sterol (plant acid) used traditionally to treat dizziness and bladder disorders, and which has now been shown to inhibit the growth of tumour cells and also to have anti-inflammatory properties (Drewes, Hall, Learmonth and Upfold 1984).

### THREATS TO BIODIVERSITY IN SOUTHERN AFRICA

As in other sub-regions of Africa, natural habitats in Southern Africa are coming under increasing pressure from expansion of agriculture and plantation forestry, human settlements, mining activities, and other commercial or subsistence activities, both inside and outside of protected areas. Individual species are threatened through habitat loss, selective harvesting, poaching, and through the spread of alien invasive organisms.

Ivory and animal skins impounded in Dar-es Salaam, Tanzania

*Sabine Vielmo/Still Pictures*

One of the greatest pressures on wild species is the trade in plant and animal products, such as ivory, horn and skins. Over the past 30 years, trade restrictions, mainly through CITES, have been used at the global level as a tool to control trade and thus help conserve populations in the wild. Implementation of regulations has had varied levels of success in Southern Africa. For example, the listing of the Black Rhino in Appendix 1 (species threatened with extinction) of CITES during the 1970s has not helped to revive the rhino population, which is still too low for breeding and multiplication in the wild. Trade controls have resulted in higher prices being paid on the black market for rhino horn which, in turn, encourages poaching of wild populations. As already mentioned, restrictions on the trade in ivory and sound conservation practices have seen significant growth in national herds of elephant in Botswana and Zimbabwe, and these countries are now pushing for limited trade in ivory to provide economic incentives for continued conservation.

Sub-regional cooperation plays a significant role in the conservation of biological resources in Southern Africa, and the Southern African Convention for Wildlife Management has been successful in regional monitoring, assessment and management of wildlife resources. However, such conservation measures need continued resourcing and support, so that benefits continue to be derived from conservation of species in the wild.

Alien invasive species of plants and animals are causing massive disturbance in natural ecosystems across Africa. In Southern Africa, the introduction of alien tree species, originally for plantation, is of greatest concern. The Catalogue of Problem Plants in Southern Africa (Wells, Balsinhas, Joffe, Engelbrecht, Harding and Stirton 1986) lists 789 species some of which, including *Acacia saligna* and *Hakea sericea*, have dominated areas to the extent that natural vegetation has been almost completely lost. Others, for example pine and eucalyptus trees, present a threat to water availability because they use greater amounts of water than the natural vegetation, and therefore reduce the amount of run-off reaching streams and rivers. Other species form dense stands that reduce the amount of light reaching the understorey, physically strangle native species and inhibit regeneration of native seeds. These impacts reduce the diversity and cover of indigenous plant species, and thus alter functioning of the ecosystem.

In South Africa, where the problem of alien invasive species has been well quantified and documented, about 180 species of trees and shrubs have invaded, covering 10 million hectares (8 per cent of the land area) (Versveld, Le Maitre and Chapman 1998). The plant diversity of the Cape Floral Region is particularly threatened by invasive species, with an estimated 33 of 70 threatened plant species being potential extinction victims of invasions of alien woody plants (Hall, De Winter and Van Oosterhout 1980).

As in other sub-regions, the water hyacinth (*Eichornia crassipes*) is a problematic invasive plant in Southern Africa, forming dense mats that block water channels, disrupting flow patterns, reducing light and nutrients reaching below the surface of the water, and thus creating an undesirable habitat for native plants and animals. Decaying mats of the weed generate unpleasant odours and lead to eutrophication of the water body. Areas afflicted by the water hyacinth include Lake Kariba and Lake Chivero (Zambia/Zimbabwe).

There is a marked lack of information available on most invertebrates, algae, bacteria and fungi in Southern Africa, including on their genetic diversity. It is, therefore, thought that many species in the sub-region (as elsewhere) become extinct before they can be named and described. Lack of knowledge of biodiversity issues has also been compounded by the fact that indigenous knowledge has not been accepted or documented by research institutes or in publications. As a result, protected areas, many of which were established more than three decades ago without consultation of local people, were set aside without accurate assessment of the biological richness within their boundaries. Thus, some areas that have little significance in terms of biodiversity are protected, while many others with significant biodiversity lack protection. In addition, farmers, who have custody of much of the sub-region's biological diversity, are rarely invited to share their knowledge it, even though that knowledge extends to crop and animal genetic diversity, and includes wild plant and animal species that serve humanity as biological resources. The lack of comprehensive knowledge of biodiversity in Southern Africa also contributes to growing discontent about unauthorized access to biodiversity and lack of reciprocity in benefit sharing, mainly on the part of the rich developed countries. For example, while acknowledging that developing drugs is costly, it is also important to attain goals of wealth creation that will provide substantial benefit to those who conserve biodiversity through a culture of bio-partnership, rather than indulging in bio-piracy.

In spite of the numerous pressures on Southern Africa's biological resources, only one species of mammal (the Blue Antelope) has become extinct in recent times, but several sub-species have been lost. The demise of the Blue Antelope has been blamed on competition for grazing from sheep farming, and on subsistence hunting. The African Wild Dog is also an endangered species in Southern Africa, surviving only in large protected areas (Ledger 1990). Similarly the Bearded Vulture has undergone serious population declines in the sub-region and is now restricted to the Drakensberg Mountain range of South Africa and Lesotho. Declines in the Bearded Vulture's population have often been blamed on reduced prey, changing animal husbandry practices and direct persecution—in Lesotho, for example, the bird is killed for its plumage which is used in traditional ceremonies. The number of threatened plant species in Southern Africa continues to grow. Estimates show that 58 species had become extinct by 1995 compared to 39 in 1980, while the number of endangered plant species grew from 105 to 250 during the same period (Hilton-Taylor 1996). The numbers of threatened species in Southern Africa are shown in Table 2b.9. In terms of area, Southern Africa has the highest concentration of threatened plant species in the world (Cowling and Hilton-Taylor 1994). A large percentage of these are in the Cape Floral Region, and are threatened by the rapid urbanization of the Cape Metropolitan Area.

### Table 2b.9 Threatened species in Southern Africa 2000

| Country | Mammals | Birds | Reptiles | Amphibians | Fishes | Inverts | Plants | Total |
|---|---|---|---|---|---|---|---|---|
| Angola | 18 | 15 | 4 | 0 | 0 | 6 | 19 | 62 |
| Botswana | 5 | 7 | 0 | 0 | 0 | 0 | 0 | 12 |
| Lesotho | 3 | 7 | 0 | 0 | 1 | 1 | 0 | 12 |
| Malawi | 8 | 11 | 0 | 0 | 0 | 8 | 14 | 41 |
| Mozambique | 15 | 16 | 5 | 0 | 3 | 7 | 36 | 82 |
| Namibia | 14 | 9 | 3 | 1 | 3 | 1 | 5 | 36 |
| South Africa | 41 | 20 | 19 | 9 | 30 | 111 | 45 | 275 |
| Swaziland | 4 | 5 | 0 | 0 | 0 | 0 | 3 | 12 |
| Tanzania | 43 | 33 | 5 | 0 | 15 | 47 | 236 | 379 |
| Zambia | 12 | 11 | 0 | 0 | 0 | 6 | 8 | 37 |
| Zimbabwe | 12 | 10 | 0 | 0 | 0 | 2 | 14 | 38 |

## TOWARDS SUSTAINABLE MANAGEMENT AND CONSERVATION OF BIODIVERSITY IN SOUTHERN AFRICA

The past 30 years have seen expansions in protected areas in Southern Africa, from just 260 areas (6 per cent of the total area) in 1989 (although this assessment excluded Tanzania, which was still associated with Eastern Africa), to a current total of 578 nationally protected areas with a combined area of over 65 million hectares (including the sizeable reserves of Tanzania) (WRI UNEP and UNDP 1990; World Bank 2001). There are also currently 44 marine protected areas in Southern Africa (see Table 2b.10).

Southern Africa is home to some of the world's largest protected areas, including: the Okavango Delta in Botswana, the largest inland delta in the world (16 000 sq. km); the Selous Game Reserve in Tanzania (52 200 sq. km); and the Namib-Naukluft National Park in Namibia (49 768 sq. km) (McCullum 2000). However, Swaziland has seen one of its three protected areas opened up for other land uses in the last 20 years, and civil wars in Angola and Mozambique have resulted in the loss of some protected areas. The war in Mozambique saw widespread habitat and species loss in the Gorongosa National Park and Marromeu Buffalo Reserve in the Zambezi Delta. The ecosystems have not yet fully recovered (Chenje 2000).

The Peace Parks initiative is based on a concept similar to that of the Biosphere Reserve, and several sites are now being established as transfrontier conservation areas. These include: the Kgalagadi Transfrontier Park, established in 1998 between Botswana and South Africa; the Maloti-Drakensberg Park between Lesotho and South Africa; the Gaza-Gonarezhou-Kruger Park between Mozambique, South Africa and Zimbabwe; and the Greater Limpopo Transfrontier Park, between Mozambique, Zimbabwe and South Africa. Internationally protected areas are shown in Table 2b.11.

Efforts have also been made to deal with invasive species. For example, the Working for Water Programme was set up in 1995 to provide sustainable employment through alien-clearing projects. There are currently more than 300 projects which have cleared more than 235 000 hectares of alien-infested land, rehabilitated a further 50 000 hectares, and employed 21 000 people in the year 2000 (Working for Water 2000). Countries affected by water hyacinth have initiated biological and chemical control programmes in addition to mechanical clearance, with some success to date (Global Water Partnership 2000).

The positive trends in biodiversity conservation in the sub-region are partly attributed to the fact that all countries in Southern Africa have ratified the Convention on Biological Diversity, the Bonn Convention and the Ramsar Convention. To a large extent, the provisions of these conventions have also been included in national policies. Furthermore, many Southern African governments have adopted structured approaches to decentralizing management of natural resources through Community Based Natural Resources Management (CBNRM) programmes. CBNRM programmes have helped to extend resource access and management rights over the past 15 years to farmers operating in some communal lands. This has resulted in rural farming communities deriving direct benefits from wildlife through returns from either safari hunting or direct sales (Cumming 2000).

Prominent CBNRM programmes in the sub-region include Zimbabwe's Communal Areas Management Programme for Indigenous Resources, hailed for raising household incomes by as much as 15–25 per cent in CAMPFIRE-designated areas since 1998 (CAMPFIRE and Africa Resources Trust 1999). Zambia's Administrative Management Design for Game Areas (ADMADE), South

### Table 2b.10 Nationally protected areas in Southern Africa

| Country | Terrestrial Number | Terrestrial Area (000 ha) | Terrestrial % land area | Marine Number |
|---|---|---|---|---|
| Angola | 13 | 8 181 | 6.6 | 4 |
| Botswana | 12 | 10 499 | 18.0 | |
| Lesotho | 1 | 7 | 0.2 | |
| Malawi | 9 | 1 059 | 8.9 | |
| Mozambique | 11 | 4 779 | 6.0 | 7 |
| Namibia | 20 | 10 616 | 12.9 | 4 |
| South Africa | 390 | 6 619 | 5.4 | 20 |
| Tanzania | 39 | 13 817 | 14.6 | 9 |
| Zambia | 35 | 6 366 | 8.5 | |
| Zimbabwe | 48 | 3 071 | 7.9 | |
| **Total** | **578** | **65 014** | | **44** |

*Data not available for Swaziland*

## Table 2b.11 Internationally protected areas in Southern Africa

| Country | Biosphere Reserves* Number | Biosphere Reserves* Area (000 ha) | World Heritage Sites Number | World Heritage Sites Area (000 ha) | Ramsar Sites Number | Ramsar Sites Area (000 ha) |
|---|---|---|---|---|---|---|
| Angola | 0 | | 0 | 0 | 0 | 0 |
| Botswana | 0 | | 1 | 1 | 1 | 6 864 |
| Lesotho | 0 | | 0 | 0 | 0 | 0 |
| Malawi | 1 | N/A | 1 | 9 | 1 | 225 |
| Mozambique | 0 | | 0 | 0 | 0 | 0 |
| Namibia | 0 | | 0 | 0 | 4 | 630 |
| South Africa | 4 | N/A | 1 | 243 | 17 | 500 |
| Swaziland | 0 | | 0 | 0 | 0 | 0 |
| Tanzania | 3 | N/A | 4 | 6 860 | 2 | 3 474 |
| Zambia | 0 | | 1 | 4 | 2 | 333 |
| Zimbabwe | 0 | | 2 | 733 | 0 | 0 |
| **Total** | **8** | | **10** | **7 850** | **27** | **12 026** |

*Some Biosphere Reserves are also World Heritage Sites or Ramsar sites

Source: Ramsar 2002, UNDP and others 2000, UNESCO 2002

Africa's Peace Parks initiative and Namibia's Living In a Finite Environment (LIFE) are other successful CBNRM programmes in the sub-region. The LIFE programme has seen local communities reap substantial benefits from sales of thatching grass, crafts, tourism and trophy hunting. In the Kunene region, game guards are employed from the local communities and Himba nomads, creating jobs and stimulating incomes whilst helping to conserve wildlife. Public-private partnerships are being developed to ensure conservation of threatened species in the Kafue wetlands of Zambia, whilst providing a source of materials and income to local communities. Despite these successes, some analysts have questioned whether these CBNRM-created incentive structures can promote a conservation ethic among rural residents. One study concluded that it was increased enforcement activities, and not any increased motivation by residents to conserve animals, which led to declines in poaching in Zimbabwe and Zambia. It appears that poaching remains a problem in these areas (Gibson 1999). Indications are that CAMPFIRE may, in some cases, have been successful in stimulating communities to secure rights to wildlife resources and may have more influence over utilization of the resource. This may have allowed communities to improve their agricultural infrastructure and equipment and to improve household conditions (Murombedzi 2001). Whether this change in the community relationship to the local authority will result in enhanced conservation of biodiversity remains to be seen.

In the Kunene region of Namibia, members of the local communities and Himba nomads are employed as game guards, thus creating jobs and stimulating incomes, whilst also helping to conserve wildlife.

UNEP

Governments and scientific institutions are slowly realizing the value of indigenous knowledge and attempts are now being made to document and record what still exists and to incorporate traditional conservation methods into modern ones. For example, more than 5 000 samples of mainly traditional food crop seed species have been collected by states that are members of the Southern African Development Community (SADC), for preservation at the sub-regional genetic resources centre in Lusaka, Zambia. In addition, a Genome Resource Banking in Southern Africa programme has been initiated, and materials from many wildlife species, including buffalo and rhino, have been banked. Plans are underway to include other species. Many computer databases—increasingly being used as tools to develop biodiversity inventories—have now been set up and placed on the Internet, including a meta-database of resources, a fish database and a database for threatened plants. The Southern African Bird Atlas (Harrison, Allan, Underhill, Herremans, Tree and Brown 1997) maps the distribution of bird records from across Southern Africa, providing a major source of information against which conservation threats can be determined or judged. Other responses to calls for greater documentation and application of indigenous knowledge include policies to assign intellectual property rights to certain countries, communities or individuals, and participation in the World Bank's Programme on Indigenous Knowledge in Africa. Indigenous knowledge is being used via this programme in the treatment of HIV/AIDS in Tanzania, through the use of medicinal plants to treat secondary infections and community support programmes that assist people living with AIDS to live positively and raise modest incomes (World Bank 2001b).

## CENTRAL AFRICA

Central Africa has a wide diversity of habitats spanning dense humid forests, savannas, semi-deserts (on the Sahelian borders), freshwater lakes, mangrove forests and coral reefs. Species diversity and endemism are high in the sub-region, mainly due to the abundant tropical forests. The forest of the Congo Basin is the second largest contiguous area of forest in the world, and the largest in Africa. It is also one of the most

biologically diverse and most poorly understood of Africa's ecosystems (IUCN, WWF and GTZ 2000). Data compiled in 1992 indicate that, of the 40 850 plant taxa so far enumerated in Central Africa, nearly 16 per cent are endemic to the area, and 175 of them are classified as rare (WRI, UNEP and UNDP 1992). There are 11 000 forest plant species in the Democratic Republic of Congo (DRC) alone and more than 3 000 of these are endemic. Cameroon has 8 000 forest plant species while the Central African Republic is home to 1 000 endemic species of plants (IUCN, WWF and GTZ 2000). Bird diversity is also high in Central Africa, with more than 1 000 species in the DRC (IUCN, WWF and GTZ 2000).

### ECOLOGICAL, ECONOMIC AND SOCIAL VALUES OF BIOLOGICAL RESOURCES IN CENTRAL AFRICA

Central Africa's biological resources are the backbone of the sub-region's economy, and support millions of livelihoods. Timber extraction is growing rapidly, and timber exports exceeded 50 per cent of all exports from Equatorial Guinea in 1993. They totalled more than 1.7 million $m^3$ in 1998 (IUCN, WWF and GTZ 2000). Up to 63 per cent of the Central African countries' population live in rural areas, and many people are dependent on forest resources such as wood for construction and fuel, plants and animals for food, medicines, clothing and household items.

The Drill *(Mandrillus leucophaeus)*—one of Africa's least known primates—is facing increasing pressure as a target for bushmeat.

*Frank W. Lane*

## Table 2b.12  Threatened species in Central Africa 2000

| Country | Mammals | Birds | Reptiles | Amphibians | Fishes | Inverts | Plants | Total |
|---|---|---|---|---|---|---|---|---|
| Central African Republic | 12 | 3 | 1 | 0 | 0 | 0 | 10 | 26 |
| Chad | 17 | 5 | 1 | 0 | 0 | 1 | 2 | 26 |
| Congo | 12 | 3 | 1 | 0 | 1 | 1 | 33 | 51 |
| DRC | 40 | 28 | 2 | 0 | 1 | 45 | 55 | 171 |
| Equatorial Guinea | 15 | 5 | 2 | 1 | 0 | 2 | 23 | 48 |
| Gabon | 15 | 5 | 1 | 0 | 1 | 1 | 71 | 94 |
| Sao Tome & Principe | 3 | 9 | 1 | 0 | 0 | 2 | 27 | 42 |
| Cameroon | 37 | 15 | 1 | 1 | 27 | 4 | 155 | 240 |

*Source: IUCN 2000a a*

### THREATS TO BIODIVERSITY IN CENTRAL AFRICA

Over the years, wildlife habitat in the sub-region has come under increasing pressure from conversion to alternative land uses, particularly cultivation of cash crops, subsistence slash-and-burn cultivation and expansion of human settlements. Weak infrastructure and lax enforcement of protection have contributed to excessive rates of deforestation. Resettlement campaigns by the French administration, together with rural-urban migration, have also left large areas of forest unpopulated and, therefore unpatrolled. In the late 1980s, it was estimated that only 50 per cent of an estimated original 404 390 000 hectares of wildlife habitat remained (McNeely, Miller, Reid, Mittermeier and Werner 1990).

The rate of forest loss is a particular cause for concern in Central Africa, with the DRC losing more than 500 000 hectares of forest per year between 1990–2000 and Cameroon losing more than 200 000 hectares (FAO 2001). Even though these are not the highest rates of deforestation in Africa—and allowing for the fact that even these large losses represent small percentages of the total forest area—the impacts on the functioning and biodiversity of the ecosystem, and the impacts on the local communities, are highly detrimental. Loss of habitat has resulted in species becoming threatened or extinct—in the DRC, for example, 40 species of mammals and 28 bird species are threatened (IUCN 2000a). The communities dependent on these species for food, construction, medicinal products or subsistence incomes are forced to find alternatives or go without. Agricultural, pharmaceutical and industrial opportunities are also lost, and economies suffer in the long term as a result. More sustainable rates of harvesting of forest products must, therefore, be adopted in order to ensure medium- to long-term supply.

In addition to the threat of habitat loss, numbers of threatened or endangered species in Central Africa are increasing, because of pollution and overharvesting of selected species for food, medicinal and commercial purposes. For example, gorillas, chimpanzees, mandrills, forest elephants, buffalo and antelopes are coming under increasing pressure for sale as bushmeat, and illegal logging is destroying large areas of their habitat. The Drill (*Mandrillus leucophaeus*) is one of Africa's least-known primates, and is one of the most endangered. Its range is limited to parts of Nigeria, Cameroon and Bioko Island in Equatorial Guinea. Drills are under severe pressure from hunting (Gadsby and Jenkins 1998). The savannas of northern Cameroon are an important wildlife habitat for the critically endangered, endemic, black rhino subspecies *Diceros bicornis longipes*. Fewer than 20 individuals remain, due to pressure from poaching. War and civil unrest are also contributing factors in the decline of these species because of resettlement of refugees in forest habitats, illegal poaching by soldiers and guerrillas, and clearing of habitat for military training. The DRC and Cameroon—richest in endemic mammals, birds and higher plants—also have the highest numbers of threatened species (see Table 2b.12). Gabon and Cameroon had the highest numbers of plants known to be threatened in the sub-region in 1997 (World Bank 2000).

## Table 2b.13 Protected areas in Central Africa

| Country | Number | Terrestrial Area (000 ha) | % land area | Marine Number |
|---------|--------|---------------------------|-------------|---------------|
| Cameroon | 18 | 2 098 | 4.4 | |
| Central African Republic | 13 | 5 110 | 8.2 | |
| Chad | 9 | 11 494 | 9.0 | |
| Democratic Rep. Congo | 15 | 10 191 | 4.3 | 1 |
| Congo | 9 | 1 545 | 4.5 | 1 |
| Equatorial Guinea | 0 | 0 | 0.0 | 4 |
| Gabon | 5 | 723 | 2.7 | 4 |
| **Total** | **69** | **31 161** | **33.1** | **10** |

*Data not available for Sao Tome & Principe*

Source: World Bank 2001

## TOWARDS SUSTAINABLE MANAGEMENT AND CONSERVATION OF BIODIVERSITY IN CENTRAL AFRICA

In response to loss of natural habitat in Central Africa, the network of protected areas has been expanded though creation of new areas and extension of existing nationally protected areas, and through the creation of protected areas of international significance, such as Biosphere Reserves and Wetlands of International Importance (Ramsar sites). Protected areas in the tropical forest of the Congo basin amount to 6 per cent of the total forest area, and include the rainforest refuge areas of the Korup National Park, Mount Cameroon and Dja Forest Reserve (Cameroon); the Crystal Mountains (Gabon); Maika National Park and Salonga National Park (DRC); and the Mayombe Forest Reserve (DRC and Congo) (IUCN, WWF and GTZ 2000). The Youndé Declaration—a 12-point resolution on the conservation and sustainable management of the forests of the Congo Basin—was signed, in 1999, by the heads of state of Cameroon, Central African Republic, Chad, Congo, Equatorial Guinea and Gabon. Under its guidance, the Sangha Park has been created, linking the protected areas of Lobeke National Park in Cameroon, the Dzanga-Sangha in the Central African Republic, the Nouabale-Ndoki Park in Congo, and production forests and hunting zones surrounding each of these areas. This is one of the first efforts at a coordinated approach to forest resource conservation, through harmonization of the laws and policies of the six countries, and coordination of patrolling for illegal activities. Nationally and internationally protected areas in Central Africa are shown in Table 2b.13 and Table 2b.14 respectively.

However, protected areas in Central Africa are reportedly still experiencing degradation, mainly as a result of poor enforcement of protection regulations. Logging activities, bushmeat poaching, agriculture, and oil exploration regularly encroach on protected areas. For example, forest concessions have been granted in Gabon in the La Lopé Wildlife Reserve, the Wonga Wongé Presidential Reserve, and in the Monkalaba and

## Table 2b.14 Internationally protected areas in Central Africa

| Country | Biosphere Reserves* Number | Area (000 ha) | World Heritage Sites Number | Area (000 ha) | Ramsar Sites Number | Area (000 ha) |
|---------|--------|---------------|--------|---------------|--------|---------------|
| Cameroon | 3 | 850 | 1 | 526 | 0 | |
| Central African Republic | 2 | 1 640 | 1 | 1 740 | 0 | |
| Chad | 0 | | 0 | 0 | 2 | 1 843 |
| Congo | 2 | 246 | 0 | | 1 | 439 |
| Democratic Rep. Congo | 3 | 283 | 5 | 6 855 | 2 | 866 |
| Equatorial Guinea | 0 | | 0 | | 0 | 0 |
| Gabon | 1 | 15 | 0 | 0 | 3 | 1 080 |
| **Total** | **11** | **3 034** | **7** | **9 121** | **8** | **4 228** |

Source: Ramsar 2002, UNDP and others 2000, UNESCO 2002

*Data not available for Sao Tome & Principe.  *Some Biosphere Reserves may also be World Heritage Sites or Ramsar sites*

Offoué Reserves, where logging activities are affecting at least 50 per cent of the protected areas (IUCN, WWF and GTZ 2000).

Patrolling of reserves in the sub-region is also desperately under-resourced, with rates as low as one guard per 35 000 hectares in Congo (IUCN, WWF and GTZ 2000). Other constraints on management of protected areas include war and civil conflict, lack of an integrated management vision, high demand for bushmeat from urban areas and the international market, inadequate staff, equipment, and infrastructure for patrolling and enforcement of regulations, and inadequate involvement or exclusion of local communities in protected area management.

Steps towards rectifying this situation include projects implemented in conjunction with WWF to protect the Black Rhino in the northern savannas of Cameroon, and to create awareness and sustainable rates of bushmeat hunting in Gabon, by establishing quotas and using local people as patrol guards. Central African countries have also ratified the Convention on Biological Diversity, confirming their commitment to protecting biological resources. They have also formulated a sub-regional Plan of Convergence for Forests of Central Africa (Plan de Convergence) which was validated by the first Conference of Ministers in Charge of Forests at its December 2000 session. In the context of this sub-regional plan, each member country was to draw up its own emergency action plan (Plan d'Action d'Urgence) for the forestry sector. Institutional and legal frameworks for conservation have also been established—for example, the National Programme for Environmental Management (Programme National de Gestion de l'Environnement). National Environmental Action Plans (NEAPs), forestry laws, and environmental management laws are other measures that countries have introduced. National Biodiversity Strategies and Action Plans (NBSAPs) have also been developed for many Central African countries, and sub-regional initiatives such as Global Environment Facility (GEF) Biodiversity projects, the Ecosystèmes Forestiers d'Afrique Centrale (ECOFAC) programme, and the Central Africa Regional Program for the Environment (CARPE) have been launched. ECOFAC has been very active in assessment of biological resources, legislation pertaining to use of the resources, traditional methods of resource management, and in making recommendations for

Mayombe rainforest reserve, Congo

*Michel Gunther/Still Pictures*

protection of various habitats under different types of management (such as recommendations for National Parks, Ramsar sites, and agro-forestry projects). CARPE is a long-term initiative funded by USAID and aimed at identifying the necessary conditions for reducing deforestation in the Congo Basin. This will be achieved through gathering of baseline information on the forest resources and threats to the ecosystem, establishing monitoring programmes, and building capacity amongst decision makers. To date a vast array of reports, maps, and briefing notes have been published.

## WESTERN AFRICA

Habitat diversity in Western Africa ranges from semi-desert and savanna to tropical forests, mangroves, freshwater lakes and rivers, and inland and coastal wetlands. The Upper Guinea forest, which extends from western Ghana through Côte d'Ivoire, Liberia, and Guinea to Southern Sierra Leone, is a biologically unique system that is considered one of the world's priority conservation areas because of its high endemism (Conservation International, 1999). Nearly 2 000 plants and more than 41 mammals are endemic to the ecosystem. Species diversity is also high, with more than 20 000 butterfly and moth species, 15 species of even-toed ungulates, and 11 species of primates.

## ECOLOGICAL, ECONOMIC AND SOCIAL VALUES OF BIOLOGICAL RESOURCES IN WESTERN AFRICA

The richness of Western Africa's biological resources has constituted the basis of survival of the sub-region's indigenous societies. The local human populations have developed knowledge systems and practised traditions which have protected and conserved plants, animals, water resources and other components of their life support systems. In Ghana, sacred groves protect biodiversity in three different ways: by protecting particular ecosystems or habitats; by protecting particular animal or plant species; and by regulating the exploitation of natural resources (Ntiamoa-Baidu 1995). Many plant species are also used in Ghana in traditional herbal medicines (Mshana, Abbiw, Addae-Mensah, Adjanouhoun, Ahyi, Ekpere, Enow-Oroc, Gbile, Noamesi, Odei, Odunlami, Oteng-Yeboah, Sarpong and Tackie 2000), and the Kakum National Park in Ghana, with its canopy walkway, attracts thousands of visitors a year, helping to boost the economy as well as awareness of environmental issues.

## THREATS TO BIODIVERSITY IN WESTERN AFRICA

Since the beginning of the last century, biological resources in Western Africa have been rapidly degraded and lost through practices such as large-scale clearing and burning of forest, overharvesting of plants and animals, indiscriminate use of persistent chemical pesticides, draining and filling of wetlands, destructive fishing practices, air pollution, and the conversion of protected lands to agricultural and urban development. These activities are the results of uncontrolled population growth and increasing poverty, as well as of economic policies and priorities. For example, economic pressures led to concessions being granted to foreign logging companies to exploit Western Africa's tropical moist forests and prices of cash crops, especially in the 1980s, resulted in clearing of large areas of natural habitat for agriculture. Benin, Côte d'Ivoire, Liberia, Mauritania, Niger, Nigeria, Sierra Leone, and Togo all have rates of deforestation of more than 2 per cent per year (FAO 2001). Remnants of forest vegetation are presently found in protected areas in coastal countries. The Upper Guinea forest extends over approximately 420 000 square kilometres, but estimates of existing forests suggest a loss of nearly 80 per cent of the original extent (Conservation International 1999). The remaining forest is highly fragmented and spread across national borders. The forest fragments that remain are under severe threat, mainly arising from slash-and-burn agriculture, which accounts for much of the sub-region's subsistence food production.

Savannas are the dominant ecosystems in Western Africa, after tropical forests. Like the forests, they also support extremely biologically diverse communities of animals and plants but persistent exploitation for food, fuelwood and other resources from the savanna has resulted in their widespread degradation. For example, the rich and extensive savanna vegetation found in the northern portions of the sub-region has been severely degraded, with resultant loss of vegetation cover, fertile top soil and wild faunal species.

Political instability in Liberia, Sierra Leone and Senegal has created large numbers of refugees that add further pressure to the threatened forests through resettlement and subsistence agriculture. Political instability also creates economic distress indirectly in the sub-region, resulting in unsustainable resource use and lack of patrolling and enforcement of protection regulations.

Another major biodiversity issue in Western Africa is the loss and degradation of wetlands. Coastal and inland wetlands in Western Africa have been regarded as wastelands constituting habitats for pests and thus representing a threat to public health. As a result of this perception, wetlands in Western Africa have been under constant threat from development activities, especially agriculture and construction of harbours. Draining or in-filling of wetlands changes hydrological regimes so that they no longer provide suitable habitats for wildlife. Untreated effluents from domestic, commercial and industrial sources in nearby settlements have polluted coastal wetlands creating a toxicity risk for flora and fauna.

Rapid urbanization in coastal areas has created a number of very large cities in the sub-region, for example Lagos (Nigeria), Accra (Ghana) and Abidjan (Côte d'Ivoire). These cities surround major coastal wetlands some of which have been degraded by pollution and eutrophication to the extent that they have now become unsightly sources of odour and are biologically unproductive (IDRC 1996). Korle Lagoon in

Accra is one such example. Degradation of wetland ecosystem in the sub-region has also been attributed to extraction of woody material for fuel and charcoal production for domestic use and for curing of fish for the market. In Senegal, *Salvinia molesta* molesta (an invasive waterweed) appeared in the Djoudj Park and delta in 1999, and has since spread to a number of man-made lakes. This is a serious threat to the delta, which is a vital habitat for many migratory species.

Habitat loss is not the only threat to wildlife in Western Africa. The demand for bushmeat is driving high rates of poaching and an international trade in endangered species and wildlife products is also flourishing. A series of surveys of endangered primates in the forest reserves of Eastern Côte D'Ivoire and southern Ghana from 1993 to 1999 document the first recorded extinction of a widely recognized primate taxon, Miss Waldron's red colobus (*Procolobus badius waldroni*). Hunting rather than habitat loss is considered to be the primary cause (McGraw, Monah and Abedi-Lartey 1998, Oates, Abedi-Lartey, McGraw, Struhsaker and Whitesides 2000).

Drills (*Mandrillus leucophaeus*) (see Central Africa) are also found in the Cross River area of Nigeria (Gadsby and Jenkins 1998). Like the critically endangered Cross River Gorilla (*Gorilla gorilla diehli*), this species is found in an area which straddles the international border between Nigeria and Cameroon, as well as the sub-regional border between Western and Central Africa (Oates 2001)

Rural people in Western Africa depend heavily on medicinal plants for their health needs. However, as a result of extensive agricultural practices and annual bush fires, many medicinal plants have been lost at a time when conscious efforts are being made in many countries to promote herbal and traditional medicine.

Other species are threatened by a few invasive species of animals and plants. The Nypa palm, for example, is threatening mangrove forests in coastal Western Africa, and the bracken fern is encroaching on savanna ecosystems. Invasive plants such as these use up available water and nutrient resources and thus deprive native species and reduce biodiversity. The threatened species in Western Africa are shown in Table 2b.15.

## Table 2b.15  Threatened species in Western Africa, 2000

| Country | Mammals | Birds | Reptiles | Amphibians | Fishes | Inverts | Plants | Total |
|---------|---------|-------|----------|------------|--------|---------|--------|-------|
| Benin | 7 | 2 | 1 | 0 | 0 | 0 | 11 | 21 |
| Burkina Faso | 7 | 2 | 1 | 0 | 0 | 0 | 2 | 12 |
| Cape Verde | 3 | 2 | 0 | 0 | 1 | 0 | 2 | 8 |
| Côte d'Ivoire | 17 | 12 | 2 | 1 | 0 | 1 | 101 | 134 |
| Gambia | 3 | 2 | 1 | 0 | 1 | 0 | 3 | 10 |
| Ghana | 13 | 8 | 2 | 0 | 0 | 0 | 115 | 138 |
| Guinea | 11 | 10 | 1 | 1 | 0 | 3 | 21 | 47 |
| Guinea Bissau | 2 | 0 | 1 | 0 | 1 | 1 | 4 | 9 |
| Liberia | 16 | 11 | 2 | 0 | 0 | 2 | 46 | 77 |
| Mali | 13 | 4 | 1 | 0 | 1 | 0 | 6 | 25 |
| Mauritania | 13 | 2 | 2 | 0 | 0 | 0 | 0 | 20 |
| Niger | 11 | 3 | 0 | 0 | 0 | 1 | 2 | 17 |
| Nigeria | 25 | 9 | 2 | 0 | 2 | 1 | 119 | 158 |
| Senegal | 11 | 4 | 6 | 0 | 1 | 0 | 7 | 29 |
| Sierra Leone | 11 | 10 | 3 | 0 | 0 | 4 | 43 | 71 |
| Togo | 9 | 0 | 2 | 0 | 0 | 0 | 9 | 20 |

Source: IUCN 2000a

## TOWARDS SUSTAINABLE MANAGEMENT AND CONSERVATION OF BIODIVERSITY IN WESTERN AFRICA

The countries of Western Africa have responded to the problems of habitat loss by placing natural areas under protection. However, the number and size of protected areas in Western Africa varies from country to country (see Table 2b.16). In 1999, Burkina Faso and Senegal had over 10 per cent of their land area under national protection, whereas in Guinea and Guinea-Bissau this was less than 1 per cent, although they do have marine protected areas (World Bank 2001a).

International efforts to conserve natural habitats have been very successful in Western Africa, mainly as a result of ratification of the Ramsar Convention, and the Convention on Biological Diversity. There are 15 Biosphere Reserves in the sub-region, 10 World Heritage Sites and 37 Ramsar sites (see Table 2b.17).

Nearly all countries within the sub-region are signatories to the Convention on Biological Diversity and the Ramsar Convention, and many have drawn up programmes and projects under these agreements. Capacity development activities are also underway, under the aegis of new institutions created to coordinate and implement them. Most notable in this area has been GEF support for biodiversity programmes and projects in the sub-region. Western Africa was the principal African recipient of GEF biodiversity funding by mid-1998, with emphasis on coastal, marine and freshwater ecosystems.

In the arid and semi-arid areas of the sub-region, emphasis is on plant genetic resources, protected area management and capacity building. At the country level, relevant legal instruments have been enacted to protect and conserve biological diversity, especially forests, fauna and wetlands. However, these are largely out of date and too under-resourced to be implemented satisfactorily. More recently, National Action Plans and Conservation Strategies for the environment in general, and for forests, wildlife and biodiversity in particular, have been formulated and implemented with external funding. For example, Sierra Leone began implementation of its Biodiversity Strategy and Action Plan in December 2001, and its government has embarked on a joint programme with a non-governmental organization (NGO), the Conservation Society of Sierra Leone, as a partnership for sustainable Biodiversity Management.

Western African states nevertheless face several obstacles to implementation, including financial and human resource constraints, lack of awareness among the general public and among decision-makers, inadequate legal structures at the national level, and ineffective cooperation between countries in the sub-region. It is critical that these efforts be strengthened and that they become sustainable, financially as well as environmentally. Developing policies with donor funds is a step in the right direction, but funding streams need to be established to ensure implementation of policies and enforcement of regulations.

In Côte d'Ivoire, the Taï National Park, a large tract of undisturbed rainforest, is a World Heritage Site but is nevertheless threatened by slash-and-burn agriculture, poaching, and illegal logging and mining activities. A long-term management plan has recently been developed for the area, by WWF in conjunction with local people, and is currently being implemented. The Gashka Gumpti National Park, in Nigeria, encompasses

## Table 2b.16 Nationally protected areas in Western Africa

| Country | Number | Terrestrial Area (000 ha) | % land area | Marine Number |
|---|---|---|---|---|
| Benin | 2 | 778 | 6.9 | |
| Burkina Faso | 12 | 2 855 | 10.4 | 1 |
| Côte d'Ivoire | 11 | 1 986 | 6.2 | 3 |
| Gambia | 6 | 23 | 2.0 | 5 |
| Ghana | 10 | 1 104 | 4.6 | |
| Guinea | 3 | 164 | 0.7 | 1 |
| Guinea Bissau | 0 | 0 | 0.0 | 2 |
| Liberia | 1 | 129 | 1.2 | |
| Mali | 13 | 4 532 | 3.7 | |
| Mauritania | 9 | 1 746 | 1.7 | 5 |
| Niger | 6 | 9 694 | 7.7 | |
| Nigeria | 27 | 3 021 | 3.3 | |
| Senegal | 12 | 2 181 | 11.1 | 7 |
| Sierra Leone | 2 | 82 | 1.1 | |
| Togo | 9 | 429 | 7.6 | 1 |
| **Total** | **123** | **28 724** | **68.2** | **25** |

*Data not available for Cape Verde*

Source: World Bank 2001a

## Table 2b.17  Internationally protected areas in Western Africa

| Country | Biosphere Reserves* Number | Biosphere Reserves* Area (000 ha) | World Heritage Sites Number | World Heritage Sites Area (000 ha) | Ramsar Sites Number | Ramsar Sites Area (000 ha) |
|---|---|---|---|---|---|---|
| Benin | 1 | 623 | 0 | 0 | 2 | 139 |
| Burkina Faso | 1 | 186 | 0 | 0 | 3 | 299 |
| Côte d'Ivoire | 2 | 1 480 | 3 | 1 504 | 1 | 19 |
| Gambia | 0 | | 0 | 0 | 1 | 20 |
| Ghana | 1 | 8 | 0 | 0 | 6 | 178 |
| Guinea | 2 | 133 | 1 | 13 | 6 | 225 |
| Guinea Bissau | 1 | 110 | 0 | 0 | 1 | 39 |
| Liberia | 0 | | 0 | 0 | 0 | 0 |
| Mali | 1 | 2 349 | 1 | 400 | 3 | 162 |
| Mauritania | 0 | | 1 | 1 200 | 2 | 1 231 |
| Niger | 2 | 25 128 | 2 | 7 957 | 4 | 715 |
| Nigeria | 1 | <1 | 0 | 0 | 1 | 58 |
| Senegal | 3 | 1 094 | 2 | 929 | 4 | 100 |
| Sierra Leone | 0 | | 0 | 0 | 1 | 295 |
| Togo | 0 | | 0 | 0 | 2 | 194 |
| **Total** | **15** | **31 111** | **10** | **12 003** | **37** | **3 674** |

*Data not available for Cape Verde. *Some Biosphere Reserves are also World Heritage Sites or Ramsar sites.*

Source: Ramsar 2002, UNDP and others 2000, UNESCO 2002

a range of habitats across a range of altitudes from 450 m to 4 000 m. Chimpanzees are among the threatened species that survive in the forests. The Nigerian Conservation Foundation and WWF are developing a plan to promote tourism in the park, stimulating the economy through conservation efforts. The Drill Rehabilitation and Breeding Centre (DRBC) in Calabar, Nigeria was set up in cooperation with the Cross River State Ministry of Agriculture, Forestry Department and Cross River National Park. This is the only purpose-built *in-situ* captive-breeding programme for an endangered African primate. The programme is achieving a high rate of live births and successful rearing of offspring. In May 2000, the Afi Mountain Wildlife Sanctuary was 'gazetted' (that is, legally protected) and will protect one of the Nigerian populations of the Cross River Gorilla, as well as drills, chimpanzees and other primate species. It also houses a new facility for the DRBC. The DRBC is also conducting community outreach programmes to bolster support for conservation activities within the community (Gadsby and Jenkins 1998, Oates 2001).

Traditional protection and conservation beliefs and practices have always existed in all countries in the sub-region, but some of them have been eclipsed by modern activities. However, African countries are now realizing the importance of documenting knowledge and implementing traditional practices. The United Nations University Institute for Natural Resources in Africa (Accra, Ghana) has, for example, initiated a project to catalogue local indigenous plants that are useful as food and for pharmaceuticals, and to promote homestead gardens (Baidu-Forson 1999). Also in Ghana, the traditional grove system has been reintroduced into conservation models, enhancing biodiversity while sustainably using the resources in surrounding buffer zones of the grove (Oteng-Yeboah 1996).

Actions at the local level include a project in Ghana to support the conservation and sustainable use of

•

*Traditional knowledge and practices must be reincorporated into regional conservation strategies, and communities and commercial companies must be provided with alternative resources or means of production. Simply excluding activities from one area intensifies the pressure on other areas and resources*

•

medicinal plants. The main aims of the project are to establish and document baseline data, build capacity and raise awareness, and relieve the pressure from harvesting of these plants in the wild. Plans are in place to establish a working example of a medicinal plant garden, and to provide training to promote others across Ghana to help people to become able to manage their own medicinal garden businesses. The World Bank's Indigenous Knowledge for Africa Programme has encouraged research and promotion of the role of traditional hunters in natural resource management in Burkina Faso (World Bank 2001b). In Mauritania, WWF has supported the Banc d'Arguin National Park (a World Heritage Site) by assisting local communities and park officials to improve patrolling and control illegal fishing.

## CONCLUSION

The range of climatic conditions and geomorphology found in Africa has created a wide diversity of habitats, which various species of flora and fauna have evolved to exploit. As a result, the region is exceedingly well endowed with diverse biological resources. The forests of Africa are particularly diverse, and support many millions of Africans by providing them with food, clothing and construction materials, medicinal products, and cultural and recreational facilities. However, in many cases, the value of these resources is undetermined, because they are not traded on the open market. Other products, such as timber are commercially traded and make substantial contributions to GDPs. Studies have estimated the value of biological resources to be millions of dollars every year.

Africa's biodiversity has been subjected to increasing pressures of habitat loss (through conversion of natural habitats to urban, industrial or agricultural uses), overharvesting (due to increasing population and rising consumption levels), pollution (from urban and industrial sources), and the introduction of alien species (which dominate and alter habitat conditions). These pressures are set to continue and intensify over the next decade or two, because of rapid population growth and extensive use of natural resources in most economic activities. At the root of this, however, is a lack of proper valuation of natural resources. In many cases, resources

are not perceived to be limited and are, therefore consumed as if they were 'free', rather than being consumed sustainably, that is, leaving sufficient resources in the system to regenerate or reproduce, and thus ensure a continued supply.

Traditionally, African societies recognized the importance of biological resources, and employed traditional conservation measures. Under colonial administrations, the perceived solution to habitat loss was to declare nationally protected areas and to ban all or most activities. The current rate of species extinctions all across the Africa region testifies that this approach is insufficient. Traditional knowledge and practices must be reincorporated into regional conservation strategies, and communities and commercial companies must be provided with alternative resources or means of production. Simply excluding activities from one area intensifies the pressure on other areas and resources. Community conservation, or community based natural resource management can take a variety of forms, and presents a range of options. It is likely to be more successful where care has been taken to understand the situation faced by each community, and to apply a strategy that best fits the circumstances (Adams and Hulme 2001). Ratification of international conventions and establishment of regional action plans is a measure of political commitment to resolving these issues, but needs to be supported with human and financial resources not just to comply with obligations, but to implement activities and projects at the national and sub-national level. It is important to take into account the inequalities and local costs that form the basis for losses of biodiversity, and to avoid the temptation of trying to rely entirely on free-market economics (Western 2001) Win-win situations must be developed in order to sustain livelihoods and economies while retaining a biological resource base which is intact and functioning properly.

Many of the most valuable biodiversity resources extend beyond national or sub-regional borders. In order to avoid conflicts, and to manage and use these resources sustainably, African countries and sub-regional groupings must cooperate in devising policies, programmes and projects that harmonize biodiversity management and conservation throughout ecologically determined regions. This calls for a sustained effort on regional integration of environmental management.

## REFERENCES

Abu Zeid, K.M. (1995). *A Geographic Information System: An Initiative for the Integrated Sustainable Development of the Nile River Basin*. CEDARE Publications, Cairo

Adams, W. M. and Hulme, D. (2001). If community conservation is the answer in Africa, what is the question? Oryx 35 (3), pp. 193–200

Alderman, C.L. (1991). Privately owned lands: their role in nature tourism, education, and conservation. Pages 89–323 in J. A. Kusler, editor. *Ecotourism and resource conservation*, Vol.1. Omnipress, Madison, Wisconsin

Baidu-Forson (ed., 1999). Africa's Natural Resource Conservation and Management Surveys: Summary of the UNU/INRA Regional Workshop. The United Nations University Institute for Natural Resources, Accra

Baquar, S.R. (1995). The Role of Traditional Medicine in a Rural Environment. In: Sindiga, I., Nyaigotti- Chacha, C., Kanunah, M.P. 1995. *Traditional Medicine in Africa*. East African Educational Publishers Ltd., Nairobi

Berger, R. and Mugo, V. (2001). Draft Report on the Consultancy on the Development of EIA Guidelines, Regulations and Procedures, and Criteria for Listing EIA Experts (Under Sections 58–68 of the Environmental Management and Co-ordination Act, No. 8 of 1999. Republic of Kenya)

Burgess, N. D., H. de Klerk, J. Fjeldså, T. M. Crowe, and C. Rahbek (2000). A preliminary assessments of congruence between biodiversity patterns in Afrotropical forest mammals and forest birds. Ostrich 71: pp. 286–291

Butynski, T.M. (2001). Africa's Great Apes. In: *Great Apes and Humans: The Ethics of Coexistence*. Beck, B., Stoinski, T., Hutchins, M., Maple, T.L., Norton, B.G., Rowan, A., Stevens, E.F., and Arluke, A. (eds). Simithonian Institution Press, Washington D.C.

CAMPFIRE and Africa Resources Trust (1999). Campfire Factsheets. Available from http://www.campfire-zimbabwe.org/campfire_factsheets.html

Chenje, M. (ed., 2000). *State of the Environment Zambezi Basin 2000*. SADC/IUCN/ ZRA/SARDC, Maseru/Lusaka/Harare

CITES (2002). Introduction to CITES. Available on www.cites.org

Conservation International (2002), www.conservation.org/xp/CIWEB/ strategies/hotspots/guinean_forest.xml

Conservation International (1999). *Conservation Priority-Setting For The Upper Guinea Forest Ecosystem, West Africa*. Conservation International, Washington D.C.

Cowling, R.M. and Hilton-Taylor, C. (1994). Patterns of Plant Diversity and Endemism in Southern Africa: An Overview. In B.J. Huntley, Botanical Diversity in Southern Africa. *Strelitzia* 1: 31–52, National Botanical Institute, Pretoria

Croal, P. (2000). The Wealth of Indigenous Peoples and Development Assistance. Paper for the ASA 2000 Conference: Participating in Development: Approaches to Indigenous Knowledge 2–5 April 2000, School of Oriental and African Studies (SOAS), University of London

Cunningham, A.B. (1993). African Medicinal Plants: Setting Priorities at the Interface between Conservation and Primary Health Care. People & Plants Working Paper 1: African Medicinal Plants. WWF

Daily Nation, 13 December 2001 (Kenya)

DEA&T (1997). White Paper on the Conservation and Sustainable Use of South Africa's Biological Diversity. Government Gazette No. 18163. Government Printer, Pretoria

Dery, B.B. & Otsyina, R. (2000). Indigenous Knowledge and the Prioritization of Medicinal Trees for Domestication in The Shinyanga Region of Tanzania. In: A.B. Temu, G. Lund, R.E. Malimbwi, G.S. Kowero, K. Klein, Y. Malende, I. Kone (eds., 2000). *Off-forest tree resources of Africa*. A proceedings of a workshop held at Arusha, Tanzania, 1999. The African Academy of Sciences (AAS)

Draz, O. (1969). The Hema System of Rangeland in the Arabian Peninsula. FAO Report number /PL/PEC/13, FAO, Rome

Drewes, S.E., Hall, A.J., Learmonth, R.A., & Upfold, U.J. (1984). Isolation of Hypoxoside from Hypoxis rooperi and Synthesis of [E]-1,5Bis [3'4'dimethoxyphenyl] pent-1-en-1-ync. *Phytochemistry*, 23; 1313–1316

Cumming, D.H.M. (2000). Drivers of Resource Management Practices – Fire in the Belly? Comments on 'Cross-cultural Conflicts in Fire Management in Northern Australia: Not so Black and White' by Alan Andersen. Conservation Ecology 4 (1): 4, [online] URL: http://www.consecol.org/vol14/iss1/art4

Emerton L., & Muramira, E.T. (1999). *Uganda's Biodiversity: Economic Assessment*. A report prepared for the National Environment Management Authority as part of the Uganda National Strategy, Biodiversity Strategy and Action Plan. Kampala

Emery, A.R. (2000). *Guidelines: Integrating Indigenous Knowledge in Project Planning and Implementation*. ILO, WB, CIDA, KIVU Nature Inc. Washington, D.C and Hull, Quebec

EPA/MEDC (1997). *The Conservation Strategy of Ethiopia – Volume 1 The Resources Base, Its Utilisation and Planning for Sustainability*. Environmental Protection Authority in collaboration with the Ministry of Economic Development and Cooperation. Federal Democratic Republic of Ethiopia, Addis Ababa

Fa, J.E., Garcia Yuste, J.E. & Castelo, R. (2000). Bushmeat Markets on Bioko Island as a Measure of Hunting Pressure. *Conservation Biology*, Vol 14, No. 6 pp.1602–1613

FAO (1995). Country Information Brief Food and Agriculture Organization of the United Nations. Ethiopia, Draft June 1995

FAO (2001). Global Forest Resource Assessment. FAO, Rome

FAOSTAT (2000). Statistics Database. FAO, Rome

Fay, D., Palmer, R., and Timmermans, H. (2000). *From Confrontation to Negotiation at Dwesa-Cwebe Nature Reserve: Conservation, Land Reform and Tourism Development on South Africa's Wild Coast*, Human Science Research Council, Pretoria

Gadsby, E.L. and Jenkins, Jr. P.D. (1998). The Drill- Integrated *in-situ* and *ex-situ* conservation. In: *African Primates: The Newsletter of the Africa Section of the IUCN/SSC Primate Specialist Group* Vol 3, Nos 1&2 1997–1998

Ghabbour, S. I. (1975). National Parks in Arab Countries. *Environ. Conserv*. 2: 45–46

Gibson, C.C. (1999). *Politicians and Poachers : The Political Economy of Wildlife Policy in Africa*. Cambridge University Press, Cambridge

Gichohi, H., Gakahu, C. and Mwangi, E. (1996). Savanna Ecosystems. In: T.R. McClanahan & T.P. Young (eds) *East African Ecosystems and their Conservation*. 1996 . Oxford University Press, Oxford

Gilpin, M. E. and Soulé, M. E. (1986). Minimum viable populations: Processes of species extinction. Soulé, M. E. Conservation Biology: The Science of Scarcity and Diversity. Sunderland, MA: Sinauer Associates, Inc.

Global Water Partnership (2000). *Southern African Vision for Water, Life and the Environment in the 21st Century ad Strategic Framework for Action Statement*. SATAC. http://www.watervision.org

GOM/ERM (1998). *National Environmental Strategies for the Republic of Mauritius: National Environmental Action Plan for the Next Decade*. Government of Mauritius and Environmental Resources Management, London. Draft Report, October 1998

Greenwire (2001). Bushmeat Trade Threat to Wildlife. *Natural Resources*; 10(9) 22 May 2001

Hall A.V., De Winter, B., and Van Oosterhout, S.A.M. (1980). Threatened Plants of Southern Africa. South African National Scientific Programmes Report 45, CSIR, Pretoria

Hall, S.J.G. and Bradley, D.G. (1995). Conserving livestock breed biodiversity. *Trends in Ecology and Evolution*, Vo. 10, No. 7, pp. 267 – 270

Hall, S.J.G. and Ruane, J. (1993) Livestock Breeds and Their Conservation: A Global Overview. *Conservation Biology* Vol 7, No. 4 pp. 815– 825

Harper, D. and Mavuti, K. (1996). Freshwater Wetlands and Marshes. In: T.R. McClanahan and T.P. Young (eds) *East African Ecosystems and their Conservation*. 1996. Oxford University Press, Oxford

Harris, L.D. (1984). The Fragmented Forest: Island Biogeographic Theory and the Preservation of Biotic Diversity (Univ. of Chicago Press, Chicago, 1984)

Harrison, J.A., Allan, D.G., Underhill, L.G., Herremans, M., Tree, A.G. and Brown, C.E. (1997). *The Atlas of Southern African Birds Vols. 1 and 2*. BirdLife South Africa, Johannesburg

Hegazy, A.K. (2000a). Intra-Population Variation In Reproductive Ecology And Resource Allocation Of The Rare Biennial Species Verbascum Sinaiticum Benth In Egypt. *Journal of Arid Environments*, 44: 185–196

Hegazy, A.K. (2000b). Reproductive Diversity And Survival Of The Potential Annual Diplotaxis Harra (Forssk.) Boiss. (Cruciferae) In Egypt. Paper presented before the International Conference on the Conservation of Biodiversity in the Arid Regions. 27–29 March 2000, Kuwait

Hegazy, A.K., Diekman, M. and Ayad, W.G. (1999). Impact of Plant Invasions on Ecosystems and Native Gene Pools. In A. K. Hegazy (ed.) *Environment 2000 and Beyond*. Printed by Horus for Computer and Printing, Cairo, Egypt. pp. 275–310

Hegazy, A.K., Fahmy, A.G., and Mohamed, H.M. (2001). Shayeb El-Banat Mountain Group On The Red Sea Coast: A Proposed Biosphere Reserve. *Symposium on Natural Resource Conservation in Egypt and Africa*, 19–21 March 2001. Cairo University, Egypt

Hilton-Taylor, C. (1996). Red Data List of Southern African Plants *Strelizia* 4: 1–117

Hulme, D. and Murphree. M. (eds., 2001). *African Wildlife & Livelihoods: The promise and performance of Community Conservation*. Heinemann/James Currey, Oxford

IDRC (1996). Ghana: The Nightmare Lagoons. Article written by Theo Andersen, Friends of the Earth, and published on: http://www.idrc.ca/books/reports/e234-13.html

IUCN (2000b). North Africa Porgramme. Available on: http://www.iucn.org/places/wcana/northafr.htm

IUCN (1997). 1997 IUCN Red List of Threatened Plants. IUCN, Gland. Available on: http://www.unep-wcmc.org.uk

IUCN (2000a). 2000 IUCN Red List of Threatened Species. Compiled by Craig Hilton-Taylor. Available on: http://www.IUCN.org

IUCN, WWF and GTZ (2000). Protected Areas Management Effectiveness Assessment for Central Africa. A development Report. Prepared by Elie Hakizumwami and others for the Forest Innovations project. Presented at the WWF 'Beyond the Trees' Conference, Bangkok, May 2000

Johnstone, R. (2001). Communing with Nature. *Travel News* No. 82, March 2001

Kassas, M. (1972). National Parks in Arid Regions. Proceedings of Second World Conference on National Parks. 15p

Kohler-Rollefson, I. and McCorkle, C. (2000). Domestic Animal Diversity, Local Knowledge and Stockraiser Rights, Paper for the ASA 2000 Conference: Participating in Development: Approaches to Indigenous Knowledge 2–5 April 2000, School of Oriental and African Studies (SOAS), University of London

Langholz, J. (1996). Economics, Objectives, and Success of Private Nature Reserves in Sub-Saharan Africa and Latin America. *Conservation Biology*, Vol. 10, No. 1 pp 271–280

Larson, P. and Bromley, D. (1991). Sudan and the Guar Gum Fiasco. World Development Journal

Laurenson, K., Sillero-Zubiri, C., Thomson, H., Shiferaw, F., Thirgood, S and Malcolm, J. (1998). Disease as a threat to endangered species: Ethiopian wolves, domestic dogs and canine pathogens. *Animal Conservation* 1 Part 4 pp. 273–280

Ledger, J. (1990). *Southern Africa's Threatened Wildlife*. Endangered Wildlife Trust, Johannesburg

Lovejoy T.E., R.O. Bierregaard, A.B. Rylands, J.R. Malcolm, C.E. Quintela, L.H. Harper, K.S. Brown, A.H. Powell, G.V.N. Powell, H.O.R. Schubart and M.B. Hays (1986). Edge and other effects of isolation on Amazon forest fragments. In: Soule, M.E. (ed.), Conservation Biology: the science of scarcity and diversity. Sinauer Associates, Sunderland, Mass., pp. 257–285

Lovett, J.C. (1998). Eastern Arc Mountain Forests: Past and Present. In Schulman, L., Junikka, L., Mndolwa, A. and Rajabu, I., (Eds) 1998. *Trees of Amani Nature Reserve*. Helsinki University Press

Low, A.B. and Rebelo, A.G. (eds., 1996). *Vegetation of South Africa, Lesotho and Swaziland*. Department of Environmental Affairs and Tourism, Pretoria

Martens, E.E. (1995). Causes of Biodiversity Loss in Coastal Ecosystems. In: L.A. Bennun, R.A. Aman and S.A. Crafter, (eds) *Conservation of Biodiversity in Africa: Local Initiatives and Institutional Roles* Proceedings of a Conference held at the National Museums of Kenya, 30 August –3 September, 1992. National Museums of Kenya, Nairobi

MBIFCT (1994). Proposal for the Establishment of the Mgahinga and Bwindi Impenetrable Forest Conservation Trust (MBIFCT) Trust Administration Unit (TAU). Mgahinga and Bwindi Impenetrable Forest Conservation Trust, Kampala

Mbora, D.N.M. and Wieczkowski, J. (2001). Impacts of Micro-and Small Enterprises on the Environmental Conservation of Fragile Ecosystems: A Case Study of the Tana River Primate National Reserve In: D.L. Manzolillo Nightingale (ed). *Micro and Small Enterprises and Natural Resource Use*. 2001. Proceedings of a workshop held at ICRAF, Nairobi, Kenya. 21–22 February 2001. Micro-Enterprises Support Programme and United Nations Environment Programme, Nairobi

McCullum (ed., 2000). *Biodiversity of Indigenous Forests and Woodlands in Southern Africa*, SADC/IUCN/SARDC, Maseru/Harare

McGraw, W.S., Monah, I.T. and Abedi-Lartey, M. (1998). Survey of Endangered Primates in the Forest Reserves of Eastern Cote D'Ivoire. *African Primates: The Newsletter of the Africa Section of the IUCN/SSC Primate Specialist Group* Volume 3, Numbers 1 & 2, 1997–1998

McNeely, J.A., Miller, K.R., Reid, W.V., Mittermeier, R.A. and Werner, T.B. (1990). *Conserving the World's Biological Diversity*. IUCN, WRI, CI, WWF-US, The World Bank, Gland

Mittermeier, R. A. and Myers, N., Gil, P.R. and Mittermeier, C.G. (2000). *Hotspots; The Earth's Biologically Richest and Most Endangered Terrestrial Ecoregions*. CEMEX and Conservation International, Washington

MoFPED (2000). *Statistical Abstracts 2000*. Ministry of Finance Planning and Economic Development. Kampala

Moyini Y. and Uwimbabazi, B. (2000). *Analysis of the Economic Significance of Gorilla Tourism in Uganda*. The International Gorilla Conservation Programme/African Wildlife Foundation

Mshana, N.R., Abbiw, D.K., Addae-Mensah, I., Adjanouhoun, E., Ahyi, M.R.A., Ekpere, J.A., Enow-Oroc, E.G., Gbile, Z.O., Noamesi, G.K., Odei, M.A., Odunlami, H., Oteng-Yeboah, A.A., Sarpong, K. and Tackie, A.N. (2000). *Traditional Medicine and Pharmacopoeia: Contribution to the Revision of Ethnobotanical and Floristic Studies in Ghana*. Scientific, Technical and Research Commission of the Organisation of African Unity (OUA/STRC)

Mugabe, J. (1998). Biodiversity and Sustainable Development in Africa In: J. Mugabe and N. Clark (eds) *Managing Biodiversity: National Systems of Conservation and Innovation in Africa*. African Centre for Technology Studies (ACTS), Nairobi

MUIENR (2000). Uganda Biodiversity Report 2000. Makerere University Institute of Environment and Natural Resources. Kampala

Murombedzi, J. (2001). Committees, Rights, Costs and Benefits: Natural Resource Stewardship and Community Benefits in Zimbabwe's CAMPFIRE Programme. In: D. Hulme and M. Murphree (Eds) 2001. *African Wildlife & Livelihoods: The promise & performance of Community Conservation*. Heinemann/ James Currey, Oxford

Myers, N. (1990). The Biodiversity Challenge: Expanded Hotspot Analysis. *The Environmentalist*, 10: 243–256

Natures Truth (2001). http://www.naturestruth.com/product_african_potatoe.htm

NEIC (1994). *State of Environment Report for Uganda 1994*. Natinal Environment Information Centre. Ministry of Natural Resources. Kampala, Uganda

Newmark, W. D. (1996). Insularization of Tanzanian Parks and the Local Extinction of Large Mammals *Conservation Biology* Vol 10 No 6 pp 1549–1556

Ntiamoa-Baidu (1995). Indigenous vs. Introduced Biodiversity Conservation Strategies: The Case of Protected Area Systems in Ghana Issues in African Biodiversity No. 1, African Biiodiversity Series. Biodiersity Support Prrogram, May 1995. (http://www.worldwildlife.org /bsp/publications/showhtml.cfm?uid=36)

Oates, J. (2001). Cross River Gorilla Workshop. *Oryx* 35: (3), 263–266

Oates, J.F., Abedi-Lartey, M., McGraw, W.S., Struhsaker, T.T. and Whitesides, G.H. (2000). Extinction of a West African Red Colobus Monkey *Conservation Biology*, Vol 14, No. 5 pp. 1526–1532

Olal, M.A., Muchilwa, M.N. and Woomer, P.A. (2001). In: D.L. Manzolillo Nightingale (ed). *Micro and Small Enterprises and Natural Resource Use*. 2001. Proceedings of a workshop held at ICRAF, Nairobi, Kenya. 21–22 February 2001. Micro-Enterprises Support Programme and United Nations Environment Programme, Nairobi

Ole Kaasha, J. M. (2001). Ecotourism Micro-Enterprise as a Tool to Conserve Natural Resources In: D.L. Manzolillo Nightingale (ed). *Micro and Small Enterprises and Natural Resource Use*. 2001. Proceedings of a workshop held at ICRAF, Nairobi, Kenya. 21–22 February 2001. Micro-Enterprises Support Programme and United Nations Environment Programme, Nairobi

Ole Lengisugi, N.A.M.O and Mziray, W.R. (1996). The Role of Indigenous Knowledge in Sustainable Ecology and Ethnobotanical Practices among Pastoral Maasai: Olkonerei– Simanjiro Experience. Paper Presented at the 5th International Congress of Ethnobiology at the Kenyatta International Conference Centre, Nairobi, Kenya, 2–6 September 1996

Olsen, J., Rodgers A. and Salehe, J. (1999). *Woodland and Tree Resources on Public Land in Tanzania II: Sustainable Management at Local Level*. Paper prepared for a Workshop on Off-forest Tree Resources , Arusha, Tanzania 12–16 July 1999. Faculty of Forestry and Nature Conservation, Sokoine University of Agriculture, Tanzania, AAS, ICRAF, FAO, IUFRO, FS, Sweden, USDA Forestry Service, USA, and GTZ

Oteng-Yeboah, A.A. (1996). Biodiversity in Three Traditional Groves in the Guinea Savanna, Ghana. In: L.J.G. van der Maesen, X.M. van der Burgt and J.M. van Medenbach de Rooy (Eds), *The Biodiversity of African Plants*. Kluwer Academic Publishers, Dordrecht

Quézel, P., Médail, F., Loisel, R. and Barbero, M. (1999). Biodiversity and Conservation of Forest Species in the Mediterranean Basin. *Unasylva No. 197 - Mediterranean Forests*. Vol: 50(2) Published by the FAO, Rome

Ramsar (2002). List of Wetlands of International Importance. 1 February 2002. Available on: www.ramsar.org

Republique de Madagascar (1997). Convention sur la Diversité Biologique, Premier Rapport National. Ministère de l'Environnement, Office National de l'Environnement (ONE), Association Nationale pour la Gestion des Aires Protégées (ANGAP). Projet GF 1200/96/59. Décembre 1997

RFIC (1998). Draft du Rapport National sur la Stratégie et le Plan d'Action en Matière de Diversité Biologique. République Fédérale Islamique des Comores, Ministère de la Production Agricole, des Ressources Marines, de l'Environnement et de l'Artisanat, Projet PNU/FEM COI/97/A/IG/99

Rodgers, A., Salehe, J. and Olsen, J. (2000). Woodland and tree resources on public land in Tanzania: National policies and sustainable use. In: A.B. Temu, G. Lund, R.E. Malimbwi, G.S. Kowero, K. Klein, Y. Malende, I. Kone (eds) 2000 *Off-forest tree resources of Africa*. A proceedings of a workshop held at Arusha, Tanzania, 1999. The African Academy of Sciences (AAS)

Science in Africa (2001). NERICA - New Rice Transforming Agriculture for West Africa. Guy Manners, WRDA, Ivory Coast. August 18th 2001

Sheldon, J.W., Balick, M.J. and Laird, S.A. (1997). *Medicinal Plants: Can Utilization and Conservation Coexist?* Advances in Economic Botany, V. 12, Charles M. Peters, Editor, The New York Botanical Garden, N.Y.

Situma, J. (2001). Mission Bulk Transfer *Swara* Vol. 24: 2, May–August 2001

Skole, D. and C. Tucker. (1993). Tropical deforestation and habitat fragmentation in the Amazon: Satellite data from 1978 to 1988. *Science* 260: 1905–09

Stephenson, D.J. Jr. (1999). The Importance of the Convention on Biological Diversity to the Loita Maasai of Kenya. In: D. Posey (ed) 1999 *'Cultural and Spiritual Values of Biodiversity'* United Nations Environment Programme, Nairobi

Ucko, P.J. and Dimbleby, G.W. (1969). *The Domestication And Expoitation Of Plants And Animals*. Gerald Duckworth & Co. Ltd., London

UNDP (2000). Human Development Report 2000. United Nations Development Programe, New York

UNDP, UNEP, World Bank and WRI (2000). World Resources 2000–2001: People and Ecosystems the Fraying Web of Life. World Resources Institute, Washington D.C.

UNEP (1999). *Western Indian Ocean Environmental Outlook*. United Nations Environment Programme, Nairobi

UNESCO (2002). World Network of Biosphere Reserves 411 Reserves in 94 Countries. United Nations Educational, Scientific and Cultural Organization, MAB Secretariat, France

UNESCO (1996). Protecting Natural Heritage in North Africa and the Middle East. United Nations Educational, Scientific and Cultural Organization, Cairo

UNESCO/UCO (1998). *Multipurpose Species in Arab African Countries*. United Nations Educational, Scientific and Cultural Organisation, Cairo

UWA (2000). *Protected Area Systems Plan for Uganda*. Uganda Wildlife Authority. Kampala, Uganda

Van Wyk, B-E., Van Oudtshoorn, B. and Gericke, N. (1997). *Medicinal Plants of South Africa*. Briza Publications, Pretoria

Versveld, D.B., Le Maitre, D.C. and Chapman, R.A. (1998). Alien Invading Plants and Water Resources in South Africa : A Preliminary Assessment. Water Research Commission, Pretoria Tt99/98

Vigne, L. (1999). High and Wild , *Swara* Volumes 22:2 & 22:3, pp. 24–27

Warren, D. M. (1995). In: L.A. Bennun, R.A. Aman and S.A. Crafter, (eds) *Conservation of Biodiversity in Africa: Local Initiatives and Institutional Roles*. Proceedings of a Conference held at the National Museums of Kenya, 30 August–3 September, 1992. National Museums of Kenya, Nairobi

WCMC (1992). *Global Biodiversity: Status of the Earth's Living Resources*. A Report compiled by the World Conservation Monitoring Centre, Editor Brian Goombridge, in collaboration with The Natural History Museum, London, and in association with IUCN, UNEP, WWF, and WRI. Chapman and Hall, London

WCMC (1991). Biodiversity Guide to Ethiopia. Report Financed by the Commission of European Communities. World Conservation Monitoring Centre, Cambridge

WCMC (2000). Global Biodiversity: Earth's Living Resources in the 21st Century. By: Groombridge, B. and Jenkins, M.D., World Conservation Press, Cambridge

Wells M.J., Balsinhas A.A., Joffe H, Engelbrecht V.M., Harding G and Stirton C.H. (1986). A Catalogue of Problem Plants in Southern Africa. Memoirs of the Botanical Survey of South Africa 53, 1–658

Western, D. (1997). *In the Dust of Kilimanjaro*. Island Press, Washington, D.C.

Western, D. (2001). Taking the Broad View of Conservation—A Response to Adams and Hulme *Oryx* 35 (3) 201–203

Westing, A.H. (1993). Building Confidence with Transfrontier Reserves: The Global Potential In: A.H. Westing, Ed. 1993 *Transfrontier Reserves for Peace and Nature: A Contribution to Human Security.* United Nations Environment Programme, Nairobi

Wilkie, D., Shaw, E., Rotberg, F., Morelli, G. and Auzel, P. (2000). Roads, Development and Conservation in the Congo Basin *Conservation Biology* Vol 14, No. 6 pp. 1614 –1622

Working for Water (2000). Annual Report. Department of Water Affairs and Forestry, Pretoria

World Bank (2000). World Development Report. The World Bank, Washington D.C.

World Bank (2001a). African Development Indicators 2001. The World Bank, Washington D.C.

World Bank (2001b). Mainstreaming Indigenous Knowledge, Africa Region. The World Bank, Washington. Update available on: http://www.worldbank.org

World Bank (1996). *The Experience Of The World Bank In The Legal Institutional And Financial Aspects Of Regional Environment Program: Potential Application Of Lessons Learned For The ROPME And PERSGA Programs.* World Bank, Washington D.C.

WRI, UNEP and UNDP (1992). *World Resources Report 1992–93,* Oxford University Press, New York and Oxford

WRI, UNEP and UNDP (1990). *World Resources 1990–91,* Oxford University Press, New York and Oxford

WWF and IUCN (1994). *Centres of Plant Diversity: A Guide and Strategy for Their Conservation.* Vol 1. IUCN Publications Unit, Cambridge

**PART C: COASTAL AND MARINE ENVIRONMENTS**

UNEP

## REGIONAL OVERVIEW

There are various definitions of the geographical, legal and functional scope of a coastal zone. According to Clark (1996), *'at a minimum, the designated coastal zone includes all the inter-tidal and supra-tidal areas of the water's edge; specifically all the coastal floodplains, mangroves, marshes and tide-flats as well as beaches and dunes and fringing coral reefs.'* Africa's 40 000 km coast consists of a narrow, low-lying coastal belt which, as shown in Figure 2c.1, includes the continental shelf and coasts of 32 mainland countries.

Coastal ecosystems and issues relating to management of the coast often cross political boundaries and frequently extend inland. Similarly, the boundaries of marine ecosystems do not necessarily correspond with those of Exclusive Economic Zones (EEZ)—the 200 nautical-mile zones from the land edge of coastal states where they have sovereign rights over natural resources and over certain economic activities. Discussion of issues affecting the coastal environment may therefore extend beyond country borders or over more than one sub-region. There may, therefore, be some overlap in the analyses presented here.

**Figure 2c.1 Map of Africa showing coastal countries, cities, and EEZ**

## ECONOMIC, SOCIAL AND ECOLOGICAL VALUE OF COASTAL AND MARINE ENVIRONMENTS

Coastal ecosystems are some of the most biologically productive in the world, occupying only 8 per cent of the Earth's surface, but accounting for 26 per cent of all biological productivity (Hare 1994). These high productivity rates result from the extreme climatic and physical conditions of coasts, and the dynamic nature of the forces acting on these zones. Opportunistic organisms have adapted to these conditions and their rapid rates of growth and reproduction when conditions are favourable have led to high productivity rates.

The African coastal zone supports a diversity of habitats and resources, encompassing mangroves, rocky shores, sandy beaches, deltas, estuaries and coastal wetlands, coral reefs and lagoons. Coral reefs and mangroves are especially important features because they protect the coastline by moderating storm and wave impacts and because mangroves stabilize sand and soils, cycle nutrients, absorb and break down waste products, provide wildlife habitat, and maintain biodiversity. Reefs and mangroves also contribute significantly to the economies of coastal countries by providing opportunities for tourism and for harvesting of resources. For example, mangrove species are used extensively by local communities for construction material, fuel, food and animal fodder, and for medicinal preparations. Mangroves extend from Senegal to Angola on Africa's west coast and from Somalia to South Africa on the east coast. Coral reefs—some of them spectacular—are abundant in the Red Sea and the Western Indian Ocean.

Periodic upwelling in African waters encourages diverse and rich production in fisheries, including crustaceans, fish, and molluscs. In 1997, total marine fish catch exports from Africa contributed US$445 million to countries' economies (FAOSTAT 2001). Fisheries in estuaries and lagoons also contribute to national economies, accounting for more than 75 per cent of fishery landings in Africa (IPCC 1998). Fisheries also provide significant amounts of employment, particularly in small islands such as Cape Verde and Seychelles where more than 33 per cent of agricultural workers are employed in the fisheries sector (FAO 1996). Artisanal fishing activities are also an important source of income for coastal communities and fish is an important source of protein for many African populations, as shown in Figure 2c.2.

Oil and gas reserves and other mineral deposits, including diamonds off the western and south-western African coast, are additional important economic resources for coastal countries. For example, in Benin, Ghana, Nigeria, Sierra Leone and Togo the majority of industries and oil and mineral mining activities are located in the coastal zone.

**Figure 2c.2 Contribution of fish to the African diet**

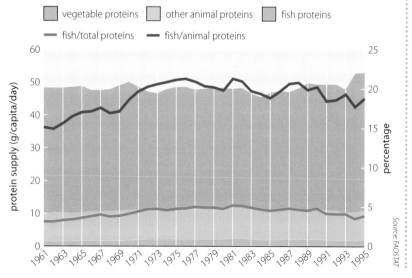

Source: FAOSTAT

The ports of Mombasa, Maputo and Durban are also major trade centres. Industrial development has expanded in the neighbouring areas to take advantage of opportunities for trade, tourism and other commercial activities opened up by the ports.

The highly attractive and diverse resources of the coastal and marine environments mean that coastal areas in Africa (as elsewhere in the world) are experiencing rapid population growth, industrial expansion and infrastructure development. Most colonial settlements in Africa were established along the coast in order to maximize trade opportunities, and this has resulted in all but three of African countries from Mauritania to Namibia having their capital cities on the coast. A large proportion of Western Africa's urban population also now lives in coastal cities. The exceptional demand for resources and infrastructure development in the coastal zone is putting pressure on fragile ecosystems and 38 per cent of Africa's coastal ecosystems are under severe threat from development-related activities (FAO 1998). The major challenges facing coastal African countries at present are coastal erosion, the potential impacts of climate change, overharvesting of resources and pollution. Although the causes of coastal erosion and climate change are entirely different the effects of climate change are so frequently and so firmly interlinked with those of erosion that they are described in the same sections in the following analyses.

## COASTAL EROSION AND CLIMATE CHANGE

Erosion and deposition are natural and dynamic processes occurring in coastal zones throughout the world. However, certain human activities, both inland and in the coastal zone itself can change patterns of erosion and deposition, and thus place a burden on the adaptive capacity of coastal ecosystems and pose a challenge to effective coastal management. Human activities including conversion of natural coastal habitats such as wetlands and mangroves to urban or agricultural uses lead to their loss or modification. For example, it is estimated that about 40 per cent of Nigeria's mangroves had been lost by 1980 as a result of clearing for development, coastal erosion and increased salinity (WRI 1990). When such ecosystems are reduced, the natural buffer they provide against

wave action and storm surges is compromised, resulting in increased erosion and worsened impacts of flooding.

Mining of coastal resources such as coral and sand also contributes to erosion by disturbing the surface and exposing the substrate to rain, rivers and wave action. Mining of coral to provide stone for construction means that the protection the coral normally gives to the shoreline is lost, and land recession results.

Damming of rivers further inland forces sediments to settle upstream allowing fewer sediments to reach the river mouth. This increases the river's scouring potential leading to higher rates of erosion in the coastal zone. For example, damming of rivers in Western Africa has accelerated rates of coastal erosion, and these have now reached as much as 30 m/yr (Chidi Ibe and Quelennac 1989). Damming of the Nile River at Aswan reduced the nutrient load reaching the Nile Delta, and this reduced sardine fishery catches from 22 618 million tonnes in 1968 to only 13 450 million tonnes in 1980, with rates still falling. Although the reservoir behind the dam created new fishing opportunities, there are doubts as to the sustainability of harvesting of these resources, and many coastal communities have lost their livelihood. (Acreman 1999). Sedimentation in ports and harbours can also interfere with activities, and dredging is costly.

Sea level rises resulting from global climate change will exert additional pressures on the coastal zone, causing inundation of low-lying areas, erosion of infrastructure, displacement of populations and contamination of freshwater sources. According to the Intergovernmental Panel on Climate Change (IPCC), the average global sea level has risen by 1–2 mm per year during the past century. The most likely cause of this is expansion of seawater and widespread loss of land ice, caused by higher mean global temperatures. The IPCC also predicts that, by 2100, the global sea level could rise by up to one metre (IPCC 2001a). The consequent flooding, and changes in salinity, wave conditions and ocean circulation will put natural habitats and human settlements at risk of flooding and accelerated erosion. The extent and severity of the impacts of storms will also increase as a result of further climatic changes, and because the buffering capacity of coral reefs and mangrove systems will have been lost. Human settlements and economic activities in the Gulf of Guinea, Senegal, Gambia, Egypt, and along the eastern

*Damming of the Nile River at Aswan reduced the nutrient load reaching the Nile Delta and this reduced sardine fishery catches from 22 618 million tonnes in 1968 to only 13 450 million tonnes in 1980, with rates still falling*

African coast, including the Western Indian Ocean, are likely to be most severely affected (IPCC 2001b). Some of these countries may be unable to cope with the financial and technical burden of implementation of mitigation measures (Leatherman and Nicholls 1995).

In addition to sea level rise, sea-surface temperatures in the open tropical oceans surrounding Africa are predicted to rise by about 0.6–0.8 °C (less than the global average). This may increase the frequency and intensity of tropical cyclones and result in bleaching of coral that will impact on the economies of countries on the eastern African coast, of the Western Indian Ocean Islands, and of the countries of the Red Sea.

### Mitigation of coastal erosion

African countries have promulgated laws and regulations requiring environmental impact studies to be carried out before development in coastal zones or hinterlands, thus regulating change of land use and expansion of human activities. Countries that have taken such action include Egypt, Gambia, Ghana, Kenya, Mauritius, Nigeria, South Africa, Swaziland, Tanzania, Uganda, Zambia and Zimbabwe. Short-term measures against coastal erosion include the construction of physical barriers such as sea walls and groynes, although these are expensive and require constant maintenance.

In addition, many African countries are adopting more integrated environmental management strategies, based on ecological management units rather than political boundaries. A holistic management approach improves the integration of different resource-use objectives and reduces competition between user groups. Coastal erosion is therefore being tackled in some countries through improved water catchment management, more conservative agricultural practices, and soil conservation programmes.

Integrated Coastal Zone Management (ICZM), discussed more fully below, is a holistic approach that has been adopted widely within Africa's coastal zone to help to reduce the causes and impacts of coastal erosion. Sub-regional and regional cooperative programmes and action plans (such as UNEP's Regional Seas Programme) provide the framework and necessary capacity building for implementation of ICZM at the national level. If the state of coastal and marine ecosystems, and the sustainability of commercial and subsistence activities dependent on them, are to be improved, continued support for such programmes is necessary, in the form of trained staff, finance, equipment for policy implementation, research, monitoring and enforcement of regulations.

## HARVESTING OF COASTAL AND MARINE RESOURCES

Marine and coastal fishery resources are extremely important in Africa, both to national economies and to the livelihoods of local communities. In the late 1990s, fishing contributed more than 5 per cent to GDP in Ghana, Madagascar, Mali, Mauritania, Mozambique, Namibia, Senegal and Seychelles, and the shrimp fishery on the Sofala Bank in Mozambique contributed 40 per cent of Mozambique's foreign exchange (FAO 1997). From 1973 to 1990, fish supplied an average 20 per cent of the animal-protein intake of the population in sub-Saharan Africa (FAO 1996). Improvements in refrigeration and transport technologies have increased the availability of fish and shellfish to inland population centres and to international markets, and this has pushed up prices, especially for species such as lobster and prawns. Population increases, both inland and in coastal centres, have also contributed to increasing demand for fish and seafood. In addition, technological developments in commercial boats and nets and in fishing techniques over the past 30–40 years have increased the potential volume of fish catches and contributed to depletion of fish stocks (Chenje and Johnson 1996). According to the most recent data from the FAO, exploitation of fish stocks increased between 1974 and 1999, by which time at least 70 per cent of fish stocks worldwide were considered fully or overexploited (FAO 2000). Figure 2c.3 shows the fish catch from African waters in the 1972–2002 period.

Certain methods of harvesting can also be destructive and a cause of depletion of marine and coastal resources. For example, dynamite fishing is still practised in the coastal zone of Eastern Africa where it damages coral reefs and has resulted in declines in the fisheries in these areas. The practice has, however, been eliminated from some protected marine areas, thanks to good conservation and educational measures (WWF

2001a). Bottom trawling is also a destructive method which disturbs benthic (seabed) communities and drags up accumulated material such as sand, rocks, plants and non-target animal species, all of which are regarded as waste and are dumped elsewhere.

Overfishing by foreign fleets is another significant factor in the decline of African fish stocks, particularly along the west African coast. For example, the European Union (EU) pays about US$234 million in subsidies for EU boats to have access to Mauritanian waters (WWF 2001b). African governments originally saw fishing by foreign fleets as an easy means of earning foreign exchange, but overfishing has made it a serious threat which depletes fish stocks and forces local small-scale fishermen to endanger their lives by fishing further and further out to sea, or to fish in protected areas such as marine national parks. The impacts of overharvesting of fishery resources include the suppression of local livelihoods, reduced capacity to meet food requirements and reduced potential for economic returns on exports. Over the past 30 years, availability of fish per capita in Africa as a whole has declined and, in some countries (for example, Ghana and Liberia), the average diet contained less fish protein in the 1990s than it did during the 1970s (FAO 2000).

An outlook study over the next ten years predicts that local supplies of fish may continue to decline in Africa (FAO 2000). The reasons for this include lack of resources for enforcing controls on fishing in overexploited multi-species fisheries, particularly those exploited by numerous companies from all over the world. In addition, any aquaculture developments are likely to focus on high-value products and, therefore, concentrate mainly on export markets.

## Sustainable harvesting of coastal and marine resources

Declining catches, together with a decrease in the mean sizes of fish caught, have led to calls for the protection of line fish stocks in some African countries. Management measures have been introduced including minimum size limits, bag limits, use of appropriate fishing gear, and closed seasons.

International agreements—between African countries, and between African and European or other international fisheries—also have an important role to play in attaining sustainable harvesting of coastal and

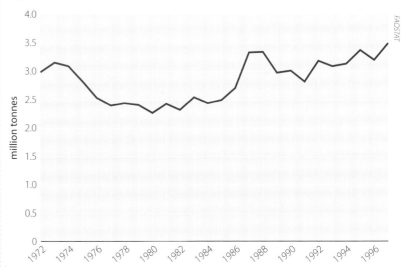

**Figure 2c.3 Marine fish catch for African coastal countries, 1972–97**

FAOSTAT

marine resources. For example, the 1982 United Nations Convention on the Law of the Sea (UNCLOS)—which came into force in 1994—defines world coastal boundaries, development rights, the extent of EEZs, and covers research and policing issues. Thirty-six African countries have signed the Convention, and are thus obliged to protect and preserve the marine environment by cooperating regionally and internationally, and to adopt policies and regulations to deal with land-based sources of marine pollution. Many of the provisions of the Convention are legally binding, and an international tribunal has been established to resolve disputes over resources in international waters.

In spite of such agreements, African countries are still experiencing exploitation of their resources by foreign fishing fleets. For example, the large stocks of small pelagic (open sea) species off the north-west and south-west coasts could be harvested at a low cost and could constitute an adequate replacement in local African diets for the exported high-value products. Countries along the Gulf of Guinea should develop joint strategies with countries in north-west and south-west Africa to exploit these stocks as a source of cheap and nutritious fish for local consumers. Existing regional fishery management organizations would provide an institutional mechanism for coordinating national policies in this area.

Uncontrolled discharge into coastal waters

*UNEP*

## POLLUTION OF AFRICA'S COASTAL AND MARINE ENVIRONMENT

Africa's coastal and marine ecosystems are under extreme pressure from pollution from both land-based and marine-based sources. Among these are: uncontrolled discharges of industrial waste and sewage from coastal settlements; refuse blown or washed out to sea from formal or informal rubbish dumps; general and toxic wastes deliberately dumped at sea; and oil spills and leaks. Effluents from fish processing plants and industries located in the coastal zone are frequently discharged directly into the sea or into surrounding watercourses and wetlands, from where contaminants are washed out to sea. This practice has been driven by lack of affordable alternative disposal facilities, often because the rate of industrial development in the coastal zone has outstripped rates of infrastructure provision, or because industries consider the tariffs for waste disposal to be too high. Residues of fertilizers are also washed down in rivers, and contribute to eutrophication of coastal waters and the development of algal blooms and toxic red tides.

In densely populated urban centres, domestic sewage is sometimes discharged directly or indirectly into the sea. This is due to the pressures of tourism and population growth and a demand for infrastructure in the coastal zone which exceeds the capacity of local authorities to supply adequate sanitation and treatment of wastes. Solid domestic waste can also be blown or washed out to sea or onto beaches when waste collection, treatment and disposal services are inadequate.

The impacts of coastal and marine pollution are widespread and affect natural habitats, human communities and economic activities. For example, contamination of shellfish through red tides, heavy metals in industrial effluents or accumulation of toxic organic compounds can result in severe economic losses. Pollution of coastal waters by sewage can expose local communities and tourists to cholera, typhoid, and hepatitis. Solid waste washed up on the shore is unsightly and a health hazard (especially sharp objects and toxic substances), and may have a deterrent effect on tourists. At sea, solid waste, especially plastics, can be mistaken for food items by dolphins, turtles, seals and sea birds. These creatures are also at risk from entanglement and poisoning from solid wastes.

The waters around Africa are major transportation routes for oil and there have been many serious accidents in recent years, including the break up of the *Apollo Sea* in 1994, and the *Treasure* in 2000, both off the Cape of Good Hope. Oil spills resulting from such accidents smother plants and animals and break down thermal insulation in sea birds and mammals. Soluble organic compounds from the spilled products get into the food web where they are toxic to wildlife and can cause behavioural changes, physiological damage or reproductive failure. Clean up and subsequent disposal of oily wastes is difficult and extremely expensive. In response to this, national and regional oil spill contingency plans have been established in many parts of Africa. However, it is not just oil spills that are a problem. Oil tankers frequently empty ballast and wash engines on the high seas and residues of degraded oil are consolidated and washed onto the shores by wind, currents and waves, resulting in tar balls. Added to this are leaks from oil drilling activities, port handling of oil and petroleum products and refineries located in coastal zones, and leaks from barges, tankers, pleasure craft and fishing boats.

## Enhancing coastal and marine environmental quality

Responses to marine and coastal pollution include the ratification and national implementation of international agreements, such as the Convention for the Prevention of Pollution from Ships (MARPOL), the Regional Convention for the Conservation of the Red Sea and Gulf of Aden Environment (Jeddah Convention), the Barcelona Convention's Mediterranean Action Plan, the Nairobi Convention, designed to ensure that resource development in Eastern Africa's coastal zone is in harmony with the maintenance of environmental quality, and the Convention for Cooperation in the Protection and Development of the Marine and Coastal Environment of the West and Central African Region (Abidjan Convention). Whilst some conventions have been implemented with significant success to date, others have experienced delays in ratification, and have received inadequate human-resource and financial support. The Nairobi Convention, for example, was signed in 1985 but only came into force in 1996. Difficulties in monitoring and enforcement of penalties for non-compliance have also been experienced, mainly because of the extensive territories requiring policing and lack of efficient surveillance systems. The governments of Africa have embarked on an African process for development and management of the coastal areas and are now committed to strengthening and reviewing the Abidjan and Nairobi Conventions. Public health legislation and municipal cleaning of coastal areas are additional responses from some governments of coastal African countries.

### *Integrated Coastal Zone Management*

Integrated Coastal Zone Management (ICZM) is a holistic management approach which considers not just individual resources but the entire coastal and marine environment, and thereby aims to overcome the fragmentation inherent in the sectoral management approach. It has become a popular mechanism for national implementation of regionally accepted

Traditional small-scale fishing in the Comoros

*Roland Seitre/Still Pictures*

management objectives and is intended to overcome the diverse nature and distribution of resources, the complex activities and sources of environmental degradation, and competition for access and use of resources between user groups. ICZM has been defined as '*a continuous and dynamic process by which decisions are taken for the sustainable use, development, and protection of coastal and marine areas and resources*' (Cicin-Sain and Knecht 1998). By recognizing the relationships between different elements of the environments and the impacts of multiple uses and pressures, sustainable harvesting levels and patterns of non-consumptive use can be achieved and maintained. According to Agenda 21 (Chapter 17), ICZM programmes should be designed to:

- identify existing and projected uses of coastal areas with a focus on their interactions and interdependencies;
- concentrate on well-defined issues;
- apply preventive and precautionary approaches in project planning and implementation, including prior assessment and systematic observation of the impacts of major projects;
- promote the development and application of methods such as natural resource and

environmental accounting that reflect changes in value resulting from uses of coastal and marine areas;

● provide access for concerned individuals, groups and organizations to relevant information and opportunities for consultation and participation in planning and decision-making.

### Towards ICZM in Africa

During the past decade, the international community has recognized the intensity and complexity of pressures on coastal and marine environments, and has demonstrated a commitment to more integrated management. An important step in this direction was the 1995 Washington Declaration, in which 110 governments recognized the importance of integrated management of coastal zones, and adopted the Global Programme of Action for the Protection of the Marine Environment from Land-based Activities (GPA).

UNEP was tasked with coordination of the GPA which is designed as a global framework and source of practical guidance for devising and implementing sustained action to prevent, reduce, control and/or eliminate marine degradation from land-based activities. Specific aims of the GPA are to identify and assess problems relating to the coastal and marine environments, establish priorities for action, set management objectives for tackling these priority issues, develop strategies to achieve these objectives, and evaluate the effectiveness of the strategies. The GPA will be implemented through UNEP's Regional Seas Programme which provides support to regions in implementing its objectives and in building capacity for sustained implementation.

The Regional Seas Programme was initiated by UNEP in 1974 and has been repeatedly endorsed as a regional approach to the control of marine pollution and the management of marine and coastal resources. In Africa, there are Regional Seas Programmes for Eastern Africa (encompassing Comoros, France (La Reunion), Kenya, Madagascar, Mauritius, Mozambique, Seychelles, Somalia and the United Republic of Tanzania), the Mediterranean (encompassing the African states of Algeria, Egypt, Libya, Morocco, and Tunisia, as well as many European states) the Red Sea and Gulf of Aden (encompassing the African states of Djibouti, Sudan and Somalia), and for Western and

Central Africa (encompassing Angola, Benin, Cameroon, Cape Verde, Congo, Côte d'Ivoire, Equatorial Guinea, Gabon, Gambia, Ghana, Guinea, Guinea-Bissau, Liberia, Mauritania, Namibia, Nigeria, Sao Tome and Principe, Senegal, Sierra Leone, Togo and Congo). Action plans have been developed under each of these programmes, as well as conventions and protocols for implementation of regional objectives at the national level.

Many African governments have realized the benefits of ICZM, and have enacted policies and legislation to put its principles into effect. However, sustained resourcing (that is, financial support, equipment, training of personnel, and monitoring of activities) is required to achieve maximum benefits to the coastal and marine environments and the economies that depend on them.

## NORTHERN AFRICA

The main issues of concern in the coastal zones of Northern Africa are the effects of rapidly developing tourism and industrial activities along the coast, combined with rapid population growth (UNEP 1997). These pressures are beginning to have impacts on the quality and stability of the physical and biological coastal environment. There are also serious concerns over the potential impacts of climate change, particularly the vulnerability of coastal settlements and natural habitats to sea level rise and saltwater intrusion.

### Economic, social and ecological value of coastal and marine environments in Northern Africa

The marine and coastal zones of Northern Africa support tourism, fishing, and petroleum industries. The significant onshore and offshore oil and gas deposits in the sub-region are the mainstays of countries' economies, providing opportunities for export and employment.

Coral reefs are widely spread and well developed in the Red Sea, with 194 species of coral and at least 450 common reef-associated species. These are also some of the most northerly located coral reefs, and have many endemic species

Mangroves are found along the southern Red Sea coast, and are important sources of molluscs, crabs, shrimp, fish, and raw materials for construction, animal fodder and fuel. They are also important nesting sites for migratory waterbirds. Sea grasses are also fairly common along the southern Red Sea, and rare or protected species such as turtles and dugongs add to the species richness and diversity that attracts an estimated one million tourists per year to the region (UNEP 1997).

Northern Africa also has economically important marine fisheries, from which total catches increased from 845 211 tonnes in 1990 to 1.1 million tonnes in 1997, an increase of about 30 per cent (FAOSTAT 2001). The trend in marine fish catches in the Northern Africa sub-region over the past 30 years is shown in Figure 2c.4.

Although fishing activity has increased in the Mediterranean over the years, its fisheries are not showing signs of overexploitation. In fact, surprisingly—as the demand for fisheries products is growing and as most countries lack formal and coordinated fisheries management—there has been an increase in production for all major species (FAO 1997).

Mediterranean states are party to the FAO's General Fisheries Council for the Mediterranean, but this is largely ineffectual and fishing by commercial fleets is virtually unregulated. At the 22nd session of the Council, member states were urged to negotiate and implement an effective management regime based on the precautionary principle and reflecting the underlying tenets of important international initiatives (WWF 1997). Artisanal fishery resources are also important in the Mediterranean and Red Seas. Shallow water species such as butterfly fish and damselfish are most common, although the area also supports rock lobster, cuttlefish, shrimp, and sea cucumber.

More than 40 per cent of the Mediterranean area's population live in coastal zones (UNEP 1996), with the greatest densities along the Nile Delta and Algerian coast (up to 500–1000 inhabitants/km² in some areas) (Blue Plan 1996). Urbanization in coastal north-west Africa has been driven mainly by oil discoveries and increased industrialization in or near the coastal areas and the associated new economic opportunities. This has led to an almost complete transformation of north-west African society and gradually increasing pressures on coastal areas.

**Figure 2c. 4  Trends in marine fish catch for Northern Africa, 1972–97**

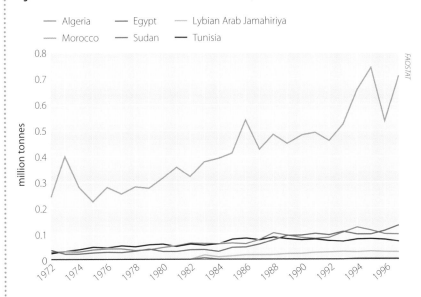

FAOSTAT

## COASTAL EROSION AND CLIMATE CHANGE IN NORTHERN AFRICA

Most lowland coastal areas in Northern Africa, especially deltas and islands, are subjected to slow tectonic subsidence that will accentuate the effects of the rise in sea level predicted as a result of climate change. For example, lowlands in Libya stretch along the Gulf of Sirte where there is an important oil shipment facility and, in Tunisia, lowlands are of great economic value for agriculture, urbanization, harbours, industry, and tourism. These would all experience accelerated coastal erosion and inundation of land. It is, however, the deltaic plain of the Nile River, the most important coastal lowland of the Mediterranean shoreline, that is expected to suffer the greatest losses. A one-metre rise in sea level would inundate 2 000 km² of land in the lower Nile Delta area and render 1 000 km² of agricultural land unusable. The costs of this are estimated at US$750 million (Khafagy, Hulsbergen and Baarse 1992). If erosion is taken into account, an additional 100 km² of land would be lost, at a cost of US$60 million (Khafagy and others 1992).

The Egyptian city of Alexandria would be severely affected, losing most of its infrastructure, the country's most popular beaches and cultural and scenic monuments. There would be associated losses of revenues from tourism and of industrial, residential and agricultural land. Alexandria's population currently

*A one-metre rise in sea level would inundate 2 000 km² of land in the lower Nile Delta area and render 1 000 km² of agricultural land unusable. The Egyptian city of Alexandria would be severely affected, losing most of its infrastructure, the country's most popular beaches and cultural and scenic monuments*

stands at three million and is predicted to grow to eight million by 2030, by which time half the residents would be at risk from inundation and displacement. However, public awareness and concern is low, with fewer than 20 per cent of residents prepared to move away, and the majority of the population regarding it as the government's responsibility to provide protection. In addition to inundation of land, sea level rise would also increase coastal erosion, flooding and saltwater intrusion into underground aquifers. The coastal zone is likely to suffer increased impacts from wave action and storm surges due to sea level rise. The average rate of coastal erosion in Alexandria is currently only 20 cm/yr, but three beaches have disappeared since the beginning of this century and eight of the other twelve are showing signs of erosion (El Raey and others 1995, IPCC 1998).

### Mitigating the effects of climate change on Northern Africa's coastal zone

A three-pronged approach is required to mitigation of sea level rise and other effects of climate change on Northern Africa's coast, including research and monitoring of potential impacts, development of appropriate responses to those impacts, and the integration of response strategies into planning and development, policies and programmes. Furthermore, this approach should take account of the ongoing coastal erosion and pollution as well as of the activities responsible for these, both in the coastal zone itself and in the hinterland. An integrated approach is therefore required, with a focus on long-term issues appropriate to managing the impacts of sea-level rise and climate change. It is, therefore, recommended that Northern African countries continue with and strengthen assessments of vulnerability of coastal areas to the consequences of climate change, as well as formulating and implementing adaptation options within integrated coastal management strategies (Nicholls and de la Vega-Leinert 2000).

### POLLUTION OF NORTHERN AFRICA'S COASTAL AND MARINE ENVIRONMENT

The Mediterranean and Red Seas are particularly sensitive to pollution because they are semi-enclosed basins with straits leading to the Atlantic and Indian Oceans respectively, and through which replacement of seawater is slow. They also have high evaporation rates and relatively low run-off from the arid lands surrounding them. Exchanges of freshwater and seawater are therefore low in these basins and dilution is limited, meaning that pollution entering the basins is likely to become concentrated over time.

In the Red Sea, there is a high risk of marine-based pollution, including: phosphate poisoning; industrial pollutants; sewage, solid waste and other contaminants related to tourism activities; dumping of general and hazardous wastes from ships; and leaks and spills of oil. More than 100 million tonnes of oil are transported through the Red Sea annually and both regulation of maritime traffic and maintenance of navigational aids are inadequate (World Bank 1996a). As a result, there have been over 20 oil spills along the Egyptian Red Sea coast since 1982 (Pilcher and Alsuhaibany 2000). Offshore oil leaks and spills from the oil industry are another threat to human and coastal and marine resources in the Northern Red Sea, and Suez Gulf area.

An emerging issue of concern is the increasing presence of Red Sea fish and algal species in the Mediterranean basin, transported via ballast water of cargo ships passing through the Suez Canal (FAO 1997). Introduction of such alien species could potentially disturb the balance of species in the Mediterranean ecosystem.

Domestic and agricultural wastes contribute to pollution in Northern Africa's coastal zone, with sewage being the major cause for concern, followed by persistent organic pollutants (POPs) from pesticide residues, then heavy metals and oils, predominantly from industrial effluents. Sewage discharge and agricultural run-off into coastal waters are causing eutrophication and algal blooms have been reported on the coral reefs of the Sudanese coast (Pilcher and Alsuhaibany 2000).

Egypt has a flourishing tourism industry with sites such as Hurghada and Sharm el Sheikh well developed and expanding rapidly. However, some reports suggest that these and other resorts, such as the Ras Mohammad National Park, are experiencing pollution and reef degradation caused by levels of tourism and infrastructure that are too intensive for these sensitive habitats (UNEP 1997).

*The Mediterranean and Red Seas are particularly sensitive to pollution; they are semi-enclosed basins with straits leading to the Atlantic and Indian Oceans, through which replacement of seawater is slow. Exchanges of freshwater and seawater are low in these basins and dilution is limited, meaning that pollution entering the basins is likely to become concentrated over time*

### Box 2c.1  Protection of the environment in the Red Sea

Recognizing the need for international cooperation in marine and coastal management in Northern Africa, an interdisciplinary research programme for the Red Sea was initiated, in 1974, under the auspices of the UNEP Regional Seas Programme. Following from this, an Action Plan for the Red Sea and Gulf of Aden was approved (1976). In 1982, the Jeddah Convention for the Conservation of the Red Sea and Gulf of Aden Environment was signed, along with the Protocol for Regional Cooperation in Combating Pollution by Oil and Other Harmful Substances in Cases of Emergency. Djibouti, Egypt, Jordan, Palestine, Saudi Arabia, Somalia, Sudan, and Yemen are parties to the Convention. In 1995, the Regional Organization for the Conservation of the Environment of the Red Sea and Gulf of Aden (PERSGA) was formally established to implement the objectives of the Convention and its protocol. Achievements to date include environmental assessments of the coast and surveys of natural habitat, collection of oceanographic data, impact assessments of shrimp and pearl industries, training workshops in combating oil pollution, environmental impact assessments, ICZM, and establishment of a marine national park. Publications include state of the environment reports, directories of capacities and legislation and an assessment of the land-based sources and activities affecting the marine and coastal environment.

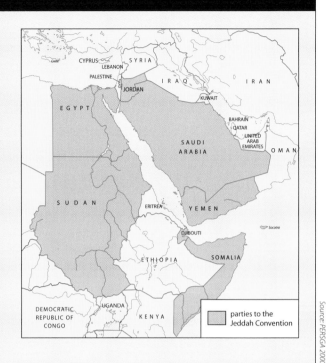

parties to the Jeddah Convention

Source: PERSGA 2000

### Enhancing coastal and marine environmental quality in Northern Africa

The most significant response from governments in Northern Africa to combat pollution of all types has been the establishment of the organization for the Protection of the Environment in the Red Sea and Gulf of Aden (PERSGA). The purpose and achievements of PERSGA are detailed in Box 2c.1. Based in Jeddah, Saudi Arabia, its main function is to implement the Regional Convention for the Conservation of the Red Sea and Gulf of Aden Environment (Jeddah Convention).

The Mediterranean Action Plan (MAP) was adopted in Barcelona, Spain, in 1975, by 16 Mediterranean States and the European Community, under the auspices of UNEP and within the framework of its Regional Seas Programme. The aim of MAP is to protect the environment while encouraging sustainable development in the Mediterranean Basin. It has six associated protocols covering coastal zone management, pollution assessment and control, protection of ecosystems and preservation of biodiversity. MAP was revised in 1995 to become more action-oriented and an instrument for sustainable

development. Activities to date include gathering of pollution-trend data and compliance monitoring, the establishment of a list of Specially Protected Areas of Mediterranean Importance and related programmes for protection and conservation of species, and regulation of the introduction of non-indigenous or genetically modified species. The MAP Coastal Areas Management Programme is a mechanism for enhancing cooperation between national and local authorities and institutions—13 projects have been implemented since 1989. A Mediterranean Environment and Development Observatory has also been established to provide information to support decision making (UNEP 2001).

There are two Marine Protected Areas (MPA) along the Gulf of Aqaba, and there is a proposal for additional protected areas. In the northern Red Sea, there are three MPAs which need support through further strengthening of regulations, and proposals for new protected areas. There is only one protected area in the Dahlak Islands (in the south-eastern Red Sea), recognized as forming unique habitats that are under increasing pressure from tourism and oil transportation. There are 14 coastal protected areas and 5 marine

Unique habitats in the south-eastern Red Sea are under increasing pressure from tourism and oil transportation in the region.

*Rafel al Ma'ary/Still Pictures*

areas along the Mediterranean coast of Northern Africa. There are no MPAs in the Gulf of Suez and the only one proposed covers only a portion at the southern end of the Gulf. There are proposals for a much wider network of protected areas all along the coast. Overall, the standards of protection on both the Mediterranean and Red Sea coasts are below those afforded by protected area mechanisms. In line with the shifting paradigm of environmental management—away from protection and towards responsible and sustainable development— Egypt Morocco and Tunisia have taken steps to develop and implement ICZM plans. However, as these are recent measures, few conclusive studies have been undertaken to measure their progress.

### Towards ICZM in Northern Africa

All countries in Northern Africa are party to either the Barcelona Convention or the Jeddah Convention (both in the case of Egypt). The objectives of the Barcelona Convention, which came into force in 1978, are to achieve international cooperation for a coordinated approach to protection and enhancement of the marine environment and coastal zone of the Mediterranean. The Jeddah Convention came into force in 1985 and sets similar objectives for the Red Sea and Gulf of Aden. These conventions and the strategies in their associated Action Plans have set the background for development of ICZM at the national level.

In Egypt, an ICZM strategy has been developed by the Egyptian Environmental Affairs Agency and is being prepared for distribution and implementation. A National Environmental Sustainable Tourism Strategy is also being developed and pilot-scale implementation is under way in the Red Sea. Egypt has also signed all regional and international agreements regarding the protection of seas and coastal zones, and a number of ecotourism projects for the Red Sea and the Mediterranean Sea are attempting to implement the principles of sustainable coastal development (CSD 1999).

In Morocco, protection of the coastal and marine environments is under the joint responsibility of the Ministries of Services (ports) Transport and Shipping, Environment, and the Interior. These ministries are responsible for compliance with regulations on environmental impact assessments and pollution and wastes, especially agricultural effluents, sewage, and flushing of ballast waters. The Ministry of Fisheries has jurisdiction over harvesting of marine and coastal resources and is responsible for ensuring sustainable use of these resources. There are also measures in place to ensure coordination between these organizations, including the National Emergency Plan for dealing with environmental accidents (1996), the National Environment Council and the Inter-ministerial Committee for Coastal Tourism (CSD 1999).

Tunisia has ratified the MARPOL Convention and has established a national plan for dealing with emergencies and accidents. Projects are also underway to monitor and rehabilitate marine fish stocks, to establish and enforce water quality guidelines, and to protect against pollution by hydrocarbons from oil processing and transport activities, giving force to the policy on sustainable use of marine resources.

## EASTERN AFRICA

The issues facing the coastal Eastern African countries are erosion and pollution of coastal and marine environments, as well as harvesting pressures resulting from rapid population growth in the coastal zone and expansion of the tourism industry. Rises in sea level and in temperature resulting from global climate change are also emerging as issues of concern.

## ECONOMIC, SOCIAL AND ECOLOGICAL VALUE OF COASTAL AND MARINE ENVIRONMENTS IN EASTERN AFRICA

The Eastern African coast supports a diversity of ecosystems, including dry coastal forests, coastal dunes, floodplains, freshwater and saltwater marshes, mangrove forests, coral reefs, lagoons, sandy beaches and rocky shores. These support a rich and diverse resource base, including fish and seafood, construction materials, energy sources, wildlife habitat and tourism opportunities, as well as industrial and transportation activities.

The Red Sea coral reefs off the coasts of Djibouti, Eritrea, and Somalia are in good, often pristine, condition, with 30–50 per cent live coral cover, and the richest diversity of coral and other reef species in the entire Indian Ocean (Pilcher and Alsuhaibany 2000).

Rich mangrove forests not only protect the shoreline from storm surges and buffeting by ocean swell, but are also breeding grounds for many species of waterbirds. Along the coast of Kenya and southern Somalia, mangroves support commercial crab, oyster, and mullet fisheries, as well as artisanal harvesting of these and other species. These are extremely important to local economies and to many communities where fish is almost the sole source of animal protein (FAO 1997).

The natural assets of the Eastern African coast make it an ideal tourist destination. Kenya's tourism industry is the country's second largest earner of foreign exchange, contributing 19 per cent to GDP (World Bank 2000). In addition, the larger scale fisheries and associated industries provide a valuable source of foreign exchange and make a significant contribution to GDP.

## COASTAL EROSION AND CLIMATE CHANGE IN EASTERN AFRICA

Erosion is a common phenomenon along the Eastern African coast (UNESCO 1997, Ngoile 1997). Human activities, including beachfront developments, and mining of sand, coral and lime, contribute to accelerated rates of erosion. Activities inland—such as clearing and draining water catchments and river basins for agriculture and aquaculture, and clear-felling of coastal and inland forests—are additional causes of accelerated erosion.

Residential areas have expanded and commercial developments such as hotels and resorts have been constructed rapidly and widely along the coast of Eastern Africa, from southern Sudan to Kenya, often without adequate planning or provision of basic services such as waste disposal and sanitation. Many developments are concentrated on beaches and sea fronts, to benefit from the clean beaches, clean lagoon waters and healthy coral reefs, and to attract as many visitors as possible. Excavation of sand, limestone, coral stone and other building materials, as well as clearance of natural habitat to make way for urban and commercial developments, contribute to coastal erosion. Coastal development in Somalia has been suspended after the collapse of the government, but as peace and stability return, it is expected that coastal development will be accelerated (Pilcher and Alsuhaibany 2000).

Erosion in one area leads to accretion of material in other areas, resulting in shifting coastal features (for example, dunes, beaches, cliffs and shoreline). For example, the Tana Delta and Sabaki Estuary (Kenya) have accreted extensive coastal plains due to coastal drift and the high sediment loads carried by the Tana and Sabaki Rivers. These high sediment loads are a result of poor land use practices in their upper catchments and high rates of soil erosion from their agricultural lands. Such a high rate of sediment discharge is threatening the ecological integrity of marine and coastal habitats such as mangroves, seagrass meadows and coral reefs. In addition, the high concentrations of silt in river water make the rivers unattractive for recreational purposes and limit the extent to which the water can be used for other purposes (UNEP 1998a). Furthermore, dams upstream on the Tana River have reduced water flows reaching the delta and lowered frequency of flooding, causing die-off in coastal forests (International Rivers Network 1995–2000).

Coastal erosion also threatens the breeding and nursery grounds of marine species, some of which are listed as endangered. Five sea turtle species recorded in the Western Indian Ocean and the Red Sea, and the cetacean *Dugong dugon* are threatened with extinction because of loss of breeding and nursery grounds (IUCN 1998). The breeding sites for turtles have been excavated, built on, or simply washed bare.

The impacts of erosion and shoreline retreat in Eastern Africa are likely to be compounded by sea level rise resulting from global climate change. Sea level rise will increase the intrusion of seawater and flooding as most of the low-lying coastal plains are just a few metres above the highest spring tide water level (Okemwa, Ntiba and Sherman 1995). A rise of one metre would result in damage to or loss of infrastructure and natural habitats, which in turn would reduce economic potential of tourism and other commercial ventures and displace hundreds of thousands of people, especially in coastal cities such as Mombasa (IPCC 1998).

Sea temperatures are also predicted to increase in certain areas under the effects of global climate change and this would stress coral reef ecosystems and the economic activities that they support (IPCC 2001b). For example, the El Niño disturbances of 1998 caused sea temperatures in the Western Indian Ocean to rise by 1–1.5 °C, and between 50 and 90 per cent of the corals off the Kenyan coast were killed through bleaching. The surviving corals had recovered significantly by 1999 but damaged areas were still not re-colonized (Obura and others 2000). More than 30 per cent of the coral in Djibouti was killed by bleaching as a result of the same El Niño event (Pilcher and Alsuhaibany 2000).

The negative impacts of aggravated coastal erosion include loss of investments, loss of agricultural lands, loss of critical habitats and loss of employment opportunities. Governments and investors have to spend large sums on mitigating processes and the potential for tourism is reduced because recreation sites are lost. Cleared mangrove swamps and destroyed coral reefs also expose coastal areas to inundation or contamination from seawater.

### Mitigation of coastal erosion in Eastern Africa

The Nairobi Convention covers, among other things, regulation and minimization of ecological damage from dredging and land reclamation, and introduces a requirement for prior assessment of environmental impacts to control and coordinate urban and industrial development and the development of tourism. All countries in Eastern Africa are party to the Convention and, now that the difficulties in its implementation (that is, legal and institutional arrangements), have been overcome, regulation of coastal development and mitigation of erosion will be priority action areas.

At the local level, private developers are putting up

defensive structures such as sea walls and groynes. However, in some cases, these have actually increased rather than slowed rates of erosion (Anon 1996) and mangrove and reef rehabilitation projects are now recognized as more effective long-term measures.

## HARVESTING OF COASTAL AND MARINE RESOURCES IN EASTERN AFRICA

Eastern Africa is unusual in that its coastal fish production is low in relation to the surface area and potential productivity of its fisheries, in spite of large amounts of nutrients made available by coastal upwelling resulting from the Somali Current. The coastal fishery yield for the entire eastern and south-eastern African coast, including the Western Indian Ocean Islands, represents less than 1 per cent of global landings and most of the coastal fish stocks of the sub-region are considered to be fully exploited or overexploited (FAO 1997).

Destructive fishing practices pose a threat to coastal fisheries and coral reefs. Use of dynamite, pull-seine nets, poisons, and heavy pressure on selected species and juveniles are widespread along the Eastern African coast, contributing to decline of the ecosystem. However, national and international pressure to ban these practices has stimulated the empowerment of local communities to monitor and manage their resources (Obura, Suleiman, Motta and Schleyer 2000).

The countries of the Eastern African coast are the main exploiters of their coastal waters, but their EEZ is being increasingly harvested by foreign fleets from Europe and Eastern Asia. Somalia is experiencing an extreme case of this, as outlined in Box 2c.2. Reported catches by foreign nations increased dramatically in the early 1990s, with the Republic of Korea, Japan, France, Taiwan and Spain playing a major role (Okemwa 1998). Shark populations are declining rapidly with consequent drops in shark-fin catches by fishermen from Yemen, Somalia, Djibouti and Sudan. Most shark fishing is illegal and also impacts on turtles, dolphins and finfish which get caught in the nets and lines. Lack of surveillance and enforcement is a contributing factor (Pilcher and Alsuhaibany 2000).

### Box 2c.2 Somali fisheries require international control

Somalia has one of the longest coastlines of any African country—around 3 300 km. Highly productive upwelling provides significant potential for the development of offshore tuna fisheries. However, the fall of Somalia's government, in 1991, left the country without a central government or control of its waters. Control has been assumed by self-promoted militia, some of whom have made controversial fishing arrangements with foreign countries, whilst others operate like pirates demanding ransoms from foreign vessels. Although these are highly dangerous waters, access to Somali fisheries is now virtually open, driven principally by foreign interests and demand for high-value tuna, shark and ray fins, lobster, deepwater shrimp and demersal whitefish. Harvesting rates are thus not known and it is not possible to determine whether marine resources are being harvested sustainably or not. Furthermore, the years of civil conflict have damaged the fisheries infrastructure, and have reduced previous oil spill response capability, aids to navigation, and search and rescue capacity.

Somalia is party to the Convention on International Trade in Endangered Species (CITES), UNCLOS, and the Nairobi and Jeddah Regional Seas Conventions. However, few of the provisions of these treaties are being implemented since the breakdown of national governance, and Somalia is looking to the international community to assist with implementation and enforcement of regulations. The current challenge is to develop a regional institutional proposal to address the situation.

Yemeni stern trawler (unflagged) fishing illegally with warps visible, 1.5 miles off Bossasso, 28 February 1998

Source and photo: Coffen-Smout 1998

### Sustainable harvesting of coastal and marine resources in Eastern Africa

Eastern African countries are party to the 1982 UNCLOS which establishes fishing rights, and to international agreements on harvesting limits and areas. However, countries in the sub-region, like many African countries, lack both the infrastructure to exploit their own territories and the capacity to police and monitor the activities of international fleets or unsustainable harvesting rates. In addition, fishing agreements with foreign countries have been poorly defined, based on the need by African states for income and foreign exchange.

In Eritrea, marine resource harvesting is being regulated as peace returns to the area, through development of formal management procedures. Similar efforts are anticipated in Somalia. In Kenya, the MPA system is managed by the Kenya Wildlife Service and extends over 5 per cent of the coastline. There are two types of protection: full protection in marine parks, and the traditional resource harvesting allowed in marine reserves. Tourism is the main activity at all sites, and plans are under way to involve local communities and other stakeholders in the management of some areas (Obura and others 2000).

### POLLUTION OF EASTERN AFRICA'S COASTAL AND MARINE ENVIRONMENT

Environmental quality in the coastal and marine environments of Eastern Africa is under increasing pressure from land- and sea-based sources of pollution. Land-derived agrochemical and municipal wastes and sea-based petroleum wastes are the major causes of pollution in the sub-region (Martens 1992, Okemwa and Wakwabi 1993). Residues of fertilizers and pesticides from agricultural inputs in the hinterland enter main drainage systems and are washed into the sea, where they have cumulative effects in the marine and coastal environment (Onyari 1981). At the same time, increased siltation resulting from deforestation in the hinterlands also impacts the coastal habitats by increasing the turbidity of the waters, and smothering habitats, flora, and fauna. Eutrophication is not yet a serious issue in the sub-region, but isolated pockets in sheltered bays (especially along the Kenya coast) are threatened with blooms of toxic algae (Wawiye, Ogongo and Tunje 2000) and phytoplanktonic bacteria (Mwangi, Kirugara, Osore, Njoya, Yobe, and Dzeha 2000).

The coastal waters of the Red Sea and Western Indian Ocean are the major sea routes for large petroleum and oil tankers supplying the world with products from the Middle East. Major shipping routes run close to the coral reefs near the port of Djibouti and Port Sudan and ships often discharge oily wastes and sewage. Ships also cause physical damage to the reefs when poor navigation brings them into collision with the reefs (Pilcher and Alsuhaibany 2000). Longshore currents and winds in the Western Indian Ocean are instrumental in the horizontal distribution and spread of pollutants, particularly in bringing oil slicks from the open sea (beyond the EEZ limit) into the coastal waters. In addition to the elevated risk of high-impact oil spills, frequent transport operations also contribute to oil pollution—oil tankers often empty ballast and wash engines on the high seas and residues of degraded oil are consolidated and washed ashore by onshore winds, currents and waves. Tar balls litter beaches with deleterious effects on wildlife and on humans that use the beaches (Munga 1981). Soluble PCBs from these products poison marine life and accumulate in the food web, causing physiological disorders in top predators.

Treated and untreated effluents from municipal, industrial and domestic sources further contribute to coastal and marine pollution. Waste treatment facilities are often poorly designed or sited, and are old and poorly maintained or overloaded. Location of dumpsites and municipal sewage outfalls is therefore critical in maintaining environmental quality (especially water quality, both inshore and offshore) when urbanization and other forms of coastal development increase. (Mwaguni and Munga 1997).

Most Eastern African coastal municipalities do not have the capacity to handle the vast quantities of sewage and solid wastes they generate every day. For example, the Mombasa Municipal Council (Kenya), can handle only 30 per cent of the waste generated (Anon 1996). In Djibouti and Somalia, sewage treatment plants are few in number and are, generally, poorly maintained (Pilcher and Alsuhaibany 2000). Large volumes of solid and liquid waste are therefore disposed of at sea or are disposed of in an unsatisfactory manner and end up by being washed or blown out to sea, where they pose a threat to wildlife and human health. It is

evident that a restructuring of waste management policies and plans is required to handle the increase in solid wastes and sewage in the coming years.

## Enhancing coastal and marine environmental quality in Eastern Africa

Governments in Eastern Africa have enacted public health legislation to regulate responsible environmental use of chemicals. There are also integrated resource management plans aimed at improving land use practices in the hinterland, thereby reducing the incidence and impacts of siltation and eutrophication. Global and regional agreements and treaties are also in place for enhanced cooperation on environmental management in the region. Furthermore, the Nairobi Convention calls for enhanced management of land-based and marine-based sources of pollution, and mitigation of their impacts.

Effective implementation, monitoring, and regulation of activities requires enhanced political commitment and coordination, as well as sustained resourcing. For example, Kenya's National Oil Spill Response Committee should be given legal status and contingency plans should be developed comprehensively. The Kenya Marine Fisheries and Research Institute also requires further financial and human resources to support its research work and monitoring of the coastal and marine environments (FAO/EAF 1999).

Djibouti and Somalia are parties to the Jeddah Convention and members of PERGSA, its implementation agency. Progress in enhancing environmental quality in these countries includes assessment, monitoring and state of the environment reporting, capacity building in oil spill response and integrated coastal management.

### Towards ICZM in Eastern Africa

ICZM and appropriate legislation and regulations on environmental management are important steps towards the sustainable use of Eastern Africa's coastal zone and resources in the hinterland. There are already national initiatives that should be encouraged by enhanced regional cooperation (Linden 1993, Ngoile 1997, PERSGA 2000). For example, ICZM plans for Kenya and Eritrea use a site-specific approach for the integration of beachfront/seafront developments and waste management in coastal urban centres. Under these

*Most Eastern African coastal municipalities do not have the capacity to handle the vast quantities of sewage and solid wastes they generate every day; large volumes of solid and liquid waste are disposed of at sea, or in an unsatisfactory manner, and end up by being washed or blown out to sea, where they pose a threat to wildlife and human health*

Source: FAO/TAF 1999

## Box 2c.3  Priorities for enhancing environmental quality in Eastern Africa

- Creation of incentives for waste recycling
- Upgrading of waste treatment plants and landfills
- Establishment of environmental impact assessment policies and practices, to reduce coastal erosion and pollution, and to design criteria for shoreline protection
- Implementation of policies that address poor agricultural practices
- Implementation of water quality guidelines and introduction of monitoring practices, together with enforcement of polluter payments
- Public awareness campaigns on improved waste management
- Support for reef restoration and mangrove planting projects
- Improvements to waste handling and implementation of cleaner technologies
- Strengthening of institutional capacity and fund-raising

plans, location of dumpsites and sewage outfalls will be properly planned and their effects closely monitored.

If the success of such initiatives is to be sustained, their institutional structures will need to be strengthened, their coordination improved, and their capacity and funding will need to be increased. Priorities include improved waste management and erosion control in the coastal zone and, in the hinterland, improvement of agricultural practices and introduction of measures such as reforestation schemes, and soil conservation programmes to prevent soil erosion (see Box 2c.3).

## WESTERN INDIAN OCEAN ISLANDS

Small oceanic islands, such as those in the Western Indian Ocean, experience problems in controlling and regulating activities in their relatively extensive EEZ, making them vulnerable to overexploitation of their marine resources, particularly deep-water fisheries. In addition, land- and marine-based sources of pollution, associated with rapidly growing coastal populations, development of tourism and oil transportation by sea, are a further cause for concern over the state of the coastal and marine environments.

The Western Indian Ocean Islands are also experiencing accelerated rates of coastal erosion resulting from poor coastal planning and development, exacerbating their vulnerability to climate-change-related sea level rise which threatens to inundate large areas of land and displace large populations. Sea temperature rises in the Western Indian Ocean, also associated with global climate change, may cause bleaching of the exceptionally biologically rich and economically important coral reefs.

## ECONOMIC, SOCIAL AND ECOLOGICAL VALUE OF COASTAL AND MARINE ENVIRONMENTS IN THE WESTERN INDIAN OCEAN ISLANDS

The oceanic islands of Comoros, Madagascar, Mauritius and Seychelles have evolved in isolation from continental Africa, and were colonized by humans only relatively recently. They are, therefore, unique in their physical characteristics, and in terms of the biological communities they support. The fringing coral reefs and the reefs surrounding the islands include the Aldabra Atoll (Seychelles), one of the most spectacular reefs in the world and a World Heritage Site.

All the islands support well-developed tourism industries which are important sources of income and foreign exchange. Coastal tourism is, for example, the mainstay of the economy in Seychelles, contributing

Mangroves provide a spawning ground for fish and shellfish, and provide resources for construction, weaving and food. They also offer protection against coastal erosion and saltwater intrusion.

*Dominique Halleux/Still Pictures*

46–55 per cent of GDP, 70 per cent of foreign income, and employing 20 per cent of the population (International Ocean Institute 2001).

There are also extensive seagrass beds, mangroves, diverse fisheries and seafood species, and lagoons. These resources mainly support artisanal fishermen (about 90 per cent of all fish landings are from small-scale operators (International Ocean Institute 2001). The majority of fish landed in the Western Indian Ocean Islands are from coastal waters, and molluscs, prawns, shrimp and lobster are important economically and in local diets. Mauritius and Seychelles are also exploiting open sea tuna fisheries on a more commercial scale, and shrimp fisheries, which have a high potential for foreign exchange earnings from export, are growing in importance in Madagascar.

## COASTAL EROSION AND CLIMATE CHANGE IN THE WESTERN INDIAN OCEAN ISLANDS

The coastlines of all the Western Indian Ocean Islands are threatened by erosion. Coastal erosion and destruction of natural habitats such as mangroves and coral reefs in many parts of the sub-region are the major threats to sustainable use of resources and development in the coastal zone. Erosion is primarily a result of uncoordinated and inappropriate developments in the coastal zone, due to rapid population growth and development of the tourism industry.

Mining of sand, coral, limestone and shells is depleting the buffer zone provided by the coral reefs and, as a result, the shores are more exposed to wave action, storm surges, and inundation. For example, in Mauritius, a million tonnes of coral sand are excavated every year, by hand, and transported with dugout canoes (Bigot, Charpy, Maharavo, Abdou Rabi, Paupiah, Aumeeruddy, Villedieu and Lieutaud 2000). Sand mining also has noticeable impacts in Comoros and is creating major problems in Madagascar.

Fishermen using dynamite, walking on the coral reef and using nets out of season contribute to a level of coral destruction that is out-pacing the coral's natural capacity for regeneration. The situation is fragile and political commitment is needed to ensure that action is taken to arrest the decline in quality of the coral reefs, to restore the damage done, and to establish effective monitoring systems.

One of the most serious impacts of coastal erosion is increased vulnerability to the rise in sea level that will result from global climate change. The IPCC predicts an average sea level rise of up to one metre by the year 2100 (IPCC 2001a). In the Western Indian Ocean Islands, this would flood natural habitats such as mangroves, and contaminate habitats further inland with saltwater, changing the ecological conditions and biological composition. Precious groundwater resources may also be lost through saltwater intrusion.

In Seychelles, a one-metre sea level rise would submerge many of the islands and result in a loss of 70 per cent of the land area (Shah 1995, Shah 1996). Mauritius is estimated to fare better, losing less than one per cent of its land. (IPCC 1995). This would, nevertheless displace around 3 000 people and even small losses of beach area could have detrimental impacts on tourism and fisheries activities.

Global climate change also brings a threat of change in sea temperature and increasing frequency and intensity of cyclones or tropical storms. In 1998, the coral reefs of the Western Indian Ocean were subjected to intensive bleaching from an El Niño-induced sea temperature rise and up to 30 per cent of the coral in Comoros and 80 per cent in Seychelles were killed (PRE/COI 1998). Initial assessments of the economic costs of this coral bleaching were between US$700 million and US$8 000 million (Wilkinson, Linden, Cesar, Hodgson, Rubens and Strong 1999), although recovery has been observed in Mauritius and Rodrigues.

Loss of habitat and food supply has caused depletion of fish populations dependent on coral throughout the sub-region and this may reduce the aesthetic quality of the environment essential for development of sustainable tourism. In addition, the decrease in income from fishing and the loss of fish in local diet may affect human health (UNEP 1999a).

### Mitigation of effects of climate change in the Western Indian Ocean Islands

In view of the predicted sea temperature rises which will become more common as global temperatures increase, mitigation efforts to protect coral reefs must include reductions in GHG emissions (Reaser, Pomerance and Thomas 2000). All of the Western Indian Ocean countries have signed the UNFCCC and Seychelles has signed the Kyoto Protocol. However, these countries contribute negligible amounts to global GHG emissions and it is, therefore, important that major GHG contributors be held to their commitments if the world's coral reefs are to be protected.

Another key component for protection against future damage to reefs is monitoring and improved understanding of the ecological system and of the natural and human influences on it. Sea level changes are being monitored carefully in the sub-region, and Madagascar and Mauritius have set up Cells for Monitoring and Analysis of Sea level (CMAS), as part of the IOC-UNEP-WMO pilot project on sea level changes and associated coastal impacts. These cells will cooperate with others in the Western Indian Ocean (from Bangladesh to Kenya), sharing data and identifying common problems. The ultimate aim is to develop a framework for data collection and analysis which will provide understanding of variability and allow assessment of trends (Sheyte 1994). Information and recommendations should be integrated into regional and sectoral development policies and action plans (CSD 1999).

## HARVESTING OF COASTAL AND MARINE RESOURCES IN THE WESTERN INDIAN OCEAN ISLANDS

The EEZ associated with the Western Indian Ocean Islands extends over more than 2 106 km$^2$, an area larger than the state of California. This not only creates difficulties for administration and protection, but also for monitoring and regulation of fishing practices and harvesting rates (Brooks/Cole 1998). International regulations are in place to protect the interests of small island states, but resources are required for their monitoring and enforcement, as evidenced by the

dramatic increase in reported catches by distant-water fishing nations during the 1990s (FAO 1997). Catches of non-target, endangered species, especially turtles, dolphins, and dugongs, are also cause for concern in the sub-region, as are the destructive practices of dynamite fishing, purse-seining and drag-netting.

Domestic fish catches in Comoros, Seychelles and Mauritius grew steadily between 1975 and 1995, but have declined recently by up to 24 per cent (see Table 2c.1). This is in contrast with catches from other low-income countries, reported to have increased by 4 per cent during the same period, and with world total catches which are reported to have grown by 8 per cent (UNEP 1999a, Commonwealth Secretariat 2000). The FAO believes that this stagnation of catch is a result of stocks being fully exploited, as there has been no reduction in fishing activities during this period (FAO 1997).

## Sustainable harvesting of coastal and marine resources in the Western Indian Ocean Islands

Control of overfishing requires elaborate marine regulatory facilities and surveillance, and lack of these has meant that compliance with regulations in the Western Indian Ocean Islands has been weak to date. However, measures have been introduced recently, including training of fishermen and provision of equipment to fish beyond the reef and in deep waters, to encourage re-growth of populations in coastal waters (UNEP 1999a). Legislation to outlaw turtle fishing and to protect the species has proved difficult to enforce and evaluation of the impact has been hindered by lack of specific data. National coral reef monitoring networks have been established, under the Indian Ocean Commission's Regional Environment Programme, with annual reporting on the state of the coral reef and its resources.

| Table 2c.1 Marine fish catch in the Western Indian Ocean Islands (thousands of tonnes) 1975–97 | | | | | |
|------|---------|------------|-----------|------------|-------------|
|      | **Comoros** | **Madagascar** | **Mauritius** | **Seychelles** | **World Total** |
| 1975 | 3 850   | 19 020     | 7 038     | 3 950      | NA          |
| 1990 | 12 200  | NA         | 14 700    | 5 400      | 86 408      |
| 1995 | 13 200  | 85 463     | 16 933    | 7 000      | 91 558      |
| 1997 | 12 500  | NA         | 13 700    | 5 300      | 93 329      |

UNEP 1999c. Commonwealth Secretariat 2000

A festival of underwater film has also been held to promote awareness of the underwater world and the need for resource conservation. There are three nationally protected marine reserves in the Western Indian Ocean Islands where fishing activities are restricted, and one World Heritage Site (the Aldabra Atoll in Seychelles).

## POLLUTION OF THE WESTERN INDIAN OCEAN ISLANDS' COASTAL AND MARINE ENVIRONMENT

Increasing quantities of domestic and industrial waste discharged without treatment into coastal waters have seriously affected coastal and marine areas. This is a priority concern because of the high level of economic dependence on tourism and other uses of coastal and marine resources. Coastal populations are increasing and, in general, most of the countries in the sub-region suffer from a lack of planning for urban development and for development of tourism. Many informal settlements have grown up as a result of this situation (UNEP 1998b). Approximately 41 per cent of the coastline of Mauritius has now been developed for urban use, tourism or industrial purposes. These developments are overloading existing wastewater and sewage treatment services, and the rate of development of new infrastructure has not kept pace with rates of population influx. For example, Seychelles treats only a fraction of its wastewater (19 per cent of domestic wastewater in 1995), and only eight hotels had sewage treatment plants in 1997 (Shah 1995,

The Western Indian Ocean Islands are home to a wide range of endemic bird species, such as this Great Frigatebird (*Fregata Minor*) found in the Aldabra Islands, Seychelles.

*Gilles Martin/Still Pictures*

Shah 1997). In Mauritius, 66 per cent of coastal residents discharge their waste into the sea and, in Comoros, there are no wastewater treatment facilities (UNEP 1999a).

Agricultural run-off, containing large volumes of silt, fertilizer and chemical residues, further contributes to pollution problems by smothering habitats, poisoning some species, and encouraging the proliferation of algal blooms which in turn cause further loss of flora and fauna (PRE/COI 1998). Smothering of seagrass beds, for example, results in loss of shelter, food, and nursery grounds for important and valuable fish, shellfish, dugong and turtles (UNEP 1999a).

Industrial developments in the coastal zone include large fish processing plants, tanneries, sugar refineries and shrimp farms. Many of these do not have wastewater treatment facilities, and they discharge their wastes directly into the ocean (UNEP 1999a).

The increasing use of motorized vessels rather than human or sail-power for fishing and pleasure craft is increasing pressure on the environment by causing oil slicks and by direct physical damage from boat propellers.

There is also a high risk of major oil spills, given that 30 per cent of all oil exports from the Near East pass through the sub-region (Salm 1996). This high level of transport traffic also exposes the marine and coastal environments to oil pollution through discharge of ballast water and oil leaks.

Commercial and domestic pollution of lagoons and coastal waters throughout the sub-region has reduced the numbers and variety of fish available for local consumption. This, in turn, has resulted in overexploitation of the remaining stocks of certain popular species, causing collapse of the coral reef ecosystems and reducing tourism opportunities (UNEP 1999a). Protected and endangered species found in the region, such as marine turtles, dugongs, whales and dolphins, are reportedly declining as a result of increasing levels of waste, notably plastics, and of overfishing and predation by humans and other predators (UNEP 1998b).

### Enhancing coastal and marine environmental quality in the Western Indian Ocean Islands

All of the Western Indian Ocean Islands have signed the Nairobi Convention, but effective actions for controlling pollution are yet to be implemented. The Indian Ocean Commission has developed a five-year action plan for a

Regional Environmental Programme that concentrates on monitoring of coral reefs and ecotoxicology, development of an environmental information system, and ICZM pilot projects. The Regional Environmental Programme has established an agreement and a framework for cooperation, as well as a proposed cooperation policy.

At the national level, Mauritius has introduced control of marine pollution into its national law, as part of its obligation's under the UNCLOS. The island has also drafted regulations to control pollution by oil, noxious liquid wastes, harmful packaged substances, sewage and garbage, thus implementing the Convention on the Prevention of Pollution from Ships (MARPOL) (UNEP 1999a). The provisions of the Convention would be strengthened by ratification by other Western Indian Ocean Islands. Comoros, Madagascar, Mauritius and Seychelles have all developed National Environmental Action Plans, and Seychelles and Mauritius have Environmental Protection Acts (UNEP 1999a). Provision of port reception facilities for oily and other wastes (as required under MARPOL) is recommended, to enhance response capacities to oil spills. Additional resources should be sought to improve equipment, training, and financial resources for emergency response activities.

At the local level, schools and NGOs have been involved in coastal protection and beach cleaning campaigns to prevent refuse, especially plastic bottles, being washed out to sea.

### Towards ICZM in the Western Indian Ocean Islands

The Indian Ocean Commission's Regional Environmental Programme has been instrumental in helping the sub-region's countries to develop and implement Environmental Management Plans and sustainable development policies. These include measures such as coordination of development in the coastal zones and requirements for assessments of environmental impacts prior to certain developments being undertaken.

Programmes to reduce coastal erosion have been implemented in Seychelles and in Mauritius, with limited success. An approach combining construction of physical defences, relocation of populations and industries, and adoption of new building designs is likely to be most effective (Commonwealth Secretariat 1997).

Comoros, Madagascar, Mauritius and Seychelles are parties to the UNEP Regional Seas Programme for Eastern Africa, for which the Regional Coordinating Unit

is in Seychelles. The mission of the Eastern African Regional Coordinating Unit is 'to provide leadership and encourage partnerships by inspiring, informing and enabling nations and people of the Eastern African Region and their partners to protect, manage and develop their Marine and Coastal Resources in a sustainable manner.' The components of an Action Plan for Eastern Africa provide a framework for comprehensive action at the sub-regional and national level and activities are designed to help the governments of the sub-region strengthen their environmental management processes (UNEP 2001).

The objectives of the Action Plan for Eastern Africa are to promote sustainable development through development of appropriate legislation, pollution prevention, protection of the sub-region's living resources, strengthening of institutional coordination and activities, capacity building, and raising awareness. Particular attention is paid to assessment of causes and impacts of coastal and marine environmental degradation, and adoption of financial arrangements for successful and sustained implementation (UNEP 2001).

## SOUTHERN AFRICA

The major issue for coastal countries of Southern Africa is the depletion of fish stocks by unsustainable levels of harvesting. There are also increasing incidences of pollution from activities on land and from oil spills and potential impacts from sea level rise including inundation of major coastal settlements with associated damage to ecosystems, infrastructure and displacement of populations.

### ECONOMIC, SOCIAL AND ECOLOGICAL VALUE OF COASTAL AND MARINE ENVIRONMENTS IN SOUTHERN AFRICA

The Southern African coastline extends from Angola on the west (Atlantic) coast to Tanzania on the east (Indian Ocean) coast. The coast is rich in fish, seafood, mangroves and coral reefs as well as oil, diamonds and other mineral deposits. The long sandy beaches and warm waters of the Indian Ocean create good opportunities for tourism, and the many deep-water ports along the Southern African coast present opportunities for industry and export.

These coastal resources are important economically, at the subsistence level and commercially. In South Africa, for example, the annual revenue from coastal resources has been estimated at more than US$17 500 million (approximately 37 per cent of South Africa's GDP). This includes revenues from transport and handling of cargo, tourism and recreation, and commercial fishing industries (DEA&T 1998). Mangrove forests along the east coast, from Tanzania to northern South Africa, support a diversity of tree species that are used for furniture, firewood, building dugout canoes, and of which the leaves are collected for animal fodder. Plants are also used for medicinal purposes—*Xylocarpus granatum*, for example, is said to cure stomach problems and hernia (Sousa 1998). Mangrove forests also provide important nursery grounds and habitats for crustaceans and fish, exploited by artisans and commercial fishermen alike. Estimates of the value of the shrimp fisheries of the Sofala Bank (Mozambique), for example, are as high as US$50–60 million per year (Acreman 1999), or about 40 per cent to the country's net foreign exchange earnings (Sousa 1997). The mangrove forests also protect the coastline from storm surges and other natural hydrological influences such as high-amplitude tidal ranges and disturbances resulting from currents (Tinley 1971).

## COASTAL EROSION AND CLIMATE CHANGE IN SOUTHERN AFRICA

Sea level rise as a result of global climate change would cause inundation of the extensive mangroves of Mozambique and Tanzania, and these would retreat, thus increasing rates of erosion of the shoreline. The coastal lagoons of Angola would also be inundated. Sea level rise is also a major threat to low-lying coastal urban centres and ports, such as Cape Town, Maputo, and Dar es Salaam. Its impacts could result in a loss of income from coastal industries and port activities throughout the sub-region, as well as loss of opportunities for development of tourism (IPCC 1998). In Tanzania, a sea level rise of 0.5 m would inundate more than 2 000 km² of land, costing around US$51 million, and a rise of 1.0 m would inundate 2 100 km² of land and erode a further 9 km², resulting in costs of more than US$81 million (IPCC 1998).

The coral reefs off the coasts of Mozambique, South Africa and Tanzania are under threat of bleaching due to sea temperature rise resulting from El Niño events and global climate change. In 1998, the El Niño induced sea temperature rise of about 1 °C caused the death of up to 90 per cent of the corals in the sub-region (Obura and others 2000).

### Mitigation of the effects of climate change Southern Africa's coastal zone

All countries of the sub-region, with the exception of Angola, have ratified the UNFCCC but, because most of them (with the exception of South Africa) contribute negligible amounts to global $CO_2$ emissions, more immediate mitigating measures are required. Construction of physical barriers is a short-term measure which has been implemented, but relocation of human settlements and industry could also be considered.

## HARVESTING OF COASTAL AND MARINE RESOURCES IN SOUTHERN AFRICA

Most of Southern Africa's coastal and marine resources are under pressure from unsustainable rates and methods of harvesting, resulting from increasing demand on marine resources for food (driven by population increase, rising demand by wealthy consumers, export markets, and tourists). Demand

Shrimp and prawn seafood fisheries are important for the local and national economies.

comes not only from the maritime countries but also from inland nations. FAO trends for marine harvest, indicate a decline in marine stocks since 1972 for most countries in the sub-region (FAOSTAT 2001). This is illustrated by Figure 2c.5.

Mangrove forests are also subjected to unsustainable harvesting pressures and are being cleared for agricultural uses, salt production and human settlements. The rate of deforestation in Mozambique has been more than 3 per cent per year in the past 18 years (Saket and Matusse 1994). Lack of monitoring and research as well as inadequate policy enforcement also contribute to overharvesting of mangrove resources by both domestic and commercial users. The slow regeneration rates of mangrove trees and mining of coastal sands are exacerbating the rate of loss of mangrove habitats and are accelerating erosion rates in the coastal zone. Loss of mangrove habitat also impacts on productivity of artisanal and commercial shrimp and crab fisheries. Elevated sediment loads in coastal waters (due to erosion in the coastal zone and upstream) can increase turbidity and cause siltation of estuaries. In the open ocean, sediments can be deposited, smothering fragile habitats such as coral reefs, and benthic habitats in sheltered bays. This not only has impacts on the ecosystems but may also affect the potential revenues from tourism.

## Sustainable harvesting of coastal and marine resources in Southern Africa

Reduced catches and a decrease in the mean sizes of fish caught have led to calls for the protection of line fish stocks by many governments in the sub-region, although controls have not always been easy to monitor and enforce. Fisheries management measures include minimum size limits, bag limits, closed seasons, and closed areas (marine reserves). For example, under the Marine Living Resources Act of South Africa (1998), all South African fish stocks must be used on a sustainable basis and overexploited populations must be allowed to recover to sustainable levels before harvesting is resumed. In December 2000, drastically reduced stocks of line fish prompted South Africa's Minister of Environmental Affairs and Tourism to declare a State of Emergency, suspending the activities of commercial, artisanal and recreational anglers until stocks regenerate (DEA&T 2000a).

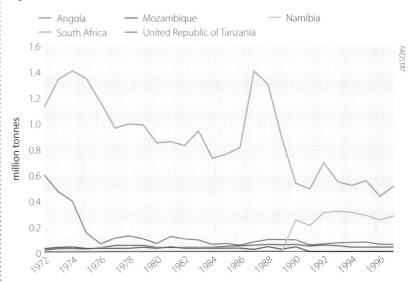

**Figure 2c.5 Trends in marine fish catch for Southern Africa, 1972–97**

— Angola  — Mozambique  — Namibia
— South Africa  — United Republic of Tanzania

Public awareness, policy directives and technological development are essential for ecosystem management and some countries have made progress in this regard. Other proposals for the management of coastal and marine resources include formulation of marine resource management plans, community-based management of mangroves and fishery resources, and institutional capacity building (Sousa 1998).

MPAs have been established in Southern Africa to limit harvesting of coastal and marine resources. There are 44 MPAs along the sub-region's coast, mostly under the jurisdiction of central or provincial governments (WCMC 1999). Where MPAs have been formally established and regulated, as in some parts of South Africa, inshore fisheries have successfully recovered (Msiska, Jiddaw and Sumaila 2000). In other areas, however, lack of education and of enforcement of regulations has hindered the success of MPAs in regulating extraction rates of some species. For example, more than 17 tonnes of Abalone (a species covered by Appendix III of the CITES) were confiscated from poachers along the coast of Cape Town during December 2001 and January 2002 (Craig Haskins, City of Cape Town, personal communication).

By contrast, informal protection has been successful in other areas. In Namibia, for instance, fishing within 200 m of the coast is illegal although there is no formal protection, and fisheries resources within this zone have been successfully protected (Msiska and others 2000). In some countries, partnerships are being formed to protect marine and coastal resources, as governments recognize the need to balance subsistence and small-scale activities with commercial demand and export revenues. This can be expected to substantially improve not only the management of resources but also relations between fishermen and authorities because subsistence fishermen will come to regard themselves as stakeholders and managers of common resources and not simply as acting under instruction from the government.

## POLLUTION OF SOUTHERN AFRICA'S COASTAL AND MARINE ENVIRONMENT

Marine and coastal ecosystems are being degraded rapidly in Southern Africa by pollution from land-based activities and dumping at sea. Land-based pollution sources include discharge of sewage, industrial effluents, stormwater run-off, wind-blown litter, suspended sediments and agro-chemicals. The increase in these types of pollution is largely a result of the rapid growth in population and in tourism in coastal centres, and of unsustainable land management practices inland. For example, in South Africa, the populations of Cape Town and Port Elizabeth, two major coastal cities, grew by 22 per cent and 24 per cent respectively during the 1990s (Macy 1999). Raw sewage is discharged in these cities, because the large

and rapidly growing population requiring sanitation facilities exceeds the capacity of municipal treatment plants. In South Africa, there are some 63 ocean outfalls along the coast, discharging approximately 800 000 m³ of sewage and industrial effluent into the sea every day. Most of the large pipelines discharge into deeper waters, but 27 older pipelines discharge above the high water mark (DEA&T 1999). This is dangerous to human health, because bathing waters are contaminated and popular seafood species such as mussels may become contaminated.

Industrial effluents in the sub-region come mainly from large fish processing plants, abattoirs, and chemical and manufacturing industries. In the case of Mozambique, for example, 126 factories in and around Maputo do not have waste treatment plants and their drains discharge toxic wastes, poisons, non-degradable substances and organic matter into their neighbourhoods (Chenje and Johnson 1996). Most of Tanzania's textile mills release dyes, bleaching agents, alkalis and starch directly into Msimbazi Creek in Dar es Salaam, from where they can easily flow into the Indian Ocean (Chenje and Johnson 1996). Agricultural run-off containing fertilizer residues and soil sediments contributes to silting of estuaries and to smothering of habitats, and has also been suspected of contributing to the occurrence of toxic algal blooms (such as red tides). Pollution of coastal and marine environments poses a threat to human health, through direct contact or through consumption of contaminated fish or seafood. It also degrades marine environments, resulting in declines in economic returns from fisheries and tourism.

Solid waste also enters the marine environment through stormwater run-off or is blown out to sea. Plastics constitute an increasing proportion of marine and coastal litter and present a particular hazard because of their endurance in the environment. Litter, especially plastics, kills many marine animals through ingestion and entanglement. It is also unsightly and a hazardous deterrent to beach users (Ballance, Ryan and Turpie 2000, Ryan 1996). Efforts have been made to reduce the volume of plastics entering the marine environment in South Africa, including regulations on the thickness of plastics used in the packaging industry and incentives for re-use of plastic bags or use of alternative materials (Ministry for Environmental Affairs and Tourism 2000).

South African fur seal, Namibia

*Klein/Hubert/Still Pictures*

Sources of pollution at sea include accidental and deliberate discharges of oil and dumping of garbage such as plastics. Dredge spoils—often rich in heavy metals such as lead, copper, zinc, mercury and cadmium—are dumped at designated sites. There have been several recent incidences of oil spills off the South African coast that have had serious adverse effects on the Africa penguin populations in the area, and on other marine life, particularly large numbers of sea birds, seals, as illustrated in Box 2c.4.

## Enhancing coastal and marine environmental quality in Southern Africa

South Africa's Prevention and Combating of Pollution of the Sea by Oil Act (1981) provides for the prevention and combating of pollution in South Africa's territorial waters. However, in 1986, responsibility for administration of the Act was divided between the Department of Transport and Environmental Affairs and the Department of Tourism and this has created problems in assigning responsibilities during emergencies and reduced the effectiveness of coordination with community groups (Trevenen-Jones 2000).

Pollution problems are further complicated by lack of awareness and weak policies and institutional frameworks, or legal instruments for enforcing national and international regulations. The most effective means of controlling coastal and marine pollution and degradation is therefore to demonstrate to industry and the public the benefits of maintaining a healthy environment—for example, improved aesthetic qualities and leisure facilities, improved harvests, and increased revenues from tourism. South Africa's Coastal Policy White Paper (from which a draft Bill is to be tabled in the country's parliament in 2002) includes measures to minimize, control, and monitor pollution of coastal waters from point and non-point sources (DEA&T 2000b).

### *Towards ICZM in Southern Africa*

Mozambique, South Africa and Tanzania participate in the UNEP Regional Seas Programme for Eastern Africa, and have ratified the Nairobi Convention. Under the convention, several capacity-building exercises have taken place in the sub-region and countries are well on their way to developing and implementing national ICZM policies and programmes. Angola and Namibia come under the UNEP Regional Seas Programme for

---

> **Box 2c.4  Oil spills and emergency responses in South Africa**
>
> In June 1994, the iron ore carrier the *Apollo Sea* broke up and sank in Table Bay. The resulting pollution caused oiling of 10 000 penguins, of which 5 000 died. The clean-up costs of this disaster were estimated at around US$1.5 million, including costs of beach cleaning, penguin rehabilitation, and interruption of port activities. Six years later, another vessel, the *Treasure*, spilled more than 1 500 tonnes of fuel oil and 20 000 penguins were oiled. In the world's largest and most successful seabird rescue operation (involving members of the public from all over South Africa) almost 20 000 birds were translocated to avoid further oiling, and oiled birds were rehabilitated and released.

Source: University of Cape Town 2001; Minister of Transport's declaration to Senate, 30 August 1994

---

West and Central Africa and have signed the Action Plan and Abidjan Convention, in 1981. South Africa has also shown interest in participating in implementation of the Abidjan Convention.

Angola Namibia and South Africa, the three countries bordering on the Benguela Current Large Marine Ecosystem, have initiated a cooperative management plan designed to overcome previous fragmentation in coastal zone management and to ensure integrated and sustainable development of coastal and marine resources. Improved understanding through research and monitoring, together with increased capacity and resources, is critical to successful implementation of the plans (O'Toole, Shannon, de Barros Neto and Malan 2001).

Countries on the east coast of Southern Africa have been actively involved in cooperation for improved coastal zone management, including a meeting of Ministers of the Environment from east-coast African states, in Seychelles in 1996. A Secretariat for Eastern Africa Coastal Area Management (SEACAM) was established at the meeting, to assist countries in implementing ICZM. Achievements under this cooperative arrangement include training workshops and publication of training manuals, development of an Eastern Africa Coastal Management database, information dissemination via newsletters, a website, distribution of documents and participation in international conferences.

At the national level, all countries in the sub-region are in the process of developing ICZM policies. For example, the Marine Research Institute in Angola currently has a number of research programmes aimed at improving understanding of the living marine resources, in order to provide management recommendations to the Ministry of Fisheries. However, lack of funding is limiting the activities of research vessels and laboratory operations. Capacity is being enhanced in coastal and marine resource management through university courses and at two Maritime Technical Schools. Mozambique developed a national ICZM policy and programme in 1998 and has since implemented environmental legislation. ICZM projects are also being conducted in Namibia for particularly sensitive regions of the coast such as the Erongo region. South Africa is expected to pass a draft Bill relevant to ICZM through its parliament in 2002 (see above), and Tanzania developed a Green Paper in 1999.

The Central African countries of Cameroon, Congo, and Gabon are among the top net oil-exporting countries in Africa.

*Chris Martin/Still Pictures*

## CENTRAL AFRICA

The total length of coastline of Central African countries is 1 789 km, the area of the continental shelf to a depth of 200 m is 66 500 km$^2$, and the EEZ extends over 537 900 km$^2$. Central Africa's coastal zone is characterized by lagoons, mangroves, seagrass beds, sandy beaches and estuarine wetlands constituting vital resources for subsistence activities and for economic development.

With large human settlements and major economic activities along its coast, Central Africa is one of the most vulnerable areas to sea level rise. Current patterns of coastal erosion are exacerbating the problem, and many of the unique coastal wetland habitats are under threat. Pollution from land- and sea-based sources is also a priority issue, particularly with the potential expansion of offshore oil and gas production activities.

### ECONOMIC, SOCIAL AND ECOLOGICAL VALUE OF COASTAL AND MARINE ENVIRONMENTS IN CENTRAL AFRICA

Central Africa's coastal belt is an area of significant commercial activity including mining, agricultural plantations and industrial developments. The belt is experiencing rapid urbanization as a result of these activities. Offshore economic resources, including petroleum and gas, are also significant, and the Central African countries of Cameroon, Congo and Gabon are among the top net oil-exporting countries in Africa. Despite political turmoil in Central Africa, the region has seen offshore and inland crude oil production rise from 650 000 barrels per day (bbl/d) in 1993 to 875 000 bbl/d in 1998. The largest increases in those years were in Equatorial Guinea, Congo and Gabon. Central Africa's proven gas reserves (about 3 per cent of the continent's total) are concentrated in Cameroon, Congo, Equatorial Guinea and Gabon. Central Africa accounted for less than 1 per cent of Africa's natural gas production in 1997, but plans are under way to increase production and utilization of gas in national electricity generation (EIA 1999). This could have implications for marine environments, particularly in terms of habitat degradation and pollution.

Commercial fisheries are also important resources for the coastal nations of Central Africa, although the FAO considers stocks of demersal fish (that is, living close to the sea bottom) to be either close to or fully exploited (FAO 1997). It has recommended that efforts to catch these species be reduced or redirected, to relieve pressure on the inshore zone and on juveniles. Similarly, catches of small pelagics (that is, open sea species) in the west and central Gulf of Guinea were considered to be fully exploited, although further south, small pelagic stocks are underexploited (FAO 1997). Pelagic fish abundance around the coast is largely controlled by variability in intensity of coastal upwelling and nutrient levels (FAO 1997). Coastal Central African countries are experiencing similar problems to those in Western Africa with exploitation of their waters by foreign fleets (FAO 1997). There is an urgent need to tighten agreements and to enforce regulations in order to protect national interests, local economies and subsistence livelihoods.

## COASTAL EROSION AND CLIMATE CHANGE IN CENTRAL AFRICA

Population densities in coastal urban centres in Central African countries are increasing under the dual pressure of population growth and migration. Major coastal cities include Douala in Cameroon (population 1.6 million in 2000) (UNCHS 2001), and Libreville in Gabon (population approximately 400 000 in 1993, around 50 per cent of the total population of Gabon) (World Bank 1997). Migration to the coast is prompted by opportunities for agriculture (fertile soils and favourable climate) and employment (large number of industries based on the coast). The resulting conversion of natural habitat to urban settlements and agricultural plantations, together with poor resource management practices inland, has accelerated rates of coastal erosion and this is now a significant problem in Central Africa. The rate of coastal erosion in Gabon, for example, is reported as having reached around 10 m per year as a result of clearance of mangrove forests (ESA-ESRIN 1996). Erosion is accelerated by construction of dams upstream of the coastal zone. Dams reduce the sediment load in rivers reaching the coastal areas and control their flow patterns, thereby increasing their erosive potential.

Development of coastal infrastructure, and poor design and management of coastal cities, lead to clearance of natural, stabilizing vegetation and increased exposure to wind and water erosion, thus contributing to destabilization of the sands and soils in the coastal zone. Mining of dunes also destabilizes the coastal zone, and enhances the potential for erosion by changing patterns of erosion and deposition. Eroded material is washed out to sea where it settles out along shipping routes which then have to be dredged to prevent grounding of ships, particularly oil tankers (ESA-ESRIN 1996).

In oil producing states, the development of canals for oil transportation is an additional modification to the coastline that has contributed to altered patterns of erosion and accretion of material.

Coastal erosion also renders Central Africa's coastal settlements and economic activities more vulnerable to sea level rise resulting from global climate change. Impacts include intrusion and contamination of freshwater sources by seawater, flooding, damage to infrastructure and displacement of populations. Cameroon and Gabon have low-lying lagoonal coasts which support large and growing populations, as well as some unique habitats for fisheries and waterfowl habitat. Sea level rise would aggravate existing problems of coastal erosion and increase the risk of saltwater intrusion into surface and groundwater resources (IPCC 1998).

### Mitigation of coastal erosion and climate change in Central Africa

Cameroon has signed the Declaration for Environmentally Sustainable Development of the Large Marine Ecosystem of the Gulf of Guinea (Accra Declaration), pledging political commitment to environmentally sustainable development in the Gulf of Guinea. One means of enhancing environmental conditions in the Gulf is to establish ICZM plans and institutions to implement policy at national level. Another is to increase existing efforts to prevent and mitigate the effects of coastal erosion and sea level rise, funded by international donor agencies and to be implemented within the framework of the Gulf of Guinea Large Marine Ecosystem Programme (see 'Enhancing coastal and marine environmental quality in Western Africa' and Box 2c.5). The Accra Declaration calls for improved sharing of information and coordination of activities between member countries (Cameroon, and the Western African States of Benin, Côte d'Ivoire, Ghana, Nigeria and Togo). The Declaration was signed in 1998, but is not legally binding and progress has been slow. Three MPAs have been established along the Central African coast, with the aim of protecting areas of natural habitat from modification, overharvesting and pollution (WCMC 1999). As elsewhere in Africa, lack of resources and weak institutional structures contribute to frequent infringement of MPAs by commercial and artisanal operations.

## POLLUTION OF CENTRAL AFRICA'S COASTAL AND MARINE ENVIRONMENT

Marine pollution is a cause for concern in Central Africa, especially in the Gulf of Guinea extending from Guinea-Bissau in Western Africa to Gabon, and in the nearshore waters of oil-producing states. Sources of pollution include offshore oil exploration, coastal industries and oil refinery activities, urban solid wastes and sewage from coastal towns, and illegal activities including dumping of wastes at sea. Marine pollution disturbs habitats, disrupts functioning of ecosystems and causes loss of biodiversity.

The significant oil reserves off the coast of Gabon have been exploited and have contributed to the country's economic growth. But they have also contributed to the risk of pollution by oil and hydrocarbons from spills, leaking valves, corroded pipelines, ballast water discharges, and production-water effluents (Chidi Ibe 1996). Diesel and other toxic chemicals are present in drilling fluids, adding to the burden of pollution from oil exploration and exploitation. Heavy metals are also associated with oil extraction. These are both toxic to marine life and, because they accumulate in the food chain, ultimately pose a threat to human health through consumption of fish and seafood (Chidi Ibe 1996). Tar balls have appeared on the beaches of Pointe Noir, the economic capital of Congo, and residents of the coastal zone have complained of pollution and an oily aftertaste in locally-caught fish (Pabou-M'Baki 1999).

Domestic sewage and agricultural effluents are additional significant sources of marine pollution. This form of pollution arises because the rates of urban and industrial development in the coastal zone exceed the capacity of municipal wastewater treatment facilities. In some cities along the coast of the Gulf of Guinea, less than 2 per cent of households have water supply and sanitation (UNIDO 2000).

## Enhancing coastal and marine environmental quality in Central Africa

The Gulf of Guinea Large Marine Ecosystem Programme has been successful in establishing regional effluent regulations and standards, and an industrial waste management programme has been piloted. Cameroon has been an active participant in the programme, establishing a National Steering Committee and adopting a regional Integrated Coastal Area Management Programme that lays the foundations of environmental policy. A waste management programme will also be implemented in Cameroon in the near future (UNIDO 2000). Additional measures to combat marine pollution in the sub-region include signing and ratification by Central African countries of the 1976 MARPOL Convention on preventing dumping of wastes at sea, and the Montego Bay Convention on the Law of the Sea.

The Abidjan Convention, to which all coastal Central African countries are party, calls for cooperation in combating pollution in cases of emergency through early warning of emergencies, and cooperation in clean up, as well as mitigation activities. As a follow up to signing of the Convention, a Regional Workshop on Oil Spill Preparedness, Response and Cooperation for West and Central Africa was held in Luanda, Angola, in

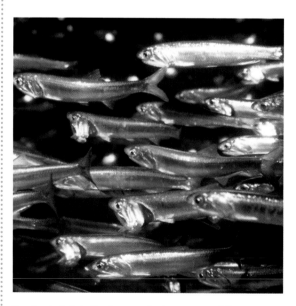

Small pelagic fish, like these anchovies, are one of the most abundant species along the Central and Western African coastlines.

*Earthviews/FLPA*

November 2000. The focus of this was to generate and facilitate communication and links within the WACAF region. A series of conclusions and recommendations were drawn up at the workshop, focusing on improving national and regional cooperation (IPIECA 2000).

### Towards ICZM in Central Africa

Cameroon, Congo, Democratic Republic of Congo, Equatorial Guinea, Gabon, and Sao Tome and Principe are all members of the UNEP Regional Seas Programme for West and Central Africa, and are party to the Abidjan Convention. Although this has facilitated the development of ICZM policies and programmes at the national level, many countries have nevertheless experienced slow progress. The institutional structures and capacity for effective coordination and holistic integration of marine and coastal ecosystem issues into development planning and environmental management remain weak. Political commitment and financial resources also need to be augmented. Programmes to raise awareness among the public and among other stakeholders have also been initiated, but funds, staff and equipment are required to implement regulations and environmental protection activities.

The Sustainable Fisheries Livelihoods Programme involves the Central African countries of Cameroon, Central African Republic, Chad, Congo, Democratic Republic of the Congo, Equatorial Guinea, Gabon, and Sao Tome and Principe. The aim of this programme is to reduce poverty in coastal and inland fishing communities. It is heavily dependent on the participation of communities and sharing of benefits among the 7 million people whose livelihoods depend directly on the use of marine and freshwater resources. Communities will be involved in environmental and economic assessments, and the programme will include extensive public awareness activities to improve communication and cooperation on sustainable development (FAO 1999).

## WESTERN AFRICA

Coastal erosion and the risk of sea level rise are the most serious issues facing coastal countries in Western Africa. However, there are also concerns over unsustainable harvesting patterns and rising levels of pollution.

Mangroves provide a natural habitat for fish, crustaceans, molluscs and water birds.

*Adrian Arbib/Still Pictures*

### Economic, social and ecological value of coastal and marine environments in Western Africa

The coast of Western Africa spans a broad range of habitats and biota, including the pristine islands of the Bijagos Archipelago and the islands of Cape Verde. Ecosystems and resources are diverse, including abundant mangrove forests, sandy beaches, lagoons, coastal wetlands and plentiful fisheries. Nearly 200 fish species were recorded in the area between 1950 and 1994, and there are a total of 22 local countries and 25 foreign fishing nations (FAO 1997). It is estimated that more than half a million people in Mauritania, Guinea-Bissau and Senegal depend directly on fisheries for incomes and food supply (IPS 2001). There are approximately 6.5 million hectares of mangroves (mainly *Rhizophora spp.*) along the coast of Benin, Côte d'Ivoire, Ghana and Nigeria, providing habitat for fish, crustaceans, molluscs and water birds (Akpabli 2000).

Storm surges are common along the coast, and patterns of erosion and accretion are highly dynamic. The protection afforded by the mangroves and other coastal wetlands is therefore vital in stabilizing the coastal zone and enabling infrastructure and development.

There are abundant oil and gas reserves off the Western African coast, especially around the Niger Delta, as well as mineral deposits (including placer minerals in Sierra Leone) and abundant sand, gravel and limestone, as well as opportunities for shipping and tourism activities.

Population pressures are among the factors that have and will continue to contribute to substantial resource degradation in the coastal zones of Western Africa. For example, in Ghana, 35 per cent of the population live on the coast, and 60 per cent of industry is concentrated in the Accra-Tema metropolis (Chidi Ibe 1996). In Nigeria, about 20 million people (22.6 per cent of the country's population) live along the coastal zone, and 13 million people live in the coastal capital of Lagos which is also the centre for 85 per cent of the country's industrial activity (UNCHS 2001, Chidi Ibe 1996). The coastal region of Dakar (Senegal) is home to about 4.5 million people (66.6 per cent of Senegal's population) and 90 per cent of the country's industries (IPCC 1998). The coastal population in Western Africa is likely to rise to about 20 million by 2020, through growth of existing coastal populations and migration from inland areas (Snrech, Cour, De Lattre and Naudet 1994).

Traditionally, opportunities for agriculture and employment in the more humid coastal areas have encouraged steady migrations from the Sudano-Sahelian area towards the coast. Much of the coastal rainforest has been cleared to make way for agricultural plantations and urban development, and what remains is decreasing at an annual rate of between 2 and 5 per cent (World Bank 1996b). Fragile coastal ecosystems, such as the stretch of coast between Accra (Ghana) and the Niger Delta (Nigeria), are under further stress because of increasing demand for resources compounded by industrial and urban development and their associated pollution loads.

## COASTAL EROSION AND CLIMATE CHANGE IN WESTERN AFRICA

Coastal erosion has been recognized as one of the most crucial issues along the coast of Western African with erosion rates of 23–30 m per year being recorded in some areas (Smith, Huq, Lenhart, Mata, Nemesova and Toure 1996). Mining of sand and gravel from estuaries,

The Akosombo hydro-electric power project, Ghana

*Ron Giling/Still Pictures*

beaches, and directly from the continental shelf, contributes to coastal erosion and shoreline retreat. In some cases, construction of ports and harbours perpendicular to the littoral zone can cause acute down-drift erosion, and this has been experienced in Benin, Côte d'Ivoire, Ghana, Liberia, Nigeria and Togo. The Bight of Benin, off the coast of Guinea, is an area particularly affected by erosion resulting from the construction of jetties and large harbours with breakwaters extending into the sea, dredging activities, and the extraction of sand for construction (Wellens-Mensah 1994). Accentuated erosion has also been documented in the Niger Delta. This is one of the impacts of offshore oil production, causing subsidence of the continental shelf and effectively raising the sea level (Chidi Ibe, Awosika, Ihenyen, Ibe and Tiamiyu 1985).

Construction of the Akosombo Dam in Ghana and the Kainji Dam in Nigeria has lowered the sediment loads in rivers that reach the coast by up to 40 per cent, making less sediment available to replace that eroded or extracted in the coastal zone (Wellens-Mensah 1994). As a result, coastal erosion east of Accra has been accelerated, reaching a rate of 6 m/yr. In Togo and Benin, coastal retreat has exceeded 150 m over the past 20 years and is threatening the potential for future development in the coastal zone (UNEP 1999b).

Climate change scenarios for the sub-region predict increases in frequency and intensity of tidal waves and storm surges which will exacerbate erosion problems by

● *Construction of the Akosombo Dam in Ghana and the Kainji Dam in Nigeria has lowered the sediment loads in rivers that reach the coast by up to 40 per cent, making less sediment available to replace that eroded or extracted in the coastal zone. As a result, coastal erosion east of Accra has been accelerated, reaching a rate of 6 m/yr* ●

moving greater amounts of coastal material (Allersman and Tilsmans 1993). Predictions also include a rise in sea level of one metre which would result in land loss of 18 000 km$^2$ along the Western African coast. Major cities such as Banjul, Abidjan, Tabaou, Grand Bassam, Sassandra, San Pedro Lagos and Port Harcourt would be inundated, with damage to infrastructure, and displacement of populations (Jallow, Barrow and Leatherman 1996, ICST 1996; Awosika, Chidi Ibe and Schroeder 1993). Inundation of Dakar, in Senegal, for which conditions are described above, would also create a significant problem of relocation and resettlement (Dennis, Niang-Diop and Nicholls 1995).

Natural habitats and resources are also at risk from sea level rise. Mangroves, for example, constitute an important resource as they stabilize coastal lands, prevent erosion and provide breeding grounds and sheltered habitats for many species and provide raw materials for medicine, food, and construction to the local communities. Inundation of these habitats would displace many species and disrupt the economic activities they support.

Sea level rise of more than one metre would flood more than 15 000 km$^2$ of the Niger Delta and force up to 80 per cent of the population to higher ground, with consequent damage to property estimated by the IPCC at US$9 000 million (World Bank 1996b, Leatherman and Nicholls 1995). The Niger Delta encompasses wetlands and lagoons that are the spawning grounds for commercial shrimp and oysters, and bait-fish for the large tuna industry. It also contains 1 300 oil wells responsible for 90 per cent of Nigeria's oil exports and foreign exchange earnings (French, Awosika and Ibe 1995). Flooding would disturb these habitats and economic activities and could worsen problems associated with oil pollution.

## Mitigation of coastal erosion and climate change in Western Africa

Coastal protection measures to date have included piecemeal construction of groynes, sea walls and other physical barriers, often at high costs and, in many cases, further contributing to the problem rather than curbing it (Wellens-Mensah 1994). An integrated, holistic framework for preventing coastal erosion is required to address the causes of erosion inland, as well as those in the coastal zone. As erosion accelerates under sea level

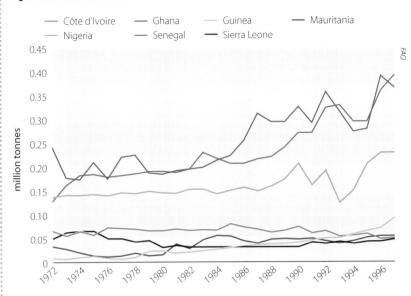

**Figure 2c.6 Marine fish catch for Western Africa, 1972–97**

— Côte d'Ivoire    — Ghana    — Guinea    — Mauritania
— Nigeria    — Senegal    — Sierra Leone

*FAO*

rise, considerable financial resources will need to be accessed and allocated, for example, to build sea walls along the Barrier Coast (near Lagos). Protection of the oil wells in the Niger Delta will be another significant cost (French and others 1995). New drilling technologies may have to be developed and harbours will require upgrading, to maintain their functionality.

## HARVESTING OF COASTAL AND MARINE RESOURCES IN WESTERN AFRICA

Small pelagic fish—herring, sardines and anchovies— are the most abundant species in the fisheries off the Western African coast and these represented almost half of all catches in 1994 (see Figure 2c.6). Catches have increased on average by 20 per cent per year since the 1950s under the impetus of national fisheries development policies. Total catches of demersal stocks on the Mauritania continental shelf fell by almost two-thirds between 1984 and 1992, which could reflect a change in fishing strategy. In Mauritania, Senegal and the Gambia, demersal resources are considered to be fully exploited or overexploited. In Cape Verde waters, the most important resource is tuna, and recent data indicate that fisheries' resources in Cape Verde are not fully exploited. In the west and central Gulf of Guinea, potential shrimp catches have been estimated at 4 700 tonnes, which is above the maximum sustainable yield (FAO 1997).

Since the late 1980s, market forces have driven foreign fleets to fish in these waters, to the detriment of catches by local countries. For example, fishing agreements between African nations and the EU have been poorly negotiated because African governments are in need of the foreign exchange and income capital. Not only have commercial fishery stocks been drastically reduced, but other species such as dolphins, sharks and turtles have also been affected and their numbers are now declining (WWF 2001b).

### Sustainable harvesting of coastal and marine resources in Western Africa

Benin, Côte d'Ivoire, Ghana, Nigeria and Togo are party to the UNCLOS, which protects national and international fishing rights and exploitation zones. However, additional resources are required to enforce these regulations and to prosecute law-breakers. Guinea-Bissau, Mauritania, and Senegal have taken additional action to protect their fisheries and Mauritania is banning all fishing except traditional, non-motorized boat fishing by local communities in the Banc d'Arguin National Park. Guinea-Bissau is establishing the Joao Viera/ Poilao National Park in the Bijagos Archipelago, as a refuge for green turtles, dolphins, sharks, rays and migratory waterbirds. Senegal is also expected to establish marine protected areas in the near future (WWF 2001b). These are in addition to the 25 existing MPAs established to reduce pressures on natural resources from overharvesting, pollution, and modification of the physical characteristics of the coastline. However, enforcement of protection regulations in most MPAs has been limited by lack of resources.

### POLLUTION OF WESTERN AFRICA'S COASTAL AND MARINE ENVIRONMENT

As Western African economies have diversified and concentrated on exports, sources of industrial pollution have developed in the coastal zone. The main sources are breweries, textile industries, tanneries, aluminium smelting, petroleum processing and edible oil manufacturing. At present, effluents are often discharged untreated into rivers, lagoons and the coastal waters of the Gulf of Guinea, and this is likely to increase with rising economic pressures to expand industrial operations (Akpabli 2000). The Korle Lagoon, a large coastal wetland in Accra (Ghana), has been severely degraded by pollution from industrial and domestic sources (WRA 1997). Agricultural pollution is widespread, with chemical residues, fertilizers and soil being washed from the surrounding cultivated areas, and causing eutrophication in coastal wetlands and estuaries.

Pollution of such environments reduces their potential to support wildlife and commercial fisheries. Polluted waters are also a risk to human health, through direct contact or contamination of drinking water sources. Pollution by sewage creates a risk of typhoid, paratyphoid, and hepatitis infections, through direct contact and consumption of contaminated seafood. Microbial and bacteriological contamination are cause for concern in the Bay of Hann, near Dakar (Senegal), in Ebrie Lagoon (Abidjan) and in Lagos Lagoon (Nigeria) (UNEP 1984).

Offshore mining and oil drilling activities are major sources of oil pollution, mainly because of leaking pipes, accidents, ballast water discharges and production-water discharges. Drilling also involves the use of heavy metals, such as vanadium and nickel, and contamination of seawater with these metals is known to affect plants and animals. Oil pollution damages coastal resources and habitats, as well as fisheries, reducing catches and incomes.

### Enhancing coastal and marine environmental quality in Western Africa

The Gulf of Guinea Large Marine Ecosystem Programme is a jointly funded, regional cooperative programme for improving environmental quality and productivity in the Gulf of Guinea (see Box 2c.5). Benin, Côte d'Ivoire, Ghana, Nigeria and Togo from Western Africa and Cameroon in Central Africa, have participated in the programme which has established a framework for sub-regional cooperation and national level, integrated coastal management plans. It also facilitated the adoption of the Accra Declaration (Declaration for Environmentally Sustainable Development of the Large Marine Ecosystem of the Gulf of Guinea), in 1998. The Declaration, aims at institutionalizing a new ecosystem-wide strategy for joint actions in environmental and natural resource assessment and management in the Gulf of Guinea. Ministers of participating countries have also called for the initiation of a second phase of the project with participation expanded to involve 10 other countries from Senegal to Angola. The Programme received

## Box 2c.5  Enhancing environmental quality in the Gulf of Guinea

The Gulf of Guinea coastal area (from Guinea-Bissau to Gabon) is one of the world's most productive marine regions, supporting 80 million inhabitants through use of the fisheries, habitat, and energy resources. Rivers and lagoons serve as important waterways for the transportation of goods and people. The Gulf is also rich in petroleum deposits, and is important globally for its marine biological diversity. Unfortunately, pollution from residential and industrial sources has resulted in habitat degradation, loss of biological diversity and productivity, and risks to human health. Coastal erosion, resulting from urbanization and clearing of mangroves constitutes a further threat to the ecosystem. Large oil and gas deposits make a major contribution to the region's economy, but also bring additional habitat modification and pollution risks.

Regional cooperation has been successful in addressing some of these environmental stresses within the framework of the Gulf of Guinea Large Marine Ecosystem Programme,

funded by UNDP, GEF, and NOAA. The programme includes the Industrial Water Pollution Control and Biodiversity Conservation in the Gulf of Guinea project, in which all of the programme countries have participated. This aims to improve the health of the coastal waters by strengthening institutional capacities for pollution prevention and remediation, developing an integrated information management system, establishing a region-wide ecosystem monitoring and assessment programme, identifying measures to prevent and control pollution, and developing policies and strategies for sustainable development of the Gulf's resources.

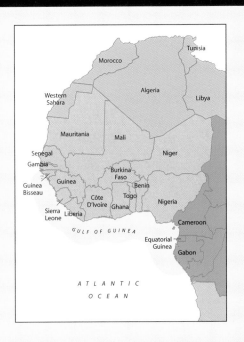

The project has been successful in building capacity within institutions, designing a regulatory policy to conserve fisheries resources, rehabilitating mangroves, establishing regional effluent regulations, and in getting all countries to adopt a regional integrated coastal areas management programme.

Source: UNEP 2002

recognition by other coastal African nations during the Pan-African Conference on Sustainable Integrated Coastal Management (PACSICOM) (Mozambique, 1998) and by the Advisory Committee on the Protection of the Sea (ACOPS) Conference (South Africa, 1998).

### Towards ICZM in Western Africa

Western African States signed the Abidjan Convention in 1981. This obligates them to place controls on land and marine-based sources of pollution, to harmonize and strengthen national policies and to cooperate with other countries in the sub-region to enhance environmental management. Parties to the Convention are also required to take steps to control and mitigate coastal erosion and its causes, and to develop contingency plans to prevent and deal with pollution arising from oil exploration and transport activities. Under the Convention, countries are also obliged to conduct environmental impact assessments prior to new developments in the coastal zone as a means of regulating uncoordinated, unplanned

developments which could accelerate pollution and erosion. Unfortunately, the Convention was unsuccessful in establishing an effective Regional Coordinating Unit, and progress has been slow. In response, UNEP has established a joint secretariat for the Abidjan and Nairobi Conventions, operational from September 2000. The new work programme of the Abidjan Convention countries includes assessments of coastal erosion and activities for improving the management of coastal ecosystems, with a special focus on mangroves and oil pollution.

## CONCLUSION

The African coastline has abundant and diverse natural resources and highly productive ecosystems that provide protection and stabilization of the physical coastline, regulation of global atmospheric gases, and nutrient cycling. The natural beauty of the coastline and its abundant resources have attracted high numbers of tourists and migrants in recent years. Local communities are heavily dependent on coastal resources such as mangrove trees for construction and medicinal and food products, and for subsistence or small-scale trade. Inland communities are now also able to access these resources and demand, as well as the price, for some food species is sufficient to support national fishing industries. These fisheries and other industries (notably oil and gas, and tourism) make substantial contributions to the national economies of coastal African countries. The coastal and marine resources therefore have great ecological, social and economic importance, both locally and for the global community.

Abundance of natural resources and economic opportunities have led to very high rates of migration and urbanization, tourism, and development in Africa over the past 30 years. Housing and urban infrastructure, industrial sites, ports, agricultural activities and hotel and leisure facilities have all also developed, and have brought with them activities such as mining of sand, limestone and coral to provide building materials. These pressures have combined to: destabilize Africa's coastal zone; increase erosion; smother habitats; deplete resources; pollute ecosystems; and reduce biodiversity. The consequence of these impacts has been a drop in economic opportunities and increasing poverty amongst coastal communities dependent on natural resources. Pollution levels are also threatening human health, directly through exposure to contaminants in coastal waters at popular resorts, and indirectly through accumulation of toxins in seafood. This pattern of overextraction and overloading with wastes is likely to continue, if not intensify, in future.

The challenge for Africa is to use its resources wisely, so that economic development can be achieved without destroying the resource base on which it is founded. ICZM is a proposed tool for doing this, and one that has been adopted, in principle, by many coastal African nations. However, implementation in many countries has been hampered by lack of human and financial resources, lack of scientific data and monitoring programmes, and by institutional fragmentation and lack of cooperative mechanisms and integrated development models. Similarly, international treaties, such as the United Nations Law of the Sea, and MARPOL, have been signed, but are ineffective without the necessary commitment to ensure implementation and enforcement of penalties for non-compliance.

## REFERENCES

Acreman, M. (1999). Water and Ecology: Linking the Earth's Ecosystems to its Hydrological Cycle. *Revista CDIOB d'Afers Internacionals*, pp. 45–46 (April 1999)

Akpabli, K.R. (2000). Integrated Management Of The Gulf Of Guinea Fight Against Coastal Pollution. Paper Presented At The Sustainable Development For Coastal Zones And Instruments For Its Evaluation. October 23–26, 2000. Bremerhaven

Allersman, E. and Tilmans, W.K. (1993). Coastal Conditions In West Africa; A Review. Ocean And Coastal Management 19: 199–240

Anon (1996). KENYA: Towards Integrated Management And Sustainable Development Of The Kenya Coast. Findings And Recommendations For An Action Strategy In The Nyali-Bamburi-Shanzu Area. Available on: http://www.ipieca.org/activities/oil/

Awosika, L.F., Chidi Ibe, A. and Schroeder, P. (eds., 1993). *Coastlines of West Africa*. A Compilation Of Papers Presented At The Coastal Zone '93 Conference, New York. American Association Of Civil Engineers

Ballance, A., Ryan, P.G. and Turpie, J.K. (2000). How Much Is A Clean Beach Worth? The Impact Of Litter On Beach Users In The Cape Peninsula, South Africa. *South African Journal Of Science* 96(5): 210_213

Bigot, L., Charpy, L., Maharavo, J., Abdou Rabi, F., Paupiah, N., Aumeeruddy, R., Villedieu, C. and Lieutaud, A. (2000). The Status Of Coral Reeds Of The Southern Indian Ocean; The Indian Ocean Commission Node For Comoros, Madagascar, Mauritius, Reunion And Seychelles. Pp 72–93 In Wilkinson, C. *Status Of Coral Reefs Of The World: 2000*. Australian Institute Of Marine Science And Global Coral Reef Monitoring Network, Townsville

Blue Plan (1996). *A Blue Plan For The Mediterranean People: From Thought To Action*. The Blue Plan Regional Activity Centre, Cairo. September 1993

Brooks/Cole (1998). Segar Introduction To Ocean Sciences. Brooks/Cole Biology Research Centre. Available on: http://www.brookscole.com/biology/member/student/oceansci/promo_feature2.html

Chenje, M. and Johnson P. (eds., 1996). *Water In Southern Africa*. SADC/IUCN/SARDC, Harare/Maseru

Chidi Ibe, A. and Quelennec, R.E. (1989). Methodology For Assessment And Control Of Coastal Erosion In West Africa And Central Africa. UNEP Regional Sea Reports And Studies No. 107. United Nations Environment Programme, New York

Chidi Ibe, A. (1996). The Coastal Zone and Oceanic Problems Of Sub-Saharan Africa. In; Benneh, G., Morgan, W.B. and Uitto, J.I., (eds.) *Sustaining The Future; Economic, Social And Environmental Change In Sub-Saharan Africa*. United Nations University Press, Tokyo

Chidi Ibe, A., Awosika, L.F., Ihenyen, A.E., Ibe, C.E. and Tiamiyu, A.I. (1985). Coastal Erosion In Awoye And Molume Villages, Ondo State, Nigeria. A Report For Gulf Oil Co. Nigeria Ltd

Cicin-Sain, B. and Knecht, R.W. (1998). Integrated Coastal And Ocean Management: Concepts And Practices. Washington, DC: Island Press. (517 pages)

Clark, J.R. (1996). *Coastal Zone Management Handbook.* Boca Raton, FL: Lewis Publishers (694 pages)

Coffen-Smout, S. (1998). Pirates, Warlords And Rogue Fishing Vessels In Somalia's Unruly Seas. Available on: http://www.chebucto.ns.ca/~ar120/somalia.html

Commonwealth Secretariat (1997). A Future for Small States: Overcoming Vulnerability. Commonwealth Secretariat, Marlborough House, Pall Mall, London

Commonwealth Secretariat (2000). *Small States Economic Review And Basic Statistics.* Commonwealth Secretariat, Marlborough House, Pall Mall, London

CSD (1999). Country Reports to the 7th Session of the Commission for Sustainable Development, New York. Available on: http://www.un.org/esa/agenda21/natlinfo

DEA&T (1998). Green Paper for a Coastal Policy for South Africa. Department Of Environmental Affairs And Tourism, Pretoria

DEA&T (1999). State Of The Environment South Africa. Department of Environmental Affairs and Tourism, Pretoria. Available on: http://www.ngo.grida.no/soesa/nsoer

DEA&T (2000a). The South African Line Fishery: Past, Present, & Future. Department Of Environmental Affairs and Tourism, Pretoria

DEA&T (2000b). White Paper for Sustainable Coastal Development in South Africa. Department Of Environmental Affairs And Tourism, Pretoria

Dennis, K.C., Niang-Diop, I. and Nicholls, R.J. (1995). SENEGAL: Senegal's Shoreline Is Already Receding And Salinity Threatens Wetlands. *Climate Alert* Volume 8, No. 2 March–April 1995

EIA (1999). Energy In Africa. US Department Of Energy/ Energy Information Administration, Washington D.C.

ESA-ESRIN (1996). Protection of mangrove forests and coastal erosion in Gabon and Nigeria. European Space Agency Publications Division *ERS-SAR Application 10*

FAO (1996). *Fisheries And Aquaculture In Sub-Saharan Africa: Situation And Outlook In 1996*. Fisheries Circular No. 922 FIPP/C922. FAO, Rome

FAO (1997). Review Of The State Of World Fishery Resources: Marine Fisheries. FAO, Rome

FAO (1998). Coastal Environments Under Threat. FAO Factfile. Available on: http://www.fao.org/news/factfile/ff9804.htm

FAO (1999). The Sustainable Fisheries Livelihoods Programme In West Africa. FAO Fisheries Department, Rome

FAO (2000). *State Of The World Fisheries And Aquaculture 2000*. FAO, Rome

FAOEAF/5 (1999). Land-Based Sources And Activities Affecting The Marine, Coastal And Associated Freshwater Environment In Comores, Kenya, Madagascar, Mauritius, Mozambique, Seychelles And United Republic Of Tanzania. FAO EAF/5 Project Office, Division Of Environmental Conventions, United Nations Environment Programme, Nairobi

FAOSTAT (2001). Fishstat: Database Of Fisheries Data. FAO, Rome

French, G.T., L.F. Awosika and C.E. Ibe (1995). Sea level rise and Nigeria: potential impacts and consequences. J. Coastal Res., special issue 14, pp. 224–242

Hare, T. (1994). *Habitats.* Macmillan, New York

ICST (Ivorian Country Study Team) (1996). Vulnerability of Coastal Zone of Côte d'Ivoire to Sea Level Rise and Adaptation Options. Report on the Côte d'Ivoire/USA Collaborative Study on Climate Change in Côte d'Ivoire

International Ocean Institute (2001). *The Coastal And Inshore Marine Environment Of The Western Indian Ocean Region At The Dawn Of The 21st Century*. Report Of The International Ocean Institute. Available on: http://www.ioinst.org/reports/report2.htm

International Rivers Network (1995–2000). The Environmental Impacts Of Large Dams. Available on: http://hww.irn.org/basics/impacts.shtml

IPCC (1995). Climate Change 1995. Impacts, Adaptations And Mitigation Of Climate Change: Scientific And Technical Analysis. Contribution Of Working Group II To The Second Assessment Report Of The Intergovernmental Panel On Climate Change. Edited By Robert T. Watson, Marufu C. Zinyowera, and Richard H. Moss. WMO and UNEP. Cambridge University Press, Cambridge

IPCC (1998). The Regional Impacts Of Climate Change. Intergovernmental Panel On Climate Change, Geneva

IPCC (2001a). Climate Change 2001: The Scientific Basis. Third Assessment Report Of Working Group 1. IPCC, Geneva

IPCC (2001b). Climate Change 2001: Impacts, Adaptation & Vulnerability. Third Assessment Report Of Working Group 1. IPCC, Geneva

IPIECA (2000). Oil Spill Preparedness And Response

IPS (2001). Three West African Nations To Ban EU Fishing Fleets. Report By Brian Kenety, 14 March 2001. IPS, Brussels

IUCN (1998). *Report of The Western Indian Ocean Turtle Excluder Device (TED) Training Workshop*. Mombasa, Kenya, 27–31 January 1997. IUCN Eastern Africa Programme

Jallow, B.P., Barrow, M.K.A. and Leatherman, S.P. (1996). Vulnerability Of The Coastal Zone of the Gambia to Sea Level Rise and Development of Response Options. *Clim. Res.*, 6, 165–177

Khafagy, A.A., Hulsbergen, C.H. and Baarse, G. (1992). Assessment of the Vulnerability of Egypt to Sea Level Rise. In: O'Callahan, J., (ed.) *Global Climate Change and the Rising Challenge of the Sea*. Proceedings of the IPCC Workshop, March 1992, Margarita Island, Venezuela. National Oceanographi and Atmospheric Administration, Silver Spring, MD

Leatherman, S.P. and Nicholls, R.J. (1995). Rising Seas Threaten Cities, Erode Beaches And Drown Wetlands In Key Developing Countries. *Climate Alert* Volume 8, No. 2 March–April 1995

Linden, O. (ed. 1993). *Workshop And Policy Conference On Integrated Coastal Zone Management In Eastern Africa Including The Island States.* Coastal Management Centre (CMC) Conf. Proc. 1, 371p. Metro Manila, Philippines

Macy, P. (1999). Water Demand Management In Southern Africa: The Conservation Potential. *Publications On Water Resources*, No.13, Department Of Natural Resources And The Environment/SIDA, Harare

Martens, E. (1992). Causes Of Biodiversity Loss In Coastal Ecosystems. Pp.69–80 In Bennun, L.A., R.A Aman, And S.A. Crafter (eds.) *Conservation Of Biodiversity In Africa; Local Initiatives And Institutional Roles*. National Museums Of Kenya, Nairobi

Ministry For Environmental Affairs And Tourism (2000). Mr Moosa Announces Penalty For Offenders Against Plastic Bag Regulations. Media Statement 23 May 2000, Pretoria

Msiska, O.V., Jiddawi, N. and Sumaila, U.R. (2000). The Potential Role of Protected Areas in Managing Marine Resources in Selected Countries of East and Southern Africa. Paper presented at the International Conference on the Economics of Marine Protected Areas, 6–7 July 2000

Munga, D. (1981). Some Observations On Petroleum Pollution Along The Kenya Coast. Pp 290–297 In *Proceedings Of The Workshop Of KMFRI On Aquatic Resources Of Kenya*. 13–19 July 1981

Mwaguni, S. and Munga, D. (1997). Land Based Sources And Activities Affecting The Quality And Uses Of The Marine Coastal And Associated Freshwater Environments Along The Kenyan Coast. (Unpublished Report)

Mwangi, S., Kirugara, D., Osore, M., Njoya, J., Yobe, A. and Dzeha, T. (2000). Status Of Marine Pollution In Mombasa Marine Park And Reserve And Mtwapa Creek, Kenya (Unpublished Report)

Ngoile, M. A. K. (1997). Coastal Zone Issues And ICM Initiatives In Sub-Saharan Africa. *Marine and Coastal Management* Vol. 37, No 3. 269–279

Nicholls, R.J. and De La Vega-Leinert, A.C. (2000). Report on the SURVAS Expert Worksop on "African Vulnerability And Adaptation To Impacts Of Accelerated Sea-Level Rise" Cairo, 5th–8th November 2000 http://www.survas.mdx.ac.uk/publica2.htm#cairo

O'Toole, M.J., Shannon, L.V., de Barros Neto, V. and Malan, D.E. (2001). Integrated Management of the Benguela Current Region: A Framework for Future Development. In B. von Bodungen and R.K.Turner (ed.) *Science and Integrated Coastal Management*. Dahlem University Press

Obura, D., Suleiman, M., Motta, H. And Schleyer M. (2000). Status Of Coral Reefs In East Africa: Kenya, Mozambique, South Africa And Tanzania. pp. 65–76 In: Wilkinson, C. *Status Of Coral Reefs Of The World: 2000*. Australian Institute Of Marine Science And Global Coral Reef Monitoring Network, Townsville

Okemwa, E. M.J. Ntiba and K. Sherman (1995). Status and Future of Large Marine Ecosystems of the Indian Ocean: A Report of the International Symposium and Workshop. Marine Conservation and Development Reports. IUCN, Switzerland

Okemwa, E. and Wakwabi, E.O. (1993). Integrated Coastal Zone Management Of Kenya. Pp193–203 In Linden, O. (ed) *Workshop And Policy Conference On Integrated Coastal Zone Management In Eastern Africa Including The Island States*. Coastal Management Centre (CMC) Conf. Proc. 1, 371p. Metro Manila

Okemwa, E. (1998). Application Of The Large Marine Ecosystem Concept To The Somali Current. In Kenneth Sherman and others (eds.), *Large Marine Ecosystems Of The Indian Ocean: Assessment, Sustainability, And Management* (Oxford: Blackwell Science, 1998) pp.73–99

Onyari, J.M. (1981). The Need For Aquatic Pollution Studies In Kenyan Inland Waters. Pp 264–289. In *Proceedings Of The Workshop Of KMFRI On Aquatic Resources Of Kenya*. 13–19 July 1981.

Pabou-M'Baki, (1999). Congo-Brazzaville—Oil Production And Sea Pollution. ANB-BIA Supplement, Issue/Edition Nr 380–15/12/1999

PERSGA (2000). The Regional Organisation For The Conservation Of The Environment Of The Red Sea And Gulf Of Aden (PERSGA) Incorporating the Strategic Action plan for the Red Sea and Gulf of Aden. Information Brochure published by PERSGA. Available on: http://www.unep.ch/seas/main/persga/brochure.pdf

Pilcher, N. and Alsuhaibany, A. (2000). Regional Status Of Coral Reefs In The Red Sea And Gulf Of Aden. Pp. 35–54. In Wilkinson, C. *Status Of Coral Reefs Of The World: 2000*. Australian Institute Of Marine Science And Global Coral Reef Monitoring Network, Townsville

PRE/COI (1998). Rapport Régional Sur Les Récifs. Programme Régional Environment, Commission De l'Ocean Indien, Quatre Bornes, Mauritius. Avril 1998

Reaser, J.K., Pomerance, R. and Thomas, P.O. (2000). Coral Bleaching and Global Climate Change: Scientific Findings and Policy Recommendations. *Conservation Biology*, 14(5): 1500–1511

Ryan, P.G. (1996). Plastic Litter In Marine Systems: Impacts, Sources And Solutions. *Plastics Southern Africa* 26(6): 20–28

Saket M. and Matusse, R.V. (1994). *Study For The Determination Of The Rate Of Deforestation Of The Mangrove Vegetation In Mozambique*. FAO/PNUD Project MOZ/92/013 National Directorate For Forestry And Wildlife, Ministry Of Agriculture And Fisheries, Maputo

Salm, R.V. (1996). The Status Of Coral Reefs In The Western Indian Ocean With Notes On The Related Ecosystems. In: The International Coral Reef Initiative (ICRI) Western Indian Ocean An Eastern African Regional Workshop Report, 29 March– 2 April 1996, Mahé, Seychelles. UNEP (Biodiversity/Water) ICRI WIO/EAF WG.1/REP. UNEP 1997

Shah, N.J. (1995). Managing Coastal Areas In The Seychelles. *Nature And Resources*, 31 (4) 16–33. UNESCO, Paris

Shah, N.J. (1996). Climate Change Poses Grave Danger To African Island States. *Splash* 12 (1) 14–15, SADC

Shah, N.J. (1997). *Country Presentation: The Seychelles*. In Brugiglio, L., (Ed) *Report Of The Workshop On Integrated Management Of Freshwater, Coastal Areas And Marine Resources In Small Island Developing States*. Malta 8–12 December 1997. University Of Gozo Centre Of Islands And Small States Institute. Foundation For International Studies, Malta, And UNEP Nairobi

Sheyte, S. (1994). IOC-UNEP-WMO Pilot Activity On Sea Level Changes And Associated Coastal Impacts. In IOC-UNEP-WMO-SAREC Planning Workshop On An Integrated Approach To Coastal Erosion, Sea Level Change And Their Impacts. Zanzibar, United Republic Of Tanzania, 17–21 January 1994. IOC Workshop Report No 96, Supplement 1, IOC, Paris

Smith, J.B., Huq, S., Lenhart, S., Mata, L.J., Nemesova, I. and Toure, S. (1996). *Vulnerability And Adaptation To Climate Change.* Environmental Science and Technology Library, Kluwer Academic Publishers

Snrech, S., Cour, J-M., De Lattre, A. and Naudet, J.D. (1994). West African Long-Term Perspective Study, Preparing For The Future: A Vision Of West Africa In The Year 2020, Summary Report. Provisional Document, CINERGY, Abidjan, Côte d'Ivoire

Sousa, M.I. (1997). 'Zona Costeira De Moçambique: Foz Do Delta Do Zambeze Até À Beira', In: *Recursos Florestais E Faunísticos Do Norte De Sofala*, Vol. 3, Informação De Base, Ministério Da Agricultura E Pescas, Maputo

Sousa, M.I. (1998). *Mangroves In Mozambique*. National Directorate For Forestry And Wildlife, Ministry Of Agriculture And Fisheries, Maputo

Tinley K.L. (1971). Determinants of Coastal Conservation: Dynamics and Diversity of the Environment as Exemplified by the Mozambique Coast. *Proceedings of a Symposium on Nature Conservation as a Form of Land Use, Gorongosa National Parks*, 13–17 September 1971, SARCCUS

Trevenen-Jones, A. (2000). Treasure Oil Spill Poses Challenge For South African Environmental Policy. 1 Ecomeme 2000

UNCHS (2001). Cities In A Globalising World; Global Report On Human Settlements 2001. United Nations Centre For Human Settlements (HABITAT) Nairobi

UNEP (1984). The Marine And Coastal Environment Of The West And Central Africa Region And Its State Of Pollution. Regional Seas Programme Reports And Studies No RSRS 46. United Nations Environment Programme, Nairobi

UNEP (1996). The State Of The Marine And Coastal Environment In The Mediterranean Region. MAP Technical Reports Series No. 100. UNEP, Athens

UNEP (1997). *Assessment Of Land-Based Sources And Activities Affecting The Marine Environment In The Red Sea And Gulf Of Aden.* Regional Seas Reports And Studies No. 166. United Nations Environment Programme, Nairobi

UNEP (1998a). Eastern African Atlas Of Coastal Resources: Kenya. United Nations Environment Programme, Nairobi

UNEP (1998b). *Water-Related Environmental Issues And Problems Of Comores And Their Potential Regional And Transboundary Importance.* TDA/SAP-WIO. Maputo

UNEP (1999a). *Western Indian Ocean Environment Outlook*. United Nations Environment Programme, Nairobi

UNEP (1999b). Overview Of Land-Base Sources And Activities Affecting The Marine, Coastal, And Associated Freshwater Environments In The West And Central Africa Region. UNEP Regional Seas Reports And Studies No. 171

UNEP (2001). UNEP Mediterranean Action Plan. June 2001. Available on: http://www.unepmap.org/

UNESCO (1997). IOC-INC-WIO IV: Reports Of Governing And Major Subsidiary Bodies. Fourth Session, Mombasa, Kenya. 6-10 May 1997

UNIDO (2000). Gulf Of Guinea: Water Pollution Control And Biodiversity Conservation (Success Story). UNIDO Media Corner. Available on: http://www.unido.org/doc/100452.htmls

University Of Cape Town (2001). A Brief History Of Penguin Oiling In South African Waters. L. Underhill, Avian Demography Unit, University Of Cape Town. Available on: http://www.uct.ac.za/depts/stats/adu/oilspill/oilhist.htm

Wawiye, P., Ogongo, P. and Tunje, E. (2000). Survey Of Potential Harmful Marine Micro-Algae In Kenyan Waters. Unpublished Report For IOC-UNESCO Contract No. 298.077.8

WCMC (World Conservation Monitoring Centre) (1999). Protected Areas Database 1999

Wellens-Mensah, J. (1994). Aspects Of Coastal Erosion In West Africa— The Case Of The Bight Of Benin. In IOC-UNEP-WMO-SAREC Planning Workshop On An Integrated Approach To Coastal Erosion, Sea Level Change And Their Impacts. Zanzibar, United Republic Of Tanzania, 17–21 January 1994. IOC Workshop Report No 96, Supplement 1, IOC, Paris

Wilkinson, C., Linden, O. Cesar, H., Hodgson, G., Rubens, J. and Strong, A.E. (1999). Ecological and Socio-Economic Impacts of 1998 Coral Bleaching in the Indian Ocean: An ENSO Impact and Warning of Future Change? Ambio 28: 188–196

World Bank (1996a). T*he Experience Of The World Bank In The Legal Institutional And Financial Aspects Of Regional Environment Program: Potential Application Of Lessons Learned For The ROPME And PERSGA Programs*. World Bank, Washington D.C.

World Bank (1996b). *Towards Environmentally Sustainable Development In West Central Africa*. Agriculture And Environment Division, West Africa Department, Africa Region. World Bank, Washington D.C.

World Bank (1997). Country Information Centre: Gabon. Available on: http://www.Ifc.org/ABN/Cic/Gabon/English/Prof.wtm

World Bank (2000). Countries: Kenya. Available on: http://www.worldbank.org/Afr/Ke2.wtm

WRA (1997). Hydrological Study Of The Korle Lagoon, Accra, Ghana. Water Resource Associates, Wallingford

WRI (World Resources Institute) (1990). World Resources 1990–1991. World Resources Institute, United Nations Environment Programme, United Nations Development Programme, The World Bank. Oxford University Press, New York

WWF (1997). Mediterranean Fisheries Face A Grim Future. Press Release 13 October 1997

WWF (2001a). Africa Programme: Oceans and Coasts. Available on: http://www.panda.org/africa/oceans.htm

WWF (2001b). West Africa Puts EU To Shame. Press Release 13 March 2001. WWF European Policy Office

PART D: FORESTS

Patryck Vaucoulon/Still Pictures

## REGIONAL OVERVIEW

Forests and woodlands have played a critical role in the survival of human populations. They have been direct providers of shelter and food for people and their livestock, and of water, medicinal plants, building materials and fuel. But forests and woodlands also regulate our environment indirectly by slowing soil erosion, controlling run-off of rainwater and storing it, and regulating its release into our rivers and lakes. Globally, they help to regulate the climate and protect coastlines. Furthermore, forests and woodlands sustain many of our cultural, spiritual and religious values as well as playing an important role in the socio-economic development of industrial countries and being a vital resource for the socio-economic stability of developing ones. Loss of forests and woodlands, therefore, means loss of a vital resource and disruption of the socio-economic activities they support. The emphasis in recent years on sustainable development has meant that the use of forests and woodlands has come under greater scrutiny in order to preserve a healthy resource base and sustain social and economic benefits.

Tracking long term trends in forest cover involves the compilation and analysis of large quantities of data that are not always consistent or comparable and the task is further complicated by different definitions of what constitutes 'forest' (IPCC 2000, UNEP 2001, Matthews, Payne, Rohweder and Murray 2001). Forests may be defined in terms of administrative categories, land use or land cover (IPCC 2000). The United Nations

Food and Agriculture Organization (FAO), in collaboration with UNEP and the UN Economic Commission for Europe (UNECE), produces an assessment of the world's forests every 10 years. These are widely cited, and much of the information presented below is taken from the 2000 FAO Forest Resources Assessment. For the FAO report, the term 'forest' means land with a tree canopy cover of more than 10 per cent and area of more than 0.5 hectares, and a minimum tree height of 5 m. A canopy cover threshold of 10 per cent can include land that might be considered tundra, wooded grassland, savanna or scrubland (Matthews 2001). The discussion in this section will, therefore, include forests, woodlands and savannas. It should be noted that natural events and human activities can affect open and closed canopy forests in different ways.

Forests cover approximately 30 per cent of the world's surface, with tropical and subtropical forests (and woodlands) comprising 56 per cent, temperate and boreal forests accounting for 44 per cent. These are all natural forests with the exception of 5 per cent of forest plantations, (FAO 2001a). It has been estimated that, since pre-agricultural times, global forest cover has been reduced by at least 20 per cent, and perhaps by as much as 50 per cent (UNDP, UNEP, World Bank and WRI 2000).

The total forest cover in Africa was estimated to be slightly less than 650 million hectares in 2000, equivalent to 17 per cent of the global forest cover, and approximately 22 per cent of Africa's land area (FAO

2001a). Africa has 14 different types of forest, in temperate and tropical climates, as shown in Figure 2d.1, although the extent of forest cover varies between sub-regions. Forests make up approximately 45 per cent of the land area of Central Africa, constituting 37 per cent of Africa's total forest cover. In contrast, only 8 per cent of the land area of countries in Northern Africa have forest cover and most of this is in Sudan (FAO 2001a).

## ECOLOGICAL, ECONOMIC AND SOCIAL VALUE OF FORESTS AND WOODLANDS

Forests and woodlands are remarkable ecosystems. They have high productivity rates—more than 800 gC/m²/yr in moist tropical forests (WCMC 2000)—and support rich and diverse animal and plant communities that, together, provide resources and opportunities sustaining livelihoods and commercial operations.

Forests and woodlands provide resources and environmental services at global, regional and local levels. At the global level, evapotranspiration and cloud cover over tropical rain forests play a role in maintaining a thermal balance in the earth's atmosphere. Forests also filter out pollution and are a sink for atmospheric $CO_2$, thus helping to mitigate global climate change. Loss of forests and woodlands can contribute to local and perhaps regional climate variability (BSP 1992, Laurance 1998, IPCC 1998) because, when forest is cleared, both the albedo (proportion of solar radiation which is reflected by the earth's surface) and local temperatures rise.

Clearing of forests can affect evapotranspiration and hydrological cycles because trees (and especially trees in tropical rainforests) recycle much of the rainwater that falls over an area. It is estimated that about 50 per cent of the precipitation in the Amazon Basin originates from evapotranspiration; the proportion of rainfall recycled in Central Africa may be as high as 75 to 95 per cent (BSP 1992, Laurance 1998). Clearing of tropical forests is implicated in reductions in local rainfall in many areas, including Côte d'Ivoire and the Gambia (Park 1992, WCMC 1992). Deforestation can also set off a chain of events that can result in intensified droughts affecting areas in other regions or sub-regions that might be more vulnerable to an increase in dry spells. For example, it has been argued that the prolonged

**Figure 2d.1 Map of African forest cover and types**

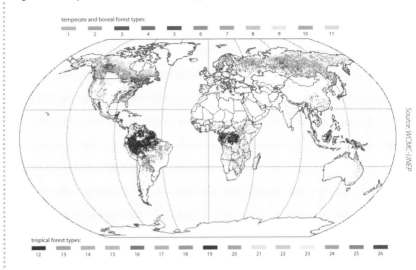

temperate and boreal forest types:
1 2 3 4 5 6 7 8 9 10 11

tropical forest types:
12 13 14 15 16 17 18 19 20 21 22 23 24 25 26

Source: WCMC-UNEP

droughts affecting the Sahelian areas of Northern Africa are caused, in part, by the destruction of forests in Western Africa (Park 1992).

Forests and woodlands also regulate soil and water quality, protecting the soil from erosion and contributing to its fertility, intercepting rainfall and channelling run-off, and maintaining the balance of elements and nutrients in the air, soil, water and organisms. They prevent silting of water downstream, and control the drought-flood cycles in rivers. Major hydroelectric schemes can suffer if these phenomena are disrupted, resulting in a lower capacity for power generation that can affect industries and their ability to provide employment.

The destruction of forests upstream of mangrove forests can harm the mangroves by causing increased sediment loads in rivers, and by contributing to global warming (Wass 1995). In coastal areas, mangrove forests protect coastlines and riverbanks by stabilizing sediments and controlling erosion. They also absorb the impact of waves and of storm floods, regulate salt intrusion inland and trap sand, preventing its moving onto the land behind beaches. In addition, mangrove forests protect coral reefs and beaches, absorb pollution from the ocean and provide a habitat for many species of commercial fish.

The value of all of these 'ecosystem functions' is not easy to estimate in monetary terms. Furthermore, the benefits accrue to the global community and not

*It is estimated that about 50 per cent of the precipitation in the Amazon Basin originates from evapotranspiration; the proportion of rainfall recycled in Central Africa may be as high as 75 to 95 per cent. Clearing of tropical forests is implicated in reductions in local rainfall in many areas, including Côte d'Ivoire and the Gambia*

View of Gatamayu
Forest, Kenya

*Christian Lambrechts*

only to local communities or users of forest resources. There is now increasing recognition of their importance and of these functions and of the need to harvest forest products on a sustainable basis in order to maintain them (FAO 1999). A recent study of forests in Madagascar has highlighted both the short-term and long-term values of different uses of forests, as detailed in Box 2d.1.

The moist tropical forests of Africa support an estimated 1.5 million species (WCMC 2000), which in turn support the local communities in terms of their food, shelter, utensils, clothing, and medicinal needs. By far the most dominant use of woodland resources is for domestic energy needs, mainly from wood and charcoal. In sub-Saharan Africa alone, traditional fuels accounted for 63.5 per cent of total energy use in 1997 (World Bank 1999). Other forest and woodland resources gathered and used by households, or traded informally amongst villagers, include meat, fruits and vegetables, construction and craft materials, medicinal products and honey. In Western and Central Africa, more than 60 wildlife species are commonly consumed and bushmeat (mainly small animals and invertebrates) harvested from forests is a traditional protein supplement to the diets of local communities (FAO 1995). The Cross River State rainforest of Nigeria is home to over 700 species of plants and animals, over 430 of which are used by local residents (CRSFP 1994). For instance, bushmeat provides 70 per cent of the animal protein in Southern Ivory Coast; 80–90 per cent in Liberia and 55 per cent in Sierra Leone (FAO 1990). In the Western African savannas, from the Gambia to

Cameroon, local residents ferment wild beans (*Parkia* sp.) to make a nutritious traditional food that provides protein and fat. The pericarp (plant ovary wall) is a source of vitamin C to children who eat it raw (FAO 1995). In South Africa, communities in woodland areas are known to have regularly used between 18 and 27 wild products from up to 300 species of plants and animals (Shackleton, Netshiluvhi, Shackleton, Geach, Ballance and Fairbanks 1999), and in Namibia wild foods provided up to 50 per cent of household food requirements in rural villages (Ashley and LaFranchi 1997). Villagers also gain benefits by using forests and woodlands as grazing areas and as sources of animal fodder, and through agro-forestry and inter-cropping. An example of a comparative valuation of such woodland resources is shown in Figure 2d.2.

### Box 2d.1  The value of forests in Madagascar

A study from Masoala National Park in Madagascar, showed that the value of forest products sustainably harvested by the local villagers could be as much as US$200 000 over 10 years. By comparison, the income that could be earned over the same period through slash-and-burn agriculture was estimated at only US$12 000. However, Madagascar could earn a total of US$90 million in foreign trade by selling the timber from the forest. When the researchers included the value of the forest in terms of global climate regulation, preserving the forest could save twice this amount of money.

*Source: Kremen Niles, Dalton, Daily, Ehrlich, Fay, Grewal*

In addition to such tangible benefits, forests and woodlands have been important for cultural, spiritual and religious purposes. The Zigua ethnic group in Tanzania, for example, protect 748 forests, which they use for burial sites and ceremonies, worshipping, traditional practices and training, *Koluhombwa* (places where people with incurable diseases are left to die), meeting places and boundaries and for water protection (Mwihomeke, Msangi, Mabula, Ylhaisi and Mndeme 1998). Some of the forests have multiple uses. Conservation of resources and biodiversity may not have been the immediate goal in protection of these forests, but there are indications that those that remain are high in rare or previously unknown species and are highly valued by the adjoining communities (Mwihomeke and others 1998). Types of use of traditionally protected forsests are shown in Figure 2d.3.

At a national level, the commercial exploitation of African forests and woodlands is an important source of income, foreign exchange and employment. For example, Cameroon, one of Africa's leading producers and exporters of tropical logs and sawn timber, earned US$436 million in 1998 from export of wood products, mainly sawnwood (FAO 2001a). South Africa is Africa's largest producer of industrial roundwood and an important producer and exporter of pulp and paper (almost exclusively from plantations). In 1998, exports of wood products totalled US$837 million (FAO 2001a). Apart from tropical hardwoods, forests provide a wide array of products that have industrial value: oils, gum, latex, resins, tannins, steroids, waxes, edible oils, rattans, bamboo, flavourings, spices, pesticides and dyes (Park 1992). Many commercial crops originated from tropical forest plants, including coffee and bananas, oranges, sugar, pineapples, rice maize and cocoa. There is concern that, as forests are degraded, genetic resources needed for the development of new food plants may also be lost (Park 1992). For example, the role of forest crop in the provision of mother (shade) trees for the establishment of cacao and coffee plantations by local communities is significant.

The enormous economic, social, cultural and environmental value of forests means that the high rates of deforestation in Africa are cause for attention and require immediate remedial action. However, it is not just clear felling but loss of certain species and processes that causes damage to forests (both natural and human-induced).

**Figure 2d.2  Comparative woodland resource valuation in Iringa villages**

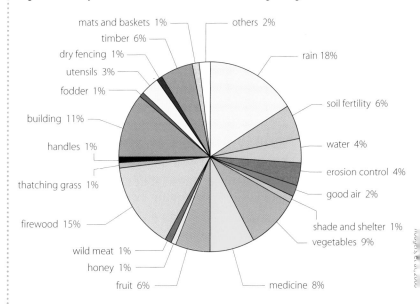

**Figure 2d.3  Types of use of the traditionally protected forests of Zigua ethnic group (by forests)**

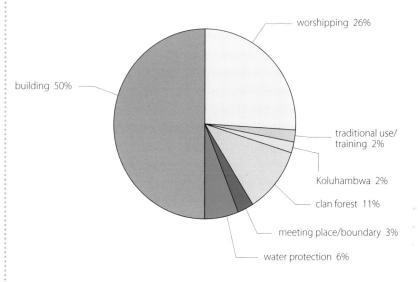

### FOREST COVER AND QUALITY

Deforestation is defined by the FAO as 'the conversion of forest to another land use or the long-term reduction of the tree canopy cover below the minimum 10 percent threshold.' Forest degradation is defined as 'changes within the forest that negatively affect the structure or function of the stand or site, and thereby lower its capacity to supply products and/or services' (FAO 2001a). In 1999, the FAO reported that 10.5 per cent of Africa's forests had been lost between 1980 and 1995, the highest rate in the developing world and in sharp contrast to the net afforestation seen in developed countries. Forest loss between 1990 and 2000 was more than 50 million hectares, representing an average deforestation rate of nearly 0.8 per cent per year over this period (FAO 2001a). As a consequence, availability of forest resources per capita declined from 1.22ha/person in 1980 to 0.74ha/person in 1995 (African Development Bank 2001). A regional comparison of changes in forest area is shown in Figure 2d.4. Deforestation can have a number of negative impacts on local agricultural production (see Figure 2d.5).

The pressures causing this decline are complex and operate at several different levels. They may be anthropogenic (caused by humans) or natural, and direct or indirect. Direct causes include commercial timber production, clearing of land for agriculture and urban expansion, and harvesting of wood for fuel and charcoal. These activities also open up forests by the construction of access roads to logging sites, fragmenting the forests and facilitating further clearance, resource extraction, and grazing by locals and commercial organizations. Forest fragmentation can lead to losses in biodiversity by cutting migratory routes for certain animal species, allowing invasion by alien species or changes in microclimate (UNDP and others). Indirect causes of deforestation include population growth, policies, agreements, legislation, lack of stakeholder participation and market factors that encourage the use of forest products, leading to loss, fragmentation or degradation (Rodgers, Salehe and Olson 2000). Other causes of forest loss include conflicts, civil wars and lack of good governance (Verolme and Moussa 1999).

Commercial logging has caused the greatest rates of deforestation in Africa in the past 30 years or more, with governments hard-pressed to earn foreign currencies and stimulate their economies. The global demand for roundwood is set to increase by 1.7 per cent per year over the next decade, and implementation of more sustainable harvesting methods is, therefore, a priority to prevent further deforestation (FAO 1999). Local processing and export of value-added products could raise export revenues and employment in timber-producing countries.

After commercial logging, consumption of wood as fuel is a major contributor to reduced forest cover and quality. Africa is the world's largest consumer of biomass energy (as a percentage of total energy consumption), which is largely wood and charcoal (see Figure 2d.6). Biomass consumption accounts for 5 per cent of total energy consumption in Northern Africa, 15 per cent in South Africa, and 86 per cent in sub-Saharan Africa excluding South Africa (EIA 1999). The demand for wood and charcoal in Africa is set to increase by over 45 per cent over the next 30 years, due to increases in population and demand for energy (FAO 2001b). Overharvesting of wood for fuel and charcoal production brings changes to the species composition of a forest or savanna. Local people are impacted by having to spend longer and search further to meet their daily fuel requirements. Development of alternative sources of energy is therefore a priority for the African region, and should be facilitated—for example, under the funding mechanisms of the Kyoto Protocol.

Slash-and-burn agriculture is another human activity contributing to declining forest cover and quality in Africa. This practice depletes the rich fertile forest soils of nutrients, and farming communities using this method have therefore traditionally shifted locations. However, the large and growing rural populations in Africa are making such shifts impossible and this breakdown in traditional management techniques is threatening the remaining forests with increasing rates of clearance and insufficient recovery time. Commercial agriculture, especially plantation agriculture, is also playing a significant part in this cycle of forest loss and soil depletion.

**Figure 2d.4  Regional comparison of change in forest area, 1980–95**

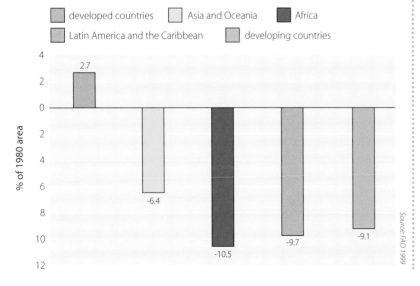

Source: FAO 1999

**Figure 2d.5  The relationship between forest clearance and crop failure**

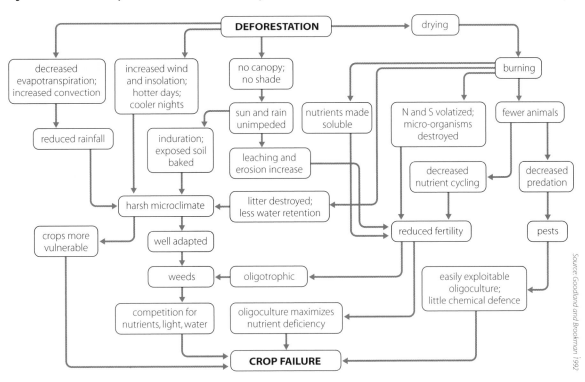

*Source Goodland and Brookman 1992*

Natural causes of forest and woodland loss, fragmentation or degradation include landslides, volcanism, fire, wind, pests and changes in water tables that change the salinity of soils (IPCC 2000, Medley and Hughes 1996). Other natural causes are river meanders cutting off areas of riverine forest from the river, after which they degrade, and large herbivores such as elephants and rhino which can change the understorey vegetation and the composition of the forest or woodland community (Medley and Hughes 1996, Western 1997). Mangrove forests can undergo change as a result of pests, sedimentation, newly formed sand dunes, high marine pollution levels, and rises in sea levels (Wass 1995).

These natural events can also be precipitated, aggravated or in some cases mitigated by human activity. For example, landslides can be caused by loss of ground cover through clearing for agriculture or cutting of timber for commercial purposes, roads or mining activities. Fires can start in other areas through human activity, and then spread to forests. On the other hand, burning can also improve regeneration of some tree species (Medley and Hughes 1996). Changes in water tables or in salinity can be caused by clearing of

trees, irrigation or interference with the hydrology of an area, including rivers flows. Large herbivores (such as elephant and rhino) may degrade their habitat if the size of continuous habitat is reduced. On the other hand, if these animals are excluded, seed dispersal and

**Figure 2d.6  Woodfuel use in Africa, 1970–2030**

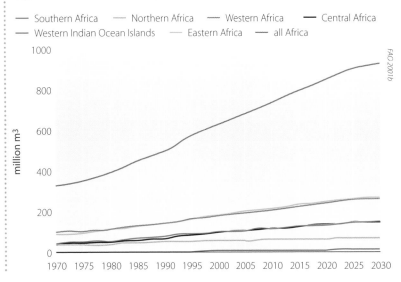

*FAO 2001b*

propagation of certain species may be interrupted (Ganzhorn, Fietz, Rakotovao, Schwab and Zinner 1999). Pests can become a problem if alien species are introduced by human activities, or if the predator-prey balance is disrupted by agrochemicals or other agents. Interestingly, some traditional activities, such as livestock grazing have been shown to control a serious pest (the bruchid beetle) on *Acacia tortilis* trees (Medley and Hughes 1996).

Fragmentation of forests exposes species adapted to the sheltered, moist, dark forest interior to greater intensities of sunlight, greater wind speeds, or increased levels of predation, reducing the number of species available to local communities, both at present and in the future. The pace and duration of the decline in species richness in forest fragments has been difficult to establish, but fragments of tropical rainforest are high priority conservation areas as they could form the basis for the regeneration of larger areas of forest (Turner and Corlett 1996).

## TOWARDS SUSTAINABLE MANAGEMENT AND CONSERVATION OF FORESTS AND WOODLANDS

Since the 1992 'Rio Earth Summit' (UNCED), the issues facing the world's forests, particularly tropical forests in developing countries, have received greater international recognition and been the object of increased local action, as illustrated by Box 2d.2. Policies aimed at improving forest management and harvesting sustainability include removal of subsidies for commercial logging and privatization of state-owned forests. Greater stakeholder participation in forest management is also emerging and, in some countries, partnerships are being formed between the state or private companies and local communities, including previously dispossessed user groups. Legal frameworks controlling the ownership and use of forests are being reviewed in many African countries (Alden Wily 2000, Alden Wily and Mbaya 2001).

The value of timber as well as non-timber forest and woodland products and the value of ecosystem

---

**Box 2d.2  The Global Workshop to Address the Underlying Causes of Deforestation and Forest Degradation (Costa Rica, January 1999)**

The Global Workshop was convened to support ongoing work of the Intergovernmental Forum on Forests and proposals for action on the underlying causes of deforestation and forest degradation.

In the African workshop it was agreed that direct causes of deforestation were logging and timber production; fuelwood consumption; forest fires; human settlements; and conversion to agricultural land. Factors hindering Sustainable Forest Management were identified as:

- poor governance and inappropriate and conflicting policies;
- inadequate macroeconomic policies and access to trade opportunities;
- inadequate institutional capacity and inappropriate technology;
- unsatisfactory tree and land tenure;
- improper valuation of forest resources;
- poverty;
- rapid population growth; and
- low levels of awareness and inadequate stakeholder participation.

Overall recommendations of the workshop included the formulation of policies to address:

1) trade and consumption patterns;
2) improving indigenous peoples, local communities and other stakeholder involvement (including issues of land tenure);
3) resolving investment and aid policies and financial flows, issues of debt servicing, incentives and subsidies, role of the private sector, governance; and
4) valuation of forest goods and services, including cultural values and ecosystem services.

Specific steps identified to combat deforestation in Africa include:

- providing an enabling policy framework;
- creating awareness;
- ensuring stakeholder participation in forest management;
- provision of adequate resources;
- ensuring equitable distribution of benefits;
- educating the public on forest values;
- reviewing Structural Adjustment Programmes;
- assessing forest resources;
- reforming economic policies; and
- encouraging good cultural practices.

*Source: Verolme and Moussa 1999, IISD 1999*

services provided by forests and woodlands are also receiving greater attention, and forest management is being reformed to include these aspects. Information on forests is becoming more accurate and widely available, with the use of technologies such as remote sensing and geographic information systems being used to document and present information. The African Timber Organization (ATO) was formed in 1976 by Angola, Cameroon, Central African Republic, Congo, Côte d'Ivoire, Democratic Republic of Congo (then Zaire), Equatorial Guinea, Gabon, Ghana, Liberia, Nigeria, Sao Tome and Principe, and Tanzania. Collectively, these countries have more than 80 per cent of the total African forest cover. At its first regional seminar (Libreville, Gabon in 1993), the ATO agreed to establish a regional sustainable forest management process with the ATO coordinating the programme and ensuring transparency and credibility. The ATO has developed principles, criteria and indicators for sustainable forest management, with assistance from the Forest Stewardship Council (FSC) and the the International Tropical Timber Organization (ITTO). In May 1996, a preliminary version of criteria and indicators for sustainable forest management was approved and, by 2000, all but nine African countries had embarked on a programme of developing and implementing criteria and indicators, either through the ATO or alternative organizations active in other sub-regions. In Southern Africa, Namibia, South Africa and Zimbabwe have criteria and indicator programmes set up through the Dry Zone Africa Process, and also have some forest areas certified by the FSC. Most countries have some forest in protected areas, although the amount varies (FAO 2001a). Some 77 per cent of Niger's forests are protected, 76 per cent are protected in Rwanda. Djibouti, Egypt and Eritrea, on the other hand, have no protected forests (FAO 2001a).

To date, sustainable forestry development in Africa has been hampered by: inadequate political commitment; weak or inappropriate institutions or policies; weak and poorly funded forestry departments; poor adoption and coordination of funding mechanisms; failure of the international community to translate forest conservation concerns into financial support; and national budgetary constraints (often worsened by Structural Adjustment Programmes). There is a need to encourage local efforts to mobilize resources for forest management in order to reduce donor dependency and to provide support for post-project financing to firmly consolidate the gains made during the life of the project. The African Development Bank is now proposing a consortium approach to remove these barriers and to improve funding for sustainable forestry development. The proposed consortium would provide a better channel for obtaining and allocating funding for policy development and sustainable development initiatives (Kufakwandi 2001).

The sub-regional analyses below highlight the priority issues in each area, and give further details on the policy and management actions in place to improve the sustainability of use of resources from Africa's forests and woodlands.

## NORTHERN AFRICA

The Northern African sub-region has an arid climate, and desert or semi-desert environments are dominant. Forests and woodlands are, therefore, not common in most of the six countries of the sub-region, except along the coast of the western Mediterranean countries, the Atlas Mountains and in the tropical zone which cuts through parts of Sudan. There are also some mangroves along the coast of the Red Sea. The total forest/woodland area in Northern Africa is estimated as 68 million hectares, and constitutes 8 per cent of the total land area and about 10 per cent of Africa's forests (FAO 2001a). Although some countries have actually shown increases in forest cover in the past 30 years (owing to establishment of plantation forestry), the major concern in the region is still loss and degradation of natural forests and wooded areas.

### ECOLOGICAL, ECONOMIC AND SOCIAL VALUE OF FORESTS AND WOODLANDS IN NORTHERN AFRICA

Non-forest types of woody vegetation in Northern Africa include forest-like stands of trees, shrubs and large bushes found in sandy *wadis* (wetlands) and depressions. The summits of coastal and inland high mountains support many wooded plants. Hydrophilous reeds occur in wetlands, and trees are cultivated as windbreaks or hedgerows around farms, along roads and canals constituting additional forest-like resources in the region. All wooded areas, although not counted in the forest area figures, are important for forest products, grazing and control of desertification through soil stabilization and regulation of hydrological systems (AOAD 1998, Hegazy 1999). No significant commercial timber production is practised or planned in the sub-region, although small plantations do exist. Local people use more than two-thirds of forest plants for food, medicinal purposes, construction, energy, and livestock rearing, and 35 per cent of plants are known to be multipurpose, that is to say useful for more than one application (FAO 1999, UNESCO/UCO 1998). In most countries in the sub-region, forestry's contribution to GDP is low. Sudan is the only country in the sub-region in which the forestry sector contributes significantly to GDP—as much as 13 per cent. Production of industrial wood is very limited with about two-thirds of the demand for industrial rounded wood and processed wood products in the sub-region being met by imports (FAO 1999).

## FOREST COVER AND QUALITY IN NORTHERN AFRICA

It is believed that much more of Northern Africa had forest cover originally, and that this has declined over several centuries because of climatic conditions and human intervention (CAMRE/UNEP/ACSAD 1996, Gilani 1997, Thirgood 1981 and AOAD 1998). However, only Sudan has experienced deforestation over the past decade, losing 10 million hectares of forest since 1990 (an average of 1.4 per cent of its forest per year) (FAO 2001a). By contrast, other countries have increased the extent of their forests and woodlands, mainly through establishment of plantations. Egypt has experienced the greatest increase—3.3 per cent per year (FAO 2001a).

The pressures resulting in loss of forest and wooded vegetation include extensive land clearing for human settlements and agricultural activities, overgrazing by livestock, overcollection of fuelwood and charcoal production. Frequent natural and human-made fires in the Mediterranean and tropical areas of the sub-region have also contributed to the reduction of natural forests, and have degraded the soils that support them, enhancing the desertification process.

The growth in Northern Africa's population has also increased the demand for forest products for energy and various domestic uses in the sub-region, especially

Dried earth, Tunisia

*Yvette Tavernier/Still Pictures*

charcoal manufacture. Although the majority of the sub-region's energy needs are met through electricity generated by burning of fossil fuels, fuelwood and charcoal are still vital sources of energy, especially for the poor. Five per cent of all energy consumed in Northern Africa comes from biomass (compared to 86 per cent in sub-Saharan Africa), and total fuelwood use is currently 58 million $m^3$ per year, predicted to increase by 20 per cent over the next 30 years (EIA 1999, FAO 2001b).

In the Maghreb countries, forests and wooded vegetation in uplands and on slopes play an important role in land stabilization, erosion control and regulation of hydrological flow. However, deforestation in these areas has resulted in increased flooding, erosion, desertification and siltation of dams (FAO 1997). Recently the construction of roads, quarrying and mining industries, and building of dams and irrigation canals, have fragmented remaining forest cover, leading to losses of biodiversity. In addition, mass tourism has increased in forest areas over the past 10 years, contributing to opening of forest canopies and disturbing natural processes and wildlife.

Large areas of the sub-region have also become infested or replaced by exotic species, introduced intentionally or accidentally. Other factors contributing to the decline in forest area and quality in Northern Africa include: ambiguity as to ownership; lack of technical personnel for research, monitoring, and enforcement of protection regulations; and lack of financial resources and development techniques.

## Towards sustainable management and conservation of forests and woodlands in Northern Africa

Concerns about forest deterioration are reflected in the substantial afforestation and reforestation programmes and various local level measures that have been introduced recently to protect and increase forest areas (see Box 2d.3). The main method used by countries in extending their forests and wooded areas has been planting of forest plantations with the aims of stabilizing sand-dunes, rehabilitating range and steppe areas, managing water catchments and protecting agricultural areas (FAO 1993). Morocco has large eucalyptus plantations (219 000 hectares), and coniferous (59 400 hectares) and dune fixing plantations (247 500 hectares). By 1990, the multipurpose green belt plantations in Algeria had mobilized a vast range of forest resources covering more than 150 000 hectares. Tunisia has reforested more than 312 000 hectares and has modernized its forest tree nurseries (Lamhamedi, Ammari, Fecteau, Fortin and Margolis 2000), and Egypt has planted about 34 000 hectares of trees. Forest plantation efforts in these countries have, however, not been able to compensate for the loss of natural forests (FAO 1996 and 1997).

Forest reserves have been set up in some countries, ranging from 19 per cent in Libya to 4 per cent in Tunisia (Egypt does not have natural forests, only plantations) (FAO 2001a). Sustainable forest management has been adopted and is being implemented by all countries in the sub-region, through the Near East and Dry Zone Africa processes in the case of Sudan (FAO 2001a). Although forest legislation has been in place in most of the countries of Northern Africa since the nineteenth century, loopholes in the laws and lack of enforcement have limited its effectiveness in terms of protecting forests and wildlife resources (FAO 1999). The situation is made more complicated by: ambiguity regarding ownership; lack of technical personnel and agricultural extension services; lack of financial resources and development techniques; poor forest management; underlying market and policy failures of forest resource pricing; and trade policies.

Reforestation and afforestation programmes are an essential part of environmental management in such an arid environment, and should be priorities for the next decade, because they will prevent or slow the rate of soil erosion, and will provide considerable benefits in terms of soil and water quality regulation.

## Box 2d.3 Community-level forest conservation in Northern Africa

Commercial opportunities for cultivation of medicinal forest plants in nurseries are being realized in several countries. In Egypt three large nurseries and 20 small ones have been established in cooperation with local Bedouins, and have played a significant role in decreasing the collection of medicinal plants from the wild.

The Tree Lovers Association (TLA) in Cairo has been instrumental in raising awareness and protection of the Wadi Degla area, a desert valley of outstanding beauty and biological diversity. In collaboration with the governmental agencies, a local women's group has created a steering committee for managing the park, including measures to guard against further removal of woody vegetation. The Mediterranean coast countries are also expected to establish a policy of replanting, improving forestry management conditions, integrating trees in urban and tourism developments, and establishing conservation areas.

## EASTERN AFRICA

The climate of Eastern Africa supports a variety of forest and woodland cover from dense tropical forests, in the humid and mountainous regions of Uganda, Burundi and Rwanda to the dry savannas of the Horn of Africa. Approximately 13 per cent of the total land area of Eastern Africa is covered in forests and woodlands, and this constitutes approximately 5 per cent of the total African forest cover. However, the percentage of forest and woodland ranges from 30 per cent in Kenya (although just 2 per cent of this is closed canopy forest), to less than 1 per cent in Djibouti (FAO 2001a, Wass 1995). There are also abundant mangrove and coastal forests in Eastern Africa. The major issue in this sub-region is conversion of natural forest to alternative land uses, predominantly cultivation and grazing, although urban encroachment is also a contributing factor.

View of Mount Kenya
National Reserve

*Christian Lambrechts*

## Ecological, economic and social value of forests and woodlands in Eastern Africa

The forests of the Eastern Arc mountain chain, running through Kenya and Tanzania, and the Albertine Rift Montane Forests of the western border of Uganda, are of particular biological importance (Rodgers 1998, Mittermeier, Myers, Gil and Mittermeier 2000, MUIENR 2000). The Eastern Arc Mountains are the oldest mountains in the sub-region and their climate, influenced by the Indian Ocean, has given rise to areas of forest which have evolved largely in isolation, given their high altitude and separation from one another. Isolation has resulted in large numbers of animals and plants being endemic to these forests and they have been identified as one of the 25 internationally recognized hotspots of biodiversity. They harbour 30–40 per cent of Tanzania's species (Iddi 1998, Mittermeier and others 2000). The Eastern Arc Mountains also form the catchment for rivers supplying the hydropower that represents 61.5 per cent of Tanzania's total electricity generation capacity (Iddi 1998).

There are few estimates of the indirect value of forests in the sub-region. From the few that exist, the water catchment protection value of the Mount Kenya Forest is reported to be about US$55 million in terms of effects on production and replacement cost (Emerton 1997). The Albertine Rift Montane Forests occupy one of the most significant geological and biogeographic regions of Africa, and support a wealth of largely endemic biodiversity. The forests on these mountains play a vital role in intercepting precipitation and channelling run-off into Africa's two largest hydrological networks (The Nile and Congo basins). The forests are also important in terms of atmospheric exchanges and global and regional climate regulation, as well as protecting and enhancing soil stability and fertility (ARCOS 2000).

Although commercial timber exploitation is limited in Eastern African countries, all forest and woodland areas are important in terms of the natural resources they provide to local communities. In Kenya, it is estimated that some 2.9 million people (more than 10 per cent of the population of the country) live within 5 km of natural forests, (Emerton 1992). The value of forest resources to these communities has been estimated at US$94 million per year, comprising fuelwood, grazing, polewood and timber (Emerton 1993). Fuelwood and charcoal supply the majority of the sub-region's energy, meeting 96 per cent of energy needs in Uganda and 75 per cent in Kenya (FAO 2001a).

Non-wood forest products are also used extensively in the sub-region. In Uganda, for example the combined value of medicines, bamboo shoots, wild foods, shea butter, oil, honey, gum arabic, curios and weaving materials has been estimated at about US$40 million per year (Emerton and Muramira 1999). The potential of medicinal plants in Eastern Africa is widely acknowledged, and they are used by Maasai, Kipsigis, Turkana and other tribes. The Maasai have a well-established pharmacopoeia for treating livestock diseases. The use of more than 60 species or subspecies of plants for ethnoveterinary purposes has been documented among the Olkonerei Maasai. These plants have been shown to act on a wide range of pathogens as well as regulating fertility, inflammation, and digestive disorders in livestock (Ole Lengisugi and Mziray 1996). Several of these species are now being documented and researched for potential commercial application, including the introduced tree species, *Azadirachta indica*, which is being researched for antimalarial properties by the Kenya Medical Research Institute's Traditional Medicine and Drugs Research Centre, Nairobi. In Uganda, the National Chemotherapeutics Research Laboratory, in Kampala, is conducting research on many indigenous plant species (Cunningham 1997).

Weaving using wood products and wood carving are important traditional crafts that contribute substantially to household incomes and local economies in Eastern Africa. It has been estimated that there are 60 000 woodcarvers in Kenya alone, with each carver generating an additional five jobs in harvesting of the wood, and sanding and polishing of the finished carvings. The annual value of exported carvings has mushroomed from around US$60 000 in the 1950s to US$20 million today (Cunningham 2001).

## FOREST COVER AND QUALITY IN EASTERN AFRICA

Between 1990 and 2000 Eastern Africa lost 9 per cent of its total forest and woodland cover (FAO 2001a). The highest rates of deforestation were experienced in Burundi (9 per cent per year), Rwanda (4 per cent per year), and Uganda (2 per cent per year) (FAO 2001a). This is not, however, a recent problem because the forests of the sub-region have been under pressure from population growth and increasing demand for fuel and for agricultural land for decades. In Uganda, for example, it is estimated that forests 'originally' (that is, around 1890) covered 45 per cent of the country but now account for only 21 per cent (MUIENR 2000, FAO 2001a). Similarly, Ethiopia's woodlands and bushland originally covered 30 per cent of the country but now represent around 4 per cent and some of the remaining forest areas are categorized as heavily disturbed and unable to produce at their full potential (EPA/MEDC 1997, FAO 2001a). A study of forest cover and quality in Ethiopia showed that up to 70 per cent of forest cover was cleared or severely degraded by human impacts between 1971 and 1997 (EIS News 1999).

Clearance of forests and woodlands for agricultural use, to feed the growing population, is perhaps the single most important cause of deforestation in Eastern Africa even though large areas of the sub-region are considered not suitable for agriculture. For example, only 29 per cent of the land area in Ethiopia is considered suitable (EPA/MEDC 1997). The percentage

is much smaller in Djibouti. To this pressure must be added the problem of declining soil fertility in cultivated areas, especially when production pressures do not allow adequate fallow periods. This causes further deforestation and encroachment by local communities and refugees, the impacts of which have been aggravated by weak policies (see Box 2d.4).

The people of Eastern Africa are largely dependent on wood for energy, with daily per capita consumption of around 1–2 kg (NEIC 1994, EPA/MEDC 1997). According to the FAO, the demand for fuelwood (including charcoal) in Eastern African countries will increase by more than 40 per cent in the next 30 years, with total demand in 2030 exceeding 271 million $m^3$/yr (FAO 2001b). Additional pressures on forests include extraction of timber for building materials, damage by forest fires and pests, and inappropriate forest policies or lack of their enforcement. Forestry departments in the sub-region have often been associated with those responsible for agriculture, water or environment and have been institutionally overshadowed. They have also often lacked the funding to implement regulations,

*The Maasai have a well-established pharmacopoeia for treating livestock diseases. The use of more than 60 species or subspecies of plants for ethnoveterinary purposes has been documented among the Olkonerei Maasai*

### Box 2d.4 Encroachment in forest reserves of Uganda

In 1975 a group of people called the Kanani Cooperative Farmers Society entered Compartment 173 of Mabira Forest Reserve. The district administration perceived them as a self-help agro-forestry project rather than as encroachers and hence supported their activities. The Forest Department consequently gave cultivation permits to 115 of its members. The permits specified the following:

- no more forested land would be cleared;
- valuable timber tree species would not be destroyed; and
- no buildings would be erected in the reserve.

However, confusion arose as to interpreatation of Uganda's 1976 Land Reform Order, which stated that 'land which is not under lease or occupation by customary tenure… is hereby specified to be land that may be occupied by free temporary license.' People might have taken this to mean that any Ugandan is free to settle anywhere, provided such land is not already occupied. By the end of 1977, 200 more encroachers had entered the reserve and the number grew to over 1800 by 1981. They degraded over 7241 hectares of the reserve.

In Mount Elgon National Park, agricultural encroachment from 1970 through the 1980s laid bare more than 25 000 hectares of what was initially virgin forest. In Kibale National Park, over 10 000 hectares of forest were cleared by encroachers. Other forest reserves that have been affected include Luung, Mubuku and Kisangi Forest Reserves in Kasese district and Kasyoha-Kitomi Forest Reserve in Bushenyi District.

Source: NEMA 1999

conservation activities, or development of trade in forest products. Several countries are now in the process of correcting these institutional weaknesses. Uganda, for example, is transforming its Forestry Department from a line ministry into a parastatal to be known as the National Forest Authority (UFSCS 2000). Forest policies are also being reviewed or revised, or new ones drafted, in Kenya, Uganda and Ethiopia.

The impacts of deforestation and degradation of wooded areas include increased potential soil erosion and loss of soil fertility, alteration of local climatic and hydrological conditions, and changes in biodiversity. Eastern Africa is home to the world's remaining population of mountain gorillas (*Gorilla beringei beringei*), in the Virunga Volcanoes. With a population of only about 320 individuals, mountain gorillas are one of two critically endangered subspecies of gorillas (the other subspecies is in Western Africa). The gorillas of the Bwindi Impenetrable National Park, previously thought to be mountain gorillas may, in fact, be a separate subspecies (Butynski 2001). The gorilla is under threat of extinction through habitat loss and disease brought about by the increasing proximity of humans resulting from the opening up of roads in forests, hunting and tourism. As a result of the conflicts in Rwanda, a massive increase in human traffic through the Virunga Volcanoes, and subsequent military presence are probably responsible for the presence of previously unidentified intestinal parasites found in the mountain gorillas (Butynski 2001). Loss of habitat combined with an increase in diseases is potentially disastrous for such a small population.

Excessive removal of wood for fuel removes vital nutrients from forests, as well as nesting material and shade for many species. Human demands on forest resources have resulted in wood deficiency leading to increased dependency on costly imports. Deficiency in fuelwood also forces people to walk further and spend longer in search of wood to meet their daily requirements and they are therefore turning to alternatives. There is, however, increasing concern over some of these. For example, agricultural crop residues and animal dung are meeting as much as 8 per cent of Ethiopia's energy requirements (EPA/MEDC 1997), but burning of these resources as fuel can give off noxious gases such as nitrogen and sulphur oxides which can cause respiratory disorders. Furthermore, these materials are principal sources of organic nutrients and as such are traditionally used as fertilizers. Removing them from the agro-forest ecosystem has direct adverse impacts on agricultural productivity.

## Towards sustainable management and conservation of forests and woodlands in Eastern Africa

In addition to national forestry policy and institutional reform, international, sub-regional and national efforts are being made to conserve forest resources and to reduce the pressures on forest ecosystems. Reforestation programmes are being implemented through funding arrangements under the Kyoto Protocol, such as the rehabilitation of degraded areas of the forests of Mount Elgon National Park and Kibale National Park (Uganda), implemented jointly by the Ugandan government and a Dutch energy utility consortium (NEMA 1999). As a response to the rising demand for fuelwood, rural electrification is being promoted in some countries such as Uganda. However, the rural poor cannot afford the investments needed for electrical appliances or the electricity tariffs. Alternative means of meeting the energy requirements include promoting the establishment of woodlots. The success of these efforts will depend largely on improvements in land tenure arrangements together with the realization that available arable land is in short supply. Agroforestry techniques are being successfully introduced to help boost agricultural productivity and

Illegal logging in the Aberdares Forest Reserve, Kenya

*Christian Lambrechts, Kenya Forests Working Group*

economic gain. These have been taken up by many Kenyan farmers, and multipurpose tree species are now providing plant nutrients, animal fodder, building poles, fuelwood and timber. Unfortunately, for a variety of reasons, including inadequacy of institutional capacity and lack of funds, not all areas are being covered. Efforts are also underway to improve the sustainability of the wood carving industry, by discouraging use of endangered species while protecting the livelihoods of millions of entrepreneurs (see Box 2d.5).

In addition to reviewing policy, legal status and regulations, some countries in the sub-region have drawn up, or are currently drawing up, Forest Action Plans or Programmes as the main frameworks for rehabilitation and expansion of forest cover. Djibouti, Ethiopia, Eritrea, Kenya, Somalia and Uganda have established criteria and indicators for sustainable forest management through the Dry Zone Africa process, and Ethiopia and Kenya now have some of their forested land under management plans. Burundi and Rwanda do not have such management measures in place as yet (because wars and civil unrest have presented barriers to these activities). However, both countries have large percentages of forest in protected areas (30 per cent and 76 per cent respectively) (FAO 2001a). The needs of some Kenyan communities for fuelwood, poles, sawnwood, wood-based panels, and pulp and paper have been met largely through afforestation and reforestation efforts at household and industrial/commercial levels. Rwanda has the largest network of forest plantations in the sub-region and these help to meet some of the timber needs of the local population.

## WESTERN INDIAN OCEAN ISLANDS

The humid, tropical climate in most parts of the Western Indian Ocean Islands is conducive to forest growth. However, frequent cyclones can cause extensive damage to forest cover, and Mauritius and Madagascar are also prone to droughts. This limits the extent of closed canopy forest but encourages savannas and thorn forests. In total, about 20 per cent of the land area of the Western Indian Ocean Islands has forest or woodland cover, less than 2 per cent of the African total (FAO 2001a). Madagascar has the most forest cover, at 20 per cent forest and 12 per cent woodland. Mauritius

> **Box 2d.5 Kenya's wood carving —options for sustainability of livelihoods and environment**
>
> Kenya's extensive and successful wood carving industry is contributing to deforestation and degradation of wooded areas through selective harvesting. The four most commonly used species are *Dalbergia melanoxylon, Brachylaena huillensis, Combretum schumannii* and *Olea africana*. Carvers are aware of the threat this causes to the environment and their livelihoods and have identified fast-growing, widely grown tree species, such as neem, jacaranda, grevillea and mango, as alternatives. These species can be harvested without environmental damage, and their use for carving generates additional incomes to farmers who grow them for fruit or other purposes. However, these species require curing before carving, and carvers require some incentive to change their current practices. The WWF has launched a campaign to raise consumer awareness and encourage tourists and local residents to buy carvings from sustainably produced woods. Certification is being explored as an option to facilitate the switch, sustaining the livelihoods of the carvers and helping to conserve the environment.
>
> *Source: Cunningham 2001*

and Seychelles each have about 7–8 per cent forest and significant woodland. Comoros, once heavily forested, now has only 4 per cent forest and 13 per cent woodland (FAO 2001a, UNEP 1999, UNDP 2000). Types of forest include lowland evergreen broadleaf rainforest, upper and lower montane forest, semi-evergreen moist forest, mangroves and savanna.

The issue of greatest concern in this sub-region is the high rate of deforestation and its environmental consequences, including soil erosion, desertification, and loss of ecosystem processes.

## ECOLOGICAL, ECONOMIC AND SOCIAL VALUE OF FORESTS AND WOODLANDS IN THE WESTERN INDIAN OCEAN ISLANDS

The Western Indian Ocean Islands constitute one of the 25 internationally recognized biodiversity 'hotspots' (Mittermeier and others 2000), and have relatively large areas of original forest habitat intact. Species diversity and endemism are extremely high for all major plant and animal groups throughout the islands, but this is especially the case for the forests and woodlands. There are also whole genera and families that are endemic to the region. For example, more than 80 per cent of the 10 000–12 000 species of flowering plants in Madagascar are endemic, as are 91 per cent of the 300

*As a response to the rising demand for fuelwood, rural electrification is being promoted in some countries, such as Uganda. However, the rural poor cannot afford the investments needed for electrical appliances, or the electricity tariffs*

reptile species. Twelve per cent of all living primate species are found in Madagascar, and all 33 species of lemur are endemic. Praslin Island (Seychelles) is home to the endemic Coco-de-Mer palm (*Lodoicea maldivica*), and some of the larger islands also have dry palm forests unique to the Seychelles. Mangroves are widespread around most of the islands of the sub-region.

Commercial timber production is limited in most of the islands, although many wood and non-wood products are used locally. Madagascar, the largest commercial producer, produces modest volumes of sawn timber, and small quantities of wood-based panels and paper. In 1998, the export value of these products was US$8 million (FAO 2001a). Important non-wood forest products in the sub-region include medicinal plants, ornamental plants, fruits, honey, essential oils, meat and animal fodder. Fuelwood is a vital resource for local communities, especially in the poorer nations of the sub-region. In Madagascar, for example, more than 90 per cent of households depend on fuelwood and charcoal, driven by increasing poverty and price inflation. By contrast, only 8 per cent of people in the

Baobab forest, Madagascar

*UNEP*

Seychelles depend on fuelwood even as a supplementary source of energy (FAO 2001a).

## FOREST COVER AND QUALITY IN THE WESTERN INDIAN OCEAN ISLANDS

It has been estimated that up to 85 per cent of the original forest cover of the Western Indian Ocean Islands has been lost (WCMC 1994). There has been a steady decline over several decades, with deforestation between 1990 and 2000 of less than 1 per cent per year for most islands. Comoros is an exception with deforestation rates exceeding 4 per cent per year (FAO 2001a). The major causes of deforestation are clearance to make way for shifting subsistence agriculture, for commercial plantations or for industry and residential areas. For example, there are now extensive coconut plantations in Seychelles, as well as a relatively large planted forest estate of Casuarina and Albizia. Other introduced species are commercially exploited (sometimes illegally) for the distillation of ylang ylang perfume or for furniture making. Forest fires are a threat in some areas and burning of forests to create cattle pasture can also cause deforestation. The risk of fire and drought may be augmented by increasingly long dry seasons resulting from climate change. Causes of forest degradation include selective harvesting of wood for fuel and for ornamental plant and animal collections.

Total fuelwood demand in the sub-region is currently estimated at 10 million m$^3$, and is projected to rise to 18 million m$^3$ over the next 30 years (FAO 2001b). More sustainable production methods must be established and affordable alternative energy sources must be made available if further deforestation and forest degradation are to be prevented.

The loss of natural forest areas has precipitated the near extinction of tree species such as Tatamaca, Ebony and Baobab, common 50 years ago. Loss of forest has also depleted natural habitats for a wide variety of plants, insects, birds and wild animals (UNEP 1999). In Madagascar, the demise of a specific group of seed dispersers may have resulted in the decline of tree regeneration in an entire forest ecosystem—one study concluded that regeneration of the dry deciduous forest with all of the primary forest tree species in the western part of the country depends on the presence of a particular species of lemur (Eulemur fulvus) (Ganzhorn and others 1999). The prolific invasion by shrubs, bushes and exotic species, such as Chinese guava, privet, liane cerf and poivre marron, is also evidence of habitat degradation throughout the sub-region's forests and woodlands. Soil erosion on steep slopes in areas that were previously mountainous forests is now causing siltation of fresh water ecosystems, lagoons and water reservoirs throughout the sub-region causing drying of watercourses. This is especially prevalent in Comoros. These changes in ecosystems are contributing to climate change and loss of cultural heritage, and are threatening local livelihoods (UNEP 1999).

### Towards sustainable management and conservation of forests and woodlands in the Western Indian Ocean Islands

Madagascar has 4 per cent of the forest area in National Parks (FAO 2001a), and Seychelles has the best-preserved area of natural forest at the Morne Seychellois National Park. Some forest sites in Mauritius that are heavily infested with alien plants have been designated as Conservation Management Areas, and removal of exotic species has been attempted. This is a lengthy, resource-intensive process, requiring the continued commitment of resources to ensure that progress is not reversed. Mauritius and Seychelles have also joined the Dry Zone Africa process to establish criteria and indicators for sustainable forest management.

Other responses in the sub-region have included attempts at reforestation, but these have been inadequate (UNEP 1999). For example, only 500 hectares of cleared forest was reforested in the Comoros during the 1990s and many of the trees planted were utilized before maturity. In Madagascar, there were attempts at reforestation during the 1970s, to provide wood for fuel, crafts, and industry but efforts were not sustained, and these areas have also been depleted (UNEP 1999). Commercial forest plantations are also limited. Seychelles has the largest area of plantations, at 11 per cent of its land area; Madagascar's plantations cover just 2 per cent of its land (UNEP 1999).

In 1998, wood production in all Western Indian Ocean Island countries was insufficient to meet domestic demand and imports of wood products exceeded exports (FAO 2001a). More needs to be done to protect the little remaining natural endemic forest and to enforce the forest protection legislation available

in all the countries of the sub-region (UNEP 1999). Several international conservation organizations have been active in the region, conducting research, mapping, and making biological inventory studies, as well as providing baseline information for the establishment of new protected areas and assisting with policy development for sustainable natural resource use. These and other efforts to bring together social and economic development imperatives and environmental issues need further strengthening and development.

## SOUTHERN AFRICA

Southern Africa has a range of forest and woodland types that follow the rainfall distribution of the sub-region. The wetter, more northern parts of the sub-region support more closed canopy forest, whilst drier countries in the south have predominantly woodlands and savannas. The total forest and woodland area of Southern Africa amounts to 32.7 per cent of the sub-region's total area, and constitutes 34 per cent of all of Africa's forests (FAO 2001a). Angola has the highest forest cover with 56 per cent of the land area under forests; Lesotho has the lowest with less than 1 per cent (FAO 2001a). There are four forest and woodland types, namely deciduous broadleaf forests (temperate forest types), lower montane forest, mangroves, deciduous/semi-deciduous broadleaf forest and savannas (tropical forest types). Southern Africa also has six regions of exceptional plant species diversity, and forest species are abundant in many of these (White 1983).

The major issues of concern in the forests of Southern Africa are degradation of forests and woodlands, and overexploitation of certain species, resulting in loss of ecosystem goods and services.

### Ecological, economic and social values of forests and woodlands in Southern Africa

Forest products are a valuable source of export earnings and revenue throughout the sub-region, and the communities living in forest or woodland areas are highly dependent on forest products for meeting everyday food and energy needs. For example, in 1998, South Africa's exports of forest products totalled US$837 million, mainly from wood pulp and paper most of which was produced in plantation forests. Zimbabwe's exports were US$42 million, mainly from sawnwood (FAO 2001a). Forests and woodlands are important to local communities, mainly as a source of domestic fuel, either wood or charcoal. For example, about 80 per cent of Mozambique's population live in rural areas and depend on wood for cooking and for heating of water for domestic use, space heating and drying of foodstuffs. The charcoal industry generates about US$30 million annually, and is the sole source of income for about 60 000 people (Kalumiana 1998). Important non-wood forest products include honey, beeswax, bamboo, reeds, mushrooms, caterpillars, fodder, wild edible plants and fruits, leaves and bark for weaving, and resins. The medicinal plant trade is extensive and profitable in Southern Africa with approximately 3 000 species (10 per cent) of Southern African plants used medicinally and around 350 species commonly and widely used (van Wyk, Van Oudtshoorn and Gericke 1997). Other species harvested from the wild contribute as much as 40 per cent to household incomes (Cavendish 1999) or between US$200–1000 per year (Shackleton, Shackleton and Cousins 2000).

## FOREST COVER AND QUALITY IN SOUTHERN AFRICA

Southern Africa has one of the fastest growing populations in the world and faces the challenge of trying to increase food supplies by some 3 per cent per year. To date, this challenge has been addressed by improvements in agricultural production, and by bringing large areas of wooded land under cultivation. The sub-region is therefore witnessing an increasingly high rate of deforestation, mainly due to human activities. Rates of deforestation over the last 10 years have ranged from 2.4 per cent per year in Malawi to 0.1 per cent per year in South Africa, whilst Swaziland has experienced a 1.2 per cent per year net increase in the its forest area (FAO 2001a). The other main pressure leading to deforestation in the sub-region is fuelwood harvesting and tree cutting for charcoal (Chenje 2000). Studies show that both rural and urban demands for wood energy have increased, and it is expected that these demands will almost double over the next 30 years due to growing populations and economic conditions (FAO 2001b). For example, worsening

poverty in some urban areas is forcing many people to turn to charcoal and fuelwood to meet their domestic energy requirements because these are cheap and fuelwood can be collected rather than purchased (World Bank 1996). In some countries, particularly those in which charcoal use is prevalent (Angola, Mozambique, Tanzania and Zambia), trading in charcoal is a major source of income for some households, contributing to the rate of wood collection and deforestation. In Zambia, for example, more than half the country's fuelwood is converted to charcoal, requiring the clearance of some 430 km$^2$ of woodland every year to produce more than 100 000 tonnes of charcoal (Chenje 2000). Commercial logging is a further cause of deforestation and, in some cases, refugees from the wars in Angola and Mozambique have resettled in wooded areas, contributing to tree clearance.

Selective harvesting of forest species is intensifying pressures on forest and woodland resources, and has the potential to degrade habitats, reduce biodiversity, and impair ecosystem functions such as water and soil quality regulation, even though clear felling does not occur. Commercial exploitation of medicinal plants from forests and woodland is an increasingly important component of selective deforestation, resulting in the localized disappearance of certain plant species. Commercialization of particular crafts, such as basket making, is also causing the disappearance of certain plant species. Another issue of concern is that species introduced for plantation forestry production are rapidly invading grassland ecosystems where they use vast amounts of water, disturb ecosystem functioning, and reduce biodiversity. Selective harvesting results in changes to the biodiversity and relative abundance of different species, as well as changing the microclimate, global climate, nutrient cycles and hydrological regimes.

Reduced forest cover and reduced forest quality accelerate soil erosion, sedimentation and siltation as well as siltation-induced flooding at points far away from the deforested areas. Soil erosion is probably the most important factor in the decline in agricultural productivity. This impacts most severely on rural communities that are most dependent on small-scale agriculture and natural resources for their livelihoods

Other impacts of forest and woodland loss or degradation include loss of resources for future exploitation, either local or commercial, for agriculture, pharmaceuticals or crafts. It is estimated that 300 000 to 400 000 hectares of natural grassland forests and plant species of medicinal value are destroyed in Tanzania each year. Tanzania has one of the largest ruminant livestock populations in Africa, many of them owned by pastoralists, and it is clear that the seriousness of this loss for the health and welfare of livestock (and therefore for that of their owners) will escalate in the near future (Ole Lengisugi and Mziray 1996).

## Towards sustainable management and conservation of forests and woodlands in Southern Africa

Most forests in Southern Africa are either state-owned or privatized with ownership by local communities being only weakly supported by laws (Alden Wily 2000). The failure to involve local people and have tangible benefits accrue to them has disassociated people from forests and even made them co-exploiters and co-destroyers of their own forests. With growing democratic governance and rapid spread of environmental awareness, community participation in forest management and land tenure reform is being promoted and is gaining popularity as a sustainable means of natural resource management and income generation. In Tanzania, for example, a 'Land Act' clearly states that forest reserves may be owned by the state, private companies, or communities and legislation in South Africa also makes provision for community ownership of reserved forests. Local community actions have been encouraged in Malawi (see Box 2d.6) and recent legislation enacted in the country recognizes traditional ownership rights in the form of Village Forest Areas. Joint Management Committees in Zambia, Local Resource Management Councils on Mozambique, and Management Authorities in Namibia, have also been established to coordinate allocation of access rights and/or distribute benefits among the local population (Alden Wiley, 2000).

The Southern African countries of Angola and Tanzania are members of the ATO, and each has participated in developing criteria and indicators for sustainable forest management (FAO 2001a). Namibia, South Africa and Zimbabwe have some forest area certified by the FSC, and all countries of the sub-region have some of their forests in protected areas. Efforts to promote sustained regional self-sufficiency in forest and wood products, and to enhance the economic and

Illegal logging of Camphor trees, Kilimanjaro Forest Reserve, Tanzania

*Christian Lambrechts, UNF/UNEP/KWS/University of Bayreuth/WCST*

environmental performance of the forestry sector are being pursued through the SADC Forestry Sector. The Forestry Sector coordinates programmes in: forestry education and training; improved forest resource management; forest resources assessment and monitoring; forest research; utilization of forest products; marketing; and environmental protection. The Sector also developed the Forestry Policy and Development Strategy, approved in September 1997. The Strategy identifies priorities for cooperation in

### Box 2d.6 Local community action in Malawi

In Malawi, a recent project has successfully trained members of local communities in agro-forestry and extension work. Member of local communities working as extension officers for the Department of Forestry have, until recently, been excluded from most training programmes, and are thus not performing their duties optimally or achieving recognition. A local development organization has worked with these people, training them in: forestry and agro-forestry practices; the Forestry Policy Act; participatory rural appraisal tools; extension techniques; and community motivation. This has boosted both their morale and performance, and has improved communication among a wide range of stakeholders.

*Source: COMPASS 2000*

building capacity in sustainable management, protection and conservation of forest resources, facilitating of trade in forest products, promoting sustained regional self-sufficiency in forest and wood products, improving public awareness of forestry, and enhancing forest research.

## CENTRAL AFRICA

Central Africa's high and reliable rainfall supports extensive forest cover throughout the sub-region, with the exceptions of the northern part of Cameroon, Chad and the Central African Republic. Forests and woodlands cover, in total, about 45 per cent of the land area of Central Africa, and constitute 37 per cent of Africa's total forest cover (FAO 2001a). Most countries in the sub-region are, therefore, well endowed with forests, with Gabon having the greatest cover (85 per cent), and Cameroon, Congo, Democratic Republic of Congo (DRC) and Equatorial Guinea having more than 50 per cent. The exception is Chad, which, because of its northerly location and arid environment, has only 10 per cent forest cover (FAO 2001a). Tropical forests are the predominant type in the sub-region, namely, lowland evergreen broadleaf rainforest, montane forests and freshwater swamp forests. Savanna cover is also significant in the northern parts of Cameroon, Central African Republic, and Chad and southern DRC.

The major issue of concern in Central Africa is rapid deforestation, mainly for commercial timber exports. The damage caused to remaining forest areas by timber extraction is an additional concern for the sub-region.

### ECOLOGICAL, ECONOMIC AND SOCIAL VALUES OF FORESTS AND WOODLANDS IN CENTRAL AFRICA

Dense, tropical forests provide essential ecosystem services such as nutrient cycling, water and soil protection, and exchange of atmospheric gases (absorbing $CO_2$ and releasing oxygen). The tropical, moist forest of Central Africa comprises the second largest contiguous area of tropical forest in the world, and thus plays a very important role in atmospheric carbon sequestration and the mitigation of potential climate change. Other benefits of forests include the extremely high levels of biodiversity, which has enormous untapped potential for agricultural, pharmaceutical and nutritional applications. There are five regional hotspots in Cameroon, noted for their plant species richness and the presence of endemic bird species.

Commercial logging is the primary source of revenue from Central Africa's forests. This is mainly carried out by foreign logging companies and, thus, ensures substantial amounts of foreign exchange for the countries of the sub-region. Cameroon, for example is one of Africa's leading producers and exporters of sawn timber and tropical logs, and ranks fifth in the world

Oil exploration in a rainforest, Gabon

*Arnaud Greth /Still Pictures*

(FAO 2001a, WRI 2001). In 1998, its exports exceeded US$436 million, approximately 5 percent of GDP (FAO 2001a, World Bank 2001). In the same year, Equatorial Guinea exported US$62 million of wood-based panels, representing 14 per cent of its GDP (FAO 2001a, World Bank 2001).

Central Africa's forest ecosystems are also home to a large number of communities from an estimated 250 ethnic groups (WRI 2001). These communities depend on wild and cultivated resources from forests, including bark, vegetables, fruits, flowers, honey, resins, fungi, medicinal plants and wildlife, for local consumption, as well as for export. Inter-cropping or agroforestry practised within these communities is a vital source of vegetables, grains and fruit, for use by households or for trade. The forests also offer opportunities for tourism, which would generate foreign exchange and stimulate rural development, although these are largely untapped at present because of low levels of infrastructure and/ or conflict in Central African countries.

## FOREST COVER AND QUALITY IN CENTRAL AFRICA

With resources of such enormous potential value and countries with such low levels of economic and social development, it is not surprising that Central Africa is exploiting forest resources and experiencing large-scale deforestation. Between 1990 and 2000, a total of 9 million hectares of forest were cleared, 4 per cent of the total area. The highest annual rates of deforestation were recorded in Cameroon (0.9 per cent), Chad and Equatorial Guinea (0.6 per cent each), while insignificant rates were recorded in Gabon and Sao Tome and Principe (FAO 2001a). The main causes of deforestation are commercial logging (including illegal activities), clearing for commercial and subsistence agriculture, and fuelwood harvesting. The improved access to forest areas, created by roads constructed for timber trucks, not only fragments the forests, but also facilitates exploitation by local communities and resettlement by refugees, who eventually introduce wildfire through their various life-sustaining activities. Previously inaccessible areas are also opened up to poachers and hunters whose activities threaten wildlife.

By far the greatest threat to forests in the sub-region is commercial logging and the unsustainable rates of harvesting practised by many companies. In

**Figure 2d.7 Logging concessions in Cameroon 1971 and 1995**

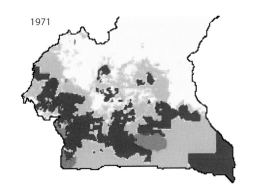

1971

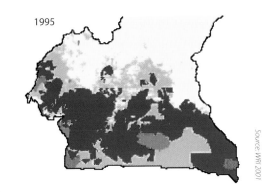

1995

*Source: WRI 2001*

Cameroon alone, the number of registered logging companies has more than quadrupled, increasing from 106 in 1980 to 479 in 1998 (see Figure 2d.7) (WRI 2001). In addition, weak protection regulations and lack of law enforcement have resulted in large numbers of logging violations, and logging concessions now encircle protected areas such as the Dja Forest Reserve, a World Heritage Site. Violations of forest regulations committed within concession areas include felling the wrong timber species, logging protected species, and cutting undersized trees. Unauthorized logging, and logging within a protected area, have also been recorded in Cameroon (WRI 2001). The need to encourage local milling and the recruitment of local skilled labour is vital in promoting joint conservation efforts.

Clearing of forests for agriculture and collection of fuelwood are other sources of forest loss and degradation in Central Africa. These pressures are set to increase over the next 20–30 years, driven by rapid population growth, poor social development, low employment opportunities, low-income levels and lack of affordable alternative sources of energy.

The impacts of the various pressures on the sub-region's forests are both positive and negative. Positive impacts include the creation of employment in the forestry sector, tax and other sources of revenue. These activities lead to more monetary exchange, the development of a road infrastructure for easier transportation and marketing of local produce, and thus the improvement of conditions for the local populations. On the other hand, the logging methods used and the scale on which they are applied exceed the natural regeneration capacity of the forests. In addition, the negative impacts of soil compaction and chemical pollution on the speed of natural regeneration could be significant. Commercial logging leaves gaps in the canopy that change the microclimate and thus species composition. In closed canopy forest, the regeneration rates are sufficiently high to close gaps relatively quickly. However, in the drier areas of Central Africa, where savanna is the predominant form of wooded vegetation, trees take much longer to grow back, and their removal therefore has a greater impact on the ecosystem. Loss of forest habitat also leads to loss of biodiversity which, in turn, reduces opportunities for commercial agricultural or pharmaceutical exploitation. Other negative impacts of forest degradation include soil degradation and reductions in productivity from agroforestry, susceptibility to alien invasive organisms, and disruption of hydrological systems. Local communities suffer by having to search further and further afield to collect wild resources such as fuelwood.

## Towards sustainable management and conservation of forests and woodlands in Central Africa

The importance of forests and forest resources in Central Africa has led stakeholders, especially governments, the international community and NGOs, to intensify measures for forest resource conservation and sustainable management. Countries in the sub-region have engaged in re-forestation programmes in places where deforestation has occurred, and in forest regeneration and rehabilitation where there has been degradation. Unfortunately, data are not available to show the extent of such actions. Further measures include the protection of forests through the establishment of protected areas, the extent of which varies from a total of 27 per cent of forest and woodland

---

### Box 2d.7 International cooperation in Central Africa

The Yaoundé Forest Summit, held in Cameroon in March 1999, was a major event in cooperation between Central African countries towards tropical forest conservation. The Summit produced the Yaoundé Declaration, signed by Cameroon, Central African Republic, Chad, Congo, DRC, Equatorial Guinea, and Gabon. The Declaration is a 12-point plan of action for sustaining Central African forests, through commitments such as the creation and extension of protected forest areas, and plans to combat illegal logging and poaching. In December 2000, the heads of state of these nations met again, and plans were put in place to establish a transborder park, between Cameroon, Central African Republic and Congo. The Sangha park, as it will be called, is the fusion of existing protected areas in the three countries, and the production forests and hunting zones that surround them, totalling one million hectares. This is the first park of its kind in Central Africa, and will help to effect harmonization of national forest policies, policing, research and monitoring.

Source: WWF 2000

---

area in Chad to 9 per cent in DRC (FAO 2001a). Regional cooperation in protection of forests has also been strengthened in recent years through the Yaounde Forest Summit and Yaoundé Declaration (Box 2d.7).

With the exception of Chad, all of the countries of Central Africa are members of the ATO, and have developed criteria and indicators for sustainable forest management under its auspices. Chad has developed criteria and indicators through the Dry Zone Africa process. Other sub-regional initiatives include the Central African Regional Environmental Programme (CARPE), a long-term initiative on the part of the USAID which aims to address the issues of deforestation and biodiversity loss in the Congo Basin forest zone. CARPE works with all governments in the sub-region, as well as American private voluntary organizations and appropriate federal agencies. Its five-year pilot programme has been designed to gather and disseminate baseline information on forest resources, and to characterize and prioritize the threats to forests and opportunities for their sustainable management. It also aims to strengthen the capacity of decision makers in sustainable forest management.

Ecosystèmes Forestiers d'Afrique Centrale (ECOFAC) is another sub-regional programme that seeks to reconcile development and conservation of the natural environment, by working closely with local communities and stakeholders. So far, significant work has been channelled into developing botanical inventories and documenting information on forest dynamics (productivity, mortality, regeneration and phenology). These data provide a baseline for understanding of how the ecosystem functions and for implementation of sustainable resource development. ECOFAC has also worked extensively with indigenous groups and has studied their interactions with the environment, as well as pressures on the forest resources exerted by various groups and activities.

Many countries in the region have revised their forest policies, and/or have developed National Environmental Action Plans or Biodiversity Conservation Strategies. However, their success is dependent on the commitment of political will, financial resources, trained staff and equipment to enforce the new regulations and protection measures. Effects of these responses include an increase in the availability of information on sustainable forest management, raised awareness of forest issues, limited reforestation efforts and the extension of protected areas.

## WESTERN AFRICA

Western Africa is characterized by a marked gradation of climate which is reflected in zones of vegetation cover. Dense rain forest and semi-deciduous forests dominate in the coastal belt. Moving northwards, there is a transition from forest to savanna, which eventually gives way to sub-Sahelian savanna in northern Mali, Mauritania and Niger. In total, there were 72 million hectares of forest in 2000, equivalent to almost 12 per cent of the sub-region's land area, and representing 11 per cent of Africa's total forest cover (FAO 2001a).

The Upper Guinea Forest, a strip of tropical moist forest that runs parallel to the coast from Guinea to Cameroon (see Figure 2d.8) is one of the world's 25 biodiversity hotspots, and ranks first in terms of mammalian species diversity (Conservation International 2001). It is estimated that only 20 per cent of the original extent still remains, and this is highly

**Figure 2d.8 Map of Upper Guinea Forest**

_Conservation International_

fragmented (Conservation International 1999). The rarest subspecies of gorilla, the Cross River Gorilla (_Gorilla gorilla diehli_) is found in fragments of this forest in Nigeria and Cameroon. With a remaining population of 200–250 individuals, this species is critically endangered (Butynski 2001). The largest continuous section of the Upper Guinean forest is in Liberia, where civil unrest threatens conservation activities. However, several initiatives are underway in Côte d'Ivoire, Equatorial Guinea, Ghana, Guinea, Liberia, Nigeria, Sierra Leone and Togo to manage the forest and protect endangered species. Forest loss and fragmentation are thus the most important issues in Western Africa.

### ECOLOGICAL, ECONOMIC AND SOCIAL VALUES OF FORESTS AND WOODLANDS IN WESTERN AFRICA

Commercial timber production is an extensive and lucrative occupation in Western Africa, contributing significant proportions of income and foreign exchange. For example, in 1998, Côte d'Ivoire exported US$228 million worth of wood products (mostly sawnwood), Ghana exported US$140 million worth (FAO 2001a). Forest and woodland products are also extremely important to local communities, and the people of Western Africa are highly dependent on forest and savanna resources for their energy needs, most of which are met from wood. In 2000, more than 175 million m³ of wood were used in Western Africa for fuelwood and charcoal production (FAO 2001b). In Gambia, 85 per cent of energy needs are met from wood and charcoal, in Niger, biomass, mainly wood, meets 90 per cent of needs (FAO 2001a). Other resources heavily used by local communities are wildlife (bushmeat), medicinal plants, wood and rattan for construction, furniture and crafts, honey, nuts and fruits, and animal fodder, gums, dyes, teas, spices and aromatics. The most widely used wild food species is the wild yam, a staple in Western African diets. Palms, together with some 35 other species, are important resources for the wine and beer making that contributes significantly to the daily incomes of rural households in Western Africa (African Ethnobotany Network 2000). Palm cabbage, made into porridge, is an important 'hungry season' food during the rainy period immediately before the rice harvest.

## FOREST COVER AND QUALITY IN WESTERN AFRICA

Deforestation, and forest degradation and the associated loss of forest products and environmental services, are serious challenges facing Western African countries. Between 1990 and 2000, a total of 12 million hectares of forest were cleared, 15 per cent of the sub-regional total (FAO 2001a). These high rates of deforestation are largely attributable to governmental concessions on commercial logging. The timber and wood products are mostly exported to earn foreign currencies and contribute a considerable amount to GDP. Forests have also been cleared for agriculture, particularly during the 1970s and 1980s when high incomes could be earned from cash crops such as coffee, cotton and sugar.

National policies for infrastructure development have also contributed to fragmentation of forests. The process starts with the construction of large access roads by logging companies which encourages the rural farming communities to settle deep inside the forest where the soils are most fertile. Other policies have played an indirect role in deforestation over the past 30 years. For example, governments have encouraged the conversion of forest reserves into agricultural lands in order to attract and resettle people from more densely populated and resource deficient areas. In addition, forest reserves have been converted into plantations to satisfy industrial demand for wood (Asibey 1990).

Collection of fuelwood contributes to forest

clearance and degradation through selective harvesting. The use of wood for fuel is likely to increase, according to predictions by the FAO (2001b), due to a rapidly increasing population and demands for energy which cannot be met by alternative sources in the short to medium term.

Current logging methods are destructive, and it has been estimated that to harvest $1m^3$ of log, about $2m^3$ of standing trees are destroyed (Serageldine 1990). This, in turn, significantly modifies the ecosystem by creating gaps in the canopy, exposing shade-adapted species to sunlight and allowing colonization by new species. Removal of vegetation has also increased the potential for soil erosion and run-off. This is especially problematic on steep slopes or in fragile ecosystems such as the mangroves and transitional marginal forests or derived savanna areas that have lower regeneration capacity than humid forests. Furthermore, it takes up to 700 years to restore a fully balanced, mature forest. Natural forests are therefore usually replaced by secondary forests which do not afford the same level of resources and services as primary forests. These secondary forest areas then typically become sources of local firewood collection, are converted to plantations, or used for grazing, activities which further degrade or destroy them.

Forest clearance and degradation are increasing in most countries in Western Africa, because of increasing population pressures for agricultural land and for energy use. Population growth is linked directly to increased demand for food and occupancy of space. However, it is the local communities that often suffer most from forest degradation as they lose vital sources of firewood, construction materials, clothing, pharmaceutical products, food, hunting accessories, cultural and religious apparatus, and grazing land for animals. Civil strife and its repercussions can fuel deforestation by creating needs for shelter and energy among refugees who will attempt to meet such needs from the forest.

Forests, especially dense tropical forests, play a vital role in carbon sequestration and, therefore, in global climate regulation, as well as in regulating local air quality and rainfall patterns. When forest cover is removed not only are these global functions lost but often—because the presence of forest reduces temperatures and evaporation, and limits the negative impacts of wind—a microclimate conducive to the development of animal life and vegetation growth is also lost. Clearing of the forest, therefore, limits the maintenance of a vibrant biological diversity.

## Towards sustainable management and conservation of forests and woodlands in Western Africa

European settlers built the foundation for modern management of forests in almost all of the countries of the sub-region. When these countries gained their independence (about 40 years ago), they inherited forest and wildlife resource management systems institutionalized in the form of state-owned protected areas. Forest in protected areas ranges from 77 per cent in Niger to 1 per cent in Guinea-Bissau and Liberia (FAO 2001a). Conservation International has been largely instrumental in developing and implementing management programmes aimed at conserving the fragmented Upper Guinea Forest hotspot. Activities have included environmental assessments and prioritization of conservation needs, building capacity in natural resources management agencies, employment creation for local communities in ranger activities, and tourism development (Conservation International 2001). Other responses include large-scale reforestation programmes, although these have encountered difficulties in implementation. New approaches built around 'grass-roots projects' have been introduced in some countries, putting more emphasis on community-level enterprises (Compaoré, Issaka and Yacouba 2000). Reintroduction and mainstreaming of indigenous forest conservation practices have also been successful in some parts of Western Africa (see Box 2d.8).

From the 1980s, the sub-region saw the appearance of projects related to the management of natural forests through the preparation and implementation of Forest Management Plans and Strategies. Also, in response to the Tropical Forest Action Plan, many Western African countries have initiated major forestry sector reviews in the past 10 years. Criteria and indicators for sustainable forest management are being developed and implemented in all Western African countries with the exceptions of Benin, Guinea, and Sierra Leone, coordinated by the ATO. Burkina Faso, Côte d'Ivoire and Togo have some of

### Box 2d.8  Indigenous forestry practices of farmers in Sierra Leone

In Sierra Leone, farmers normally leave tree cover at a distance of 3–5m from footpaths in order to maintain a continuum of shade for those using the paths during the peak of the hot season from March to April. Also, the clearing of a farm site normally excludes forest fringe vegetation in order to encourage wildlife and fish reproduction, and also to define a boundary between common property and farmers' fields. These practices could form the basis for the use of traditional knowledge in forestry in development modern strategies for conservation.

*Source: E.K Alieu, Director of Forests Ministry of Agriculture, Forestry and Marine Resources, Sierra Leone*

their forests under management plans (FAO 2001a). In October 2000, the ATO announced its intentions to coordinate a common system of certification for timber and forest products, to ensure customers that the timber has come from sustainably managed forests.

Policy measures to reduce the pressures on forests and woodlands from fuelwood collection and charcoal production include the development and expansion of energy generation from renewable resources (such as hydropower and solar power), as well as through centralized power generation from fossil fuels using cleaner technologies. Under the Kyoto Protocol, developing countries in Africa will receive funds for development and adoption of cleaner technologies as well as for tree-planting and afforestation programmes.

## CONCLUSION

Approximately 22 per cent of the African region is covered with forests, ranging from open savanna to closed tropical rainforest. Forests provide a great many goods and services which benefit local communities and national economies, as well as providing international environmental benefits. Commercial forest products include timber for construction and paper, but forest resources provide much more to local communities including food, construction materials, grazing areas for livestock, cultural and medicinal products, sites for religious practices and leisure activities, and fuel for cooking, heating and lighting. Forests protect and stabilize the soil, recycle nutrients to maintain soil quality and regulate water quality and flow. Forests are also vast sinks of $CO_2$, and thus play a critical role in mitigating global climate change, the impacts of which are predicted to be most severe for African countries and other developing nations. By protecting soils and by regulating temperatures, rainfall and hydrological systems, forests provide basic support systems for agriculture and industry and, therefore, for the economies of African nations.

Natural forests and woodlands in Africa have been drastically reduced in size over the past century but particularly so since independence, as countries have struggled to improve their economies through exploitation of natural resources. Deforestation for commercial timber sales and clearance for agricultural and urban developments are the most intensive pressures, as well as overharvesting of wood for fuel, medicinal products, and construction materials. Remaining forests have also been degraded as a result of clear felling, fires, selective harvesting, and encroachment. Impacts of this degradation include losses of biodiversity, radically increased rates of soil erosion, reductions in water quality and increased risk of flooding in surrounding areas, and loss of livelihoods for local communities. The global community has also lost potential pharmaceutical and food products.

Pressures remain, even though many countries in Africa have protected forest reserves, initiated reforestation and afforestation programmes, and developed policies and programmes for sustainable forest management. And those pressures are set to increase with increasing populations and demand for forest resources. The developing countries are struggling to commit the necessary financial and human resources to their efforts in order to be able to turn away much needed foreign exchange proposed in return for further logging concessions. If they are to succeed, the true value of the variety of forest resources and services on which local communities and the global community depend needs to be made explicit and either incorporated into the price of forest products or traded on the open market. Economic and social development priorities need to be integrated into forest conservation measures, so that local communities can share in the management of the resource and in the benefits of trade in their products. This requires further

understanding of the types of forests, the complexity of the issues, and establishment of clear objectives and commitment of resources to implementation of sustainable development policies. Essential elements that must be taken into account include recognition of the intrinsic features of the local physical environment, impacts of climate change, deforestation factors, skills in forest management, scarcity of forest resources, deep-rooted traditions and human impacts, economic forces and political events. There must also be concerted efforts to provide alternative sources of energy, more efficient forms of energy utilization (such as mudstoves) and to provide income to people relying on natural forests. The major solutions for forest problems in the region include reforestation of the original and more prosperous areas and afforestation of multipurpose forests that can be used for grazing, wood production and other traditional uses. Establishment of multipurpose forests will diversify outputs, an important asset to avoid overuse of one-purpose forest.

## REFERENCES

African Development Bank (2001). *Gender, Poverty and Environmental Indicators on African Countries 2001–2002*. African Development Bank, Adbijan

African Ethnobotany Network (2000). Review of Ethnobotanical Literature for West and Central Africa. Bulletin No. 2, August 2000

Alden Wily, L.A. (2000). Democratising the Commonage: The Changing Legal Framework for Natural Resource Management in Eastern and Southern Africa with Particular Reference to Forests. Paper presented at the 2nd CASS/PLAAS Regional Meeting, 16–17 October 2000. Workshop theme: Legal aspects of Governance in CBNRM. University of the Western Cape, South Africa

Alden Wily and Mbaya, S. (2001). Land, People and Forests in Eastern and Southern Africa at the Beginning of the 21st Century. The Impact of Land Relations on the Role of Communities in Forest Future. Nairobi, IUCN –EARO

AOAD (1998). Study on Suitable Modern Technologies for Forest Wealth Development in the Arab Region. AOAD, Khartoum

ARCOS (2000). Albertine Rift Conservation Society: Special Focus Albertine Rift Montane Forests. Available on: http://www.africanconservation.com/arcos.html

Ashley, C. and LaFranchi, C. (1997). Livelihood Strategies of Rural Households in Caprivi: Implications for Conservancies and Natural Resource Management. DEA Research Discussion Paper 20. Windhoek: DEA

Asibey, E.O.A. (1990). Development of Private Forest Plantation to Take Pressure off Natural Forest in Sub-Saharan Africa. AFTEN World Bank, Washington D.C.

BSP (1992). *Central Africa: Global Climate Change and Development— Synopsis.* Biodiversity Support Program – WWF, The Nature Conservancy and WRI. Corporate Press, Landover, Maryland

Butynski, T.M. (2001). Africa's Great Apes. In: *Great Apes and Humans: The Ethics of Coexistence.* Beck, B., Stoinski, T., Hutchins, M., Maple, T.L., Norton, B.G., Rowan, A., Stevens, E.F. and Arluke, A. (eds). Simithonian Institution Press, Washington D.C.

CAMRE/UNEP/ACSAD (1996). *State of Desertification in the Arab Region and the Ways and Means to Deal With It.* ACSAD Publications, Damscus

Cavendish, W. (1999). Empirical Regularities in the Poverty-Environment Relationship of African Households. WPS 99–21

Chenje, M. (ed. 2000). *State of the Environment Zambezi Basin 2000,* SADC/IUCN/ ZRA/SARDC, Maseru/Lusaka/Harare

Compaoré A., Issaka, M. and Yacouba, M. (2000). *Synthesis of the Regional Workshop on Natural Resources Management: Evolution and Perspectives.* April 2000, FRAME-ing the Future of Sub-Saharan Africa's Natural Resources, Frame Contact Group meeting, 2–5 May 2000, Saly Portugal

COMPASS (Community Partnerships for Sustainable Resources Management in Malawi) (2000). COMPASS Success Story: Agroforestry and Extension Capacity Building Training for Forestry Patrolmen (Forestry Extension Agents). June 2000. Available on: http://www.tamis.dai.com

Conservation International (2001). West Africa: The Guinean Forest Hotspot. Available on: http://www.conservation.org Conservation Regions: Africa

Conservation International (1999). *Conservation Priority-Setting for the Upper Guinea Forest Ecosystem, West Africa.* Conservation International, Washington D.C.

Cross River State Forestry Project (CRSFP) (1994). Technical Report of the Overseas Development Administration, UK

Cunningham, A.B. (2001). Ecological Footprint of the Wooden Rhino: Depletion of Hardwoods for the Carving Trade in Kenya. *People and Plants Online.* Available on: http://www.rbgkew.org.uk/peopleplants/ regions/kenya/ hardwood.htm

Cunningham, A.B. (1997). Review of Ethnobotanical Literature from Eastern and Southern Africa *The African Ethnobotany Network Bulletin* No.1, November 1997

EIA (1999). *Energy in Africa.* U.S. Department of Energy- Energy Information Administration, Washington D.C.

EIS News (1999). *Ethiopia: Extent and Dynamics of Deforestation in Ethiopia.* Advisory Assistance to the Forest Administration Study by Matthias Reusing, Addis Ababa

Emerton, L. (1997). *An Environmental Economic Assessment of Mount Kenya Forest.* Report prepared for EU by the African Wildlife Foundation, Nairobi

Emerton, L. (1993). *The Value of Kenya's Indigenous Forests to Adjacent Households.* Forest Department, Nairobi

Emerton, L. (1992). *A District Profile of Kenya's Gazetted Indigenous Forests.* KIFCON/Forest Department. Ministry of Environment and Natural Resources. Nairobi

Emerton, L. and Muramir, E.T. (1999). *Uganda Biodiversity: Economic Assessment.* A report prepared for the National Environment Management Authority as part of the Uganda National Strategy, Biodiversity Strategy and Action Plan. Kampala

EPA/MEDC (1997). *The Conservation Strategy of Ethiopia – Volume 1 The Resources Base, Its Utilisation and Planning for Sustainability.* Environmental Protection Authority in collaboration with the Ministry of Economic Development and Cooperation. Federal Democratic Republic of Ethiopia, Addis Ababa

FAO (2001a). *Forest Resources Assessment 2000.* FAO, Rome

FAO (2001b). Broadhead, J., Bahdon, J. and Whiteman, A. (in prep). *Past Trends and Future Prospects for the Utilization of Wood for Energy.* Global Forest Products Outlook Study Working Paper GFPOS/WP/05 FAO, Rome

FAO (1999). *State of the World's Forests 1999.* FAO, Rome

FAO (1997). FAO *Provisional Outlook for Global Forest Products Consumption, Production and Trade to 2010.* Forestry Policy and Planning Division, Forestry Department, FAO, Rome

FAO (1996). *Forestry Policies of Selected Countries in Africa.* FAO, Rome

FAO (1995). *Non-Wood Forest Products For Rural Income And Sustainable Forestry.* FAO, Rome

FAO (1993). *Forestry Policies in the Near East Region: Analysis and Synthesis.* FAO, Rome

FAO (1990). *The Major Significance of 'Minor' Forest Products: The local use and value of forests in the west African humid zones.* FAO, Rome

Ganzhorn, J.U., Fietz, J., Rakotovao, E., Schwab, D. and Zinner, D. (1999). Lemurs and the Regeneration of Dry Deciduous Forest in Madagascar. *Conservation Biology* Vol. 13 No. 4 794–804

Gilani, A. (1997). Soil Degradation and Desertification in Arabic Countries. *Journal of Water and Agriculture,* 17: 28–55

Hegazy, A.K. (1999). Deserts of the Middle East. In M. A. Mares (ed.), *Encyclopaedia of Deserts.* Norman: University of Oklahoma Press

Iddi, S. (1998). Eastern Arc Mountains and their National and Global Importance. *Journal of East African Natural History* 87: 19–26 (1998)

IISD (1999). A Summary Report of the Underlying Causes of Deforestation and Forest Degradation. Sustainable Developments 21(1), January 1999

IPCC (2000). *Land Use, Land-use Change, and Forestry.* Intergovernmental Panel on Climate Change, Cambridge University Press, Cambridge

IPCC (1998). *The Regional Impacts of Climate Change—An Assessment of Vulnerability*. R.T. Watson, MC. Zinyowera and R.H. Moss (eds). IPCC, Geneva

IUCN (2001). The North Africa Programme. Available on: http://iucn.org/places/wcana/Egypt.htm

Kalumiana, O.S. (1998). *Woodfuel Sub-Programme of the Zambia Forestry Action Programme*. Ministry of Environment and Natural Resources, Lusaka

Kremen, C., Niles, J.O., Dalton, M.G., Daily, G.C., Ehrlich, P.R., Fay, J.P., Grewal, D. and Guillery, R.P. (2000). Economic Incentives for Rain Forest Conservation Across Scales. *Science*, June 9 2000; 1828—2832

Kufakwandi, S. (2001). *Consortium Funding for Sustainable Forestry Management: African Perspectives and Priorities*. African Development Bank, Abidjan

Lamhamedi, M.S., Ammari, Y., Fecteau, B., Fortin, J.A. and Margolis, H. (2000). Development Strategies for Forest Tree Nurseries in North Africa. *Cahiers d'études et de recherches francophones / Agricultures*. Vol. 9, Issue 5, September—October 2000

Laurance, W.F. (1998). A Crisis in the Making: Responses of Amazonian Forests to Land Use and Climate Change *Trends in Ecology and Evolution* Vol. 13 No. 10 pp. 411—415

Matthews, E. (2001). Understanding the FRA 2000, Forest Briefing No. 1, World Resources Institute. Available on: http://www.wri.org

Matthews, E., Payne, R., Rohweder, M. and Murray, S. (2000). Pilot Analysis of Global Ecosystems: Forest Ecosystems. World Resources Institute, Washington D.C.

Medley, K.E. and Hughes, F. M.R. (1996). Riverine Forests. In: McClanahan, T.R. and T.P. Young (eds) *East African Ecosystems and their Conservation* , 1996 Oxford University Press, Oxford

Mittermeier, R. A., Myers, N., Gil, P.R. and Mittermeier, C.G. (2000). *Hotspots; The Earth's Biologically Richest and Most Endangered Terrestrial Ecoregions*. CEMEX and Conservation International, Washington

MUIENR (2000). Uganda Biodiversity Report 2000. Makerere University Institute of Environment and Natural Resources. Kampala

Mwihomeke, S.T., Msangi, T.H., Mabula, C.K., Ylhaisi, J. and Mndeme, K.C.H. (1998). Traditionally Protected Forests and Nature Conservatioin in the North Pare Mountains and Handeni District, Tanzania. *Journal of East African Natural History*, Vol 87, 279—290, (1998)

NEIC (1994). *State of Environment Report for Uganda 1994*. National Environment Information Centre. Ministry of Natural Resources. Kampala, Uganda

NEMA (1999). *State of Environment Report for Uganda 1998*. National Environment Management Authority. Kampala

Ole Lengisugi, N.A.M.O and Mziray, W.R. (1996). The Role of Indigenous Knowledge in Sustainable Ecology and Ethnobotanical Practices among Pastoral Maasai: Olkonerei—Simanjiro Experience. Paper Presented at the 5th International Congress of Ethnobiology at the Kenyatta International Conference Centre, Nairobi, Kenya, 2—6 September 1996

Park, C. C. (1992). Tropical Rainforests. Routledge, London and New York

Rodgers, A., Salehe, J. and Olson, J. (2000). Woodland and Tree Resources on Public Land in Tanzania: National Policies and Sustainable Use. In: A.B. Temu, G. Lund, R.E. Malimbwi, G.S. Kowero, K. Klein, Y. Malende, and I.Kone, eds. 2000 *Off-forest Tree Resources of Africa*. A proceedings of a workshop held at Arusha, Tanzania, 1999. The African Academy of Sciences (AAS)

Rodgers, W.A. (1998). An Introduction to the Conservation of the Eastern Arc Mountains. *Journal of East African Natural History* 87: 1—7 (1998) Special Issue: Eastern Arc Mountains

Serageldine, I. (1990). La Protection des Forêts Tropicales Ombrophiles de l'Afrique, Conference sur la Conservation des Forêts Tropicales Ombrophiles de l'Afrique Occidentale et Centrale, 5—9 Septembre 1990

Shackleton, C.M., Netshiluvhi, T.R., Shackleton, S.E., Geach, B.S., Ballance, A. and Fairbanks, D.H.K. (1999). Direct use Values of Woodland Resources from Three Rural Villages. Unpubl. Report No ENV-P-I 98210. Pretoria: CSIR

Shackleton, S.E., Shackleton, C.M. and Cousins, B. (2000). The Economic Value of Land and Natural Resources to Rural Livelihoods: Case Studies from South Africa. In Cousins, B., (ed) *At the Crossroads: Land and Agrarian Reform in South Africa into the 21st Century*. NLC and PLAAS. Cape Town, University of the Western Cape

Thirgood, J.V. (1981). *Man and the Mediterranean Forest: A History of Resource Depletion*. Academic Press., New York

Turner, I.M. and Corlett, R. T. (1996). The Conservation Value of Small, Isolated Fragments of Lowland Tropical Rain Forest *Trends in Ecology and Evolution* Vol. 11 No. 8 pp 330—333

UFSCS (2000). *Uganda National Forestry Policy*. Uganda Forestry Sector Coordination Secretariat. Kampala

UNDP (2000). *Human Development Report 2000* Oxford University Press, Oxford

UNDP, UNEP, World Bank and WRI (2000). World Resources 2000—2001: *People and Ecosystems: The Fraying Web of Life*, World Resources Institute, Washington D.C.

UNEP (1999). *Western Indian Ocean Environment Outlook*. United Nations Environment Programme, Nairobi

UNEP (2001). *An Assessment of the Status of the World's Remaining Closed Forests*, Early Warning and Assessment Technical Report. United Nations Environment Programme, Nairobi

UNESCO/UCO (1998). *Multipurpose Species in Arab African Countries*. UNESCO, Cairo

Van Wyk, B-E., Van Oudtshoorn, B. and Gericke, N. (1997). *Medicinal Plants of South Africa*. Briza Publications, Pretoria.

Verolme, H.J.H. and Moussa, J. (1999). Addressing the Underlying Causes of Deforestation and Forest Degradation—Case Studies, Analysis and Policy Recommendations. Biodiversity Action Network, Washington, D.C.

Wass, P. (1995). Kenya's Indigenous Forests: Status, Management and Conservation. P. Wass, Editor. IUCN Forest Conservation Programme. IUCN, Gland and ODA, Cambridge

WCMC (1992). *Global Biodiversity: Status of the Earth's Living Resources*. A report compiled by the World Conservation and Monitoring Centre, B. Groombridge (ed.), Chapman and Hall, London

WCMC (2000). *Global Biodiversity: Earth's Living Resources in the 21st Century*. Groombridge, B. and Jenkins, M.D., World Conservation Press, Cambridge

WCMC (1994). *Biodiversity Source Book*. World Conservation Press, Cambridge

Western, D. (1997). *In the Dust of Kilimanjaro* Island Press, Washington D.C.

White, F. (1983). The Vegetation of Africa: a Descriptive Memoir to Accompany the UNESCO/AETFAT/UNISO Vegetation Map of Africa. *Natural Resources Research* 20. UNESCO, Paris

World Bank (2001). *African Development Indicators*. The World Bank, Washington D.C.

World Bank (1999). World Development Indicators Database. Available on: http://www.worldbank.org

World Bank (1996), Development in Practice, Toward Environmentally Sustainable Development in Sub-Saharan Africa, A World Bank Agenda, World Bank Publication, World Bank, Washington, D.C.

WRI (2001). Global Forest Watch: Cameroon's Forests. Available on: http://www.globalforestwatch.org

WWF (2000). The Yaoundé Forest Summit Update. Available on: http://www.wwfcameroon.org/forestsummit/summit.htm

**PART E: FRESHWATER**

UNEP

## REGIONAL OVERVIEW

### AVAILABILITY OF FRESHWATER

Freshwater availability is one of the most critical factors in development, particularly in Africa. Some 71 per cent of the Earth's surface is water. However, less than 3 per cent is freshwater, and most of that is either in the form of ice and snow in the polar regions, or in underground aquifers.

Africa's share of global freshwater resources is about 9 per cent, or 4 050 km³/yr (Shiklomanov 1999, UNDP, UNEP, World Bank and WRI 2000). These freshwater resources are distributed unevenly across Africa, with Western Africa and Central Africa having significantly greater precipitation than Northern Africa, the Horn of Africa and Southern Africa. The wettest country, the Democratic Republic of Congo (DRC), has nearly 25 per cent of average annual internal renewable water resources in Africa, with 935 km³/yr. By contrast, the driest country, Mauritania, has just 0.4 km³/yr, or 0.01 per cent of Africa's total (UNDP and others 2000). Average water availability per person in Africa is 5 720 m³/capita/year compared to a global average of 7 600 m³/capita/year, but there are large disparities between sub-regions, as shown in Figure 2e.1

**Figure 2e.1 Annual renewable water availability per capita (1995)**

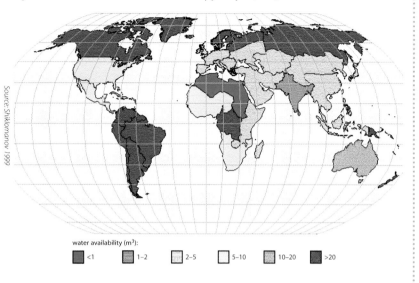

*Source: Shiklomanov 1999*

water availability (m³):
<1  1–2  2–5  5–10  10–20  >20

Problems with freshwater availability in Africa are further complicated by highly variable levels of rainfall. As a result, large numbers of people are dependent on groundwater as their primary source of freshwater. In Algeria, for example, more than 60 per cent of all withdrawals are from groundwater and, in Libya, 95 per cent of all withdrawals are from groundwater (UNDP and others 2000). Some countries, including Algeria, Egypt, Libya, Mauritius0, Morocco, South Africa and Tunisia, make use of desalinated water to assist in meeting their withdrawal requirements (UNDP and others 2000, GOM/ERM 1998).

Most African countries also experience extremes of rainfall (periodic flooding or drought). There is some evidence that both droughts and floods have increased in frequency and severity over the past 30 years (OFDA 2000). In particular, the Sahelian zone has experienced a continued decline in rainfall compared to pre-1960s averages, and Lake Chad has shrunk to 5 per cent of its size 35 years ago (NASA Global Earth Observing System 2001). Severe droughts were experienced in 1973 and 1984, when almost all African countries suffered reduced rainfall, and several million people in the Horn of Africa and the Sahelian zone, and in Southern Africa, were affected.

The freshwater lakes of Africa have a total volume of 30 567 km³, covering a surface area of 165 581 km². Not only are they important in water flow regulation,

flooding control and water storage, but they are also important for meeting human needs. Lake Tanganyika alone could supply water to 400 million people through the annual extraction of less than 1 per cent of its volume (Khroda 1996). With the exception of Lake Tana in Ethiopia, all African lakes are shared across international borders, which makes international cooperation a necessary condition for equitable use and development of lake resources. Wetlands cover about 1 per cent of Africa's total surface area, and are found in almost every country (WCMC 1992). The largest wetlands include: the Zaire swamps; the Sudd in the upper Nile River; the Lake Victoria basin; the Chad basin; the Okavango delta; the Bangweulu swamps; and the floodplains and deltas of the Niger and Zambezi Rivers.

Africa has the highest rate of population growth in the world, and is also one of the regions that is most vulnerable to climate change. The Intergovernmental Panel on Climate Change (IPCC) predicts that average run-off and water availability will decline in the countries of Northern and Southern Africa, impacting on freshwater ecosystems, and advancing desertification in the Sahelian zone and in Northern Africa (IPCC 2001). In addition, increased frequency of flooding and drought will also stress freshwater systems and pressurize water supply networks. As a result, 25 African countries are expected to experience water scarcity or water stress over the next 20–30 years, as shown in Figure 2e.2 (UNEP 1999a).

All the above factors combine to create enormous challenges for water storage, supply and distribution, as well as for water treatment (purification and wastewater treatment). Despite considerable efforts to develop water storage and infrastructure, particularly

### Box 2e.1 Africa's Water Vision

Water is a precious resource. In Africa, it can be a matter of life and death. It can also be a matter of economic survival. Yet it can be both an instrument and a limiting factor in poverty alleviation and economic recovery, lifting people out of the degradation of having to live without access to safe water and sanitation, while at the same time bringing prosperity to all in the region. A radical change in approach is required if water is not to become a constraint, but an instrument, to socio-economic development in Africa.

*Source: World Water Council 2000*

**Figure 2e.2 Countries expected to experience water stress or scarcity in 2025**

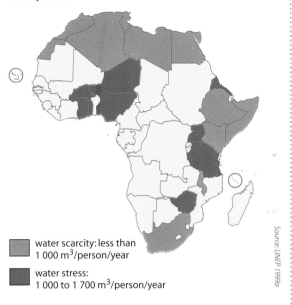

Source: UNEP 1999a

■ water scarcity: less than
1 000 m³/person/year

■ water stress:
1 000 to 1 700 m³/person/year

in the past 30 years, large numbers of people remain without access to water for domestic use, and many farmers do not have access to water for irrigation. By contrast, some industrial, agricultural and domestic users have access to subsidized water supplies, and have no incentive to use water carefully, or to reuse or recycle water. Access to water resources is thus a priority issue for the countries of Africa, together with rising concerns over water quality, due to excessive water withdrawal and declining availability, and pollution from a variety of sources.

## ACCESS TO FRESHWATER RESOURCES

Water stress (less than 1 700 m³/capita/yr) or water scarcity (less than 1 000m³/capita/yr) is already observed in 14 of the 53 African countries (WRI 2000). The high demand for water is driving unsustainable practices, and competition for water resources between sectors, communities, and nations. On occasion, it has been the cause of strained relations and hostility. For example, in Libya, annual withdrawals of groundwater are more than 500 times the rate of replenishment (UNDP and others 2000). In Egypt, 90 per cent of all freshwater resources are derived from the Nile River and, because this is a shared watercourse with nine

other nations, securing access and usage rights has been a contentious process.

With such limited, variable or unevenly distributed freshwater resources in Africa, it is not surprising that access to water is a major factor in social and economic development. Over the past 30 years, the response to this has been to dam or to modify large rivers in order to provide water for agricultural, domestic and industrial use, as well as to supply hydropower (see Box 2e.2). Low investment in water supply and infrastructure maintenance, increasing demand from all sectors and inequitable access policies have added further strain to the situation, resulting in highly skewed access to water resources.

As shown in Figure 2e.3, the major user of freshwater resources is the agricultural sector, which accounted for 63 per cent of all withdrawals in Africa in 1995 (Shiklomanov 1999), and which irrigates more than 12 million hectares (ha) of farmland (FAOSTAT 2001). The more arid countries of Northern and Southern Africa are more dependent on irrigation than those in Western and Central Africa, however and, thus, securing water for irrigation is a high priority for economic

*• Low investment in water supply and infrastructure maintenance, increasing demand from all sectors and inequitable access policies have added further strain to the situation, resulting in highly skewed access to water resources •*

### Box 2e.2  Large dams in Africa

There are more than 1 200 dams in Africa, more than 60 per cent of which are located in South Africa (539) and Zimbabwe (213). Most of these were constructed during the past 30 years, coincident with rising demands for water from growing populations. The overwhelming majority of dams in Africa have been constructed to facilitate irrigation (52 per cent) and to supply water to municipalities (20 per cent), although almost 20 per cent of dams are multipurpose. Although only 6 per cent of dams were built primarily for electricity generation, hydroelectric power accounts for more than 80 per cent of total power generation in 18 African countries, and for more than 50 per cent in 25 countries. Only 1 per cent of African dams have been constructed to provide flooding control.

Besides providing these benefits and services, however, dams have had several negative impacts, including: large-scale displacement of people; altered patterns of erosion and flooding; loss of land, due to flooding; loss of income from downstream fisheries; and changes to sedimentation rates. These ecological and social concerns—together with additional stresses on water resources, which are expected in many parts of Africa as a result of increasing population pressures and global climate change—are changing attitudes towards large dams. The construction of smaller dams and the development of micro-hydropower facilities are being investigated as more sustainable means of supplying water and power.

Source: World Commission on Dams 2001

**Figure 2e.3  Water use by sector in Africa 1990–2025**

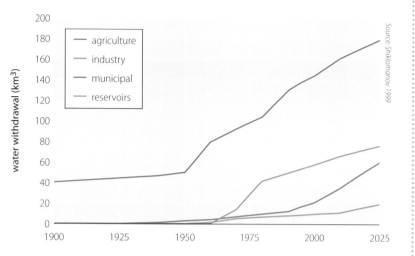

*Source Shiklomanov 1999*

Africa (Hinrichsen, Robey and Upadhyay 1997). Per capita usage in Africa (on average 47 litres/person/day) is also far below that of other countries (85 litres/person/day in Asia, 334 litres/person/day in the UK and 578 litres/person/day in the US) (Hinrichsen and others 1997).

There are large differences in access to water between countries, between urban and rural areas, and between population groups. Inadequate water supply and sanitation, particularly in the cramped living conditions of urban informal settlements, increases the risk of waterborne diseases and infections. Diarrhoeal infections are particularly prevalent amongst children, and constitute a major cause of preventable deaths.

Consumption of water consumption also varies, as well as access to water. Typically, the greater the access to water, the greater the consumption. By contrast, people who spend a large proportion of their day fetching and carrying water are most conservative with its use. A study in Eastern Africa, for example, showed that urban dwellers use more water than rural households, and households with piped water use more than three times the amount of water used by households without piped water (IIED 2000). In South Africa, extreme disparities exist in water consumption between different population groups, with residents of urban informal settlements using less than 50

development and stability. However, few countries can afford the financial investment in efficient irrigation systems, and water losses through leaking pipes and evaporation are as high as 50 per cent in South Africa alone (Global Water Partnership 2000). Notable exceptions are Mauritius and Northern African countries, where drip irrigation systems are in place and faulty pipes are being replaced. With increasing population and consumption patterns, the demand for food is increasing across Africa, and freshwater withdrawals for agriculture are predicted to rise by more than 30 per cent over the next 20 years (Shiklomanov 1999).

Compared to the agricultural sector, the domestic sector in Africa uses little water. However, domestic use has also increased over the past 30 years and is set to continue this trend (see Figure 2e.3). In 1950, the domestic sector was responsible for less than 3 per cent of all water withdrawals, but this had risen to 4.4 per cent by 1995, and is predicted to account for 6 per cent of all withdrawals by 2025 (Shiklomanov 1999). It is important to note, however, that only 62 per cent of the African population received piped water from municipal sources in 2000 (WHO/UNICEF 2000). Therefore, actual withdrawals for domestic consumption are higher than the estimated percentage from municipal withdrawals. Compared to domestic consumption in other regions of the world, Africa's domestic sector is a moderate user of water. For example, in Europe, the domestic sector accounts for 13 per cent of all withdrawals, more than twice that in

**Figure 2e.4  Water supply coverage in Africa in 2000**

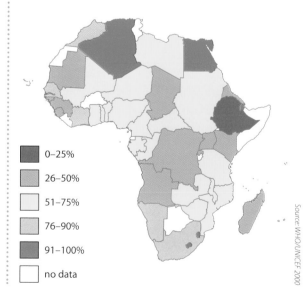

- 0–25%
- 26–50%
- 51–75%
- 76–90%
- 91–100%
- no data

*Source WHO/UNICEF 2000*

litres/person/day, and residents of middle-upper class suburbs consuming 750 litres/person/day—thirteen times more (Napier 2000). Losses from domestic water distribution systems also account for significant wastage, for example, up to 50 per cent of domestic consumption in the case of Mauritius (Government of Mauritius 1994).

## Improving access to freshwater resources in Africa

One of several international efforts to improve access to drinking water was the United Nations International Drinking Water Supply and Sanitation Decade (1981–90). This initiative was approved at the United Nations Water Conference in Argentina in 1977, following recommendations that governments 'develop national plans and programmes for community water supply and sanitation' and that 'national development policies and plans should give priority to the supply of drinking water for the entire population and to the final disposal of wastewater' (United Nations Mar del Plata Plan of Action 1977). However, population growth and the proliferation of unplanned settlements in Africa negated much of the progress and, by 1990, only marginal improvements had been made (WHO/UNICEF 2000). In the following decade (1991–2000), some of these gains had been lost (WHO/UNICEF 2000). The World Health Organization's (WHO) Africa Regional Office developed an Africa 2000 initiative in 1994. This is an international effort to expand water supply and sanitation services in Africa, through changing attitudes towards water and sanitation development. At the second regional consultation of African Health Ministers in 1998, recommendations were made for expanding the Africa 2000 initiative, on the strength of results achieved so far.

In the coming decades, major revisions to water policies and pricing in Africa are required. In the semi-arid countries of Northern and Southern Africa, these are immediate priorities, if the additional stresses of climate change and population growth are to be offset. Until recently, there have been few attempts at demand management. On the contrary, water sector policies have focused on meeting increasing demand, as a means of encouraging economic and social development. However, some countries have revised their water policies and pricing mechanisms, as

Village on the Nile River

*Photodisc Inc.*

measures to manage demand and to encourage more conservative water use. Recycling of wastewater as irrigation water, and upgrading of reticulation networks, are other vital measures which are in place in some countries. Africa's Water Vision—developed through consultative processes in 1999 and 2000, and presented at the 2nd World Water Forum in The Hague in 2000—stresses the need to change attitudes towards water supply and use, and proposes a framework for building on these achievements to date. The ten key elements of Africa's Water Vision are as follows:

- There is sustainable access to safe and adequate water supply and sanitation in order to meet the basic needs of all.
- There is sufficient water for food and energy security.
- Water for sustaining ecosystems and biodiversity is adequate in both quantity and quality.
- Water resources institutions have been reformed, in order to create an enabling environment for the effective and integrated management of water in national and transboundary water basins, including management at the lowest appropriate level.
- basins serve as a basis for regional cooperation and development, and are treated as natural assets for all within such basins.
- There is an adequate number of motivated and highly skilled water professionals.

- There is an effective and financially sustainable system for data collection, assessment and dissemination for national and transboundary water basins.
- There are effective and sustainable strategies for addressing natural and human-made water resources problems, including climate variability and change.
- Water is financed and priced to promote equity, efficiency, and sustainability.
- There is political will, public awareness and commitment among all for sustainable water resources management, including the mainstreaming of gender issues and youth concerns, and the use of participatory approaches (World Water Council 2000).

The proposed mechanism for attaining Africa's Water Vision is comprised of four key components, namely:

- strengthening governance of water resources;
- improving our understanding of water-related issues;

- meeting urgent water needs; and
- providing financial resources for the development and management of water resources in the future.

Africa's Water Vision stresses the need for enhanced regional cooperation, and for a new model for water resources management (World Water Council 2000).

The significant movement towards Integrated Water Resources Management (IWRM) in several countries is, in part, a reflection of the thinking consolidated in Africa's Water Vision. The principles of IWRM include the management of water resources at the basin level, rather than within politically defined boundaries, and with all stakeholder groups having roles and responsibilities in managing the resource. This integrated approach also ensures that social, economic, environmental and technical considerations are taken into account in the management and development of water resources.

Public-private partnerships in water resources management and water supply programmes are also gaining popularity in many countries, as a means for ensuring a sustained supply and development of infrastructure. Such partnerships have been successful in providing services to large urban centres, as well as to districts and rural communities (Sharma 1996). International agreements and protocols have also been established, either as proactive measures or in response to escalating conflict over shared watercourses. The Nile Basin Initiative (NBI), the Regional Programme for the Sustainable Development of the Nubian Sandstone Aquifer (NSA), and the Southern African Development Community (SADC) Protocol on Shared Water Courses are successful examples of transboundary cooperation for the sustainable use and development of water resources.

Hydrological Cycle Observing Systems (HYCOS) projects are in operation in the Mediterranean zone, and in the Southern Africa, Western Africa and Central Africa sub-regions. Similar systems are being established for the Inter-Governmental Authority on Development (IGAD), Congo Basin and Nile Basin areas. These systems have been established primarily as monitoring and assessment stations, building capacity in data collection and information dissemination. They monitor hydrological conditions, and have provided valuable information, such as early warning of rainfall fluctuations and extreme events (WMO 2001).

Bella nomad girls collect water in Burkina Faso

*Mark Edwards/Still Pictures*

## WATER QUALITY

Not only is the availability of freshwater a major constraint to development but, in all sub-regions of Africa, the quality of freshwater is also raising concerns. Some of the manifestations of poor water quality are: eutrophication in lakes and dams; contamination of groundwater with nitrates and salts; and loss of aquatic habitats and biodiversity.

Eutrophication is the proliferation of algae due to raised levels of nutrients in the water body. Poor agricultural practices, and the unregulated discharge of untreated or partially treated sewage, are contributing factors. Algal blooms change the composition and functioning of the natural biota. They can be hazardous to human health, they make the environment less attractive for recreation and sport, and they increase the cost of water treatment. Raised levels of nutrients in water bodies can also encourage the spread of water weeds, and many lakes and wetlands—including the Caprivi wetland system (Namibia), Lake Victoria (East Africa) and Lake Kariba (Zambia/Zimbabwe)—are experiencing problems of infestation by the water hyacinth, *Eichornia crassipes*. This weed forms dense mats that block water channels, disrupting flow patterns and presenting a hazard for boat traffic. Decaying mats of the weed generate bad odours and lead to eutrophication of the water body. Affected countries have initiated biological and chemical control programmes, in addition to mechanical clearance, with some success to date (Global Water Partnership 2000).

Groundwater pollution is a particular concern in the more arid countries, which are highly dependent on underground aquifers for drinking water supplies. The main pollutants are nitrates and phosphates, together with chemical residues from seepage of agricultural runoff, and from inappropriate discharge of industrial and domestic wastewater. In low-lying areas, salinization of groundwater aquifers is a further concern, through the intrusion of saltwater and seepage of irrigation water. Expected sea level rise under climate change scenarios will cause further widespread contamination of aquifers. Therefore, it is critical that action is taken now, in order to ensure the sustained protection and sustainable use of underground water resources.

Freshwater lakes, wetlands and dams are also experiencing loss of biodiversity, as a result of industrial pollution and contamination of water sources by acid-mine drainage. Lake Chivero in Zimbabwe, for example, experienced massive fish deaths in 1970, 1991 and 1996, due to high levels of ammonia and heavy metals, and low oxygen levels (Gumbo 1997). The Korle Lagoon in Ghana has been severely degraded, due to nearby development and associated effluents, dumping of solid waste, and run-off from surrounding areas, such that almost all wildlife has disappeared (Accra Sustainable Programme 2001). These threats to wetlands, and the requirements of the 1971 Ramsar Convention on Wetlands of International Importance have encouraged many countries to develop Wetlands Conservation Strategies, including Ghana, South Africa and Uganda.

Contamination of freshwater habitats, such as lagoons, wetlands, lakes and rivers, not only threatens ecological health, but the health of humans as well. Declining water quality results in an increased risk of water-related disease outbreaks. Diseases such as diarrhoea, ascariasis, dracunculosis, hookworm, schistosomiasis and trachoma are on the increase, particularly in urban informal settlements, where levels of water supply and sanitation provision are low. Lack of water of sufficient quality also reduces agricultural sustainability and potential output which, in turn, force increased importation of foodstuffs and agricultural products. Similarly, poor water quality limits economic development options, such as water-intensive industries and tourism.

### Improving water quality in Africa

Measures to control water quality have been implemented in many countries, including: establishment and enforcement of potable water and wastewater standards; and rehabilitation of existing wastewater treatment facilities. The Polluter Pays Principle has been adopted in many policies and legislation, but enforcement has been often hampered by a lack of resources for effective monitoring and prosecution. Other responses include: schemes for improving drainage; purification and decontamination of freshwater systems; and public awareness campaigns. Although only recently implemented, these responses have had localized success in improving access to potable water and raising awareness. Public-private partnerships have also been successful in implementing large urban sanitation projects (Sharma 1996).

●

*Groundwater pollution is a particular concern in the more arid countries, which are highly dependent on underground aquifers for drinking water supplies. The main pollutants are nitrates and phosphates, together with chemical residues from seepage of agricultural runoff, and from inappropriate discharge of industrial and domestic wastewater.*

●

Increasing water shortages and declining water quality are compounding factors limiting Africa's development. As freshwater availability decreases, it increases the concentration of pollutants, lowering the water quality. And as water quality declines, water shortages are exacerbated, through increased costs of water treatment and increased time spent in collecting water. Therefore, the challenges are: to ensure equitable access to clean water; to manage the demand for water for domestic, industrial and agricultural purposes; and maintain healthy ecosystems, in order to ensure sustained supplies of good quality water.

The African Initiative on Land and Water, funded by the Global Environment Facility (GEF), was set up to address in an integrated manner complex environmental challenges impacting on land and water quality. It provides a framework for coordinating bilateral and multilateral funding mechanisms, as well as for harmonizing existing sustainable development initiatives. Thus, a consolidated national and sub-regional sustainable development framework is proposed, in order to control the direct and indirect impacts of land degradation and desertification, including: sedimentation; siltation; and pollution of national and international water bodies. In addition, the benefits accruing from sound environmental management will be equally distributed between stakeholders. A catchment- or watershed-based approach is considered appropriate, taking into account natural environmental variability, in terms of rainfall and biomass production, and linking closely to biodiversity conservation objectives. The initiative will use monitoring and assessment information to support the development of flexible and adaptive policies and management strategies at the catchment or watershed level (UNEP 2002).

The following sections provide further information on specific freshwater issues relating to each sub-region in Africa.

## NORTHERN AFRICA

The Northern Africa sub-region is dominated by arid conditions and extensive deserts, with the exception of parts of southern Sudan and an intermittent narrow strip along the Mediterranean shoreline, where the climate is more humid. The major issue of concern is, therefore, freshwater availability for domestic, agricultural and industrial consumption. Although most people have access to water resources, as a result of high levels of infrastructure development, demand management and integrated water resources management are priorities for improving adequacy and equity in supply. Water quality is an emerging issue, particularly with regard to salinization from poor irrigation methods, and pollution from industrial and domestic wastewater disposal.

### AVAILABILITY OF FRESHWATER IN NORTHERN AFRICA

The average total annual precipitation in Northern Africa is estimated at 1 503 km$^3$/yr, equivalent to 7 per cent of the total precipitation for Africa (FAO 1995). Distribution of this precipitation varies dramatically, with almost 75 per cent falling in Sudan (the average is 436 mm/yr, but it ranges from 20 mm/yr in the north to more than 1 600 mm/yr in the south) and just 3 per cent in Egypt (about 18 mm/yr) (FAO 1995). Only 5.6 per cent of the precipitation is available for renewing stream flows and for recharging shallow groundwater aquifers. The rest is mainly lost by evaporation, transpiration and seepage. Per capita water availability ranges from 26 m$^3$/yr in Egypt to 1 058 m$^3$/yr in Morocco (UNDP and others 2000).

Further disparity is evident when the countries of Northern Africa are compared to those in sub-Saharan Africa. The total internal renewable water resources for Northern Africa represent 2.5 per cent of the African total, but Northern Africa's withdrawals represent 46 per cent of the total African withdrawals. This disparity partially reflects the harsh climatic conditions, and partially indicates an enhanced level of water resources development. It is the effectiveness of such management schemes, and heavy dependence on transboundary supplies, that has facilitated the growth of the sub-region's population and economies.

Renewable groundwater resources are in the form of shallow alluvial aquifers, recharged from the main rivers (for example, the alluvial aquifer beneath the Nile delta in Egypt) or from precipitation (along the north African Mediterranean coast). In the Sahara desert, the major water resources are the combined NSA and the Continental Intercalaire non-renewable aquifer, which extend from Egypt to Mauritania. Current annual rates of groundwater withdrawal in the sub-region are 407 per cent of the recharge rate in Egypt, and 560 per cent in Libya (UNDP and others 2000). Exploitation of groundwater resources over the past ten years has led to a reduction in water pressure levels at the oasis of the western desert. Overextraction from the delta shallow aquifer has led to increased water salinization and a rapid inland advance of the saltwater interface.

The NSA is a huge fossil water resource, located in the eastern Sahara desert in northeastern Africa (see Figure 2e.5). It is shared among four countries—Chad, Egypt, Libya and Sudan—and contains an estimated 150 000 km$^3$ of groundwater (CEDARE 2000). The total current extraction from the NSA is estimated at 1 500 million m$^3$/yr. The Centre for Environment and Development for the Arab Region and Europe (CEDARE) is developing a regional strategy for the sustainable utilization of the aquifer, to be adopted by the four sharing countries. This strategy will consider sustainability of the resource, as well as the development dimension in each country, based on current and future needs.

Global warming and regional climatic change impose an additional possible threat to the already scarce freshwater resources in Northern Africa. The Nile Basin

**Figure 2e.6 Water use by sector in Northern Africa 1900–2025**

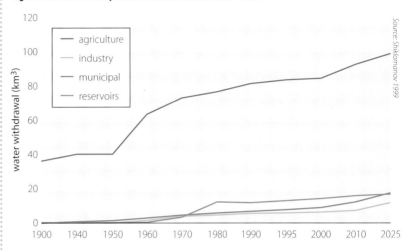

Source: Shiklomanov 1999

has a low run-off efficiency index and a high dryness index, rendering it highly susceptible to climate change (IPCC 1998). Run-off is likely to decrease with global warming, even if rainfall increases, because evaporation rates are so high. Scenarios for the future range from a 30 per cent increase in river flow to a 78 per cent decrease in river flow (IPCC 1998), presenting even greater challenges for international cooperation in water resources management. Northern Africa is already frequently affected by cycles of droughts and flooding and, with climate change, these are expected to increase. In the dry lands, which dominate most of the sub-region, population growth will push people onto marginal land, which is highly vulnerable to desertification, thus exacerbating the impacts of climate change.

### Access to freshwater in Northern Africa

The water supply coverage in the sub-region is generally high (ranging from 72 per cent in Libya to 95 per cent in Egypt), although urban areas are better supplied than rural areas. In rural areas, water supply coverage varies (from 58 per cent in Morocco to 94 per cent in Egypt) (WHO/UNICEF 2000). Access to sanitation is also generally good but, again, urban areas are better supplied than rural areas (WHO/UNICEF 2000). This is because of the high levels of water resources development, which have been implemented by the governments of Northern African countries. However, the demand for water from all sectors is rapidly increasing, thus placing enormous pressure on the ecosystems which supply and regulate water availability and quality (see Figure 2e.6).

**Figure 2e.5 Map of the sandstone aquifer in the Nubian Basin**

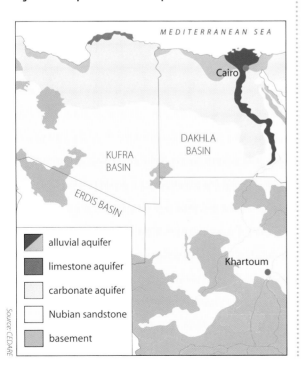

Source: CEDARE

Agriculture, particularly the production of cash crops, is heavily dependent on irrigation, and consumes more than 80 per cent of withdrawals, while industry and the domestic sector consume 5.6 per cent and 10 per cent of withdrawals respectively (UNDP and others 2000). Irrigated areas represent 100 per cent of cultivated areas in Egypt, 26 per cent in Sudan and 9 per cent in Tunisia (FAOSTAT 1997). In response to the rising demand for water and increasing water scarcity, some countries supplement their water resources through desalination and the reuse of treated municipal wastewater. Tunisia, for example, supplements surface and groundwater extraction with 8.3 million m$^3$/yr of desalinated seawater, and 20 million m$^3$/yr of treated wastewater (FAOSTAT 1997). In Egypt, 8 per cent of the total water used comes from the reuse of agricultural or municipal wastewater, and less than 1 per cent comes from the desalinisation of seawater, although the flourishing tourism industry is pushing for the severalfold increase of the current desalination capacity by 2010 (FAOSTAT 1997).

## Improving freshwater availability and access in Northern Africa

The interdisciplinary nature of water problems requires new methods to integrate the technical, hydrological, economic, environmental, social and legal aspects into a coherent framework of IWRM (McKinney and others 1999). Since the 1990s, most Northern African countries have realized that the business-as-usual scenario of dealing with water management and water security issues is no longer suitable to cope with future challenges, and they are starting to implement IWRM strategies. Countries have created enabling environments for IWRM by developing legal and institutional frameworks for effective operation. For example Egypt, Morocco and Tunisia have developed national policies and master plans for water management, based on IWRM principles, and the Algerian government has created five Basin Authorities for IWRM. In Tunisia and Egypt, groundwater recharging schemes have been adopted, together with equitable models of water use between user groups. Water use associations have also been formally established in Morocco and Tunisia. Also under the banner of IWRM, a national drought management plan has been put in place in Morocco, and a national programme for the

### Box 2e.3 Overcoming conflicts associated with transboundary water resources

Egypt's annual water consumption demands are met by the Blue Nile River (86 per cent), which flows from Lake Tana in Ethiopia, and by the White Nile River (14 per cent), which flows from Lake Victoria in Uganda. Egypt presently uses the majority of the River Nile's flows, and could potentially suffer crippling water reductions if other countries, such as Ethiopia, begin to utilize their share of the Nile's waters for hydroelectric power. Water rights to the Nile have thus become an important issue for the ten Nile riparian countries (Burundi, Democratic Republic of Congo, Egypt, Eritrea, Ethiopia, Kenya, Rwanda, Sudan, Tanzania and Uganda.).

The Nile Basin Initiative, launched in 1999, is joint programme of action between the ten Nile riparian countries, with the objectives of ensuring: sustainable resource development; security; cooperation; and economic integration. The first meeting of the International Consortium for Cooperation on the Nile (ICCON) took place in June 2001 in Geneva, where the Nile Council of Ministers put forward their Strategic Action Plan to the donor community. So far US$140 million has been pledged in support of the region's commitment to peace and sustainable resource development.

Source: EIA 2000, ICCON 2001

abatement of wastewater discharge into the Nile River has been successfully instituted in Egypt. However, continued support for IWRM is required in terms of: ongoing capacity building; institutional coordination; improved information exchange and processing; and sustainable funding and political commitment.

The Nile Basin Initiative represents an excellent example of cooperation within the framework of IWRM between the ten Nile riparian countries (see Box 2e.3). The Joint Authority for the Nubian Sandstone Aquifer is another example of cooperation between Sudan, Libya, Egypt and Chad, in order to formulate and to monitor strategies for the rational utilization of the NSA.

## WATER QUALITY IN NORTHERN AFRICA

Major water quality problems in Northern Africa include: salinization; pollution; and eutrophication of surface and/or groundwater. Increased salt content of surface and groundwater is mainly due to irrigation return flows and saltwater intrusion along the Mediterranean coast. Poorly sited or illegal waste dumping and seepage from landfills, together with localized sanitation problems, are the major causes of surface and groundwater pollution; and eutrophication of wetlands occurs as a result of high nitrate and phosphate levels in run-off from agricultural areas. Extensive algal blooms and mats of water hyacinth (*Eichornia crassipes*) in Sudan are also a result of raised nutrient levels in water bodies. These mats constitute problems for water quality, because they release toxins into the water, cause discolouration and reduce dissolved oxygen levels, thus creating an unsuitable habitat for many beneficial freshwater species, and increasing the costs of water treatment. Land erosion and dam reservoir sedimentation are also identified as major environmental problems in the sub-region. Reservoir sedimentation in Morocco was estimated at 50 million $m^3$/yr in 1994, and was expected to reach 100 million $m^3$/yr by the onset of the 21st century. The resulting annual loss of storage capacity is equivalent to the annual water requirements for irrigating 6 000 ha of cultivated land (FAO 1995).

Between the Aswan High Dam and Cairo, the Nile River receives domestic wastewater from 43 towns and 1 500 villages, and 35 major factories (with a combined wastewater discharge of 125 $m^3$/yr), as well as 2 300 million $m^3$/yr of irrigation return flows (containing chemical residues, sediments and nutrients) (Myllyla 1995). The health risks associated with such water quality problems are high, especially for children. In Morocco, the incidence of diarrhoea amongst children is highly correlated to lack of access to treated piped or treated well water (Kelly, Khanfir, David, Arata and Kleinau 1999). Tunisia's largest reservoir, the Sidi Salem, and Algeria's Mitidja and Saida aquifers, are showing deteriorating quality, in particular, elevated nitrate levels originating from industrial effluents and agricultural run-offs (World Bank 1996). This is of particular concern, because these are the main water supplies for the capital of Algeria, Algiers.

#### Improving water quality in Northern Africa

Special development programmes devoted to enhancing drinking water quality and sanitation services are expected to be in place by 2010 in Northern Africa. Implementation has already begun in some areas, such as Cairo, where the gigantic Greater Cairo Wastewater Project was implemented in the 1990s in order to convey and to treat the effluents from 7 million inhabitants of Cairo City. Industrial pollution abatement and pollution prevention programmes were prepared by the Egyptian Environmental Affairs Agency, and successful prevention of untreated industrial effluent discharge into the Nile River was confirmed by the Ministry of State for Environmental Affairs by 1998. National water monitoring programmes have also been launched by the National Research Centre and the Nile River Institute, but there is lack of coordination in their efforts, and insufficient technical and financial resources are additional barriers. The World Bank (WB) is investing more than US$2 000 million through the GEF for water-related projects in Egypt, and the US Agency for International Development (USAID) is another sponsor, funding water supply and sanitation schemes in Cairo (Myllyla 1995). The Egyptian Environmental Affairs Agency has passed a new environmental protection law, and Environmental Impact Assessments (EIAs) will be required for new big industrial developments.

In Tunisia, wastewater management began in the 1970s with the launch of a national water policy, and the establishment of the National Wastewater Management Agency. In recent years, the focus of the National Wastewater Management Agency has been in providing services for rural areas, implementing the State Environmental and Public Health Policy, and maintaining infrastructure related to sewage in urban and industrial areas. Environmental water requirements are considered, and Tunisia is seeking to supplement domestic, industrial and agricultural demands through the reuse of treated wastewater (Citet 2001).

### EASTERN AFRICA

Eastern Africa experiences high variability in rainfall over time and space, including frequent episodes of flooding or drought. There is also competition for access to water resources between user groups and between countries. Some of the countries are not only dependent on freshwater for domestic, agricultural and

•

*Between the Aswan High Dam and Cairo, the Nile River receives domestic wastewater from 43 towns and 1 500 villages, and 35 major factories (with a combined wastewater discharge of 125 $m^3$/yr), as well as 2 300 million $m^3$/yr of irrigation return flows (containing chemical residues, sediments and nutrients)*

•

industrial consumption, but also for hydropower generation. Hence, freshwater availability and access is a priority issue for the sub-region. Concerns have been raised in recent years about declining water quality and, in particular, about the infestation of Lake Victoria with water hyacinth (*Eichornia crassipes*).

## AVAILABILITY OF FRESHWATER IN EASTERN AFRICA

Eastern Africa, on the whole, is fairly well endowed with freshwater, with total average renewable freshwater resources amounting to 187 km$^3$/yr (UNDP and others 2000). Uganda has the largest share of this, with 39 km$^3$/yr (1 791 m$^3$/capita/yr), whilst Eritrea has the least, with 2.8 km$^3$/yr (data on per capita resources are not available) (UNDP and others 2000). The amount and distribution of rainfall varies across Eastern Africa, with annual averages ranging from 147 mm for Djibouti to more than 1 000 mm for Uganda, Rwanda and Burundi (FAOSTAT 2000). Intra-annual variations are also high, ranging from: 50–300 mm for Djibouti; 250–700 mm for Somalia; 750–2 000 mm for Uganda; and 100–2 400 mm for Ethiopia (FAOSTAT 2000). These intra-annual variations determine, to some extent, water availability. For example, more than 75

per cent of Ethiopia's rainfall occurs in intense downpours over a period of 3–4 months, whilst conditions are relatively dry for the rest of the year (Ministry of Water Resources 1998). The intensity of these rains and the lack of vegetative cover cause most rainfall to be lost as run-off or evaporation, with only a small percentage available to recharge underground aquifers. Surface water, therefore, dominates freshwater resources in Eastern Africa (the groundwater resources of Ethiopia and Eritrea, for example, are just 2.6 km$^3$ of the total resources for these countries) (FAOSTAT 1996). Surface water resources are also important in power generation (see Box 2e.4).

The drier countries in the Horn of Africa (Ethiopia, Eritrea and Somalia) frequently experience drought, and have been devastated by drought-induced famine on several occasions over the past 30 years. The largest freshwater source in Eastern Africa is Lake Victoria, the second largest lake in the world. Lake Victoria provides freshwater to the populations of Uganda, Kenya and Tanzania directly and, through the Nile River, to Sudan and Egypt. It is also the life and livelihood support of millions of people living around the lake, providing: fish; irrigation water; tourism and recreation; communications; and transport. Other major freshwater lakes in Eastern Africa include: Lake

*Annual freshwater withdrawals are a small percentage of the total available, ranging from less than 3 per cent of the total resources available in Burundi to 12 per cent in Rwanda*

### Box 2e.4  Hydropower development in Eastern Africa

The world's largest storage dam is the Owen Falls Dam on the River Nile in Uganda. The 162-megawatt (MW) capacity hydroelectric station at the dam supplies most of Uganda's electricity requirements, and exports 30 MW each year to Kenya. In 1999, however, increased domestic demand in Uganda resulted in a drop in supply to Kenya. Kenya's own hydropower stations supply 78 per cent of the country's electricity (670 MW in 1999). Ethiopia's hydropower potential has been estimated at 15 000–30 000 MW, although less than 2 per cent of this had been exploited by 1993, and 90 per cent of all energy consumed is derived from biomass. With such dependency on hydroelectric power, the Eastern African countries are vulnerable to power shortages during times of low rainfall, as experienced during 1999 and 2000 by Kenya and Ethiopia. This, in turn, adversely impacts on the economy, as a result of losses in industrial

productivity, commercial activities, and transport and communication networks. The government of Kenya is subsequently promoting the development of diesel and geothermal power plants.

*T. de Salis/Still Pictures*

Source: Bermudez 1999, Hailu 1998

Tanganyika in Tanzania; Lake Edward, Lake George, Lake Kyoga and Lake Albert in Uganda; Lake Turkana in Kenya; and eleven freshwater lakes in Ethiopia.

Annual freshwater withdrawals are a small percentage of the total available, ranging from less than 3 per cent of the total resources available in Burundi to 12 per cent in Rwanda (UNDP 2000). However, variability in rainfall results in frequent bouts of water scarcity and, during these times, demand exceeds supply. Human settlement patterns also influence, and are influenced by, freshwater availability. For example, in Kenya, only 33 per cent of the land area has adequate and dependable water, but this area is home to 70 per cent of the population.

With Eastern Africa's population growing rapidly, demand for freshwater is already becoming a problem. Figure 2e.7 shows the current and expected sectoral water use for Eastern Africa. Demand for freshwater in the domestic sector is also rising because of increasing per capita water usage. In 1980, per capita urban water usage in Uganda was 90 litres/day; and this was expected to almost double by the year 2000 (NEMA 1999).

Eastern Africa is home to a sizeable population of pastoralists. One of the main environmental problems associated with pastoralism is overstocking of animals, leading to depletion of drinking water sources and degradation of vegetation. Currently, Uganda's livestock freshwater demand is reported to be 81 million m$^3$/yr, and is projected to increase to 233 million m$^3$/yr by 2010 (NEMA 1999). Compounded by increasing demand from the domestic sector, this will present huge challenges to water resources management and water supply services. In addition, more land is being brought under cultivation in many countries, as part of strategies to increase food production and security. Ethiopia's potentially irrigable land area is 3.7 million ha, of which only 160,000 ha have been developed (Ministry of Water Resources 1998). Ethiopia is considering expanding irrigation activities into the Shebelle and Genale river valleys, with some of the irrigation plans also calling for diversion of streams. This could have adverse effects on downstream water users, and could potentially disrupt hydrological systems and aquatic ecosystem health.

IPCC predicts that rainfall will decrease in the already arid areas of the Horn of Africa, and that

**Figure 2e.7 Water use by sector in Eastern Africa, 1900–2025**

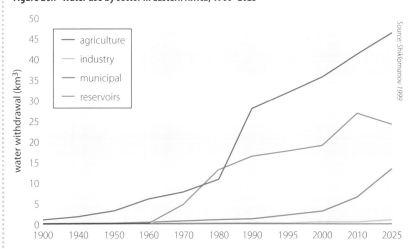

Source: Shiklomanov 1999

drought and desertification will become more widespread (IPCC 2001). As a result of the increasing scarcity of surface freshwater, groundwater aquifers are being mined. Wetlands areas are also being used to obtain water for humans and livestock, and as additional cultivation and grazing land. This alters hydrological cycles, leaving the surrounding area more prone to flooding.

View of the southern end of Lake Turkana in Kenya

Frants Hartmann/FLPA

## Box 2e.5  Water resources management in Ethiopia

Until the early 1990s, there were eight public agencies involved in the development and provision of water resources in Ethiopia. The National Water Resources Commission was responsible for irrigation. The Water Resource Development Agency was responsible for the design, implementation and operation of large- and medium-scale irrigation projects. The Irrigation Development Department within the Ministry of Agriculture was entrusted with the planning and construction of small-scale irrigation. Feasibility studies and planning of irrigation schemes was undertaken by the Ethiopian Valleys Development Studies Authority and the Water Well Drilling Agency, which took over from the Valleys Development Agency and the Development Projects Studies Authority. The Ethiopian Water Works Construction Agency constructed all water project infrastructure, and the Water Supply and Sewerage Authority supplied water services and sanitation for urban and rural settlements. Not surprisingly, there was often a great deal of duplication of effort and wastage of resources among these myriad, autonomous and semi-autonomous agencies. There was also incomplete geographical coverage of water service provision. In 1994, the Ministry of Water Resources was established as a single, unified public agency responsible for water development in Ethiopia.

Source: Rahmato 1999

### Access to freshwater in Eastern Africa

Even where freshwater is potentially available, it is not always accessible. In Ethiopia, for example, despite an extensive network of lakes and rivers, only 24 per cent of the country's population has access to clean water, and only 13 per cent in the rural areas (WHO/UNICEF 2000). In Djibouti, by contrast, water supply and sanitation coverage is 100 per cent in both urban and rural areas (WHO/UNICEF 2000). A significant barrier to the effective development of water storage, treatment and supply has been the fragmentation of responsibilities amongst government departments, as illustrated by the Ethiopian example (see Box 2e.5).

### Improving availability and access to freshwater in Eastern Africa

The major international programmes for water resources management in the sub-region are the Lake Victoria Environmental Management Programme (LVEMP) and the Nile Basin Initiative. The LVEMP was established in 1995, with funding from GEF. The focus was primarily on: fisheries management; pollution control; control of invasive weeds; and catchment land use management. The achievements to date include: bio-control of water hyacinth (*Eichornia crassipes*); involvement of local communities in fisheries research and management; and afforestation in the surrounding catchment.

At the national level, policy responses include: revision of water resources development policies; improved reticulation and treatment; and greater involvement of stakeholders in water management and supply. In Kenya, water and sanitation schemes have been commercialized in pilot areas in Kericho, Eldoret and Nyeri. These pilot studies will test whether privatization contributes to meeting the goals of the Kenyan Water Act (Cap. 372), namely, to enhance the provision, conservation, control, apportionment and use of water in Kenya. The international donor organization, SIDA, is also providing support to the government of Kenya to produce a new water policy that will support the rights of villages to own and run their own water systems (SIDA 2000).

Uganda's long-term goal for the water sector is a system of full cost-recovery for services provided, but with the provision of cross-subsidized safe water services for low-income groups. The responsibility for water supply in urban areas has been decentralized away from the national Water Development Department, to local authorities. Safe water and sanitation facilities in Uganda have been improved, although rural supply rates (less than 50 per cent of the population) lag behind urban supply rates (60–75 per cent) (NEMA 2001). A National Wetlands Policy was formulated and passed by the Ugandan government in 1994, calling for, amongst other things: capacity building for wetlands management; public awareness; and wetlands resource assessment. Where housing construction design permits, rainwater is also being harvested.

In 2001, Ethiopia initiated a process to develop a sectoral strategic action plan for the realization of the objectives of the national water policy. The strategy prioritizes the interest and roles of different stakeholders, who are invited to make inputs to the strategy development.

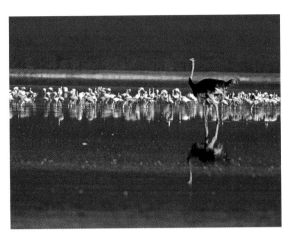

Ostrich and flamingos, Lake Nakuru, Kenya

M and C Denis-Huot/Still Pictures

## WATER QUALITY IN EASTERN AFRICA

A further environmental issue concerning the freshwater resources of Eastern Africa is pollution from domestic, agricultural and industrial sources. Low levels of water supply and sanitation, and overburdening of municipal services, create situations where untreated sewage is discharged directly into watercourses, particularly in areas surrounding informal settlements. For example, in Mombasa, the domestic sewerage system was designed to serve about 17 per cent of the current population. The outfalls for both domestic sewage and stormwater run-off are located in the Kilindini and Tudor Creeks, and sludges from septic tanks and pit latrines are usually disposed of at the Kibarani dumpsite on the shores of Makupa Creek (UNEP 1998).

Poor agricultural practices contribute to the pollution of freshwater sources in two ways. Firstly, the increasing use of agrochemicals contributes to the pollution of both surface water and groundwater, through run-off. These chemicals either cause eutrophication (by increasing the nitrogen and phosphorous loads in water bodies), or are toxic to flora and fauna. Unsustainable farming practices and overgrazing increase the susceptibility of the soil to erosion. Increased sediment loads in rivers, in turn, contribute to siltation in dams and lakes, and the smothering of habitats, flora and fauna. Lake Victoria, for example, has become the recipient of increased concentrations of nitrogen and phosphorous, washed

down from the surrounding plantations of tea and coffee. This has led to the invasion and rapid domination of the water hyacinth, *Eichornia crassipes*, which has formed dense mats, blocking navigation channels and choking boat engines. Other adverse impacts of the weed include modification of aquatic and wetland environments, through changing the concentrations of nutrients and dissolved oxygen, as well as reducing the light penetrating below the surface of the lake and releasing toxins. It also reduces the quality of water for drinking and domestic use, and provides an ideal habitat for disease carriers, such as mosquitoes. The Lake Victoria Environmental Management Programme is implementing biological, chemical and mechanical means of controlling the spread of the weed, and the Kisumu Innovation Centre-Kenya (KICK) has initiated marketing a number of crafts made from the weed (Olal, Muchilwa and Woomer 2001).

In 2001 Lake Nakuru and Lake Bogoria in Kenya experienced large numbers of deaths among the flamingos for which the lakes are famous. Pollution by heavy metals is suspected to be the primary cause of these deaths, resulting from contamination of the lakes by sewage, industrial effluent and organochlorines, which are present in agricultural run off (Environment News Service 2001). Industrial developments, particularly in urban centres and along the coast, have inadequate or expensive wastewater treatment facilities and, therefore, discharge of effluents directly into rivers and estuaries is common. The growing level of industrialization in greater Mombasa, for example, is causing considerable concern as a result of effluent discharge. Complex organic compounds and heavy metals may be contained in effluents and spillages from commercial industries and textile mills, which are located along the coast and banks of the Mombasa, Kilindini, and Port Reitz creeks (Mwangi and Munga 1997). A decline in the quality of freshwater creates even greater stress for water supply schemes. It can also limit industrial, domestic and agricultural water use, and it contributes to the prevalence of waterborne diseases.

### Improving water quality in Eastern Africa

Responses aimed at improving water quality and at mitigating the environmental impacts of poor quality water include the promulgation by several countries in Eastern Africa of water quality standards and effluent

*In 2001 Lake Nakuru and Lake Bogoria in Kenya experienced large numbers of deaths among the flamingos for which the lakes are famous. Pollution by heavy metals is suspected to be the primary cause of these deaths, resulting from contamination of the lakes by sewage, industrial effluent and organochlorines, which are present in agricultural run off*

## Box 2e.6  Wetland rehabilitation in Uganda

Wetlands in Kampala are used by the city council as dumping sites for municipal waste. The associated pollution has rendered these areas unsuitable for wildlife or human settlement. Resultant flooding in the city has led to: the interruption of transport and communications services; loss of property; and outbreaks of waterborne diseases. Kampala is also experiencing increasing numbers of problematic Marabou storks. These birds are wetland species, which have been forced into the city as scavengers, due to the loss of their natural habitat. The storks live on the top of buildings, and constantly dirty the city's streets and roads.

Education and improved town planning are recommended as solutions to the growing problems of wetland destruction and its consequences, together with improved wetland protection and management. Additional measures include the 1971 Ramsar Convention on Wetlands of International Importance, and cooperation in environmental pollution control under the auspices of IGAD.

Source: NEMA 1999

controls. A Water Action Plan, providing a framework for the protection and development of Uganda's water resources, was prepared in 1994, based on the guiding principles of IWRM established at the United Nations Conference on Environment and Development (UNCED) in 1992. Uganda's Water Statute, incorporating many of the recommendations from the Water Action Plan, was promulgated in 1995, and regulations for the control of Water Resources, Water and Sewerage and Wastewater Discharges were issued in 1998. A draft National Water Policy is currently being reviewed prior to submission to The Ugandan government for approval (NEMA 1999). Wetland rehabilitation is also underway, in order to improve water quality and ecological functionality (See Box 2e.6).

Ethiopia's water supply and sanitation sector shows an encouraging participation of communities, for example, through the contribution of beneficiary communities in the development and construction of water supply schemes and sanitation facilities (WHO/UNICEF 2000). Under the Africa 2000 initiative, launched in 1996, a study on low-cost latrine technology options was carried out in 1997, and a study on the impact of water and sanitation-related diseases on development is also nearly complete. Several health education trainers have completed their training and are being deployed.

Kenya's government has formulated a comprehensive water supply, demand and management policy, which was approved by the government in 2001. The policy outlines methods to improve water supply, methods for regulation, pricing and ways to enlist the participation of other stakeholders.

## WESTERN INDIAN OCEAN ISLANDS

Availability of freshwater in the Western Indian Ocean islands is a priority issue, as a result of: variable rainfall and high run-off; increasing domestic, agricultural and industrial consumption; and a lack of storage capacity. Pollution from a variety of sources is contributing to restricted availability of water for human use, and is contributing to environmental degradation. Therefore, implementation of integrated water resource management strategies is a priority in the sub-region.

### AVAILABILITY OF FRESHWATER IN THE WESTERN INDIAN OCEAN ISLANDS

Access to freshwater in the islands of the Western Indian Ocean Islands is complicated not only by variability in rainfall, but also by disparities between the abundance of water and the distribution of the population. Only Madagascar has abundant internal freshwater resources: a total of 337 km$^3$/yr (UNDP and others 2000). However, these resources are unevenly distributed across the country due to the terrain, and poor infrastructure development means that large sectors of the population, especially in rural and coastal areas, lack access to water. Mauritius has just 2.21 km$^3$/yr, or 1 970m$^3$/capita/yr, of freshwater, which is close to the water stress threshold of 1 700 m$^3$/capita/yr (UNDP 1998). This poses a challenge for the supply of water to urban centres.

Although the climate in the Western Indian Ocean is humid maritime, most of the rainfall occurs during a few months—December to April. Therefore, many islands experience periods of water shortages. The southwest region of Madagascar is the driest part of the Western Indian Ocean Islands, and experiences periodic drought (UNEP 1999b). In addition, monsoon rainfall and cyclones are common, and much of the water is lost through run-off. Comoros has an almost complete absence of surface water, because of the high permeability of its soil (UNEP 1999b). Deforestation and the clearing of land has occurred in all the islands, and has contributed to high rates of run-off.

There are many rivers (92 in Mauritius alone), natural and human-made lakes, and considerable groundwater resources in Madagascar and Mauritius, although these have not been evaluated or exploited thoroughly (UNDP and others 2000, UNEP 1999b). Wetlands are also important habitats on all of the islands in the sub-region, providing breeding grounds for large numbers of waterfowl. However, these are coming under pressure for land development, especially in the smaller islands, where tourism and population growth are driving the demand for housing and industry. To date, Madagascar has declared two wetlands as Wetlands of International Importance under the Ramsar Convention, and Mauritius and Comoros have each designated one Ramsar site.

## Access to freshwater in the Western Indian Ocean Islands

Data for total annual withdrawals are not available for Comoros or Seychelles but, in Madagascar and Mauritius, withdrawals by all sectors represent a small percentage of annual renewable resources (6 per cent and 16 per cent respectively) (UNDP and others 2000, UNEP 1999b). The agricultural sector is by far the greatest water user, accounting for 99 per cent of all withdrawals in Madagascar and 77 per cent in Mauritius (UNDP and others 2000, UNEP 1999b). This reflects the importance of agriculture in economic development and livelihood subsistence and, in Mauritius, particularly the production of sugar cane, yields of which are among the highest in the world (FAOSTAT 1995). The area under irrigation grew from 12 000 ha in 1970 to 17 500 ha in 1995, most of which was sugar cane (FAOSTAT 1995). In Seychelles, water is used to irrigate orchid farms, as well as food crops, in times of water shortages. Groundwater is largely used for this purpose, as it was on Mauritius until the costs of pumping became prohibitively high (FAOSTAT 1995).

Details of water consumption are not available for the sub-region. However, in Mauritius, the second largest user of freshwater is the domestic sector which accounts for 16 per cent of all withdrawals (UNEP

1999b). This is largely to support the other economic mainstay of Mauritius, namely, the tourism industry, and it has been supported by significant investment in water supply infrastructure. Consequently, 100 per cent of the population, in both rural and urban areas, has access to piped water and improved sanitation (WHO/UNICEF 2000). By comparison, only 47 per cent of the population in Madagascar has access to piped water (although, in urban areas, rates are as high as 85 per cent), and 70 per cent of urban poulations and 30 per cent of rural populations have access to improved sanitation (WHO/UNICEF 2000). In Comoros, only 33 per cent of people have safe sanitation (UNEP/COI 1997).

With growing populations, and expansions in tourism and other industrial sectors, the demand for freshwater is expected to increase over the next 25 years in all the Western Indian Ocean states. Supplies per capita in Mauritius are falling, and are predicted to be within water stress levels by 2025 (Johns Hopkins 1998). In Comoros, per capita annual water resources are expected to fall to 760 m$^3$, creating a situation of water scarcity (UNDP 1998). In Seychelles, water shortages were so severe during 1998 (due in part to the El Niño Southern Oscillation phenomenon) that brewing and fish canning industries were temporarily closed (UNEP 1999b). This had such an impact on the local economy that the government has since commissioned a water desalination plant, in order to augment supplies. Losses from old water pipes further compound problems of current and future supply, accounting for up to 50 per cent of domestic withdrawals in Mauritius alone (Government of Mauritius 1994). Low storage capacity by households and in the public system aggravates the situation (Government of Mauritius 1994). The dry spells and periodic droughts which are common in Mauritius force water rationing in the north of the island, and increase the risk of contamination through the infiltration of untreated surface water.

The islands of the Western Indian Ocean are extremely vulnerable to climate change, not only in terms of sea level rise and associated loss of land and infrastructure, but also in terms of reduced availability of freshwater. Potential inundation of groundwater aquifers could foreclose options to exploit these resources, and alterations in rainfall quantities and

Sugar cane plantation: young crop with mature crop behind.

*Duncan Smith/Holt Studios*

distribution patterns, together with temperature and evaporation changes, may also reduce surface water availability (IPCC 2001).

### Improving availability and access to freshwater in the Western Indian Ocean Islands

These conditions of water scarcity and stress will continue to restrict economic and social development in the sub-region through curtailment of industry, agriculture, tourism and subsistence, unless a dramatic shift in policy is made towards IWRM. Major challenges include: the shortage of freshwater; improvement in freshwater management; and balancing the competing demands of population, agriculture and tourism. At present, an adequate framework of integrated policy is lacking. Water pricing does not reflect the true cost of production and distribution, and the capital cost of increasing supply. Lack of sensitivity to water conservation by industrialists, tourists and local residents exacerbates the problem (UNEP 1999b). Therefore, the priority is demand management, through public awareness and education, as well as through economic incentives. Reforms to date in Mauritius

include: the diversification of crops, away from the heavily irrigation-dependent sugar cane; and conversion to more efficient drip, pivot, guns and sprinklers. Localized irrigation from small impoundments is being encouraged, instead of reliance on large dams that incur higher evaporative losses (FAOSTAT 1995).

Seychelles is looking to increase its crop production through improved irrigation, but also through the use of hydroponics, a technique that is currently under development (FAOSTAT 1995).

## WATER QUALITY IN THE WESTERN INDIAN OCEAN ISLANDS

Problems of water availability are worsened by the contamination of existing supplies by pollution from various sources. Sugar cane production for example, as well as consuming a high proportion of local water, is also a major polluter of water aquifers, as a result of heavy use of chemical fertilizers. Irrigation of other crops also contributes to the contamination of surface water and groundwater resources, through chemical residues, increased silt loads and higher salt concentrations. As rivers and estuaries are drying up, and as nitrate and phosphate levels increase, so they are engulfed by algal blooms and dense mats of water hyacinth (*Eichornia crassipes*). Sewage, wastewater and solid waste dumping are significant contributors to pollution of groundwater, especially in Mauritius (Institute for Environmental and Legal Studies 1998). Some aquifers have nitrate levels of 50 milligrams/litre (mg/l), a level considered a risk to health (Jootun, Bhikajee, Prayag and Soyfoo 1997). This is ecologically detrimental, removing nutrients and dissolved oxygen from the water, and elevating the level of toxins in the water so that it becomes unfit for consumption.

Saline intrusion from the sea is also contaminating underground freshwater supplies in various parts of the sub-region. Sources of industrial pollution include: textile factories; fish processing plants; breweries; and tanneries. In Port Louis in Mauritius, industrial effluent is only pretreated before it is discharged into the sea at only 800 metres (m) from the shore (Institute for Environmental and Legal Studies 1998). Just along the coast, at the Coromandel industrial zone, untreated effluent is discharged only 600 m out to sea, and has caused the death of the coral reef opposite Pointe aux Sables (Institute for Environmental and Legal Studies 1998). In Madagascar, industrial wastewater receives little treatment, and pollution control legislation does not exist (ONE and INSTAT 1994, ONE 1997). There is no wastewater treatment system in Comoros, and most effluent eventually ends up in the sea (UNEP 1999b).

## Improving water quality in the Western Indian Ocean Islands

Improving water availability and quality is now a priority for the sub-region. Water desalination is practised in Seychelles and Mauritius, and Mauritius has a series of projects geared to increasing the volume and quality of water supply in its National Environmental Action Plan (GOM/ERM 1998). Projects for increasing reservoirs and sewage treatment works, and the rehabilitation of old, leaking water systems, are in hand. Metering and charging for water use by volume is being introduced in the region. Legislation for environmental and water protection is in place in most countries. Recycling domestic wastewater for commercial users is projected. Education and sensitization programmes with schools and the media have been introduced (UNEP 1999b). It has yet to be seen whether these separate efforts will be effective in the absence of an integrated sub-regionwide policy framework. The extent of the worsening situation appears to have been underestimated.

## SOUTHERN AFRICA

Southern Africa is mostly semi-arid, and experiences variation in rainfall, both over time and between countries. This sub-region is also expected to experience further variability in rainfall, reduced precipitation and increased evaporation, as a result of climate change. With a rapidly growing population, and demands from the domestic, agricultural and industrial sectors for water, freshwater availability is a priority concern for the sub-region. Discriminatory access policies and pricing systems have also skewed the distribution of access to water resources across population groups. An additional concern in the sub-region is declining water quality due to domestic and industrial pollution, and eutrophication and salinization due to agricultural pollution.

## AVAILABILITY OF FRESHWATER IN SOUTHERN AFRICA

Southern Africa's annual average surface water resources are approximately 534 $km^3$/yr, but they are distributed unevenly due to: frequently low and variable rainfall; terrain; evaporation rates; and vegetation and soil cover. For example, Angola, the wettest country in the sub-region, has average annual internal water resources of 184 $km^3$/yr (14 000 $m^3$/capita/yr), and Mozambique and Zambia have 100 $km^3$/yr and 80 $km^3$/yr respectively (5 000 $m^3$/capita/yr and 8 700$m^3$/capita/yr). By contrast, the driest countries, Botswana and Namibia, have just 2.9 $km^3$/yr and 6.2 $km^3$/yr respectively (1 700$m^3$/capita/yr and 3 500$m^3$/capita/yr respectively) (UNDP and others 2000).

The areas of low rainfall are in many cases also coincidental with areas of highest evaporation potential, and variability in rainfall can result in periodic episodes of severe and prolonged droughts, particularly in the southwest. In these areas, groundwater resources are particularly important (see Box 2e.7). By contrast, the northern and eastern areas are subject to occasional floods, the most recent example being in 1999–2000. The excessive rains of this season affected Mozambique, Botswana, Zambia, Zimbabwe and South Africa. Some 200 000 ha of cropland were flooded and more than 150 000 families were affected. The estimated cost of recovery is millions of US dollars (Mpofu 2000). Although the SADC Early Warning

---

**Box 2e.7  Importance of groundwater in Southern Africa**

Groundwater is the main source of water for about 60 per cent of both rural and urban residents throughout Southern Africa. A large part of the sub-region is characterized by small towns, villages and dispersed rural settlements. Thus, access to reticulated surface water resources has been limited because of the high costs and long distances that need to be covered in order to establish infrastructure for formal water services. The sustainable management of groundwater is, therefore, important for rural livelihoods in the sub-region, even though the proportion of water coming from groundwater sources is relatively small.

*Source: Chenje 2000*

System was able to predict the heavy rains, most countries were ill-prepared for the magnitude and duration of the floods, stimulating investigations and revisions of response strategies.

The sub-region's largest freshwater lake, and the third largest lake in Africa, is Lake Malawi, with a surface area of 31 000 km². The lake is important in terms of fishing activities and the tourist industry which it supports. Lake Malawi has the largest number of fish species of any lake in the world, estimated at more than 500 species, of which 90 per cent are thought to be unique to the lake. The most important species biologically, and in terms of local livelihoods, are the 400 or more cichlid species, of which all but five are endemic to Lake Malawi (Ribbink, Marsh, Ribbink and Sharp 1983). Because of this extraordinary biodiversity, the southern part of the lake is registered as a national park, and was established as a Natural World Heritage Site in 1984. Major rivers include the Zambezi River, whose basin is shared by eight Southern African countries and is home to 40 million people. The Zambezi basin also supports many local communities, as well as commercial agriculture and forestry, manufacturing and mining, conservation and tourism (Chenje 2000).

A study of the potential impacts of climate change on freshwater resources in Southern Africa predicts an overall reduction in rainfall, by as much as 10 per cent across the whole sub-region, and up to 20 per cent in parts of South Africa (WWF 2000). Evaporation rates will increase by 5–20 per cent, as a result of raised temperatures, which will reduce run-off, and decrease water security and agricultural potential. Coincident with this will be increases in the frequency and intensity of flooding and drought (WWF 2000). In addition to variability and long-term decline in precipitation, water resources development is further complicated by the uneven distribution of population—particularly, high population densities in arid areas. Per capita average annual water resources in Angola, for example, exceeds 14 000 m³/capita/yr whereas, in South Africa, each person has only slightly more than 1 000 m³ per year (UNDP and others 2000).

Environmental degradation is a further contributor to decline in water availability, through loss of vegetation, and the disruption of microclimates and hydrological cycles. Dense stands of alien vegetation in Southern Africa are particularly disruptive in this regard, because they use much larger amounts of water than indigenous species (see Box 2e.8). In some areas of Southern Africa, up to 50 per cent of wetlands have been transformed, and the Caprivi wetland system (Namibia) has been reduced to almost 25 per cent of its original size. This has been due to: draining of wetlands, for agricultural or infrastructure development; reduced flows; aquatic weeds choking water courses; increasing use of pesticides; and overextraction of reeds, wood and other materials for construction, weaving and crafts (Chenje 2000). Wetlands act as sponges, absorbing excess water in times of heavy rainfall and, thus, buffering the effects of flooding. To prevent further degradation of these important habitats, two countries of the sub-region, South Africa and Zambia, are parties to the 1971 Ramsar Convention, and have designated certain areas as Wetlands of International Importance.

*Lake Malawi has the largest number of fish species of any lake in the world, estimated at more than 500 species, of which 90 per cent are thought to be unique to the lake. The most important species biologically, and in terms of local livelihoods, are the 400 or more cichlid species, of which all but five are endemic to Lake Malawi*

---

**Box 2e.8  Alien invasive vegetation and water use in Southern Africa**

Alien invasive vegetation is a significant and growing problem in many areas of Southern Africa, especially in riparian zones. The water hyacinth (*Eichornia crassipes*) is choking many water bodies in South Africa, Swaziland, Malawi and Zimbabwe, and exotic timber species, such as pines (*Pinus* sp.) wattles (*Acacia* sp.) and eucalypts (*Eucalyptus* sp) take up more water than indigenous species, thereby reducing mean annual run-off. Zimbabwe and South Africa have initiated biological and chemical control programmes for the water hyacinth,

with some success to date. South Africa's Working for Water Programme, launched in 1995, is a nationwide alien plant control programme, building capacity, and generating employment and incomes among some of the country's poorest communities. The programme uses mechanical, chemical and biological control methods and, during 2000, the teams cleared 238 000 ha and rehabilitated 51 000 ha of land which was infested with alien vegetation. The activities of that year employed 21 000 people.

*Sources: MacDonald 1989, Chenje 2000, MacDonald and Richardson 1986, Global Water Partnership 2000, Working for Water 2000*

## Access to freshwater in Southern Africa

With such intense and varied pressures on freshwater ecosystems, Southern Africa faces a serious water supply challenge. This is exacerbated by rapid population growth, high rates of urbanization, and imperatives of economic development and social equity, causing an increase in demand for agricultural, industrial and domestic water uses. At present, agriculture accounts for 74 per cent of Southern Africa's total water use, domestic users account for 17 per cent and industry accounts for 9 per cent (UNDP and others 2000). The rapid rise in demand for water is best illustrated in South Africa where, although 4.3 million households do not have water services, the increase in domestic demand over the past four decades has been four times greater than that of the agricultural sector. The domestic demand in South Africa is projected to increase to 23 per cent of the total by 2030, an increase of more than 200 per cent since 1996 (DEA&T 1999). Current projections estimate that serious shortfalls in water provision will occur within the next 10–20 years. Figure 2e.8 shows the current and predicted trends in water use by all sectors in Southern Africa.

Some urban water distribution networks in Southern Africa are poorly maintained and highly inefficient. In South Africa, for example, the supplier Rand Water estimates that up to 70 per cent of water is lost every year from the Soweto supply, due to leakages, costing US$100 800 per day (UNCHS/UNEP 2001). Similarly, irrigation is said to be less than 50 per cent efficient, with the majority of farms being irrigated by wasteful flooding and overhead sprinkler systems, while only 10 per cent use the more efficient microjet and drip irrigation systems (Chenje 2000). Mozambique has larger water resources than other countries in the sub-region, yet access to potable and irrigation water is more limited than in the drier countries, mainly due to infrastructure damage from the long civil war (IUCN 2000).

## Improving availability and access to freshwater in Southern Africa

The response to this situation over the past 30 years has been to dam or to modify nearly all the water courses in the sub-region, in order to meet demand. Southern Africa has the highest concentration of dams and interbasin transfer schemes anywhere in Africa

**Figure 2e.8 Water use by sector in Southern Africa, 1900–2025**

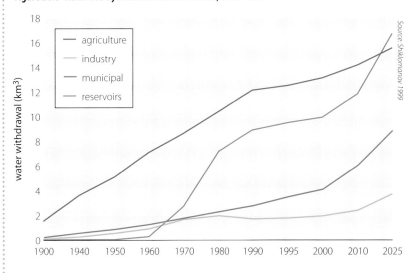

(World Commission on Dams 2001). The Kariba Dam, for example, located on the Zambezi River, and shared between Zambia and Zimbabwe, has a capacity of 180 $km^3$. With a surface area of some 5 500 $km^2$, it is one of the largest dams in the world. It was constructed primarily to provide hydropower to the two countries, and has a generation capacity of 1 320 MW (Soils Incorporated (Pty) Ltd and Chalo Environmental and Sustainable Development Consultants 2000).

Significant strides have been made towards the development of infrastructure for water supply and sanitation services. Access to safe drinking water improved from an average figure of 52.6 per cent in 1990 (WRI, UNEP and UNDP 1990) to 58 per cent in 1998 (WRI, UNEP, UNDP and World Bank 1998). Such access, however, varies between rural and urban areas, with the latter enjoying better access. There are also disparities between countries, with the greatest access to water in Botswana (95 per cent) and the least access in Angola (38 per cent) (WHO/UNICEF 2000). Water policies need revising to redress this issue, combining demand management (using economic incentives to control usage) with meeting basic needs (through improved equity in access and supply). South Africa has enacted such legislation, in the form of the National Water Act, Act 36 of 1998. In Windhoek, Namibia, water tariffs were increased in 1996 by 30 per cent, and household water consumption exceeding 60 $m^3$/month was penalized with even higher rates, with the effect of reducing average water consumption by 25 per cent (Eales, Forster and Du Mhango 1996).

*Some urban water distribution networks in Southern Africa are poorly maintained and highly inefficient. In South Africa, for example, the supplier Rand Water estimates that up to 70 per cent of water is lost every year from the Soweto supply, due to leakages*

In addition, countries of Southern Africa have made considerable efforts to cooperate in the management of shared water resources (see Box 2e.9). At the national level, countries have also made a shift towards IWRM and true cost pricing, in order to control demand. Instruments invoked to effect this include the removal of subsidies, and cost recovery from consumers. In Namibia, for example, subsidies—including those on water—accounted for 1 per cent of GDP in 1993 (IUCN 2000), but have gradually been lifted, while, in Zimbabwe, most large-scale and smallholder farmers used not to pay for irrigation water at all. The reuse of wastewater is being increasingly considered as a step towards IWRM in the region. The city of Windhoek already treats and reuses water to supply 19 per cent of the 42 510 m$^3$ that the city uses every day (IUCN 2000).

## WATER QUALITY IN SOUTHERN AFRICA

Freshwater resources in Southern Africa are also under pressure from pollution, and water quality is a growing problem, particularly in urban areas and close to industrial centres. Inadequate sewage treatment and disposal facilities are the main causes of urban water pollution, resulting in localized incidences of high faecal coliform counts. In the Victoria Falls town in Zimbabwe, for example, faecal coliform counts were found to be 350 times the level in unpolluted water (Feresu and van Sickle 1990). Khami Dam, near the city of Bulawayo, receives a daily discharge of up to 700 m$^3$ of effluent (Chenje 2000). These bacteria cause intestinal infections and diarrhoeal epidemics, if allowed into drinking water supplies. Diarrhoea is one of the main causes of death in Africa, particularly among children, and the main causes of diarrhoea are poor water quality and sanitation. Contamination of water supplies also creates problems for irrigation and industrial uses, requiring additional processing, which is expensive.

Agricultural run-off is causing accumulation of excess nutrients, such as phosphates and nitrates, in water bodies (eutrophication), as well as sedimentation. This changes the composition and functioning of the natural biota; makes the environment less attractive for recreation and sport; releases toxic metabolites, and taste and odour-causing compounds; and complicates water treatment. In 1996, Lake Chivero, near Harare, experienced algal blooms, infestations of water hyacinth and fish deaths, as a result of high levels of ammonia and low levels of oxygen in the lake (Gumbo 1997). The most likely sources of contaminants are: sewage from nearby settlements; agricultural run-off; and industrial discharges (Gumbo 1997). The Kafue River in Zambia is the recipient of almost all of the 93 000 tonnes (t) of waste from mining, chemical, fertilizer and textile industries (Chenje 2000). Mining activities also contribute to declining water quality and degradation of freshwater habitats. The rivers in Greater Soweto, for example, are heavily polluted by mining drainage water, and show evidence of low pH values, high electrical conductivity and high sulphate values (GJMC 1999). A further source of pollution is from solid waste which has not been properly disposed of, from which pollutants are leached into surface and groundwater. Metals and batteries were found to be prevalent in solid wastes from the town of Kariba, constituting a major input of metals and heavy metals into the Zambezi River (Chenje 2000).

### Improving water quality in Southern Africa

The heads of state of the Southern African countries recently made a decision to bring forces together, in order to develop a holistic regional strategic approach to IWRM and development in the region. This resulted in the formulation of the Regional Strategic Action Plan

---

**Box 2e.9 Mechanisms for international cooperation over shared water resources in Southern Africa**

Rising demand for water resources has the potential for conflict between user groups and between countries where water resources are shared. To ensure close cooperation for the judicious, sustainable and coordinated use of shared watercourses, the SADC Protocol on Shared Watercourse Systems was developed and came into force in 1998. In addition to the protocol, there are also several bilateral agreements to facilitate sharing of water resources. These include:

- the Zambezi River Action Plan, between Zambia and Zimbabwe;
- the Permanent Okavango River Basin Water Commission;
- the Joint Permanent Technical Commission Over the Lesotho Highlands Water Project; and
- the Permanent Joint Technical Commission on the Cunene River.

Cooperation has also been developed in the areas of research and hydro-meteorological data, through SADC-HYCOS. Whilst there is commitment and cooperation at the sub-regional political level, further capacity needs to be built, in order to effectively implement the tenet of shared water management at grassroots level.

*Source: SADC 2000*

in 1997, and represents a significant commitment towards meeting the challenge of providing adequate water service and supply in the sub-region, as well as protection of the environment. The Regional Strategic Action Plan reaffirms the importance of the sub-region's water resources, and its influences on all aspects of the region's economic and social performance. Other initiatives that have occurred in the sub-region in the water sector, in a bid to achieve regional integration, include the Water Weeks, instituted by SADC. The Water Weeks comprise a series of national workshops aimed at informing key stakeholders about SADC-wide water initiatives, and to involve them in discussions about the implementation of these initiatives.

Effluent water standards have been established and implemented in South Africa and Namibia, as measures to control water quality. In South Africa, in addition to a General Effluent Standard, there is a Special Effluent Standard, designed to protect mountain streams that can support trout (Water Act 54 of 1956). A National River Health Programme has also been implemented, in order to provide regularly updated information on the biological health of river systems in South Africa, and to provide recommendations for IWRM at the catchment level. The National River Health Programme is also responsible for a series of 'State of the Rivers' reports, which present the information in a user-friendly manner.

## CENTRAL AFRICA

Central Africa rarely experiences problems of water availability, because rainfall is high and generally predictable. However, there is growing competition between user groups, and access to water, particularly in rural areas, is seen as a priority for development. Localized problems of water quality are being raised, especially in coastal areas, where industrial, agricultural and domestic wastewater discharges are high, and where there is an additional threat of saltwater intrusion.

### AVAILABILITY OF FRESHWATER IN CENTRAL AFRICA

With the exceptions of the deserts of northern Chad and the Sahelian parts of northern Cameroon and central Chad, the Central Africa region is well-endowed with

water resources. The total annual renewable resources for the region in 2000 were estimated at 1 775 km$^3$/yr (UNDP and others 2000). There were significant variations between countries however, with the DRC having the greatest water resources (935 km$^3$/yr or 18 000 m$^3$/capita/yr), and Chad being the driest country (15 km$^3$/yr or 1961 m$^3$/capita/yr) (UNDP and others 2000). Although water use has been on the increase, water deficit is generally unknown in the region. The Congo River basin is the largest in Africa, covering 12 per cent of the region, and is shared by nine countries. The water resources of this area are vital to support livelihoods and economic development, in particular, by providing irrigation water for the cultivation of cash crops. There is some inter-annual variation in rainfall in Central Africa, where flooding is more common in the humid zone. Only in Chad and northern Cameroon is drought a serious threat, and drought frequency has increased over the past 30 years (IPCC 1998). In 1973, drought killed 100 000 people in the Sahel, and even countries in the humid zone suffered lowered rainfall and reduced crop yields (Gommes and Petrassi 1996).

The Lake Chad basin is a depression of the seven countries grouped around it, forming a freshwater lake (the Conventional Basin), which is shared by Cameroon (9 per cent), Chad (42 per cent), Niger (28 per cent) and Nigeria (21 per cent). Lake Chad is an important source of water and economic activities, including agriculture and fisheries. Satellite images show that the lake has shrunk considerably over the past 30 years, and is now 5 per cent of its former size, due to persistent low rainfall in the region. IPCC predicts reduced rainfall and run-off, and increased desertification, risks in the Sahelian belt (IPCC 2001), which could mean further reductions in the size of the lake. This would threaten the livelihoods of fishermen and farmers in the region, and poses a challenge for IWRM and international cooperation to ensure sustainability of the resource.

### Access to freshwater resources in Central Africa

In 1998, the annual withdrawal of freshwater for Central Africa was estimated to have been less than 1 per cent of the total available. However, the uneven distribution of water resources, with respect to time and population distribution, has created challenges for water supply. The traditional response to this challenge has been to dam rivers and to distribute water to the people, rather than resettling people closer to water resources. The DRC has 14 dams, Cameroon has nine, and Gabon and Congo each have two. Their main functions are to provide water for domestic consumption, and to provide hydroelectric power, although it is estimated that only half the potential for hydroelectric power is exploited (WRI, UNEP and UNDP 1992). Because of the relatively high reliability of rainfall in this sub-region, irrigation is not always required, and the agricultural sector only consumes 33 per cent of all withdrawals, whereas the domestic sector accounts for more than 50 per cent of all withdrawals (UNDP and others 2000, Shiklomanov 1999). Despite these efforts to provide water to municipalities in Central Africa, there are significant shortfalls, for example, in Chad, where only 27 per cent of the population has access to improved water sources (WHO/UNICEF 2000).

The reasons for the relatively low access to water in the sub-region include poor economic growth and, thus, low investment in infrastructure development and maintenance, as well as rapid population growth and migration to urban centres. The demand for water for domestic use is predicted to increase fivefold over the next 25 years, due to population growth and increases in per capita consumption (see Figure 2e.9). Significant improvements to the existing infrastructure and supply networks are thus required, in order to improve access to potable water over the next 20–30 years. Increasing demand from other sectors is also expected, as agricultural and industrial developments expand to meet economic growth imperatives.

## Improving availability and access to freshwater resources in Central Africa

The increases in demand for water in Central Africa are unlikely to lead to conditions of water stress or water scarcity, because the estimated withdrawals are still small compared to the available resources. However, reductions in rainfall related to climate change are expected in parts of northern Cameroon and Chad, and this will exacerbate already inadequate water supply systems. Therefore, localized problems of water supply may be exacerbated. In addition, further studies are required to investigate the ecological consequences of damming rivers, diverting flows and abstracting water, before ecologically and socially acceptable standards can be established.

Cameroon, Chad, Niger and Nigeria formed the Lake Chad Basin Commission in 1964, and were later joined by Central African Republic in 1999. The objective of the Lake Chad Basin Commission is to ensure the most rational use of water, land and other natural resources, and to coordinate regional development. In March 1994, the commission approved the Master Plan for the Development and Environmentally Sound Management of the Natural Resources of the Lake Chad Conventional Basin. This document consisted of 36 projects relating to: water resources; agriculture; forestry; biodiversity management livestock; and fishery developments. Funding was made available through the GEF, in order to coordinate these projects by means of: the establishment of joint regulations on the development and use of resources; information collection and sharing; and joint actions and research programmes. An international campaign to save Lake Chad was also launched. In July 2000, the Lake Chad Basin Commission states met, and agreed on a proposal to establish the whole of Lake Chad as a transnational Ramsar site. Other projects agreed to include the development of an Inter-basin Water Transfer Scheme, and strengthening of Joint Patrol Teams (Lake Chad Basin Commission 2000).

## WATER QUALITY IN CENTRAL AFRICA

Freshwater quality in parts of Central Africa is declining, as a result of: pollution from industrial and sewage outflows; agricultural run-off; and saltwater intrusion.

**Figure 2e.9 Water use by sector in Central Africa, 1900–2025**

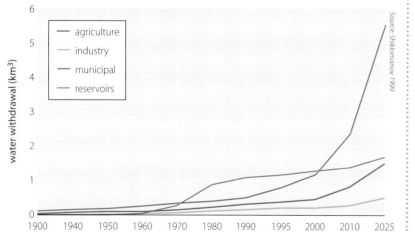

Source: Shiklomanov 1999

The Kouiloo River, Congo—one of many rivers in the coastal area of Central Africa facing contamination by industry and siltation from excessive soil erosion.

*Michel Gunther/Still Pictures*

The Congo River has been recognized as one of the cleanest in the world, due to the absence of industry, large urban settlements and agriculture along its banks (Johnson 1999). However, coastal areas are experiencing run-off and effluent from neighbouring agro-plantations and industries, which contaminate rivers, marshes and groundwater reserves. Clear felling in the hinterland also contributes to high rates of soil erosion, and to the siltation of rivers and estuaries. In particular, the Lobe and Kienke Rivers (Cameroon) are especially high in sediment, which is deposited in the coastal zone and prevents the development of reef-building coral (World Bank 2000). Salinization occurs where freshwater sources come into contact with brackish or saltwater. The coastal industrial town of Douala in Cameroon has problems with the salinization of drinking water from aquifer resources, because rising sea levels facilitate permeation of underground aquifers by seawater. There is also inadequate provision of sanitation for the city's 1.4 million inhabitants, and raised nutrient levels are recorded in the estuary, as result of sewage discharge (Gabche and Smith 2001).

Coastal freshwater resources are further threatened by sea level rise, which the IPCC predicts could be as much as one metre by 2100 (IPCC 2001).

The impacts of declining water quality are both ecological and social, with the contamination of freshwater habitat resulting in: losses of biodiversity; regulation of water flow; and reduced availability of exploitatable water for domestic, industrial, agricultural and recreational functions. There are also heightened risks of waterborne diseases associated with lack of potable water and sanitation, and costs of water treatment are increasing at a time when investment in this sector is already inadequate. For example, the drinking water supply in Yaounde, Cameroon, experiences occasional shortages and interruptions, driving many people to fetch water from alternative sources, such as springs and wells. A recent study of the bacteriological status of fifteen of these wells and springs revealed high densities of faecal bacteria and, thus, high risk of disease (Nola, Njine, Monkiedje, Sikati, Foko, Djuikom and Tailliez 1998). Several wetlands in the region have been degraded, due to the construction of dams and irrigation schemes, which divert water, and thus alter hydrological regimes. The reduction in water flow, and the increased silt and nutrient levels from agricultural return flows, have resulted in loss of favourable habitat and breeding grounds, and several wetland species have been lost.

## Improving water quality in Central Africa

In response to pressures affecting water quality, many programmes and actions are under way in Central Africa. These include: schemes for the drainage, purification and decontamination of freshwater systems; water management programmes; public awareness campaigns; ratification of relevant regional and transfrontier conventions for water resources protection and management; and efforts to implement water quality standards and control. National and international organizations have been instrumental in water supply and sanitation projects at the local level, and vehicles for international cooperation have been established for management of shared water resources. The Gulf of Guinea Large Ecosystem Project, for example, includes the management of land-based activities which contribute to declining water quality. The GEF-funded Reversal of Land and Water

## Box 2e.10  Wetland rehabilitation in Cameroon

Each year, the Logone River (Cameroon) floods an area of about 6 000 km². This wetland supports large herds of giraffe, elephant, lions and various ungulates (including topi, antelope, reedbuck, gazelle and kob). Part of the floodplain has been designated as the Waza National Park, which attracts thousands of tourists per year, and which serves as a fish nursery in the flood season. Livestock grazing is also an important activity, supporting the livelihoods of many local communities. However, the wetland has been significantly reduced in size since the construction of a barrage across the floodplain in 1979.

This barrage created Lake Maga, which supplies water for the irrigation of nearby farmland. Fish stocks fell by 90 per cent, and grazing potential has decreased, impacting on local livelihoods. Plans to rehabilitate the wetland were established in 1993, and the embankments along the river were modified at an estimated cost of US$5 million over eight years. Stakeholders and local community members were involved in the planning and design of the project, which reinundated several thousand hectares of land. Small-scale fishing activities have recommenced, and clean water has been supplied to 33 villages from 37 wells. Together with training in health and sanitation, this has succeeded in lowering the incidence of diarrhoea by 70 per cent.

Degradation project has been initiated through the WB, in order to improve the management of resources in the Lake Chad basin. Gabon now has three Ramsar sites, following ratification of the Ramsar Convention in 1987 (Ramsar 2001), and Cameroon is undertaking the rehabilitation of wetlands that were compromised due to inappropriate development policies (see Box 2e.10).

Between 1990 and 2000, the percentage of the population with access to potable water supply increased from 52 per cent to 62 per cent in Cameroon, and from 59 per cent to 60 per cent in Central African Republic (WHO/UNICEF 2000). Access to sanitation has shown very slight increases, and rates are still far below other countries in Africa. Continued increased investment in this area is required, in order to mitigate pollution and the outbreak of diseases arising from inadequate sanitation and wastewater treatment.

## WESTERN AFRICA

Most Western African countries are well-endowed with freshwater, except for those bordering on the Sahel, which frequently experience drought. Accessibility to freshwater and its integrated management nonetheless remain major concerns in Western Africa. Inappropriate management of freshwater, and competition between user groups, limit efforts by governments and the international community to encourage economic development and to improve the standard of living in Western Africa. There are also rising concerns over freshwater quality, in terms of pollution from domestic effluents and industrial wastewater, particularly in the coastal zone.

## AVAILABILITY OF FRESHWATER IN WESTERN AFRICA

With the exception of Cape Verde, all the countries in the sub-region share surface water resources with one or more other countries. The sub-region is drained by three major basin systems. The Niger basin drains an area of 2 million km² (33 per cent of the total surface area of the sub-region), and involves 9 of the 16 sub-regional countries, including Cameroon and Chad. Other important basins are: the Senegal basin, shared by four countries; the Gambia basin, shared by three countries; the Bandama basin in Côte d'Ivoire; the Comoe basin, shared by four countries; and the Volta basin, shared by five countries. The sub-region's freshwater resources are unevenly distributed between countries. Liberia, for example, has internal renewable water resources of more than 63 000 m³/capita/yr, and Mauritania has only 150 m³/capita/yr (UNDP and others 2000). Temporal variation in rainfall is common throughout the sub-region, but only those countries in the northern Sahelian zone (Mali, Mauritania, and Niger) regularly experience drought, whilst countries in the wetter coastal belt are periodically affected by floods.

Three major types of groundwater aquifers are observed in the region, namely: basement aquifers; deep coastal sedimentary aquifers; and superficial aquifers. The availability of groundwater varies considerably from one type of substrate to another, and according to the local levels of precipitation and infiltration, which determine the actual recharge. In Mauritania, for example, internal renewable groundwater resources are estimated at 0.3 km³/yr (FAOSTAT 1997), and these are important sources of water for domestic use, irrigation and livestock watering. About 400 000 people live in the 218 oases, and are dependent on 31 400 wells, extracting the water manually (FAOSTAT 1997). The water is used to irrigate 4 751 ha of palm trees, with 244 ha of annual crops under them (FAOSTAT 1997).

Six Western African countries are expected to experience water scarcity by 2025, namely: Benin; Burkina Faso; Ghana; Mauritania; Niger; and Nigeria (Johns Hopkins 1998). Climate change is predicted to bring about reduced rainfall and increased evaporation in the areas to the north, advancing the rate of desertification in the Sahel (IPCC 2001). Countries in the coastal zone may experience more intense rainfall and increased run-off. Combined with existing high rates of deforestation and degradation of vegetation cover, this could have serious consequences for soil erosion and agricultural productivity.

### Access to freshwater resources in Western Africa

Demand for water has been steadily increasing in all sectors, as a result of population growth; commercial agricultural expansion; and industrial development. Current total withdrawal of water for domestic, industrial and agricultural consumption is 11 km³/yr, and demand for water from all sectors is expected to increase to some 36 km³/yr by 2025, as shown in Figure 2e.10 (UNDP and others 2000, Shiklomanov 1999).

The Volta and Niger Rivers have been dammed to supply water for irrigation and domestic consumption, as well as to generate hydroelectric power. However, this has created problems of accelerated erosion in the coastal zone, as well as marginalization of pastoralists, who are dependent on seasonal floods (for example, Acreman 1999). Despite agriculture being the largest water user, accounting for 70 per cent of all withdrawals in the sub-region in 1995 (Shiklomanov 1999), irrigation potential remains largely untapped, especially in the Sudano-Sahelian zone, where only 16 per cent (5.3 million ha) of potentially irrigated lands have been developed (Falloux and Kukendi 1988).

Access to piped water and sanitation also remains low, despite considerable improvements during the International Drinking Water Supply and Sanitation Decade (1981–1990). In 2000, total water supply coverage was highest in Senegal (78 per cent) and lowest in Sierra Leone (23 per cent), with most urban areas in the sub-region better supplied than rural areas (WHO/UNICEF 2000).

**Figure 2e.10 Water use by sector in Western Africa, 1900–2025**

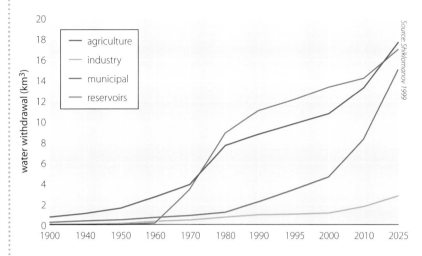

Source: Shiklomanov 1999

### Improving availability and access to freshwater resources in Western Africa

IWRM is recognized as a priority for Western Africa. However, its implementation has been constrained by a number of factors. The main obstacles are limited technical skills in the sub-region, coupled with the complexity of the integrated approach. In addition, the increasing demand for water, freshwater pollution and disruptions to the hydrological cycle have complicated water resources management. The deterioration of most of the infrastructure for hydrometeorological and hydrogeological data collection, compilation, analysis and dissemination, as well as lack of funding, have led to poor maintenance. This situation is aggravated by political instability, due largely to armed conflicts, ethnic strife and inadequate grassroots participation in decision making. However some countries have adopted strategies on IWRM and development, or have enacted water laws and established institutions to enforce the laws. Economic incentives for water demand management have been successfully implemented in some areas, for example, Conakry, Guinea (see Box 2e.11).

A recent project to improve water supply to 90 villages in Mauritania has demonstrated the importance of the involvement of community members in the design, establishment and maintenance of facilities. This UNDP-funded programme created a private network for the construction and maintenance of water

### Box 2e.11  Private sector involvement in water services in Western Africa

In 1998, 50 per cent of all the water pumped in Conakry (Guinea), was unaccounted for, and only 10 per cent of water bills were collected. Since a private company took over the management of water services in the city in 1989, water losses have been halved, and 85 per cent of bills are now paid. In Lomé (Togo), the public water utility has recently undergone improvements. Losses are now only 20 per cent, almost all water bills are paid and water subsidies are not required.

*Source: Menard and Clarke, 1996*

supply equipment, providing employment to the community, and ensuring back-up supplies of spare parts. A water-point committee was also established in each village, in order to take responsibility for management and maintenance costs, and local people were trained to repair the equipment. On completion of the project, more than 90 per cent of the villages were independently funding the maintenance of their pumps, with help from the people trained during the project (UNDESA 2000).

Women collecting water from a village well in Burkina Faso

*Glen Christian/Still Pictures*

There is also a growing tendency in the countries of Western Africa to intensify monitoring, assessment and policy reform for the water resources sector. New measures are being introduced for water conservation and the development of infrastructure, in order to improve the integration of water and land management. Furthermore, during the past three decades, numerous intergovernmental organizations responsible for the development and management of shared rivers in the region have emerged (in particular, for the management of the Senegal, Gambia and Niger Rivers), although their actual performance remains far below their potential.

## WATER QUALITY IN WESTERN AFRICA

Rising levels of pollution of surface and groundwater resources in the sub-region compounds inadequate access to freshwater. The primary sources of pollution are sewage and industrial effluents. Agricultural run-off is an additional burden, through contamination with chemical residues and silt, and increased nutrient levels. This is particularly noticeable in the coastal areas of Western Africa, where most industries and commercial agricultural plantations are located. Coastal wetlands, which are particularly sensitive to pollutants, are experiencing severe degradation. The Korle lagoon in Accra, Ghana, for example, is one of the most heavily polluted wetlands in Western Africa, receiving effluents from solid waste dumps, industries, sewage outfalls, and agricultural run-off. Effluents washed out to sea are also having a negative impact on coastal and marine habitats, and are posing a health risk to humans (IDRC 1996). In recent years, 36 wetlands in Western Africa have been designated as Wetlands of International Importance under the 1971 Ramsar Convention, including six sites in Ghana and six in Guinea (Ramsar 2001).

Degradation of water quality inland also creates problems for domestic, industrial and agricultural users. In Yeumbeul, Senegal, 7 000 households are dependent on a groundwater source that was recently found to harbour nitrate concentrations well above WHO's guideline of 50 mg/l. Two sources of contamination were identified, namely: shallow nearby latrines; and waste organic matter carried in groundwater (Tandia, Diop and Gaye 1999). Diarrhoea, ascariasis (roundworm), dracunculiasis (guinea worm) and schistosomiasis (bilharzia) are among the most prevalent waterborne diseases, and are associated with standing water, and lack of safe drinking water and sanitation.

### Improving water quality in Western Africa

The Africa 2000 initiative of the WHO Africa Regional Office has seen considerable gains in water supply and sanitation in Western Africa. For example, in Ghana, Drinking Water Quality Guidelines are the same as WHO guidelines, and disinfection of drinking water has 100 per cent coverage in Accra. A new policy on water and sanitation provision, introduced in 1993, encourages user communities to become more self-reliant, especially in rural areas, where the provision of services is through community ownership and management of the facilities. In Mauritania, gains under the Africa 2000 initiative include: a research programme to optimize water treatment with the use of appropriate technology; and training of water treatment plant operation and maintenance technicians. Training of hygiene workers in water quality analysis, and the promotion of a hygiene education programme in villages, schools and urban areas, has resulted in several latrines being built in schools and individual households.

A number of Western African countries are members of the African Union of Water Distributors, namely: Benin; Burkina Faso; Côte d'Ivoire; Ghana; Guinea; Liberia; Mali; Mauritania; Niger; Senegal; and Togo. The African Union of Water Distributors has been instrumental in improving water exploitation, distribution and pricing among member states. Western African states have also formed a water research organization, the Inter-African Committee for Hydrology Studies, based in Ouagadougou. This committee deals with extension work to spread new technologies and information concerning water in Africa (Biémi 1996).

Countries of Western Africa would benefit from developing and enforcing effective water pollution prevention and control programmes which combine strategies for reducing pollution at source, prior to EIAs of new developments. Criteria for monitoring the biological, physical and chemical quality of water bodies need to be established, and mechanisms need to be put in place to address deviations from standards. Above all, water resources require integrated management, with the participation of all stakeholders in decision making and management, and with the inclusion of the environment as a legitimate water user. International water-quality monitoring and management programmes—such as the Global Water Quality Monitoring Programme, the UNEP Environmentally Sound Management of Inland Waters and the Food and Agriculture (FAO) Regional Inland Fishery Bodies—provide support and coordination for such activities.

## CONCLUSION

Africa's water resources are characterized by extreme variability, both over time and space. Rainfall varies from less than 20 mm/yr in the arid countries encompassing the Sahara desert to more than 1 000 mm/yr in the humid tropical belt of Western, Eastern and Central Africa. Inter-annual variations can be extremely high, and drought and/or flooding is common in most African countries. The cost of such extreme events runs to millions of dollars every year, a price that many African countries cannot afford either to incur or to prevent.

Natural climatic variations, and the location of major urban and industrial centres in dry or water-stressed areas, presents a formidable challenge for water service providers. With anticipated increases in unpredictability and variation in precipitation due to global climate change, and increases in demand from a rapidly growing population and economic developments, the way in which water resources are managed in Africa needs to be radically reformed in order to meet the goal of equitable access to sufficient water of acceptable quality, and to facilitate agricultural and industrial development. (See Box 2e.12.)

Increases in demand for freshwater are anticipated in all African countries, in all sectors (domestic, agricultural and industrial), over the next 10 years. In some countries, demand is projected to double within the next 30 years. This arises not just from an increase in the number of people requiring water, but also from increasing consumption patterns, especially among the wealthier communities. Likewise, industries rarely pay the true price of water and are, therefore, not encouraged to recycle or to reduce their consumption. Surface and groundwater

*Increases in demand for freshwater are anticipated in all African countries, in all sectors (domestic, agricultural and industrial), over the next 10 years. In some countries, demand is projected to double within the next 30 years*

## Box 2e.12  Attaining Africa's water vision

In order to achieve Africa's water vision of sustainable access to clean, safe water for all, various reforms to water policies and management strategies are required. The most important of these, which will be monitored according to a series of milestones over the next 25 years, are: political commitment and support at the grassroots level, together with openness and accountability in decision making; enhanced information gathering and dissemination; regional cooperation and decisive action; and sustainable financing and cost recovery methods.

*Source: World Water Council 2000*

supplies have been exploited, and some countries have turned to alternative sources of water, in order to attempt to meet this increasing demand, including desalinization of seawater and recycling of wastewater.

Most African countries cite declining water quality as a priority issue for the environmental and developmental agenda. The most common sources of freshwater pollution are sewage, industrial effluents and agricultural run-off. They are due, in large part, to inadequate wastewater treatment facilities, as well as to ineffective pollution control.

The principles of IWRM have been widely adopted in Africa, as a systematic approach for tackling the technical, hydrological, economic, environmental, social and legal aspects of water quality and supply problems. A central tenet of IWRM is to move away from supply management methods, such as the construction of dams, towards demand management methods, such as removal of water subsidies and the enforcement of polluter payments. IWRM also requires the joint management of water resources by all users, from local communities to water service providers, municipalities, industries and agricultural organizations. Although adopted in principle in the region, implementation of IWRM has, to date, been impeded by: capacity constraints; lack of financial resources; institutional fragmentation; poor availability of information; and lack of commitment by various partners.

●

*Most African countries cite declining water quality as a priority issue for the environmental and developmental agenda. The most common sources of freshwater pollution are sewage, industrial effluents and agricultural run-off. They are due, in large part, to inadequate wastewater treatment facilities, as well as to ineffective pollution control*

●

## REFERENCES

Accra Sustainable Programme (2001). Reaching Consensus Through a City Consultation. Available on: http://www.unchs.org/scp/cities/accra.htm

Acreman, M. (1999). Water and Ecology Linking the Earth's Ecosystems to its Hydrological Cycle. In *Revista Cidob d'Afers Internacionals*, 45–46, April 1999

Bermudez, O. (1999). The Mineral Industry in Kenya. In U.S. *Geological Survey Minerals Yearbook 1999*. USGS, Reston, VA

Biémi, J. (1996). Water Crises and Constraints in West and Central Africa: The Case of Côte D'Ivoire. In *Water Management in Africa and the Middle East: Challenges and Opportunities* (eds. Rached, E., Rathgeber, E. and Brooks, D. IDRC, Ottawa

CEDARE (2000). Programme for the Development of a Regional Strategy for the Utilisation of the Nubian Sandstone Aquifer System. Draft report, CEDARE, Cairo.

Chenje, M., (ed.) 2000. *State of the Environment Zambezi Basin 2000*. SADC/IUCN/ZRA/SARDC, Maseru, Lusaka, Harare

Citet (2001). Water and Waste Water Management. Tunis International Center for Environmental Technologies. Available on: http://www.citet.nat.tn/english/water/assainissement.html

DEA&T (1999). *State of the Environment South Africa. Department of Environmental Affairs and Tourism*, Pretoria, South Africa. Available on: http://www.environment.gov.za

Eales, K., Forster, S. and Du Mhango, L. (1996). Strain, Water Demand, and Supply Directions in the Most Stressed Water Systems of Lesotho, Namibia, South Africa, and Swaziland. In *Water Management in Africa and the Middle East: Challenges and Opportunities* (eds. Rached, E., Rathgeber, E. and Brooks, D.). IDRC, Ottawa

EIA (2000). Country Analysis Briefs: Egypt: Environmental Issues. United States Energy Information Administration. Available on: http://www.eia.doe.gov/emeu/cabs/egypt.html

Environment News Service (2001). Kenya's Pink Flamingos Weighed Down by Heavy Metals. Environment News Service 16 July 2001. Available on http://ens.lycos.com/ens/jul2001/2001L-07-16-04.html

Falloux, F. and Kukendi, A.. (eds.) (1988). Lutte contre la Désertification et Gestion des Ressources Sahélienne et Soudanienne de l'Afrique de l'Ouest, Document Technique no. 70F

FAO (1995). Irrigation in Africa in Figures. Water Report 7. FAO, Rome

FAOSTAT (1995). AQUASTAT Database: Country Profiles for Mauritius, Seychelles. FAO, Rome

FAOSTAT (1996). AQUASTAT Database: Country Profile for Ethiopia. FAO, Rome

FAOSTAT (1997). AQUASTAT database: Country Profiles for Egypt, Mauritania, Sudan, Tunisia. FAO, Rome

FAOSTAT (2000). AQUASTAT Database. FAO, Rome

FAOSTAT (2001). FAOSTAT Database: Agriculture Data. FAO, Rome

Feresu, S.B. and van Sickle, J. (1990). Coliforms as a Measure of Sewage Contamination of the River Zambezi. In *Journal of Applied Bacteriology* 68, pp. 397–403.

Gabche, C.E. and Smith, S.V. (2001). Cameroon Estuarine Systems. In *Land-Ocean Interactions in the Coastal Zone International Project Biogeochemical Modelling Node*. Available on: http://www.nioz.nl/loicz/

GJMC (1999). State of the Environment Johannesburg. Available on: http://www.ceroi.net/reports/johannesburg/csoe/navIntro.htm

Global Water Partnership (2000). *Southern African Vision for Water, Life and the Environment in the 21st Century and Strategic Framework for Action Statement*. SATAC. Available on: http://www.watervision.org

GOM/ERM (1998). *Mauritius NEAP II: Strategy Options Report.* Government of Mauritius and Environmental Resources Management

Gommes, R. and Petrassi, F. (1996). Rainfall Variability and Drought in Sub-Saharan Africa Since 1960. In FAO *Agrometeorology Working Paper* No. 9. FAO, Rome

Government of Mauritius (1994). *National Physical Development Plan*. Ministry of Housing, Lands and Country Planning, Port Louis, Mauritius

Gumbo, B. (1997). Integrated Water Quality Management in Harare. 23rd WEDC Conference: Water and Sanitation for All: Partnerships and Innovations. Durban, South Africa

Hailu, S.S. (1998). Hydropower of Ethiopia: Status, Potential and Prospects. In *Ethiopian Association of Civil Engineers Bulletin* vol. 1, no. 1

Hinrichsen, D., Robey, B. and Upadhyay, U.D. (1997). Solutions for a Water-Short World. In *Population Reports*, series M, no. 14, December 1997. Johns Hopkins School of Public Health, Population Information Program, Baltimore

ICCON (2001). Nile Basin Initiative website. Available on: http://www.nilebasin.org/ICCON1.htm

IDRC (1996). Ghana: The Nightmare Lagoons. Article written by Theo Andersen, Friends of the Earth. Available on: http://www.idrc.ca/books/reports/e234-13.html

IIED (2000). *Drawers of Water II: Thirty years of Change in Domestic Water Use and Environmental Health in East Africa*. International Institute for Environment and Development, London

Institute for Environmental and Legal Studies (1998). Mauritius and its Environment; Water: Resources, Uses and Pollution. Available on: http://www.intnet.mu/iels/index.htm

IPCC (1998). T*he Regional Impacts of Climate Change*. Intergovernmental Panel on Climate Change, Geneva

IPCC (2001). IPCC Third Assessment Report of Working Group 1: C*limate Change 2001: Impacts, Adaptation and Vulnerability: Summary for Policy Makers*. IPCC, Geneva

IUCN (2001). Waza-Logone Floodplain Restoration, Cameroon. Available on: http://www.iucn.org/themes/wetlands/waza.html

IUCN (2000). *Water Demand Management: Towards Developing Effective Strategies for Southern Africa*. IUCN, Harare

Johns Hopkins (1998). Solutions for a Water-Short World. In Population Report, vol. XXVI, no. 1, September 1998. Johns Hopkins Population Information Program, Baltimore, Maryland. Available on: http://www.jhuccp.org/popreport/m14sum.stm

Johnson, D. (1999). Congo River Called one of World's Cleanest. 3 December 1999. Africana.com On-line News

Jootun, L., Bhikajee, M., Prayag, R. and Soyfoo, R. (1997). Report on Inventories of Land-based Sources of Pollution in Mauritius. University of Mauritius, Mauritius

Kelly, P., Khanfir, H., David, P., Arata, M. and Kleinau, E.F. (1999). Environmental and Behavioural Risk Factors for Diarrhoeal Diseases in Childhood: A Survey in Two Towns in Morocco. Activity Report No 79 Prepared for USAID Mission to Morocco under Environmental Health Programme Activity No. 526CC

Khroda, G. (1996). Strain, Social and Environmental Consequences, and Water Management in the Most Stressed Water Systems in Africa. In *Water Management in Africa and the Middle East: Challenges and Opportunities* (eds. Eglal Rached, Eva Rathgeber and David B. Brooks). IDRC, Ottowa

Lake Chad Basin Commission (2000). Final Communique of the Tenth Summit, July 2000. Available on: http://www.ramsar.org/w.n.chad_summit_e.htm

Macdonald I.A.W. (1989). Man's Role in Changing the Face of Southern Africa. In *Biotic Diversity in Southern Africa: Concepts and Conservation*. (ed. B.J. Huntley) pp. 51–77. Oxford University Press, Cape Town

Macdonald, I.A.W. and Richardson, D.M. (1986). Alien Species in Terrestrial Ecosystems of the Fynbos Biome. In The *Ecology and Management of Biological Invasions in Southern Africa* (eds. Macdonald, I.A.W., Kruger, F.J. and Ferrar, A.A.) pp. 77–91. Oxford University Press, Cape Town

McKinney, D.C., X. Cai., M.W. Rosegrant, C. Ringler, and C.A. Scott. (1999). Modeling Water Resources Management at the Basin Level: Review and Future Directions. In *SWIM Repot 6*. Colombo, Sri Lanka. International Water Management Institute.

Ménard, C. and Clarke, G. (1996). A Transitory Regime: Water Supply in Conakry, Guinea. Université de Paris, France/World Bank, Washington D.C. USA

Ministry of Water Resources (1998). *Ethiopia Water Resource Management Policy*. Federal Democratic Republic of Ethiopia

Mpofu, B. (2000). *Assessment of Seed Requirements in Southern African Countries Ravaged by Floods and Drought 1999/2000 Season*. Commissioned by SADC/GTZ, published on the SADC Regional Remote Sensing Unit's website: http://www/FANR.SADC.net

Mwangi, S. and Munga, D. (1997) Kenya Chapter of the Strategic Action Plan for Land-based Sources and Activities Affecting the Marine, Coastal and Associated Fresh Water Environment in the Eastern African Region. Report prepared by the Food and Agriculture Organization of the United Nations project for the Protection and Management of the Marine and Coastal Areas of the Eastern African Region (EAF/5)

Myllyla, S. (1995). Cairo—A Megacity and its Water Resources. Presented at the Third Nordic Conference on Middle Eastern Studies: Ethnic Encounter and Culture Change. Joensuu, Finland, 19–22 June 1995

Napier, M. (2000). Human Settlements and the Environment. Report prepared for the Department of Environmental Affairs and Tourism, Pretoria, South Africa. Available on: http://www.environment.gov.za/soer/index.htm

NASA Global Earth Observing System (2001). A Shadow of a Lake: Africa's Disappearing Lake Chad. GSFC On-line News. Available on: http://www.gsfc.nasa.gov/gsfc/earth/environ/lakechad/chad.htm

NEMA (1999). State of the Environment Report for Uganda 1998. National Environment Management Authority, Kampala

NEMA (2001). State of Environment Report for Uganda 2000/2001. National Environment Management Authority, Kampala

Nola, M., Njine, T., Monkiedje, A., Sikati Foko, V., Djuikom, E. and Tailliez, R. (1998). Bacteriological Quality of Spring and Well Water in Yaounde, Cameroon. In Cahiers D'études et de Recherches Francophones/Santé vol. 8, issue 5, pp. 330–6.

OFDA (2000). Statistics Database. United States Office for Disaster Assistance

Olal, M.A., Muchilwa, M.N. and Woomer, P.A. (2001). Water Hyacinth Utilisation and the Use of Waste Material for Handicraft Production in Kenya In: Micro and Small Enterprises and Natural Resource Use. (ed. D.L. Manzolillo Nightingale) 2001. Proceedings of a workshop held at ICRAF, Nairobi, Kenya. 21–22 February 2001. Micro-Enterprises Support Programme and United Nations Environment Programme, Nairobi

ONE (1997). Rapport Sur L'Environment Urbain. Cas de la Zone d'Antannarivo. Edition 1997. Banque Mondial IDA 2125 MAG. Office National Pour L'Environment, Antannarivo

ONE and INSTAT (1994). Rapport sur l'Etat de l'Environnement. Office National de l'Environnement et Institut National de la Statistique. PNUD (UNDDSMS/MAG/91/007). Banque Mondiale (IDA/2125 MAG), Edition 1994, Antananarivo

Rahmato, D. (1999). Water Resource Development in Ethiopia: Issues of Sustainability and Participation. Forum for Social Studies Discussion Paper, June 1999. Addis Ababa

Ramsar (2001). The Ramsar List of Wetlands of International Importance. Available on: http://www.ramsar.org/index_list.htm

Ribbink, A.J. Marsh, A.C., Ribbink, A.C. and Sharp, B.J. (1983). A Preliminary Survey Of The Cichlid Fish Of The Rocky Habitats Of Malawi. In African Journal of Zoology 18(3), pp. 149–310

Sharma, N. (1996). Summary Report of Water Resources Management in Sub-Saharan Africa Workshops. World Bank, Washington D.C.

Shiklomanov, I.A. (1999). World Water Resources: Modern Assessment and Outlook for the 21st Century. Federal Service of Russia for Hydrometeorology and Environment Monitoring, State Hydrological Institute, St Petersburg

Soils Incorporated (Pty) Ltd and Chalo Environment and Sustainable Development Consultants (2000). Kariba Dam Case Study. Prepared as an input to the World Commission on Dams, Cape Town. Available on: www.dams.org

Tandia, A.A., Diop, E.S. and Gaye, C.B. (1999). Nitrate Groundwater Pollution in Suburban Areas: Example of Groundwater from Yeumbeul, Senegal. In Journal of African Earth Sciences, 29(4), pp. 809–822

United Nations Mar del Plata Plan of Action (1977). Report of the United Nations Water Conference, Mar del Plata. (United Nations publication, sales number E.77.II.A.12), Chapter I.

UNCHS/UNEP (2001). Managing Water for African Cities: Demonstration Project in Johannesburg. Available on: http://www.un-urbanwater.net/cities/joburg.html

UNDESA (2000). Water-supply in Rural Areas: Feedback from Experiments in 90 villages in Mauritania. Available on: http://www.un.org/esa/sustdev/success/cb3.htm

UNDP (1998). Human Development Report 1998. United Nations Development Programme. Oxford University Press, New York

UNDP, 2000. Human Development Report 2000. United Nations Development Programme. Oxford University Press, New York.

UNDP, UNEP, World Bank and WRI (2000). World Resources 2000–2001: People and Ecosystems: The Fraying Web of Life. World Resources Institute, Washington D.C.

UNEP (1998). Eastern Africa Atlas of Coastal Resources. United Nations Environment Programme, Nairobi

UNEP (1999a). Global Environment Outlook. United Nations Environment Programme, Nairobi

UNEP (1999b). Western Indian Ocean Environment Outlook. United Nations Environment Programme, Nairobi

UNEP (2002). African Land and Water Initiative of the Global Environment Facility: Towards An African Strategic Partnership On Sustainable Development For The New Millennium. Available on: http://gef-forum.unep.org/Background%20Papers/Land-water/bg_land-water.htm

UNEP/COI (1997). Waste Management in Small Island Developing States in the Indian Ocean Region. Report of Regional Workshop, Proceedings and Papers Presented. Quatre Bornes

World Conservation Monitoring Centre (WCMC), 1992. Global Biodiversity: Status of the Earth's Living Resources. Chapman and Hall, London

WHO/UNICEF (2000). Global Water Supply and Sanitation Assessment 2000 Report. World Health Organization, Geneva

WMO (2001). Status of WHYCOS Components. Available on: http://www.wmo.ch/web/homs/status.html#aoc-hycos

Working for Water (2000). Annual Report. Department of Water Affairs and Forestry, Pretoria

World Bank (1996). From Scarcity to Security: Averting a Water Crisis in the Middle East and North Africa. The World Bank, Washington D.C.

World Bank (2000). Chad/Cameroon Petroleum Development and Pipeline Project. The World Bank, Washington D.C.

World Commission on Dams (2001). Africa: Irrigation and Hydropower Have Been the Main Drivers for Dam Building. Dams and Water: Global Statistics. Available on: www.dams.org

World Water Council (2000). The Africa Water Vision for 2025: Equitable and Sustainable Use of Water for Socio-economic Development. Presented at the 2nd World Water Forum in The Hague, The Netherlands, March 2000. Available on: http://watercouncil.org/Vision/Documents/AfricaVision.PDF

WRI (2000). *Pilot Analysis of Global Ecosystems: Freshwater Ecosystems.* World Resources Institute, Washington

WRI, UNEP and UNDP (1992). *World Resources Report 1992–9.* Oxford University Press, New York and Oxford

WRI, UNEP and UNDP (1990). *World Resources 1990–91.* Oxford University Press, New York and Oxford

WRI, UNEP, UNDP and World Bank (1998). *World Resources 1998–99.* Oxford University Press, New York and Oxford

WWF (2000). Climate Change and Southern Africa. Available on: http://www.livingplanet.org/resources/publications/climate/Africa_issue/page1.htm

PART F: LAND

## REGIONAL OVERVIEW

Stretching 7 680 km north to south and 7 200 km from east to west, Africa is the second largest region in the world, accounting for 20 per cent of the world's land mass (2 963 313 000 hectares (ha)) (FAOSTAT 2001). Some 66 per cent of Africa is classified as arid or semi-arid, and the region experiences extreme variability in rainfall (UNEP 1999a). Climatic regimes are roughly symmetrical around the equator, and are mirrored in the pattern of vegetation. Thus, dense tropical forests are found in the high-rainfall equatorial belt, with a gradient to both north and south, through savannas, grasslands and deserts. Approximately 22 per cent of Africa's land area is under forest (650 million ha), 43

per cent is characterized as extreme deserts (1 274 million ha), and 21 per cent (630 million ha) is suitable for cultivation (FAO 2001a, Reich, Numbem, Almaraz and Eswaran 2001, UNEP 1999a). By 1999, it was estimated that about 200 million ha (32 per cent of the suitable area) had actually been cultivated (FAOSTAT 2001). At the same time, it was estimated that 30 per cent of the total land area (892 million ha) were being used as permanent pasture (FAOSTAT 2001).

Africa has abundant natural terrestrial resources and potential for economic, social and environmental development. Patterns of land use in Africa are equally diverse and complex, extending beyond agriculture (which, in this context, refers to both cultivation of crops and rearing of livestock). However, many of these have

•

*The contribution of agriculture to the formal economy and to employment in many African countries, although substantial, does not take account of the significant contribution of small-scale cultivation and livestock production to livelihoods*

•

been discussed in other sections of this chapter, and the key issues identified for this section are land quality and productivity, and land tenure, as they relate to food production systems and food security. Figure 2f.1 shows the different types of land cover and patterns of land use in Africa.

The people of Africa are largely rural, and they traditionally practise small-scale cultivation, or pastoralism, which is more common in the more arid areas of Northern, Eastern and Southern Africa. Pastoralists herd cattle, camels, sheep and goats, and migrate according to the seasonal abundance of fodder for the livestock. Although men and women both play important roles in agriculture, the production and preparation of food for the household is the main activity for most rural women (FAO 2001b). Cultivation is also an important means of supplementing diets and incomes in urban areas, and urban agriculture is growing faster in Africa than in any other region of the world (Asomani-Boateng and Haight 1999, Mougeot 1998). Other land resources are also widely used, both at the household level and commercially. These include: medicinal products; raw materials for construction and crafts; bushmeat; and wood for fuel. Together, they contribute up to 50 per cent of household food requirements and up to 40 per cent of household incomes (for example, Ashley and LaFranchi 1997, Cavendish 1999). This direct dependency of most Africans on land, and the heavy economic dependence of many African countries

on agricultural (as well as mineral) resources, create a unique regard for land in Africa, as well as unique production pressures and competition for resources.

## THE REGIONAL IMPORTANCE OF CULTIVATION AND LIVESTOCK PRODUCTION

The contribution of agriculture to the formal economy and to employment in many African countries, although substantial, does not take account of the significant contribution of small-scale cultivation and livestock production to livelihoods. Nor does it represent the cultural value of terrestrial resources (particularly livestock), or the ways in which traditional practices have shaped the environment. In many countries, for example, cattle provide the primary source of dietary protein, are traded for cereals, and contribute to the ploughing and fertilization of cultivated areas (FAO 1997a). Thus, the total value of terrestrial resources is much greater than reflected in the figures that follow below.

Agriculture employs the largest number of workers, and generates a significant share of gross domestic product (GDP) in many African countries. In 1990, the agricultural sector accounted for 68 per cent of the workforce in sub-Saharan Africa, and 37 per cent of the workforce in Northern Africa. This compared with industry, which accounted for 9 per cent and 25 per cent of the workforce respectively (World Bank 2001). In 1999, agriculture contributed more than US$64 484 million to the economy of sub-Saharan Africa (18 per cent of GDP for 1999), and US$26 188 million to Northern Africa (13 per cent of GDP in 1999) (World Bank 2001).

The main commercial crops grown in the region are: cereals; cocoa; coffee; cotton; fruit; nuts and seeds; oils; rubber; spices; sugar; tea; tobacco; and vegetables. These contribute significantly (alongside minerals and metals) to exports and foreign exchange earnings. In 2001, for example, Africa produced 67 per cent of the world's cocoa, 16 per cent of the world's coffee, and 5 per cent of world's cereal production (FAOSTAT 2001). Livestock and livestock products contribute about 19 per cent to the total production value from agriculture, forestry and fisheries in sub-Saharan Africa (FAO 1997a). Animal production is highly skewed towards the drier regions, with Ethiopia, Kenya, Nigeria, Sudan and Tanzania accounting for 50 per cent of Africa's animal production (FAO 1997a).

**Figure 2f.1 Map of land cover and use in Africa**

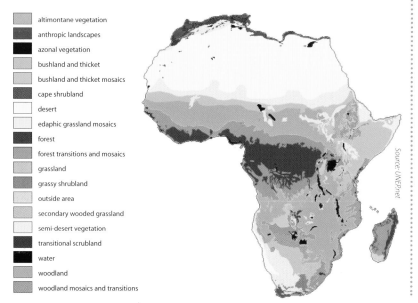

altimontane vegetation
anthropic landscapes
azonal vegetation
bushland and thicket
bushland and thicket mosaics
cape shrubland
desert
edaphic grassland mosaics
forest
forest transitions and mosaics
grassland
grassy shrubland
outside area
secondary wooded grassland
semi-desert vegetation
transitional scrubland
water
woodland
woodland mosaics and transitions

*Source: UNEP.net*

This dependency on agriculture was highlighted by a World Bank (WB) study in 1993, which found that a 1 per cent growth in agricultural production in Africa can stimulate 1.5 per cent overall economic growth, due to positive impacts on industry, transport and other services (World Bank 1993a). Agriculture has, therefore, been the focus of strategies and hopes for economic development over the past three decades (for example, African Development Bank 1998). However, the full potential for development in agriculture has not been realized.

Most agriculture in Africa is rain-fed (except in Northern African countries and the Western Indian Ocean Island states, where irrigation potential has been well developed), and most African countries experience large inter-annual and intra-annual variations in rainfall, with frequent extremes of flooding or drought. Thus, vulnerability to crop failure translates into economic insecurity. In addition, African farmers have been heavily taxed through price-fixing, export taxes and taxes on agricultural inputs, whilst countries outside the region have enjoyed massive subsidization (Wolfensohn 2001, World Bank 1994, Oyejide 1993). Furthermore, many African economies are dependent on a narrow range of agricultural commodities, creating greater vulnerability to failure (resulting from outbreaks of pests, climatic variations, price fluctuations and so on). For example, the same nine commodities (banana, cocoa, coffee, cotton, groundnut, rubber, sugar, tea and tobacco) accounted for 70 per cent of the region's agricultural exports over the period 1970–1995 (Oyejide 1999). Between the 1960s and the mid-1990s, a number of economic shocks hit Africa's agricultural base, with the result that market share for a number of commodities declined sharply, including: cocoa (for which market share declined by 50 per cent); coffee (42 per cent); and cotton (35 per cent). Groundnuts suffered the most dramatic decline in market share, from 70 per cent to just 2 per cent over this period (Oyejide 1999).

However, over the past decade and, in particular, over the last half of the past decade, there have been signs of an economic recovery in several African countries (World Bank 2001), and agricultural value added in Africa grew by 4 per cent between 1990 and 1997 (Oyejide 1999). Continued agricultural development and global competitiveness requires the liberalization of domestic pricing, and improved export market access and, therefore, participation in world trade negotiations is critical for the region.

## EXTENT AND PRODUCTIVITY OF CULTIVATION AND LIVESTOCK PRODUCTION SYSTEMS

In response to growing domestic populations in Africa, and to policies directed at economic growth through agricultural export, the extent of land under cultivation has risen steadily from 166 million ha in 1970 to 202 million ha in 1999. The area of land under permanent pasture increased rapidly during the latter half of the 1980s, and then declined sharply to 892 million hectares in 1999 (approximately the same as in 1970) (FAOSTAT 2001).

Absolute productivity over the past 30 years has also increased for both crops and livestock (see Figure 2f.2 and Figure 2f.3). However, the gains in cultivated area and productivity have been outweighed by rapid population growth and, hence, increased demands on

*Absolute productivity over the past 30 years has increased for both crops and livestock; however, the gains in cultivated area and productivity have been outweighed by rapid population growth and, hence, increased demands on food supply*

**Figure 2f.2  Crop production indices for Africa, 1970–2000 (total and per capita)**

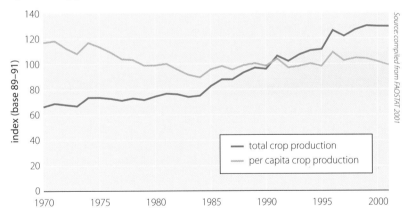

Source: compiled from FAOSTAT 2001

**Figure 2f.3  Livestock production indices for Africa, 1970–2000 (total and per capita)**

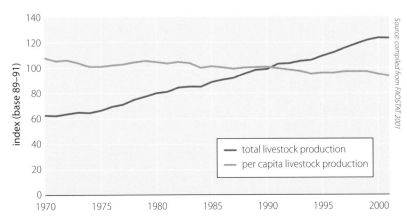

Source: compiled from FAOSTAT 2001

food supply. For example, despite increases in crop production of 2.3 per cent in 1998 and 2.1 per cent in 1999, population growth averaged 2.5 per cent and, therefore, per capita crop production declined (FAO 2001c). Average per capita arable land which is actually cultivated has also fallen, as a result of population growth, from 0.5 ha/capita in 1965 to 0.3 ha/capita in 1990 (Cleaver and Schreiber 1994).

Limited economic growth, and myriad land tenure policies and management practices (including the increasingly widespread poverty and marginalization of subsistence farmers) have also contributed to limited gains in nutritional status, and to increasing dependence on food aid. Although average per capita daily calorie consumption has increased slightly, the number of undernourished people in Africa has doubled since 1970 (FAO 2000a). The region is a net importer of cereal crops for domestic consumption, and the ratio of imports to exports is escalating. Recurrent drought has resulted in crop failures and an inability to accumulate food reserves, and civil wars have restricted food distribution. Consequently, over the past 30 years, millions of people have faced food shortages. In 2000, these totalled 28 million in sub-Saharan Africa, in at least 16 countries (FAO 2001d).

## LAND QUALITY AND PRODUCTIVITY

As a result of pressures to increase production: marginal land is being brought under cultivation or grazing; fertilizers and pesticides are widely used by commercial operations (although applications of organic matter in small-scale practices is declining); and fallow periods are being reduced. These activities, although designed to increase productivity, can result in exhaustion of the production capacity of the land, manifested as: declining yields; vegetation and soil degradation and loss; and, in extreme cases, desertification. Climatic variability and change, and inappropriate land use or land tenure policies, add to the pressures and magnify the impacts. The current situation is that approximately 22 per cent of vegetated land in the region (494 million ha) has been classified as degraded, and 66 per cent of this are classified as moderately, severely or extremely degraded (UNU 1998). Africa is not unique in experiencing this problem, as shown in Figure 2f.4, but the effects on food security and the anticipated impacts of climate change make land degradation a priority issue for African leaders.

Land degradation and reduced productivity can be categorized as: hydrological and chemical degradation; physical degradation; or biological degradation. Hydrological and chemical degradation encompasses: waterlogging; salinization; sodication; and chemical pollution. It is caused by the use of low quality water for irrigation, and environmental pollution. Chemical degradation—resulting from pollution from industrial, household and medical refuse, and mining wastes—also occurs in selected sites. Physical degradation includes deterioration of soil structure and the occurrence of compacted layers, and can be due to: overstocking; inappropriate use of machinery; mining and quarrying activities; frequent waterlogging; and exposure to erosion. Biological degradation refers to the loss of nutrients and micro-organisms vital for maintaining healthy productive crops, and is due to the exhaustion of soil fertility, as a result of: intensive cropping; removal of crop residues; nutrient deficiencies; and insufficient organic matter.

The causes of land degradation in Africa are climatic variability and management practices, in addition to physical factors, such as the slope of the land and soil structure. Although most often associated with cultivation or grazing areas, degradation can also affect: forests, woodlands and savannas; urban and peri-urban areas; and protected areas. For example, clearance of forest vegetation (by fire, drought or overgrazing, or for alternative land uses) leaves soils

**Figure 2f.4  Regional comparison of land degradation**

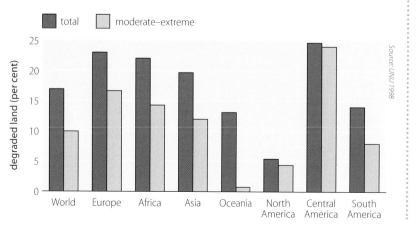

degraded land (per cent)

Source: UNU 1998

more susceptible to erosion by wind and rain, especially when on steep slopes, in high rainfall zones or when combined with poor management, such as overgrazing. Draining of wetlands (for cultivation or urban development) disrupts the hydrological cycle and renders the surrounding areas more prone to flooding, and may no longer provide suitable habitat for wildlife, or raw materials for construction and crafts. Conversion of natural habitats to cultivation or grazing in unsuitable areas can also set off negative feedback systems of degradation, as detailed in Box 2f.1

Over the past 30 years, soil structure has been damaged, nutrients have been depleted and susceptibility to erosion has been increased, as a result of: increasing application of chemicals; use of inappropriate equipment and technologies; and commercial mono-specific plantations. Irrigation in areas of high evaporation, and by inappropriate technologies, for example, increases the rate of salinization of the soil, because the water is rapidly lost, leaving behind a crust of once-dissolved salts. Likewise, intense grazing, especially in marginal areas and by a single species of livestock, can degrade vegetation, and can lead to soil compaction and accelerated erosion. One study found that most soil degradation in Africa was attributable to overgrazing (50 per cent), followed by: poor agricultural management practices (24 per cent); vegetation removal (14 per cent); and overexploitation (13 per cent) (WRI 1992). Between 1980 and 1995, Africa's permanent pasture declined slightly, indicating either conversion to cultivation, or abandonment due to extreme degradation, or a combination of both (UNU 1998).

A visible impact of land degradation is soil loss. Information regarding rates of soil loss in Africa are fragmented and country-specific, with estimates ranging from 900 t/km$^2$/yr to 7 000 t/km$^2$/yr (Rattan 1988). Likewise, studies of the economic impacts of soil loss are localized and varied, but are estimated to reach up to 9 per cent of GDP (UNU 1998). Loss of soil not only impairs productivity for future cultivation, but also causes: sedimentation in dams and rivers; smothering of aquatic and coastal habitats; and eutrophication. This, in turn, leads to reduced biodiversity in, and productivity of, these systems. Ultimately, these effects are felt in the lowered economic and nutritional status of African people.

### Box 2f.1 Demographic change and land quality

In traditional agro-pastoral systems, when land productivity declined through overcultivation or overgrazing, the farmers shifted to a new area whilst the former area recovered. However, with increasing population and economic pressures, production systems have changed, and the rate of conversion of natural habitat is faster. The rate of degradation is also faster, because inputs are minimized and fallow periods are reduced, in order to maximize production over the short term. The overall result is that more land is brought under grazing or cultivation, in order to counteract losses due to degradation. This impacts on natural habitats, biodiversity and ecological functioning, and means that food production requirements are not met.

Certain techniques have all been used successfully to improve productivity and over the long term, whilst maintaining a health resource base. These techniques include: crop rotation; increasing crop diversity; using livestock manure and crop residues as fertilizers; the construction of windbreaks; and agroforestry

With appropriate agricultural practices, rates of soil loss can be reduced, and soil fertility and productivity can be restored, as recently shown in Ethiopia, Kenya, Malawi, Senegal, Somalia and South Africa (Nana-Sinkam 1995, Hoffman and Todd 2000). A Soil Fertility Initiative for sub-Saharan Africa was established in the 1990s, and launched at the 1996 World Food Summit, in response to growing concerns over soil degradation and loss. This is a participatory initiative, with technical partners including: the International Fertilizer Industry Association; the International Food Policy Research Institute (IFPRI); the International Centre for Research in Agroforestry (ICRAF); the International Fertilizer Development Centre; the Food and Agriculture Organization (FAO); and the WB. The approach combines policy reform and technology adaptation, aimed at conserving natural resources and improving farmer's livelihoods through the design and implementation of integrated plant nutrient management programmes, which use a combination of available organic sources of nutrients, supplemented by mineral fertilizers. Many countries are currently preparing National Soil Fertility Action Plans as part of this programme (Maene 2001).

Desertification describes an extreme form of degradation in dryland areas, caused by climatic and management factors, where the land is no longer productive. Some 66 per cent of Africa is classified as desert or drylands and, currently, 46 per cent of

*Over the past 30 years, soil structure has been damaged, nutrients have been depleted and susceptibility to erosion has been increased, as a result of: increasing application of chemicals; use of inappropriate equipment and technologies; and commercial mono-specific plantations*

Africa's land area is vulnerable to desertification, with more than 50 per cent of that under high or very high risk (Reich and others 2001). The most vulnerable areas are along desert margins, as shown in Figure 2f.5. These areas account for 5 per cent of the land area, and are home to an estimated 22 million people (Reich and others 2001). Climate change is predicted to reduce rainfall, to increase evaporation, and to increase the variability and unpredictability in rainfall for many areas of Africa (IPCC 1998, IPCC 2001). This, in turn, will lead to greater vulnerability to drought and desertification. In combination with continuing pressure for economic growth, and the rapid population growth rates, across the region, this will further threaten food security, unless coherent land tenure and management policies are established and enforced.

## Improving land quality and productivity

In recognition of their vulnerability to declining land quality and desertification, African countries were largely instrumental in establishing the United Nations Convention to Combat Drought and Desertification in Countries Experiencing Serious Drought and/or Desertification, Particularly in Africa (UNCCD) in 1992 (UNCCD 2000). Since then, most African countries have embarked on National Action Plans, together with awareness-raising campaigns and, by 2001, 17 countries had completed and formally adopted their programmes (UNCCD 2001). Action plans have also been developed at the sub-regional level: in Northern Africa by the Arab Maghreb Union (AMU); in Western Africa by the Permanent Inter-State Committee for Drought Control in the Sahel (CILSS); in Eastern Africa by the Intergovernmental Authority on Development (IGAD); and for Southern Africa by the Southern African Development Community (SADC) (UNCCD 2001). A Regional Action Programme is also being developed, and will be coordinated by the African Development Bank (ADB) in Abidjan. Desertification, poverty, development pressures and climatic factors interact in a complex manner to influence food security. It is, therefore, essential that desertification be tackled within a development framework, and in a participatory manner. The approach must combine: political and legal reform; economic and social development strategies; land tenure reform; international partnerships; capacity building; and financial sustainability.

**Figure 2f.5  Vulnerability to desertification in Africa**

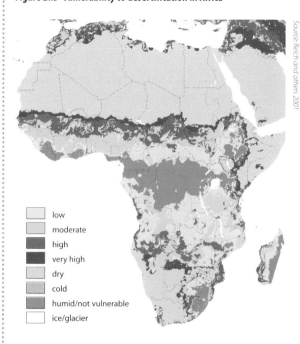

- low
- moderate
- high
- very high
- dry
- cold
- humid/not vulnerable
- ice/glacier

Source: Reich and others 2001

## LAND RIGHTS

The issue of land rights in Africa is a highly complex and sensitive social and political issue, closely linked with poverty and land degradation issues. The following terminology is applied throughout this review to avoid ambiguity or confusion:

- 'land rights' include rights over access to land and resources;
- 'land tenure' refers to the conditions and institutional arrangements under which land is held;
- 'land reform' encompasses reform of land rights and policies, as well as land redistribution or the physical reallocation of land; and
- 'land restitution' refers to the restoration of land to indigenous people, following alienation under colonialism or apartheid.

Traditional land tenure systems in Africa were developed in accordance with variations in physical conditions and cultures, although they were largely centred on communal access to resources and sharing of benefits. Tenure was rarely recorded or registered, and land rights were largely allocated through

### Box 2f.2  Women and land tenure

In most parts of Africa, both modern and traditional laws favour patriarchal ownership of land and control of resources. Women may be granted access to resources through their fathers or husbands on a temporary basis but, if they become widowed, they may be forced to leave their land. There is also a financial bias towards male ownership, as women tend to have lower incomes than men and, therefore, may not be able to afford to purchase land, or to acquire access to credit schemes. Women are also discriminated against in the quality of land they have access to, and often have marginal and remote land.

However, with land reform efforts progressing in many African countries, access to land, ownership and registration systems are helping to shift the balance of equity in recognition of the rights of vulnerable groups, including women and minority, ethnic or nomadic groups.

inheritance or other regulatory and distributive mechanisms. Traditional systems offer more security of tenure than is often recognized by supporters of individual tenure systems, although women generally have lower status than men (Cleaver and Schreiber 1994). The insecurity of women in this regard is illustrated in Box 2f.2. In pastoral communities, land may be under tenure or may be open access. Here, access is regulated according to seasonal abundance of resources, and is determined by traditional governance systems. The mobility of pastoral communities buffers against climatic variability and provides them with greater food security.

However, colonial regimes and newly independent governments perceived traditional tenure and access systems to be insecure and poorly suited to commercial, settled agricultural development and conventional economic growth (Cleaver and Schreiber 1994, Toulmin and Longbottom 1997). Thus, land largely became the property of the state, and was then redistributed with discrimination along lines of wealth, race or gender (Cleaver and Schreiber 1994). Resulting conflicts between traditional and contemporary tenure and access policies have frequently led to the mismanagement of resources and to conflicts between

user groups. In the Western Indian Ocean state of Comoros, for example, there are three types of land tenure system in place: traditional, colonial and Islamic (RFIC 1998). The best land is reserved for commercial crops which are mainly for export, whilst small-scale and subsistence farmers are left with less productive, even marginal areas, such as steep slopes. Without proper terracing, these areas are prone to soil erosion (COMESA 2001). Extreme inequity in land distribution is seen in South Africa, where access to land averages just over 1 ha/capita for black farmers and 1 570 ha/capita for white farmers (SARIPS 2000). A further complication affecting access rights and food security across Africa is that of population growth, resulting in greater and greater land fragmentation under the traditional system of inheritance and sub-division. As each successive generation inherits a smaller and smaller parcel of land, average farm size per household is declining, thus contributing to food insecurity.

Over the past few decades, governments have realized that centralized management of land resources is also inappropriate, and processes of land reform are underway in many countries. Individual land titling may be appropriate, for example, in urban areas but, in pastoral areas, communal ownership may be a more viable alternative. Legal recognition of traditional land-use practices could be a means of avoiding the overexploitation of resources associated with lack of ownership and accrual of benefits (the 'tragedy of the commons'). Mozambique is one of the countries where demarcation of community lands is being undertaken and, in Tanzania, the Village Land Act provides for collective land ownership for pastoralists (DFID 1999). Protection of sacred forests through community ownership and management is also emerging as a successful model of forest conservation (Alden Wiley 2000). Uganda's 1998 Land Act attempts to integrate traditional and contemporary tenure and access policies, allowing: legal registration of traditional land rights; representation of traditional authorities in dispute resolution; and the formation of communal land associations. However, there is still confusion among the public over their rights, and implementation has been restricted through lack of resources (DFID 1999).

Another key element in the discussion of land reform processes is the effect of different tenure systems on investment in resource management and on

*Legal recognition of traditional land-use practices could be a means of avoiding the overexploitation of resources associated with lack of ownership and accrual of benefits*

productivity. Evidence of subsistence level production being less efficient is still inconclusive, and comparisons are complicated by existing market failures and by inequities in market access. The role of the private sector needs to be explored further, and a clear strategy of land use and production needs to be developed and understood by all participants before implementation. Otherwise, there is a risk of collapse in production during a transition period (DFID 1999).

Other issues for concern and for continued discussion include: the means for developing effective, legitimate institutions for the management of land rights; the implementation of market-based instruments for the redistribution of land; the question of land restitution; and the continuing marginalization of women, indigenous peoples, and pastoralists or hunter-gatherers (DFID 1999). The role of the state in facilitating or administrating land reform is also under the spotlight, particularly in Zimbabwe, where the land issue has become the central campaign issue between political parties (Moyo 2002).

Delays in reform have been considerable, and governments continue to hold legally defined de jure ownership rights over much of rural Africa. By contrast, and often as a consequence, rural communities and individuals exert de facto rights to land and resources, based on claims to traditional land rights, and in protestation over slow reforms (Cousins 2000). In some countries, most notably, Southern African countries, where there was extreme inequity in access to and ownership of land and resources, the process of reform has been catalysed by demonstrative and, in some cases, violent action. In Zimbabwe, for example, there have been several violent clashes over illegal land occupation (for example, Drimie and Mbaya 2001).

Land reform is a highly contested and sensitive issue, requiring an appropriately sensitive approach. Policies must consider country-specific situations and objectives, within an overall development context. Policies will necessarily be developed through several iterations, with the involvement of all stakeholders, particularly those most marginalized (Drimie and Mbaya 2001).

The following sections give further details of specific issues and policies relating to land resources in each of the African sub-regions.

## NORTHERN AFRICA

Due to the extreme aridity, a major issue in Northern Africa is scarcity of arable land (or land that is suitable for cultivation). Average annual precipitation is just 7 per cent of Africa's total, and there are large inter-annual and intra- annual variations (FAO 1995b). Distribution of rainfall between the countries is also varied, with more than 70 per cent falling in Sudan, and just 3 per cent in Egypt, where more than 90 per cent of precipitation is lost through evaporation or transpiration (FAO 1995b). These harsh climatic conditions, and the predominance of shallow, highly erodible soils, make cultivation a precarious occupation. Arable land represents 26.4 per cent of the total land area, and 18.7 per cent is currently cultivated, although the extent of cultivated area ranges from 2.6 per cent in Egypt to 77.4 per cent in Morocco (FAOSTAT 2001). Rangelands currently occupy about 13 per cent of the total land area (mostly in Algeria and Sudan) although, over the past 50 years, half of these have been reclaimed for cultivation (AOAD 1998, Le Houerou 1997).

Despite severe physical limitations, agricultural and pastoral activities contribute significantly to national economies and traditional lifestyles. Thus, land cultivation in the sub-region is becoming increasingly dualistic in nature, with a high technology agribusiness sector developing alongside traditional smallholder agriculture. In almost all countries, some farmers still harvest their crops by hand, whilst commercial agriculture is heavily mechanized, employing highly efficient irrigation systems, tractors, multi-furrow ploughs and combine-harvesters. There is a pressing need to integrate the two sectors, and to combine the wisdom from each (Lycett 1987).

### IMPORTANCE OF CULTIVATION AND LIVESTOCK PRODUCTION IN NORTHERN AFRICA

In 1990, agriculture employed about 37 per cent of the workforce in the Arab countries, and 69 per cent in Sudan. This was a steep decline for the Arab countries, from 51 per cent in the previous two decades (World Bank 2001). The major crops produced in the sub-region include: cereals (wheat, barley, rice and sorghum); fruit (citrus, dates and olives); vegetables (beans); sugar (beet and cane); and nuts and seeds

A Bedouin nomad diverts water to his crops in a small-scale irrigation project.

*Nigel Dickinson/Still Pictures*

(sesame and groundnuts). Total agricultural exports for Northern Africa (excluding Sudan) were US$2 451 million in 1997, and value added in agriculture was 13 per cent of GDP in 1999 (World Bank 2001). Commercial farming is heavily dependent on irrigation and fertilizer use. For example, in Egypt, all cultivated land is irrigated although, in other countries, the percentage is lower (FAO 1995b). Over the past decade, Northern African countries have used between 1 million t and 1.5 million t of fertilizer per year, approximately 45 per cent of Africa's total (World Bank 2001).

The population of Northern Africa doubled between 1970 and 2000 (from 85 million to 174 million people), and is continuing to grow at an average of 2 per cent per annum (UNPD 1996, World Bank 2001). This, together with increasing consumption and demand for luxury foods (Miladi 1999), has been responsible for rising demands on agricultural production and for pressures on the land resources. Responses to meet this rising demand have included: enhancing cropping intensity; extending the area of land under cultivation; and intensive irrigation and use of chemicals and other inputs (FAOSTAT 2001).

## EXTENT AND PRODUCTIVITY OF CULTIVATION AND LIVESTOCK PRODUCTION SYSTEMS IN NORTHERN AFRICA

The total area under cultivation in the sub-region has grown from approximately 35 million ha in 1970 to more than 45 million ha in 1999 (FAOSTAT 2001). Sudan has witnessed the most dramatic expansion (from 11 million ha in 1970 to 17 million ha in 2000), although there are still large areas of land with potential for cultivation, which have not yet been exploited (FAOSTAT 2001). For other countries, most or all arable land is under cultivation, due to significant technological and engineering developments. These include: the construction of the Aswan High Dam on the Nile River in Egypt; the Al-Salam canal development, which diverts water from the Nile River to Sinai, Egypt; and the great covered human-made river in Libya (Hegazy 1999). In other countries, the potential for expansion has been limited by: availability of water resources; inequity of land resource distribution; availability of capital investment; human resources; supply of energy; and prevailing soil characteristics.

The percentage of cultivated land that is irrigated has also expanded considerably over the past 30 years, from 6 million ha in 1970 to nearly 8 million ha in 1999 (FAOSTAT 2001). Irrigated area as a percentage of agricultural area varies from 100 per cent in Egypt to about 15 per cent in Morocco and Sudan, where rain-fed agriculture is more reliable (FAOSTAT 2001). This expansion has been outstripped by population growth, however, as shown by declining area of cultivated land per capita, which fell from 0.59 ha/capita in 1970 to 0.33 ha/capita in 1990 (calculated from FAOSTAT 2001). Total crop and livestock production has also increased over the past decade but, again, the rapid increase in population has resulted in a net decline in per capita production, as shown in Figure 2f.6 and Figure 2f.7 (FAOSTAT 2001).

Intensification and expansion of food production have contributed to food self-sufficiency and food security in the sub-region. However, imports far exceed exports, and the gap between imports and exports is growing (Miladi 1999). Northern Africa has high self-sufficiency in vegetables, fruits and tuber crops (more than 97 per cent) but, for sugar, cereals and plant oils, self-sufficiency rates are 33.2 per cent, 57.3 per cent and 62.7 per cent respectively (Arab League 1999).

**Figure 2f.6  Crop production indices for Northern Africa, 1970–2000 (total and per capita)**

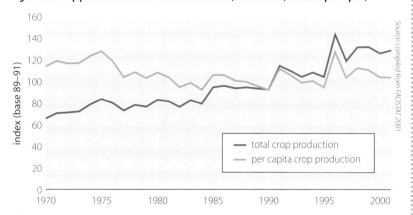

*Source: compiled from FAOSTAT 2001*

**Figure 2f.7  Livestock production indices for Northern Africa, 1970–2000 (total and per capita)**

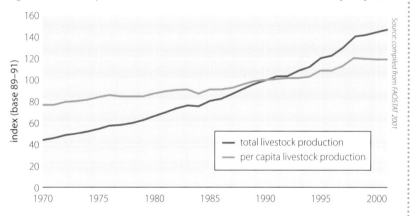

*Source: compiled from FAOSTAT 2001*

## LAND QUALITY AND PRODUCTIVITY IN NORTHERN AFRICA

Land quality and productivity are declining in cultivated areas, rangelands and forests, as a result of a number of pressures. These include: rapid population growth; climatic stresses and drought; overgrazing; clearing of forests and woody vegetation; cultivation of marginal lands; unfavourable land tenure; irrational agricultural production; and trade policies (El Bagouri 2000).

Forests and wooded areas are rapidly being degraded as wood is cut for fuel, and to bring more land under cultivation. Not only does this result in reduced natural habitat and natural resources available from forests, but loss of soil cover and protection increases the risk of erosion by wind and rain, as well as increasing the risk of flooding and landslides in these areas. In Egypt, cultivation occupies a narrow strip along the Nile River (3.5 per cent of the total land area of the country), surrounded by desert on either side. In these desert margins, wind erosion is a threat, and may affect agriculture production on the edges of cultivated areas. To slow down the effect of wind erosion, the government of Egypt has a long-term plan to establish a protective belt of perennial green vegetation cover (FAO/AGL 2000).

Rain-fed areas suffer particularly from prolonged and/or frequent dry spells and droughts, and from inappropriate farming systems which, in turn, have been encouraged by macroeconomic policies and pressures. In Sudan, for example, the so-called 'suitcase farmers' have caused widespread loss of productivity, through uncontrolled, low-input sorghum farming (Lusk 1987). High prices for this cereal encouraged extensive monoculture, with inadequate fallow or rest periods for the soil to recover, or alternative cropping to return nutrients removed from the soil by sorghum. This has caused: serious losses of soil, through wind and water erosion; loss of soil fertility; and loss of biodiversity, through degradation of natural plant cover and deforestation. Fertility decline in Egypt is managed by using mineral fertilizer—mainly nitrogen and phosphorous fertilizers—at quite high rates in comparison with other countries in Africa. The use of organic matter and crop residues is not common in Egypt (FAO/AGL 2000).

Areas under irrigated cultivation are showing signs of chemical degradation, such as: increased soil salinity; water-logging; and pollution. Irrigation in areas with naturally high evaporation rates and inadequate drainage, for example, leads to the accumulation of high levels of soluble salts in the topsoil. Salt-affected or salinized soils have become widespread over the past 30 years, as a result of the use of saline drainage water and brackish water in irrigation (Goossens, Ghabour, Ongena and Gad 1994, FAO 1994, Framji 1974). In Egypt, 30–40 per cent of the cultivated irrigated land is now salt-affected (mostly in the northern part of the country, along the Nile Delta), and this accounts for 15 per cent of total production losses (FAO/AGL 2000). This region is now considered a global 'hot spot' of land salinization, together with the most fertile river basins of Asia and Latin America (Scherr and Yadav 1996 in FAO 2000b).

Responses to curb salinization and the hydrological degradation of soils include: improvements to leaching and drainage of irrigation water; land levelling (to avoid collection of irrigation water in pools); applications of gypsum, organic matter and farm manure; and the selection of salt-tolerant crops during the reclamation programmes (FAO/AGL 2000).

Rangelands are subject to: overstocking and overgrazing; mismanagement of land, often associated with poorly defined or inappropriate tenure policies; and the conversion of rangelands to cultivation. The environmental impacts of land degradation include: reduced water quality from sedimentation and pollution, thus leaving water unsuitable for human use or irrigation; impaired navigation of watercourses; and increased costs for flood control and dam maintenance. Ultimately, productivity is reduced, affecting food security in the long term and the short term. According to the FAO (FAO/AGL 2000), only 5 per cent of Egypt's cultivated lands are in excellent condition, 40 per cent are classified as good, and the remaining 55 per cent are classified as either medium or poor. Furthermore, the FAO estimates that soil degradation in Egypt results in an overall production loss of 53 per cent (FAO/AGL 2000).

According to Reich and others (2001), the areas most vulnerable to desertification are along the desert margins and, within these areas, where population densities are high, and where land management practices are insensitive to climatic or physical restraints. Thus, large areas of Northern Africa are highly vulnerable to desertification, along the borders of the Sahara desert and the arid Sahelian region. In Algeria, for example, 85 per cent of the land area is desert and, of the remaining 15 per cent, 14 per cent is moderately, highly or very highly vulnerable to desertification (Reich and others 2001). In Morocco, 52 per cent of the land area is vulnerable to desertification, and 45 per cent is either moderately, highly or very highly vulnerable (Reich and others 2001).

### Improving land quality and productivity in Northern Africa

The consequences of declining productivity could include: reduction in return from capital investment and labour inputs; lower incomes, especially for small-scale farmers; and increasing rural-urban migration, in search of more secure livelihoods. Therefore, combating degradation is of paramount importance for sustainable development. Northern African countries are carrying out varied activities in order to assess and to monitor degradation, and to test techniques to mitigate desertification.

Steppe and alfa grass, Mergueb Reserve, Algeria

*Arnaud Greth/Still Pictures*

All the countries of the Northern African sub-region have ratified the UNCCD. Tunisia has produced a National Action Plan, Morocco has established a website for information dissemination, and most countries have produced national reports. In 2000, the Union du Maghreb Arabe (AMU) produced a sub-regional action plan for the Arab countries of Northern Africa (UNCCD 2001). Greater efforts are also being exerted to reduce the rate of population increase, especially in Tunisia and Egypt. In Morocco, a decentralization process is under way, in order to encourage rural development and to improve food and economic security in rural areas and in agricultural lifestyles. Participatory implementation tools have been piloted, alongside a new legal framework for the development of rain-fed agriculture. In Sudan, the government passed a decentralization law in 1998, transferring responsibility for revenue generation and expenditure to local authorities.

### LAND RIGHTS IN NORTHERN AFRICA

Current land tenure policies in Northern Africa have been derived from pre-colonial and colonial tenure systems, largely influenced by the type of land use most appropriate for various areas, and economic pressures of different periods. Although they may have once been optimal for enhancing productivity under the prevailing social and demographic situations, tenure policies have not always kept pace with changing demographic patterns, demand for resources or national priorities. Thus, resource overexploitation is apparent in some areas, contrasting with inefficient use of resources in other areas. Lack of tenurial security has been reported as a major constraint to land development (FAO 1992) and, throughout the sub-region, much land remains under poorly defined ownership. Much of it is 'open access', although sometimes called 'common property', and large proportions are state-owned. Until recently, 20 per cent of Tunisia's total land area (3 million ha) was under communal ownership, and 6 per cent was state-owned. However, in 1991, the government privatized 1.2 million ha of agricultural land and 600 000 ha of rangelands (World Bank 2000a). This reflects a change in economic and social thinking towards more commercial production. Access to land resources is a primary determinant for development, and is considered a major factor responsible for the failure of some governments to engage the people in land development programmes. In some countries, such as Libya, the state assumes ownership of any land which is not assigned to sedentary populations, that is, they have nationalized lands held in some sense collectively by local communities (Harbeson 1990).

Land tenure reform processes in the sub-region attempt to optimize land use, minimize degradation and stem the rapid rates of urbanization. Many of the governments in the sub-region have also adopted liberalization and structural reform policies, that is, marketing, pricing and trading produced food and agricultural commodities with the removal or minimization of subsidies for inputs. Policies have been adopted to enhance exports and foreign trade. Measures have been taken to encourage intra-regional trade, through bilateral trade agreements among the countries of Northern Africa, and the establishment of free trade zones. Research efforts to enhance sustainable development have also been encouraged. Greater participation of stakeholders through farmers and users associations is receiving greater attention, and gender issues are being discussed.

In Tunisia, for example, land reform has focused on the settlement of land claims through the registration and certification of ownership, and through a dual plan for promoting economic stability and environmental protection (World Bank 2000a). However, despite the government's attempts to encourage registration, half of all eligible lands have not been registered and, even amongst registered lands, successive land transfers often go unrecorded. The process of privatizing state

and collectively owned lands has also been slow, as has the emergence of land markets. Fragmentation of land and resources remains a principal concern, which the government is attempting to redress by means of: legislative modifications; extension programmes; and improved administrative procedures (Gharbi 1998). In Morocco, until recently, land was owned by the state, by religious orders or by communities. The government has encouraged land reform through incentives for: streamlining registration; limiting fragmentation; the standardization of tenant contracts; and the privatization of state-owned and church-owned lands. A recent report concludes that private tenureship provides greater security and greater incentives for investment in land improvements, such as access to financing (Taleb 1998).

## EASTERN AFRICA

The Eastern African sub-region is characterized by two fragile ecosystems, namely: mountainous and hilly areas (predominantly in Burundi, Rwanda, Uganda, Kenya and Ethiopia); and semi-arid or arid (dryland) areas (predominantly in Djibouti, Eritrea, Ethiopia and Somalia). These areas support most of the sub-region's population (with densities of more than 200 people/km$^2$), and are the centres of crop cultivation. For example, the highlands of Ethiopia (above 1 500 masl) constitute about 45 per cent of the total land area, and are inhabited by 80 per cent of the population and by 75 per cent of the country's livestock (EPA/MEDC 1997). The dryland areas have low rainfall, and are extremely vulnerable to drought and desertification, especially in the Horn of Africa (rainfall in Djibouti, for example, is only 147 mm/yr). The Horn of Africa experienced severe droughts in 1972–73 and 1984–85, in which millions of people lost their lives, their homes, their livestock or their livelihoods (FAO 2000a). In Djibouti, Eritrea and Somalia, less than 5 per cent of the land area is under cultivation (FAOSTAT 2001). Ethiopia and Kenya have 10 per cent and 8 per cent respectively of land under cultivation. In Burundi, Rwanda and Uganda, higher and more predictable rainfall facilitates relatively extensive cultivation (42 per cent, 35 per cent and 45 per cent respectively) (FAOSTAT 2001). All countries in the sub-region, except

Coffee picking: worker harvesting ripe cherries in Tanzania

*Nigel Cattlin/Holt Studios*

Uganda, have extensive pastures (FAOSTAT 2001). In the drier regions, these are used by nomadic herders, because livestock production is preferable to the risks of cultivation.

Most of the population in Eastern Africa (more than 70 per cent) is rural, practising subsistence agriculture (WHO/UNICEF 2000). More than 95 per cent of Ethiopia's agricultural output is generated by small-scale farmers, who use traditional farming practices (FAO 2000a). Rapid population growth and increasing demand for food, combined with high variability in rainfall and frequent drought, is putting pressure on farmers to clear more natural vegetation, and to cultivate more and more marginal land. Shortening of fallow periods and high intensity of rainfall contribute to creating conditions which are conducive to land degradation, soil erosion and desertification (NEMA 2000). Thus, the main issues of concern are population growth, agricultural practices and food security.

## IMPORTANCE OF CULTIVATION AND LIVESTOCK PRODUCTION IN EASTERN AFRICA

Although most people in the sub-region are involved in subsistence agriculture, commercial agriculture is also an economic mainstay, contributing significantly to employment, GDP and exports. In Burundi and Rwanda, for example, more than 90 per cent of the workforce

have been employed in agriculture for the past three decades (World Bank 2001). In Ethiopia, Kenya, Somalia and Uganda, rates of employment in this sector have been between 80 and 95 per cent (World Bank 2001). The main crops are: bananas; beans; cassava; coffee; cotton; maize; millet; rice; sesame; sisal; sorghum; sugar; tea; and wheat. Economic contributions from agriculture have been significant over the past 30 years, including an average of 45 per cent of GDP in Burundi, Ethiopia and Uganda, and 21 per cent of GDP in Kenya (World Bank 2001). The value of agricultural exports from the sub-region is also substantial, reaching US$526 million in Ethiopia and US$1157 million in Kenya during 1997 (World Bank 2001). Considering that these are among the poorest countries in the world, the value of agriculture, and the precariousness of depending on so few rain-fed crops, cannot be underestimated.

## EXTENT AND PRODUCTIVITY OF CULTIVATION AND LIVESTOCK PRODUCTION SYSTEMS IN EASTERN AFRICA

The total area under cultivation has increased over the past three decades for all countries, except for Burundi, which has experienced a slight decline (FAOSTAT 2001). Food production has also climbed over the past 30 years, although with considerable inter-annual variation. However, population growth has exceeded increases in production. The drier countries in the Horn of Africa, where climatic variation and drought are more common, show the greatest overall decline in per capita food production, and the greatest variation between years, as shown in Figure 2f8 and Figure 2f.9.

This has resulted in declining food security, and decreased per capita food intake. Daily per capita calorie intake in Kenya, Uganda, Rwanda and Burundi was less in 1997 than in 1970 and, in Ethiopia, almost 50 per cent of the population is undernourished (UNDP 2000, FAO 2001c). Over the same period, there were declines in per capita supply of protein and fats in the same countries (UNDP 2000). Malnutrition is also of serious concern, and deficiencies in iodine and vitamin A are common among children below the age of six years in Ethiopia (FAO 2001c).

Declining per capita food production and per capita food intake is causing the countries of Eastern Africa to

**Figure 2f.8  Crop production indices for Eastern Africa, 1970–2000 (total and per capita)**

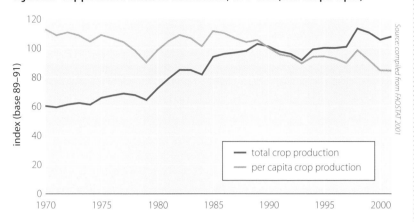

Source compiled from FAOSTAT 2001

**Figure 2f.9  Livestock production indices for Eastern Africa, 1970–2000 (total and per capita)**

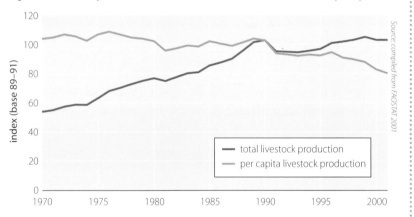

Source compiled from FAOSTAT 2001

become more and more dependent on food imports and food aid. For example, Ethiopia has been a food-deficit country for several decades, and average cereal food aid during the period 1984–99 was 14 per cent of total cereal production (FAO/AGL 2000). These severe droughts and food shortages catalysed the need for a drought monitoring facility for Eastern and Southern Africa. In March 1985, the leaders of the meteorological centres of the region met in Nairobi, and agreed to establish the present Drought Monitoring Centres (DMCs) in Nairobi (Kenya) and Harare (Zimbabwe), within the respective meteorological departments. The main objective of the Drought Monitoring Centres is to provide early warning of drought, based on meteorological information, thus preparing countries and alleviating the devastating impacts of drought. A regional project, 'Drought Monitoring for Eastern and Southern Africa', was set up in January 1989, with 21 participating countries, namely: Angola; Botswana; Burundi; Comoros; Djibouti; Ethiopia; Kenya; Lesotho;

Madagascar; Malawi; Mauritius; Mozambique; Rwanda; Seychelles; Somalia; Sudan; Swaziland; Tanzania; Uganda; Zambia; and Zimbabwe. The second phase of the project included Eritrea, Namibia and South Africa (DMC 2002). Other responses include the United Nations Inter-Agency Task Force strategy for improving food security, which focuses on the underlying causes of chronic food insecurity in the seven countries. It proposes diversification of livelihoods away from the traditional dependence on agriculture, as well as means of enhancing resilience to climatic variation within agricultural practices. Macro issues, such as market reforms, improving access to trade and information, and environmental protection, are also central to the strategy (UN 2000).

## LAND QUALITY AND PRODUCTIVITY IN EASTERN AFRICA

Population and climatic pressures have taken their toll on land and resources in Eastern Africa, and the sub-region is experiencing some of the most rapid degradation rates in Africa (Henao and Baanante 1999). It has been estimated that 2 million ha of Ethiopia's highlands have been degraded beyond rehabilitation, and an additional 14 million ha severely degraded (EPA/MEDC 1997). The same study estimated that more than 25 per cent of the country is experiencing desertification, and the annual rate of topsoil loss is reported to be 1 900 million tonnes (EPA/MEDC 1997). As farms have been sub-divided over the years, decreasing farm size has led to shorter fallow periods and, in some places, to continuous cropping, in order to sustain productivity levels. Crop residues are rarely ploughed into the soil, and applications of other organic matter have been low, resulting in higher requirements for inorganic fertilizers, with consequent problems of salinization and pollution (FAO 2001c). In the early 1990s, Rwanda and Burundi experienced nutrient depletion rates of more than 100 kg ha/yr of nitrogen, phosphorous and potassium (NPK) (Henao and Baanante 1999). Almost 80 per cent of Kenya's total land area is classified as arid or semi-arid and, in these areas, removal of vegetation cover (through overgrazing and for charcoal production) exposes the soil to wind and water erosion. Soil compaction occurs in areas where there is excessive

## Box 2f.3 Extent and quality of grazing areas

Quality of grazing lands refers not only to the quantity of vegetation cover, but also to the quality of the grazing available, as well as other factors, such as water availability, mineral content and parasite load. A good illustration of this is provided by the interactions of Maasai cattle with wildlife in the pastoral areas of Kenya. It has been suggested that, when the Maasai were excluded from their traditional grazing areas under British colonial administration, wildebeest moved in to graze these high-quality short grass plains and, consequently, their numbers increased. Exclusion of the Maasai means that their cattle are now restricted to lower quality areas, which the herdsmen would traditionally have avoided due to their high parasite load.

*Source: Hempwood and Rodgers 1991*

trampling by animals and, in cultivated areas, soil fertility is declining, as a result of the exhaustion of soils by mono-specific cropping and reduction of fallow periods (FAO/AGL 2000). Colonial land policies have also contributed to degradation through the marginalization of pastoralists, as shown by the example in Box 2f.3. In irrigated areas of Kenya, approximately 50 per cent of the soils are affected by salt, as a result of the poor management of irrigation (FAO/AGL 2000). In Uganda, the estimated proportion of degraded land ranges from 20 per cent to 90 per cent, with soil degradation, resulting from overgrazing and soil compaction, a common feature in the major cattle-rearing areas (FAO/AGL 2000).

In Djibouti, where 85 per cent of the land area is dryland or desert, the remaining 15 per cent of the land has been classified as moderately, severely or extremely vulnerable to desertification. In Eritrea, 42 per cent of the area not already classified as desert is vulnerable. In Ethiopia, this is 26 per cent and, in Kenya, it is 35 per cent (Reich and others 2001). Soil fertility has also declined under cultivation with little organic inputs and short fallow periods. Other causes of land degradation are: inappropriate crop production practices; overgrazing; land fragmentation; deforestation; uncontrolled bushfires; and inefficient mineral exploration techniques. According to Slade and Weitz (1991), the annual cost of soil erosion in Uganda

is in the order of US\$132–396 million. Soil loss estimates in Ethiopia range from US\$15 million (FAO 1986) to US\$155 million (Sutcliffe 1993), equivalent to between 1 per cent and 5 per cent of GDP. Cumulative economic losses, as a result of lowered productivity, could reach US\$3 000 million for Ethiopia (Bojo 1996).

The impacts of soil degradation include: increased risk of flooding; sedimentation in rivers, lakes and dams; smothering of coastal habitats; and eutrophication. For example, in Uganda, uncontrolled flooding was experienced in 1997–98, as a result of the El Niño rains, and because of extensive vegetation removal, landslides were experienced (NEMA 1999). In April 1999, thousands of fish deaths were observed in the Albert Nile River, most likely caused by eutrophication, resulting from soil erosion and increased levels of fertilizers in the river. Soil degradation also contributes to rising rural poverty and food insecurity, because productivity is reduced, and subsistence farmers are less and less able to accumulate reserves of grain. Ultimately, rural-to-urban migration and encroachment into gazetted natural reserves occur, as are now extensively experienced in Uganda (NEMA 1999).

### Improving land quality and productivity in Eastern Africa

Responses to control and to reduce land degradation, erosion and desertification in the sub-region include: the ratification of international agreements; the development and implementation of regional action plans; reform and development of national policies; and effecting local level soil conservation practices. All governments in Eastern Africa, except Somalia, have signed and ratified the UNCCD. Uganda, Djibouti and Ethiopia have produced national reports and National Action Plans, and have held awareness-raising meetings. Rwanda, Burundi and Kenya have produced national reports. The Intergovernmental Authority on Development (IGAD) has also produced a sub-regional action plan for the countries in the Horn of Africa (UNCCD 2001).

IGAD was created in 1986 by Djibouti, Ethiopia, Kenya, Somalia, Sudan and Uganda (and later joined by Eritrea), in order to coordinate development in the Horn of Africa. The IGAD policy on food security and environment aims to establish, amongst other activities: a regional integrated information system for drought

*The annual cost of soil erosion in Uganda is in the order of US\$132–396 million. Soil loss estimates in Ethiopia range from US\$15 million to US\$155 million, equivalent to between 1 per cent and 5 per cent of GDP*

Erosion, Kenya

*Charlotte These/Still Pictures*

monitoring and early warning; diversification of household energy sources; and implementation of the UNCCD. Research activities are also focusing on the promotion of the sustainable production of drought-tolerant, high-yielding crop varieties, and capacity building programmes have been established to improve integrated water resources management (IWRM). Efforts to remove barriers to trade and economic growth include the coordination and harmonization of policies in trade, industry, tourism, communications infrastructure and telecommunications.

The Ethiopian government has made it a priority to improve food insecurity following the tragic famine of 1983–84, and has been successful in avoiding famine, despite droughts in 1991–92 and 1993–94. This success has been facilitated by: early warning of drought by IGAD; the establishment of an Emergency Food Security Reserve; and substantial assistance from international agencies and donors (FAO 2001c). Vulnerability analysis and mapping by the United Nations (UN) World Food Programme (WFP) has also helped in providing advance information on crop yields. The Ethiopian government is also implementing a Sectoral Investment Programme, focusing on: crop diversification; strengthening early warning; improving access to facilities for the very poor (such as extension services and financial resources); and small-scale irrigation development (FAO 2001c). The capacity of soil and water conservation training institutions is being increased at the various universities and agricultural colleges. The programme is accompanied by a population policy aimed at reducing fertility, in an

attempt to address one of the major pressures in land management—population growth (FAO 2001c).

In Uganda, a national land use policy has recently been formulated, although a new law, the Land Act 1998, was enacted two years ago (Moyini 2000). Additionally, the Ugandan government is approaching the population-agriculture issue in three ways. First, in a direct approach, a National Population Policy was formulated in 1995. Second, in an indirect approach, universal primary education was established, in the expectation that a literate population will better appreciate the benefits of family planning (NEMA 1999). Third, the government's Plan for Modernization of Agriculture is expected ultimately to shift the rural population away from subsistence agriculture into agroprocessing industries, leaving few and more efficient farmers (MoFPED/MAAIF 2000).

At the community level, soil management practices in semi-arid eastern Kenya include: strip cropping; intercropping; cover cropping; agroforestry; crop rotation; ridging; mulching and application of manure; terracing; and cutting of drains. National afforestation programmes are also being established as long-term measures to conserve land and soil resources (FAO/AGL 2000).

## LAND RIGHTS IN EASTERN AFRICA

Land tenure in Eastern Africa is a sensitive and complex issue. At independence, the countries in the sub-region established quite different tenure reforms, all aimed at improving productivity. For example, in Ethiopia, all land became public land, with leasing or sale of land being forbidden (FAO 2001c) whereas, in Kenya, the government pursued private ownership (Bruce, Subramanian, Knox, Bohrer and Leisz 1997). In both Kenya and Ethiopia, fragmentation of land parcels through subdivision has reduced the average farm size to less than 1 ha in many areas. This has the result that fallow periods have been reduced or are omitted altogether, in order to produce sufficient quantities to meet the needs of the family. In spite of these policies, the countries of the sub-region have all suffered impediments to large-scale agricultural development, and the majority of the population are small-scale farmers (Bruce and others 1997).

The nomadic herdsmen of the Horn of Africa have suffered extreme marginalization and reduced food security since the colonial governments seized control of all central rangelands and, later, reallocated them for mechanized farming, expansion of irrigated agriculture and declaration of wildlife reserves (see Box 2f.4). Thus, the herdsmen have greatly reduced access to fodder, and are frequently also denied access to crop residues in farming areas, except in return for a fee. Cropping has extended steadily, woodcutting in former pasture areas has been sanctioned, nomadic routes have been disturbed and watersources have not been maintained (DFID 1999). The lack of institutional support for nomadic pastoralists has further excluded their participation in decision making or land use planning (DFID 1999).

Governments are recognizing that central control of land and agricultural resources is limited by capacities and resources, and that land policy reform is needed to encourage the formation of farms of viable size, for sustainability and growth of agricultural output (FAO 2001c). In addition, just as state ownership has not yielded the anticipated growth in agricultural production, private ownership has also shown little benefit to increasing production, largely as a result of market failures. Therefore, market reform must go hand in hand with tenure reform (Bruce and others 1997). Policy-makers are also reforming attitudes towards communal land tenure and access, and realizing that, under certain conditions, communal systems provide security of tenure, environmental and production sustainability, and conflict avoidance (Bruce and others 1997). However, this transformation has been slow, and is still experiencing opposition in some countries. In Kenya, for example, individual titling is still regarded as the political and social ideal and, therefore, claims to communally owned land are often thrown out of court. This has led to land grabbing, or illegal occupancy in some areas, notably in urban areas and state forests (DFID 1999). Means for strengthening the voice of community groups include the decentralization of political power and the formation of natural resource use councils, comprised of community members (DFID 1999). In Uganda, the new Land Act (1998) combines objectives of agricultural productivity and equity by promoting democratization and good governance with some redistribution of land rights. Implementation of the Land Act (1998) has been hindered by lack of an overall land policy, and by insufficient strategic

---

**Box 2f.4  Conflicts in land use due to land policy failures**

Many of Kenya's major wildlife reserves are in traditional pastoral areas (for example, Maasailand and Samburu). The livestock belonging to the indigenous pastoralists (the Maasai and Samburu tribes) are excluded from the parks, because conservation areas were established under colonial rule, and the prevailing philosophy was to preserve and to protect the land from human activities. However, this results in restrictions on important grazing areas (including springs and other water sources) and disrupts traditional management practices. At the same time, the parks are not fenced and the wild animals are not herded. Therefore, they are able to leave the reserves at certain times of the year and to graze in the same areas as livestock. This means that the areas outside of the reserves incur additional pressure – from the livestock that can no longer migrate into the neighbouring reserves, and from the game that migrates out of the reserve. The impact of the wildlife is also greater than that of the cattle, as they have 'extended grazing hours', feeding throughout the night, whereas cattle are kept in enclosures

*Source: Sindiga 1999, personal communication, Ole Kamuaro Ololtisatti, Purko Maasai*

---

planning, limited resources and capacity, and widespread corruption (DFID 1999).

A further consideration in land reform is the issue of gender. Although women are responsible for most household and commercial agricultural production (FAO 2001b), their rights to own land are severely diminished, being largely through husbands or fathers (Bruce and others 1997). Governments' recognition of women's rights, and the issue of gender reform, have not progressed as far in Eastern Africa as in Southern Africa, although Burundi, Eritrea and Ethiopia are starting to encourage the inheritance of land by women, and the allocation of land to couples to create household holdings (Bruce and others 1997).

## WESTERN INDIAN OCEAN ISLANDS

Madagascar dominates the sub-region in terms of land mass, occupying nearly 600 000 km$^2$, the fourth largest island in the world. The remaining countries are made up of archipelagos of between 3 and 115 islands, with a combined size of 2 000 km$^2$. Large parts of the Western Indian Ocean Islands are mountainous, rugged and dry, and unsuitable for cultivation. Only Madagascar is large enough to support a significant amount of permanent pasture (41 per cent of the land area) and livestock production (UNEP 1999b). The dominant land use in the sub-region is cultivated crops in Comoros and Mauritius (40 and 48 per cent of the land area respectively), whilst all countries have significant cover of forests and woodlands, which are widely used for grazing and gathering of wild resources (UNEP 1999b). Most of the islands experience monsoon rains from November to April, and total rainfall varies from island to island, within the range 700 mm/yr to more than 2 000 mm/yr. Cyclones are also common in some of the islands, and this can be highly erosive to exposed soils. By contrast, dry spells and droughts are not uncommon, especially in southern Madagascar.

## IMPORTANCE OF CULTIVATION AND LIVESTOCK PRODUCTION IN THE WESTERN INDIAN OCEAN ISLANDS

Agriculture is an important activity in the Western Indian Ocean islands, both at the subsistence level and at the commercial level. The major commercial crops are: bananas; cassava; cloves; coffee; copra; onions; potatoes; rice; sugar; sweet potatoes; tea; vanilla; and ylang ylang, a perfume which is unique to Comoros. In 1970, agriculture employed 83 per cent of the workforce in Comoros, 84 per cent in Madagascar and 34 per cent in Mauritius. This fell slightly in all countries over the following two decades, with Mauritius showing the greatest decline, to 17 per cent by 1990 (World Bank 2001). Over the past 30 years, the agricultural sector has also contributed significantly to the economies of the sub-region's countries, with the result that the best land has been reserved for the commercial production of luxury commodities, and the countries are net importers of cereals and staples.

In Comoros, the contribution of agriculture to the economy has ranged from 35 per cent to 38 per cent of GDP between 1980 and 1999 (World Bank 2001). In Madagascar, the contribution to GDP has been slightly lower, at between 24 and 28 per cent (World Bank 2001). In the Seychelles, agriculture's contribution to GDP has fallen from nearly 10 per cent in 1980 to less than 5 per cent in 1999, as tourism has grown in income contribution and foreign exchange earnings (World Bank 2001). Exports from agriculture in 1999 were estimated at US$92 million from Madagascar, and US$405 million from Mauritius (World Bank 2001).

## EXTENT AND PRODUCTIVITY OF CULTIVATION AND LIVESTOCK PRODUCTION SYSTEMS IN THE WESTERN INDIAN OCEAN ISLANDS

Subsistence agriculture is practised on all islands in the sub-region, especially in Madagascar and Comoros, where slash-and-burn agriculture is a common means of supplementing household food requirements and incomes. However, due to economic pressures for agricultural exports and foreign exchange earnings, the best land is often reserved for commercial crop production (especially sugar, copra, vanilla, coffee and ylang ylang) (UNEP 1999b). As a result of rising population pressure and of increasing demand for land for subsistence farming, more natural habitat is being converted to cultivation, and the soils are becoming degraded, especially in the marginal areas, where low-input subsistence agriculture is practised. As a result, there are signs of reduced productivity compared to levels 50 years ago (UNEP 1999b).

These same pressures have resulted in increased area of land under cultivation in Comoros (from 90 000 ha in 1970 to 120 000 ha in 1999) and in Madagascar (from 2 300 ha in 1970 to 3 100 ha in 1999). In Mauritius and Seychelles, the area of land under cultivation has remained fairly constant (at 100 000 ha and 5 000 ha respectively) and, in Reunion, the area of cultivated land has actually declined slightly, from almost 60 000 ha to less than 40 000 ha (FAOSTAT 2001).

Although absolute food production indices have also climbed over the past 30 years, as a result of population growth and largely export-driven markets, food production per capita has declined in all countries, as shown in Figure 2f.10 and Figure 2f.11 (FAOSTAT 2001). In Madagascar, where there is significant rearing of livestock, production indices have also risen, by almost 50 per cent of the output in 1970 but, again, this has been outstripped by population growth, and per capita production rates have fallen by almost 40 per cent (FAOSTAT 2001). In Mauritius, livestock rearing and processing, principally chicken farming and fish farming, has produced an increase in food production (UNDP 2000).

The gap between food production and population growth, and the emphasis on commercial agriculture rather than production for domestic consumption, have resulted in a food deficit, particularly in cereals and other staples (FAO/GIEWS 1998). As a result, some countries, such as Mauritius and Seychelles, import significant quantities of cereals, whereas Comoros and Madagascar are poorer countries, which cannot afford expensive imports and, therefore, are often dependent on food aid to make up some of the deficit. In 1998, both food imports and aid were high for Comoros and Madagascar, due to an outbreak of brown locusts, which caused crop damage. The locust outbreak may have been linked to the El Niño event, which caused hot, dry conditions (FAO/GIEWS 1998).

## LAND QUALITY AND PRODUCTIVITY IN THE WESTERN INDIAN OCEAN ISLANDS

The heavy pressures on land in the sub-region have resulted in: the degradation of, or the conversion of, natural vegetation; clearing of forests; loss of productivity; and soil erosion. Deforestation rates are high, because land is cleared for commercial cultivation, as well as for urban and industrial developments. This exposes the soil to wind and water erosion, and alters the capacity to regulate soil and water quality, flow regulation and flooding control. The rainforest area of Madagascar, for example, is experiencing escalating rates of clearance for cultivation and cattle rearing. Intense grazing pressures are, in turn, affecting areas of grassland and permanent pastures, leading to: loss of vegetation cover; soil compaction; and erosion.

Lack of erosion control techniques, exacerbated by frequent, severe tropical cyclones, leaves the soils exposed and vulnerable to extensive erosion (World Bank 1995). The central highlands of Madagascar have been classified as 'hotspots' of soil erosion, due to intense pressures from low-input agriculture and, in the southwest of the country, shortening of fallow periods has caused nutrient depletion of the soils, rendering them more vulnerable to the impacts of drought (UNEP 1999b). In Comoros, soil fertility has declined and soil structure has deteriorated, such that enormous soil losses are experienced with each monsoon (UNEP 1998). In Reunion, the steep sides of the volcano have been cleared of vegetation, and they experience highly erosive rainfall, resulting in landslides, which are effective barriers to sustainable agriculture. In these small islands, the coastal zone is greatly threatened by soil losses of this magnitude, because sediments which are washed

**Figure 2f.10  Crop production indices for the Western Indian Ocean Islands, 1970–2000 (total and per capita)**

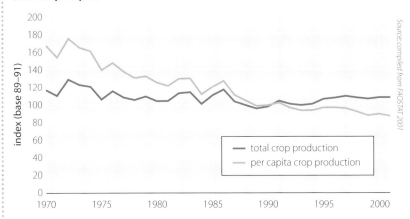

*Source: compiled from FAOSTAT 2001*

**Figure 2f.11  Livestock production indices for the Western Indian Ocean Islands, 1970–2000 (total and per capita)**

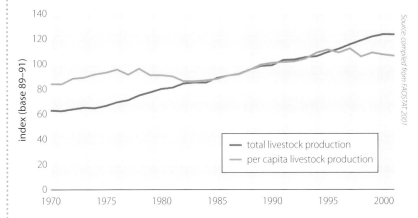

*Source: compiled from FAOSTAT 2001*

out to sea smother fragile and economically important habitats, such as coral reefs and mangroves. This is a serious problem in Seychelles (Shah 1997), where additional coastal erosion occurs due to wave action and sea level rise. Land reclamation projects in Seychelles have produced 200 ha, with a further 395 ha under development, despite damage to the coral (Shah 1997).

Rates of soil loss, and the economic impacts of this, have not been quantified for this sub-region, but it has been estimated that, in Madagascar, approximately 45 per cent of the land is under moderate, high or very high risk of desertification. This follows the general pattern seen in Africa of highest vulnerability in areas of low-input agriculture combined with high population, or along desert margins (Reich and others 2001). Madagascar's National Environmental Action Plan (NEAP) places the costs of land degradation at 15 per cent of GDP, or US$290 million (USAID 1988). This was

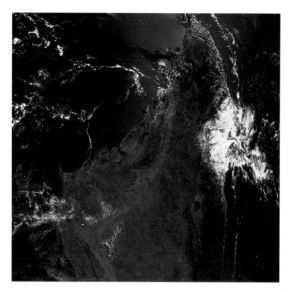

Satellite image showing soil erosion and deposition in the ocean off Madagascar

*Jacques Descloitres, MODIS Land Rapid Response Team, NASA/GSFC*

mostly attributed to deforestation, although soil erosion was estimated to cost between US$4.9 million and US$7.6 million in any particular year. The costs included: reductions in agricultural productivity; raised infrastructure maintenance and investment costs; and damage to coastal ecosystems (USAID 1988). Whilst this was a small fraction of agricultural GDP that year, the cumulative costs of soil degradation over time could be significantly higher, as shown by estimates from other countries (UNU 1998).

A further issue of concern, particularly in Mauritius, is the extensive use of inorganic fertilizers, herbicides and pesticides in commercial sugar plantations. Average annual fertilizer applications have been estimated at 600 kg/ha, more than five times the world average of 113 kg/ha (Government of Mauritius 1991, WRI, UNDP, UNEP and World Bank 1998). This excessive exposure to agrochemicals poses a health risk for farmworkers, and run-off reaching rivers and coastal waters is poisoning wildlife and contributing to eutrophication (UNEP 1999b).

### Improving land quality and productivity in the Western Indian Ocean Islands

Climate change and increasing populations (both resident and tourist) are likely to increase the pressures on land resources, in order to meet the growing demand for food amid more variable climatic conditions. In response, all countries in the sub-region have ratified the UNCCD, although none of the countries have yet produced National Action Plans. Madagascar, the country most affected by drought, organized a National Awareness Seminar in 1997, and has established a national focal point for coordinating the National Action Plan (UNCCD 2001).

All countries have also produced NEAPs, in addition to a Regional Environment Programme coordinated by the Indian Ocean Commission, Mauritius (IOC). Each of these contains action programmes to combat degradation, and implementation is well underway in Madagascar and Mauritius. In Madagascar, environmental awareness has increased since adoption of a national conservation strategy in 1984, and several successful conservation schemes have been completed. The NEAP complements these activities by focusing on: biodiversity protection and management; the creation of a national environmental fund for environmental improvement projects; additional research, including land mapping and management; environmental education and training; and institutional support.

### LAND RIGHTS IN THE WESTERN INDIAN OCEAN ISLANDS

Land tenure arrangements vary within and between countries, which often creates conflicts over land ownership, access to resources, land use and distribution of benefits. These, in turn, influence investment patterns and land management practices, and can lead to land degradation. The richer parts of the region have established more equitable distribution of land between their differing communities, with effective protection of land rights. In the poorer islands, however, much needs to be done, and the lack of equitable land policies and practices present direct threats to sustainable land quality. In Mauritius, 90 per cent of land is privately owned, and more than 80 per cent of families own their own homes with government-registered deeds. By contrast, in Madagascar, land is generally owned through inheritance without title deeds, resulting in conflicts over land rights, and in short-term exploitation of soil (FAO 1997b). In Comoros, three types of land tenure laws apply:

colonial law, customary law and Islamic law. As in Madagascar, lack of registration in Comoros creates confusion over ownership and access rights (RFIC 1998). In Seychelles, 70 per cent of the land is state-owned and is then leased to smallholders (Republic of Seychelles 1997, UNDP 1997). Since 1993, however, land reform processes are underway in Seychelles, in order to transfer land back to original owners (UNEP 1999b). In Madagascar, a land reform study was started in 1993 to improve understanding of the relationship between land tenure and degradation, and the likely future consequences of rising population and demand for land and resources. Additional studies are required to determine the role of traditional property rights in sustainable land use and to make recommendations for short-term local action programmes, coupled with reforms in land tenure and land use, as well as development policies (World Bank 1993b).

## SOUTHERN AFRICA

Southern Africa has a total land area of 6.8 million km², of which almost 33 per cent is covered by forest, 21 per cent is desert, and the remaining natural habitat is largely savannas and grasslands. Rainfall in the sub-region ranges from 50mm/yr in the arid deserts of Botswana, Namibia and South Africa, to more than 1000mm/yr in the equatorial forests of Angola, Malawi, Mozambique and northern Zambia. In most areas, rainfall is largely seasonal, falling over a period of just a few months, often in the form of intense thunderstorms or showers. Where vegetation cover is reduced, this can lead to higher rates of soil erosion. Likewise, most of the sub-region experiences high variability in rainfall, and frequent or prolonged periods of flooding and drought. Grazing lands currently cover 49 per cent of the area, predominantly in savannas and grasslands and, especially, in the drier countries where forest cover is lower (FAOSTAT 2001). Permanent crops and arable lands cover slightly less than 6 per cent of the land area, and are predominantly rain-fed, except in South Africa, where irrigation potential is relatively well developed.

### THE IMPORTANCE OF CULTIVATION AND LIVESTOCK PRODUCTION IN SOUTHERN AFRICA

The proportion of the Southern African population employed in agriculture in 1970 was 71 per cent. In 1980, it was 64 per cent and, by 1990, it was 60 per cent (World Bank 2001). Proportions varied, however, from 87 per cent in Malawi to 14 per cent in South Africa (World Bank 2001). The major crops include maize, wheat, tobacco, tea, cashew nuts, sugar cane, coffee and cotton, and these contribute significantly to GDP and to exports. In Tanzania, agriculture contributes up to 50 per cent of GDP and 50 per cent of export earnings (Government of Tanzania 2001). Livestock production is particularly important, accounting for approximately 30 per cent of agricultural earnings. South Africa's agricultural exports totalled US$2 464 million in 1997, and Zimbabwe's reached US$1 157 million (World Bank 2001). In Malawi and Mozambique, agriculture accounts for approximately 35 per cent of GDP, but less than 10 per cent in mineral-rich countries, such as Botswana and South Africa (World Bank 2001).

Small-scale agriculture and pastoralism are widely practised in Southern Africa, although the value of these practices is not reflected in national accounts. For example, in Tanzania, approximately 3.8 million households practise small-scale farming, and roughly 10 per cent of these practise pastoralism or agro-pastoralism (Government of Tanzania 2001). Although cattle dominate, sheep and goats are extremely important sources of protein, accounting for about 12 per cent of the national meat supplies (Government of Tanzania 2001). In Botswana, approximately 70 per cent of the population lives in rural areas and is dependent on agricultural activities—both rain-fed cultivation of crops and livestock rearing—for their livelihoods (Botswana Agricultural Census Report 1993).

### EXTENT AND PRODUCTIVITY OF CULTIVATION AND LIVESTOCK PRODUCTION SYSTEMS IN SOUTHERN AFRICA

Over the years, the sub-region has seen some expansion in both cropland and permanent pastures, as a response to rising population and demand for food, as well as to policies aimed at increasing exports. The total cultivated area has grown from 32 million ha in 1970 to 39 million ha in 1999, whereas the extent of permanent pastures has remained almost constant, at 332 million ha (FAOSTAT 2001). In some instances, these areas are inappropriate for cultivation or grazing, as a result of low or variable rainfall, or unsuitable topography and soil quality.

Absolute production of crops and livestock has increased since 1970, but has lagged behind population growth and, therefore, the per capita production indices show a decline (see Figure 2f12 and Figure 2f.13). Countries have been dependent on imports of grain over the past three decades and, on occasion, have required food aid, particularly during times of flooding or drought. In addition, per capita calorie intake in many countries is now lower than it was in 1970, and protein intake has declined quite considerably in Malawi, Zambia and Zimbabwe (FAOSTAT 2001).

Current levels of nutrition in the sub-region are, on average, 2 231 calories/capita/day. However, they vary from 1 782 calories/capita/day in Mozambique (which is recovering from civil war) to 2 956 calories/capita/day in South Africa, one of the wealthiest countries in the sub-region (Trueblood, Shapouri and Henneberry 2001). Climate variability contributes significantly to fluctuations in production, and food supplies tend to be in surplus or in deficit. Mozambique and Angola have also been most heavily dependent on imports and food aid, especially since 1980. For example, between 1963 and 1965, Angola imported 9.5 kg/capita of grain but, between 1993 and 1995, the figure had increased to nearly 50 kg/capita (Trueblood and others 2001). In Lesotho, imports grew even more dramatically, from 19 kg/capita to 98 kg/capita, whereas, in Botswana, food imports remained constant, and exports grew from nothing to 3 kg/capita (Trueblood and others 2001).

## LAND QUALITY AND PRODUCTIVITY IN SOUTHERN AFRICA

One of the challenges facing Southern Africa is to feed the growing population through increased agricultural production. Such production pressures have led to resources being overexploited, and vegetation and soil degradation is a major concern. Studies indicate that more than 50 per cent of Southern Africa's land degradation is caused by overgrazing from cattle, sheep and goats, some of which are bred in unsuitable areas (UNEP 1999a). The timing and density of stocking affect the productivity of grazing land, particularly in areas where rainfall is limited or variable (Erskine 1987). Drought is a further contributing factor, as is the land tenure type, because traditional communal access grazing systems have conflicted with commercial production pressures, population growth, and colonial and post-colonial tenure systems, resulting in less willingness to share input costs, and greater incentives to exploit land for individual gains. Because the extent of suitable land is limited, there is increasing pressure to open up forests for grazing land and irrigation, and to cultivate marginal lands—a situation which exacerbates land degradation. Deforestation rates in the region average 0.5 per cent per year (equivalent to the loss of 13 000 km² per year), mainly due to the expansion in agricultural land (World Bank 2000b). Removal of

**Figure 2f.12  Crop production indices for Southern Africa, 1970–2000**

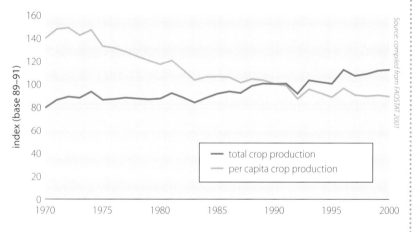

Source: compiled from FAOSTAT 2001

**Figure 2F.13  Livestock production indices for Southern Africa, 1970–2000**

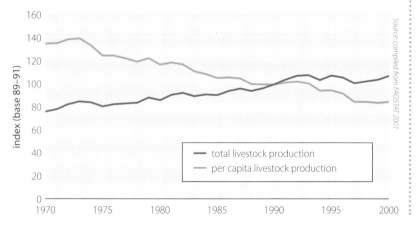

Source: compiled from FAOSTAT 2001

vegetation cover, poor agricultural practices, loss of wetlands and overgrazing are mainly responsible for increasingly dramatic impacts of flooding in Southern Africa in recent years (UNDHA 1994).

Some of the visible manifestations of land degradation in Southern Africa include: rapid soil erosion; declining crop yields; siltation of rivers and reservoirs; and deterioration of grazing lands. In Zambia, for example, it is estimated that the country loses 3 million tonnes per year (t/yr) of topsoil from cultivated land while, in Swaziland, the rate of soil loss is 50 000 (t/yr) and, in South Africa, it is between 300–400 million (t/yr) (Chenje and Johnson 1994). Knock-on effects of vegetation and soil loss include: changes in the abundance and diversity of species; and impaired water quality and air quality regulation.

Declining soil fertility is largely brought about by continuous cultivation without the application of fertilizer or manure. Fertilizer use in the sub-region has increased for some countries whilst, for others, the removal of government subsidies has lead to increasing costs of fertilizers and, thus, their use is declining. Many farmers can no longer afford them, or apply sub-optimal amounts and, therefore, soil fertility further declines.

## Improving land quality and productivity in Southern Africa

There have been a number of responses in Southern Africa to combat loss of vegetation cover and soil fertility, and to rehabilitate land resources. Some of the programmes are in the form of community-based natural resources management projects (CBNRM), or of district and national environmental action plans or biodiversity strategy action plans, under the auspices of regional and international conventions. For example, the SADC has established an erosion hazard mapping programme, aimed at defining seriously affected areas, as well as assisting in the design of appropriate conservation strategies, by providing guidance in regional planning, environmental monitoring and land utilization programmes. The SADC also produced a Sub-Regional Action Programme to Combat Desertification in Southern Africa, in line with UNCCD. All the countries of Southern Africa are party to this convention, and Lesotho, Malawi, Swaziland, Tanzania and Zimbabwe have also produced National Action Plans (UNCCD

Drought-stricken maize crop, Tanzania

*Nigel Cattlin/Holt Studios*

2001). The sub-regional action plan also identifies priority issues to be dealt with in curbing desertification, including strengthening of early warning systems and the development of alternative sources of energy (Chenje 2000). An environment education programme aimed at establishing an SADC network for environmental education is also being implemented, with support from international donors. The Regional Environmental Education Centre coordinates the activities, which include the development of environmental education policy, and training of trainers.

Following the Inter-African Conference on Soil Conservation and Land Utilization in Goma (Democratic Republic of Congo) in 1948, several regional committees for the conservation and utilization of soil have been established, including the Southern African Regional Commission for the Conservation and Utilisation of the Soil (SARCCUS). Through the Sub-committee for Land-use Planning and Erosion Control, SARCCUS drew the attention of its member countries to negligible levels of soil conservation in the region. In 1974, soil erosion surveys were undertaken throughout Southern Africa. However, owing to a lack of a standard procedure meeting the specific needs of the region, no overall assessment of the soil erosion situation was carried out. Following several attempts over a long period, a classification system eventually materialized, with several countries participating, culminating in a meeting in Gaborone in 1981 (SARCCUS 1981).

## LAND RIGHTS IN SOUTHERN AFRICA

Access to land and resources in Southern Africa is perhaps the most socially and politically sensitive issue, now being entangled in environmental and political agendas. Minority individual tenure and state-based conservation practices were imposed on land which was traditionally owned by indigenous people, and most of the black populations were removed and confined to areas insufficient in extent or quality to meet production requirements, as detailed in Box 2f.5. Although colonial and post-colonial apartheid policies have largely been replaced, or are in the process of transformation, ownership and access to resources is now largely determined by economic status, with commercial farmers occupying the best farmland and contributing most visibly to the economies. For example, in Zimbabwe, a minority of 4 500 mainly white commercial farmers control more than 33 per cent of the country's prime agricultural land.

Land inequities were, until recently, most extreme in South Africa, where some 70 000 white farmers owned 87 per cent of the arable land, and 2 million black subsistence farmers were restricted to 13 per cent of the land (Moyo 1998). The imposition of state-controlled institutions significantly undermined the traditional and cultural institutional structure for resource management, thereby alienating the indigenous people from their cultural and governance aspirations. Traditional tenure security was effectively eroded, and CBNRM came under threat. Inequitable access to land underlies the food and agricultural problems facing Southern Africa today, and their impact on poverty. Botswana's Tribal Land Act (1968) facilitated the conversion of tribal land to individual lease for residential, arable or grazing purposes (DFID 1999). This has led to the expansion of commercial cattle ranching, which contributes to vegetation and soil degradation, and to the marginalization of traditional hunter-gatherer communities, such as the Bushmen. The traditional authority of the chiefs has been replaced by a Land Board and its Board Secretariat, a Technical Committee and Land Officers (DFID 1999). Land Rights issues in Southern Africa are manifest in increasing pressure on resources, land-based conflict, and pressure for rural development and land reform. In Zimbabwe, the conflict over land encompasses ancestral claims, claims by veterans of the independence war and gender imbalances.

### Box 2f.5 Colonial influences on land rights in Southern Africa

Colonial policies on land tenure and access influenced patterns of land use and management in several ways in Southern Africa. Shifting cultivation was seen as destructive to forests, for example, and legislation creating forest reserves was passed, leaving farmers with little option but to intensify production from existing cultivated or grazing areas. The traditional communal land tenure system was perceived to be insecure and a further cause of environmental degradation and, therefore, land was either leased from the state or privatized. It has since been acknowledged that state or private ownership can be just as harmful to the natural resource base, and extensive land tenure reforms are underway, with a greater recognition of indigenous rights and practices, as well as a greater appreciation of the role of women in agriculture.

Source: Annersten 1989

In response, most countries of Southern Africa are developing new policies, through reorganization and transformation, in order to address the needs of previously disadvantaged masses. A number of strategies were adopted, in order to achieve the objectives of land reform, including land redistribution and resettlement programmes. For example, in Zimbabwe, the government plans to acquire 5 million ha of the total 11.3 million ha of land belonging to the commercial sector, in order to complete its resettlement programme. Following land identification and planning, selected groups are to be resettled according to six different models, covering mixed farming, specialized farming and ranching (Government of Zimbabwe 1998). However, there have been significant delays and, of the targeted 160 000 families aimed for resettlement, only 60 000 had been resettled by 1988 (African Development Bank 1993). Other policy instruments that have been used include five-year national development plans to reorganize communal areas by agricultural potential. These plans presented options for sustainable and viable agricultural production, in an attempt to alleviate fears of falling agricultural productivity levels and economic recession.

*Land inequities were, until recently, most extreme in South Africa, where some 70 000 white farmers owned 87 per cent of the arable land, and 2 million black subsistence farmers were restricted to 13 per cent of the land*

In South Africa, the land reform process has attempted to provide for tenure, use and access rights that are either individual or group-based, with existing or new community organizations qualifying for such rights on the basis of demonstrable public support. Land Rights Officers are proposed, in order to ensure the participation of all stakeholders in decision making, and Land Rights Boards will arbitrate in the event of disputes and will make recommendations to the Minister (DFID 1999). Implementation is in the early stages, however, and assessments of effectiveness would currently be premature. In Mozambique, the 1998 Land Law is beginning to be implemented, following an extensive public awareness and discussion programme. Surveys to register land rights have commenced in certain areas, and verbal testimony to tenure under customary law is sufficient to register tenure rights under the new law (DFID 1999).

In many countries, land reform processes have been strengthened by the creation of central agencies, such as government departments for land, agriculture, local government and resource development. These institutions provide land, credit facilities, and a range of technical and professional services.

The current mixture of land tenure systems allows varying degrees of access to resources by women. Under the private freehold system, women have rights to access land, but very few of them have the resources to purchase such land on the open market. On the other hand, communal land held under the traditional or customary system allows women secondary access through marriage but, as soon as the marriage breaks, they lose the right to cultivate lineage land (SARDC-WIDSAA 2000). However, through processes of land reform, liberalization and improved status of women, women are slowly beginning to control a sizeable proportion of rented, purchased and allocated land.

## CENTRAL AFRICA

Central Africa is predominantly covered in forest and savanna. The coastal humid belt, with high and relatively constant rainfall, supports dense tropical forests, whereas the northern parts of Cameroon, Central African Republic and Chad are drier, with more variable rainfall, and the dominant vegetation is savanna. Land use in the sub-region is sensitive to climatic and vegetation characteristics, with forestry and commercial plantation agriculture largely found in the humid zones (where rainfall reaches up to 4000 mm/yr), and livestock rearing, with some subsistence cultivation, in the semi-arid zones (where rainfall averages 500 mm/yr). The semi-arid zone is also highly vulnerable to climatic variations and drought, which limit agricultural expansion. Soils are highly vulnerable to erosion, because most of the rainfall occurs in intense heavy storms, and because the clay and silt content makes the soils prone to crusting when exposed (Njinyam 1998).

Approximately 8 per cent of the total area is currently used for arable and permanent crops (with Cameroon having the largest share, at 15 per cent of its land area), and 16.5 per cent is used as permanent pasture (FAOSTAT 2001). Irrigated agriculture is limited, partly because the fertile soils and the high, reliable rainfall in the humid zone are conducive to rain-fed agriculture, and partly because the infrastructure development required to establish irrigated cultivation in the semi-arid zone has so far been prohibitively expensive. Despite these favourable conditions in Central Africa, large-scale agricultural development has been limited by national market failures and international trade barriers. Shifting cultivation (or slash-and-burn agriculture) was a traditional means of coping with variability, but this practice is no longer sustainable, because there are much larger populations now requiring land. The priority issues in Central Africa are, therefore: improving food security, through enhanced production and distribution of resources; and reducing the pressures that shifting cultivation has on forests and woodlands.

## IMPORTANCE OF CULTIVATION AND LIVESTOCK PRODUCTION IN CENTRAL AFRICA

In 1970, 81 per cent of Central Africa's labour force was employed in agriculture, with Chad having the highest at 92 per cent. By 1980, the percentage had fallen to 74 and, in 1990, the average was 68 per cent (World Bank 2001). The reasons for this decline include population growth exceeding agricultural expansion, and industrial development. Pastures account for much of the agriculturally productive land in the Sahel, and

*Through processes of land reform, liberalization and improved status of women, women are slowly beginning to control a sizeable proportion of rented, purchased and allocated land*

pastoralists and agropastoralists are integral parts of local and regional economies. It was estimated in 1995 that there were more than 404 000 pastoralists in Chad (about 15 per cent of the country's population), with pasture areas covering about 55 percent of the national territory (FEWS 1995).

The major crops in the sub-region include: cassava; cocoa; coffee; cotton; groundnuts; maize; millet; palm oil; rubber; and sorghum. In 1980, the total value of agricultural exports for the region was US$1 148 million, although Cameroon took the lion's share at US$699 million. In 1990, exports fell to just US$909 million, mainly as a result of commodity price fluctuations (World Bank 2001). By 1997, however, markets had recovered in most countries, although the war in Democratic Republic of Congo had an enormous impact on its agricultural exports and, as a result, totals

for the sub-region were just US$796 million (World Bank 2001). Value added in agriculture in 1980 ranged from 40 per cent of GDP in Chad to just 12 per cent in Gabon (World Bank 2001). In 1990, the percentage for Chad fell to 24 whilst, in Democratic Republic of Congo, the percentage contribution of agriculture to GDP has climbed from 27 in 1980 to 58 in 1999, mainly due to other economic activities having been disrupted due to the war (World Bank 2001).

## EXTENT AND PRODUCTIVITY OF CULTIVATION AND LIVESTOCK PRODUCTION SYSTEMS IN CENTRAL AFRICA

Land development for crop cultivation increased rapidly between 1970 and 1985 (from 18 million ha to 21 million ha), but then the rate of expansion slowed, until 1999, when the total cultivated area was 21.5 million ha (FAOSTAT 2001). The period of rapid increase must have come as a result of the boom in export crop prices, especially from coffee and cocoa sales, followed by the economic crises of the 1990s, during which land use for export cash crops decreased. Over the same period, the extent of permanent pasture has shown little change, and it stands at approximately 80 million ha (FAOSTAT 2001).

Absolute production for crops has increased steadily over the past 30 years, but crop production per capita has declined, due to population growth rates exceeding food production capacity (see Figure 2f.14) (FAOSTAT 2001). Interestingly, absolute livestock production indices for the sub-region have increased quite significantly, but livestock production per capita has remained relatively constant (see Figure 2f.15) (FAOSTAT 2001).

As a result of declining yields and civil wars, calorie intake in Cameroon, Central African Republic and Democratic Republic of Congo in 1999 was lower than in 1970. For the other countries, it had improved, but was still, on average, just 2 292 cal/capita/day (FAOSTAT 2001). To compensate for this, many countries have imported large amounts of cereals and, during times of drought, Cameroon, Democratic Republic of Congo and Chad have requested food aid. The Democratic Republic of Congo is normally an importer of cereals but, in 1998, livestock production suffered from the civil strife (more so than crop

**Figure 2f.14  Crop production indices for Central Africa, 1970–2000 (total and per capita)**

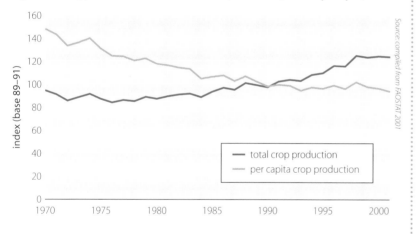

Source: compiled from FAOSTAT 2001

**Figure 2f.15  Livestock production indices for Central Africa, 1970–2000 (total and per capita)**

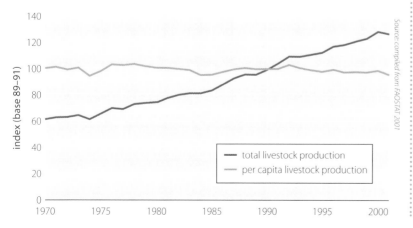

Source: compiled from FAOSTAT 2001

production), and food production shortfalls were estimated at 118 000 tonnes. Due to the disruption of trading activities, particularly in Brazzaville, commercial food imports were reduced, and 46 000 tonnes of cereal were required to meet the population's needs. A variety of coping mechanisms were enacted, including increasing effort in alternative food production areas (fishing and hunting, and short-cycle crops), as well as food aid for vulnerable groups (FAO/GIEWS 1998).

## LAND QUALITY AND PRODUCTIVITY IN CENTRAL AFRICA

Land degradation, defined as the deterioration in the quality and productive capacity of land (Benneh, Agyepong and Allotey 1990), has been identified as one of the major environmental challenges facing the Central Africa sub-region. The main contributors to land degradation in this sub-region are erosion and soil compacting, as a result of extensive removal of vegetation, and exposure of the soils to heavy rainfall, increased evaporation and wind action. The main reasons for vegetation removal are commercial logging and tree cutting to provide domestic fuel, as well as clearance of forests for commercial or subsistence cultivation.

The rate of forest loss in Central Africa is a cause for concern in terms of its impacts on biodiversity, atmospheric change and hydrological cycles, in addition to the concerns regarding soil erosion (UNU 1998). Chemical degradation also occurs, because of: intensive cultivation of marginal areas without sufficient fallowing; use of chemical rather than organic fertilizers; and salinization, through inundation with saltwater or irrigation with poor quality water. For example, the Congo Basin has lost more than 1 million ha of original forest cover, contributing to soil erosion and the sedimentation of waterbodies (WRI 2001). The Lake Chad Basin has also suffered severe vegetation loss, and potential for soil loss and desertification is high (WRI 2001).

Declining productivity and soil structure in the Sahelian zones of Chad and Cameroon are exacerbated by unpredictable rainfall and drought, resulting in extreme degradation and desertification. Chad is currently experiencing the greatest vulnerability to desertification, with 58 per cent of the area already

Goat herd overgrazing fragile pasture, Chad

*Vincent Dedet/Still Pictures*

classified as desert, and 30 per cent classified as highly or extremely vulnerable (Reich and others 2001). There is a large area in Democratic Republic of Congo (64 per cent), which is classified as hyper-arid, and sands of the Kalahari desert have encroached on the savanna vegetation (Reich and others 2001). Central African Republic and northern Cameroon have also been experiencing desertification since the severe drought of 1972–73 (Njinyam 1998).

In the coming decades, the threat of desertification will increase, as a result of climatic changes, such as: increased evaporation; reduced rainfall and run-off; and increased frequency and severity of drought (IPCC 2001). In addition, civil unrest or conflict can result in vast movements of refugees, many of whom are settled in marginal or fragile areas. Such social and environmental pressures were clearly demonstrated in 1997, when Central African Republic (having to cope with internal disputes) received more than 50 000 refugees from Sudan, Chad, Democratic Republic of Congo, Congo Brazzaville and Rwanda. The arrival of these displaced persons put a visible strain on the already stressed food security situation (Njinyam 1998).

The consequences of land degradation, and of soil erosion and compaction, are manifest as a result of the declining ability to support natural or domesticated plant and animal production. Ultimately, this translates to reduced nutritional status of the population and to reduced export revenues. In addition, communities which are dependent on wild produce—such as fruits, nuts, animals and mushrooms, and wood for fuel—have to search further and further afield to meet their needs, and may experience food shortages or even famine during drought years. Extreme reductions in productivity may result in people abandoning their farms and migrating to urban centres, in search of improved security.

### Improving land quality and productivity in Central Africa

Political and economic development policies, as well as conflicts and civil unrest, have also played a role in declining food security in parts of the sub-region, as shown by the example of Cameroon in Box 2f.6. A comprehensive, integrated approach to improving food security and land quality is, therefore, a current environmental and developmental priority for Central Africa. To this end, countries of the sub-region have ratified UNCCD. Chad and Cameroon are the only countries to have so far produced national reports,

however, and Chad has also produced a National Action Plan. The CILSS, which encompasses Chad, has developed a sub-regional action plan to combat desertification (UNCCD 2001).

Cameroon and Chad have also developed NEAPs, which provide an overall framework for: improvement of land use; harmonization of land use policies; and environmental management. Implementation of these plans needs strengthening through additional resources and institutional arrangements. In Cameroon, the government has also embarked on a tree-planting programme, aimed at stopping the advancing desert.

## LAND RIGHTS IN CENTRAL AFRICA

One of the most important issues related to land and natural resources management in Central Africa is the form of land tenure and the system of access rights. These policies have a direct effect on people's security and on their investment in land and resources management which, in turn, affect productivity and land quality. The procedures and conditions of land attribution are cumbersome and complex, and there are a number of disparities between traditional and modern rights to land ownership. Whereas traditional land rights are granted through inheritance, and are centred on communal access to resources, statutory laws encourage state and private ownership, with the emphasis on commercial production rather than on household production. Differences between customary and statutory law produce conflicting situations on the ground, and can lead to disputes over access to resources. As elsewhere in Africa, the interests of certain population groups with lower levels of political recognition (including pastoralists, women and certain castes) are often marginalized (IIED 1999). Rapid population growth and migration in the sub-region have added to the pressures on land resources, especially on commercial production in the more fertile coastal zones. The influence of Islamic law in some areas, and interventions by development projects, add further complexity to the allocation of resource distribution and access rights (IIED 1999).

Where there is insecurity in land tenure, or competing interests for land use, incentives to invest in adequate land management are reduced, thereby increasing vulnerability to degradation and conflict. An

---

### Box 2f.6  Agricultural development in Cameroon

In 1972, the government of Cameroon embarked on a 'Green Revolution', whereby agricultural production and export was promoted, in order to boost the economy. Monocropping was encouraged, in order to increase the production of cacao, coffee, cotton and rubber, which were fetching high prices on the international markets. Fertilizers and pesticides were subsidized (by up to 65 per cent and 100 per cent respectively), which attracted many farmers to this style of input-dependent agriculture.

However, in 1986, the value of cacao and coffee fell drastically, plunging the country into economic crisis. Subsidies were completely removed, and the use of agrochemicals declined, together with soil fertility. Not only did farmers suffer from lost revenue for their products but, because they had been encouraged to grow crops for export, there was a national deficit of cereal staples. The country had to import food and to request food aid, which has subsequently disrupted local production and food security. For example, imported rice was sold at a lower price than local farmers could produce it, making rice farming unviable.

Consequently, many farmers abandoned their land and migrated to urban centres, seeking greater income and food security. In 1990, the government embarked on a comprehensive research and development programme to improve food security and land management. The approach was for agricultural extension workers to engage with local farmers, and to develop affordable means of improving productivity, food security and soil fertility. A number of local level projects have been successfully implemented, encouraging farmers to use organic matter, such as animal dung, instead of expensive inorganic fertilizers, and to encourage crop diversification and rotation, and agroforestry. Some farmers have been able to produce sufficient quantities to sell in local markets, and assistance in marketing is provided through the non-governmental organization (NGO) network.

example of this is the large area around Lake Chad (northern Cameroon, Chad and Central African Republic), which is used primarily by migratory herdsmen. Their livestock often compete with settled cultivators for land and water, sometimes resulting in violent clashes (Njinyam 1998). The wetter areas in the coastal zone of Central Africa are not suitable, due to the widespread occurrence of tsetse flies (which carry trypanosmiasis) and other diseases. Thus, the arid savannas of the north are vital resources, and assured access to dry-season grazing reserves and crop residues in villages is necessary, in order to ensure the continued significant contribution of livestock production to local communities and national economies (IIED 1999).

In Cameroon, the declaration of conservation areas under the colonial administration has alienated local communities from land and resources which they owned under customary laws. Their claims to the land are not recognized, and any occupation or use of such areas must be negotiated with the relevant responsible state department (Vabi and Sikod 2000). This situation has resulted in the communities exercising de facto user rights in a manner of open access, and these activities are rarely incorporated into conservation areas management plans. Better communication and planning between conservation authorities and local communities is required, in order to meet the objectives of both parties in a sustainable way (Vabi and Sikod 2000).

In response to the various impacts at national level, most countries have elaborated land-zoning plans which indicate the boundaries for various land uses. Registration and titling have also been used in some countries, as a means of improving tenure security, in order to encourage investment in agricultural inputs. However, this has been hampered by the enormous administrative burden of reclassification and recording, and it has not been able to effectively promote access rights for women or pastoralists (IIED 1999). Land reform is also challenged by the existing overlaps and contradictions between traditional customary laws, land use practices and statutory laws. These need to be resolved as a first step in delivering a more equitable and workable system of land rights.

In Chad, the government passed a Forest Code in 1989, confirming the dominant role of the state in ownership of forest land. However, this code simply replaced the customary laws, and placed sole responsibility for administration with the central government. Decentralization of government to local authorities may relieve the administrative burden of central ministries, thus improving the effectiveness of the system and speeding up the process of reform (IIED 1999). To improve effectiveness, stakeholder dialogue and voluntary participation need to be encouraged, and policy reform needs to be complemented with initiatives that target the underlying causes of land degradation, and promote food self-sufficiency and economic development (IIED 1999). In addition, the lessons learned from the implementation of CBNRMs could be incorporated into law reform, such as: options for the co-management of forest and wildlife resources; and improving conflict resolution mechanisms (IIED 1999). CILSS, the sub-regional resource management organization, provides a useful structure for encouraging such debate and sharing of experiences between countries with similar challenges.

## WESTERN AFRICA

Land cover and land use in Western Africa are largely determined by climate, and a dramatic gradation is seen from north to south in rainfall and vegetation cover. In the north, average annual rainfall is 350–850 mm/yr and savannas are the dominant ecosystems along the southern border of the Sahel (Mali, Mauritania, Niger and northern Senegal). Here, climate variability is greatest, and drought is common, and often severe. Cultivation is limited, and the dominant agricultural activity is pastoral livestock rearing. For example, in Mali and Niger, cultivation represents just 4 per cent of the land area and, in Mauritania, it is less than 1 per cent (FAOSTAT 2001). By contrast, permanent pasture accounts for 25 per cent of the land area in Mali, and nearly 40 per cent in Mauritania (FAOSTAT 2001). In the equatorial and coastal zone, rainfall is higher, with greater inter-annual and intra-annual reliability, ranging from 1 000 mm/yr to 4 000 mm/yr, although periodic flooding occurs (FAOSTAT 2001). In 2000, forest cover totalled 72 million ha (almost 12 percent of the land area), although it is highly fragmented, and under increasing threat from charcoal production and collection of wood for fuel, commercial logging, and plantation and slash-and-burn agriculture (FAO 2001a). Nearly 11 per cent of the total area of Western Africa is currently cultivated. Most of it is rain-fed agriculture, and cultivation mostly occurs in the equatorial belt. Togo and Nigeria have the largest percentage of land under cultivation (42 per cent and 33 per cent respectively), followed by Côte d'Ivoire and Ghana (23 per cent each) (FAOSTAT 2001).

Most of Western Africa's population depends on the land for their subsistence, as well as for the production of cash crops. However, this dependency, rising economic pressure and population growth have resulted in increased demands on productivity of the land over the past 30 years. As a result, forest and agricultural lands have undergone rapid degradation and reductions in productivity.

## IMPORTANCE OF CULTIVATION AND LIVESTOCK PRODUCTION IN WESTERN AFRICA

Although subsistence agriculture is widespread, commercial agriculture also contributes significantly to the economy of the sub-region with, on average, 65 per cent of the workforce employed in this sector over the past three decades. Burkina Faso, Mali and Niger had the highest rates of employment in agriculture—more than 90 per cent—in 1990 whereas, in Cape Verde and Nigeria, employment in agriculture reached only 31 and 42 per cent respectively (World Bank 2001).

There is a wider variety of crops grown in Western Africa than in some other sub-regions, although the

Harvesting cocoa pods, Ghana

*Ron Giling/Still Pictures*

most important are: cotton; coffee; cocoa; cassava; groundnuts; maize; millet; palm oil; rubber; sorghum; and yams. International price fluctuations have heavily influenced income from agricultural produce over the past 30 years, but it has remained one of the mainstays of GDP. For example, in Benin and Burkina Faso, the value of agricultural exports grew steadily between 1980 and 1997 (from US$55 million to US$198 million and from US$80 million to US$119 million respectively) (World Bank 2001). Côte D'Ivoire, Ghana and Nigeria experienced declines in export values between 1980 and the mid-1990s, but have seen gains since then, whereas Gambia, Liberia, Senegal and Sierra Leone have not recovered so well. Mali, Mauritania and Guinea have seen fairly constant export values (World Bank 2001). In Mauritania, value added in agriculture was US$306 million (compared to GDP of US$1 252 million) in 1999, up from US$160 million in 1980 (when GDP was US$753 million).

## EXTENT AND PRODUCTIVITY OF CULTIVATION AND LIVESTOCK PRODUCTION SYSTEMS IN WESTERN AFRICA

Total land area under crop cultivation in the sub-region has grown from 51 million ha in 1970 to 66 million ha in 2000 (FAOSTAT 2001). Absolute crop production has also increased, whereas per capita production was the same in 2000 as it was in 1970, having declined until 1985, and then recovered, as shown in Figure 2f.16. Livestock production has also increased in absolute terms, although per capita production increased between 1975 and 1985, and then declined to below 1970 levels (see Figure 2f.17).

Whilst there has been an overall increase in the daily per capita calorie supply in Western Africa (from an average of 2 252 calories/capita/day in 1970 to 2 612 calories/capita/day in 1999), Liberia and Sierra Leone have seen declines (FAOSTAT 2001). This is probably due to supply and distribution disruptions, as a result of the civil wars in these countries. Conflicts and climatic variability have also been the cause of Western African countries being food-deficit countries, and among the most food insecure in the world (Staatz, Diskin and Estes 1999). These countries have been dependent on food imports and aid, and are likely to continue to be so for the foreseeable future. For example, Sierra Leone was a net exporter of rice in the 1960s but, by 2001, was importing it at an approximate cost of US$22 million a year (Verheye 2001). The food situation in 2001 was also particularly severe in Burkina Faso, Liberia and Niger, where the FAO was warning of the need for food aid (FAO 2001d).

Efforts to improve food security have been instigated at the sub-regional and national levels. For example, the US Agency for International Development (USAID) is currently developing a Western Africa Strategy, in order to determine priority needs and areas of assistance (Staatz and others 1999). A further sub-regional response has come from the CILSS and the Club du Sahel. The Club du Sahel is a donor group of the Organization for Economic Cooperation and Development (OECD), which interfaces with CILSS. It was established after the 1972–73 drought, in order to ensure that the disaster was not repeated. Together, these institutions developed a Charter for Food Aid to the Sahel in 1990. The charter, the first of its kind, provided guidelines for improving food aid practices, and integrating foreign assistance into long-term food security objectives. The three main focus areas are: enhancing understanding of the food situation; coordination of donations; and provision of food aid (OECD/ Club du Sahel 2001). Mali and Niger have established well-functioning early warning systems (EWS), whereas other countries are experiencing start-up problems. Lack of coordination between donors, and confusion over requests for aid, have also been experienced in some countries (OECD/ Club du Sahel 2001).

## LAND QUALITY AND PRODUCTIVITY IN WESTERN AFRICA

Land degradation is identified as a major issue in many Western African countries, specifically with regard to: the degradation of forest cover; intensive cultivation practices; and natural disasters, such as droughts and desertification. The current state is a reflection of increasing pressures on land resources over the past 30 years, due to: population growth; rising demand for production and energy; and worsening poverty. Policies in support of foreign exchange earnings are additional pressures, which have intensified land degradation in both the dry savanna and the wet forest zones of the sub-region.

**Figure 2f.16  Crop production indices for Western Africa 1970–2000**

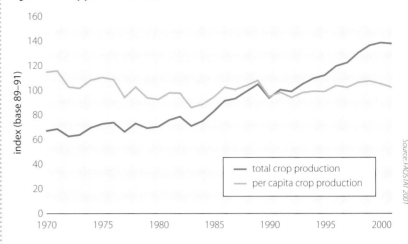

Source: FAOSTAT 2001

**Figure 2f.17  Livestock production indices for Western Africa 1970–2000**

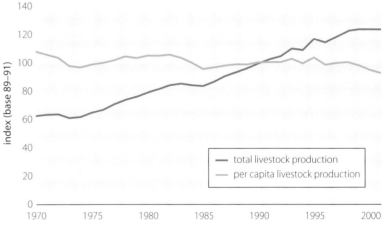

0Source: FAOSTAT 2001

Loss of soil fertility and soil erosion are the obvious manifestations of land degradation in Western Africa. In the northern Sahelian zone, where animal husbandry is the dominant agricultural practice, the main agent of soil erosion is the wind, which is supported by dry climatic conditions. Overgrazing and trampling reduces the vegetative cover and causes compacting of the soil, which is then vulnerable to erosion. In the wet forest zones of the sub-region, sheet and gully erosions are predominant, due to loss of vegetative cover. In commercial plantations and subsistence farms alike, soil fertility is declining, as a result of the prohibitive costs of inorganic fertilizers. In Ghana, for example, water erosion is the major problem, with more than 220 000 $km^2$ of land affected by sheet and gully erosion, of which 50 000 $km^2$ is severely affected (FAO/AGL 2000). Nutrient depletion is widespread in

Ghana, especially in the coastal areas, due to high losses of organic matter caused by: reduction of fallow; intensive cropping; and inadequate applications of fertilizers (FAO/AGL 2000). The long term on-site consequence of erosion and nutrient depletion is the reduction in crop yield, and annual productivity losses of 2.9 per cent in all crops and livestock have been recorded in Ghana (FAO/AGL 2000). The impacts of this on Ghana's economy have been estimated at 2–5 per cent of agricultural GDP. Furthermore, additional losses and costs may be incurred, for example, through eutrophication of water bodies, resulting from sedimentation and accumulation of fertilizer residues. Soil degradation also has consequences at the household level, affecting incomes and food intake, as well as encouraging rural-to-urban migration.

In the Sahelian belt of Western Africa, there is an increasing risk of erosion and desertification (FAO 2000b, Reich and others 2001). In Niger, for example, only 19 per cent of the country is non-desert, and most of this is highly or very highly vulnerable to desertification (Reich and others 2001). Mauritania is similarly affected, with 93 per cent of the country classified as hyper-arid, and the remaining 7 per cent at moderate to very high risk of desertification (Reich and others 2001). The IPCC predicts that rainfall and run-off will decline, and that evaporation will increase, in this zone, further contributing to desertification pressures in future (IPCC 2001).

### Improving land quality and productivity in Western Africa

All the countries of Western Africa have ratified the UNCCD, and Benin, Burkina Faso, Cape Verde, Gambia, Mali, Niger and Senegal have produced National Action Plans. The Economic Community of Western African States (ECOWAS) and the CILSS have jointly produced a Sub-regional Action Plan (UNCCD 2001).

Following an upsurge of general environmental awareness as a result of the Stockholm Conference on the Human Environment in 1972, many countries in Western Africa have formulated environmental and land use policies aimed at stopping land degradation. NEAPs, National Conservation Strategies (NCSs) and other such strategy frameworks have been adopted to address these land issues. In most instances, these policy responses have been initiated or sponsored by

external agencies and institutions, which provided the bulk of the funding. National and sub-regional institutions have also been created to address specific issues, including: the development of drought-resistant crop varieties; collation of reliable climatic data; and soil conservation measures. Population policies have also been adopted in most of the countries, and relevant institutions are in place to implement these policies and related programmes. However, family planning programmes adopted in countries in the sub-region since the 1970s have not reduced the population growth rate, which still averages 3 per cent (ECOWAS 2000).

Local programmes to improve land management include: stone bunding, in order to control erosion on slopes in the northern savanna region; and applying organic matter to cultivated areas, in order to replace nutrients. Composting, agroforestry and the establishment of woodlots are additional land and water conservation measures in operation in Ghana (FAO/AGL 2000). In Niger, the Keita Integrated Development Project was launched in 1984, in order to address declines in agricultural production caused by drought, desertification and population growth. Women have been the main participants in the project, and have been involved in: dune stabilization; land reclamation; soil and water conservation; reforestation; crop production; rural engineering; training; and setting up credit and alternative incomes (Carucci 2000).

## LAND RIGHTS IN WESTERN AFRICA

Countries of Western Africa, as in other sub-regions, are bound by a complex interaction of land tenure policies, which often create insecurity of tenure and exploitation of production in the short-term, and which lay the foundations for conflict over resources (Toulmin and Longbottom 1997). Under customary law, chiefs are the dominant land owners, through whom members of the community obtain access to resources via ownership, sharecropping, tenancy and pledging. The countries inherited tenure laws from the colonial government and, at independence, state-ownership became the norm, and land was then either sold or leased privately. Thus, customary land tenure practices have been weakened, and are no longer recognized by the state or rural communities (DFID 1999). Attempts to reconcile

Agricultural plots , shaded by trees, in the Dogon region, Mali

*J.P.Delobelle/Still Pictures*

differences in land tenure policies, and the influence of Islamic laws and development ideals have, in many cases, created greater confusion and conflicts between land users (Toulmin and Longbottom 1997).

Conflicts over land and resource access rights have most frequently been between farmers and herders, migrant farmers and governments. An example is the establishment of cultivation in areas used by seasonal herders, such as the northern savanna areas and seasonal floodplains (Toulmin and Longbottom 1997, Maltby 1986). The impacts of inappropriate land tenure are economic, environmental and social (Ouedraogo and Toulmin 1999). Economic impacts are loss of livelihoods and increasing poverty, as well as unrealized potential of commercial agriculture. Environmental impacts include disincentives for investment in land care or land improvement, and social consequences include deterioration of community cohesion and rising levels of conflict.

Many Western African countries are undergoing land policy reform, either in the preparatory discussion phase, or in the pilot phase, testing different legislative approaches. The main objectives of these reforms are: clarification of rights and access; harmonization of existing legislation; creation of incentives for enhanced land management; raising productivity and economic development; improving dialogue between stakeholders; and diffusing potential conflicts (Ouedraogo and Toulmin 1999). The main type of reform has been to convert greater areas of land to private ownership, particularly in favour of large-scale agricultural developers (Ouedraogo and Toulmin 1999). Côte d'Ivoire's Rural Land Plan, which was implemented in 1988, is a programme of mapping and recording the boundaries of landholdings and existing rights of individuals and groups. This is done through consultation with local inhabitants and validation by village committees, thereby integrating customary laws into a new rural land tenure code (DFID 1999).

## CONCLUSION

Land and terrestrial resources in Africa have unparalleled economic, social and environmental value. Traditionally, African societies are agrarian or pastoral, depending directly on subsistence farming to meet their daily needs. Commercial agriculture holds an equally important position, employing the largest share of the workforce in most countries, and contributing significantly to national economic growth, export earnings and foreign exchange. However, national and household dependency on agricultural output has been a significant factor in limited economic growth over the past three decades. Climatic instability has caused significant and frequent variability in production, and narrow crop diversity, and national and international market failures, have facilitated recurrent economic losses.

Rapid population growth, and policy pressures to increase production, have forced the cultivation of greater and greater areas of land in all sub-regions, and the extension of cultivation and grazing to marginal areas. Combined with limited application of organic or inorganic fertilizers, reductions in fallow periods, restrictions on crop diversity, inappropriate irrigation, and an increasing use of herbicides and pesticides, this has resulted in the physical, chemical and biological degradation of vegetation and soil. Soil erosion and desertification rates are increasing as a result, and declines in productivity have been noted.

More than 20 per cent of Africa's vegetated lands are classified as degraded, and 66 per cent of this is moderately to severely degraded. The worst affected areas are along desert margins, and the problem is

*Rapid population growth, and policy pressures to increase production, have forced the cultivation of greater and greater areas of land in all sub-regions, and the extension of cultivation and grazing to marginal areas*

likely to intensify over the next 30 years, as a result of population growth and increasing climate variability. Land degradation impacts are felt most keenly by the poor, because they are forced to cultivate marginal lands, such as desert margins, which get degraded more rapidly. Hence, productivity losses are more rapid, and affected households become increasingly food insecure. Environmental consequences of erosion include: sedimentation; pollution; eutrophication of waterbodies; smothering of aquatic habitats; and changes to biodiversity.

Responses have included: ratification of international conventions; and implementation of awareness raising and action planning at national level. Sub-regional cooperation has also been initiated, through national and sub-regional organizations, such as CILSS, ECOWAS, IGAD and SADC, which have developed food security strategies, and strategies for improving the condition and management of resources. NEAPs have provided overall frameworks for enhancing planning and development of land resources, and EIAs have facilitated the implementation of the tenets held in the NEAPs. Local level responses have perhaps seen the most rapid and dramatic results, because they have not experienced the institutional and logistical constraints of actions attempted at national and sub-regional levels. Local level measures have included: diversification of cropping; increased fallow; crop rotation; creation of wind breaks; application of manure and other organic wastes; and responsible and conservative irrigation schemes.

Land tenure and access to land resources are complex issues in Africa, impacting on food security, environmental sustainability and social security. Colonial legislation often conflicted with traditional tenurial systems, and was aimed at increasing commercial production, often at the expense of household production. With independence, African countries attempted to redress the issue of traditional access rights, whilst maintaining control of resources, especially economically important ones. Realizing the inadequacies, overlaps and contradictions in existing land policies, governments embarked on further processes of land reform. However, these are fraught with tensions between user groups and different land uses. Women, generally, have fewer ownership rights although, in many countries, women are involved in

most food production activities. Political and environmental refugees frequently place additional burdens on issues of tenure, and on the resources themselves. Refugees who are settled in marginal or sensitive areas cause extensive degradation, whilst further potential conflicts exist with neighbouring communities.

Improved security of tenure can greatly improve land management practices, and rural development programmes should focus on greater inputs to farming, freer trade and higher value addition. This will ensure that: greater income is earned from production; greater food security is awarded to the household producer; and expansion of agriculture into marginal areas is controlled. Poverty alleviation schemes are also required, in order to keep people from abandoning their farms, and to ensure that they are able to afford inputs, such as fertilizers and efficient irrigation systems (UNU 1998).

Policy reforms are another essential component of resolving conflicts over resources and of ensuring greater investment in sound resource management activities. However, these reforms have to be made carefully, with the full participation of all stakeholder groups. This is a process which will take time, but which will avoid further conflicts in rights or their administration. Legislation is also only a part of what governments can contribute towards land reforms. Other actions include: decentralization of administrative power; more comprehensive land use planning and management frameworks; more effective participation by stakeholders; and economic diversification, in order to relieve some of the dependence on commercial agriculture and forestry. The involvement of women and children is particularly important, because the spread of HIV/AIDS will mean that, in future, more households are headed by these groups. Pastoralists must be involved in management plans for grazing areas, in order to maintain livestock production at the household level and to maintain its contribution to national economies.

## REFERENCES

African Development Bank (1993). *Economic Integration in Southern Africa*. Vol. 3, Oxford University Press, Oxford

African Development Bank (1998). *African Development Report*. African Development Bank, Abidjan

Alden Wiley, L. (2000). Land Tenure and the Balance of Power in Eastern and Southern Africa. In *Natural Resource Perspectives* no. 58 (June 2000). Overseas Development Institute, London

AOAD (1998). *Arab Agriculture Statistics Yearbook*. Arab Organization for Agricultural Development, Khartoum

Arab League (1999). Unified Economic Report (Arab World). Arab League General Secretariat, Cairo

Ashley, C. and LaFranchi, C. (1997). Livelihood Strategies of Rural Households in Caprivi: Implications for Conservancies and Natural Resource Management. DEA Research Discussion Paper 20. Department of Environmental Affairs, Windhoek

Asomani-Boateng, R. and Haight, M. (1999). *Reusing Organic Solid Waste in Urban Farming in African Cities: A Challenge for Urban Planners*. IDRC, Ontario

Benneh, G., Agyepong, G.T. and Allotey, J.A. (1990). *Land Degradation in Ghana*. Commonwealth Secretariat, London and University of Ghana

Bojo, J. (1996). The Costs of Land Degradation in Sub-Saharan Africa. In *Ecological Economics* 16, pp. 161–173

*Botswana Agricultural Census Report 1993*. Agriculture Statistics Unit, Government Printers, Gaborone

Bruce, J., Subramanian, J., Knox, A., Bohrer, K. and Leisz, S. (1997). Land and Natural Resource Tenure on the Horn of Africa: Synthesis of Trends and Issues Raised by Land Tenure Country Profiles of Eastern African Countries, 1996. GTZ, Sahara and Sahel Observatory, United Nations Economic Commission for Africa Sub-Regional Workshop on Land Tenure Issues in Natural Resources Management in the Anglophone Eastern Africa with a Focus on the IGAD Region, August 1997

Carucci, R. (2000). Trees Outside Forests: An Essential Tool for Desertification Control in the Sahel. In *Unasylva* no. 200. FAO, Rome

Cavendish, W. (1999). *Empirical Regularities in the Poverty-Environment Relationship of African Households*. WPS 99–21

Chenje, M. (ed.) (2000). *State of the Environment Zambezi Basin 2000*. SADC/IUCN/ ZRA/SARDC, Maseru/Lusaka/Harare

Chenje, M. and Johnson, P. (eds) (1994). *State of the Environment Southern Africa*. SADC/IUCN/SARDC, Harare/Maseru

Cleaver, K. and Schreiber, G. (1994). Reversing the Spiral: The Population, Agriculture and Environment Nexus in sub-Saharan Africa. Directions in Development Series of the World Bank, Washington, D.C. In *Findings, Africa Region*, no. 28, December 1994

COMESA (2001). Comoros Agricultural and Natural Resources. Common Market for Eastern and Southern Africa. Available on: http://www.comesa.int/

Cousins, B. (2000) Tenure and Common Property Resources in Africa. In *Evolving Land Rights, Policy and Tenure in Africa* (eds Toulmin, C. and Quan, J.) 2000. Department for International Development/International Institute for Environment and Development/Natural Resources Institute, London

DFID (1999). Land Rights and Sustainable Development in Sub-Saharan Africa: Lessons and Ways Forward in Land Tenure Policy. In *Report Of A Delegate Workshop On Land Tenure Policy In African Nations*, 16–19 February 1999. U.K. Department for International Development, London

DMC (2002). Drought Monitoring Centre Background and History. Available on: http://www.meteo.go.ke/dmc/

Drimie, S. and Mbaya, S. (2001). Land Reform and Poverty Alleviation in Southern Africa: Towards Greater Impact. Conference Report and Analysis for the Southern African Regional Poverty Network. Human Sciences Research Council, Pretoria. Available on: http://www.hsrc.ac.za/corporate/conferences/sarpn/20010604Report.html

ECOWAS (2000). *Annual Report 2000*. Economic Community for Western African States, Abuja

El Bagouri, I.H. (2000). Marginal Lands in the Arab Region: Potentials and Constraints. Paper presented before the Regional Workshop on Degradation and Rehabilitation of Marginal Lands in the Arab Region, 2–4 July 2000. CEDARE, Cairo

EPA/MEDC (1997). *The Conservation Strategy of Ethiopia—Volume 1 The Resources Base, Its Utilisation and Planning for Sustainability*. Environmental Protection Authority in collaboration with the Ministry of Economic Development and Cooperation. Federal Democratic Republic of Ethiopia, Addis Ababa

Erskine, J.M. (1987). *Ecology and Land Usage in Southern Africa: A Survey of Present-day Realities, Problems and Opportunities*. Africa Institute of South Africa, Pretoria

FAO (1986). *Highlands and Reclamation Study: Ethiopia*. Final Report Vols I and II. Ministry of Agriculture and FAO, Addis Ababa

FAO (1992). *Key Aspects of Strategies for the Sustainable Development Of Drylands*. FAO, Rome

FAO (1994). *A System Perspective for Sustainable Dryland Development In the Near Eastern Region*. FAO, Cairo

FAO (1995a). *Modules on Gender, Population and Rural Development With a Focus on Land Tenure and Farming System*. November 1995. FAO, Rome

FAO (1995b). *Irrigation in Africa in Figures*. Water Report 7. FAO, Rome

FAO (1997a). *Agriculture, Food, and Nutrition for Africa—A Resource Book for Teachers of Agriculture*. FAO, Rome

FAO (1997b). *Agricultural Production and Trade Opportunities in the Indian Ocean*. Findings of a Mission to Madagascar, Mauritius and Seychelles 18 May–9 June 1997. FAO Commodities and Trade Division, 11 July 1997

FAO (2000a). *The State of Food and Agriculture 2000*. FAO, Rome

FAO (2000b). *Agriculture Towards 2015/30*. Technical Interim Report, April 2000. FAO, Rome

FAO (2001a). *Forest Resources Assessment 2000*. FAO, Rome

FAO (2001b). Gender and Food Security Fact Files. Available on: http://www.fao.org/Gender/en/agri-e.htm

FAO (2001c). *The State of Food and Agriculture 2001.* FAO, Rome

FAO (2001d). Seventeen Countries Are Facing Exceptional Food Emergencies in Sub-Saharan Africa- FAO Concerned About Deteriorating Food Situation in Sudan, Somalia and Zimbabwe. Press Release 01/48

FAO/AGL (2000). Management of Degraded Soils in Southern and Eastern Africa (MADS-SEA-Network). Land and Plant Nutrition Management Service of the FAO, Rome. Available on: http://www.fao.org/waicent/faoinfo/agricult/agl/agll/madssea/intro.htm

FAO/GIEWS (1998). Food Supply Situation and Crop Prospects in Sub-Saharan Africa. Global Information and Early Warning System Report No 1/98. Part III: Crop Prospects and Food Supply Position in Individual Countries (situation as of early March 1998). FAO, Rome

FAOSTAT (2001). Statistics Database of the United Nations Food and Agriculture Organization. FAO, Rome

FEWS (1995). FEWS Bulletin 23 December 1995. FAO, Rome

Framji, K.K. (ed.) (1974). Irrigation and Salinity: A Worldwide Survey. International Commission on Irrigation and Drainage, New Delhi

Gharbi, M. (1998). Private, Collective and State Tenure in Tunisia. Land Reform; Land Settlement and Cooperatives. In *Sustainable Dimensions*. FAO, Rome. Available on: http://www.fao.org/sd/LTdirect/Irindex.htm

Goossens, R., Ghabour, T. K., Ongena, T. and Gad, A. (1994). Waterlogging and Soil Salinity in the Newly Reclaimed Areas of The western Nile Delta of Egypt. In *Environmental Change in Drylands* (eds. Millington, A.C. and Pye, K) pp. 365–377. John Wiley and Sons Ltd

Government of Mauritius (1991). *State of the Environment in Mauritius*. Ministry of Environment and Quality of Life, Government of Mauritius, Port Louis

Government of Tanzania (2001). Tanzania Agriculture and Livestock Profile. Available on: http://www.tanzania-online.gov.uk/agriculture.htm

Government of Zimbabwe (1998). Land Reform and Resettlement Programme. Draft for Discussion October 1998. Available on: http://www.worldbank.org.zw/complement.pdf

Harbeson, J. W. (1990). Post-Drought Adjustments Among Horn Of Africa Pastoralists: Policy And Institution-Building Dimensions. In *Sub-Regional Seminar On The Dynamics Of Pastoral Land And Resource Tenure In The Horn Of Africa*. Sustainable Dimension, FAO, Rome

Hegazy, A.K. (1999). Deserts of Middle Eastern. In *Encyclopedia of Deserts* (ed. Mares, M.M.) pp. 360–364. University of Oklahoma Press, Norman

Henao, J. and Baanante, C. (1999). Nutrient Depletion in the Agricultural Soils of Africa. In *2020 Brief* October 1999. International Food Policy Research Institute, Washington D.C.

Hoffman, M.T. and Todd, S.W. (2000). A National Review of Land Degradation in South Africa; The Influence of Socio-Economic Factors. In *Journal of Southern African Studies* 26(4), pp. 743–758

Homewood, K.M. and Rodgers, W.A. (1991). *Maasailand Ecology*. Cambridge University Press, Cambridge

IIED (1999). Land Tenure and Resource Access in Western Africa: Issues and Opportunities for the Next Twenty-Five Years. IIED Drylands Programme. Executive Summary Available on: www.iied.org/drylands/lt_exec.html

IPCC (1998). The *Regional Impacts of Climate Change: An Assessment of Vulnerability*. IPCC, Geneva

IPCC (2001). *Climate Change 2001: Impacts, Vulnerability, and Adaptation*. IPCC Working Group II, Geneva

Le Houerou, H. N. (1997). Biogeography of the Arid Steppeland northern of the Sahara. In *Reviews in Ecology: Desert Conservation and Development* (eds. Barakat, H.N. and Hegazy, A.K.). Metropole, Cairo

Lusk, G. (1987). Arab Agriculture—A Country By Country Survey (Sudan). In *Arab Agriculture 1987*. Falcoln Publishing, Bahrain

Lycett, A. (1987). Arab Agriculture 1987—Buoyant Prospects for Expansion Despite Revenue Slump. In *Arab Agriculture 1987*. Falcoln Publishing, Bahrain

Maene, L. (2001). Promoting Integrated Plant Nutrition: Farmer Field Schools in Zambia and Other Case Studies. Sustainable Developments International, Edition 1. Available on: http://www.sustdev.org/agriculture/articles/edition1/index.shtml

Maltby (1986). Waterlogged Wealth; Why Waste the Worlds Wet Places? International Institute for Environment and Development. Earthscan Publishers, London

Miladi, S.S. (1999). Changes in Food Consumption Patterns in the Arab Countries. In *International Journal of Food Sciences and Nutrition* vol. 49, pp. 23–30

MoFPED/MAAIF (2000). *Plan for Modernisation of Agriculture: Eradicating Poverty in Uganda—Government Strategy and Operational Framework*. Ministry of Finance, Planning and Economic Development/Ministry of Agriculture, Animal Industry and Fisheries, Kampala

Mougeot, L.J.A. (1998). Urban Agriculture Research in Africa: Enhancing Project Impacts. Cities Feeding People Report Series, Report No 29. IDRC, Ontario

Moyini, Y. (2000). *Report on the First Stakeholders Workshop on the Formulation of a National Landuse Policy for Uganda*. Prepared for the Ministry of Water, Lands and Environment and the National Environment Management Authority, Kampala

Moyo, S. (1998). Land Entitlements and Growing Poverty in Southern Africa. In *Southern Review*. SAPEM, Harare

Moyo, S. (2002). Land Reform and Transition in Zimbabwe. Paper presented at EDGA Conference, 21 Jauary 2002, Cape Town

Nana-Sinkam, S.C. (1995). Land and Environmental Degradation and Desertification in Africa. Joint ECA/FAO Agriculture Division Publication, FAO, Rome

NEMA (1999). *State of Environment Report for Uganda 1998*. National Environment Management Authority, Kampala

NEMA (2000). *Draft National Soils Policy for Uganda*. National Environment Management Authority, Kampala

Njinyam, S. (1998). Diminishing Surface Water and Soil Erosion, Major Handicaps to Sustainable Agricultural Development in the Semi-Arid Zones of UDEAC. The World Bank/ WBI CBNRM Initiative. Available on: http://srdis.ciesin.org/cases/cameroon-003.html

OECD/Club du Sahel (2001). The Food Aid Charter in the Sahel. Available on: http://www1.oecd.org/sah/activities/prevent/prevent-e/histocha.htm

Ouedraogo, H. and Toulmin, C. (1999). Tenure Rights and Sustainable Development in Western Africa: A Regional Overview. Paper Prepared for the DFID Workshop On Land Tenure, Poverty And Sustainable Development in Sub-Saharan Africa, 16–19 February 1999

Oyejide, T.A. (1993). Effects of Trade and Macro-Economic Policies on African Agriculture. In *The Bias Against Agriculture; Trade and Macro-economic Policies in Developing Countries* (eds. Bautista, R.M. and Valdes, A). ICEG/IFPRI, ICS Press, San Francisco

Oyejide, T.A. (1999). Agriculture in the Millennium Round of Multilateral Trade Negotiations: African Interests and Options. Paper Prepared for the World Bank's Integrated Programme of Research and Capacity Building to Enhance Developing Countries Participation in the WTO 2000 Negotiations. Presented at the Conference on Agriculture and the New Trade Agenda in the WTO 2000 Negotiations, 1–2 October 1999, Geneva, Switzerland

Rattan, L. (1988). Soil Degradation and the Future of Agriculture in Sub-Saharan Africa. *Journal of Soil and Water Conservation*, 43; 444–51

Reich, P.F., Numbem, S.T., Almaraz, R.A. and Eswaran, H. (2001). Land Resource Stresses and Desertification in Africa. In *Responses to Land Degradation. Proc 2nd International Conference on Land Degradation and Desertification* (eds. Bridges, E.M., Hannam, I.D., Oldeman, L.R., Pening, F.W.T., de Vries, S.J., Scherr, S.J. and Sompatpanit, S.). Khon Kaen, Thailand. Oxford Press, New Delhi, India

*Republic of Seychelles* (1997) (eds. Shah, N.J., Payet, R. and Henri, K.) Seychelles Biodiversity Strategy and Action Plan. Republic of Seychelles Ministry of Environment with UNEP and IUCN, December 1997

RFIC (1998). *Draft du Rapport National sur la Strategie et le Plan d'Action en Matiere de Diversitie Biologique*. Repbulique Federale Islamique des Comores, Ministere de la Production Agricole, des Resources Marines, de l'Environement et de l'artisanat, Projet PNUD/FEM COI/97/A/IG/99

SANE (1997). The SANE Process in Cameroon. Sustainable Agricultural Networking and Extension. Available on: http://www.cnr.berkeley.edu/~agroeco3/sane/monograph/

SARCCUS (1981). *A System for the Classification of Soil Erosion*. Department of Agriculture and Fisheries, Pretoria

SARDC-WIDSAA (2000). *Beyond Inequalities: Women in Southern Africa*. SARDC, Harare

SARIPS (2000). *SADC Human Development Report: Challenges and Opportunities for Regional Integration*. SAPES Books, Harare

Shah, N.J. (1997). Country Presentation: The Seychelles. In *Report of the Workshop on Integrated Management of Freshwater, Coastal Areas and Marine Resources in Small Island Developing States* (ed. Brugiglio L) Malta, 8–12 December 1997, University of Gozo Centre of Islands and Small States Institute, Foundation for International Studies, Malta and UNEP Nairobi

Staatz, J.M., Diskin, P. and Estes, N. (1999). Food Aid Monetization in Western Africa: How to Make it More Effective. MSU Agricultural Economics Food Security II Cooperative Agreement Policy Syntheses No 49

Sutcliffe, J.P. (1993). Economic Assessment of Land Degradation in the Ethiopian Highlands: A Case Study. National Conservation Strategy Secretariat, Ministry of Planning and Economic Development, Addis Ababa

Taleb, B.K. (1998). Towards Private Land Ownership: The State's Role in the Modernization of Land Tenure in Morocco. Land Reform; Land Settlement and Cooperatives. In *Sustainable Dimensions*. FAO, Rome. Available on: http://www.fao.org/sd/LTdirect/lrindex.htm

Toulmin, C. and Longbottom, J. (1997). Managing Land Tenure and Resource Access in Western Africa. Report from a Regional Workshop held in Goree, Senegal, 18–22 November 1996

Trueblood, M.A., Shapouri, S. Henneberry, S. (2001). Policy Options to Stabilize Food Supplies: A Case Study of Southern Africa. In ERS *Agriculture Information Bulletin* No. 764, April 2001. Economic Research Service, US Department of Agriculture

UN (2000). The Elimination of Food Insecurity in the Horn of Africa: Summary Report. Inter-Agency Task Force on the UN Response to Long Term Food Security, Agricultural Development and Related Issues. ECA, FAO, IFAD, UNDP, UNEP, UNICEF, WFP, WHO, WMO and the World Bank

UNCCD (2000). Combating Desertification in Africa. Fact Sheet 11. Published by the Secretariat of the United Nations Convention to Combat Drought and Desertification. Available on: http://www.unccd.int/publicinfo/factsheets/showFS.php?number=11

UNCCD (2001). Action Programme to Combat Desertification: Africa. Updated list November 2001. Available on: http://www.unccd.int/actionprogrammes/africa/africa.php

UNDHA (1994). *First African Sub-Regional Workshop on Natural Disaster Reduction*, 28 November to 2 December 1994, United Nations Department of Humanitarian Affairs, Gaborone

UNPD 1996. *World Urbanization Prospects: The 1996 Revision*. United Nations Population Division, New York.

UNDP (1997). Development Cooperation Report: Seychelles. United Nations Development Programme, UNDP, Port Louis

UNDP (2000). *Human Development Report 2000*. Oxford University Press, Oxford

UNEP (1998). *Water-Related Environmental Issues and Problems of the Comores and Their Potential Regional Transboundary Importance*. TDA/SAP-WIO, Maputo

UNEP (1999a). *Global Environment Outlook*. United Nations Environment Programme, Nairobi

UNEP (1999b). *Western Indian Ocean Environment Outlook*. United Nations Environment Programme, Nairobi

UNU (1998). Land Degradation and Rural Poverty in Africa: Examining the Evidence. United Nations University/ Institute for Natural Resources Assessment Lecture Series 1. UNU/INRA, Accra

USAID (1988). Madagascar Environmental Action Plan: Volume 1 General Synthesis and Proposed Actions. US Agency for International Development, Washington, D.C.

Vabi, M.B. and Sikod, F. (2000). Challenges of Reconciling Informal and Formal Land and Resource Access Tenure: Evidence from WWF-supported Conservation Sites in Cameroon. Paper presented at the 2nd Pan African Symposium on the Sustainable Use of Natural Resources in Africa, Ouagadougou, Burkina Faso, July 2000, coordinated by IUCN

Verheye, W. (2001). Food Production or Food Aid: An African Challenge. In Findings no. 190, September 2001. The World Bank, Washington D.C.

WHO/UNICEF (2000). *Global Water Supply and Sanitation Assessment 2000 Report*. World Health Organization/United Nations Children's Fund

Wolfensohn, J.D. (2001). Putting Africa Front and Center. Remarks to the United Nations Economic and Social Council, 16th July 2001, Geneva, Switzerland

World Bank (1993a). A Strategy to Develop Agriculture in Sub-Saharan Africa and a Focus for the World Bank. Technical Report, Dept Africa Region. The World Bank, Washington D.C.

World Bank (1993b). Shifting Cultivation, Population Growth, and Unsustainable Agriculture: A Case Study of Madagascar. Africa Technical Department, The World Bank, Washington D.C.

World Bank (1994). *Adjustment in Africa: Reforms, Results and the Road Ahead*. The World Bank, Washington D.C.

World Bank (1995). *Towards Environmentally Sustainable Development in sub-Saharan Africa: A World Bank Agenda*. Draft for Discussion, Africa Technical Department, the World Bank, Washington D.C.

World Bank (2000a). A Review of Rural Land Tenure Issues and Consolidation Policy in Tunisia. Land Policy Network Regional Perspectives. Available on: http://wbln0018.worldbank.org/Networks/ESSD/icdb.nsf/d4856f112e805df4852566c9007c27a6/4421fae238d4c8e9852567ec005a0162/$FILE/TunisiaExeSum.pdf

World Bank (2000b). *African Development Indicators 2000*. The World Bank, Washington D.C.

World Bank (2001). *African Development Indicators 2001*. The World Bank, Washington D.C.

WRI (1992). *World Resources 1992–3*. Oxford University Press, Oxford and New York

WRI (2001). Factors Contributing to Watershed Degradation. Global Topics. Available on http://www.igc.org/wri/watersheds/degrade.html

WRI, UNDP, UNEP and World Bank (1998). *World Resources Report 1998–1999: A Guide to the Global Environment: Environmental Change and Human Health*. Oxford University Press, Oxford

**PART G: URBAN AREAS**

## REGIONAL OVERVIEW

Thirty-eight per cent of Africa's population, that is to say 297 million people, live in urban areas. By 2030, this is expected to grow to approximately 54 per cent of Africa's projected population of around 1405 million (UNCHS 2001a). The level of urbanization in Africa is on a par with that in Asia, but lower than the global figure of 47 per cent, and well below the European and North American levels of more than 70 per cent (UNCHS 2001a).

However, it must be borne in mind that the definition of what constitutes an urban area differs from one African country to another. For example, in Uganda a settlement with a population of more than 100 is classified as urban, whereas in Nigeria and Mauritius an urban area has a population of more than 20 000 (UNCHS 2001b). There are also difficulties in defining a city, because cities are not only defined on the basis of population size but also of administrative or legislative functions. Large cities, however, are generally those with populations exceeding 1 million, and mega-cities have populations of more than 10 million (UNCHS 2001b).

Africa's rate of urbanization of 3.5 per cent per year is the highest in the world, resulting in more urban areas with bigger populations, as well as the expansion of existing urban areas (UNCHS 2001a). There are currently 40 cities in Africa with populations of more than 1 million, and it is expected that by 2015, 70 cities will have populations of 1 million or more. Lagos, with its current population of 13.4 million is the largest city in Africa, and the sixth largest in the world. Cairo, Africa's second largest city, has a population of 10.6 million and ranks nineteenth in the world (UNCHS 2001b). The growth of Africa's urban population is shown in Figure 2g.1.

Northern Africa is Africa's most urbanized sub-region with, on average, 64 per cent of its population living in urban centres. Libya is the most urbanized country, with 87.6 per cent of the population living in urban areas (UNCHS 2001a). Central Africa and the Western Indian Ocean Islands are also considerably urbanized, with average urban populations of 48 per cent each, followed by Western Africa (38 per cent) and Southern Africa (36 per cent). The least urbanized sub-region is Eastern Africa (26 per cent), and Rwanda is the least urbanized country, with an urban population of just 6.2 per cent (UNCHS 2001a). However, Eastern Africa has the highest average rate of urbanization for any of Africa's sub-regions, averaging 4.5 per cent per year. Malawi has the highest rate of all African countries (6.3 per cent per year).

## THE SIGNIFICANCE OF URBAN AREAS

The reasons for rapid growth of urban populations include overall high population growth rates, and 'pull factors', such as opportunities for employment and education, and improved access to health care, which attract people from urban areas. African cities account for 60 per cent of the region's GDP, and are important centres for education, employment and trade (UNCHS 2001b).

The colonial influence on development resulted in many of Africa's urban centres and national capitals being located on the coast, maximizing access to trade, international travel and development. However, there are also many social challenges associated with urbanization, such as the influx into urban areas of people forced out of rural areas by declining agricultural

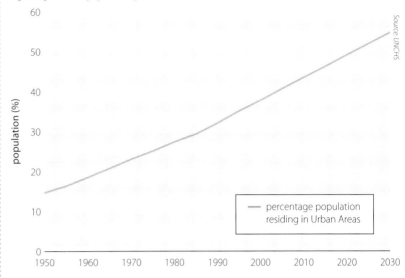

**Figure 2g.1  Urban population growth in Africa, 1950–2030**

*Source UNCHS*

yields, and who come to the urban areas in the hope of employment and greater income security. In many urban areas, rates of economic growth and infrastructure development have lagged behind urbanization rates, resulting in high levels of unemployment, inadequate standards of housing and services, and impacts on human health and development. Environmental disasters and conflicts have also caused many people to flee rural areas and to seek refuge in urban centres. In Mozambique, about 4.5 million rural people were displaced to urban areas because of civil strife in the 1980s (Chenje 2000) and the third largest settlement in Sierra Leone is a camp for displaced persons (UNCHS 2001b).

## URBAN AREAS AND THE ENVIRONMENT

Urban centres, and cities in particular, have developed from administrative and transport centres to commercial hubs, and centres of education and technology, manufacturing and processing, trade and employment. Urban dwellers have lifestyles that contrast starkly with those of their rural forebears or contemporaries, not least in their interactions with the environment. In the majority of nations, cities generate the lion's share of economic activity, ultimately consume most of the natural resources, and produce most of the pollution and waste. These problems are usually associated with unplanned or unserviced settlements (slums) where predominantly poor

•

*African cities account for 60 per cent of the region's GDP, and are important centres for education, employment and trade*

•

inhabitants do not have access to adequate housing, water supply, sanitation, waste disposal or electricity. These are also highly visible impacts, and they affect the health and well-being of many millions of people. However, environmental degradation is caused as much by the excessive consumption of resources (especially water) and generation of non-biodegradable wastes prevalent in affluent urban areas as it is by illegal dumping and burning of sewage and solid wastes in informal settlements (for example, Napier 2000). The key requirements for sustainable urban growth are:

● coherent, integrated planning;
● development that is environmentally and socially sensitive;
● security of tenure and financing;
● sufficient investment in infrastructure to keep pace with the rate of growth of urban populations, and their demands for adequate shelter, services, and security;
● integrated and innovative demand management with respect to resources and services; and
● rural development programmes to assist in slowing the rate of urban population growth.

## UNPLANNED SETTLEMENTS

The high rate of growth of urban populations in Africa has resulted partially from an increase in the number of urban households, brought about by changes in standards of living and attitudes towards familial dependence. The rate of growth in number of households across Africa averaged 3.1 per cent between 1985 and 2000, and is set to continue at this rate until 2030 (UNCHS 2001a). The consequent increase in demand for basic housing and services for urban populations, as well as skewed distribution of investment towards affluent suburban developments, has resulted in the rapid expansion of illegal or unplanned and unserviced settlements, with unhealthy living conditions and extreme overcrowding. For example, in 1993, about 55 per cent of Nairobi's population lived in informal settlements (USAID 1993) and, in South Africa, nearly half the population did not have adequate housing in the late 1990s (DEA&T 1999). In Monrovia, Liberia, 42 per cent of households were reportedly living as squatters in 1998 (UNCHS 2001c). Furthermore, only 60 per cent of urban dwellings in

*In spite of their contribution to national economies, African municipalities receive only 14 per cent of GDP in revenue—an average of US$14 per capita per year and 200 times less than the revenue of municipalities of high income countries*

Unplanned settlements often lack basic services.

*Mark Napier*

Africa are considered permanent, and almost half fall short of compliance with regulations (UNCHS 2001b). In African cities the average person has just 8 $m^2$ of floor space, indicating conditions of extreme overcrowding. By comparison, residents of Asian cities have 9.5 $m^2$; in industrialized countries the average is 34.5 $m^2$; and the global average is 13.6 $m^2$ (UNCHS 2001b). Overcrowding exacerbates rates of transmission of infectious diseases, such as gastro-intestinal infections, and respiratory diseases such as tuberculosis, commonly associated with poor ventilation and air pollution.

Low revenue in African municipalities, and consequently low spending, has led to development and maintenance of infrastructure being severely curtailed. In spite of their contribution to national economies, African municipalities receive only 14 per cent of GDP in revenue—an average of US$14 per capita per year and 200 times less than the revenue of municipalities of high income countries (UNCHS 2001b). They spend only US$12 per capita per year. This situation has been further compounded by slow economic growth over the past three decades and a bias amongst donor organizations in favour of development projects in rural areas.

An additional concern is that the poor often pay higher prices for housing and associated services. In African cities, people spend approximately 40 per cent of their incomes on rent, with the Arab nations paying the most at 45 per cent. This is more than twice the amount paid by residents of higher income countries, and indicates the high demand for housing and slow rates of provision of rented accommodation in Africa (UNCHS 2001b). Local governments are attempting to address this issue by increasing the production of low-cost housing stocks, and introducing housing subsidies for low-income groups (Department of Housing 2000, UNCHS 2000). However, tremendous backlogs still exist, as housing development falls short of population growth and urbanization rates. The residents of informal urban settlements often pay more for access to water (see Figure 2g.2), whereas water rates to planned suburban developments and industrial and agricultural users can be heavily subsidized (UNCHS 2001b).

Access to water supply, sewerage, electricity and telephone in African cities ranks the lowest in the world (UNCHS 2001b). Overall access to improved water sources in urban areas in Africa has increased marginally, from 84 per cent in 1990 to 85 per cent in 2000. However, 'improved water' includes household connection, public standpipes, boreholes, protected dug wells or springs, and rainwater collection (WHO/UNICEF 2000). Similarly, 'improved sanitation' includes connection to a public sewer or septic system, a pour-flush toilet, a simple pit latrine or a ventilated pit latrine (WHO/UNICEF 2000). The overall figure for access to sanitation in African towns and cities declined marginally from 85 to 84 per cent in the same period (WHO/UNICEF 2000). Poor water supply and sanitation create conditions conducive to high rates of water-related diseases, such as cholera, dysentery, scabies and eye infections. Diarrhoea and dysentery are among the most prevalent childhood diseases and a significant cause of under-five mortality. They correlate strongly with lack of availability of clean water (UNCHS 2001b). In addition, health and education facilities in unplanned settlements are often under-resourced or unaffordable, and unemployment and poverty are rife.

Unplanned settlements can, however, also be important economic centres, contributing to the informal economy through provision of services, such as food provision and sale, domestic services, car and household maintenance, hairdressing and childminding, as well as trade and recycling (for example, Napier, Ballance and Macozoma 2000). A good example of the contribution of the informal sector is the 'Zaballeen', a group of informal waste collectors in Egypt, who

**Figure 2g.2 Regional cost of water**

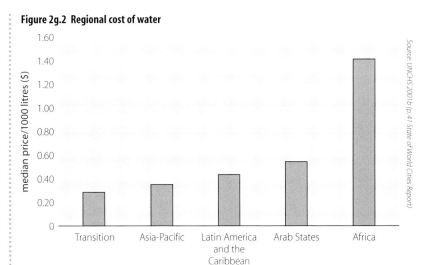

Source: UNCHS 2001b (p.41 State of World Cities Report)

separate and recycle different types of waste, thereby providing a valuable service to the urban environment whilst also making a living (UNCHS 1999).

Environmental impacts of rapid, unplanned urban growth include: loss of natural habitats; changes in, and sometimes loss of, biodiversity; and alterations to ecological functions such as hydrological cycles and atmospheric exchange. Fragile environments, such as delicate slopes, natural drainage waterways and flood-prone areas, are at risk from rapid urbanization. In unplanned settlements—where space is at a premium and shelters are erected on steep slopes, on wetlands or in flood zones—not only are residents at risk from flooding and subsidence, but the ecosystems are also vulnerable to pollution and physical degradation. Residents of Johannesburg's Alexandra Township were devastated by flooding and suffered loss of property and outbreaks of cholera in early 2000. In Cairo, the risk of earthquake puts millions of people in dense and insufficiently stable urban settlements in danger of loss of life and property. Improved planning, reductions in the backlog of low-cost housing, and improved disaster early warning and preparedness schemes are required to reduce this threat.

## Management of pollution and waste

Pollution and waste are often concentrated in urban centres, due to the high population densities and generally higher consumption patterns relative to rural areas. This presents a visible challenge for urban planners and managers in selecting landfill sites so as to minimize impacts on the environment and human health. Solid waste production and management, and air pollution, are the issues of greatest concern in African towns and cities.

Urban centres in developing countries are experiencing high and growing levels of air pollution due to their rapid rates of growth and to industrial development. The major sources of air pollution are: vehicle exhausts; industrial emissions; and domestic use of wood, coal, paraffin or refuse for heating and cooking. Lead pollution from vehicles is also causing great concern and, in some cities, it combines with dust and sand or gets trapped near the ground due to temperature inversions. Some cities have introduced incentives for conversion to non-leaded fuels, or have encouraged the sale of newer, cleaner vehicles. In dense, informal settlements, the use of traditional fuel sources contributes to rising ambient levels of sulphur dioxide, nitrogen oxides, carbon monoxide, ozone and suspended particulate matter. Exposure to these pollutants is associated with increased risk of acute respiratory infections (ARIs), particularly among children. These conditions are compounded by inadequate affordable health care for poorer urban residents. Electrification of households, promotion of low-smoke fuels and improved ventilation of houses are some of the measures that have been adopted to reduce health risks.

In the late 1990s, each person in the developing countries was generating, on average, 200 kg of solid waste per year (UNCHS 2001b), but this is increasing with increasingly affluent lifestyles, as well as (to a lesser extent) population growth. The amount of waste generated far exceeds the capacities for waste collection, treatment and disposal of most municipalities, and only one-third of the waste generated in African cities is disposed of via formal disposal routes (UNCHS 2001b). Lack of suitable landfill sites, and rapid filling of existing ones, are problems experienced by many municipalities. As a result, solid waste is often dumped or burned formally by the municipality. In Comoros, for example, domestic waste is dumped directly on the beaches (UNEP 1998) and, in Kampala (Uganda), the municipality has designated several wetlands as dumping grounds (NEMA 1999). Inappropriate dumping of solid waste poses a threat to human health, through contamination

An old quarry is used as a municipal rubbish dump in Zanzibar town.

*Andy Crump/Still Pictures*

of water supplies by harmful leachates. It also causes blockages in drainage channels exacerbating the effects of flooding, and is hazardous to wildlife. Burning of solid waste contributes to ambient air pollution by the emission of toxic fumes.

Lack of integrated waste management policies, inadequate funds and low access to appropriate, affordable technologies, and lack of incentives for alternative waste treatment and disposal, have all contributed to this situation. For example, in 1999, only seven of the 90 garbage trucks in the city of Harare, Zimbabwe were operating, because there were insufficient funds for training personnel and for equipment maintenance (UNCHS 2001b).

Waste recovery and recycling in Africa is limited to 2 per cent of all waste generated in the region (UNCHS 2001b), due to lack of economic incentives and markets for recycled materials. The most commonly recycled materials are paper, textiles, glass, plastic and metal. However, opportunities for income generation through recycling in the informal sector are gaining recognition, and several African cities have projects under way for waste collection, sorting and sale to commercial composting and recycling operations. Some cities have piloted industrial waste recycling by, for example, using wastes and effluents for electricity production. Furthermore, a large percentage of urban waste is organic matter that can be used as compost for urban agricultural development (Asomani-Boateng and Haight 1999).

## Towards sustainable urban settlements

The two most influential movements over the past 30 years for improving conditions in urban settlements have been the creation of the United Nations Commission for Human Settlements (Habitat) (UNCHS) and Local Agenda 21.

Habitat was established in October 1978 to provide leadership and coordination for the activities of the United Nations in the field of human settlements. Habitat's mission is to promote the socially and environmentally sustainable development of human settlements, and the achievement of adequate shelter for all. The Habitat Agenda was adopted as the global plan of action for achieving this mission at the Habitat II Conference, held in Istanbul, Turkey in June 1996. In the lead up to the Conference, African Ministers responsible for housing and urban development met in Kampala, Uganda. They recognized the need for urbanization in relation to economic development, but stressed the need to improve rural facilities and rural-urban linkages to slow the rate of urbanization. They also endorsed the decisions of the Dakar Declaration to involve stakeholders, build capacity in urban authorities, establish enhanced urban environmental management, and develop and implement national environmental action plans.

The other framework for action, the Local Agenda 21 programme, is expressly called for in Agenda 21 (Chapter 28), which encourages local authorities to enter into dialogue and to develop a process that is specifically designed to help achieve sustainable development at the local level, in the same way as Agenda 21 promotes this at the national and global level.

Although implementation of the Habitat Agenda and of Agenda 21 has been constrained, largely by lack of funding, many African cities have made remarkable achievements (see Box 2g.1). These include the revision or formulation of constitutions and national legislation to promote the right to adequate shelter—

almost 80 per cent of African countries now legally recognize this right (UNCHS 2001b). Some 13 out of 29 African countries recently studied recognize women's rights to own property, and more are revising or developing policies to this effect (UNCHS 2001b). Environmental policies are also being overhauled, and many African countries now require environmental impact assessments (EIAs) to be conducted prior to new developments such as housing, commercial development or road construction. Integrated water policies and waste management strategies are being developed, and municipal services are also being privatized in many urban centres, in an effort to improve coverage and maintenance. Effluent standards and tighter controls on waste management are also being developed and implemented to encourage responsible waste disposal. Housing programmes, subsidies for low-income families, poverty alleviation programmes, and decentralization strategies are additional measures being used to relieve the social, economic and environmental burden currently concentrated in urban areas.

In 1995, officials from concerned African cities met and formed the African Sustainable Cities Network (ASCN), to build capacity in participatory environmental planning within local authorities. The members started the process with an environmental management needs assessment, followed by a phased programme of

---

### Box 2g.1 African best practices in urban development

African cities have been setting standards internationally for their progress in achieving sustainable growth and improving living conditions. The 1998 Global 100 Best Practices List, drawn up by the United Nations Centre for Human Settlements to recognise various elements of urban development, contained eight African projects, and two award winning projects, one from Angola and one from Sudan. These best practices encompassed environmental management, children and youth centres, infrastructure development, sustainable community design and architecture, and poverty eradication.

*Source: UNCHS 2000*

education and training, establishing a framework for implementing new national environmental policies, and creating local sustainable development strategies. By June 2000, 31 African cities had joined the network and had embarked on activities (ICLEI 2001).

## NORTHERN AFRICA

With 64 per cent of its population living in towns and cities, Northern Africa is the most urbanized sub-region in Africa (UNCHS 2001a). Within the sub-region, urbanization rates vary between countries—36 per cent in Sudan to 95 per cent in western Sahara (UNCHS 2001a). Most of this growth has occurred in the past two decades, although urban populations have increased steadily since the colonial phase. In 1990–98, the average rate of urbanization reached 2 per cent per year (World Bank 2001a), and it is predicted that this will continue over the next 15 years, which means that, by 2015, 70 per cent of Northern Africa's population will live in cities (UNCHS 2001a).

Urban growth in Northern Africa is partially the result of rural-urban migration, but natural urban growth and reclassification account for more than 70 per cent of urban development (World Bank 1995). Furthermore, the new structural adjustment programmes (SAPs) adopted by most countries in the sub-region have brought new frontiers in industrial development in the last 10 years. By 1990, Alexandria (Egypt), Algiers (Algeria), Cairo (Egypt), Casablanca and Rabat (Morocco), Tunis (Tunisia), Tripoli (Libya) and Khartoum (Sudan) had populations of more than 1 million (UNCHS 2001a).

### THE SIGNIFICANCE OF URBAN AREAS IN NORTHERN AFRICA

Old and well-established urban centres, such as Cairo, Casablanca, Alexandria, and Tripoli continue to thrive, and have retained their character despite various economic, social, cultural and political changes. However, urban agglomerations and 'mega-cities' are important features of recent urbanization trends, as well as heavy industrialization. Most infrastructure, and societal services are centred in these cities, and they make the greatest contribution to national products. As a consequence, they play a vital part in economic development by providing opportunities for investment and employment. They have also gained a key political and administrative role.

### URBAN AREAS AND THE ENVIRONMENT IN NORTHERN AFRICA

Despite the gains outlined above, rapid urbanization in Northern Africa has also given rise to significant environmental and social problems, and is characterized by increasing urban poverty, emergence of informal settlements and slums, shortage in basic urban services, and encroachment on agricultural land. Between 15 and 50 per cent of city residents in the sub-region are urban poor living in squatter settlements, illegal subdivisions, sub-standard inner-city housing, custom-built slums and boarding houses (World Bank 2000). It has also been indicated that cultural heritage sites are especially at risk from uncontrolled development and environmental degradation in urban areas. While there have been national and international efforts to conserve this heritage, greater efforts are required to ensure its protection.

Water supply and sanitation rates are higher in Northern Africa than in some other sub-regions, but they are nonetheless variable, and many of the poorer residents of urban centres are without reliable services. In 2000, most urban residents had access to improved water supply (ranging from 72 per cent in Libya to 98 per cent in Algeria) and sanitation (87 per cent in Sudan and 100 per cent in Morocco) (WHO/UNICEF 2000). Casablanca's wastewater network, however, cannot cope with the volume of wastewater produced, despite recent upgrading by the municipality. As a consequence, large quantities of wastewater are discharged into the sea (WHO/UNICEF 2000).

### Unplanned settlements in Northern Africa

Cairo has several large informal settlements on the city periphery, and these accounted for 84 per cent of urban growth between 1970 and 1981 (ABT 1983). In 1992, there were 400 shantytowns in Casablanca, housing more than 53 000 families and, in Algiers, 6 per cent of the population are reportedly squatters (UNCHS 2001c). These settlements suffer from a shortage in basic infrastructure and, hence, have major health problems. Their illegal status compounds the problem, and residents often have to be moved to more appropriate sites, where local authorities or municipalities can provide them with basic urban services, including water supply, transport and health care (CEDARE 1997, Hamza 1994). More investment is required to provide these services to the whole urban population (Larsen 1995).

There has been land use planning and zoning in most of the cities in the sub-region. However, this has not, in some cases, prevented chaotic expansion and densification of cities. It is now the norm to find residential zones next to industrial sites or industries enveloped by housing estates, with all of the potential risks to the environment and to health which that entails.

In 1979, the Egyptian government implemented a strategy for improving the living conditions of urban slum dwellers in Cairo, by moving them to alternative locations. Residents of Eshash El-Torgman and Arab El-Mohamady were moved to El-Zawia El-Hamra, Ain Shams or Madinat El-Salam. Their original residences were close to the city centre and, thus, to the job market, whereas the resettled areas were far out of town. Reports of the negative social impact of this move prompted the government to change its approach to informal settlement upgrading, and to adopt a new twofold strategy: clearance for state land (squatter settlements) and upgrading for private land (informal settlements). As a result of social pressure, however, no areas have been cleared, but provision of infrastructure to informal settlements has made significant strides (Manal El-Batran from UNCHS 2001c). In Morocco, settlements have been upgraded in Agadir (see Box 2g.2) and, in Casablanca, the state issued a plan, in 1992, to build 200 000 houses for low-income people. Progress has been slow, however, because the extent of state-owned land has declined sharply, and privately owned land is very expensive, making subsidized housing unviable (UNCHS 2001c). Over the past three decades, the government of Libya has made housing provision a priority, and general housing, investment housing, agricultural housing and low-income housing projects have been implemented. Nearly 400 000 units were constructed in various locations, and efforts are well under way to deliver a further 60 000 (UNCHS 2001c).

## Management of pollution and waste in Northern Africa

Cities tend to be areas of high population concentration, high economic status and high levels of activity and, therefore, also consume natural resources from their vicinity and from distant sources. In doing so,

### Box 2g.2 Settlement upgrading in Agadir

The Moroccan National Shelter Upgrading Agency's project in Agadir has been recognized as one of Habitat's Best Practices for Human Settlements. Agadir was devastated by an earthquake in 1960, and housing for lower-income families has been insufficient ever since. In 1992, there were 77 separate shanty areas with 12 500 households (13 per cent of Agadir's population). The Upgrading Agency's project has provided approximately 13 000 serviced housing lots, housing units, apartment units or lots for apartment units; has provided utilities connections for an additional 3 600 families; and has created 25 000 jobs per year in construction. Reasons for the project's success include: active participation of the clients in project design, implementation and monitoring; recognition by local authorities of the clients' rights to adequate shelter; open dialogue between the parties; and integration of former squatter areas into cosmopolitan neighbourhoods.

*Source: UNCHS 2000.*

they generate large amounts of waste that are disposed of within and outside the urban areas. In Egypt, Morocco and Tunisia, solid waste generation in 1993 averaged 0.5–0.6 kg/capita/day (UNCHS quoted in WRI, UNDP, UNEP and World Bank 1998).

The effectiveness of management of municipal waste in the sub-region varies from country to country and, in some places, large amounts of wastes are not deposited in sanitary landfill sites, but are dumped in the open. In Tripoli, for instance, 65 per cent of waste is dumped in the open; in Ismailia (Egypt) 80 per cent is dumped; and in Casablanca, 90 per cent is dumped (UNCHS 2001c). With extremely warm temperatures, such waste tends to decompose rapidly, causing serious health risks and nuisance (Kanbour 1997). Open burning of waste is an informal disposal method used where waste collection services are inadequate, and it contributes to the problem of urban air pollution. Unmanaged disposal of waste affects human health, causes economic losses, and damages the physical and biological environments. For example, it has led to problems such as pollution of water sources and offensive odours, as well as an increase in disease vectors, pests and scavengers. Health risks also translate into economic losses, with sick employees staying away from their jobs and increasing the burden on public health care.

There are, however, successful examples of hygienic dumping and recycling of waste in many countries in the sub-region. For example, Tripoli recycles 20 per cent of its solid waste, and Tunis 5 per cent (UNCHS 2001c). In Egypt, waste is also recycled through the informal sector, by the 'Zaballeen'—informal sector garbage collectors who have been providing a door-to-door waste collection service since the 1960s covering wide areas, especially in Cairo. They collect mainly domestic waste and recycle up to 80 per cent of what they collect. Their activity has recently been somewhat formalized by the government (UNCHS 1999). In rural Egypt, there are organizations that promote environmental awareness among women, distributing plastic bags so they can collect their garbage, while youths are given the task of taking them to a designated collection point. Such schemes can have equal success in urban areas.

The high content of organic matter in waste has also created interest among some municipal governments in the sub-region, and several composting plants are already in operation, producing compost fertilizers and soil conditioners (Kanbour 1997). Whilst there are still no effective legal and institutional frameworks for implementation of solid waste management at national level, the issue has been given national priority in most countries in the sub-region. Privatization of solid waste management has recently been introduced in the sub-region, as a way of overcoming the lack of finance and capabilities of local governments and, in many areas, this has been more efficient and cost effective in delivering services, especially in Tunisia, Egypt and Morocco.

Crowded cities in Northern Africa have also seen a significant increase in the number of motor vehicles and traffic congestion, creating additional problems of air pollution. In large cities, the situation is compounded by the presence of heavy industries surrounding residential areas and adding to the types and quantities of pollutants. In Tunis, many factories responsible for the worst air pollution were transferred to the outskirts of the city, and air quality has improved although, at peak traffic times, smog is still a problem (UNCHS 2001c). In Egypt, environmental legislation introduced in 1994

imposed strict regulations and penalties for polluting industries. Conversion to unleaded fuel for vehicles was also initiated. Within one year, the level of airborne lead in some of Cairo's most polluted areas had decreased by up to 88 per cent (USAID 2001). Urban transport reform is still a priority in the region, however, not only in terms of promoting more fuel efficient and cleaner-fuelled vehicles, but also of upgrading roads and public transport systems to reduce congestion.

## Towards sustainable urban development in Northern Africa

Meeting the challenges raised by the urban environment requires concerted government action at all levels, in partnership with non-governmental organizations (NGOs), private enterprises, communities and citizens. Institutional structures, together with policies and frameworks, have been established in Northern African countries. For example, the high population growth and urbanization rates experienced by Libya in the 1970s and 1980s prompted the establishment of a Public Authority for Environment, a centralized agency with presence in all major centres for protection and cleansing of the environment. Solid waste disposal sites have been established and are managed by the Public Authority, and some waste is now recycled and used as inputs to industrial processes. Attention is also focused on the provision of drinking

Aerial view over Cairo, Egypt

NRSC/Still Pictures

water and maintenance of air quality (UNCHS 2001c). In Tunis, a department dealing with environmental issues has been created within the municipal administration. It is critical that such efforts be maintained, in order to continue to improve the social and environmental conditions in urban areas, in the face of rapid population growth, urbanization, and development.

## EASTERN AFRICA

Eastern Africa is the least urbanized sub-region in Africa, with 26 per cent of its population living in towns and cities (UNCHS 2001a). Within the sub-region, Djibouti is the most urbanized country with 83 per cent of the population in urban areas; Rwanda is the least urbanized at 6 per cent, although the recent violent conflicts in Rwanda and Burundi are likely to have impacted enormously on demographic patterns (UNCHS 2001a). Eastern Africa does, however, have the highest rate of urbanization of Africa's sub-regions (expected to average 4.5 per cent over the next 15 years), so the pattern is changing rapidly (UNCHS 2001a). Only a few cities in the sub-region currently have populations in excess of 1 million, namely: Addis Ababa (Ethiopia), with 2.6 million; Kampala (Uganda), with 1.2 million; Mogadishu (Somalia), with 1.2 million; and Nairobi (Kenya), with 2.3 million (UNCHS 2001a).

### THE SIGNIFICANCE OF URBAN AREAS IN EASTERN AFRICA

Rapid urban growth since the 1960s has resulted from growth of existing urban populations and from migration to urban centres of people either seeking refuge from poverty-stricken rural areas and declining agricultural productivity, or coming in search of employment, and improved security of income and housing tenure. This pattern has created a high demand for housing and urban services. However, economic growth has lagged behind growth of population and urbanization, recording on average a slight negative trend in the 1990–98 period, although with wide differences between countries. Uganda, for example, reported net economic growth of 3.5 per cent in the period, whereas Burundi reported a net loss of 4.2 per cent (UNDP 2000). Negative economic growth has

resulted in fewer funds being available for development and maintenance of infrastructure, in increased unemployment, and in people being less able to afford basic housing and services.

### URBAN AREAS AND THE ENVIRONMENT IN EASTERN AFRICA

Proliferation of slums and unplanned settlements in urban areas of Eastern Africa has been accompanied by inadequate provision of water supply and sanitation. Water supply is not only insufficient, but is also intermittent in unplanned and low-income residential areas. On average, access to clean water in the sub-region's urban areas is 80 per cent, ranging from 100 per cent in Djibouti to 60 per cent in Eritrea (WHO/UNICEF 2000[1]). Access to sanitation in urban areas ranges from 99 per cent in Djibouti to 12 per cent in Rwanda, with an average of 72 per cent (WHO/UNICEF 2000). In both cases, this represents a slight improvement on the situation in 1990 (WHO/UNICEF 2000). Residents in unplanned settlements also tend to pay more for water consumed. For instance, in Nairobi, residents of unplanned settlements paid between 30 US cents and 70 US cents for a 20-litre container of water, compared to 17 US cents per 20-litre container paid by consumers with water meters (USAID 1993). Inadequate water for cleansing and for sanitation creates opportunities for disease and pests to breed and spread.

Discharges of raw sewage into rivers and lakes have also increased, creating a toxic environment for plant and animal communities, as well as for humans. Lake Victoria and the Indian Ocean coast are particularly affected (as evidenced by frequent episodes of eutrophication) owing to the concentration of towns and cities in these areas. Governments in the sub-region have yet to put in place effective policy measures and regulations to arrest the situation. However, a programme to privatize municipal services in Eastern Africa was launched by the United Nations Centre for Human Settlements in the second half of the 1990s (UNCHS 1998). Improvements in water supply and sanitation have been made in Kenya and Uganda, although collection of payments is a serious limitation to maintenance of infrastructure and viability of continued operations (UNCHS 1998).

*Residents in unplanned settlements tend to pay more for water consumed. For instance, in Nairobi, residents of unplanned settlements paid between 30 cents and 70 cents for a 20-litre container of water, compared to 17 cents per 20-litre container paid by consumers with water meters*

View over the city of Kampala, Uganda

*Ron Giling/Still Pictures*

## Unplanned settlements in Eastern Africa

Unplanned settlements have mushroomed in Eastern Africa, with negative impacts on the social and biophysical environment. In Nairobi, for instance, about 55 per cent of the population were living in unplanned settlements in 1993 (USAID 1993), and approximately 5 per cent of residents in Entebbe and 16 per cent in Jinja (Uganda) are squatters (UNCHS 2001c). Balbala, an unplanned settlement of 240 000 residents in the city of Djibouti, accounted for the largest share of the city's growth during the past decade (UNCDF 1998). Population densities in these settlements are usually very high, which encourages the transmission of diseases such as parasites and respiratory diseases.

In response to this situation, governments have taken measures to enact housing policies that will support an environmentally sustainable housing sector. For example, Uganda formulated a Housing Policy in 1978 that influenced the upgrading of Namuwongo low-cost housing and Masese Women's Self-Help housing projects. The National Housing Strategy was completed in 1992, a Land Tenure Policy was prepared, and services and planning have been decentralized. Capacity building for land use planning is still urgently required. In Kenya, the government drafted a new housing policy in 1999, giving priority to upgrading of slum areas with minimal displacement, to allow for proper planning and provision of necessary infrastructure and related services. Between 1971 and 1991, the city of Nairobi received World Bank funding to provide urban housing and water supply and sanitation. An evaluation of the projects revealed that water supply had kept pace with the urban population growth, and tariff restructuring had helped in keeping water affordable (World Bank 1996). About 65 per cent of the urban population now has water-borne sewerage, and significant environmental improvements have been noticed. A development project in the Balbala quarter of Djibouti brought improvements to living conditions for 1 164 families and established a development model setting standards for future infrastructure, equipment, and housing development (UNCDF 1998). Additional responses in Eastern African cities include the development and/or implementation of local environmental plans in Bujumbura (Burundi), Addis Ababa, Mombasa (Kenya), Kigali (Rwanda), and Entebbe (UNCHS 2001c).

## Management of pollution and waste in Eastern Africa

The volume of solid wastes generated by urban centres in Eastern Africa has been increasing, mainly as a result of growing urban population, concentration of industries in the urban areas, consumption patterns of residents, and inadequate finances and facilities to manage waste collection and disposal. There is thus a wide variation in solid waste generation. Recent data were not available for all cities in the sub-region but, in 1993, solid waste generation ranged from 6 kg/capita/day in Kampala to 1.4 kg/capita/day in Bujumbura and 0.6 kg/capita/day in Kigali (UNCHS quoted in WRI, UNDP, UNEP and World Bank 1998). Percentages of households benefiting from garbage collection services also varied, from 47 per cent in Nairobi to 20 per cent in Kampala, and just 2 per cent in Addis Ababa (UNCHS quoted in WRI, UNDP, UNEP and World Bank 1998).

Proper collection and disposal of solid waste is necessary in order to avoid environmental pollution problems such as pollution of surface water bodies and of the ambient atmosphere, through burning or decomposition. In Kenya, an increasing amount of plastic materials has been observed among solid wastes. This is a problem, as plastics block the drainage systems leading to frequent flooding in urban areas. In Bujumbura, 33 per cent of solid wastes are deposited in open dumps, 27 per cent is burned openly, and only 15 per cent finds its way to sanitary landfill (UNCHS 2001c). A similar pattern is observed in Addis Ababa where 10 per cent is deposited in sanitary landfill, 21 per cent is dumped openly and the remainder is unaccounted for (UNCHS 2001c). In Kigali a staggering 84 per cent of solid waste is burned openly, giving rise to unpleasant and sometimes toxic fumes (UNCHS 2001c). On the outskirts of Kampala, large volumes of wastes are dumped at Lweza and Lubigi, which were originally wetlands, and Uganda's only recognized landfill, at Mpererwe, is actually managed as a dumpsite rather than as a designated sanitary landfill (Ngategize and others 2000).

Policy failures in the past have contributed to poor waste collection and management, because most urban authorities would not allow private sector involvement in waste management. There are also inadequate facilities for recycling materials such as paper, aluminium and glass, and inadequate measures in the re-use of recycled containers that may have contained harmful substances. Moreover, there is an overall lack of awareness and limited community participation in the management of solid wastes (Ngategize and others 2000). In response to this, the government of Uganda enacted the Environment Statute (1995) and regulations (1999) to address solid waste management. However, there is still no national waste management policy in Uganda, although a draft action plan for municipal solid waste management was developed in 1999 (NEMA 2001). The city of Kampala privatized its solid waste management functions recently and, in June 2000, eleven companies were operating and their contribution to waste management was increasing steadily (Ngategize and others 2000). Nairobi City Council opted to privatize its solid waste management in 1997, when it was recognized that solid waste generation was outstripping capacity to collect and treat the waste (UNCHS 1998). By 1998, improvements to waste management had been noted, although some areas were still inadequately serviced or unserviced (UNCHS 1998).

Long-term data on levels of air pollution in urban centres in Eastern Africa are not available. However, emissions from vehicles, industries and domestic consumption of coal, wood, charcoal and other fuels cause localized smog and pose risks to human health. Additional sources of air pollutants include waste dumps which emit methane, sulphur oxides and $CO_2$. There are growing numbers of motorized vehicles in the major urban centres, and many of the vehicles on the roads are old and inefficient (MoWTC 2000). In addition, leaded petrol is still widely used, and traffic congestion is a common occurrence because transport networks have not been upgraded to keep pace with the increasing numbers of vehicles (NEMA 2001). Vehicular emissions account for 95 per cent of atmospheric lead concentrations in Kampala, and residents of urban areas have higher levels of lead in their blood than those exposed to less traffic (NEMA 2001). Air quality standards for all major pollutants have been established for most countries in Eastern Africa, but lack of resources renders enforcement below optimum levels.

## Towards sustainable urban development in Eastern Africa

Entebbe, Gulu, Iganga, Jinja, Kampala, Masindi District and Njeru in Uganda, and Nakuru and Thika in Kenya, are members of the African Sustainable Cities Network, and have been working towards capacity building for improved environmental planning and management, and implementation of sustainable urban development strategies in partnership with stakeholder groups.

Jinja, Uganda is a highly industrialized and agricultural centre on the shores of Lake Victoria. As a result of these developments, and the attraction of the source of the Nile River, Jinja has experienced rapid population growth, with consequent pressure on the environment. Inadequate housing, waste collection and disposal and electricity supply result in pollution of air, water and soil, excessive vegetation removal, and eutrophication and siltation of waterways. The local environmental action plan, developed through the ASCN programme, has encouraged greater public participation in local issues and programmes. Through the system of decentralization, considerable power and decision making has been devolved to local communities, who now have the right to be involved in planning and managing their own affairs (ICLEI 2001).

The town of Nakuru in Kenya has developed a Local Agenda 21 framework to build a long-term vision in partnership with stakeholders, and to combine urban development with sustainable environmental practice. Nakuru is situated between two environmentally sensitive areas, the Menengai crater and Lake Nakuru National Park, and is a rapidly growing town with growth based on agricultural development and services, tourism and administration. There are also plans to strengthen the functions of the Municipal Council and to stimulate innovative partnerships. Plans are to be implemented on the basis of partnerships between local communities and the municipality (ICLEI 2001).

In Addis Ababa, a programme is underway to build capacity in urban planning and management and to reduce urban poverty. The programme is jointly implemented by the Ethiopian Ministry of Works and Urban Development, Addis Ababa City authorities, Ethiopian Civil Service College, UNDP, and UNCHS. The objective is to develop an urban development strategy and operational guidelines for municipalities, focusing on urban management, housing development, integrated infrastructure development, municipal finance management and upgrading of institutional capacity (UNCHS/ROAAS 2001).

## WESTERN INDIAN OCEAN ISLANDS

Some 48 per cent of the population of the Western Indian Ocean Islands lives in urban areas. The least urbanized countries are Madagascar (30 per cent) and Comoros (33 per cent) (UNCHS 2001a). Seychelles is the most urbanized, at 63 per cent, followed by Mauritius at 41 per cent (UNCHS 2001a). Annual urban population growth rates in the sub-region in the past 30 years have ranged from 1 per cent in Mauritius to more than 5 per cent in Comoros and Madagascar (UNDP 2000). Urban growth rates over the next 15 years are expected to average 2.8 per cent per year, ranging from 4.5 per cent in Madagascar and 4.1 per cent in Comoros to less than 2 per cent in Mauritius and Seychelles (UNCHS 2001a).

### THE SIGNIFICANCE OF URBAN AREAS IN THE WESTERN INDIAN OCEAN ISLANDS

The major factors encouraging rural-urban migration are the search for improved standards of living, employment and educational opportunities, and for access to communications and trade. In Mauritius, for example, economic growth averaged 7.7 per cent per year between 1985 and 1989, and 5.2 per cent per year between 1990 and 1999, with yearly growth in exports of 15.5 per cent and 6.4 per cent over the same periods (World Bank 2001a). The port and capital city, Port Louis, has facilitated this development through harbour operations and, as a consequence, has attracted numerous industries. In Seychelles, 90 per cent of the population lives on the principle island of Mahe, with the majority (71 000 people) living in the capital Victoria where the country's main economic activity, tourism, is based.

### URBAN AREAS AND THE ENVIRONMENT IN THE WESTERN INDIAN OCEAN ISLANDS

The high rate of urbanization means that urban environments and infrastructure providers are under extreme pressure to deliver goods and services. In addition, many urban centres are located on the coast, where there are risks of environmental disturbance, erosion, and pollution. The large and growing number of tourists coming to the islands is also pushing up the demand for housing and infrastructure, and land is being reclaimed and coastal wetlands drained. In Seychelles, for example, coastal sand dunes are now being used for construction and land is being reclaimed from the sea with irreparable damage to the reefs, wetlands, marine and land ecosystems (UNEP 1999).

### Unplanned settlements in the Western Indian Ocean Islands

Data on adequacy of shelter are only available for Mauritius, where most residents are owners of their properties, and dwellings are of conventional type (UNCHS 2001c). The majority of towns and cities in the sub-region are growing faster than the provision of infrastructure and there are shortfalls in water supply, sanitation, waste disposal, roads and communications infrastructure, as well as in health and educational facilities. However, the Western Indian Ocean Islands fare better than many African countries, with all urban residents in Mauritius and nearly all in Comoros enjoying access to improved water and sanitation (WHO/UNICEF 2000). In Madagascar, 85 per cent of the urban population has access to water supply and

70 per cent has access to some form of improved sanitation (WHO/UNICEF 2000).

Where these services are lacking, the urban population relies on septic tanks and pit latrines. If not properly managed (especially in informal settlements), these can contaminate groundwater, and untreated domestic sewage—together with storm water run-off from impermeable surfaces in urban areas—can threaten drinking water supplies and the quality and safety of coastal waters. The result is a risk to biodiversity and human health from enhanced breeding and transmission of intestinal parasites, pests and bacteria. Many of the urban developments—grown from villages and concentrated along the coastlines—discharge their liquid and solid wastes directly into the sea. For example, there are no waste management facilities in Comoros (UNEP 1998).

## Management of pollution and waste in the Western Indian Ocean Islands

The growing populations in urban centres in the Western Indian Ocean Islands, together with the growing number of visitors and patterns of increased consumption, are producing greater and greater volumes of solid waste. In Mauritius more than 1 000 tonnes of solid waste are generated every day, mostly in the capital, Port Louis, and other urban centres (UNEP 1998). However, solid waste collection and disposal services in urban areas in the sub-region are somewhat dysfunctional. For example, in Antananarivo, Madagascar, it is estimated that only 25 per cent of solid waste is collected. Lack of waste collection encourages illegal dumping of wastes by households and municipalities, and uncontrolled roadside dumping, or dumping on the beaches, as in Comoros (UNEP 1999, UNEP 1998). Lack of suitable landfill sites, especially on the smaller islands complicates the problem, and burning in the open is common. Burning presents a threat to public health and air quality. Water quality is also threatened when solid waste is dumped in or near surface and groundwater resources (UNEP 1998). Some of the waste generated in the sub-region is recycled. For example, Madagascar and Mauritius recycle paper, textiles and metal. In Mauritius glass, precious metals and plastics are also recycled (UNEP 1998).

Intensive animal rearing, especially the pasturing of goats around urban areas, creates pressures on the environment and human health if the waste is not disposed of properly. Contamination and eutrophication of water courses lower the ability of the natural systems to provide vital functions, such as water quality regulation and nutrient cycling, as well as impacting on biodiversity and providing breeding sites for parasites and bacteria.

## Towards sustainable urban development in the Western Indian Ocean Islands

The establishment of urban planning regulations, waste management and raising of public awareness are improving the situation in some parts of the Islands. But more needs to be done, especially in islands where urbanization and population are growing rapidly and where residential and industrial developments combine their pressures on the urban environment. In old towns and cities, the road, water, sanitation and traffic management infrastructure is in urgent need of radical renewal and attention is being given to decentralizing the population through better spatial planning for industry, government and residential population areas.

The Cities Alliance, to which Antananarivo, Madagascar is a partner, held a special session in February 2001, at which Madagascar's Minister for Regional and Urban Development reported that significantly more investment is required to keep pace with a rapidly urbanizing population. The Government of Madagascar has conducted extensive consultations with local authorities, civil society and international partners, and has formulated a Poverty Reduction Strategic Programme, focused at the local level and, in particular, on provision of infrastructure and creation of employment. Activities include the development of planning and monitoring tools, and a campaign to develop municipal awareness and capacities.

Also in Madagascar, the World Bank has funded a slum upgrading and city development strategy for Antsirabe, Antsiranana, Mahajanga and Toamasina, as well as an integrated municipal programme for poverty reduction in Fianarantsoa and Tulear. Community participation has been a significant component of both projects, and replicable methodologies have been established (UNCHS/ROAAS 2001).

## SOUTHERN AFRICA

Southern Africa's urbanization levels are currently just below the average for the Africa region, with 36 per cent of the sub-region's population living in urban areas (UNCHS 2001a). South Africa and Botswana are the most urbanized countries, with urban populations of 50 per cent each, and Malawi is the least urbanized with just 24 per cent of its population living in urban centres (UNCHS 2001a). This is considerably different from the situation 30 years ago, when just 11.2 per cent of the Southern African population lived in towns and cities (WRI, UNEP and UNDP 1992).

The current rate of urbanization is also high, and is predicted to average about 3.5 per cent over the next 15 years, although there are wide differences between countries. For example, South Africa, one of the most urbanized countries in the sub-region, has the lowest rate of urbanization, at 1.2 per cent per year, whereas Malawi, currently the least urbanized, has estimated urban growth rates of more than 6 per cent per year (UNCHS 2001a). South Africa has the biggest and most numerous urban areas in the sub-region, including Southern Africa's largest urban agglomeration, Johannesburg (population approximately 4 million, 791 000 households, area 1384 km$^2$, 720 suburbs) (UNCHS and UNEP 1997, GJMC 1999).

### THE SIGNIFICANCE OF URBAN AREAS IN SOUTHERN AFRICA

The high urbanization rates in Southern Africa are due to rural-urban migration and high population growth rates. Migration is a result of the pull factors of urban settlements—such as perceived job opportunities, and better infrastructure and housing—in addition to 'push factors' from rural areas, such as shortage of land and declining returns from agriculture. In Angola and Mozambique, urbanization has been driven largely by civil conflict which forced many rural residents to flee to relatively safer urban areas. About 4.5 million Mozambicans were displaced to urban areas during the 1980s (Chenje 2000).

Despite considerable economic development and employment opportunities in Southern Africa's urban centres, the rate of population growth and urban migration has exceeded the ability to create jobs and to raise standards of living for urban residents. However, the informal sector has grown rapidly in urban areas, and many cities now have considerable hidden economies. One such informal activity is urban agriculture. Mainly instituted at the subsistence level to counteract low levels of employment and income, urban agriculture has become an integral part of the urban economy for the poor. Up to 37 per cent of urban households in Mozambique were engaged in subsistence agriculture in 1996, and 45 per cent of low-income urban households in Zambia grew crops or raised livestock (UNDP 1996).

### URBAN AREAS AND THE ENVIRONMENT IN SOUTHERN AFRICA

The rapid growth and expansion of urban areas in Southern Africa are causing an unprecedented level of localized depletion of natural resources, discharge of unprocessed wastes into the environment and massive demands for urban services. Most Southern African municipalities have not been able to keep pace with the demand for basic services such as housing, roads, piped water, sanitation and waste disposal. Provision of health and education services and facilities has also lagged far behind urban population growth. The overall result is that the environment has become hazardous to human health through rapid spread of water-borne and respiratory diseases, and this situation is compounded by a lack of health facilities, low levels of education and employment opportunities and, hence, a reduced ability to afford improvements to living conditions.

#### Unplanned settlements in Southern Africa

The gap between urbanization rates and rates of housing and service provision—together with colonial and post-colonial apartheid development policies—has created a wide range of settlement types, with stark inequities in terms of tenure, access to land and shelter, and provision of services. Large informal settlements have developed on the periphery of towns and cities, furthest from economic opportunities, transport networks and urban amenities.

In general, these informal settlements are characterized by insecure (or illegal) tenure, unstable structures, inadequate water supply and sanitation, lack of waste disposal facilities and poor electricity

Housing development project, South Africa

*Mark Napier*

supplies. Meagre and overburdened health, education and social services compound these problems, creating unhealthy living conditions and social problems, such as high crime rates, prostitution and drug abuse (for example, Napier 2000, GJMC 1999, DEA&T 1999). Gauteng, South Africa's most urbanized province, has more than 24 per cent of its population living in informal settlements (Statistics SA 1998). In Dar es Salaam, Tanzania, about 70 per cent of the 3 million residents live in unplanned settlements, with marginal access to piped water, sanitation, drainage or basic social services (UNCHS and UNEP 1997).

Environmental impacts of poor and insufficient housing include uncontrolled urban development in fragile areas such as flood zones, on steep slopes, or in wetlands and other unique natural habitats. This, in turn, poses a threat to the residents, who are at greater risk of flooding, landslides and outbreaks of diseases, such as cholera or vector-borne diseases, such as malaria. For example, in Johannesburg's Alexandra township (South Africa) there were some 3 800 households living on an infill site, and 5 500 households living on the banks of the Jukskei River in 1999 (Alexandra Renewal Project 2001). Unusually heavy rains during December 1999 and January 2000 precipitated flooding of the Jukskei River and 120 households were washed away (Disaster Relief 2000). There have also been recent cases of cholera reported in Alexandra, and there is now a plan to move residents to alternative, more sanitary

conditions (Alexandra Renewal Project 2001).

Provision and upgrading of urban infrastructure have thus become priorities for municipalities of the sub-region, and great strides have been made in recent years to improve urban living conditions. These include the award-winning Luanda Sul Self-Financed Urban Infrastructure Programme in Angola, which aimed to provide satisfactory shelter for a significant proportion of urban migrants who fled the conflict during the 1980s. Since 1994, 2 210 dwellings have been built by the 16 702 people they now shelter, together with 12 km of power lines, 70 km of clean water pipes, 23 km of drainage and 290 000 $m^2$ of paved roads (UNCHS 2000). Lesotho's Urban Upgrading Project—through involvement of local authorities and communities—has housed 267 families, of which 134 were female-headed households (UNCHS 2000). In Namibia, a National Housing Policy was approved in 1991 and, together with a National Shelter Strategy, has facilitated production of more than 3 400 housing units, improved women's access to shelter, and is assisting an additional 1 300 families per year (UNCHS 2000). South Africa's efforts have included the production of more than 1 million low-cost houses in the past six years (Department of Housing 2000), representing an addition of at least 17 per cent to the national stock of formal housing (based on Statistics SA 1998). However, this is still short of the massive backlog, which stood at 5 million units in 1994 (Everatt 1999). Problems of inadequate housing have been fundamentally complicated by lack of access to land and financing. For instance, the housing demand in Zimbabwe now stands at more than 1 million units, from 670 000 units in 1995, with the greatest shortages being experienced in major cities, where the annual population growth rate is between 3–6 per cent. The private and public sectors are currently able to produce approximately 18 000 units per year, which is far below the required levels (SARDC 1999).

Water supply and sanitation rates in Southern Africa have, on average, improved over the past decade (the United Nations International Drinking Water and Sanitation decade), although millions of urban residents still do not have clean water or adequate sanitation (see Box 2g.3). Access to clean water is highest in urban areas in Botswana (100 per cent) and lowest in Angola (34 per cent) (WHO/UNICEF 2000). Access to sanitation is much higher in general, with Malawi, Namibia, South

### Box 2g.3 Improving water supply services in Southern Africa

Municipalities and local authorities have identified access to clean water and sanitation as priorities for development, and several local-level projects have been established, with funding from national development budgets, as well as with the help of international donors. Many countries in the sub-region have also reformed their legislation to reflect the right of all citizens to adequate sources of clean water. In Zambia, a National Water Supply and Sanitation Council has been established to coordinate supply by local utilities. Following a recent outbreak of cholera, Mozambique received a US$36 million credit from the World Bank to upgrade sanitation and to chlorinate drinking water in the cities of Maputo and Beira.

Source: World Bank 1998

● *Privatization of waste collection services has been encouraged by Structural Adjustment Programmes in an attempt to raise additional funds. Unfortunately, this has, to an extent, worsened the problem of improper waste dumping, because companies sometimes avoid using designated dumping sites, where they are supposed to pay a fee*

●

Africa, Tanzania, Zambia and Zimbabwe, having more than 95 per cent access in urban areas (WHO/UNICEF 2000). In areas where water supply and sanitation are inadequate (mainly informal settlements), there are higher risks of water-borne diseases such as dysentery, cholera, typhoid, parasitic worms and flukes, as well as skin and eye infections that can be cured by enhanced levels of hygiene. Pools of stagnant water are also ideal breeding habitats for disease vectors such as mosquitoes. Environmental impacts of poor water supply and sanitation include the risk of contamination of groundwater as well as surface water, eutrophication and changes to biodiversity. For example, the town of Victoria Falls (Zimbabwe), which has a population of more than 3 000, discharges 8 000 m³ of wastewater into the Zambezi River, including raw sewage, because the town's sewage treatment facilities are overloaded and subject to frequent breakdowns (Chenje 2000). Nitrate and phosphate levels in the effluent exceed Zimbabwean standards, and may be a contributing factor to the spread of water hyacinth in Lake Kariba, which lies downstream. Total coliform counts in the effluent are also high, and this poses a health risk to downstream communities and to the town's 32 000 annual visitors (Chenje 2000).

## Management of pollution and waste in Southern Africa

Per capita solid waste generation averages 0.7 kg/day in Zimbabwe, while, in Tanzania, it is 1 kg/day (Chenje 2000). More people living in urban areas means greater levels of waste generation. Because municipalities cannot cope with this, large quantities of solid wastes are not collected, are not treated or are not disposed of in designated sanitary landfills. On average, less than two-thirds of urban households in the sub-region have access to garbage collection services. The situation is most extreme in Lesotho, where only 7 per cent of urban households have garbage collection facilities (UNCHS quoted in WRI, UNEP, UNDP and World Bank 1998). As a consequence, large quantities of solid waste are illegally dumped or burned, resulting in increased pollution of both the air and soil. By contrast, Windhoek in Namibia recycles 4.5 per cent of waste, 3 per cent is incinerated, and the rest is disposed of in sanitary landfill (UNCHS 2001c). In Gaborone (Botswana) and Maputo (Mozambique), nearly all solid waste is disposed of in an open dump, rather than a sanitary landfill (UNCHS 2001c). This gives rise to unhygienic conditions where the proliferation of disease vectors is facilitated, and risks of surface and groundwater contamination are high. Methane emissions may also be dangerous to nearby residents and contribute to atmospheric pollution and global climate change.

The major constraint on proper management of solid waste in cities of Southern Africa is inadequate finance (UNCHS 1996). Privatization of waste collection services has been encouraged by Structural Adjustment Programmes (SAPs) in an attempt to raise additional funds (see Box 2g.4). Unfortunately, this has, to an extent, worsened the problem of improper waste dumping, because companies sometimes avoid using designated dumping sites where they are supposed to pay a fee.

The slow pace of development of urban infrastructure in Southern Africa has resulted in increasing traffic congestion throughout the sub-region, with central business districts of most major cities having inadequate public transport networks and parking space. However, the sub-region has also witnessed a significant rise in the number of cars, currently translating as 51 persons per car (WRI, UNEP, UNDP and World Bank 1998) compared to 197 persons

## Box 2g.4 Improving waste management in Dar es Salaam

In Dar es Salaam (Tanzania), less than 3 per cent of the city's solid waste is collected due to lack of appropriate vehicles and inadequate cost recovery (UNCHS and UNEP 1997). As part of a new Environmental Planning and Management Process, waste management services have been privatized, an emergency clean-up operation was launched, and management of disposal sites has improved. New City Council bylaws were introduced to facilitate collection of revenues, and innovative fiscal incentives for rehabilitating the fleet of vehicles were also established.

Source: UNCHS and UNEP 1997.

per car 20 years ago (WRI, UNEP and UNDP 1992). The advanced age of most of the vehicles, and heavy dependence on leaded fuel and diesel contribute to high levels of smog and particularly lead pollution.

Other urban sources of air pollution include industrial emissions and smoke arising from domestic consumption of coal, wood and charcoal. Human health impacts include elevated incidence of acute respiratory infections such as asthma and bronchitis. Buildings are corroded and surrounding plant communities are subjected to toxic pollutants.

Most cities in the sub-region have established air quality standards and have monitoring programmes in place. In South Africa, where the problem is greatest because of the highest density of industries and vehicles, an Atmospheric Pollution Prevention Act (1965) stipulates the latest technologies to be used in controlling noxious and offensive emissions. The same act, which is being updated, makes provision for control of smoke from industrial operations by establishing grounds for prosecution of offenders, and empowers the Minister of Transport to make regulations regarding the exhaust emissions from vehicles operating on public roads. However, enforcement of this regulation is weak owing to lack of resources for monitoring and prosecution. Some municipalities (including Cape Town and Durban) also have bylaws preventing the burning of refuse, but again, these are not always adhered to.

### Towards sustainable urban development in Southern Africa

In addition to the specific responses listed above, a number of policies and strategies have been put in place at national level. In Malawi, a national housing policy has been developed while, in South Africa, an urban development strategy was recently launched for discussion. In Zimbabwe, a policy of decentralization encourages the development of rural service centres, commonly known as growth points, to relieve the pressures of population and pollution on major urban centres. However, the policy has not met with much success, as the service centres continue to be shunned by many, probably because of the complexity of push and pull factors that contribute to urban growth rates.

Legislation requiring environmental impact assessment prior to development is another recent feature in most Southern African municipalities, and local development plans taking into consideration the priorities for integrated environmentally-sensitive and socially-acceptable development have been widely adopted. Maseru (Lesotho), Lilongwe (Malawi), Maputo (Mozambique), Windhoek (Namibia), East Rand (South Africa), Bulawayo, Chegutu, Gweru, Harare and Mutare (Zimbabwe) have all put in place local environmental plans and implemented them to some extent (UNCHS 2001c). The City of Cape Town approved an Integrated Metropolitan Management Policy in 2001 and implementation of a number of strategies for tackling air pollution, litter and illegal dumping, and for creating open space and managing environmental education has begun (City of Cape Town 2001). Funding is still the major restriction for many municipalities, and alternative strategies are required to boost local authorities' capacities.

## CENTRAL AFRICA

The average level of urbanization in Central Africa in 2000 was 48 per cent, ranging from 81 per cent in Gabon to 24 per cent in Chad (UNCHS 2001a). This represents rapid growth, over the past decade in particular, with urbanization rates of more than 3 per cent for all countries, and reaching 5 per cent or more in Equatorial Guinea and Gabon. However, urbanization rates were even higher during the previous decade, when economic growth encouraged employment in urban areas. The largest cities in Central Africa are Kinshasa (5 million) in Democratic Republic of Congo (DRC), and Douala (1.6 million) and Yaoundé (1.4 million) in Cameroon (UNCHS 2001a).

## THE SIGNIFICANCE OF URBAN AREAS IN CENTRAL AFRICA

Central Africa has a long history of urbanization. Cities or urban centres started as commercial, administrative and mining towns and as seaports during the colonial period. As a result many major or capital cities are on the coast (including Libreville and Douala) where access to trade, travel and international communications is enhanced. Libreville is the economic and administrative capital of Gabon, concentrating 50 per cent of the population, 50 per cent of employment and more than 80 per cent of GDP. Douala is Cameroon's most important city and the man industrial centre. It is a port city with industries such as aluminium smelting, brewing, textiles manufacture, and processing of wood and cocoa. Kinshasa is also a port, on the Congo River estuary, and serves as the political, administrative, and industrial capital of DRC (UNCHS 2001c). Although cities have become the main catalysts of economic growth in Central Africa, urbanization has caused massive problems of poverty and environmental degradation.

## URBAN AREAS AND THE ENVIRONMENT IN CENTRAL AFRICA

Environmental consequences of the rapid urbanization and urban population growth include intensifying pressures on natural habitats and resources to satisfy the growing demand for space, housing and water for drinking and sanitation. Municipalities and utility companies are unable to provide housing and infrastructure quickly enough to meet this demand and sub-optimal services are therefore provided, with sub-optimal environmental standards and conditions.

Dualism between customary and modern tenure laws is also apparent in the urban centres and, together with high land and property prices, is one of the causes of unplanned or illegal construction. Rapid development of informal settlements or shantytowns ensues, characterized by overcrowding, unstable or unhealthy housing, inadequate water supply and sanitation, and lack of electricity supply and waste collection.

## Unplanned settlements in Central Africa

In Yaoundé, Cameroon, the majority of urban residents are squatters or tenants. There are conflicting pressures on residents to purchase properties, and rents are high. However, property prices have also risen recently, while incomes have declined because of devaluation of the currency (UNCHS 2001c). By contrast, land prices in Douala are lower, and there is a much higher proportion of house owners and no squatter settlements (UNCHS 2001c). In DRC, security of tenure is also complicated by disparities between modern legislation and traditional laws. For example, DRC's 1973 Land Act stipulated that 'land is the exclusive inalienable and unprescriptable property of the state', but the acquisition of land remains subject to the consent of the land chief.

The city of Libreville, Gabon, is experiencing uncontrolled urban development resulting from a shortage of serviced plots, an absence of planning tools and instruments, and a lack of urban space control. The demand for housing stands at approximately 6 000 units per year (in a city with a population of 500 000), and available land for development is minimal (there are 14 hectares of 'green spaces' per 10 000 hectares) (UNCHS 2001c). The results are that more than 50 per cent of the population lack proper housing; there is rapid development of unplanned, inadequately serviced and often unsafe settlements; and illegal occupation is as high as 85 per cent (UNCHS 2001c).

One of the most important environmental impacts of uncontrolled urbanization in Central Africa is its spread into fragile ecosystems, including delicate or highly erodible slopes, natural drainage waterways or valleys, and areas that are subject to flooding. Due to the intense competition for space in urban areas, green spaces are rapidly disappearing and areas usually deemed unsuitable for housing are the only refuges available for the urban poor, who are then vulnerable to flooding, landslides, and outbreaks of pests and diseases. Although planning regulations are in place, they are poorly monitored and enforced. Development in and modification of green areas results in changes to biodiversity, risks of pollution of soil and water, changes to soil fertility and stability and, especially in wetland areas or areas where there is standing water due to lack of sanitation, high risk of disease transmission. Dense, unstable and poorly sited settlements are also vulnerable to the impacts of floods, landslides, and fires.

Water supply and sanitation provision has also fallen behind rates of urban growth in many Central African cities, largely due to lack of municipal funds and capacities. On average, 59 per cent of the urban population has access to clean water (more than 80 per cent in Cameroon and DRC), whilst 54 per cent has access to sanitation (but only 14 per cent in Congo and 25 per cent in Gabon) (WHO/UNICEF 2000). Inadequate water supply and sanitation pose a threat to human health via exposure to pathogens such as cholera and intestinal parasites. They also pose a threat to the surrounding environment if sewage and wastewater are discharged untreated. Untreated discharges contaminate soil and water bodies, creating a risk to human health via transmission of disease vectors or toxic elements, and threaten biodiversity through effects on the ecosystem such as eutrophication and contamination with heavy metals and inorganic compounds.

## Management of pollution and waste in Central Africa

Solid waste generation in Central Africa was recorded in 1993 as 0.6 kg/person/day in Brazzaville, 0.8 kg/person/day in Yaoundé, and 1.2 kg/person/day in Kinshasa (UNCHS quoted in WRI, UNDP, UNEP and World Bank 1998). Relatively few households had access to solid waste collection, however, (none in Kinshasa, 25 per cent in Bangui (Central African Republic) and 44 per cent in Yaoundé). Residents of Douala and Brazzaville fared better, with 60 per cent and 72 per cent respectively (UNCHS quoted in WRI, UNDP, UNEP and World Bank 1998).

The lack of capacity to collect, treat and dispose of solid waste stems from inadequate municipal budgets and insufficient sites. In many cases, refuse collection is restricted to high-income areas. Where refuse is collected, it is often dumped at the edge of the city in open sites such as wetlands or water courses, rather than in regulated landfill sites. Whereas 62 per cent of Douala's solid waste and 70 per cent of Libreville's waste is disposed of in sanitary landfill, the figure for Brazzaville is less than 1 per cent, and 40 per cent goes to open dumps (UNCHS 2001c). Here, it poses a health risk to rubbish pickers and to the water supply system, through leaching of toxic and decomposing material and blocking of drainage channels. Furthermore, 38 per cent of Brazzaville's waste and 32 per cent of Kinshasa's waste are burned in the open, posing a further risk to human and environmental health though the emission of toxic fumes (UNCHS 2001c). Very little is recycled, as infrastructure and services for collection and recycling have not been established and there are few markets for recycled materials.

The responses adopted by countries in the sub-region to address waste management problems include: improved urban planning and management; establishment of programmes for waste disposal; enforcement of existing regulations; urban environmental education, monitoring, and community participation; and involvement of the private sector in urban sanitation. Libreville municipality has launched a 'Clean City' campaign, and local environmental plans have been implemented to improve environmental and social conditions. However, lack of legislation empowering local authorities and lack of assistance in transfer of state funds to municipalities remain barriers to their effectiveness (UNCHS 2001c).

Air pollution is an emerging issue in urban centres of Central Africa, although few data are available at present to provide quantitative analysis of the sources or impacts. Sources of pollution include increasing vehicular traffic and advanced age of many vehicles, industrial activity and household consumption of traditional fuels. Lack of connection to electricity supply forces many households to burn coal, wood or even refuse to provide energy for heating and cooking. These fuels give off toxic fumes and are dangerous to human health, causing respiratory disorders.

## Towards sustainable urban development in Central Africa

In Cameroon, the government has embarked on a decentralization process, and there has been increased community participation, especially in implementation of projects aimed at improving living conditions (UNCHS 2001c). At the recent 25th Special Session of the United Nations General Assembly and Istanbul+5 Conference in June 2001, Cameroon also reported that it had put in place an environmental management strategy for urban development. Additional national programmes recently developed include a programme for poverty reduction, environmental protection, governance, and health, fertility and nutrition. However, no assessment of the performance of these actions was available at the time of writing. The Congolese delegation reported that the human settlement development strategy has four major components, namely security of tenure, adequate housing for all, promotion of equality in access to credit and provision of essential services. However, attention so far has been focused on reconstruction and post-conflict development, and there is still significant work to do to achieve the goals of this strategy. Gabon noted that a national Habitat committee had been established, partnerships between national and local government and civil society had been forged, and community infrastructure projects had been implemented with financial assistance from the World Bank.

## WESTERN AFRICA

Approximately 38 per cent of the population of Western Africa lives in urban areas—on a par with the average for Africa as a whole (UNCHS 2001a). Cape Verde is the most urbanized country with 62 per cent of the population living in urban areas, and Burkina Faso the least, with just 18.5 per cent of its population in urban areas (UNCHS

2001a). The predicted average rate of urbanization for 2000–2015 ranges from more than 5 per cent in Burkina Faso and Niger to 3 per cent in Cape Verde (UNCHS 2001a). Thirty years ago, only one Western African city (Lagos) had a population of more than 1 million. By 2000, cities with populations exceeding 1 million included Ouagadougou (Burkina Faso), Abidjan (Côte d'Ivoire), Accra (Ghana), Conakry (Guinea), Bamako (Mali), Ibadan (Nigeria), Lagos (Nigeria) and Dakar (Senegal). Lagos, the largest city in Africa, and the sixth largest in the world, has an estimated current population of 13.4 million people, and this is expected to grow to more than 23 million by 2015 (UNCHS 2001b).

## THE SIGNIFICANCE OF URBAN AREAS IN WESTERN AFRICA

The growth in urban populations in Western Africa is the result of a combination of high overall population growth, and migration. Migration, in turn, is a composite of rural push factors and urban pull factors. The population pressures, climatic variability, and fragmentation of tenure and traditional systems contribute to degradation of soil and vegetation, diminishing yields and worsening food insecurity in rural areas. Prevailing educational systems are also more oriented towards training people in urban occupations than to improving agriculture or animal husbandry in rural areas.

The lure of employment and the perception of improved quality of life in urban areas are increasingly attracting rural populations to the cities. The concentration of amenities (such as health care, educational and recreational facilities) in capital cities such as Abidjan, Accra, Dakar, Lagos and Bamako, and minimum wage legislation, also makes urban centres more attractive and employment more favourable than the insecure dividends of subsistence farming. Although several countries of the sub-region (including Côte d'Ivoire, Senegal and Nigeria), experience internal migrations, especially urban-urban migration, a high percentage of migrants from other countries is adding to the growth of towns and urban areas in Western Africa. The percentage of foreign residents in Côte d'Ivoire increased from 23.3 per cent in 1965 to 35.9 per cent in 1990 (Toure and Fadayomi 1992).

Freetown slum, Sierra Leone

*Edgar Cleijne/Still Pictures*

## URBAN AREAS AND THE ENVIRONMENT IN WESTERN AFRICA

Urban growth rates in Western Africa exceed the capacities of municipalities to provide adequate housing and services such as water supply, sanitation, waste disposal, communications and transport infrastructure, health services and education. High unemployment in urban areas also contributes to widespread poverty and poor living conditions.

Several coastal wetlands in Western Africa have come under increasing threat in recent years from industrial pollution, eutrophication due to sewage and contamination with solid wastes. Wildlife habitats, especially waterfowl habitats, have also been destroyed (IDRC 1996). In response, some sites have been declared Wetlands of International Importance under the 1971 Ramsar Convention, and are now protected (Ramsar 2001).

### Unplanned settlements in Western Africa

Rapid urban growth—complicated by poor urban planning and control of land use, lack of financial resources and inadequate investment in environmental management—has led to the proliferation of urban slums in Western Africa. Although it is difficult to quantify the number of people living in slums, it has been reported that 42 per cent of the population in Liberia's capital, Monrovia, are squatters and, in Nouakchott (Mauritania), approximately 12 per cent of the city's area is taken up with slums (UNCHS 2001c). In Abidjan, Côte d'Ivoire, a number of schemes have been implemented to improve the living conditions of the 20 per cent of the population estimated to be living in slums. These include the Programme for Institutional Support to Settlement Policy, creation of an Agency for Land Management and the creation of a housing bank (UNCHS 2001c). In Ghana and Nigeria, improved security of tenure and enhanced gender-equality in tenure have been reported in recent years, resulting from sound municipal governance and the evolution of democracy. Although there is no legislated right to adequate housing, the governments aim to provide protection from forced eviction by documenting ownership of properties (UNCHS 2001c). Similar efforts are underway in Senegal (see Box 2g.5).

In 2000, an average of around 70 per cent of Western Africa's urban population had access to

---

### Box 2g.5  Settlement upgrading in Senegal

Unplanned developments accounted for 25 per cent of Senegal's urban areas in 1987 when the Dalifort Settlement Upgrading Pilot Project was launched. The programme aimed to assist squatters in improving their own living conditions, whilst adhering to environmentally sound practices. To date, 500 inhabitants of Dalifort have secured land titles, and water supply, electricity, garbage collection and individual sanitation has been provided for the 7000 inhabitants. This success has sparked interest from a further nine informal settlements (with a total of 100 000 inhabitants), and a presidential revolving fund has been established to provide financial sustainability.

Source: UNCHS 2000

---

improved water supply and sanitation, although there were wide differences across the sub-region (WHO/UNICEF 2000). In Côte d'Ivoire and Senegal, for example, more than 90 per cent of urban populations had access to improved water supply whereas, in Guinea-Bissau and Sierra Leone, less than 30 per cent of the urban population had access (WHO/UNICEF 2000). More than 90 per cent of the people in urban areas in Cape Verde, Guinea, Mali and Senegal had access to improved sanitation whereas, in Sierra Leone, just 23 per cent of the urban residents had sanitation (WHO/UNICEF 2000). Data are not available for Liberia.

Water pollution, arising from inadequate water supply and sanitation, is both a public health risk and an environmental problem in many cities. Diarrhoeal diseases are amongst the most prevalent preventable diseases, especially among children, and have been correlated with inadequate sanitation. The World Health Organization (WHO) reports, for example, that washing hands with soap and water can reduce the incidence of diarrhoea by one-third (WHO/UNICEF 2000). Other water-borne diseases include parasitic worms and flukes, and skin and eye infections. Provision of adequate, clean water supply can reduce the incidence of these diseases by up to 70 per cent (WHO/UNICEF 2000). Furthermore, pollution of surrounding habitats and water sources can render them unsuitable as wildlife habitats or can even cause poisoning of wildlife and loss of biodiversity, as well as reducing their productivity and posing further health risks to local residents.

## Management of pollution and waste in Western Africa

Waste disposal is one of the most pressing and most visible environmental issues in cities in Western Africa, as elsewhere. Lack of investment in waste collection and disposal facilities, lack of sufficient suitable landfill sites, and growing volumes of waste resulting from increasing consumption, changes in packaging and increasing urban populations have all contributed to a breakdown in solid waste management. For example, in Accra it has been reported that, in 1992, just 11 per cent of the 1.4 million inhabitants had waste collection facilities, and the remainder of residents were dumping their refuse informally (Songsore 1992). In 1989, Abidjan, Côte d'Ivoire reported that none of its solid waste went to sanitary landfill, 72 per cent was disposed of in open dumps, 15 per cent was burned and 3 per cent was recycled (UNCHS 2001c). During the same year, the city of Dakar, Senegal reported that all of its solid waste was disposed of in open dumps (UNCHS 2001c). In Nouakchott, Mauritania's capital, an estimated 600–800 tonnes of household refuse are generated every day, of which 500 tonnes are collected (UNCHS 2001c). This results in large amounts of waste being blown onto streets and into drainage channels (where it causes blockages and creates a risk of flooding), or being burned, giving off toxic fumes and thus threatening human health.

Leachates from solid wastes can contaminate surface and groundwater sources, adding to the problem of inadequate access to safe water and sanitation, and increasing the risk of disease transmission. Responses have included privatization of waste collection, sorting and disposal, in an attempt to improve coverage and treatment. In Nouakchott, for example, private operators have been collecting and transporting garbage since 1997. However, they are linked to the municipality by contract, and thus still receive inadequate funding.

Reuse of organic waste in urban agriculture is an activity that is rapidly gaining popularity as a means of waste management, and one that is changing the perception of organic material—no longer seen as merely waste but as a useful input to another process.

A sizeable proportion of solid waste is organic (85 per cent in Accra, Ghana), consisting of food leftovers, vegetable matter, leaves, fruit and bones (Asomani-Boateng and Haight 1999), and urban agriculture is widely practised as a means of supplementing incomes and food supply. The organic compost or fertilizer from organic waste is a valuable addition. Despite this seemingly win-win opportunity, municipal authorities appear reluctant to encourage and commercialize this activity, because it would require changes to urban planning and changes in attitudes towards urban agriculture (Asomani-Boateng and Haight 1999).

Air pollution is another issue of concern in urban areas of Western Africa, arising mainly from vehicle and industrial emissions, and domestic use of coal, oil, wood or other 'dirty' fuels. Many vehicles in the cities of Western Africa, as elsewhere in Africa, are old and still run on leaded fuel or diesel, and have inefficient combustion. Compounding this is the rapid growth in urban populations, which is now over-burdening the transport infrastructure and road networks, and the high tariffs placed on importing new (and therefore more efficient) vehicles. Congestion and air pollution are thus common in urban areas, and concerns are being raised over the impacts on human health, in particular, the incidence of respiratory diseases. Dakar (Senegal) and Ouagadougou (Burkina Faso) are among the many Western African cities experiencing increased motorization rates. Currently, there are no air quality standards, and little has been done so far to monitor pollution levels or their impacts on the environment or the health of residents (World Bank 2001b). However, a World Bank approximation of air pollution costs in Dakar and in Ouagadougou amounted to 2.7 per cent and 1.6 per cent of GDP respectively (World Bank 2001b). Policies in place to tackle this problem include promotion of newer, less polluting vehicles and cleaner fuels, reorganization of mass transit, traffic regulation and land use planning in urban centres (World Bank 2001b). The World Bank also launched the Clean Air Initiative in sub-Saharan African Cities, in 1998, to raise awareness of the impacts of air pollution and to identify and implement strategies for improving air quality. A conference held in November 2001, in Abuja, Nigeria identified options for phase out of leaded gasoline, and provided a framework for development of action plans.

## Towards sustainable urban development in Western Africa

In addition to programmes by local authorities to improve water supply, sanitation, and waste management, many municipalities in Western Africa have enacted legislation to assist with improving urban planning and environmental management. The cities of Cotonou, Parakou, Porto-Novo (Benin) Ouagadougou (Burkina Faso), Abidjan, Yamoussoukro (Cote d'Ivoire), Banjul (Gambia), Accra, Kumasi (Ghana), Conakry (Guinea), Monrovia (Liberia), Bamako (Mali), Nouakchott (Mauritania), Ibadan, Lagos (Nigeria), Bignona, Thies (Senegal), Lomé and Sokode (Togo) have all developed, institutionalized and implemented local environmental plans (UNCHS 2001c). Many of these and other Western African countries also have national legislation or National Environmental Action Plans (NEAPS). Sustained investment in urban upgrading and development of infrastructure is a priority for Western African countries, as for all African countries, in order to keep pace with the rates of urban growth and to improve the living conditions of millions of urban residents.

Primary school children, Ghana

*Ron Gilling /Still Pictures*

## CONCLUSION

Throughout history, urbanization has been associated with economic and social progress, the promotion of literacy and education, the improvement of the general state of health, greater access to social services, and cultural, political and religious participation. When properly planned and managed, urban areas are capable of promoting human development, whilst effectively managing the extent of their impact on the natural environment.

However, population growth and rural-to-urban migrations have increased steadily, particularly in developing countries, placing enormous pressures on urban infrastructure and services, both in terms of provision and maintenance. The most serious problems confronting cities and towns and their inhabitants include inadequate financial resources, and planning skills and capacities. Coupled with the rapid growth in demand for shelter and services, as well as rising consumption, this has culminated in many social and environmental problems. These factors strain governments and local authorities' budgets, and their

capacities to achieve the goals of sustainable development. Responses urgently required include the eradication of rural poverty and improvement of the quality of living conditions, as well as creation of employment and educational opportunities in rural settlements, regional centres and secondary cities (United Nations Conference on Human Settlements 1996). Additional resources need to be committed to environmental protection and management at the national and international levels, to ensure a sustained supply of fertile and stable soils, good quality water, healthy fresh air, medicinal and nutritional products, and space for recreation.

Improved planning and design of settlements is required to make use of existing sites, thereby containing urban sprawl and limiting the extent of encroachment on natural habitat or agricultural land. Efforts should also be made to encourage the use of recycled or sustainably produced building materials, and innovations are required in building design to reduce environmental impact (for example, through improved energy efficiency). EIAs need to be better integrated with urban planning and management, so that environmental impacts are monitored and managed continuously and within a context of holistic development.

In addition to this, improvements to infrastructure provision and maintenance are a priority for most African countries, although this is not exclusively an urban issue. A significant proportion of the burden of disease in Africa could be alleviated through improvements in water supply and sanitation, waste disposal, education and community-based, preventive healthcare. Reducing air pollution in urban centres by regulating the industrial and transport sectors, as well as reducing domestic air pollution through improved access to electricity and ventilation, would also make significant inroads in reducing the disease burden in Africa.

'Recognizing the global nature of these issues, the international community, in convening Habitat II, has decided that a concerted global approach could greatly enhance progress towards achieving these goals. Unsustainable patterns of production and consumption, particularly in industrialized countries, environmental degradation, demographic changes, widespread and persistent poverty, and social and economic inequality can have local, cross-national and global impacts. The sooner communities, local governments and partnerships among the public, private and community sectors join efforts to create comprehensive, bold and innovative strategies for shelter and human settlements, the better the prospects will be for the safety, health and well-being of people and the brighter the outlook for solutions to global environment and social problems.' (United Nations Conference on Human Settlements, 1996).

## REFERENCES

ABT (1983). Informal Housing in Egypt. Report by ABT Associates, Dames and Moor, to USAID

Alexandra Renewal Project (2001). Available on: http://www.alexandra.co.za

Asomani-Boateng, R. and Haight, M. (1999). *Reusing Organic Solid Waste in Urban Farming in African Cities: A Challenge for Urban Planners*. IDRC, Ontario

CEDARE (1997). CEDARE: From Vision to Action, Meeting the environmental Challenges of the Region. Third National Focal Point Meeting of CEDARE

Chenje, M. (ed., 2000). *State of the Environment Zambezi Basin 2000*. SADC/IUCN/ZRA/SARDC, Maseru/Lusaka/Harare

City of Cape Town (2001). Integrated Metropolitan Environmental Policy. Available on: http://www.capetown.gov.za/imep

DEA&T (1999). *State of the Environment South Africa 1999*. Published by the Department of Environmental Affairs and Tourism, Pretoria. Available on: http://www.environment.gov.za

Department of Housing (2000). South African Country Report to the Special Session of the United Nations General Assembly for the Review of the Implementation of the Habitat Agenda. Department of Housing, Pretoria

Disaster Relief (2000). As Death Toll Climbs, More Rain Expected in Deluged Southern Africa. 16 Feb 2000. Available on: http://www.disasterrelief.org/Disasters/000216SAfloods2/

Everatt, D. (1999). *Yet Another Transition? Urbanization, Class Formation, and the End of National Liberation Struggle in South Africa*, Woodrow Wilson International Centre for Scholars, Washington D.C.

GJMC (1999). State of the Environment Johannesburg. Greater Johannesburg Metropolitan Council. Available on: http://www.environment.gov.za

Hamza, A. (ed., 1994). *Towards a National Strategy for Potable Water in Egypt*. The Centre for Environment and Development for the Arab Region and Europe (CEDARE), Cairo

ICLEI (2001). African Sustainable Cities Network. Available on: http://www.iclei.org/la21/ascn/index.cfm

IDRC (1996). Ghana: The Nightmare Lagoons. Article written by Theo Andersen, Friends of the Earth. Available on: http://www.idrc.ca/books/reports/e234–13.html

Kanbour, F. (1997). General Status on Urban Waste Management in West Asia, paper delivered at the UNEP Regional Workshop on Urban Waste Management in West Asia, Bahrain, 23–27 November 1997

Larsen, B. (1995). Natural Resource Extraction, Pollution, Intensive Spending and Inequities in the Middle East and North Africa. Working Paper Series. World Bank, Washington D.C.

MoWTC (2000). Vehicle Database. Ministry of Works, Transport and Communications. Kampala

Napier, M, Ballance, A. and Macozoma, D. (2000). 'Predicting the Impact of Home-Based Enterprises on Health and the Biophysical Environment: Observations from Two South African Settlements' in proceedings of the CARDO International Conference on 'Housing, Work and Development: the role of home-based enterprises', 26–28 April 2000, University of Newcastle upon Tyne

Napier, M. (2000). *Human Settlements and the Environment*. Report prepared for the Department of Environmental Affairs and Tourism, Pretoria, South Africa. Available on: http://www.environment.gov.za/soer/index.htm

NEMA (1999). *State of the Environment Uganda 1998*. National Environment Management Authority, Kampala

NEMA (2001). *State of the Environment Uganda 2000*. National Environment Management Authority, Kampala

Ngategize, P. and others (2000). Draft Strategic Plan for Solid Waste Management for Mpigi District. Environmental Monitoring Associates Ltd., Kampala

Ramsar (2001). Contracting Parties to the Ramsar Convention on Wetlands (as of August 7 2001). Available on: www.ramsar.org/key_cp_e.htm

SARDC (1999). Need For More Urban Housing in Southern Africa. Southern African News Features. Southern African Research and Documentation Centre, Harare. Available on: http://www.sardc.net/editorial/sanf/1999/10/15-10-1999-nf3.htm

Songsore, J. (1992). Review of Household Environmental Problems in Accra Metropolitan Area, Ghana. Stockholm Environment Institute (SEI), Sweden

Statistics South Africa (1998). The People of South Africa: population census, 1996. Report No. 03-01-11 (1996). Pretoria: Statistics South Africa

Toure, M. and Fadayomi, A. (1992). Politique de Population Urbanization, Migration en Côte d'Ivoire. In *Population, Urbanization and Rural Crisis in Africa*. CODESRIA, Dakar

UNCDF (1998). Project Evaluation Summary: Community Housing in Balbala. Project Numbers DJI/86/C04 and DJI/89/005 of the United Nations Capital Development Fund, New York

UNCHS and UNEP (1997). *Environmental Planning and Management (EPM) Source Book*. United Nations Centre for Human Settlement (Habitat) and United Nations Environment Programme, Nairobi. Available on: http://www.unchs.org/uef/cities/home.htm

UNCHS (1999). *State of the World Cities Report 1999*. United Nations Centre for Human Settlements, Nairobi

UNCHS (1996). *An Urbanizing World: Global Report on Human Settlements 1996*. United Nations Centre for Human Settlements (Habitat), Nairobi

UNCHS (1998). *Privatization of Municipal Services in East Africa: A Governance Approach to Human Settlements Management*. United Nations Centre for Human Settlements (Habitat), Nairobi, Kenya

UNCHS (2000). Best Practices Database. Available on: http://www.unesco.org/most/bpunchs.htm

UNCHS (2001a). *Cities in a Globalizing World: Global Report on Human Settlements 2001*. HABITAT (United Nations Centre for Human Settlements), Nairobi

UNCHS (2001b). *The State of the World's Cities 2001*. HABITAT (United Nations Centre for Human Settlements), Nairobi

UNCHS (2001c). Global Urban Observatory Statistics Database (Year of Reference 1998). United Nations Centre for Human Settlements, Nairobi. Available on: http://www.unchs.org/guo/gui/data_gui.asp

UNCHS/ROAAS (2001). UNCHS Regional Office for Africa and Arab States. Available on: http://www.unchs.org/roaas/start.htm

UNDP (1996). *Balancing Rocks: Environment and Development in Zimbabwe*. UNDP, Harare

UNDP (2000). *Human Development Report*. United Nations Development Programme, New York

UNEP (1998). *Water-Related Environmental Issues And Problems Of Comores And Their Potential Regional And Transboundary Importance*. TDA/SAP-WIO. Maputo

UNEP (1999). *Western Indian Ocean Environment Outlook*. United Nations Environment Programme, Nairobi

United Nations Conference on Human Settlements (1996). The Habitat Agenda

USAID (1993). *Nairobi's Informal Settlements: An Inventory*. Working Paper No PN-ABH-741; United States Agency for International Development

USAID (2001). Protecting Egypt's Environment. Perspectives From the Field. Available on: http://www.usaid.gov/regions/ane/newpages/perspectives/egypt/egenv.htm

WHO/UNICEF (2000). Global Water Supply and Sanitation Assessment 2000

World Bank (2001b). Transport Air Pollution Impact Assessment in Two Sub-Saharan African Cities: Dakar and Ouagadougou, Two Tales of the Same Story. World Bank Technical Note, Washington D.C.

World Bank (1995). Middle East and North Africa Environment Strategy. Report No 13601-MNA. The World Bank, Washington, D.C.

World Bank (1996). Kenya — Development of Housing, Water Supply and Sanitation in Nairobi. Impact Evaluation Report Number 15586. The World Bank, Washington, D.C.

World Bank (1998). First World Bank Water Project in Mozambique: US$36 Million Loan To Become Effective Ahead Of Schedule To Help Alleviate Cholera Outbreak. News Release No. 98/1646/AFR, The World Bank, Washington D.C.

World Bank (2000). *World Development Report 2000/2001*. The World Bank, Washington D.C.

World Bank (2001a). *African Development Indicators 2001*. The World Bank, Washington D.C.

WRI, UNDP, UNEP, World Bank (1992). *World Resources 1992–1993*. Oxford University Press, New York and Oxford

WRI, UNDP, UNEP, World Bank (1998). *World Resources 1998–99: A guide to the Global Environment; The Urban Environment*. Oxford University Press, New York and Oxford

## THE STATE OF AFRICA'S ENVIRONMENT AND POLICY ANALYSIS

# CONCLUDING SUMMARY

## ATMOSPHERE

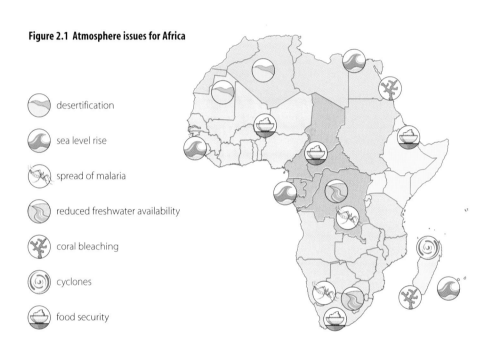

Figure 2.1 Atmosphere issues for Africa

- desertification
- sea level rise
- spread of malaria
- reduced freshwater availability
- coral bleaching
- cyclones
- food security

Africa is extremely vulnerable to climate variability and climate change. Climate variability means the seasonal and annual variations in temperature and rainfall, and their distribution within and between countries. Climate change refers to long-term changes in global weather patterns, resulting from changes in the composition of the atmosphere brought about by emissions of greenhouse gases (GHG).

Africa's climate is characterized by considerable variability and by extreme climatic events such as droughts, floods and cyclones. These have particularly serious consequences for Africa because of the difficulties experienced in the region in predicting their occurrence and in mitigating their effects, and also because many African countries lack the financial resources to make adequate and timely recovery before the next climatic event. The countries of the Horn of Africa and the Sahel are most prone to drought; those of Western and Central Africa experience flooding periodically. Cyclones occur regularly in the islands of the Western Indian Ocean.

Climate change is likely to bring increased frequency and severity of flooding and drought to those areas already experiencing variability in rainfall. Additional concerns are increased risk of desertification in Northern Africa and on the southern border of the Sahel, and reduced availability of freshwater in Southern Africa, resulting in lowered food security and the spread of infectious diseases (notably malaria) to new areas. The islands and low-lying coastal areas of Central and Western Africa are most vulnerable to the sea level rise resulting from climate change. Coral bleaching due to sea temperature rise is the biggest threat to the Western Indian Ocean Islands and to the coasts of the countries of Eastern Africa.

Africa's contribution to global GHG emissions is negligible, with the exception of emissions from South Africa and the countries of Northern Africa, which together account for the majority of the Africa region's GHG emissions. However, activities such as deforestation, inappropriate coastal development and poor land management throughout Africa contribute to worsening of possible impacts of climate change such as drought, desertification, flooding and sea level rise.

Low air quality is also emerging as an issue in many African countries, particularly the more urbanized and industrialized. In large cities, populations are at risk from respiratory infections caused by emissions from industry and vehicles and—in both urban and rural areas—from the use of wood, coal, oil, paraffin and other such fuels used for domestic consumption, and creating unhealthy indoor conditions.

Urgent action is required to develop alternative, clean and renewable sources of energy for Africa to avoid increasing GHG emissions and to stem widespread deforestation. The Kyoto Protocol makes provision for funding streams to facilitate such development. Figure 2.1 shows the major issues relating to atmosphere for Africa.

## BIODIVERSITY

**Figure 2.2  Threats to Africa's biodiversity**

- bushmeat trade
- forest loss
- encroachment of protected areas
- conflicts between wildlife and pastoralist
- loss of indigenous knowledge
- overharvesting of rare and endangered species
- alien invasive organisms
- habitat loss/conversion to other uses

Africa is endowed with rich and diverse biological resources. These have enormous value for indigenous populations and commercial enterprises, and for development of tourism. These resources are, however, declining rapidly under the pressures of habitat loss, overharvesting of selected species, the spread of alien species, and illegal activities. In Western and Central Africa, the main issues are loss and fragmentation of forest habitat and poaching of endangered species to meet the growing demand for bushmeat. In Eastern Africa, encroachment of human settlements into protected areas and pastoral areas outside of reserves and their cultivation are priority concerns. In Southern Africa, loss of indigenous knowledge and inadequate protection of intellectual property rights are hampering conservation measures, as are overharvesting (legal and illegal) of medicinal plant species, rare and endangered plants, and 'trophy' animals and exotic pets. Overharvesting of certain species, particularly medicinal plants is also the primary cause of biodiversity loss in Northern Africa. Alien invasive organisms are a widespread problem throughout the region, particularly in closed ecosystems including Lake Victoria and the Western Indian Ocean Islands.

Formal protection of Africa's biodiversity at both national and international levels has been strengthened over the past 30 years, but the paradigm of conservation is now shifting away from protection and preservation, and focusing more on sustainable use and sharing of benefits. If this is to be achieved in Africa, wider involvement of stakeholders is required, together with additional research and documentation. Biological resources can and should be used to enhance economic growth opportunities, but consumptive and non-consumptive uses need to be managed with a long-term rather than a short-term view, and a fair and protective framework needs to be established to ensure that the benefits of resource use accrue to African communities and nations, rather than to international companies. Figure 2.2 summarizes the main threats to Africa's biodiversity.

## COASTAL AND MARINE ENVIRONMENTS

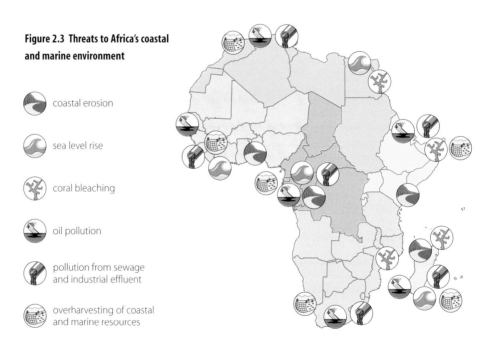

**Figure 2.3 Threats to Africa's coastal and marine environment**

- coastal erosion
- sea level rise
- coral bleaching
- oil pollution
- pollution from sewage and industrial effluent
- overharvesting of coastal and marine resources

Africa has a long and varied coastline, supporting a diversity of habitats, resources, and economic activities. There has been rapid urban and industrial development and growth in tourism in many coastal zones without adequate infrastructure planning, protection or provision. As a result of this—and of phenomena in the hinterland such as deforestation and soil erosion—coastal erosion is a growing concern, particularly in Western Africa, Eastern Africa and the Western Indian Ocean Islands. Sea level rise due to climate change is a real and serious threat to the Western Indian Ocean Islands and to low-lying coastal settlements, particularly in Northern, Western and Central Africa. Sea temperature rise has the potential to cause coral bleaching which would have damaging impacts on the economies of the countries bordering the Red Sea and the Western Indian Ocean.

Coastal and marine environments are also under pressure from pollution from land-based and sea-based sources. Oil pollution is a major threat, with high levels of oil transportation threatening the Eastern and Southern African coast and islands, and oil drilling and processing activities causing problems along the Northern, Western and Central African coast.

Domestic, industrial and agricultural effluents are problems in most sub-regions, especially around large urban or industrial coastal centres, and in Western and Central Africa where commercial plantations are common in the coastal zone. Overharvesting of coastal and marine resources is a priority concern for countries of Southern and Western Africa and for the island states, because of their relatively large area of Exclusive Economic Zone (EEZ).

Most coastal African countries are party to relevant international conventions, and have national policies and regulations for sustainable coastal development and use of marine resources. In particular, Integrated Coastal Zone Management (ICZM) plans have been developed by many countries, showing commitment to conservative use of the coastal zone and marine resources. These plans are, however, wide-ranging in nature and require considerable resources, such as trained personnel, equipment and financial resources, and more effective policing, monitoring, administration and enforcement. Lack of coordination between government departments and between countries is an additional constraint on their effectiveness. Coordination should be strengthened. Figure 2.3 shows the threats to Africa's coastal and marine environment.

## FORESTS

**Figure 2.4 Threats to Africa's forests**

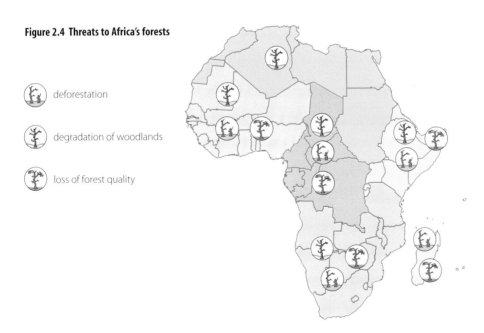

deforestation

degradation of woodlands

loss of forest quality

Only Western and Central Africa have abundant resources of closed canopy forests, although Madagascar and the wetter parts of Southern and Eastern Africa have equally important but smaller areas of forest. Savanna and woodlands are predominant in Africa's more arid countries. These ecosystems are very different from closed canopy forests, but are also rich in diverse natural resources. Both forests and woodlands play vital roles in supporting national economies and local communities through harvesting of resources, provision of habitats and services derived from the ecosystem. Since the majority of these resources and services have not been quantified in monetary terms, they are often undervalued and threatened by overexploitation.

Africa has the fastest rate of deforestation in the world. Competing land uses (agriculture and human settlements mainly) are contributing to the decline of forest and woodland areas, and the rising demand for fuelwood and charcoal is also a major cause of deforestation. In addition to ecological impacts, local communities are suffering from loss of livelihood and from loss of vital energy sources. Although policies and mechanisms for enhancing sustainable use of forests are in place in many countries, and regional cooperative arrangements are well under way, implementation and enforcement of regulations are weak, because economic forces are pressurizing governments and communities into unsustainable practices for short-term profits. Political commitment to protection of indigenous forests, sustainable harvesting practices and community ownership needs to be strengthened and development of alternative energy sources is a priority. Figure 2.4 shows the major threats to Africa's forests.

## FRESHWATER

**Figure 2.5 Freshwater issues in Africa**

 rainfall variability

 freshwater pollution

 access to water and freshwater availability

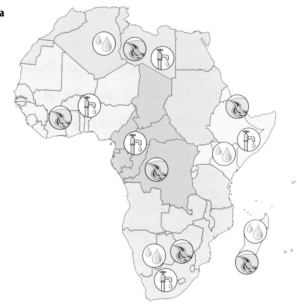

Lack of availability and inadequate quality of freshwater are the two most limiting factors for development in Africa, constraining food production and industrial activities, and contributing significantly to the burden of disease.

Western and Central African countries have relatively abundant freshwater resources and fairly predictable rainfall. However, the distribution of human populations in these countries is such that the various user groups nevertheless experience difficulties and disparities in access to water resources. For example, the rural poor have to walk long distances to collect water for domestic consumption, whereas agricultural and industrial users have access to subsidized resources.

Most other countries, particularly those in Northern Africa, the Horn of Africa and in the Sahel, experience extreme variability and unpredictability in rainfall, as well as frequent drought. User groups experience disparities in access to water resources, with Southern African countries showing the widest disparity in terms of both access and consumption.

Almost all African countries experience problems of water quality and are struggling to upgrade water treatment and wastewater processing plants. At present, large quantities of industrial and domestic wastewater are discharged untreated into watercourses and coastal waters, and the resulting pollution poses a risk to human and aquatic life. Pollution of freshwater also compounds the existing problems of water availability by raising the costs of treatment and supply.

Integrated Water Resources Management (IWRM) is a new, multi-stakeholder approach to meeting the challenges of water supply by curbing demand from certain user groups and encouraging re-use and recycling, as well as management of aquatic ecosystems and making provision for environmental water requirements. Although there is increasing recognition of the need for, and benefits of, adopting such an approach in African countries, implementation will only be effective if it is supported by adequate finances, and by trained personnel with adequate facilities. Coordination between government departments responsible for water supply and use must be brought in line with the common goal of sustainable use of resources. Figure 2.5 shows the main freshwater issues for Africa.

# LAND

**Figure 2.6  Land issues in Africa**

 degradation due to commercial agriculture

 desertification

 competition between land uses

 salinization and chemical pollution

 land tenure

The land and its resources are the cornerstones of Africa's culture and development. Pastoralism and subsistence agriculture are the traditional practices, but commercial plantation agriculture has been promoted over the past 30 years, particularly in Western and Central Africa, as the foundation of economic growth. However, Africa's soils are not generally suited to cultivation, rainfall is variable in much of Africa and irrigation has been developed in only a few countries. Dependence on agriculture (in particular a narrow range of crops) has therefore had economic and environmental drawbacks.

National development policies and international trade agreements and/or restrictions have also affected agricultural development and natural resource quality. Soils and vegetation are being degraded largely as a result of increasing use of inorganic chemicals, reduction of fallow systems, increased monoculture and cultivation of marginal areas. In Eastern and Southern Africa, conflict between user groups over land and resources is the priority issue, resulting from competition between agriculture, pastoralists and conservation areas, and from complex and inappropriate land tenure policies.

Desertification is another serious concern in Africa, particularly in the more arid zones, where climate variability and poor land management practices combine to threaten the sustained productivity of soil and vegetation.

Land tenure reform, international cooperation and integration of land resource management with development goals are required. Monitoring of climatic patterns and strategies to alleviate the pressures that economic growth places on terrestrial resources are additional priorities if Africa is to achieve sustainable development and protect its resource base. Figure 2.6 illustrates the land issues facing Africa.

# URBAN AREAS

**Figure 2.7 Impacts of urban development in Africa**

 urban issues, including:
- proliferation of slums
- inadequate waste management
- poor water supply and sanitation coverage
- urban air pollution

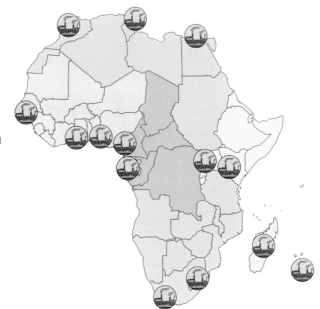

Although most Africans currently live in rural areas, rates of urbanization in the region are among the highest in the world. While urban areas offer considerable opportunities for employment, education and social services, and contribute substantially to national GDPs, they also have their own environmental problems. In Africa, poor economic growth and low investment in infrastructure have left provision of housing and basic services in urban areas lagging far behind the rate of population growth, resulting in a proliferation of urban slums. Slums often lack access to many essential services, such as water, sanitation and waste disposal; concentrate the pressures of pollution on the environment; and are detrimental or even hazardous to human health. Emissions from industry and vehicles are additional threats to human health in the larger urban areas.

Investment in urban infrastructure is essential to mitigate existing problems and to prevent exacerbation of the pressures exerted on urban areas by rapid population growth and migration. However, sustained economic growth and strategies aimed at job creation and alleviation of poverty are of pivotal importance if such investments are to be made.

Many African countries have implemented air quality standards and pollution abatement regulations, but integrated planning and investment in infrastructure development remain priorities for urban areas. Figure 2.7 illustrates the impacts of urban development in Africa.

## THE COMPLEX RELATIONSHIPS BETWEEN ENVIRONMENTAL ISSUES

Although environmental issues have been presented here by theme, it is important to recognize that they are inter-linked, and that they therefore need to be addressed with an integrated environmental management approach. Figure 2.8 illustrates links between stresses on land and water resources, others are highlighted below.

Deforestation and loss of vegetation cover increase the risk of soil erosion which, in turn, impacts on aquatic and coastal ecosystems through sedimentation and smothering of habitats. This reduces water availability and quality and contributes to coastal erosion, exacerbating the impacts of climate change on freshwater resources and of sea level rise. Deforestation also affects climatic conditions reducing rainfall and run-off at the micro level (thereby also impacting on freshwater availability and aquatic ecosystems) and contributes to global warming and its impacts at the macro level.

Reductions in rainfall (natural or induced through changes such as deforestation) also impact on infrastructure and urban development, particularly on electricity generation in many parts of Africa, where hydro-electric power is well developed. Power shortages and poorly developed central power supplies contribute to deforestation by increasing the population's dependence on wood and charcoal for fuel.

Loss of natural habitat because of urban and agricultural sprawl contributes to declines in biodiversity, loss of economic development potential and support for livelihoods. Endangerment and extinction of certain species (terrestrial, aquatic and marine) put pressure on other resources as substitutes are sought to meet people's requirements. Changing patterns of resource distribution also affect life cycles of pests and diseases, and may in turn impact on human health, and crop and livestock yields.

Alien invasive plants cause changes to freshwater availability and alter the diversity and abundance of species within ecosystems. Aquatic weeds also change water quality by adding or removing nutrients and, thus, affect drinking water quality and human health.

**Figure 2.8  Interlinked stresses on land and water resources**

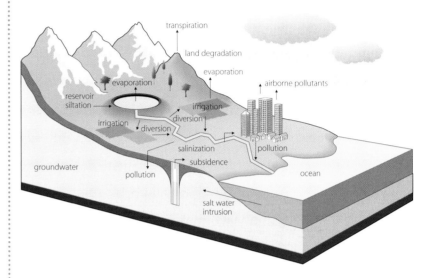

### WHAT IS CAUSING ENVIRONMENTAL CHANGE IN AFRICA?

The causes of environmental change in Africa are both natural and human-made, resulting from events and activities both within the region and external to it. For example, climate variability is a natural phenomenon that has impacts such as flooding and drought, but human activities such as deforestation and cultivation of marginal areas contribute to its impacts on human lives, livelihoods, economic growth and on the environment. Climate change is a result of anthropogenic GHG emissions, largely from countries outside of Africa, but the impacts of climate change on Africa need to be mitigated to conserve African resources (human, economic and environmental).

In many cases, it is not just one force that causes a single environmental change but a complex interaction of forces, often seemingly unconnected in time and/or space. It is also clear that some forces have impacts that are particularly relevant to the themes discussed in this chapter and, as such, should be addressed as priorities. The most important of these are listed here and are discussed more fully below:

● population growth, urbanization and increasing consumption;

- economic pressures;
- food production systems;
- energy needs and transport systems; and
- loss of indigenous knowledge and practices.

**Population growth**, urbanization and increasing consumption are putting pressure on the environment through growing demand for resources, such as freshwater, the produce of agriculture, forestry and fisheries, energy and space for housing and recreational facilities. These forces also exert pressure on the environment in terms of the large and growing volumes of waste and pollution they give rise to. In urban areas, environmental impacts are concentrated in small areas, making the surrounding ecosystems, resources and vast numbers of people highly vulnerable. Development and implementation of population policies is, therefore, a priority to ensure that population growth and urbanization patterns are in line with economic growth, resource constraints, employment opportunities and infrastructure development.

**Economic pressures** include government policies to bring more land under cultivation, to grant more logging concessions and fishing agreements, and to promote industrial development, thereby encouraging economic growth and earnings of foreign exchange. These practices are driving environmental change because they encourage unsustainable rates of resource use or extraction. For example, Africa's forests are being felled and timber exported at a rate that exceeds the forest's natural rate of regeneration. Forest areas are declining rapidly as a result. Similarly, fishing rights have been granted to foreign fishing companies that are harvesting resources faster than the natural population-replacement rate, and fish stocks are declining as a result. These activities need to be managed in a more sustainable manner so that benefits can continue to accrue to African countries in the long term. To achieve this, the economic base needs to be broadened to lessen the pressure on natural resources.

**Food production systems** are driving environmental change as an indirect result of population and economic pressures. Because there are more people to feed, and because there are pressures for economic growth through export of agricultural produce, commercial food production systems are being transformed to low input, monospecific cultivation, with one or two harvests per year. This rapidly exhausts the soil of nutrients and lowers productivity in the long term. At the subsistence level, farmers are forced to cultivate or graze more and more marginal land because there is greater competition for land from different users. This causes yields to decline, and makes resources more vulnerable to the impacts of climate variability and change, environmental degradation and desertification. The development of population policies, broadening of the economic base, and promotion of soil and vegetation conservation measures would alleviate the pressure from food production systems on the environment and increase food security.

**The energy and transport sectors** in Africa are also significant drivers of environmental change in a number of areas. For example, the transport sector is a major contributor of emissions that are harmful to human health, as well as to the atmosphere and to terrestrial ecosystems. Centralized power generation using fossil fuels is a problem in South Africa and in Northern African countries, contributing to atmospheric pollution and climate change. In Eastern and Western Africa, hydropower is one of the main sources of electricity. This has a different set of environmental problems associated with it, including inundation of land and displacement of people, disruption of ecosystems and exacerbation of flooding. A much more widespread problem associated with energy sources in Africa is the use of wood, charcoal and other traditional fuels at the household level. This not only results in high rates of deforestation but also raises the risk of acute respiratory diseases from inhalation of sulphur oxides, nitrogen oxides, carbon monoxide and soot particles. Alternative energy sources need to be developed in the near future to reduce dependence on traditional fuels and to keep emissions and other environmental impacts to a minimum.

**Loss of indigenous knowledge and practices** is causing environmental change because it results in changes in patterns of land and resource use, obscuring of conservation measures, and foreclosing of opportunities for commercial development of resources. For example, knowledge of the medicinal properties of Africa's plant species could be used in commercial pharmaceutical production, but lack of recognition or documentation of this knowledge is impeding this. Likewise, indigenous plant species that are drought tolerant could be bred and used in commercial agriculture, and traditional practices of agro-forestry and inter-cropping could be adopted to promote soil conservation. Further resources need to be channeled into research and documentation of indigenous knowledge in order to realize its potential in Africa's development.

## WHAT ARE THE CONSEQUENCES OF ENVIRONMENTAL CHANGE IN AFRICA?

An issue that emerges from this analysis of the environment in Africa over the past 30 years is that of human vulnerability. The pressures described above contribute to climate change, declining yields, the spread of diseases and pests, and reduced economic potential, among other impacts. These, in turn, threaten food production and security, human health, economic growth, lives, livelihoods and infrastructure. Human vulnerability is a highly complex issue, involving many and varied social and economic factors, and further research is required to fully understand the links between it and environmental change. These are explored further in Chapter 3 of this report, but the following conditions and trends are apparent from the foregoing analysis:

- Africa is highly dependent on rain-fed agriculture (both crop production and livestock rearing) for national economic growth, and for subsistence of large sectors of the population. Many countries, especially those in the Horn of Africa and the Sahel, are likely to experience increasing variability in rainfall, and increasing frequency and intensity of drought over the next few decades. Food production systems in these areas are already under stress from degradation of soil and vegetation and spreading desert margins and countries are already experiencing difficulties in accumulating food reserves. There is, therefore, a danger that millions of people's livelihoods may be affected and national economic growth will stagnate, as a result of repeated crop and livestock losses. Early warning and response systems need to be strengthened urgently, together with long-term means of alleviating economic dependence on cultivation and livestock rearing. The subject of early warning and response is discussed in full in the 'Atmosphere' and 'Land' sections of this chapter.

- Environmental change can be conducive to outbreaks of pests and disease that impact on human health and economic development. For example, monospecific crop cultivation is more vulnerable to destruction by pests such as locusts; rising average temperatures may encourage the spread of malarial mosquitoes to previously unaffected areas; and reduced freshwater availability and quality increase the risk of infection by gastro-intestinal parasites and bacteria, and of skin and eye infections. Issues of human health require a two-pronged approach involving both preventive and curative measures. Preventive measures include maintaining a healthy environment and applying public hygiene standards. Curative measures include investment in health care services. (See 'Atmosphere', 'Freshwater' and 'Urban Areas' sections for a fuller discussion).

- Public and private infrastructure is threatened by sea level rise, and increasing frequency of floods and cyclones. These will affect communications and transport networks, administrative and commercial activities, and families and communities. Although loss of infrastructure will result in enormous economic losses for all concerned, it is the poorer communities and the poorer nations that will suffer the most, because they do not have the financial reserves to relocate and rebuild. The sooner action is taken to mitigate these impacts—through construction of physical defences, relocation of populations, and innovative design and construction measures—the more effective such actions will be. (See 'Atmosphere' and 'Coastal Areas' sections for full discussion.)

- The rising costs of water treatment, food imports, medical treatment and soil conservation measures are not only increasing the vulnerability of food and health, but they are also draining African countries of their economic resources in the long term and reducing the effective growth of GNP. Preventive measures need to be implemented fully and immediately in order to reverse this trend and improve security in the long term.

- Loss of biological resources translates into loss of economic potential and options for commercial development in the future. For example, deforestation and biodiversity loss, together with inadequate protection of intellectual property rights, are contributing to the loss of resources as well as to a loss of understanding of their applications and, hence, of commercial potential, thus diverting possibilities for economic growth away from African nations. Loss of these resources is also compromising the health of African populations and rural livelihoods. Conservation practices need to be strengthened, and strategies for sustainable use of resources need to be developed urgently to preserve and capitalize on this potential. (See 'Forests' and 'Biodiversity' sections for a fuller discussion).

## RESPONSES TO ENVIRONMENTAL CHANGE IN AFRICA

African countries have responded to the challenges of environmental change by ratifying international agreements, participating in regional and sub-regional initiatives and programmes, and by developing National Environmental Action Plans (NEAPs), Integrated Resource Management Plans (IRMP), Environmental Impact Assessment (EIA) regulations and other relevant legislation. Such responses are important because they set the framework at the highest level for integration of environmental issues into national governance, and they establish the boundaries of and administrative capacity for action at national and sub-national levels. African countries have some of the most advanced and coherent policies for environmental management in the world, and some exemplary models of public participation in policy formation and decision making. However, Africa's nations must not be allowed to rest on their laurels. They should be encouraged by the international community to put such policies and models into practice, to monitor their effectiveness, and to enforce their objectives. This will require the allocation of financial resources, equipment, trained personnel, monitoring systems, administrative frameworks and institutions, and the enforcement of regulations through economic incentives and disincentives.

Regional responses include development and signature of the Lagos Plan of Action, the Lusaka Agreement, the Abidjan and Nairobi Conventions, and the Accra Declaration. Although some of these have experienced initial difficulties in implementation, they nonetheless testify to the political recognition of the need to improve environmental conditions, and to commitment to such improvement. Initiatives aimed at

sustainable development of shared resources—such as the Nile Basin Initiative, the Lake Chad Basin Commission and the Lake Victoria Environmental Management Plan—are evidence of the foresight of governments and of their commitment to equitable access and use of resources and to sound environmental management. Furthermore, these actions set precedents for additional cooperative arrangements in the areas of research, management, and mitigation of adverse impacts.

In addition, environmental education and awareness-raising campaigns have been successfully implemented in many African countries, as evidenced by the growing prominence given to environmental issues in decision making at national and community level. These are positive steps, in that greater awareness in all sectors of the population facilitates mainstreaming of environmental issues into all activities and decisions. Creation of incentives, such as markets for recycled materials and subsidizing of unleaded fuel, are urgently required to sustain this momentum and to turn awareness into action.

## PRIORITIES FOR FUTURE ACTION

Priorities for action include strengthening of existing mechanisms by investment in equipment and personnel, as well as enhancement of coordination between government departments and other related organizations, so that environmental and development issues are integrated into all decision making at the national and sub-regional level. Chapter 5 offers a more comprehensive and integrated analysis of policy options and makes specific recommendations. However, some broad outlines for progress can be extracted from the analysis presented in this chapter. These include:

- Using climate change fora to exert pressure on the international community to implement the proposed mechanisms for emissions trading, reforestation schemes and cleaner development measures from which everyone stands to gain economically and environmentally.

- Implementation of national and sub-regional action plans developed under the United Nations Convention to Combat Desertification in those Countries Experiencing Serious Drought and/or Desertification, Particularly in Africa (UNCCD).

- Strengthening of hazard monitoring programmes and early warning and response systems to effectively mitigate and cope with the impacts of extreme events, to increase food security and to maintain healthy ecosystems.

● Diversification of economies to alleviate dependence on agricultural production, forestry, fisheries and other natural resources. In particular, new economic activities should focus on value-addition to the natural resources that are exported, to ensure that maximum benefit accrues to Africa (in terms of employment, skills development, income and foreign exchange).

● Advancing trade negotiations to create a more equitable and operational international market for African products.

● Radical reforms to the energy and transport sectors, including: development and adoption of appropriate clean technologies to alleviate dependence on wood and fossil fuels without increasing emissions of GHGs; removal of subsidies; promotion of unleaded fuels and conversion to cleaner fuels; and upgrading of public transport systems.

● Integration of natural resources management programmes, such as ICZM and IWRM programmes. In particular, development of administrative frameworks is required to improve coordination between departments responsible for coastal planning, development and use of resources. Measures to manage demand also need to be established and implemented so that rates of consumption of resources are brought down to sustainable levels. This can be done through financial incentives such as true-cost pricing. In addition, an effective system for enforcing regulations is required.

● EIAs should be conducted as part of proactive and holistic development planning. These should be integrated into an overall development plan for a region or area, so that cumulative impacts of all developments are considered, and habitat loss and fragmentation are minimized. Highest priority should be given to improved urban planning, provision and maintenance of basic infrastructure and services, and to innovations in design to meet the demand for housing sustainably.

● Valuation of biodiversity and ecosystem services, particularly in forests and woodlands, and

economic reforms to the forestry sector, so that the true value of forests and woodlands is reflected in the prices, and so that harvesting rates are adjusted accordingly.

● Sustained participation by all stakeholders in management of natural resources, through open and transparent governance systems, involvement in community-level projects, and representation in regional resource management programmes, such as river basin management or management of protected areas.

● Strengthening of national efforts to meet the objectives of the United Nations Convention on Biological Diversity, including establishment of strategies for sustainable resource use, from within and outside of protected areas. In addition, indigenous knowledge and intellectual property rights need to be protected so that the development potential of African resources can be realized with benefits accruing to African countries and communities. Tighter trade restrictions need to be enforced to sustain the progress made in reducing poaching and illegal trade.

● Reform of tenure and access policies is a priority for nearly all African countries. Improving security of tenure encourages sound environmental management practices and sustainable production levels. Tenure reform will also raise the status of women and other disadvantaged groups, giving them equal opportunities and raising social development standards.

● Above all, these activities must be integrated into development and poverty-alleviation strategies, so that economic, social and environmental development priorities do not compete with one another, but are rather considered simultaneously and are addressed through mutually beneficial means.

# CHAPTER 3

## HUMAN VULNERABILITY TO ENVIRONMENTAL CHANGE

# CHAPTER 3

## HUMAN VULNERABILITY TO ENVIRONMENTAL CHANGE

### INTRODUCTION

Three decades ago, in 1972, the international community adopted the Stockholm Declaration, following the Stockholm Conference on the Human Environment. Principle 1 of the Declaration highlighted a healthy environment as a fundamental human right, explicitly stating: 'Man [sic] has the fundamental right to freedom, equality and adequate conditions of life in an environment of a quality that permits a life of dignity and well-being, and he bears a solemn responsibility to protect and improve the environment for present and future generations...' Since then, the Organization of African Unity (OAU) African Charter on Human and People's Rights, and dozens of relatively new African national constitutions, have enshrined a healthy environment as a fundamental human right.

Of particular interest to Africa in the Stockholm Declaration was the condemnation in Principle 1 of apartheid, racial segregation, discrimination, colonial and other forms of oppression, and foreign domination. While these socio-political issues have virtually been eliminated in the region, the environmental objectives have been compromised in many ways.

Over the past 30 years, the environment in Africa has continued to deteriorate, resulting in environmental change which is making more and more people in the region vulnerable due to increased risk and inadequate coping capability. Such deterioration has been acknowledged at various fora, and the World Commission on Environment and Development (WCED) reported in 1987: 'Today, many regions face risks of irreversible damage to the human environment that threaten the basis for human progress' (WCED 1987).

The undervaluing of the environment is a major factor in terms of overexploitation of the environment (see Box 3.1).

Human vulnerability to environmental change is complex; it may, in fact, be as complex as ecological processes, where some cause and effect linkages are still not fully understood despite centuries of scientific research. Human vulnerability to environmental change has global, local, social and economic dimensions. It is not synonymous with disasters, even though such and media interest (see Box 3.2).

### Box 3.1 Environmental concerns a priority

The advocates of sustainable development have not yet succeeded in raising environmental concerns to a high priority in all countries. The perception remains in some quarters that environmental protection is something that can and should be addressed only when a country is rich enough to do so, and that it is a 'low rate of return' activity. Yet the evidence is mounting that local environmental destruction can accelerate the poverty spiral not only for future generations, but even for today's population. It is obvious that countries which recklessly deplete their natural resources are destroying the basis of prosperity for future generations, but few policy makers have been able to persuade their constituents that, as forests disappear and water is exhausted or polluted, it is the poor of today, especially children and women, who suffer most.

Source: CSD 1997

**Box 3.2 Woman gives birth during flood disaster**

Sofia Pedro made world headlines in March 2000 when she gave birth to a daughter in a tree as the furious and raging waters of the flooded Limpopo River gushed below, laying to waste surrounding areas and devastating the lives of hundreds of thousands of her Mozambican compatriots. The Mozambican floods killed 700 people and left millions more homeless.

Perhaps the birth of Sofia Pedro's daughter—Rosita Pedro—brought to reality the juxtaposition of the birth of a new human life and the death of others, and the struggle humanity faces today in dealing with the challenges of a merciless, changed environment whose devastation grows in intensity and impact. Often, the impact of episodes such as the Mozambican floods in early 2000 is hidden behind a string of statistics: the number of confirmed deaths, the numbers injured, the livelihoods lost, the infrastructure destroyed, the habitats lost and the damage caused. As the news headlines bombard people with such figures, the human face is lost, reducing people to a footnote of another disaster event.

However, Sofia Pedro refused to be a footnote of that devastating episode of nature in Mozambique, but a living symbol of the human spirit and resilience in the storm of an increasingly unforgiving hostile environment, which has changed dramatically over the past three decades. In those swift-flowing muddy waters

below her were many people who were not so lucky. There were also deadly snakes, wild animals, livestock, and tonnes of soil on which millions of people in the Limpopo River basin depended for agriculture and food security. A way of life was swept away to the Indian Ocean to be drowned under masses of water. Left behind was human misery and people whose resilience had been compromised.

Sofia Pedro's story not only exemplifies just how people have become more vulnerable to environmental change, but also that ultimately disasters have the greatest impact on the personal level. Her story has been played out countless times since time immemorial in different regions, countries, communities and homes. Many Sofia Pedros have been rescued in floods, droughts, earthquakes, landslides and avalanches—but even more have perished and continue to do so. The threats to human life today lurk in sudden and intense events such as earthquakes and landslides, and also in more insidious and slow-setting events such as droughts, ozone layer depletion and global warming.

Long after the devastation of the Mozambican floods, Sofia Pedro's story lingers in the mind—a constant reference point not only of the fury of a river in flood, but also of the increased frequency and intensity with which the environment can unleash such terror.

## UNDERSTANDING HUMAN VULNERABILITY

The environment is always in a state of flux and has, therefore, always impacted on people and the way they live. The history of humankind abounds with examples of environmental change which have affected civilizations, or which have provided lasting lessons as to how people have been impacted by such change.

In an early example, the pioneering organized food-producing systems in the Nile Valley, under the civilization which ruled ancient Egypt for two and a half millennia from about 5 000 years ago, collapsed after its population peaked. A decline in food production was experienced due to environmental change as a result of massive flooding of the River Nile, which was catastrophic downstream. The river was also transformed over 500 years as heavy rains in the

upper catchment area produced more vegetation, reducing erosion and sediment carried downstream. This led to the reduction of the floodplain, and quantities of plant foods declined. 'The levels to which the human population had soared could not be sustained, and the pressure on resources mounted inexorably. Competition for food intensified, doubtless provoking conflict of which the massacre at Jebel Sahaba is probably an extreme example' (Reader 1997, 1998).

A more recent example involves climate variability, an issue highlighted by the WCED, which reported that: 'human overuse of land and prolonged drought threaten to turn the grasslands of Africa's Sahel region into desert. No other region more tragically suffers the vicious cycle of poverty leading to environmental degradation, which leads in turn to even greater poverty' (WCED 1987).

Africa's biophysical and socio-economic characteristics, and the complexity of its cultural diversity, are some of the factors or driving forces which contribute to environmental change which, in turn, impacts on human vulnerability and security. The Africa region is characterized by diverse patterns of elevation, geology, climatic variability and vegetation types. Millions of people in most parts of Africa are directly dependent on natural resources of the physical environment. They are, therefore, more vulnerable to environmental change than people in other regions of the world. It is important to note that people in all regions in the world are vulnerable in one way or another to environmental change, but that their coping capacity is different. For example, in 1999, between two and three times as many disaster events were reported in the United States than in India or Bangladesh. However, there were 14 times more deaths in India and 34 times more deaths in Bangladesh than in the United States in that year. Equally surprising is the fact that lightning causes more deaths in an average year in the United States than do floods, forest fires or tornadoes.

Environmentally unsustainable and inappropriate practices, such as unsuitable agricultural methods, deforestation and water pollution, are the major human-induced causes of vulnerability to environmental change. These are exacerbated by the impacts of climatic variations and interacting with the unique biophysical dynamics, thereby reducing coping capacities for most of the people who are already living in environmentally fragile areas.

For the purpose of this chapter, human vulnerability/security (see Box 3.3) is considered as a

Flooded village in the Tana River valley, Kenya, Nairobi

*Glynn Griffiths/Christian Aid/Still Pictures*

continuous variable, whereby vulnerability is the negative part of the continuum and security is the positive part. The two major constituent themes of human vulnerability are exposure to environmental hazards (or contingencies, shocks and stresses) and the coping capability of people which assures them of security.

People who have more capability to cope with extreme events or stresses are at lesser risk and are, therefore, more secure. The stresses to which an individual or household, or a broader social or geographic sub-region or region, are subjected are reflected in the state of helplessness or the lack of means to cope with risks, shocks, stresses or demands (Edralin undated). For example, many African countries in arid and semi-arid areas depend on food aid during some parts of the year. In 2000, for example, 8 million people in Ethiopia faced severe food shortages and had to depend on food aid (ELCA 2000). This was mainly due to adverse weather conditions and the impact upon food production.

Human vulnerability/security is a continuum which is characterized by situations which range from the undesirable state of vulnerability and its characteristics to the desirable state of security and its characteristics, as depicted in the Figure 3.1.

The human vulnerability/security continuum shows how vulnerability and security are defined in terms of coping capacity. Coping capacity increases as you move

## Box 3.3 Concept of human security

In 1994, the UN Human Development Report introduced the concept of human security, predicating it on the dual notion of, on the one hand, safety from chronic threats of hunger, disease and repression and, on the other hand, protection from sudden and hurtful disruptions in daily life. Environmental insecurity became shorthand for the dimension of human insecurity induced by the combined effects of natural disasters and mismanaged environmental endowment.

*Source: Geisler and de Sousa*

from the state of vulnerability towards security and vice versa, along the continuum. People, as individuals or as a community, will be at different stages of the vulnerability/security continuum depending on the socio-economic situation of each individual or group.

Individuals or groups within the vulnerability/ security continuum can be classified, in very simple terms, as falling under one of four categories along the gradations of the continuum:

- high risk and low coping capacity;
- high risk and high coping capacity;
- low risk and low coping capacity; or
- low risk and high coping capacity.

Most African countries fall under the category of high risk and low coping capacity. This is because most countries in Africa over the past 30 years have been at high risk of, for example, floods, earthquakes, lava flows, fires, droughts, civil strife, and armed conflicts and wars, which have increased poverty, exacerbated serious health problems and resulted in hunger. These disasters have displaced populations across national borders and internally, contributing to further environmental degradation, and leading to more vulnerability and insecurity. The impacts have mostly affected the poor, who have low coping capacities.

The high risk and high coping capacity scenario is very rare in Africa. The United States falls under this scenario. Some areas of the United States are at high risk from earthquakes, for example, but impacts, particularly in terms of human casualties, are low. Only pockets of Africa fall under the low risk and low coping capacity. Where natural causes of risks are absent, human-induced changes will be present to make people vulnerable. In any circumstances in Africa for now, the scenario of low coping capacities exists. The low risk and high coping capacity scenario is the ideal.

Vulnerability is also a reflection of human capacity to cope with risks or shocks. Those who are least vulnerable cope the best and enjoy security, while the opposite applies to those households, communities or broader populations who are most vulnerable and who stand to lose the most from the effects of environmental change and other risks, shocks or stresses. Coping strategies have many dimensions, from the traditional to the scientific. Traditional communities in Africa have, for millennia, adapted to environmental change in different ways, including shifts in livelihood activities

**Figure 3.1 Human vulnerability/security continuum**

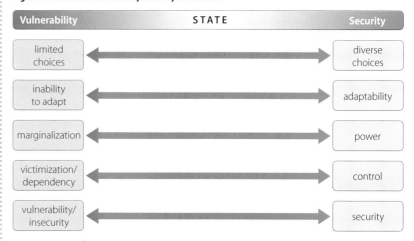

according to seasons and changes in the environment. They have also managed resources in a sustainable manner, adopting various management regimes to avoid overexploitation and to enhance their own food security (see Box 3.4).

Human vulnerability/security is a complex phenomenon with many interacting dimensions with respect to environmental change, and the resulting human responses and ability to cope with the impacts of such change. For example, desertification and drought are directly linked to poverty, food and water shortages, conflict and mass migration. They increase the risk of fire, decrease the availability of fuel and limit access to

---

**Box 3.4  Cultural value of the environment**

The fisherman who throws some of his catch into the sea after a fishing expedition in Ghana is expressing the responsibility which he has, as a member of the community, in ensuring that the fish population in the sea is not depleted. He, therefore, throws some of the live fish back into the sea so that they may continue to breed. And so each time he goes fishing, there will be fish in the sea.

At the same time, the fisherman is expressing gratitude to *Bosompo*, the divinity of the sea, for giving him some of his fish. If the fisherman does not give back some of his catch to *Bosompo*, he will feel that he has been negligent of an important cultural value—gratitude. The fisherman's action is based on the proverb: '*Bosompo ankame wo nam a, wo nso wonkame no abia*'—'If the divinity of the sea does not begrudge you of his fish, you do not begrudge him of your catch.'

And so, while the fisherman is expressing gratitude in conformity with a cultural value dating from antiquity, he is also expressing a concern for the environment by ensuring that there continue to be fish in the ocean and by acknowledging that human beings are responsible for their environment.

Source: Opoku 1999

*Interventions addressing human vulnerability to environmental change must be translated into integrated responses which reflect the inter-sectoral nature and processes of the causes and states of vulnerability.*

health care. Health effects can include malnutrition, the failure of babies to develop properly, iron and Vitamin A deficiency, infections, blindness and anaemia (Diallo 2000). Women and children are particularly vulnerable. In Africa, 49 per cent of the deaths among children less than 5 years of age are associated with malnutrition. A WHO estimate put the number of such deaths in Africa at 3.8 million in 1999 (WHO 2000a). As water sources dry up, people are forced to use heavily polluted water, leading to severe epidemics. In particular, desertification and droughts can increase water-related diseases such as cholera, typhoid, hepatitis A and diarrhoeal diseases (Menne 2000).

The dimensions of human vulnerability analysed in this chapter also include social and economic aspects, that is, poverty, food security, health, civil strife/conflicts, economic dimensions and governance. These complex, interacting dimensions of vulnerability to environmental change can act either as the constituent elements of vulnerability or, depending on the coping capacities (see Box 3.5) and resilience of an affected population, can result from, or be exacerbated by, environmental change. Interventions addressing human vulnerability to environmental change must be translated into integrated responses which reflect the inter-sectoral nature and processes of the causes and states of vulnerability. Because human security depends on the effectiveness of sustainable environmental management and the reduction of human vulnerability to environmental change and

threats, responses aimed to address disasters should be quick, adequate and coordinated (UNDP 1994).

The poor are especially vulnerable to degradation of natural systems. Both the global and the local consequences of environmental damage directly affect poor people. Global concerns, such as changes in the Earth's atmosphere, are critical to the livelihoods of poor people, and their consequences last longer than first assumed. For example, a rate of climate change is likely to cause widespread economic, social and environmental degradation over the next century. Therefore, the poorest people in Africa and other developing regions are certain to suffer the most due to failing harvests, growing water shortages and rising sea levels.

## ENVIRONMENTAL CHANGE: IMPACTS ON PEOPLE

The environment is life, supporting people and other living things. Environment is widely recognized as a 'pillar' of sustainable development. It provides essential goods and services which contribute to meeting basic human needs, and is essential to human development and quality of life. It provides services to ecosystems, including water catchments which protect freshwater resources, wetlands, riverbank environments, biodiversity habitats and ecologically functioning landscapes. The environment is also a sink of the wastes generated from different human activities.

In Africa, there is a high dependency on agro-sylvo-ecological systems, which are very sensitive to the impacts of the state of the environment and environmental change. The root causes of environmental change have both natural and human-made factors, and include interactions between them.

The environmental changes which have occurred in Africa since 1972 have been highlighted in Chapter 2, which provides a comprehensive overview of the key issues facing the region today. The changes highlighted in Chapter 2 do not only have regional dimensions, but also have sub-regional and national implications. Global processes also influence environmental change in Africa, for example, greenhouse gas emissions and their impact on climate change. The vulnerability of people in the region to environmental change, therefore,

---

### Box 3.5  Coping capacities and sustainability

Coping capacities are critical within the concept of sustainability, which is defined as encompassing:
- the ability to cope with and recover from shocks and stresses;
- economic effectiveness, or the use of minimal inputs to generate a given amount of outputs;
- ecological integrity, ensuring that livelihood activities do not irreversibly degrade natural resources within a given ecosystem; and
- social equity, which suggests that promotion of livelihood opportunities for one group should not foreclose opportunities of other groups, either now or in the future.

*Source: UNDP 1999b*

Refugees from a degraded agricultural land living in a slum in Nairobi, Kenya

*Mark Edwards/Still Pictures*

manifests itself—to a lesser or greater extent—at these different levels, and is a major factor in terms of sustainable development in Africa.

One of the impacts of human vulnerability to environmental change is the forced movement of people, creating what has come to be known as environmental refugees. The notion of environmental refugees describes a new insight on an old phenomenon—large numbers of the world's least secure people seeking refuge from insecure biophysical environments (Geisler and de Sousa 2000). Although the phrase 'environmental refugee' is controversial among advocates of the classical definition of refugees (political and social), it has gained in popular usage. It has been estimated that, globally, there were 25 million environmental refugees in 1994, more than half of whom were in Africa (Myers 1994).

## IMPACTS OF THE STATE OF THE ENVIRONMENT

In sub-Saharan Africa, 61 per cent of the population lives in ecologically vulnerable areas characterized by a high degree of sensitivity and low degree of resilience (IDS 1991). This is not necessarily by choice, but by force of circumstance, because other options are either unavailable or have been exhausted.

Perhaps one of the major threats of the state of the environment is malaria—a major killer in the region. Between 300 million and 500 million cases of malaria—which involve mostly the poor—are recorded in Africa annually. They cause between 1.5 to 2.7 million deaths, of which more than 90 per cent are children under 5 years of age (World Bank 2000, Nchinda 1998). Malaria slows economic growth in Africa by up to 1.3 per cent each year and, according to statistical estimates, the gross domestic product (GDP) of sub-Saharan Africa would be up to 32 per cent greater if malaria had been eliminated 35 years ago (WHO 2000b).

It is estimated that, by 1999, malaria had cost Africa about US$100 000 million in lost economic opportunities—or nearly five times more than all development aid provided to the region in 1999 (IRIN 2001). According to a report by the World Health Organization (WHO), Harvard University, and the London School of Hygiene and Tropical Medicine, malaria slows economic growth in Africa by up to 1.3 per cent each year. This slowdown in economic growth due to malaria is over and above the more readily observed short-term costs of the disease. With a GDP of about US$300 000 million, the short-term benefits of malaria control in sub-Saharan Africa are estimated at between US$3 000 million and US$12 000 million per year (WHO 2000b). According to UNICEF, the average cost for each nation in Africa to implement malaria control programmes is estimated to be at least US$300 000 a year. This amounts to about 6 US cents (US$0.06) per person for a country of 5 million people.

Some of the causes of malaria are summarized in Box 3.6, which also indicates the areas on which health programmes could focus in order to fight the disease.

The health crisis situation in Africa has been summarized in Table 3.1, which indicates the high percentage of the population of the region who are undernourished, and who have HIV/AIDS, malaria and tuberculosis. During 1992, cholera affected almost every

## Box 3.6 Malaria: factors related to human vulnerability

A number of factors, many of them relating to various dimensions of human vulnerability, appear to be contributing to the resurgence of malaria:

- the rapid spread of resistance of malaria parasites to chloroquine and other quinolines;
- frequent armed conflicts and civil unrest in many countries, forcing large populations to settle under difficult conditions, sometimes in areas of high malaria transmission;
- migration of non-immune populations—for reasons of agriculture, commerce and trade—to areas where malaria transmission is high;
- changing climatic conditions, especially rising temperatures and rainfall patterns;
- water development projects, such as dams and irrigation schemes, which create new mosquito breeding sites;
- adverse socio-economic conditions, leading to a much-reduced health budget and gross inadequacy of funds for drugs;
- high birth rates, leading to a rapid increase in the susceptible population of those younger than 5 years of age; and
- changes in the behaviour of the vectors, particularly in biting habits, from indoor to outdoor biters.

(N chinda 1996)

country in the region of the South African Development Community (SADC), claiming hundreds of lives.

Climate change, and human activities which transform habitats and create conditions suitable for parasites and disease organisms to breed, have a significant impact on the distribution and prevalence of vector-borne diseases (VBDs) in Africa. Climate change affects vector survival primarily through minimum temperatures, impacting the latitude and elevation of distribution, as well as the length of season permissive to transmission of VBDs (IPCC 1998). Meteorological variables, subject to climate variability and global atmospheric change, can therefore create conditions conducive to the spread of disease or, in the case of flooding or drought, clusters of outbreaks.

## IMPACTS OF ENVIRONMENTAL CHANGE

Human-induced environmental change, brought about by rapid population growth and overexploitation of natural resources, is considered to be the overriding cause of natural resource degradation, deepening poverty and increasing food insecurity in sub-Saharan Africa (FAO 1998). Such situations have forced farmers and others dependent on natural resources and agro-ecological systems to move into low-potential ecosystems, where the resulting damage can worsen and become irreversible. The degradation of natural resources which are essential for future production extends across the agro-ecological system, from the depletion of soil nutrients to overgrazing, overfishing and deforestation (FAO 1998).

Environmentally unsustainable and inappropriate practices are the major human-induced causes of vulnerability. These are exacerbated by the impacts of climatic variations and by interacting with the unique biophysical dynamics, resulting in reduced coping capacities for most of the people already living in environmentally fragile areas.

## Table 3.1  Health crisis and challenges in Africa

| Sub-region (countries) | Undernourished people (% of total population) 1996–98 | People living with HIV/AIDS | | | Malaria cases (per 100 000 people) 1997 | Tuberculosis cases (per 100 000 people) 1998 |
|---|---|---|---|---|---|---|
| | | Adult (age 15–49) 1999 | Women (age 15–49) 1999 | Children (age 0–14) 1999 | | |
| Northern Africa (6) | 6.8* | 0.24* | – | – | 1 321* | 51.5 |
| Western Africa | 22.9 | 3.5 | 176 600 | 15 303 | 9 275.7 | 69.4 |
| Central Africa | 34.8 | 5.8 | 160 937 | 13 254 | 3 240 | 98 |
| Eastern Africa | 48 | 7.59 | 534 978 | 48 439 | 6 759 | 167 |
| Southern Africa | 35.7 | 19 | 525 818 | 34 409.1 | 16 838 | 301 |
| IOC | 23 | 0.12 | 5800 (Madagascar) | 450 (Madagascar) | | |

*Note: More than 50 per cent of undernourished and more than 95 per cent of all HIV/AIDS and malaria cases for North Africa are from the Sudan.

Source: IES-Preparation WSSD 2001

Some environmental policies also contribute to human vulnerability, for example, the creation of national parks and protected areas without the necessary Environmental Impact Assessment (EIA), which takes into account social, ecological and economic aspects. Such policies, which also include the construction of reservoirs, have been described as 'ecological expropriation'. The growth of protected areas in Africa has been used as an indicator in the generation of 'other environmental refugees'. This category of environmental refugees is different from those impacted by, for example, storms, floods, droughts, fire and El Niño effects. The number of protected areas in Africa grew from 443 (88 662 000 hectares (ha), or 3 per cent of the land area) in 1985 to 746 (154 043 000 ha, or 5.2 per cent) 12 years later. Long-lasting conservation has been devastating for hundreds of thousands of Africans (Geisler and de Sousa 2000). This is mainly as a result of displacement and exclusion. However, there have been positive biodiversity conservation results.

## Impacts of land degradation

The conversion of natural habitats, such as forests or wetlands, for agriculture and cultivation of marginal areas has not only contributed to land degradation, but has also impacted on people's livelihood options. Desertification processes reportedly affect 46 per cent of Africa and, of that, 55 per cent is under high or very high risk. A total of about 485 million people are impacted (Reich and others, 2001). The projected effects of climate change in Africa include expanding arid areas and increasing variability in rainfall. This threatens to add to the current rate of desertification, a crop yield decline of 10–20 per cent in parts of the region, and food insecurity.

With wind and water erosion extensive in many parts of Africa (25 per cent of the land is prone to water erosion and about 22 per cent to wind erosion (Reich and others, 2001)), millions of peasant farmers are constantly at risk of food insecurity. Soil erosion not only reduces productivity of the land, but also requires farmers to utilize more and more fertilizers and other chemicals. While comprehensive data for soil loss rates in Africa are unavailable, estimates from the past three decades range between 900 and 57 000 tonnes/km$^2$/annum (t/km$^2$/a) (Rattan 1988). For example, more than 800 million ha of soil—representing 60 per cent of the total land areas of the semi-arid Sahel region—are affected or threatened by human-induced degradation. Some 224 million ha in the Sahel, in which the Lake Chad basin is located, are already severely degraded. Major causes of soil degradation include cropping activities such as reduced fallow periods, inadequate replenishment of soil nutrients, poorly managed irrigation and overgrazing. During the 1980s, Ethiopia's highlands were reported to have lost 3 million tonnes of topsoil annually (MacKenzie 1987). The cost of erosion in Zimbabwe was estimated to be US$20–50/ha/a on arable land and US$10–80/ha/a on grazing lands. In South Africa, estimates in 1992–93 reached US$237 million, some 15 per cent of gross national agricultural product (Mackenzie 1994).

The dependence on rain-fed agriculture increases the risk of food and economic insecurity, especially in areas of high climate variability. Restricted access to foreign markets, heavy agricultural subsidies in OECD countries and limited processing before export add to Africa's vulnerability to international price fluctuations and, therefore, to its failure to realize the full potential of its land resources. OECD countries spend more than US$300 000 million annually—roughly equivalent to the entire GDP of sub-Saharan Africa—on agricultural subsidies (Wolfensohn 2001).

●

*OECD countries spend more than US$300 000 million annually—roughly equivalent to the entire GDP of sub-Saharan Africa—on agricultural subsidies (Wolfensohn 2001).*

●

Wind and water erosion is extensive in many parts of Africa, placing millions of peasant farmers at risk of food insecurity.

*UNEP*

Conflicts over land in Africa have occurred for centuries, but have become more frequent in recent years, most notably since independence from European colonialism. These are complex issues, with conflicts over land resources between races, the state and other stakeholders, and conflicts within and between families and communities. Colonial rule imposed artificial boundaries on African populations, with little or no regard to the community fabric or wildlife habitats, and with damaging effects upon issues such as migration routes of animals. The designation of conservation areas also forced communities off their ancestral land, without compensation. During the past decade, there have been a number of land grabs and retrospective claims against governments, largely due to landlessness and displacement.

## Impacts of freshwater mismanagement and pollution

In terms of freshwater, at least 13 countries suffered water stress or scarcity (less than 1 700 m$^3$/capita/a and less than 1 000 m$^3$/capita/a respectively) in 1990. The number is projected to double by 2025 (PAI 1995). While 62 per cent of people in the region had access to an improved water supply in 2000, rural Africans spend much time searching for water as water sources become more distant and harder to locate. A total of 28 per cent of the global population without access to improved water supplies live in Africa (WHO and UNICEF 2000).

Freshwater and groundwater pollution is a growing concern in many areas, further limiting access to safe water. Poor water quality leads not only to water-related diseases, but also reduces agricultural production, meaning that more foodstuffs and agricultural products must be imported. Poor water quality also limits economic development options, such as water-intensive industries and tourism, a situation that is potentially disastrous to developing countries in Africa.

Droughts are probably the most serious factor in terms of human vulnerability to environmental change. Eastern Africa, for example, has suffered at least one drought per decade for the past 30 years (CRED-OFDA 2000). In the 1970s in Ethiopia, drought killed 400 000 people, and about 1.2 million others were displaced. About a decade later, in 1984–85, a total of 7.8 million Ethiopians were affected, causing 1 million

Droughts are probably the most serious factor in terms of human vulnerability to environmental change.

*UNEP*

deaths. For a region which is dependent on rain-fed agriculture, drought is an ill omen. In times of drought, for example, a drop in water levels in dams and rivers may affect the concentration of sewage and other effluent in rivers, resulting in outbreaks of diseases such as diarrhoea, dysentry and cholera. Reduced water flows during droughts also decrease the capacity of rivers, streams and swamps to dilute agro-chemicals and fertilizers in fields, adversely affecting soil ecosystems and potential agricultural production. These drought-related problems are likely to increase under projected climate change, although vulnerabilities and control measures will affect the impact (IPCC 1998).

The poor are usually the most affected by flood or drought-induced crop failure. Malnutrition and famine have resulted from both droughts and floods, and food imports and dependency on food aid associated with this have contributed to limiting the economic growth of affected countries. For example, in the Horn of Africa, where 76 per cent of the population is classed as agricultural, there is an important connection between food security and poverty (FAO 2001) because, for the majority of the poor, agriculture is the main source of livelihood. Several processes of environmental vulnerability are at play in the Horn of Africa: drought, which causes widespread periodic famine in the region; localized floods; and the threat of locust swarms (FAO

2001). Power generation, which drives economic activity, is also affected by episodes of drought, leading to load shedding or power outages.

Floods also contribute to the vulnerability of people in Africa. In Southern Africa, for example, the devastating floods of 1999–2000 affected more than 150 000 families (Mpofu 2000). Mozambique alone lost US$273 million in physical damage, US$247 million in lost production, US$48 million in lost exports and US$31 million in increased imports as a result of the flooding (Mozambique National News Agency 2000). Degradation of wetlands, such as the Kafue wetlands in Zambia, damming of rivers, deforestation and overgrazing lowered the environment's ability to absorb excess water, and magnified the impact of the floods (Chenje 2000, UNDHA 1994). Poor water supply and sanitation led to high rates of water-related diseases such as ascaris, cholera, diarrhoea, dracunculosis, dysentery, eye infections, hookworm, scabies, schistosomiasis and trachoma. About 3 million people in Africa die annually as a result of water-related diseases (Lake and Souré 1997). In 1998, 72 per cent of all reported cholera cases in the world were in Africa.

### Impacts of habitat and biodiversity loss

As has already been highlighted, loss and degradation of habitats have been widespread in Africa. Some 0.78 per cent of forest area was lost in 1990–2000, representing a total loss of some 5.2 million ha. Increasing demand for fuelwood, which accounted for more than 90 per cent of roundwood produced in 1990, has accelerated the destruction of forests, which now cover about one-third of the Africa region. Every year, an area of about 4 million ha (roughly the size of Switzerland) is deforested. The situation has been particularly alarming in the Lake Chad basin, as a result of the demand for fuelwood. In addition to losing populations of various species, the cost of deforestation to people in Africa may be massive, particularly considering the fact that many populations depend on wildlife for food and other goods. Deforestation is a threat to people's livelihood options, contributing to habitat loss, soil and wind erosion, and general land degradation. For a region which virtually depends on fuelwood for its energy needs, deforestation also has social and economic costs. In many sub-regions, fuelwood makes up 61–86 per cent of primary energy consumption; some 74–97 per cent of this is for household consumption. Charcoal production, which is widespread in the region and which provides many jobs, is a US$1 000 million activity (Amous undated). Unsustainable forest harvests threaten this industry, exacerbating the problem of unemployment.

Habitat and biodiversity loss can also affect tourism in the region, contributing to poor economic performance. This has serious impacts on revenue and jobs, particularly in countries in Eastern and Southern Africa, which are heavily dependent on wildlife tourism.

Wild food plays an important role in food security for rural people and is also, increasingly, a commercial commodity which is traded nationally and regionally. In many urban areas, meat from wild animals commands a significantly higher price than meat from domestic animals. Very large quantities of meat are involved. It has been estimated that, in the moist forests of Central Africa alone, as much as 1 million tonnes of wildlife (primarily forest antelope, wild pigs and primates) may be killed for food each year.

Rural and urban populations across Africa depend largely on medicinal plants, often collected from the wild, for their health needs, due to preference or a lack of affordable alternatives. Some species, such as the montane tree *Prunus africana* and the southern African devil's claw *Harpagophytum* spp., are also exported in significant quantities. Overharvesting, agricultural encroachment and unregulated burning are believed to be contributing to the decline of many species in the wild. Unsustainable activities may foreclose such livelihood options, making more and more people vulnerable.

### DISASTERS

Human mismanagement of environmental resources and processes significantly exacerbates the impacts resulting from disasters, and their effects on natural resources. Box 3.7 highlights some recent environmental disasters in Africa.

East Africa is exposed to seismic hazards due to the presence of the Rift Valley system. Earthquakes have been identified as a major threat in the area, which covers about 5.5 million km$^2$ and holds more than 120 million people (Midzi and others, 1999). The vulnerability of East African populations to seismic events has been underscored by a recent study which

*Every year, an area of about 4 million ha (roughly the size of Switzerland) is deforested.*

*It has been estimated that, in the moist forests of Central Africa alone, as much as 1 million tonnes of wildlife (primarily forest antelope, wild pigs and primates) may be killed for food each year.*

## Box 3.7 Vulnerability to natural disasters

Extreme droughts have resulted in exceptional food emergencies in Burkina Faso, Chad, Ethiopia, Kenya, Niger, Rwanda, Somalia, Sudan, Tanzania and Uganda (FAO 2001).

People in Africa are also vulnerable to floods, such as those which submerged more than 79 000 ha of planted land, severely affecting the livelihoods of nearly 120 000 farm families in central Mozambique in February and March 2001 (FAO/TCOR 2001). Farmers lost their crops, livestock, food and seed reserves, and hand tools. Simultaneously, the cyclone Dera hit the provinces of Nampula and Tete in March 2001, devastating about 2 000 families which depend on fisheries. The floods and cyclone also caused severe damage to infrastructure. As a result of the damage and loss of livelihoods due to these extreme events, the Food and Agriculture Organization (FAO) launched an appeal for US$8.71 million in assistance for Mozambique.

In North Africa (Egypt and Algeria), 22 earthquakes killed 14 405 people, and affected another 106 150 people, between 1980 and 1998 (EM-DAT undated). In mid-January 2002, lava flowing from the erupting Mount Nyiragongo destroyed half of the city of Goma in the eastern Democratic Republic of Congo, together with crops and biodiversity. More than 400 000 people fled to take refuge in neighbouring villages and in Rwanda.

advises that the region's capacity in earthquake preparedness and hazard mitigation needs to be improved 'significantly' (Midzi and others 1999).

In Central Africa, Mount Cameroon has erupted twice in the past 40 years, pouring out tonnes of lava and destroying farms and biodiversity. The last eruption was in the year 2000, and earth tremors occur every three to four years on average. Explosive emissions of toxic gases from Lake Nyos and Lake Mounoun, both crater lakes in the mountainous west of Cameroon, killed thousands of people, livestock and wildlife in 1986. Box 3.8 gives examples of disasters which have struck Nigeria in the recent past.

Although natural disasters cannot be prevented, sustainable utilization and management of the environment can increase coping capacities at

## Box 3.8 Disasters in Nigeria

Four main categories of disaster have occurred in Nigeria over the past three years, and they have had significant environmental consequences.

*Industrial accidents*
- On 10 July 2000, a pipeline in Nigeria exploded, killing about 250 villagers. Fires burned out of control about 20 km from the town of Jesse.
- On 27 January 2002, at least 600 people were drowned and thousands were rendered homeless after multiple bomb explosions at a Nigerian military armoury, triggered by an accidental fire. Mass panic ensued.

*Civil strife and conflicts (some related to community land resource ownership)*
- On 4 June 1999, ethnic clashes flared up in Nigeria's southern oil industry hub of Warri. Dozens of people were reported dead in six days of fighting.
- On 19 July 1999, at least 60 people died in clashes between Hausa and Yoruba tribes near Lagos.
- On 26 July 1999, troops were sent to Kano after at least 60 people were killed in renewed ethnic clashes in northern Nigeria.

*Property rights and unequal sharing of benefits from natural resources*
- On 1 January 1999, at least 19 people died in clashes in Nigeria's oil region after an ultimatum to oil firms to leave ethnic Ijaw areas.
- On 3 June 1999, local youths set fires at four separate points on the Warri-Kaduna products pipeline near the village of Adeje, after police arrested suspected product thieves. The number of dead was undetermined.

*Rural and urban poverty which lead people to take deadly risks to get money*
- On 18 October 1998, fire engulfed more than 2 000 villagers who were scrambling for petrol near a ruptured pipeline in Jesse, outside Warri. Nearly 1 000 people were killed.
- 14 March 1999, at least 50 villagers who were scooping up gasoline from a broken products pipeline at Umuichieichi-Umungbede village in Abia state were burned to death after an explosion.

community level. An effective management regime would include: economic policies which encourage smallholder agricultural production; the enforcement of relevant laws and regulations; incentive measures which encourage agricultural and biodiversity conservation at community level; and integrated and coordinated planning. The integration and implementation of sustainability strategies at community and national levels would enhance coping capacities by reducing dependency on aid. Sustainability strategies would develop and support an enabling environment for households and communities, and would improve forecasting abilities which could forestall the potential of environmental events to translate into environmental disasters.

## SOCIAL DIMENSIONS TO HUMAN VULNERABILITY

Changes in the structure and distribution of populations can be both a cause and an effect of vulnerability to environmental change. As land degradation spreads and food insecurity grows, as a result of climatic variability and global warming, more and more people are forced to migrate to urban areas, for example, in search of work and other opportunities. People with the most diversified livelihoods, education, power, adaptability and security, among other factors, are most likely to cope with adverse

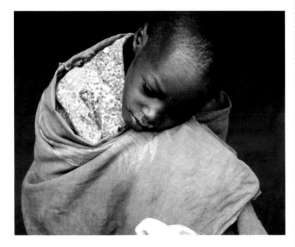

Up to 65 per cent of urban dwellers in some African regions live in poverty, with little or no access to social and urban services which constitute decent living conditions.

environmental change, because they are more secure. The reverse is true for those lacking such endowments.

### POVERTY

Environmental changes almost always have a greater impact on those who live in poverty. People in Africa, the majority of whom are poor, depend directly on what they can grow, catch or gather. Poverty can be exacerbated by environmental change in complex ways, particularly in natural resource-based African economies. Land degradation, deforestation, lack of access to safe water and loss of biodiversity, compounded by climatic variability, all contribute to a general reduction in environmental quality and to an increase in vulnerability for populations which are dependent on natural resources (World Bank 1997). Degradation of resources reduces the productivity of the poor, who mostly rely on such resources. It makes poor people even more susceptible to extreme events, such as drought or floods, economic fluctuations and civil strife (World Bank 1997). Poverty severely impedes recovery from these events, and weakens social and ecological resistance, especially as the poor are unable or do not have the opportunity to invest in natural resource management.

Table 3.2 averages the poverty (human development) indicators for the various sub-regions of Africa, using three levels of evaluation: high human development; medium human development; and low human development. The table also indicates: the number of countries in each sub-region at each level of evaluation; the number of people, on average, who live on US$1 a day; and the number of people living below the national poverty line.

Macro-economic crises affect the poor in several ways, all of which contribute to their vulnerability to environmental change. The situation is such that: living standards are reduced; the ability of the poor to grow out of poverty is constrained; malnutrition rates among children increase, as do school dropout rates; and household effects are sold at depressed prices (World Bank 2000). The situation only helps to perpetuate chronic poverty and to reduce overall economic growth. In the process of livelihood reduction, people become more vulnerable to environmental change because of diminished coping capacity, and the potential to adopt environmentally hazardous coping

**Table 3.2 Poverty indicators for African countries**

| Sub-regions (number of countries in each) | number of countries in each category | % population below income poverty line US$1 a day (1993 PPP USD) 1983–99 ave. | National poverty line 1984–99 ave. |
|---|---|---|---|
| A. HIGH HUMAN DEVELOPMENT: No African country in this category! | | | |
| B. MEDIUM HUMAN DEVELOPMENT: | | | |
| Northern Africa (7) | 5 | 23 | 19.8 |
| Western Africa (14) | 2 | 39 | 31 |
| Central Africa (6) | 4 | xx | 40 |
| Eastern Africa (7) | 1 | 27 | 42 |
| Southern Africa (11) | 6 | 32.2 | 28.5 |
| IOC (4*—Reunion excl.) | 3 | – | 10.6 |
| C. LOW HUMAN DEVELOPMENT: | | | |
| Northern Africa (7) | 2 | 29 | 75 |
| Western Africa (14) | 12 | 47.1 | 49 |
| Central Africa (6) | 2 | 67 | 64 |
| Eastern Africa (7) | 8 | 29 | 45.5 |
| Southern Africa (11) | 4 | 51 | 70 |
| IOC (4*—Reunion excl.) | 1 | 63 | 70 |

Source: Summarized from JES-Preparation WSSD 2001

strategies—such as settling in floodplains or on unstable slopes—is enhanced.

Much of the rural poverty problem is characterized by the existence of:

- increasing landlessness;
- a large number of farmers with very small holdings;
- a lack of resources to invest in agricultural improvement;
- the 'urban pull' effect;
- low levels of rural education;
- migration to urban areas of those who do achieve an acceptable level of education;
- overvalued exchange rates, which discourage agricultural exports; and
- poverty amongst women.

The rural poor are particularly vulnerable to stresses, such as extremes of temperature and rainfall (climate variation which results in drought and floods), general financial shortage, and persistent illness and bereavement. They are even more vulnerable to shocks, including famine, floods, epidemics and major changes in markets.

About 40 per cent of the region's poor live in urban areas and, depending on the countries and urban settlements, between 15 and 65 per cent of African urban dwellers live in poverty, with little or absolutely no access to social and urban services which constitute decent living conditions (Soumaré and Gérard 2000). Rapid rates of urbanization in Africa can be attributed to the effects of colonialism, rural-to-urban migration, weak rural economies and a poor industrial base which cannot absorb unskilled labour from rural areas. The result of rapid urbanization is the radical transformation of the structure of cities, accompanied by complex social, economic and environmental changes (Rabinovitch 1997). There is strong evidence to suggest

A child carries water across an open drain in a village in Ghana; water pollution and poor levels of sanitation frequently lead to a predominance of water-borne diseases in the Africa region.

*Ron Gilling / Still Pictures*

that urban environmental hazards—such as biological pathogens and various pollutants—are a major cause of or contributor to urban poverty and, for much of the urban poor population, environmental hazards are the main causes of ill-health, injury and premature death (Satterthwaite 1999).

Poor people, especially in urban areas, often settle in fragile zones with high population densities. This increases the overall impact of exposure to risk under conditions of heightened vulnerability, as the case study in the Box 3.9 illustrates.

## HEALTH

Environmental damage—whether it is water or air pollution, or waste and sanitation—has serious consequences for human health. Generally, most countries in Africa face high environmental threats to health (WRI and others, 1998), a situation which poses a profound challenge to the region.

Pollution of water and air, and their impact on human health, is of immediate concern. Water pollution and contamination impact on people in the region, resulting in the predominance of water-borne diseases. Air pollution—from industrial and car exhaust emissions, and the burning of traditional fuels in homes—kills a large number of people each year. People die from respiratory damage, heart and lung diseases, and cancer. Urban air pollution causes close to 1 million premature deaths worldwide every year, primarily due to respiratory diseases (World Bank 2000) affecting mostly the poor. About 4 million people die annually due to overcrowding, and from indoor pollution caused by burning biomass fuels for cooking and heating (WRI and others, 1998). As many as 25 million poor agricultural workers in the developing world (11 million in Africa alone) are poisoned by pesticides every year, and hundreds of thousands die.

In Africa, human vulnerability is exacerbated by poor health caused by unmitigated or heightened exposure to disease, malnourishment and undernourishment, and weak public health institutions

---

### Box 3.9  Vulnerability in Manchieyt Nasser, Cairo

The Manchieyt Nasser township, located at the heart of Cairo, developed from a long-standing limestone quarry and dump site—activities which moulded the rugged contours of the site to what is it today. The township now represents the biggest informal squatter settlement in Egypt, with between 350 000 and 500 000 people living on a mountainous site of 7.27 km$^2$.

Most of these people are exposed, on a daily basis, to a plethora of natural and human-made environmental hazards. At the same time, they face the continuous possibility of legal charges for illegal occupation of state-owned land. Most of them lack the choice to reduce their economic and health vulnerability as a result of the environmental hazards. More shanty-town dwellers were moved into the area in 1960 and, in 1972, garbage deposition was relocated to the area.

The infrastructure and social services in the township are scanty and inadequate—and some have become obsolete. The construction of dwellings on environmentally fragile and risky spots has been commonplace. Pollution in the area results from informal manufacturing activities, leaking raw sewage, the open burning of garbage and mounting heaps of solid wastes. The area has also become a refuge for outlaws and illegal practices, due to its virtual inaccessibility to security forces. These changes have only worsened an already bad environmental situation, rendering the inhabitants more and more vulnerable. Thus, the area was the most badly damaged by an earthquake which struck Cairo in October 1992. Today, a considerable number of inhabitants still live in very dangerous spots in the area.

*Osama Salem unpublished case study 2001*

and interventions. Linked to this is poverty and the coping capacities of the population at risk to effectively reduce their vulnerability to infectious diseases, poor or irregular nutrition, and the myriad of health conditions that are associated with poverty in urban and rural areas. The survival indices which are a reflection of the general health situation in Africa are summarized in Table 3.3.

Poor health, lack of or inadequate access to health services, low or skewed investment in health services (concentration in few urban centres) and dysfunctional health policies all contribute to lower life expectancy and high mortality rates in African countries, as indicated in the table. Health-related interventions may also contribute to vulnerability by, for example, contributing to the evolution of drug-resistant organisms, or by exposing food webs and people, through the process of bioaccumulation, to toxins such as DDT.

The effects of infectious diseases—such as HIV/AIDS, tuberculosis and malaria—are felt throughout society and, if unchecked, they damage the social fabric of the community, diminish agricultural and industrial production, undermine political, social and economic stability, and contribute to regional and global insecurity (WHO 2000a). The linkages between environmental change, poverty, reduced quality of health and vulnerability are complex, and their individual impacts and causal effects are not easy to isolate. However, the lack of scientifically proven evidence of such causal changes in Africa does not mean that these changes do not exist; rather, it may reflect the lack of available epidemiological data as a result of poor or absent surveillance and health

Infectious diseases such as HIV/AIDS, tuberculosis and malaria can thrive in poor communities, and the limited access to health-care services means mortality rates are high in many regions.

*UNEP*

information systems (IPCC 1998). There is, consequently, a pressing need for scientific investigation on the assumptions about environmental change and its impacts on human health and vulnerability.

The HIV/AIDS epidemic has spread with devastating speed. It is among the leading causes of death in sub-Saharan Africa (World Bank 2001), where 2.4 million adults and children are estimated to have died due to HIV/AIDS in 2000 alone (UNAIDS/WHO 2000). The HIV/AIDS epidemic is not only the most significant public health problem affecting large parts of sub-Saharan Africa; it is also an unprecedented threat to the region's development (World Bank 2000). More than 95 per cent of the 36 million people in the world living with

| Table 3.3 Health progress and setbacks in African countries | | | | | | |
|---|---|---|---|---|---|---|
| | Life expectancy at birth (years) | | Infant mortality (per 1 000 live births) | | Under-five mortality rate (per 1 000 live births) | |
| Sub-region | 1970–75 | 1995–2000 | 1970 | 1999 | 1970 | 1999 |
| Northern Africa (6) | 52 | 66.1 | 123.8 | 38.7 | 190.7 | 51.2 |
| Western Africa (16-1) | 42.7 | 49.7 | 161.3 | 106.3 | 273.3 | 174.2 |
| Central Africa (8-1) | 43.7 | 49.1 | 139.6 | 103.6 | 233.3 | 163.2 |
| Eastern Africa (8-1) | 44.7 | 50.1 | 133.6 | 94.7 | 212 | 147.9 |
| Southern Africa (11) | 47.7 | 46.2 | 127.6 | 91.3 | 205.1 | 142.2 |
| IOC (4.1) | – | – | – | – | – | – |

*Source: IES-Preparation WSSD 2001*

HIV/AIDS are in developing countries, and 25.3 million of them are in sub-Saharan Africa (UNAIDS 2001, UNAIDS/WHO 2000). In Africa, HIV/AIDS is largely a rural/urban poor issue, where a matrix of socio-economic, cultural and gender-related vulnerabilities indicate that the links between AIDS, food insecurity and poverty are strong and deadly (FAO undated).

HIV/AIDS is a threat to sustainable agriculture and rural development through its systemic impact (FAO/UNAIDS 1999). At the household level, HIV/AIDS can result in labour shortages and declining productivity, reduced income, increased expenditure on medical treatments and an increase in the dependency ratio due to the rise in the number of dependents relying on a smaller number of productive family members. Smallholder agriculture is a vital sector for rural households and national economies in many African countries. HIV/AIDS is affecting agricultural production through the decimation of household labour, the disruption of traditional social mechanisms, and the forced disposal of productive assets to meet the costs of medical care and funerals. The disease also results in the loss of traditional farming methods, inter-generational knowledge, and specialized skills, practices and customs.

## MARGINALIZATION OF TRADITIONAL SYSTEMS

One of the most destructively persistent historical legacies of Africa's past has been the subversion and destruction of indigenous coping strategies due to foreign military, political, administrative and economic interference. Colonial dispossession of the richest traditional pastoral and agricultural lands from Africans, and the commodotization of agriculture production for export purposes, marked two of the most far-reaching colonial interventions which have significantly contributed to Africa's current state of vulnerability. The loss of pasturage and farmland to colonial settlement/private ownership, and insecure land tenure for the native population, have further undermined traditional coping strategies. In Eastern and Southern Africa, in particular, indigenous Africans were confined to marginal, and increasingly degraded and unproductive, lands due to the impacts of settler agriculture, colonialism and subsequent changes in traditional land tenure which exacerbated negative environmental change after independence. In Ethiopia,

for example, foreign resource conservation measures introduced by the government between 1971 and 1985 not only eroded indigenous processes of resource conservation, but also led to soil erosion and impact on crop production (Singh 2000).

Environmental stress, such as deforestation, can cause community-level conflict.

*UNEP*

## CIVIL STRIFE AND ARMED CONFLICT

A total of 26 armed conflicts erupted in Africa between 1963 and 1998, affecting 474 million people in Africa, or 61 per cent of the population. Some 79 per cent of people were affected in Eastern Africa; 73 per cent in Central Africa; 64 per cent in Western Africa; 51 per cent in Northern Africa; and 29 per cent in Southern Africa (ECA 2001). Another impact of armed conflict is the creation of refugees. In 2001 in Central Africa and the Horn of Africa, for example, a total of about 9.6 million people were either refugees or internally displaced as a result of armed conflict (US Committee for Refugees 2001). Refugee settlements often result in environmental degradation which, in turn, increase human vulnerability, limiting livelihood options and exposing the refugees to health risks.

Environmental change due to environmental stress has an indirect impact on the outbreak of conflict. Environmental stress—including deforestation, land degradation and scarce supply of freshwater—alone, and in combination with high population density, increases the risk of low-level conflict.

Armed conflict over resources can also spill over national borders. In 1977–78, deforestation and soil

*HIV/AIDS leads to labour shortages, decreased productivity, reduced income and an increasing number of dependents. In turn, traditional farming methods are often lost together with, inter-generational knowledge, and specialized skills, practices and customs.*

degradation, in conjunction with rapid population growth, forced Somali pastoralists to migrate to Ethiopia, resulting in conflict between the two countries (Molvoer 1991). Overgrazing induced widespread deforestation and desertification in Somalia, prompting the large migrations of Somali pastoralists into Ethiopian territory. The migrations brought the Somali pastoralists face to face with local Ethiopians who were dependent on the same resources. The bitter competition between these groups fuelled cross-border tensions which eventually found an outlet in armed conflict between the two nations.

Conflicts in the region are partly attributed to disputes over environmental resources. For example, Liberia/Sierra Leone/Guinea conflicts are partly attributed to contest over the resources of the Manor River basin. The DRC/Rwanda conflicts, the Sudanese conflict, and tribal conflicts and wars in many African countries are attributed, in part, to contest over natural resources. Environmental problems which are exacerbated by civil strife, armed conflicts and wars are threatening the survival of large numbers of people in the Africa region, and these problems are becoming increasingly serious. The types of environmentally related conflicts include:

- Simple scarcity conflicts which may arise over three major types of resources: river water, fish and agriculturally productive land. These renewable resources are likely to spark conflict because they are rapidly becoming scarce in some regions, they are essential for human survival and they can be physically seized or controlled.

- Group identity conflicts, which are likely to arise from the large-scale movements of populations brought about by environmental change, for instance, the earlier Ethiopia/Somalia example.

The situation described above is vividly illustrated by a case study on natural resources scarcity and conflicts, summarized in the Box 3.10.

The analysis of the relationship between environmental change (especially that which leads to scarcity), violent conflict and security has highlighted both positive and negative social effects (Matthew 2000). The negative social effects of environmental change include:

- Decreased agricultural production and productivity, which may arise as a result of the effects of deforestation, as well as of the degrading and decreasing of available agricultural land.

- Depressing economic performance as a result of environmental degradation, leading to further impoverishment in the affected countries.

- Population displacement, compelling people to migrate in search of livelihood opportunities.

- Intensifying group identity tensions, forcing people onto marginal lands and promoting resource capture by social subgroups—all of which may generate diffuse and persistent misery, frustration and resentment.

- Rendering individuals and groups increasingly vulnerable to natural and human-made disasters.

- Disruption of legitimized and authoritative institutions and social relations.

---

**Box 3.10  Natural resources scarcity and conflicts**

The semi-arid land of northern Uganda is usually referred to as Karamoja. It is home to pastoralists called karimojong, made up of several tribes which depend on livestock for food, payment of bride price and other cash needs.

Karamoja is characterized by low/unreliable rainfall. Scarcity of water for human and animal needs, and inadequate pasture for grazing, results in overstocking of livestock in the area in relation to the carrying capacity of the limited pasture. The groundwater resources on which the population depend has been reducing because the water table in the area has been falling

since 1960, as a result of the effects of drought and other aspects of environmental degradation. Also, the rate of livestock loss is high, due to the effects of drought and disease. Furthermore, about 50 per cent of Karamoja is a protected biodiversity conservation area, where the government prohibits any human activities.

The explosive situation described above has led to internal armed conflicts and cattle raiding between the different tribes, and also to external armed conflicts with people from neighbouring countries with the same resource scarcity problems.

## Box 3.11 Conflicts and the environment

The civil war in the Democratic Republic of Congo has been devastating to the country's wildlife, killing thousands of elephants, gorillas (which are among the world's most endangered animals, with only a few hundred surviving in the wild today) and other endangered species. After three years of fighting in the Congo, the number of okapis, gorillas and elephants has dwindled to small populations. Participants in the many-sided conflict have plundered resources to fuel the fighting. Soldiers have slaughtered elephants for meat and ivory, and buffalo for meat.

In Garamba Park in northeastern Congo, an area controlled by Ugandan troops and Sudanese rebels, nearly 4 000 out of 12 000 elephants were killed between 1995 and 1999. In other parks and reserves, including Kahuzi-Biega park, the Okapi reserve and Virunga park, the situation is equally grave. In Kahuzi-Biega park, a zone controlled by the Rwandan and Rwandan-backed rebels, just two out of 350 elephant families remained in 2000—

the rest must have fled of their own accord or may have been killed, because two tonnes of elephant tusks were traced in the Bukavu area late in 2000. The war has made both humans and wildlife vulnerable.

**Park rangers with impounded ivory, Central African Republic**

*Photo: Mathieu Laboureur / Still Pictures*

*Source: United Nations 2001*

Armed conflicts, in addition to exacerbating environmental degradation and increasing human vulnerability, also cause a lot of damage to invaluable environmental resources, especially wildlife and biodiversity, as illustrated in the Box 3.11. This situation is the same with all the armed conflicts which have taken place or which continue to take place today in Africa.

Armed conflict not only contributes to the degradation of the environment, but also contributes to the breakdown of legal and institutional frameworks which are critical to environmental management. In Mozambique, the war which ended in 1992 resulted in the fragmentation and collapse of the management of protected areas (Chenje and Johnson 1994). It also foreclosed livelihood options for millions of people who were forcibly displaced to relatively safe areas but which, however, had more limited livelihood options (see Box 3.12).

One of the results is that some large communities are forced to survive on food handouts or are forced to overexploit their immediate environment in order to survive. This becomes a vicious circle, where the poor overexploit their resources, limiting the environment's ability to recover. As the state of the environment deteriorates, the people's livelihood options also become limited, worsening their poverty and

vulnerability. For example, the 1999 UN secretary-general report on the war in Angola, said that among the immediate consequences of the war were the higher level of malnutrition, especially among young children, and the dismal sanitation and health conditions which seriously increased the risk of epidemics (UN 1999). Box 3.12 also provides additional information on how armed conflicts impact people.

## Box 3.12 War causes health problems in the Democratic Republic of Congo

The number of people in critical need of food in the Democratic Republic of Congo remains at an estimated 16 million, or roughly 33 per cent of the country's population. The uprooting of rural populations and isolation from their traditional food sources, together with the declining economic situation, continue to be the underlying causes of this troubling situation, which is aggravated at Kinshasa, where about 70 per cent of the population of 7 million live on less than US$1 per day for food. Some 18 per cent of children in the inner city, and more than 30 per cent in the outskirts, suffer from chronic malnutrition. Less than 47 per cent of the population is estimated to have access to safe drinking water.

*Source: UN 2000*

## ECONOMIC DIMENSIONS
## TO HUMAN VULNERABILITY

The economy is both a pressure and a victim of environmental change. The overexploitation of resources for economic growth may cause environmental change, and such change may, in turn, negatively impact economic performance. The 1991–92 drought which hit most of Southern Africa forced the Zimbabwe stock market to decline by 62 per cent, causing the International Finance Corporation (IFC) to describe the country as the worst performer out of 54 world stock markets. The country's manufacturing sector declined by 9.3 per cent in 1992. Research shows that drought caused a 25 per cent reduction in the volume of manufacturing output, and a 6 per cent reduction in foreign currency receipts. In South Africa, a model developed by the Reserve Bank of South Africa indicated that the 1991–92 drought had a net negative effect of at least US$112.4 million on the current account of the balance of payments. It is estimated that about 49 000 agricultural and 20 000 jobs in non-agricultural sectors were lost as a result of the 1991–92 drought (Benson and Clay 1994).

There are both direct and indirect economic implications of human vulnerability to environmental change which involve costs. Direct costs are dramatically illustrated when losses resulting from the impacts of floods, earthquakes, wind storms or fires on the infrastructure and property of the affected communities are evaluated and calculated against the expenditure required to rebuild or repair lost capital assets, and to provide aid and basic services to affected people. The economic impacts of overharvesting natural resources, such as fish or timber, can result in vulnerability as these resources, on which people depend for their livelihoods, become scarce. Because the affected population has become vulnerable and has lost coping capacities, such costs are usually taken up by governments, relief agencies, donors and, in many cases, nearby sympathetic communities usually give assistance, especially in kind.

At the micro-economic level, the impacts of adverse environmental change on human vulnerability, as individuals and households become economically insecure, result from the following: reduced productivity and production; reduced income; reduced reserves; reduced purchasing power; increased demand for subsidies, aid and assistance; reduced capability to pay taxes; increased indebtedness; and poverty, food insecurity and health problems.

At the macro-economic level, the impacts of adverse environmental change follow on directly from the micro-economic impacts. They include: reduced taxes to the treasury; increased budget deficits; decreased social spending; increased foreign aid dependency; decreased debt repayment; decreased competitiveness; decreased foreign exchange; and overall poor economic performance.

Sub-Saharan Africa has low overall economic performance, including low industrial performance. The factors which contribute to the poor performance of Africa and which render most of the population vulnerable to environmental change have been summarized as follows:

- low levels of private investment due to macro-economic instability, inadequate legal systems and conflict;
- high tax and import duty rates which discourage foreign investment;
- bad governance and corruption;
- high levels of debt and dependence on foreign assistance;
- low rates of return on capital and labour;
- low overall productivity rates;
- over-valued exchange rates;
- poor infrastructure; and
- insufficient competition and monopolistic structures. (World Bank 2001)

It must be noted that the results of the above are not linear, because they involve complex economic and social processes.

The data in Table 3.4 illustrate the situation with regard to dependence on foreign assistance, foreign investment and debt servicing, which decrease export earnings.

Many African countries have persistently faced social and economic difficulties since independence. Although some countries in the region depend on mineral resources for their foreign exchange earnings, most of the countries rely principally on agriculture which has, and continues to be, the single largest employer of the population. Economic growth for most African countries has been sluggish or negative and, in most instances, has impacted heavily on the welfare of the people.

**Table 3.4  Aid, private capital and the debt crisis in Africa**

| Sub-region | Official development assistance | | | | Net foreign direct investment as % GDP | | Total debt service as % exports | |
| | Total US$m | Per capita US$ | As % GDP | | | | | |
| | 1999 | 1999 | 1990 | 1999 | 1990 | 1999 | 1990 | 1999 |
|---|---|---|---|---|---|---|---|---|
| Northern Africa | 473.5 | 14.6 | 5.3 | 1.5 | 0.6 | 1.4 | 27.9 | 18.7 |
| Western Africa | 247.5 | 51.9 | 17.7 | 11.5 | 0.6 | 1.5 | 19.3 | 16.2 |
| Central Africa | 154.2 | 32.1 | 14.9 | 6.4 | 1.2 | 3.5 | 15.3 | 9.9 |
| Eastern Africa | 314.5 | 36.7 | 20.7 | 12.4 | 0.0 | 1.2 | 37.3 | 24.1 |
| Southern Africa | 331.5 | 35.6 | 15.5 | 7.3 | 1.2 | 7.3 | 18.6 | 16.8 |
| IOC | 140.4 | 32.8 | 11.5 | 7.2 | 0.6 | 1.1 | 18.7 | 14.3 |

Source: JES-Preparation WSSD 2001 and Assessment of Progress on Sustainable Development in Africa Since Rio 1992, UNEP

The international community has a considerable and varied involvement in social and economic development in Africa. Most sub-Saharan African countries have been affected by macro-economic disequilibrium, with inflation and unsustainable current account deficits. In most countries, the balance of investment has shifted to the towns, while the productive sectors of their economies have continued to depend largely on the rural areas, because of their dependency on agriculture.

While micro-economic and macro-economic problems contribute to environmental change and human vulnerability, the situation is exacerbated by unresponsive governance. Inappropriate domestic spending and bureaucratic inefficiencies have added to the economic burden on people and the environment.

## STRUCTURAL ADJUSTMENT PROGRAMMES

Most sub-Saharan African countries have been affected by the International Monetary Fund (IMF) and World Bank Structural Adjustment Programmes (SAPs) of economic reform, with varying results for their economies and varying effects on the poorest people. Adjustment policies have been characterized by:

- removal by governments of subsidies on essential services, such as education, health and transport;
- reduction of the civil service labour force through retrenchments;
- removal of subsidies on agricultural inputs; and
- liberalization of commodity prices.

Most of the policies introduced as a result of SAPs have tended to impact negatively on the economic situation, human livelihood conditions, human vulnerability and coping capacities in Africa, but at different scales, and with considerable differentiation between countries, affected sectors and populations. The WCED noted in 1987 that: 'most of the world's poorest countries depend for increasing export earnings on tropical agricultural products that are

Oasis in the Sahara Desert, Libya

UNEP/Vesselin Voltchev

vulnerable to fluctuating or declining terms of trade. Expansion can often be achieved at the price of ecological stress' (WCED 1987). In Southern Africa, for example, SAPs have resulted in major job losses in Malawi, Mozambique, Tanzania, Zambia and Zimbabwe, increasing poverty levels and pressure on the environment (Chenje 2000). Overall, people in these countries have become more vulnerable, both socially and economically, because degraded environments are producing fewer resources.

SAPs also increase vulnerability to disaster in urban areas which are not necessarily disaster-prone by nature. The SAPs contribute to accelerated urbanization, population movement and population concentration, which make low-income urban dwellers more vulnerable to the impacts of disasters (Hamza and Zetter 1998). Liberalization of pricing systems and intensified exports have translated into forest losses in Côte d'Ivoire, where rising cocoa production has been the primary source of decreased forest cover—which has been reduced from 12 million ha in 1960 to 3.9 million ha at present. Similarly, in Cameroon, the IMF encouraged the government to reduce export taxes on forest products, to devalue the local currency and to cut government jobs. The lack of financial resources and personnel to enforce forest protection, combined with few incentives for appropriate forest management and land use, resulted in a 49.6 per cent increase in lumber exports between 1995 and 1997, and the wholesale destruction of one of Cameroon's most valuable environmental resources (FOE 2000).

## DEBT BURDEN

The persistent dependence on external financing by African countries has made it difficult for them to remain on the sustainable development track. However, many other factors have also contributed to economic stagnation; these include governance, civil strife and disasters. Regional engagement by multilateral financial institutions such as the World Bank and the IMF comes at a price for recipient countries. Sub-Saharan Africa's total external debt rose from US$176 874 million in 1990 to US$216 359 million in 1999 (World Bank 2001). Some 48 of the 52 African countries spend about US$135 000 million every year repaying debts to rich foreign creditors. The region's massive foreign debt

burden has been described as 'a new form of slavery, as vicious as the slave trade' (Colgan 2001). African countries have been receiving financial assistance from the developed countries in order to balance their development budgets in the midst of growing financial difficulties due to poor commodity prices, globalization and other factors (UNDP and others, 2000). For example, sub-Saharan countries exported goods and services worth about US$96 584 million in 1999, while total external debt amounted to US$216 359 million in the same year (World Bank 2001).

Saddled with heavy debt commitments, many African countries are unable to build and maintain economic reserves. The overall result is: further erosion of regional coping capacities; greater vulnerability to both internal and external stresses and shocks; indebtedness; increasing reliance on foreign aid; and reduced competitiveness in the face of economic globalization. Africa's heavy international debt burden contributes substantially to human vulnerability and

### Box 3.13 Africa's indebtedness

Africa's official debt at the end of 1999 stood at approximately US$170 000 million (World Bank 2000). In sub-Saharan Africa, debt service consumes between one-quarter and one-third of foreign exchange earnings, diverting resources from productive investments. Africa paid back US$1.31 in debt service for every US$1 received in aid grants in 1996 (Jubilee 2000).

Large debt service payments mean that vital social services must be sacrificed to meet debt payments, which makes the poor in HIPCs even worse off and, by implication, more vulnerable. For example, the United Nations Children's Fund (UNICEF)—which supports calls for the cancellation of debt owed by African countries—says that forest loss in HIPCs was 50 per cent greater than in non-HIPC countries between 1990 and 1995. Sub-Saharan Africa is forced to spend more on servicing its external debts than on the health and education of the region's 306 million children—while child mortality averages one-third more than other regions of the world (IPS 1999). In addition, heavy debt burdens result in increased pressure on the environment because of the growing need by HIPCs to generate foreign exchange to meet debt obligations.

**Table 3.5 Africa—total external debt**

| Sub-region | Annual average (US$ millions—current prices) | | |
| --- | --- | --- | --- |
| | 1975–84 | 1985–89 | 1999–MR |
| Northern Africa | 48 632 | 102 600 | 109 804 |
| Western Africa | 22 766 | 58 308 | 76 661 |
| Central Africa | 9 254 | 19 745 | 31 854 |
| Eastern Africa | 6 042 | 16 636 | 25 046 |
| Southern Africa | 11 084 | 26 133 | 63 237 |
| IOC | 1 721 | 4 350 | 6 074 |
| | | | |
| Total Africa | 97 717 | 228 409 | 302 655 |
| SSA | 54 892 | 136 754 | 209 816 |
| SSA Excl. S.A. | 54 892 | 136 751 | 195 025 |
| SSA Excl. S.A. & Nigeria | 46 377 | 110 830 | 163 455 |

*Source: Compiled from World Bank 2001*

security in the region. The servicing of debt consumes resources which could be spent on development, poverty alleviation and increasing coping capacities. Allied with the question of debt is that of SAPs, which often include obligations to reduce state spending, especially on social development and environmental management—a situation which tends to exacerbate the vulnerability of the poor and other marginal groups to environmental, economic and social stresses. Heavily Indebted Poor Countries (HIPCs) in Africa are also increasingly resorting to unsustainable exploitation of the region's natural resource base in order to boost foreign exchange earnings to service debt.

Table 3.5 gives summary data on the situation described in the Box 3.13.

Sub-Saharan African economic and financial problems were made much worse in the 1970s and the 1980s by a combination of:

- an investment in growth and development which failed to earn the expected rewards;
- the international debt crisis, increases in oil prices and rising interest rates, plus the inadequacy of aid programmes which were meant to provide relief;
- repeated drought, crop failure and widespread famine;
- the failure of agricultural production to contribute

significantly to growth, and the increased dependence on imported food;
- widespread warfare and civil unrest;
- the fact that SAPs have so far only been partially successful, and the success recorded has been mainly in terms of systems rather than people, who would have hoped to have their livelihood problems solved through the SAPs;
- reduced competitive ability;
- loss of intellectual property rights; and
- economic policy failures.

## FOOD SECURITY

Hunger is the most extreme manifestation of the multi-dimensional phenomenon of poverty, and the eradication of hunger is instrumental to the eradication of other dimensions of poverty. Persistent widespread hunger impedes progress in other aspects of poverty reduction, and weakens the foundation for broad-based economic growth. Hunger also represents an extreme instance of market failure, because the people who are most in need of food are the least able to express this need in terms of effective demand (FAO 2001).

The FAO defines 'food security' as a state of affairs where all people at all times have access to safe and nutritious food which enables them to maintain a healthy and active life. Food security, therefore, implies the provision of safe, nutritious, and quantitatively and qualitatively adequate food, as well as access to it by all people (NFSD 1996). Food security has three dimensions:

● availability of sufficient quantities of food of appropriate quality, supplied through domestic production or imports;

● access by households and individuals to appropriate foods for a nutritious diet; and

● optimal uptake of nourishment, thanks to a sustaining diet, clean water and adequate sanitation, together with health care.

Households seeking to preserve food security levels may resort to a number of coping strategies to gain access to food. These include: maintaining normal income-generating patterns; adaptation by means of innovative use of available resources or some divestment of liquid assets; divestment of productive assets, such as stock or land; and out-migration and destitution (USAID 1999).

Agriculture, of which 85–90 per cent is rain-fed in sub-Saharan Africa, accounts for 35 per cent of the region's gross national product (GNP), 40 per cent of exports and 70 per cent of employment (World Bank 2000). Year-to-year swings in GDP can be as high as 15–20 per cent, largely as a result of the effects of fluctuations in rainfall on agricultural production (World Bank undated). With the greatest part of African agriculture being rain-fed crop farming, food insecurity is largely caused by variability of rainfall (Khroda 1996). Moreover, about one-third of the region has a mean annual rainfall of less than 700 mm, which is too little to sustainably support rain-fed crop production.

Agricultural production varies from one sub-region to the other and is projected, due to the impact of climate change, to significantly reduce production in the tropics and sub-tropics, areas where food insecurity and

## Box 3.14 Consequences of environmental degradation of the Lake Chad basin

Over the years, the effects of drought and unsustainable human activities have continued to degrade the Lake Chad basin in general and the lake itself in particular, in terms of water availability and biodiversity. Water in the lake and its surrounding water systems is decreasing at an alarming rate. This persisting degradation of the entire ecosystem is compromising the performance of agriculture, fishery and livestock, which totally depend on the water and biological resources. The net effect has been the vulnerability of the population in the zone which is occupied mainly by nationals of Cameroon, Chad, Niger and Nigeria. The problem is expected to worsen in the coming years as population and irrigation demands continue to increase.

In 1960–63, the lake's surface area was 25 000 km², today, it is only 1.350 km². It has been calculated that 25 per cent of the decrease in surface area took place between 1966 and 1975, as a result of drought and excessive evaporation. The persisting low rainfall in the Lake Chad basin has affected its

water systems regime. This persisting/chronic degradation of the run-off of the lake's river systems has been compared to 'disease of water' or 'hyper-draining' by some experts. The increase in agricultural water use and the loss of water due to drier climate are exacerbating the massive decrease of water in the lake. Groundwater is abundant, but is difficult to exploit.

The prolonged rainfall deficits and the generalized low levels of the river systems will gradually affect the stocks of groundwater in the region. A total of 11 million people live in the Lake Chad basin. The population is projected to reach 23–30 million in 2020, increasing informal settlements and causing acute water shortage. Demographic pressure, drought, bush fires, unsustainable farming activities and deforestation are exerting pressure on water and soil. The lake is shrinking, and fisheries are decreasing. The consequences of this situation have been drastic reductions in food production in the basin and increased vulnerability as a result of food insecurity.

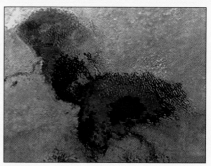

Satellite photographs showing the disappearing Lake Chad in Central Africa

*Goddard Space Flight Centre, 2001*

Source: Summarized from Nami, B (2002) Impacts of Environmental Change on Food Production in the Lake Chad Basin (unpublished)

hunger are already a problem (IPCC 1998). Box 3.14 describes the situation in the Lake Chad Basin, where the adverse climatic factor of drought, coupled with unsustainable human activities, have contributed to a reduction in the volume of water in the lake and its biodiversity.

Table 3.6 indicates that, taking the period of 1989–91 as base year = 100, the index of food production per capita in Africa was declining during the years indicated for all sub-regions except for North Africa and, to a small extent, West Africa. The same situation holds for the negative average annual percentage growth in food production for all sub-regions except for the same two sub-regions. This explains why most African countries have depended on food aid, and confirms their state of vulnerability in terms of food security.

Food insecurity results in: malnutrition and increased infant mortality rates; a heightened risk of contracting infectious diseases; ecologically and environmentally destructive coping strategies; migration; heightened dependency on aid; and stunted economic growth (IPCC 1998, USDA/ERS 2000). Estimates of undernourishment indicated that, in 1996–98, 792 million people in the developing world

and 34 million in the developed world were undernourished. The data in Table 3.7 indicate the level of undernourishment in Africa.

In sub-Saharan Africa in 1996–98, more than 34 per cent of the region's population was undernourished; some 185.9 million people experienced an average food deficit of 291 kcal/day. In this region, GNP/capita was US$297 against US$1 205 in the developed world. Sub-Saharan Africa has been identified as the region which is most vulnerable to food security, and the only region which shows increases in all indicators of food insecurity. Furthermore, the high incidence of HIV/AIDS in sub-Saharan Africa is expected to reduce agricultural production and productivity, and constraints in financial resources will limit commercial imports, leading to declining per capita consumption and, ultimately, to further undernourishment and the risk of famine (USDA/ERS 2000).

In sub-Saharan Africa, domestic food production accounts for about 80 per cent of consumption (USDA/ERS 2000), in a region where 4 out of every 10 Africans live in conditions of increasing poverty (ECA 1999). Farmers and pastoralists are vulnerable to food insecurity because they produce too little, and do not have enough food reserves. They usually have meagre

## Table 3.6 Food production per capita in Africa

| Sub-region | Index ( average 1989–91 = 100) | | | | | Average annual % growth | | |
| | 1980 | 1990 | 1993 | 1996 | 1999 | 75–84 | 85–89 | 90–MR |
|---|---|---|---|---|---|---|---|---|
| Northern Africa | 93.3 | 96 | 101.2 | 122 | 115.3 | −1.15 | 0.82 | 1.8 |
| Western Africa | 96.3 | 97 | 100.5 | 102.1 | 104.3 | −0.8 | 2.5 | 0.32 |
| Central Africa | 116.4 | 98.5 | 98.1 | 99 | 99.3 | −1.6 | −0.1 | −0.4 |
| Eastern Africa | 98.2 | 101.2 | 89 | 87.9 | 86.3 | −0.7 | −0.1 | −1.2 |
| Southern Africa | 114.2 | 100.2 | 94.2 | 95.9 | 92.2 | −2.6 | −0.2 | −1.2 |
| IOC | 114.5 | 99.3 | 96.8 | 104 | 93.8 | −1.9 | −1.7 | 0.1 |
| | | | | | | | | |
| Total Africa | 101 | 100 | 96 | 99 | 97 | −1.4 | 0.6 | −0.3 |
| SSA | 105 | 100 | 96 | 98 | 94 | −1.5 | 0.5 | −0.5 |
| SSA Excl. South Africa | 105 | 100 | 96 | 98 | 94 | −1.5 | 0.5 | −0.4 |
| SSA Excl. S.A. & Nigeria | 105 | 100 | 96 | 98 | 94 | −0.6 | 0.4 | −0.6 |

*Summarized from: The World Bank (2001): African Development Indicators 2001*

### Table 3.7 Food and nourishment in Africa (1996–98)

| Sub-region (countries) | Food availability | Prevalence of undernourishment | | Depth of undernourishment |
| --- | --- | --- | --- | --- |
| | Ave. per capita dietary energy supply (kcal/day) | Proportion of population undernourished (%) | Number of undernourished (millions) | Average food deficiency (kcal/person) |
| Northern Africa (6) | 3 055 | 8* | 10.7* | 183 |
| Central Africa (6) | 1 898 | 50 | 38.5 | 344 |
| Eastern Africa (7) | 1 833 | 42 | 52.2 | 359 |
| Southern Africa (9) | 1 736 | 45 | 28.6 | 302 |
| Western Africa (14) | 2 570 | 16 | 33.0 | 238 |
| IOC (2)* | 2 475 | 23* | 3* | 245 |

*Note: Data for the undernourished people in Northern Africa have doubled because of the poor situation in Sudan, and the same is applicable in the IOC with respect to Madagascar.
Source: Extracted from Report of FAO Committee on World Food Security—Assessment of the World Food Security Situation, 2000

savings and few other possible sources of income. They are more vulnerable to environmental change.

Macro-economic stresses, such as the transition to cash economies, and the penetration by global markets into local economies and the attendant structural changes, further serve to weaken the efficacy of traditional coping mechanisms, and exacerbate vulnerability to food insecurity. Natural hazards and armed conflict present two of the greatest obstacles to achieving necessary coping objectives; that is, increasing agricultural output while seeking additional security through alternative forms of income and stability (FAO 2000).

## ADDRESSING HUMAN VULNERABILITY TO ENVIRONMENTAL CHANGE

Increasing human vulnerability due to environmental change is a threat to sustainable social, economic and environmental development. Governments and institutions in Africa have adopted various measures in the past 30 years to deal with issues which contribute to environmental change. These have ranged from political and social measures to economic and environmental measures. At the political level, the Organization of African Unity (OAU) spearheaded the

decolonization of the region, facilitating independence to many countries and, in so doing, making natural resources more accessible to millions of people. For example, the elimination of apartheid in South Africa in 1994 helped to reduce the marginalization of most of the people to the abundant resources in their own country. As a result of apartheid, the white farmers, who make up only 5 per cent of the population, own 87 per cent of the land (Moyo 2000). The mean amount of land held per person in South Africa is slightly more than 1 ha for blacks and 1 570 ha for whites (SADC/UNDP/SAPES 2000). This situation has to change if sustainability is to be achieved. The South African government is committed to buying land and resettling the landless.

Other measures have been adopted to enhance sustainable development and to help to reduce the vulnerability of the population, including: investment in human resources; trade liberalization; review of outdated laws; and strengthening institutions at different levels. Such measures have been supported at national, sub-regional, regional and global levels. In terms of environmental management, together with other regions, African countries have adopted *Agenda 21*—the blueprint for sustainable development—and various multilateral environmental agreements (MEAs). Box 3.15 outlines some *Agenda 21* principles which are

---

**Box 3.15 *Agenda 21* principles which are relevant to human vulnerability**

A number of the principles of Agenda 21 relate to the issue of human vulnerability:

- Principle 6: The special situation and needs of developing countries, particularly the least developed and those most environmentally vulnerable, shall be given special priority. International actions in the field of environment and development should also address the interests and needs of all countries.

- Principle 8: To achieve sustainable development and a higher quality of life for all people, states should reduce and eliminate unsustainable patterns of production and consumption and promote appropriate demographic policies.

- Principle 15: In order to protect the environment, the precautionary approach shall be widely applied by States according to their capabilities. Where there are threats of serious or irreversible damage, lack of full scientific certainty shall not be used as a reason for postponing cost-effective measures to prevent environmental degradation.

- Principle 18: States shall immediately notify other States of any natural disasters or other emergencies that are likely to produce sudden harmful effects on the environment of those States. Every effort shall be made by the international community to help States so afflicted.

---

for 'a systematic combination of initiatives to develop a coherent environmental programme ... It is also recognized that a core objective of the (NEPAD) environment initiative must be to help in combating poverty and contributing to socio-economic development in Africa. It has been demonstrated in other parts of the world that measures taken to achieve a healthy environmental base can contribute greatly to employment, social and economic empowerment, and reduction of poverty' (NEPAD 2001).

NEPAD targets a number of areas for action. These include: combating desertification; wetland conservation; global warming; environmental governance; and financing.

## COMBATING DESERTIFICATION

The region has been in the forefront in terms of pushing for an international response to drought and desertification. The result was the adoption and ratification of the Convention to Combat Desertification (CCD). Since the CCD came into effect in 1996, countries (see Box 3.16 on the example of Mauritania) and sub-regions have adopted action programmes to combat desertification. In Southern Africa, for example, where general overdependence on natural resources

relevant to human vulnerability.

The latest initiative adopted by African leaders in 2001 is the New Partnership for African Development (NEPAD), whose long-term objective is to 'eradicate poverty ... and to place African countries, both individually and collectively, on a path of sustainable growth and development and thus halt the marginalization of Africa in the globalization process.' Under NEPAD, African political leaders have pledged to work both individually and collectively for peace, security, democracy, good governance, human rights and sound economic management, all of which are the conditions for sustainable development.

The African leaders also recognize that the range of issues necessary to nurture the region's environmental base is vast and complex. They emphasized the need

---

**Box 3.16 Fighting sand dunes in Mauritania**

In parts of Mauritania, villagers face two major environmental hazards: encroaching sand dunes which take over and destroy vegetation, causing desertification; and lack of water for household consumption and irrigation for agricultural production. Both situations render populations vulnerable by increasing poverty, and leading to food security and health problems, and they can cause civil unrest and strife.

The government has adopted policy measures to address the situation through the Mauritania-Agricultural Rehabilitation Programme, supported by external funding from the International Fund for Agricultural Development (IFAD), and these are yielding good results. Sand dunes, which are engulfing planted areas and which are threatening to engulf villages, are being brought under control in order to reduce environmental impacts and to make the population more secure.

The acute problem of water availability for household needs and agricultural irrigation is also being solved within the same project. Wells are being sunk where water is available in order to satisfy household and irrigation needs, and water is also being pumped from the Senegal River for irrigation. The implementation of the policy is helping to alleviate poverty, and to improve the food security problem and health of the population, thus increasing their coping capacities.

**Figure 3.2 Main causes of desertification by region**

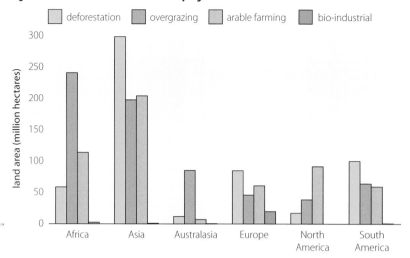

*Desertification does not refer to the moving forward of existing deserts, but to the formation, expansion or intensification of degraded patches of soil and vegetation cover.*

has been identified as the root cause of desertification, the action programme involves strengthening environmental capacities, enhancing public awareness and mobilizing their active participation in combating desertification, land degradation and the effects of drought (SADC-ELMS 1997). Figure 3.2 illustrates the main causes of desertification.

## POVERTY ALLEVIATION PROGRAMMES

Countries throughout the region have adopted various initiatives and programmes to alleviate poverty. As already highlighted, poverty is a major factor in increased human vulnerability to a stressed environment. Many national, regional and international organizations are involved in poverty alleviation programmes in Africa. For example, the African Development Bank (ADB) has developed a five-year programme which makes poverty alleviation and lasting development in Africa priority issues. The ADB's objective is to reduce poverty by half by 2015. According to ADB statistics, 40–45 per cent of the region's population live in absolute poverty, and 30 per cent are classified as extremely poor, including 70 per cent of women. The ADB will focus action on several major areas:

- agricultural and rural development, since the majority of the people in Africa live in rural areas;
- human capital development, through addressing health and education problems;
- 'transversal sectors', which include the

environment, gender equality and governance;

- regional integration, which the bank emphasizes as the only means to secure African integration in the world economy and among other groupings; and
- the promotion of the private sector as an engine of growth, job creation and exports. (Kabbaj 2000).

Recently, African environment ministers issued a statement, following an October 2001 African Preparatory Conference for the World Summit on Sustainable Development. In the statement, poverty eradication was identified as a priority area for action (see Box 3.17).

---

### Box 3.17 Eradication of poverty

Although Africa is an indispensable resource base which has been serving humanity for many centuries, poverty in Africa stands in stark contrast to the prosperity of the developed world. The process of globalization has further marginalized Africa, and this has contributed to the increasing incidence of poverty in the region. It is in this regard that the New African Initiative calls for the reversal of this abnormal situation by changing the relationship which underpins it. Achieving the poverty reduction goals of the Millennium Declaration is a joint responsibility of the North and South. It requires the adoption of a comprehensive approach which addresses the key priority areas, including:

- the removal of obstacles which prevent access to exports from developing countries to the markets of developed countries;
- debt reduction/cancellation;
- a review of the conditionalities of the IMF and the World Bank;
- promoting industrial growth, especially through small- and medium-sized enterprises;
- ensuring access to sources of energy at affordable prices, particularly in rural areas;
- promoting micro-finance;
- enhancing access to basic health services;
- sustainable rural development;
- agricultural development and food security;
- greater access to safe water and sanitation;
- reducing the vulnerability of our people to natural disasters and environmental risks; and
- access to and improved standards of education at all levels. (AMCEN 2001)

## EARLY WARNING SYSTEMS

One of the key responses to reduce the vulnerability of people is early warning. Various early warning initiatives have been implemented in the region. The Famine Early Warning System (FEWS) is perhaps one of the more widely known initiatives in Africa. The 1985 famine in Ethiopia galvanized African countries to establish FEWS, with funding support from the United States Agency for International Development (USAID). The main objective of FEWS is to lower the incidence of drought-induced famine by providing timely and accurate information to decision makers regarding potential famine conditions. The monitoring and response to famine has different phases, as shown in Figure 3.3.

Satellite data collected and processed by the US National Aeronautics and Space Administration, (NASA) and the National Oceanic and Atmospheric Administration (NOAA) are used to monitor vegetation conditions and rainfall across Africa. USAID has established a FEWS Network (FEWS NET), which is an information system designed to identify problems in the food supply system which can potentially lead to famine, flood or other food-insecure conditions in sub-Saharan Africa. FEWS NET is a multi-disciplinary project which collects, analyses and distributes regional, national and sub-national information to decision-makers about potential or current famine or flood situations, allowing them to authorize timely measures to prevent food-insecure conditions in these nations. Countries with FEWS NET representatives are: Burkina Faso, Chad, Eritrea, Ethiopia, Kenya, Malawi, Mali, Mauritania, Mozambique, Niger, Rwanda, Somalia, (southern) Sudan, Tanzania, Uganda, Zambia and Zimbabwe (USGS undated).

At a sub-regional level, in 1996–98, countries in Southern Africa established, with the support of FAO technical assistance: the Regional Early Warning System (REWS); the Regional Early Warning Unit (REWU); the Regional Remote Sensing Unit (RRSU); and the National Early Warning System (NEWS) (Chopak 2000).

## VULNERABILITY ASSESSMENT IN AFRICA

One practical example of vulnerability assessment is the in-depth vulnerability analysis mapping in Mozambique which has been used to classify the country into different food production systems. The project has entailed a

### Figure 3.3 Monitoring and response sequence

**During normal times: presence of risk of food insecurity**

*Prevention of food insecurity*
- Increase food production and supplies
- Diversify income sources
- Improve management of natural resources
- Strengthen food markets
- Identify vulnerable groups: baseline (FSVP) *

*Preparedness*
- Routine monitoring of selected indicators for early warning.
- Identify vulnerable groups: current year (CVA) *
- Develop capacity and strengthen partner institutions *
- Develop contingency planning institutions and procedures **

**When famine threatens: high probability of specific famine occurring**

*Early warning*
- Monitor and analyse indicators of food insecurity based on:
  - immediate needs and capacities of food insecure *
  - context and environment *
  - source of risk *
- Report results; issue alerts *
- Activate contingency planning groups **

**When famine sets in: occurrence of famine**

*Impact and needs assessment*
- Assess impact *
- Conduct emergency relief needs assessments (ENAs) *
- Determine objective(s) of response(s) **
- Determine capacities of households, communities and governments to respond
- Determine information need(s) of response(s) **

**As the famine continues: mitigate impact**

*Provide relief*
- Select and plan emergency responses (to rescue and provide relief to vulnerable groups) **
- Provide food relief
- Provide non-food relief (goods, inputs, services, technical assistance)
- Provide information to support response(s) **
- Monitor impact of relief operations on vulnerable groups **

**As the famine wanes: initiate recovery**

*Facilitate development to avert recurrence*
- Select and plan rehabilitation activities (restore livelihoods, foster sustainable recovery) **
- Implement longer-term response(s) and programme(s)
- Evaluate response(s) and lessons learned **

* Activities undertaken at present by FEWS FFRs in many instances
** Expanded activities for FEWS FFRs                    Source: Hutchinson 1992

## Box 3.18 Vulnerability assessments in Africa

The International Geosphere-Biosphere Programme's core project on Biospheric Aspects of the Hydrological Cycle (IGBP-BAHC) adopted a vulnerability assessment approach to develop strategies for sustainable management of groundwater resources. This approach included:

- regional integrated assessments of groundwater abstraction and recharge;
- integration of human vulnerability dimensions (particularly relating to the limited adaptive capacities of poor communities to groundwater scarcity in arid and semi-arid environments);
- stakeholder participation to identify social and environmental stresses;
- strategies for integrated water resource management which lessened pressure on the resource, promoted welfare, enhanced adaptive capacity and reduced vulnerability; and
- development of consolidated databases and continuous monitoring.

*Source: Hoff 2001*

multidisciplinary group, involving, among others: the Ministry of Health (Department of Nutrition); the Ministry of Planning and Finance (Department of Social Development, Poverty Alleviation Unit); and the National Early Warning Unit within the Department of Agriculture. Some of the preliminary mapping products include: flood risk maps; Normalized Differential Vegetation Index (NDVI) identification of drought risk areas; food systems maps; land use maps; market access maps; and health and nutritional profiles. Collection, analysis and presentation of food security and nutrition has been institutionalized within government, becoming a tool for local development, service delivery and monitoring, as well as scientific inquiry.

One of the major challenges for the region will be to design interventions which identify and target the various interacting dimensions which characterize vulnerability. In broad terms, this implies developing vulnerability assessment methodologies based on multi-disciplinary, integrated and coordinated strategies.

Human vulnerability and security assessment is a valuable tool for integrating environmental concerns into evaluations of livelihood security and sustainable development in the region. In Africa, with major populations exposed to adverse environmental change in both rural and urban areas, vulnerability assessment, IEM and disaster early warning systems can be integrated into a powerful tool for planning towards sustainable policies, plans and development. There are already examples of vulnerability assessments taking place in Africa (see Box 3.18).

## COMMUNITY-BASED NATURAL RESOURCE MANAGEMENT PROGRAMMES

Wildlife management in the region has undergone many changes over the past 30 years, from the colonial policies of protectionism of wildlife at the expense of communities, to sustainable utilization, which supports community involvement. Countries in different sub-regions have implemented or are implementing community-based natural resource management (CBNRM) programmes not only to address the issue of biodiversity conservation, but also to generate income for communities and to help to reduce rural poverty. One of the CBNRM programmes which has been introduced is the *Tchuma Tchato* ('Our Wealth') initiative in Mozambique's northern Tete province (see Box 3.19).

Poaching is one issue being addressed in an increasing number of community-based natural resource management programmes in some regions; such programmes aim to conserve biodiversity whilst at the same time generating income for communities to help reduce rural poverty.

*Gilles Nicolet/Still Pictures*

## Box 3.19 Community takes pride in its wealth

*Tchuma Tchato* ('Our Wealth'), a CBNRM project, has helped to restore the people's control over wildlife and natural resources in northwestern Mozambique. Traditional hunting, on which the community depended for both food and income, was declared 'poaching' in 1989 by a private landowner to whom the central government had leased hunting rights. Now, with restoration of both control of and the right to derive benefits from the area's natural resources, the community is more inclined to protect those resources, resulting in better conservation of wildlife and habitats.

The *Tchuma Tchato* project region is in the biodiversity-rich 'Tri-Nations Corner', where the borders of Mozambique, Zambia and Zimbabwe meet. The project encompasses six villages in the northwestern part of Tete province, in the mopane woodland, part of a vast, savannah-covered plateau encompassing more than 30 per cent of Mozambique's 799 380 $km^2$. Tete is one of the least populated of Mozambique's 10 provinces. The

project area, which is spread over 2 500 $km^2$ adjacent to the Zambezi river, has a population density of fewer than 5 people per 1 $km^2$, compared to about 2 590 people per 1 $km^2$ in Maputo, Mozambique's capital.

Initiated in 1994, *Tchuma Tchato* was started, in large part, to undo problems created by the two-phase (1964–75 and 1976–92) Mozambican civil war. The conflict destroyed social structures, displaced millions of people, and devastated wildlife management structures and institutions. The Tete region, as well as other parts of the country, became a large, uncontrolled hunting area. Elephant populations were decimated as combatants on both sides hunted the animals for meat and ivory.

*Tchuma Tchato*'s purpose is to make communities aware of the link between their economic welfare and the region's wild animals and biodiversity. The project stresses the advantages of the community acting as caretakers of the animals and the region, which means stopping poaching and

overexploitation of resources. Since the project began, the elephant population has increased and is flourishing. The region's biodiversity provides a solid basis for ecotourism. For this reason, a seven-chalet campsite, controlled by the Provincial Directorate of Agriculture and Fisheries through the *Tchuma Tchato* project manager, has been built on the bank of the Zambezi river.

Before the project started, the community derived its income primarily from hunting, subsistence agriculture and fishing. The project created new sources of income by employing some villagers to work as staff at the chalet complex and as game scouts. Revenue from both game hunting and the chalet operation is split three ways: 35 per cent to central government, 32.5 per cent to regional government and 32.5 per cent to the *Tchuma Tchato* project. The six villages involved in the project have established natural resource management councils to help manage the project on their behalf and to decide how revenues are used.

## DEBT-FOR-NATURE SWAPS

Pioneered by Kenya in 1988, debt-for-nature swaps (DNS) were aimed at curbing the runaway debt burden of the region. DNS, which involves the cancellation of external debt in exchange for the debtor government's commitment to mobilize domestic resources to fund conservation programmes, were meant to achieve two outcomes at once: to reduce external debt and to enhance environmental management. Between 1987 and 1998, more than 30 developing countries, including some in Africa, had benefited from DNS, which generated US$100 million in funding for the environment (UNSO 1998). According to the United Nations Sudano-Sahelian Office (UNSO), for African countries affected by deforestation, bilateral debt is the most promising source of DNS. This is because, in 1998, commercial debt represented 2 per cent of external debt for these countries and, also, multilateral debt had not been the subject of debt swap

transactions. The hope is that since multilateral creditors have accepted the principle that multilateral debt can be written off via the HIPC Initiative, DNS involving multilateral debt could take place. Of the 41 countries which may be eligible for the HIPC Initiative, 15 HIPC-eligible countries in Africa are severely affected by desertification (UNSO 1998).

A recent analysis of DNS concluded: 'The debt-for-nature swaps mechanism has been very modestly used until now partly because of the threat it poses on macroeconomic stability of the indebted country, partly because of the complicated dealings involved and finally because of free-riding of northern countries in environmental terms.

'Nevertheless, in its latest version, DNS have become simpler and more significant both in terms of the amount of debt involved and the financial support it provides for environmental projects without inducing macroeconomic instability in the countries involved.

The elephant is a 'main attraction' in the growing market for ecotourism—just one of many reasons why communities need to be aware of the link between their own economic welfare and their region's wild animals and biodiversity.

*UNEP*

Still, DNS has a more important role to play in the environment rather than the debt arena.' (ECLAC 2001)

DNS is only one way of tackling Africa's debt burden, and this has been recognized by African leaders, who want to see the region achieve and sustain an average GDP growth rate of more than 7 per cent per year for the next 15 years (NEPAD 2001). One of the goals is to implement national strategies for sustainable development by 2005, in order to reverse the loss of environmental resources by 2015.

## INDIGENOUS KNOWLEDGE SYSTEMS

Indigenous knowledge systems (IKS), which were discouraged during colonialism, are being revived throughout the region in order to promote human-nature linkages and to help communities to adapt to their changing environment. The Hausa of northern Nigeria, for example, developed a wealth of indigenous knowledge to cope with vulnerability to drought and famine in the sub-humid to arid regions of the Sahel (Milich 1997). These included: inter-cropping with nitrogen-fixing legumes; intensive manure application; soil conservation works; poly-cultural production of different cereals to cope with variable soil-moisture regimes; and exploiting different ecological niches to support wet and dry season production respectively. Other household and communal coping strategies included: wild food substitutes; increasing petty trading by women; selling livestock; craft production; out-migration to find work; and a strong communal ethic of sharing food with the hungry.

Pastoralists, such as East Africa's Maasai, exploit the opportunities of the natural system by migrating with their flocks to areas where pasturage has grown following rain. Other coping tactics include splitting herds to minimize risk and setting aside areas as grazing reserves. One important coping strategy throughout Africa is 'communal action based on social capital'. In this strategy, traditional societies in rural Africa draw on the collective strength of the weak to cope with stress in order to decrease vulnerability and insecurity, for example, through informal communal institutional processes, such as barter and trade.

In Southern Africa, for example, various activities have been undertaken to raise awareness among the people. There is a growing realization by governments and civil society that IKS has a lot to offer in natural resource management (Matowanyika 1999). The role of indigenous communities is recognized in Agenda 21, which acknowledges their role in developing 'a holistic traditional scientific knowledge of their lands, natural resources and environment' (UN 1992). Advocates of IKS say that the future of the region depends on people acknowledging the important role of their culture in development (see Box 3.20).

### Box 3.20 IKS foundation of Africa's future

*'If we want a strong Africa in the future we must lay its foundations on our indigenous knowledge in all areas of our lives. We will borrow ideas and skills from others since we live in an interdependent world, but if we are in possession of our minds, what we borrow will come to enrich and embellish what we already have and not supplant it.'*

Source: Opoku 1999

## INTELLECTUAL PROPERTY RIGHTS

Related to the issue of IKS is intellectual property rights (IPR), which has assumed increasing importance in terms of conservation, management, sustainable utilization and benefit sharing of genetic resources. The plunder of African intellectual property rights contributes to human vulnerability, because the region will be unable to derive benefit from its resources, particularly after patenting.

African countries which are particularly rich in genetic resources, traditional knowledge and folklore have an interest in the role of IPR, in the sharing of benefits arising from the patenting and use of biological resources and associated traditional knowledge (WIPO 2001). For example, currently, at least 97 per cent of all patents are held by nationals of countries belonging to the Organization for Economic Cooperation and Development (OECD), and at least 90 per cent of all technology and product patents are held by Northern-based global corporations (UNDP 2000, Singh 2001). Similarly, at least 70 per cent of all patent royalty payments are made between the subsidiaries of parent enterprises, and 75 per cent of the 76 000-odd patent applications filed to WIPO in 1999 came from just five OECD countries (RAFI 2000, Singh 2001). Between

---

**Box 3.22 African model law on intellectual property rights**

The African model law on intellectual property rights urges OAU member states to:

● examine ways and means of raising awareness about the protection of genetic resources, indigenous knowledge and folklore, taking into account the need to protect the rights of local communities;

● identify, catalogue, record and document the genetic and biological resources and traditional knowledge, including expressing of folklore held by their communities, within the framework of national laws.

● exchange information and experiences, and continue, within the framework of the OAU, with the search for joint solutions to problems of common concern, and with the efforts aimed at developing common positions, policies and strategies in relation to these issues.

*Source: WIPO 2001*

---

**Box 3.21 IPR statistics**

● Some 86 per cent of known higher plants, 99 per cent of the world's indigenous people and 96 per cent of the world's farmers live in the South (Africa, Asia and Latin America).

● Some 83 per cent of known diversity and *in-situ* knowledge is held in the South.

● The South's share of species diversity ranges from a low of 52 per cent of known fish species to a high of 91 per cent of reptile species. The South has 87 per cent of the global diversity of higher-order plant species and at least 83 per cent of all forests (tropical and temperate) (RAFI 1996).

● Some 75 per cent of *ex-situ* resources and technology are in the North (most of which originated in the South), and it is currently beyond the reach of the Convention on Biological Diversity (CBD) (RAFI 1996).

● Only 22 per cent of all crop gene banks are in the North, but 55 per cent of all seed accessions and 62 per cent of all crop species are in the North's collection. The large gene banks of the Consultative Groups on International Agricultural Research (CGIAR), which are located in the South, are controlled by Northern boards and funding agencies. If the CGIAR collections are deducted from the South's *ex-situ* holdings, its share of banked crop seed will plummet to about one-third of the global total (RAFI 1996).

● Some 83 per cent of recent bio-prospecting projects focus on the South's terrestrial biodiversity, 11 per cent on international waters and only 6 per cent exclusively on the North (RAFI 1996).

*Source: Singh 2001*

1980 and 1994, the share of global trade involving high-tech, patented production rose from 12 per cent to 24 per cent, and now accounts for more than half of the GDP of OECD countries if intellectual property rights on plants and livestock are included (UNDP 1999, Singh 2001). The total revenue from all patents grew from US$15 000 million in 1990 to US$100 000 million in 1998, and is expected to increase to US$500 000 million dollars by 2005 (RAFI 2000). The number of cases of intellectual appropriation of the South is growing steadily. Box 3.21 gives some statistics on IPR.

The OAU has produced model legislation to regulate traditional medicine in African countries, and member states have already adopted it. Box 3.22 highlights some of the provisions of the model law, which will be used as a basis for finalizing uniform national laws aimed at integrating African economies.

## CHALLENGES OF REDUCING HUMAN VULNERABILITY

In order to effectively reduce human vulnerability and increase security in anticipation of, during or after adverse environmental change, policies need to be adopted which adequately address environmental issues at national, sub-regional, regional and global levels, as well as enhance their implementation. These policies will:

● raise incomes and standards of living to alleviate/ eradicate poverty and to improve on livelihoods;

- improve economic performance by restructuring SAPs to focus also on people, not only on institutions, and effecting better debt management;
- assure food security by technologically transforming rain-fed agriculture to more reliable irrigation systems;
- drastically reform land ownership laws to assure security of land tenure;
- take measures to improve health;
- reduce, if not eliminate, civil strife, armed conflicts and warfare;
- develop and rationalize human resources utilization, bring demographic pressures under control and ensure the participatory approach in environmental management;
- rationalize the exploitation of natural resources at sustainable levels;
- integrate environmental concerns into general development planning;
- establish vulnerability assessments and strengthen early warning systems; and
- institute and enforce governance, and ensure a general enabling environment for sustainable development.

African countries are party to various policy instruments at global, regional, sub-regional and national levels. However, compliance and implementation with these instruments of environmental management and policy are, in many instances, ineffective (see Chapter 5). This is an indication of policy failures which have to be addressed if Africa is to move towards sustainable development. Policy failures can contribute to increased human vulnerability to environmental change. This can be due to inaction on the part of authorities, unsustainable policies, poor implementation of existing policies or insufficient human and financial resources to give effect to policies. Lack of expertise, an absence of political will, other state spending priorities and the so-called 'brain drain' can all contribute to non-effectiveness of policy.

Consequently, there is absolute need to strengthen policy implementation measures at community, national, sub-regional and regional levels. This certainly calls for support at global level, but the responsibility is that of governments, through the necessary political will and commitment, and in cooperation with their sub-regional and regional organizations.

African governments should assume their responsibilities to reduce human vulnerability and increase security by lowering risk and enhancing coping capacities. This will require:

- capacity building at community and national levels;
- the development of information and communication technology;
- the development and promotion of science and technology;
- the integration of environmental management in development planning; and
- the enhancing of governance and accountability, in order to create an enabling environment for sustainable development.

## CONCLUSIONS

Environmental change is making people in Africa more vulnerable, with the increase in exposure to risk and inadequate coping capabilities. Many factors impact exposure to risk and coping capabilities, and they have social, economic and environmental dimensions. Poor economic performance, and weak institutional and legal frameworks, as well as overexploitation and other processes, contribute to increased human vulnerability. With models suggesting that human vulnerability is set to get worse in the future (see Box 3.23), the region has to adopt strategic measures to mitigate such vulnerability and to improve human security.

The likely impacts of increasing human vulnerability in the region over the coming decades include:

### Box 3.23 Impacts of climate change

Climate change-induced famine may result in more than 50 million environmental refugees in Africa alone by 2060. Globally, sea level rise and agricultural disruption will not be the only causes of human migration. Severe water problems may affect three billion people globally by 2015, and this would encourage mass migrations. Deforestation, soil erosion and desertification may also lead to large movements of people.

*Source: Myers 1993*

- Increasing poverty, and the targets set under NEPAD to reduce poverty by 2015 will remain just that—a target.
- Governments in Africa are likely to pay less attention to the environment as they try to address the basic needs of their people. Environmental policies and institutions are likely, therefore, to remain weak.
- Women and children will continue to bear the brunt of environmental change, particularly in the region.
- Migration—illegal or otherwise—will continue to be a major factor, with millions of people risking their lives to migrate to urban areas and developed regions.
- Overexploitation of the environment will continue (and even increase), making it difficult to break the poverty circle and vulnerability to environmental change.

   Policy makers cannot afford to ignore the need to improve environmental management if the human vulnerability/environmental change circle is to be broken. Sustainable development means taking into account social, economic and environmental issues at the same time, not one at a time.

# REFERENCES

AMCEN (2001). *African Ministerial Statement*. African Preparatory Conference for the World Summit on Sustainable Development, Nairobi

Amous, S. (undated). *The Role of Wood Energy in Africa.* Forestry Department, FAO, Rome. Available on: http://www.fao.org/docrep/x2740e/x2740e00.htm

Benson, C. and Clay, E. (1994). *The Impact of Drought on Sub-Saharan African Economies: A Preliminary Examination*. Working Paper 77. London, Overseas Development Institute

Chenje, M. (ed., 2000). State of the Environment—Zambezi Basin 2000. SADC/IUCN/ZRA/SARDC/SIDA. Maseru/Harare/Lusaka, Lesotho/Zimbabwe/Zambia

Chenje, M. and Johnson, P. (eds., 1994). State of the Environment in Southern Africa. SADC/IUCN/SARDC. Maseru/Harare, Lesotho/Zimbabwe

Chopak, C. (2000). *Early Warning Primer: An Overview of Monitoring and Reporting*. FEWS NET. Available on: http://www.fews.net/resources/gcontent/pdf\1000006.pdf

Colgan, A-L. (2001). Africa's Debt—Africa Action Position Paper. Africa Action. Available at: http://www.africaaction.org/action/debtpos.htm

CRED/OFDA (2000). The OFDA/CRED International Disaster Database. Université Catholique de Louvain, Brussels, Belgium. http://www.cred.be/emdat

CSD (1997). *Report of the High-level Advisory Board on Sustainable Development for the 1997 Review of the Rio Commitments*. UN Economic and Social Council, New York. Available on: http://www.un.org/documents/ecosoc/cn17/1997/ecn171997-17add1.htm

Diallo, H. A. (2000). In *WHO: Human Health Severely Affected by Desertification and Drought*. UNCCD/WHO, Bonn/Copenhagen

ECA (1999). Economic Report on Africa 1999: The Challenges of Poverty Reduction and Sustainability. Available on: http://www.un.org/Depts/eca/divis/espd/ecrep99.htm

ECA (2001). *Transforming Africa's Economies. Economic.* Commision for Africa. Addis Ababa, Ethiopia

ECLAC (2001). *Debt for Nature: A Swap Whose Time Has Gone?* UN Economic Commission for Latin America, Mexico City

EM-DAT (undated). Natural Disaster Profiles—Region: North Africa. OFRA/CRED International Database, Université Catholique de Louvain, Brussels. Available on: http://www.cred.be/emdat/profiles/regions/nafr.htm

FAO (undated). Focus: AIDS—A Threat to Rural Africa. Available on: http://www.fao.org/Focus/E/aids/aids2-e.htm

FAO (1998). Agricultural Policies for Sustainable Use and Management of Natural Resources in Africa. Report presented at 20th regional conference for Africa, 16–20 February 1998, Addis Ababa, Ethiopia

FAO (2000). The Elimination of Food Insecurity in the Horn of Africa: A Strategy for Concerted Government and UN Agency Action—Summary Report. Available on: http://www.fao.org/DOCREP/003/X8530E/X8530E00.htm

FAO (2001). Committee on World Food Security—The World Food Summit Goal and the Millennium Development Goals. 27th session, 28 May-1 June 2001, Rome. Available on: http://www.fao.org/docrep/meeting/003/Y0688e.htm

FAO/TCOR (2001). Special Appeal for Mozambique: Relief and Rehabilitation Assistance to the Agricultural, Livestock and Fisheries Sectors. Available on: http://www.fao.org/reliefoperations/appeals/mozambi/moz2001.htm

FAO/UNAIDS (1999). *Sustainable Agriculture/Rural Development and Vulnerability to the AIDS Epidemic*. UNAIDS, Geneva

FOE (2000). *The IMF: Selling the Environment Short.* Friends of the Earth, Washington D.C.

Geisler, C. and de Sousa, R. (2000). Africa's Other Environmental Refugees. In Africa Notes, September 2000. Available on: http://www.einaudi.cornell.edu/africa/notes/September2000Notes.pdf

Hoff, H. (2001). Vulnerability of African Groundwater Sources. In *IHP Update, Newsletter of the International Human Dimensions Programme on Global Environmental Change* 2/01

Hutchinson, C. (1992). Early Warning and Vulnerability Assessments for Famine Mitigation. In *Chopak, C. (2000). Early Warning Primer: An Overview of Monitoring and Reporting*. FEWS NET. Available on: http://www.fews.net/resources/gcontent/pdf/1000006.pdf

IDS (1991). Poverty and the Environment—Micro Analytical Issues: People in Places. Institute of Development Studies, Brighton. Available on: http://www.ids.ac.uk/eldis/envids/chap31.html

Integrated Regional Information Networks (2001). IRIN World Health Day Special: Taking Africa's pulse. Available on: http://www.reliefweb.int/IRIN/webspecials/health/index.phtml

IPCC (1998). *The Regional Impacts of Climate Change: An Assessment of Vulnerability*. IPCC, Geneva

IPS (1999). RIGHTS-AFRICA: UNICEF Renews Call for Debt Cancellation. Press release, 16 June 1999. Available on: http://www.oneworld.org/ips2/june99/16_48_058.html

Jubilee 2000 (1999). Africa: Debt Background Paper. Available on: www.africapolicy.org/docs99/dbt9903b.htm

Kabbaj, O. (2000). Poverty Alleviation and Lasting Development in Africa on Top of ADB Vision. Arabic News.com. Available on: http://www.arabicnews.com/ansub/Daily/Day/000513/2000051332.html

Khroda, G. 1996. Strain, Social and Environmental Consequences, and Water Management in the Most Stress Water Systems in Africa. In *Water Management in Africa and the Middle East: Challenges and Opportunities* (eds Rached, E., Rathgeber, E. and Brooks, D. M.). IDRC Books, Ottawa

Lake, W.B. and Souré, M. (1997). Water and Development in Africa. International Development Information Centre. Available at: http://www.acdi-cida.gc.ca/express/dex/dex9709.htm

Mackenzie, C. (1994). Degradation of Arable Land Resources: Policy Options and Considerations Within the Context of Rural Restructuring in South Africa. Policy Paper No 11. Land and Agriculture Policy Centre, Johannesburg

MacKenzie, D. (1987). Can Ethiopia be Saved? *New Scientist*, 115 (1579); 54-8

Matowanyika, J. Z. Z. (1999). *Hearing the Crab's Cough: Perspectives and Emerging Institutions for Indigenous Knowledge Systems in Land Resources Management in Southern Africa*. SADC-ELMS/ IUCN, Maseru and Harare

Matthew, R. A. (2000). The Relationship Between Environment and Security. Keynote address to IUCN World Congress, 5 October 2000, Environment and Security: A Strategic Role for IUCN, Interactive Session 7, Amman

Menne, B. (2000). In WHO: *Human Health Severely Affected by Desertification and Drought*. UNCCD/WHO, Bonn/Copenhagen

Midzi, V., Hlatywayo, D. J., Chapola, L. S., Kebede, F., Atakan, K., Lombe, D. K., Turyomurugyendo, G. and Tugume, F. A. (1999). Seismic Hazard Assessment in Eastern and Southern Africa. In *Annali di Geofisica*, GSHAP Special Volume

Milich, L., 1997. Food Security in Pre-colonial Hausaland. Available on: http://ag.arizona.edu/~lmilich/afoodsec.html

Molvoer, R. K. (1991). Environmentally Induced Conflicts? *Bulletin of Peace Proposals*, 22: 175-88

Moyo, S. (2000). The land question and Land reform in Southern Africa. In Tevera, D. and Moyo, S. (eds). *Environmental Security in Southern Africa*. Sapes Trust, Harare

Mozambique National News Agency (2000). *AIM Reports*, Issue No. 194, 6 November 2000

Mpofu, B. (2000). *Assessment of Seed Requirements in Southern African Countries Ravaged by Floods and Drought 1999/2000 Season*. SADC Food Security Programme, Food, Agriculture and Natural Resources. http://www.sadc-fanr.org.zw/sssd/mozcalrep.htm

Myers, N. (1993). Total Environment Refugees Forseen. In *BioScience* vol. 43 (11), December 1993. Available on: http://www.greenpeace.org/~climate/database/records/zgpz0401.html

Myers, N. (1994). *Environmental Refugees*. Available on: http://www.gcrio.org/ASPEN/science/eoc94/EOC2/EOC2-10.html

Nchinda, T. C. (1998). Malaria: A Reemerging Disease in Africa. In *Emerging Infectious Diseases* vol. 4 no. 3

NEPAD (2001) The New Partnership for Africa's Development. NEPAD, unpublished memorandum, October 2001

NFSD (1996). Food Security for a Growing World Population: 200 Years after Malthus, Still an Unanswered Problem. Available on: http://www.foundation.novartis.com/food_security_population.htm

Opoku, K. A., Hearing the Crab's Cough: Indigenous Knowledge and the Future of Africa. In Matowanyika, J. Z. Z. (1999). *Hearing the Crab's Cough: Perspectives and Emerging Institutions for Indigenous Knowledge Systems in Land Resources Management in Southern Africa*. SADC-ELMS/ IUCN, Maseru and Harare

PAI (1995). *Sustaining Water: An Update*. Population Action International. Washington D.C.

Rabinovitch, J., 1997. UNDP's Networks and Urban Poverty, International Forum on Urban Poverty Governance and Participation: Practical Approaches to Urban Poverty Reduction, Florence 10-13 November 1997. Available on: http://magnet.undp.org/Docs/urban/Urbpov.htm

RAFI (Rural Advancement Foundation International) (1996). The Geopolitics of Biodiversity: A Biodiversity Balance Sheet. http://www.rafi.org/web/allpub-one.shtml?dfl= allpub.db& tfl=allpub-one-frag.ptml&operation=display&ro1= recNo&rf1=82&rt1=82&usebrs=true

RAFI (Rural Advancement Foundation International) (2000). In Search of Higher Ground? URL: http://www.rafi.org/web/allnews-one.shtml?dfl=allnews.db&tfl=allnews-one-frag.ptml&operation=display&ro1=recNo&rf1=120&rt1=120&usebrs=true

Rattan, L. (1988). Soil Degradation and the Future of Agriculture in Sub-Saharan Africa. *Journal of Soil and Water Conservation*, 43; 444-51

Reader, J. (1997, 1988) *Africa – A Biography of the Continent*. Penguin Books. London

Reich, P .F., Numbem, S. T., Almaraz, R. A. and Eswaran, H. (2001). Land resource stresses and desertification in Africa. In Bridges, E. M., Hannam, I. D., Oldeman, L. R., Pening, F. W. T., de Vries, S. J., Scherr, S. J. and Sompatpanit, S. (eds). *Responses to Land Degradation. Proceedings of the 2nd International Conference on Land Degradation and Desertification, Khon Kaen, Thailand*. Oxford University Press. New Delhi

SADC-ELMS (1997). *Sub-regional Action Programme to Combat Desertification in Southern Africa*. SADC Environment and Land Management Sector, Maseru

SADC/UNDP/SAPES (2000). SADC Regional Human Development Report 2000. Southern African Regional Institute for Policy Studies. Harare, Zimbabwe

Satterthwaite, D. (1999). The Links Between Poverty and the Environment in Urban Areas of Africa, Asia and Latin America. UNDP/IIED, New York

Singh, M. (2000). Environmental (In)security: Loss of Indigenous Knowledge and Environmental Degradation in Africa. In *Environmental Security in Southern Africa, 2000* (eds. Tevera, D and Moyo, S.). SARIPS, Harare

Singh, A. (2001). Human Vulnerability to Environmental Change. Draft chapter for *Global Environment Outlook 3*. In press

Soumaré, M. and Gérard, J. (2000). Reducing Urban Poverty in Africa: Towards a New Paradigm? In *Habitat Debate—Regional Perspective* vol. 6 no 4. Available on: http://www.unchs.org/unchs/english/hdv6n4/need_toilet.htm)

UN (1992). *Agenda 21*. Available at: http://www.un.org/esa/sustdev/agenda21.htm

UN (1999). *Report of the Secretary-General on the United Nations Observer Mission to Angola* (MONUA): S/1999/202. United Nations, New York. Available on: http://www.un.org/Docs/sc/reports/1999/s1999202.htm

UN (2000). *Report of the Secretary-General on the United Nations Mission in the Democratic Republic of the Congo:* S/2000/1156. United Nations, New York. Available on: http://www.un.org/Docs/sc/reports/2000/1156e.pdf

UNAIDS (2001). AIDS Epidemic Update; December 2001. Joint United Nations Programme on HIV/AIDS (UNAIDS). Available at: http://www.unaids.org/worldaidsday/2001/Epiupdate2001/Epiupdate2001_en.pdf

UNAIDS/WHO (2000). Aids Epidemic Update: December 2000. Available on: www.unaids.org/wac/2000/wad00/files/WAD_epidemic_report.htm

UNDHA (1994). *First African Sub-Regional Workshop on Natural Disaster Reduction*, 28 November to 2 December 1994, United Nations Department of Humanitarian Affairs, Gaborone

UNDP (1994). Human Development Report 1994: New Dimensions of Human Security. Oxford University Press, Oxford

UNDP (1999). Sustainable Livelihoods—Rural and Urban Poverty: Similarities and Differences. Available on: http://www.undp.org/sl/Documents/General%20info/Rural_poverty/rural.htm

UNDP (2000). Overcoming Poverty – UNDP Poverty Report 2000. Available on: http://www.undp.org/povertyreport/

UNDP/UNEP/World Bank/World Resources Institute (2000). World Resources 2000–2001. People and Ecosystems: the Fraying Web of Life. World Resources Institute, Washington D.C.

UNEP (1992). *World Atlas of Desertification*. Arnold. London

UNSO (1998). *Debt-for-Environment Swaps*. UNDP Office to Combat Desertification and Drought, New York. Available on: http://www.undp.org/seed/unso/pub-htm/swap-eng2.htm

USAID (1999). *FEWS Current Vulnerability Assessment Guidance Manual—Introduction to Current Vulnerability Guidelines*. USAID, Washington D.C.

US Committee for Refugees (2001). More than Half-Million Newly Uprooted People in Central and Horn of Africa in 2001. Available on: http://www.refugees.org/news/press_releases/2001/100301.cfm

USDA/ERS (2000). Food Security Assessment GFA12. ERS Outlook report, December 2000. Available on: http://www.ers.usda.gov/publications/GFA12/

USGS (undated). *Africa Data Dissemination Service*. Available on: http://edcintl.cr.usgs.gov/adds/adds.html

WCED (1987). *Our Common Future: The World Commission on Environment and Development*. Oxford University Press, Oxford

WHO (2000a). *HIV, TB and Malaria—Three Major Infectious Disease Threats*. Available on: http://www.who.int/inf-fs/en/back001.html

WHO (2000b). Economic Costs of Malaria are Many Times Higher than Previously Estimated. Press release, WHO/28, 25 April 2000. Available on: http://www.who.int/inf-pr-2000/en/pr2000-28.html

WHO and UNICEF (2000). *Global Water Supply and Sanitation Assessment 2000 Report*. World Health Organization and United Nations Children's Fund. Geneva and New York, Switzerland and USA http://www.who.int/water_sanitation_health/Globassessment/Glassessment6.pdf

WIPO (2001), *Global Intellectual Property Issues.* WIPO, Geneva, Switzerland. Available on: http://www.wipo.org/africa/en/activities/activities/oua_draft.htm

Wolfensohn, J. D. (2001). *Putting Africa Front and Centre*. The World Bank Group, Geneva, Switzerland, 16 July

World Bank (1997). Towards Environmentally Sustainable Development in Sub-Saharan Africa. In *Findings: Africa Region*, 78. World Bank, Washington D.C.

World Bank (2000). Regional Brief: Sub-Saharan Africa. Available on: http://www.worldbank.org/afr/overview.htm

World Bank (2001). In *Findings: Africa Region*, 185. World Bank, Washington D.C.,USA. Available on: http://www.worldbank.org/afr/findings/english/ find185.htm

WRI, UNDP, UNEP and World Bank (1998). *1998-99 World Resources: A Guide to the Global Environment—Environmental Change and Human Health*. Oxford University Press, New York/Oxford, USA/UK

# CHAPTER 4

## OUTLOOK 2002–2032

# CHAPTER 4

## OUTLOOK 2002–2032

### VISUALISING THE FUTURE

Chapter 2 describes the current state of the environment in Africa, outlining its physical attributes in terms of: the atmosphere; land; biodiversity; forests; freshwater, marine and coastal areas; and the urban environment. The chapter demonstrates that the African environment is under constant pressure, primarily from people and from nature. Given the rate at which natural resources are currently being degraded in the region, it is easy to conclude that the African environment is a highly threatened environment, and one in which the future seems mostly uncertain. And so it is not only desirable, but pertinent, that Africans should create a sustainable environment for their current and future needs. It is necessary, therefore, to redefine sustainable development, actively focusing on the reconciliation of the environmental and socioeconomic objectives of development with an adequate consideration of the needs of present and future generations.

An equally pressing issue, addressed in Chapter 3, concerns the fact that vast proportions of the population in various African countries are vulnerable to disasters resulting from the use of the environment and from social tensions which often develop when multitudes of people live together. There is a general belief that it is the poorest people who are most susceptible to these disasters. The poor are also unable to cope with the misfortunes that follow such disasters because coping mechanisms are generally poorly developed in many countries. The need to protect people who are vulnerable to disasters should be a critical component of decision-making processes.

The current state of the African environment, and the vulnerability of large populations in the region to environmental and human-induced disasters, pose an additional challenge to studies of the environment. Much of the challenge relates to the creation and maintenance of a sustainable environment whilst, at the same time, recognizing the need to develop the economic and social resources of nations. The creation of a sustainable development strategy requires insight both into the present and into the future. Since the future is, by definition, unknown, there is a need to develop mechanisms and methodologies which facilitate an understanding of the future.

This chapter will explore, compare and contrast different scenarios regarding the development of the African environment in the future. It will show which of these scenarios has the greatest promise for the region over the next 30 years.

### AMCEN's MANDATE AND SUB-REGIONAL SCENARIO DEVELOPMENT

The African Environment Outlook (AEO) is an initiative of the African Ministerial Conference on the Environment (AMCEN), with technical support from the Division of Early Warning and Assessment (DEWA) of the United Nations Environment Programme (UNEP). AMCEN was established in 1985 as an environmental intergovernmental body in Africa, with a mandate to provide policy guidance, and to identify priorities for the implementation of environmental treaties and goals at national and sub regional levels. One of AMCEN's goals is to ensure that the quality of the environment is maintained, and that Africans derive quality of life through the environment's provision of food, shelter and natural resources, which are needed to generate employment. AMCEN has made considerable progress in imparting environmental awareness to African governments. It has also facilitated the establishment of ministries of environment in many countries, as well as the enactment of environmental acts and bills which are being implemented in many countries.

AMCEN's eighth session, held in Abuja, Nigeria in April 2000, marked a turning point for the organization as the conference adopted a medium-term programme which clearly spelled out the need to provide an AEO. An AEO was identified as one of the tools which could be used to respond to Africa's persistent and severe economic and environmental problems in a sustainable way. It will provide the opportunity for AMCEN to look into the future, and to assess the various environmental and sustainable development policy options for the next 30 years, and then to identify which steps can meaningfully be taken at national, sub-regional and regional levels. Such a task involves a clear conceptualization of what the future environment should be like. It also requires great vision to create these possible future scenarios, to outline their characteristics and to show in what areas preferences lie.

Few scenarios have been developed to evaluate the environmental sustainability of Africa as a whole. However, many scenarios have been devised at sub-regional levels, which provide some general perspectives regarding the future of the region. Recent reviews by *GEO-2000* (UNEP 2000) and Raskin (2000a) describe attempts to develop scenarios for different sub-regions of Africa by: the International Institute for Environment and Development (IIED 1997); the World Bank (WB 1996); the Southern African Research and Documentation Centre (SARDC 1994); *Club du Sahel* (OECD 1995); *Beyond Hunger Study* (Achebe and others, 1990); and *Blue Plan* (UNEP 1989). Figure 4.1 outlines these scenarios. In general, the timescales covered by these scenarios range from 18 years (IIED) to almost 70 years (*Beyond Hunger*).

At least two contrasting scenarios were proposed for each region by most of the review bodies. For example, the IIED proposed 'Doomsday' and 'Sustainable Future' scenarios. The WB proposed 'Current Trends' and 'Desired Future' scenarios, for environmentally sustainable development in sub-

**Figure 4.1 Six scenarios present a general outlook for the future of the sub-regions**

**Study: IIED 1997**
**Horizon: 2015**
**Scenarios:**

1. *Doomsday scenario*
   Rapid population growth, economic stagnation, conflict and ineffective governance lead to deteriorating environmental conditions.
2. *Building a sustainable future*
   New development approach based on sustainability principles and cooperation leads to improved socio-economic conditions, peace and a clean environment.
3. *Real future*
   A compromise with elements of both of the above.

**Study: World Bank (1996)**
**Horizon: 2025**
**Scenario:**
*Looking 30 years ahead*
Africa joins the information society, but land and food shortages cause population migration and shift environmental burdens to urban centres.

**Study: SARDC (1994)**
**Horizon: 2020**
**Scenarios:**

1. *Current trends*
   Environmental conditions deteriorate in the context of economic stagnation.
2. *Desired future*
   Policies mobilize the region's physical and human resources through vigorous research education and institution building.

**Study: Club du Sahel (OECD 1995)**
**Horizon: 2020**
**Scenarios:**

1. *Laissez faire*
   Trade continues to lead to cheap imports not economic diversification as social inequity rises, international support wanes, and social and political stability is threatened.
2. *Orthodox growth*
   Good governance guides market towards development of new competitive sectors, international investments, and support.
3. *Regional integration*
   Supports local development of small enterprises modest economic growth, but stronger regional ties and less tension and conflict.

**Study: Beyond Hunger (Achee and others 1990)**
**Horizon: 2057**
**Scenarios:**

1. *Current perspective*
   Slow economic growth with increasing population leads to uncertain environmental conditions.
2. *Big lift*
   .A distictively Afro-centric development process leads to economic independence and a clean environment.

**Study: Blue Plan (UNEP 1989)**
**Horizon: 2025**

1. *Trend scenarios*
   Focus on macro-economic success and a laissez-faire policy towards population growth jeopardizes economic and social development and leads to environmental deterioration.
2. *Alternative scenarios*
   Goal-oriented development policies focus on domestic objectives leading to a cleaner environment.

Saharan Africa. Club du Sahel proposed *'Laissez-faire'*, 'Orthodox Growth' and 'Regional Integration' scenarios. *Beyond Hunger* proposed 'Conventional Wisdom' and 'Big Lift' scenarios. *Blue Plan* proposed 'Trend Scenarios' and 'Alternative Scenarios'. In almost all cases, comparisons were made between pairs of scenarios.

The 'business as usual' scenario, which comprises scenarios such as *'Laissez-faire'*, 'Orthodox Growth' and 'Current Trends', is driven by demographic change, particularly population growth and migration, and by lacklustre economic development. The 'Doomsday' scenario provides the worst-case scenario, given a *laissez-faire* approach to environmental change. Some scenarios, such as the 'Current Trends' scenario and UNEP's *Blue Plan*, look at realistic futures for specific sub-regions of Africa, and *Blue Plan* attempts to chart a desirable course for African development. There are also discussions on alternative scenarios, which feature a desirable and sustainable vision for the region.

Perhaps the most positive of the scenarios is *Beyond Hunger* (Achebe and others, 1990). This provides a vision of Africa some 100 years after the independence of Ghana, which was the first colonial territory in Africa to become independent in 1957. *Beyond Hunger*, produced before South Africa imbibed multiracial politics in 1994, already envisioned an African state of Azania for South Africa. It also envisioned the institution of mechanisms that would lead Africa to an economic development strategy which emphasized autonomy or integration into global markets. *Beyond Hunger* can be said to constitute the roots of the 'African Renaissance'— the resurgence of African culture, human resource development, outreach programmes and public participation in the development process.

## METHODOLOGY:
## THE SCENARIO APPROACH

[This section and the next section contain text drawn from Paul Raskin (30 June 2000; 1 September 2000 (revised); 16 October 16 2000 (second revision)) and *GEO-3 Scenarios: Preliminary Framework*.]

AMCEN's mandate to assess long-range environmental issues poses significant methodological challenges. Because the time horizon expands from years to decades, the long-range future cannot be extrapolated or predicted due to three types of indeterminacy: ignorance, surprise and volition. First, insufficient information on both the current state of the system and on the forces governing its dynamics leads to a classical statistical dispersion over possible future scenarios. Second, even if precise information were available, complex systems are known to exhibit turbulent behaviour, extreme sensitivity to initial conditions and branching behaviours at various thresholds. Therefore, the possibilities for novelty, surprise and emergent phenomena make prediction impossible. Finally, the future is unknowable because it is subject to human choices that have not yet been made. In the face of such indeterminacy, scenarios offer a means for examining the forces shaping the world, the uncertainties that lie ahead and the implications for tomorrow of today's actions.

A scenario is a story, told in words and numbers, concerning the manner in which future events could unfold, and offering lessons on how to direct the flow of events towards sustainable pathways and away from unsustainable ones. Development scenarios are alternative stories about the future with a logical plot and narrative. Scenarios usually include images of the future—snapshots of the major features of interest at various points in time—and an account of the flow of events leading to such future conditions.

Scenarios draw both on science—on our understanding of historical patterns, current conditions, and physical and social processes—and on the imagination in order to conceive, articulate and evaluate alternative pathways of development and the environment. In so doing, scenarios can illuminate the links between issues, the relationship between global and regional development, and the role of human actions in shaping the future. It is this added insight—leading to more informed and rational action—that is the foremost goal of scenarios, rather than prediction of the future.

Although scenarios certainly can offer quantitative insight, they are not primarily modelling exercises. The qualitative scenario narrative plays a critical role in giving voice to key aspects such as cultural influences, values, behaviours and institutions, which are not quantifiable. Thus, scenarios can provide a broader perspective than model-based analyses, while at the same time making use of various quantitative tools, such as accounting frameworks and mathematical simulation models. Quantitative analysis offers a degree of structure, discipline and rigour. Scenarios can offer texture, richness and insight.

Figure 4.2 illustrates the major features that govern the dynamics of change associated with scenarios of combined human and environmental systems. The 'current state' of the system is the outcome of an historical process, which is driven forward by a set of 'driving forces'. Moreover, the capacity of human beings to imagine alternative futures and to act intentionally means that images of the future can act as 'attractive forces' and 'repulsive forces' in shaping a scenario. In addition, there is the possibility that surprising and extreme occurrences—called 'sideswipes'—could affect development. Many unexpected events could be involved (for example, a breakdown of the climate system, a world war, cheap fusion power, a major natural disaster or a rampant global epidemic such as HIV/AIDS), but probabilities cannot be assigned, nor can all the possibilities be imagined.

Some key issues to consider in the formulation of scenarios include: the boundary; the current state; the definition and determination of driving forces; the narrative, or storyline; and images of the future.

- The boundary of the scenario is specified in several senses: spatially (for example, global, regional, sub-regional, national or local); thematically (for example, coverage of sectors and issues); and temporally (the time horizon of the analysis).

- The current state covers a range of dimensions: economic, demographic, environmental, institutional and so on. In the Global Environmental Outlook (GEO), as well as in the AEO, the dimensions are land, atmosphere, forests, freshwater, coastal and marine resources, biodiversity and urban environment.

- The important driving forces are those which condition and change the system. These are described in the GEO and AEO as demographics, economics, social issues, culture, technology, environment and governance.

- The narrative, or storyline, provides the plot by which the scenario stories unfold (often, quantitative indicators are used to illuminate aspects of the scenarios).

- An image of the future paints a picture of conditions at one or more points in time.

**Figure 4.2 Scenario dynamics**

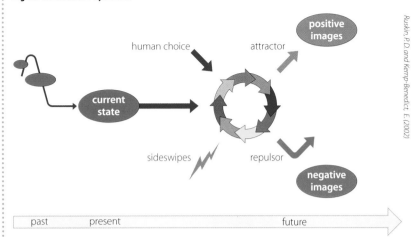

*Ruskin, P.D. and Kemp-Benedict, E. (2002)*

## AFRICA REGIONAL FUTURES: SCENARIO DEVELOPMENT IN THE AEO

The development of scenarios in the AEO followed the pattern described in the GEO. As with the *GEO-3* report, scenarios are based on the work of the Global Scenarios Group (GSG) (Gallopin and others, 1997). The GSG uses a two-tier hierarchy, described as 'classes' and 'variants', to categorize scenarios. Classes are distinguished by fundamentally different social visions, while variants reflect a range of possible outcomes within each class. The three broad classes are 'Conventional Worlds', 'Barbarization' and 'Great Transitions'. These are characterized, respectively, by: essential continuity with today's evolving development patterns; fundamental, but undesirable, social change; and fundamental and favourable social transformations. For each of the three classes, two variants are defined, giving a total of six scenarios. Thus, in the 'Conventional Worlds' class, the two scenarios that are emerging are 'Conventional Development' and 'Policy Reform'. The two scenarios in the 'Barbarization' class are 'Breakdown' and the 'Fortress World'. In the 'Great Transitions' class, the two scenarios are 'Eco-communalism' and the 'New Sustainability Paradigm'.

The 'Conventional Worlds' class envisions the global system of the 21st century evolving without major surprises, sharp discontinuities or fundamental transformations in the basis for human civilization. The future is shaped by the continued evolution, expansion and globalization of the dominant values and socioeconomic relationships of industrial society. In contrast, the 'Barbarization' and 'Great Transition' classes relax the notion of the long-term continuity of dominant values and institutional arrangements. Indeed, these scenarios envision profound historical transformations in the fundamental organizing principles of society over the next century, perhaps as significant as the transition to settled agriculture and the industrial revolution.

Within the 'Conventional Worlds' class, the 'Reference' variant incorporates mid-range population and development projections, and typical technological change assumptions. The 'Policy Reform' variant adds strong, comprehensive and coordinated government action, as called for in many policy-oriented discussions of sustainability, in order to achieve greater social equity and environmental protection. In this variant, the political will evolves for strengthening management systems and for the rapid diffusion of environmentally friendly technology. Whatever their differences, the two 'Conventional Worlds' variants share a number of premises: the continuity of institutions and values; the rapid growth of the world economy; and the convergence of global regions toward the norms set by highly industrialized countries. Environmental stress arising from global population and economic growth is left to the self-correcting logic of competitive markets. In the 'Policy Reform' variant, sustainability is pursued as a proactive strategic priority.

The 'Barbarization' variants envision the grim possibility that the social, economic and moral underpinnings of civilization deteriorate, as emerging problems overwhelm the coping capacity of both markets and policy reforms. The 'Breakdown' variant leads to unbridled conflict, institutional disintegration and economic collapse. The 'Fortress World' variant features an authoritarian response to the threat of breakdown. In this scenario, ensconced in protected enclaves, élites safeguard their privileges by controlling an impoverished majority and by managing critical natural resources while, outside the fortress, there is repression, environmental destruction and misery.

Further reflections indicate the need to reclassify these into four distinct categories. The four categories, adopted from the GEO, are: 'Conventional Development', later renamed 'Market Forces', 'Policy Reform', 'Fortress World' and the 'Great Transitions'. These are the four scenarios that are used in the AEO. Figure 4.3 shows the sketches of the behaviour over time for six descriptive variables on these four scenarios, namely: population growth; economic scale; environmental quality; social and economic equity; technological change; and the degree of social and geopolitical conflict. The curves are intended as rough illustrations only of the possible patterns of change.

The characteristics of the four scenarios may be

**Figure 4.3 Scenarios structure with illustrative patterns of change over time**

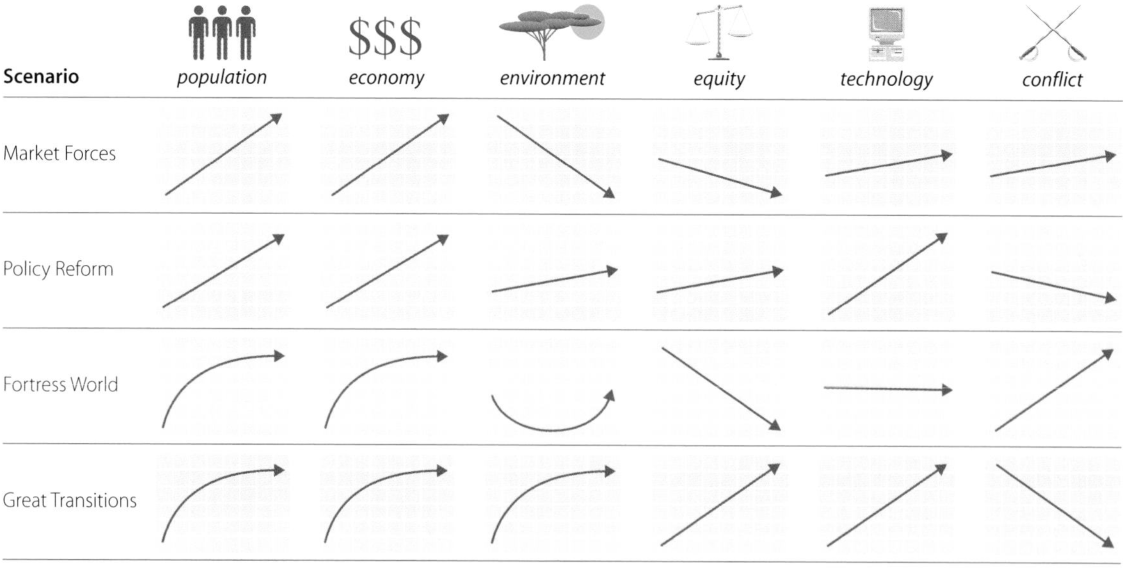

summarized as follows:

- Market Forces scenario: market-driven global development leads to convergence toward dominant values and development patterns;

- Policy Reform scenario: incremental policy adjustments steer conventional development towards environmental and poverty-reduction goals;

- Fortress World scenario: as socio-economic and environmental stresses mount, the world descends toward fragmentation, extreme inequality and widespread conflict; and

- Great Transitions scenario: a new development paradigm emerges in response to the challenge of sustainability, distinguished by pluralism, planetary solidarity, and new values and institutions.

  Scenario development proceeds in one of two directions. In the first case, one begins with the current position and then proceeds to make projections into the future. Such a strategy may be described as 'forecasting'. On the other hand, one can begin with the desirable future, and seek to manipulate variables and resources to achieve this future. Such an approach is described as 'back-casting'(see Figure 4.4). Two of the scenarios described above (the Market Forces scenario and the Fortress World scenario) may be achieved by methods of forecasting, while the other two scenarios (the Policy Reform scenario and the Great Transitions scenario) are best achieved by methods of backcasting, a procedure adopted in this study.

## QUANTITATIVE EXPRESSIONS OF THE SCENARIOS

The generation of quantitative expressions for the different scenarios requires the use of models, or input from experts. The applications described in this work were developed at the regional scale, and were organized with the help of the SEI's PoleStar software system. The goal of the PoleStar project is to give operational meaning to the notion of sustainable development, a challenging task which requires a broad view—across issues, over a range of spatial scales and over long timescales. The PoleStar software system has been applied at the global level by the SEI for UNEP's *GEO-3* report, and also for the scenarios of

**Figure 4.4 Forecast and backcast**

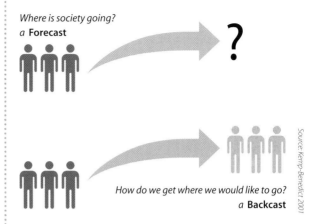

*Where is society going?*
*a* **Forecast**

**?**

*How do we get where we would like to go?*
*a* **Backcast**

*Source: Kemp-Benedict 2001*

the GSG. Scenario development in the AEO benefited from the good assemblage of data undertaken by the SEI, but has been modified in different ways. Some of the major driving variables—Gross Domestic Product (GDP), population and urban population—differ from those of the *GEO-3* report or the GSG scenarios. In part, this reflects an update of the data for Africa, using more recent information published in Cities in a Globalizing World.

For illustrative purposes, six possible outputs are examined, which represent the four dimensions of: demographics; economy and society; agriculture and forestry; and environment. The demographic dimensions consist of the total population and the urban population figures for Africa, and for each of its sub-regions. The economy and society dimension consists of the GDP measured at the purchasing power parity rates, with 1995 as base. The third dimension, agriculture and forestry, consists of severely degraded cropland. The environment dimension consists of water use and urban household water pollution. The quantitative expressions of the scenarios for these variables are presented as Figures 4.5–4.10.

Populations grow at mid-range levels in the Market Forces scenario. The spread of education for women, poverty-reduction programmes and widespread use of birth-control methods leads to a significant slowing in population growth rates in the Policy Reform scenario, and there is a radical reduction in growth rates in the

**Figure 4.5  Populations in the scenarios, by sub-region**

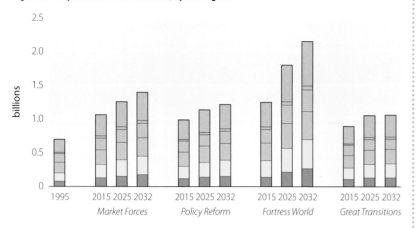

**Figure 4.6  Urban populations in the scenarios, by sub-region**

**Figure 4.7  GDP (Purchasing Power Parity (PPP)) in the scenarios, by sub-region**

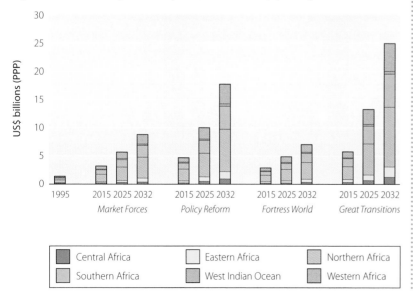

Legend:
- Central Africa
- Southern Africa
- Eastern Africa
- West Indian Ocean
- Northern Africa
- Western Africa

*UNEP 2002*

Great Transitions scenario. In the Fortress World scenario, social collapse, poverty and the failure of educational systems precipitate a massive expansion of the population, reaching, in 2032, four times the population in 1995.

Urban populations follow similar trends to population as a whole. However, the differences between the Fortress World scenario and the Great Transitions scenario are even more pronounced. Widespread alternatives to urban and periurban settlement expansion in the Great Transitions scenario lead a barely perceptible growth in total urban population. In the Fortress World scenario, the loss of farmland to élite interests, land degradation and the collapse of urban markets for rural agricultural produce drives billions of people toward the already overcrowded cities.

Patterns of GDP growth are nearly the inverse of the patterns of population growth. Growth in the Market Forces scenario is rather robust compared to recent experience in the region, but it is slow compared to other developing regions as the new millennium begins to unfold. In the Policy Reform scenario, the gap is much smaller, because several African countries see rapid growth in GDP. The collapse in society, polity and infrastructure in the Fortress World scenario underlies very sluggish growth in per capita income, which grows just barely over the rate of population growth over the scenario period. In the Great Transition scenario, with an international environment that is focusing on equity through more vigorous connections with other southern regions and in a supportive domestic environment, output per person expands rapidly, reaching an average level for the region as a whole that is close to the average of the industrialized countries in 1995.

Cropland degradation rates are highest in the Market Forces scenario and the Fortress World scenario, and the amount of cropland degraded adds to the total that must be converted from other land-use types.

Water use grows under each of the scenarios, as populations expand and incomes grow. In the Policy Reform scenario and the Great Transitions scenario, especially, rising demand for water-intensive end uses is offset by greater efficiency. However, despite relatively greater improvements in water efficiency in the Great Transitions scenario compared to the Policy Reform scenario, the rapid economic growth in the Great

Transitions scenario leads to comparable levels of water withdrawals. At the same time, water-use intensity remains well below that of the industrialized regions. In the Fortress World scenario, relatively modest water withdrawals—below those of any of the other scenarios—hide a disturbing picture of collapsing water-delivery infrastructure and degraded water sources, leading to severe water shortages and declines in the population with access to clean water.

Total water pollution from urban households depends on both the size of the urban population and the water-treatment infrastructure. The high average incomes in the Policy Reform scenario and the Great Transitions scenario, combined with low urban populations, lead to very low total water pollution loadings. In the Great Transitions scenario, urban water pollution falls after 2015, as urban population levels off and treatment continues to improve. The poor infrastructure and rapid urban population growth in the Fortress World scenario leads to a massive increase in total water pollutant loadings from urban households.

## THE DRIVING FORCES OF THE SCENARIOS

Driving forces are the mechanisms that allow change to occur. They can be thought of as clusters of shifts within society so great that they cause other significant shifts to take place. Driving forces set the initial course for development, but the complex regional system can rapidly change direction at critical thresholds of extreme turbulence and instability. Understanding the nature and interplay of driving forces is essential to developing scenarios. Driving forces are the departure points for looking at the future. They may operate with different magnitudes and directions than those of the initial stage, and they may emerge or disappear as circumstances dictate.

Current trends, on the other hand, are reflected in Chapter 2 of this report. Although they are not inevitably persistent, but evolve over time, they certainly condition the initial direction of economic, social and environmental change, and they may strongly influence even the long-term future. The driving forces control trends which are themselves influenced by social, economic and environmental conditions (Gallopin and others, 1997).

**Figure 4.8 Severely degraded croplands in the scenarios, by sub-region**

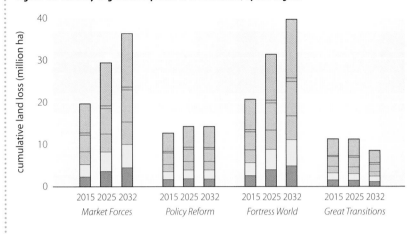

**Figure 4.9 Water use in the scenarios, by sub-region**

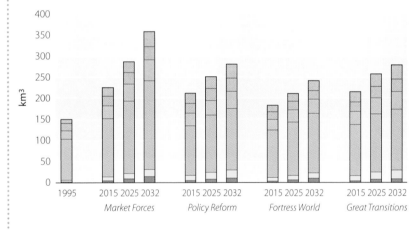

**Figure 4.10 Urban household water pollution in the scenarios, by sub-region**

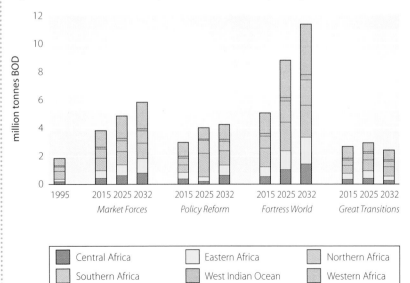

Legend:
- Central Africa
- Southern Africa
- Eastern Africa
- West Indian Ocean
- Northern Africa
- Western Africa

## THE DRIVING FORCES

### Demographics

Africa witnessed dramatic population increase, from 221 million in 1950 to 785 million in 2000 (see Figure 4.11). Despite the fact that population growth rates have declined since the mid-1980s, Africa remains the world's fastest growing region, at an estimated 2.4 per cent per annum. However, future growth rates are expected to be lower. The region will attain an estimated population of 1 406 million by the year 2030 (UNDP 2000) (see also figure 4.11). Rapid urbanization is also a main driving force, which is causing stresses in many African economies. With an average annual growth rate of 3.71 per cent (see Figure 4.12), Africa is the fastest urbanizing region of the world. Nevertheless, Africa is still very largely rural and agricultural. In 2000, the urbanization level was only 37.9 per cent, and it is projected to reach 54.5 per cent by 2030. Urban population is expected to grow from 297 million in 2000 to 766 million in 2030 (UNPD 1999).

Nonetheless, the problem of population in Africa is not related only to the population size, because Africa remains underpopulated by world standards. Rapid population growth entails challenges for the African countries to improve standards of living and to provide essential social services, including housing, transport, sanitation, health, education, job opportunities and

security. It also limits the capacity of African countries to deal with the problem of poverty. Furthermore, rapid population growth rates are leading to political and social conflicts among different ethnic, religious and social groups.

The population age structure is heavily skewed towards young people, which generates tremendous demographic momentum. About 43 per cent of the population is below the age of 15 years, about 52 per cent is between the ages of 15 and 60 years, and 5 per cent are aged 60 years or older (UNDP 2000). The 15–24 age group numbered 149 million in 1998, constituting about 20 per cent of the total African population. This workforce bulge can be the basis for more investment, greater labour productivity and rapid economic development (Makinwa-Adebusoye, 2001). With such high population momentum, reflected in the rather high fertility rates, and in improving health and medical situations, it can only be expected that the population of African countries will rise to phenomenal heights and will continue to impact the environment in significant ways. Left unattended, these trends lead to great strains and undesirable consequences on the environment.

However, the rate of population increase is not uniform. Some areas experience higher than average rates of population increase, due both to higher intrinsic rates of growth and to immigration from other areas

**Figure 4.11  Annual change in total, urban and rural populations, 1950–2030**

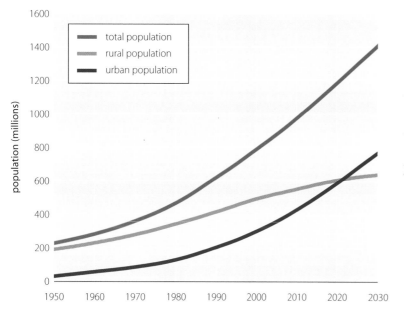

**Figure 4.12  Annual rate of change in total, rural and urban populations**

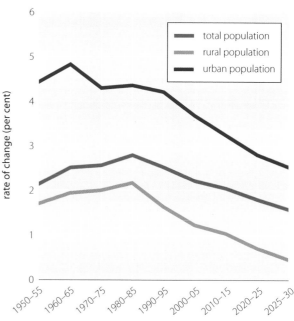

within a country, sub-region or the whole region. Such areas include town and cities, coastal regions, and the vicinity of lakes and rivers. Of these, the impact of population growth in urban areas is particularly important, because of the changes in lifestyle, consumption patterns and waste production that accompany urbanization. Conversely, some areas experience lower than average rates of population growth, due to below-average intrinsic rates of population increase and to emigration. Such areas include those in which there are conflicts or where there is severe environmental degradation.

## Economics

In relation to its size, population and relative abundance of natural resources, Africa is grossly underdeveloped, contributing, in 1999, only 1.1 per cent of global total GDP (WB 2000). Furthermore, the economies of most African countries show a marked dualism, with a relatively small commercial sector, based mostly on the extraction and export of natural resources (minerals, agriculture, other natural products and tourism), and a relatively large subsistence and informal economic sector. This dependence on the extraction and export of natural resources makes the economies of these countries particularly vulnerable to global economic fluctuations, especially in mineral and commodity prices.. The level of industrialization is low, and relatively little value is added within the region in the case of minerals and agricultural commodities. This means that industrial emissions of greenhouse gases is low, both overall and per capita, and emissions per unit of industrial and manufactured output are relatively high, because of the relatively old and inefficient equipment and technologies used by industry in Africa.

Nonetheless, over the past four decades, the economies of African countries have undergone considerable change, as they have moved from a colonial phase economy to a post-colonial economy. For instance, the development of modern, commercialized colonial agriculture was geared to the growth of cash crops for export. Although agriculture has remained a major economic sector, there have been key transformations, especially in the industrial sector, which has grown rapidly and on a large scale in some countries, to include heavy industries.

Nevertheless, the economic underdevelopment of African countries reflects, in part, their history of economic and political colonization and, partly, the economic and other policies adopted by governments following independence. These include: wage and price controls; widespread subsidies of basic commodities; a burgeoning civil service; fixed currency exchange rates, which lead to overvaluation of currencies; high tax rates; and disincentives for potential external investors. Many countries have had to accept economic Structural Adjustment Programmes (SAPs), often as a condition for being granted loans by the WB and the International Monetary Fund (IMF). The features of these programmes vary somewhat from country to country, but common elements include: strict controls on public expenditure; reforms of the structure and functioning of the civil service; reductions in barriers to trade; the removal of domestic subsidies; opening up of the economy to external investment; and allowing the value of the national currency to be determined by market forces. Nevertheless, these changes are proving to be socially disruptive and difficult to implement, with rising unemployment, sharply rising prices of even basic commodities and increasing inequity, among others.

For a long time, the promised benefits of SAPs have not been realized. Instead, these countries have drifted further away from development and their debt burdens have increased yearly (Vavi, 1999). Debt has been a stumbling block for many African nations, which have had to spend more on serving their debt than on providing basic social services. A combination of internal and external factors has precipitated Africa's debt problem. Africa's total debt stock stood at US$320 000 million in 1998, having increased by about 13 per cent from 1990.

Africa's share in the world trade is small and declining, and is being met with fierce competition from the other regions of the world that have faster and more sustained economic growth. Nevertheless, exports and imports have significant influence on the region's economy, with exports alone accounting for 25 per cent of the regional GDP, and imports providing 20 per cent of the domestic supply (ECA 1999). Africa has become severally marginalized in the global economy as it continues to face formidable barriers to northern markets. Most of the wealth generated by trade liberalization has flowed to developed countries. The rules governing world trade—set largely by the

*Africa's share in the world trade is small and declining, and is being met with fierce competition from the other regions of the world that have faster and more sustained economic growth.*

industrialized countries over the course of the 1986–94 Uruguay Round of agreements—have only contributed to Africa's economic woes.

Poverty remains endemic in Africa. In the late 1990s, more than 46 per cent of, or 290 million, people in sub-Saharan Africa lived on less than US$ 1 per day, up from 217 million people a decade earlier. Much of the poverty is based in rural areas and is, therefore, difficult to address, because of relative inaccessibility of much of the population and the limited economic opportunities for rapid development. This continuing poverty and associated vulnerability to all sorts of shocks, together with the inability of the relatively weak African economies to create conditions that would help to reduce poverty, has meant that countries in the region remain dependent on external aid.

## Social issues

Although improvements in social indicators have been sustained across the globe, Africa has lagged behind other world regions. Half of the population in Africa lacks access to health services. In rural Africa, about 65 per cent of the population are without access to adequate water supply, and 73 per cent are without access to adequate sanitation. In urban areas, about 25 per cent and 43 per cent of the population are without access to adequate water supply and sanitation, respectively (World Commission on Water 2000). The average adult illiteracy rate stands at 42 per cent. Life expectancy, which has steadily increased over the past decade to an average of 54 years, is expected to fall by 17 years as a result of the HIV/AIDS epidemic in many countries (ECA 2000). Health challenges are monumental in a region with the highest rates of fertility, maternal and childhood mortality, malnutrition, two-thirds of the world's known AIDS cases and 90 per cent of the world's annual malaria fatalities, and where half of the female population is illiterate.

Social development in African countries manifests in many dimensions, which cut across poverty and the declining standards of living, as well as issues of access to facilities and social security. On the one hand, there are issues of the provision and maintenance of basic human services and, on the other hand, issues of the consequences of the failure to do these successfully. When they became independent, most African countries inherited a system where government was absolutely responsible for providing these services and amenities, at almost no direct cost to consumers. Over the years, the ability of governments to meet the demands for providing basic services and utilities has decreased tremendously, and the effect has been one of aggravating social conditions in which nationals live. According to the latest UN figures, African nations rank, on average, lower than any other continent on the Human Development Index (HDI). Furthermore, more than 70 per cent of the countries in Africa rank 'low' on this index. In some cases, the HDI has suffered decreases, rather than improvements, over the years. Moreover, there are also great inequalities in the distribution of income, opportunity and social welfare.

Many countries have programmes which directly address these inequalities, while some have formulated development strategies designed to alleviate the problems of poverty and inequities. These inequalities remain today and, in some cases, manifest themselves in more serious forms than in the early days of independence. Compared with the rest of the world, social indicators show Africa as the least developed region, even though there are instances where improvements have been made, for example, in the areas of health, birth and deaths, fertility and educational development. In the area of educational development, the level of literacy has risen considerably in the region, and the quality of trained personnel has equally improved. African countries have witnessed the phenomenon of 'brain drain', whereby the best-trained of its manpower resources migrate into developed countries.

Poverty remains both a cause and an effect of environmental degradation, and is the main social and economic challenge for Africa. Poverty can be reduced, either by increasing economic growth or by reducing inequity. For Africa to cut poverty in half by 2015, it will require an average annual GDP growth rate of 7 per cent. Although there has been significant progress in education in Africa over the past two decades, there is much to be done. Primary school enrollment in 16 countries is less than 60 per cent, and there are more children aged 6–11 years out of school than was the case in the 1990s.

Although African women have made tremendous progress over the past four decades, there is still a gap between rhetoric and action in order to maintain the momentum of this progress in Africa. Economic and legal barriers, associated with social discrimination, continue to prevent women in Africa from improving their status and productivity, and from achieving their full potential.

## Culture

Culture and the natural environment are strongly linked. Cultural practices, values, norms and language often reflect the diversity of natural resources upon which people depend for their livelihoods, and the strategies they use to extract these resources. Africa holds many strongly traditional cultures. They can serve both as bulwarks against outside influences and as conduits through which new ideas are assimilated, through translation and representation in established idioms. Cultural norms and values shape people's perceptions, aspirations and attitudes and, thereby, their actions. Cultures often include mechanisms which regulate access to, and patterns of use and consumption of, natural resources. Most of all, culture strongly influences the choices that people make. Among the many elements comprising a culture, religious and linguistic elements, and ethnic groups living within that culture, can—and do, at times—cause changes in its total configuration. As in some other parts of the world, religion in Africa serves as a strong unifying force in some areas, and as a potentially divisive one in others. Ethnic tensions, driven by historical animosities—themselves often exacerbated by religious, economic and social tensions—are also potentially divisive and inhibitory to development. In contrast, religion can promote social coherence and support within communities, thereby encouraging self-reliance and development.

With a plurality of peoples and languages, Africa has a rich traditional culture. For many peoples of Africa, the systems of social governance, provision of services, maintenance of social cohesion and even economic development have been based on the norms that these cultures have allowed. The advent of other cultures in the region has led to some form of competition between cultures, and the assimilation of different cultures into a dominant culture. Traditional African cultures have been most affected, while the imported Islamic culture has merely suffered some damage. Today, traditional support systems, which served as social securities for all members of society, including the aged, the homeless, the sick and the poor, have more than crumbled and have not been replaced by efficient structures. Of course, traditional systems of governance have disappeared almost completely. They are being replaced by 'democratic ethics', which are being intermingled with tribal and ethnic sentiments and loyalties, and religious strives. Many Africans today are entangled in the influences of these cultures, but the influence of westernization has become very pervading.

Culture is not static, especially in this era of increasing globalization. People around the world are being exposed more and more to the norms and values of other cultures, sometimes creating tensions with their own culture and, in many instances, resulting in substantial modification or replacement of some of its elements. Consumption patterns in many parts of Africa are increasingly being modelled on a western-style consumer culture. This influences both trading and investment patterns, particularly by creating a demand for imported consumer goods while, at the same time, serving as an incentive for some of the multinational corporations to enter local markets directly through investment, partnerships or takeovers. A shift from self-sufficiency to a cash economy, and new patterns of accumulation of wealth, can bring unforeseen changes. This, in turn, can affect cultural values, as communities lose contact with the resources that were so important in determining many of their livelihoods and practices. These factors (changes in traditional patterns of access and management, as well as a diminishing natural resource base) can become sources of conflict, as well as culture change.

Western life styles are becoming increasingly commonplace. So, too, are western values, which continue to overshadow existing African cultures. However, dogmatic following of the consumer culture will reach a peak, and people will begin to see the difference between needs and wants. Effective utilization and channelling of these ideas and ideals will lead to healthier lifestyles and to the promotion of alternatives to traditional families ties, which are breaking down. The emphasis on meeting human needs rather than wants will lead to the elimination of religion and ethnicity as divisive forces in African development. The introduction of a culture of peace will bring lasting solutions to various forms of conflicts in Africa. This cultural revival will constitute a crucial fulcrum for the African renaissance, and its effects will soon become noticeable on the environment.

## Technology

Before the colonization of the region, African economies were able to survive large-scale environmental degradation, for a number of reasons. First, the population was small, and the demands on the economy were small. More importantly, the technology was appropriate and adequate, because the African people had learned over centuries to adapt systems of extraction of natural resources to be commensurate with the dictates of the environment. Things have since changed. Modern economic practices have introduced increased demand on human and natural resources, and so that technology has proved inadequate. For instance, attempts to improve agriculture has resulted in the importation of strange varieties of food crops, and the introduction of chemicals and additives to soils, plants and vegetation.

For a large part of the 20th century, Africa's role in the development of science and technology was marginalized. To a large extent, the colonial powers inhibited the development of indigenous technology in Africa, and destabilized some of the existing processes of technical growth. Indigenous manufacturing capability was deliberately undermined in order to facilitate European exports, and captive markets were created. In addition, colonial powers deprived Africa of its historical credit in contributing to advancements in science, technology and medicine (Emeagwali 2000). Furthermore, Africa has not been only a user of technologies developed in the west, but has also been a dumping ground for obsolete technologies abandoned in the west. More recently, Africa has became a testing ground for biotechnology and genetically modified (GM) products.

African countries have also discovered that modern technology is needed to manufacture the goods and services that have become imperative to sustain modern life. Caught in this cycle, dependence on imported technology, with all its consequences, has increased, and the effects on the environment have become severe. Many African countries will readily offer their land to locate even the most obnoxious industries—despite their negative impacts on the people and the environment—provided they will bring in some monetary relief or economic growth.

The craving today is for modern information technology (IT), which is at the heart of so much development. And yet it can be said that Africa has not even begun to establish itself in this arena. Because of its poor economic base, the region has not succeeded in manufacturing computers, the basic hardware required for IT development. Other impediments include: the low number of currently installed telephone lines; the limited provision of facilities such as electricity; and the high levels of illiteracy in the region.

The emergence of micropower technologies begins to revolutionize the exploitation of alternative sources of energy to meet the ever-growing needs of the region. While African countries continue to see modern IT and industrialization as principal agents for economic development, they nonetheless recognize that the timing for the application and integration of IT into national development processes will vary from country to country. With the introduction of cleaner fuels, swift transition to renewable resources and greater concern for the environment, the impact of industrialization on the environment is reduced to the barest minimum.

The use of biotechnology in agriculture and in related primary production industries has the potential to influence food and fibre production, food security and management. There are both benefits and risks associated with the release of genetically modified organisms (GMOs). The search for new pharmaceuticals, based on chemical compounds found in plants, can serve as a stimulus to conserve biodiversity. It also has the potential, albeit possibly a minor one, to serve as a source of income for local people.

## Environment

The environment, as a key driving force for change, is interlinked with other driving forces, and is, at the same time, impacted by other driving forces. Africa faces a series of environmental threats, including: water scarcity; quality and water management issues; land degradation; loss of biodiversity; the rise of infectious diseases; solid waste; degradation of coastal areas; marine pollution; urban air pollution; and climate change. Poverty, as a cause and effect of environmental degradation, has captured significant attention in Africa.

Most basic human needs (food, water and shelter) are linked to the environment. The capacity of the environment to support the demand for food depends both on the intrinsic features of that environment

*The capacity of the environment to support the demand for food depends both on the intrinsic features of that environment (climate, soils and seasonality) and the extent to which natural constraints on production can be alleviated by management.*

(climate, soils and seasonality) and the extent to which natural constraints on production can be alleviated by management. Demand for food grows both in relation to increasing populations and to changes in lifestyles. Food production is influenced by: the introduction of improved varieties of crops; improved agricultural land management practices; and climate variability. Food security is compromised by droughts and floods, and by the degradation of agricultural lands (declines in fertility, soil loss and salinization) through unsustainable farming practices. Depletion and degradation of the natural resource base intensifies competition for land and natural resources, creating potential conflict and causing human migration to less stressed areas.

Competition for limited resources also arises as demand for those resources exceeds supply. Demand increases both as a result of more people and their domesticates needing access to that resource, and of changes in the patterns of consumption. For example, across much of Africa, water is an increasingly limited resource. This is the result of ongoing degradation of water sources, such as: siltation of dams and rivers; increased demand, due both to increasing numbers of people, and to progressive urbanization and more profligate use of water; and increased demand by the agricultural and emerging industrial sectors. Increased competition for limited water resources has the potential to fuel conflict within countries, among different user groups and between countries sharing a common major water source. Conflicts are also possible between upstream and downstream users in the same watershed.

The African environment has, for several reasons, been healthier than the environments of other continents. For instance, Africa has lower levels of air pollution by carbon dioxide ($CO_2$) than all the other continents, and the ozone layer above Africa is in a better state. Furthermore, there are greater levels of biodiversity in Africa than elsewhere. Nevertheless, there are more and more onslaughts on the environment as bushes are increasingly cleared for agricultural development, and as erosion worsens as a result of putting marginal lands to use for mining and agricultural development. African countries remain highly vulnerable to loss of biodiversity and environmental degradation.

## Governance

It is impossible to achieve economic development without good governance. Since independence, African countries have been subjected to various forms of government, including: communism and socialism in Tanzania and Ghana; various forms of democracy in Côte d'Ivoire, Kenya and Nigeria; and quasi-military rule in many other countries. Chazan and others (1992) provided a six-fold classification, with the following categories of governance: administrative hegemonic; pluralist; party mobilizing; party centralist; personal coercive; and populist. Table 4.1 indicates the countries which fell into these groups. The 1960s, 1970s and early 1980s were periods of serious unrest in many African countries, and times when the role of governance was seriously called into question. The near-total breakdown of services, the increasing pauperization of the people, and the complete breakdown of law and order in some countries are continuing evidence of this unrest.

Post-colonial changes in government were achieved through military interventions and through various forms of elections, many of which were merely symbolic and not designed to effect serious changes in the structure of governance. Thus, some involved competitions for office within a single party, while others involved limited multi-party participation. Only in very rare cases were elections designed to provide a mechanism for the simultaneous turnover of both leaders and regime.

Military interventions in governance led to various forms of instability, and to the rise of insurgencies, riots, and ethnic rife and rivalries. Military expenditures increased, and accounted for significant per centages of the GDP of these countries (UNIDIR 1991). The lack of focus in governance led to a breakdown of many government institutions and to wasteful duplication of efforts in the development process.

The international community intervened in many ways, including attempts to sanitize national economies through suggestions and recommendations by the IMF and the WB. There was also the influence of multinational and transnational companies, some of which are financially stronger than many African national economies. The effects of these organizations were so pervading that, in some countries, they seemed to run parallel governments. There is no doubt today

**Table 4.1  Typology of regimes in Africa, 1951–90**

| | Regime type | Examples |
|---|---|---|
| 1 | Administrative hegemonic | Kenya, Zaire, Togo, Côte d'Ivoire, Cameroon, Zambia, Malawi, Morocco, Nigeria |
| 2 | Pluralist | Botswana, Gambia, Mauritius, Senegal |
| 3 | Party mobilizing | Ghana (Nkrumah), Mali (Keita), Guinea (Sekou Toure), Zambia, Algeria (Boumedienne), Tanzania, Zimbabwe |
| 4 | Party centralist | Angola, Mozambique, Ethiopia, Guinea-Bissau, Congo, Benin |
| 5 | Personal coercive | Uganda (Amin), Central African Republic (Bokassa), Equatorial Guinea (Nguema) |
| 6 | Populist | Ghana (Rawlins), Libya (Qaddafi), Burkina Faso (Sankara) |

Source: Chazan and others, 1992a

that African economies need some streamlining or restructuring. What most people do not agree about is not if and when this will be done, but how and by whom it will be done.

Good governance in Africa is challenged by various issues, including the collapse of the state in countries where governance has already been weakened by strife, and where governments hardly have the capacity to govern and to maintain law and order. In spite of the fact that African leaders have adopted democracy as a key element of their agendas over the past decade, the democratic process remains challenged. Narrow political considerations, personalized power and corruption have undermined the process of democracy and responsive governance. Inequity in social, economic and political systems, including gender inequity, has been a barrier to achieving good governance. This has resulted in notably increasing disparities between rich and poor in terms of income and capabilities, and in the marginalizing of women in governance.

Poverty remains widely spread in Africa. As such, poverty alleviation represents the greatest challenge to good governance. There is also the challenge of how to manage effectively financial and natural resources, promoting decentralization based on trust, transparency, accountability and capacity.

Corruption and the theft of public funds are some of the problems which have blighted the region's governance record. Many Africans wonder how much of their stolen funds, hidden somewhere in western banks, could make a difference in terms of debt repayments

and financing the region's renaissance. But, in the 1990s, the winds of change blew across the region, as the majority of people demanded greater accountability from their elected leaders. They called for transparency and respect for human rights.

Institutional reforms emerge as people begin to develop new forms of social consciousness and as they move to prevent the violation of human rights. When this happens, civil societies emerge in large numbers, and their influence increases and serves as a check on the excesses of national governments. Local participation in decision making also increases considerably. Efforts are made to reduce conflict—in countries where conflicts presently exist—by assisting with the provision of basic services and by breaking the poverty trap.

## SCENARIOS AND THEIR ENVIRONMENTAL IMPLICATIONS

### THE MARKET FORCES SCENARIO

#### Introduction

The Market Forces scenario assumes that world development evolves without major discontinuities, changes in values or other structural ruptures from the position as it existed at the end of the 20th century. However, in this scenario, the world becomes increasingly more integrated, both economically and culturally. Globalization of product and labour markets continues apace, catalysed by free trade agreements, unregulated capital and financial flows, and information technology. A number of important initiatives pave the way. The World Trade Organization (WTO) provides the legal basis for the global trading system. Barriers to trade and capital movements gradually vanish, as protectionism becomes a thing of the past. New institutional instruments promote market openness and global competition. Virtually all national governments advance a package of policy adjustments, which include: modernization of financial systems; investment in education to create a workforce that is competitive in the emerging global market; privatization; reduced social safety nets; and, in general, reliance on market-based approaches.

In the context of Africa, the Market Forces scenario is based on the assumption that African countries will adopt, willingly or otherwise, the range of policy reforms promoted by the WB and the IMF since the late 1980s. These reforms aim to improve the economic performance of developing countries by encouraging them to restructure their economies through a combination of tight fiscal and monetary policies. The objective is to limit budget deficits and to allow market-determined interest rates, more free trade, capital flows and unencumbered foreign direct investment. The reforms also include: privatization of state enterprises; extension and consolidation of private property rights; and a shift in public expenditure away from subsidies and administration towards infrastructure development and support for sectors of the economy—such as primary health care and education—which are likely to provide ultimately greater economic returns and more equitable income distribution. From a trade perspective, the main outcomes of this strategy are: global integration of commodity markets; opening up of investment markets; more mobile labour markets; and the application of global standards and regulations.

The assumptions of the Market Forces scenario may be summarized as follows:

- The dominant western model of development prevails, with the spread of consumerism/materialism and individualism. The world economy converges to this mode.
- Policies promoted by international financial institutions are adopted, either willingly or otherwise, and they are found to have positive impacts on aggregate growth as the scenario progresses.

- The most effective poverty-reduction strategy is growth promotion. Growth will tend to be broad-based and will trickle down.
- Effective institutions will emerge and spread.
- Economic growth will automatically contribute to recover the environmental damage incurred as a result of development.
- An active policy-making environment is in place. However, although policies are implemented, they tend to be market-based.

## The narratives

In the Market Forces scenario, market forces are the main determinants of: economic and social relations; the distribution of welfare, goods and services; the pattern of investment; and the development of human resources and institutions. Barriers to trade and to the movement of capital gradually diminish. In the emerging free-market economy, markets for goods, services and labour become increasingly interconnected and more integrated; unfettered financial flows and free trade facilitate this. Competition for markets and private investment stimulate initiative and become the main engines of economic growth. To promote market openness and global competition, new institutional arrangements are forged. Consumerism, individualism and self-centredness spread. Inequalities in the distribution of, and access to, the benefits of globalization grow at all levels: among individuals, communities, regions and nations. Overall, these patterns largely continue the global trends that have shaped much of the world economy over the past few decades and more.

Social and environmental stresses are resolved largely through incremental market and policy adaptations. Economic growth is expected to pay for the maintenance and repair costs of environmental damage incurred in the course of economic development. As such, policy makers are not encouraged to design mutually supportive environmental and trade policies, with the result that there is little or no strategic planning of environmental policies to mitigate the negative and to enhance the positive effects of rapid economic development.

In the Market Forces scenario, as the economic and social patterns and behaviours of developing countries converge towards those of the industrialized countries, no fundamental novel transformations, surprises or sharp discontinuities occur. Africa's future becomes increasingly shaped by the evolving patterns of globalization, and influenced by the dominant values, and social and economic relationships, of industrialized societies. Countries in Africa become increasingly integrated—nationally, regionally and internationally— as advances in communication systems, IT and the electronic media shorten both distance and time. Relations between people in different parts of Africa and across the world become more direct and immediate. Africa becomes part of the 'global village', connected through flows of knowledge and information, and stimulated by this to become more aware of, and active in, the global marketplace.

Despite increasing globalization, the nation state remains the dominant unit of governance, though national decision making is somewhat circumscribed by international treaties, agreements and obligations. Regional economic groupings emerge to pursue common development objectives. Examples are the establishment of economic trading blocks, such as: the Arab Magreb Union (AMU) in north Africa; the Economic Community of West African States (ECOWAS) in west Africa; the Economic Community of Central African States (ECOAS) in central Africa; the South African Development Community (SADC) in southern Africa; the Indian Ocean Community for the West Indian Ocean Community (IOC); and the Common Market for Eastern and Southern Africa (COMESA). There are other economic groupings, such as the formation of the Southern African power pool.

The participation of people, the private sector and non-governmental organizations (NGOs) in development continues to increase, however, as governments are encouraged to become more open, transparent, tolerant and democratic. Most national governments advance policy packages, which include steps to modernize the economies and financial systems of their countries, and to increase investment in education, in order to develop human resources and to improve their competitiveness in the global markets.

Many governments also privatize state enterprises in order to improve efficiency productivity and to reduce public expenditure. Nevertheless, pockets of resistance to these changes remain, and are characterized by corruption, nepotism and economic failure. As the welfare of populations is increasingly determined by the extent of their integration into a liberalized global market, countries which fail to embrace these changes fall further behind, both economically and in other ways.

In the Market Forces scenario, pressures of a growing population are not just confined to land-based resources. The demand for water increases—a consequence both of more people overall, and greater demand for water for agricultural, industrial and urban domestic uses. In an attempt to meet these demands, more dams are built and more groundwater is extracted. This localizes the distribution of surface water, and increases the cost of provision and distribution. Groundwater levels decline, amplifying the cost of extraction. Water pricing is introduced in order to recoup the costs of supply and to establish a mechanism for the economically efficient allocation of a limited resource. As water becomes more valuable, water recycling is encouraged. Pollution of water bodies by agricultural, industrial and domestic effluents is gradually regulated, though this takes some time to work efficiently. Despite these advances, however, a majority of people in many of the drier regions of the region remain short of water.

Given the continued prevalence of poverty in Africa, both as a result of the slow spread of economic benefits of globalization and the continued growth in the number of people in the lower socioeconomic strata, many countries in Africa continue to rely on foreign aid and external economic assistance. Foreign debt continues to increase, creating more pressure for export-led economic development, based primarily on the extraction of natural resources. The prospect of debt relief encourages economic structural adjustment and increased democratization, but the sustainability of these changes remains an open question.

The close juxtaposition of an expanding global culture and the largely conservative, traditional cultures of the region heightens social stresses. This is particularly the case between the younger members of society, who are more exposed to global culture, and their elders, for whom such changes seem to be a threat. These stresses are somewhat exacerbated by the fact that authority within communities is exercised largely by the elders in circumstances of change which, unlike the young, they resist embracing. Changes in social relationships within families, resulting from differences in education, lifestyle and location, cause the traditional support systems to break down.

## Environmental consequences
### Atmosphere
• *Africa*

Large areas of Africa are characterized by stable atmospheric conditions during the dry winter months. These result in marked thermal inversions, which trap emissions in the lower atmosphere. This natural phenomenon, coupled with a rise in emissions, results in a substantial decline in air quality. In the Market Forces scenario, despite the economic gains from urban and industrial development, most people on the region continue to rely on wood fuels for domestic energy. This is coupled with a rapid growth in industrial and vehicle emissions and, in some parts of the region, in the growth of emissions from coal-fired power stations. The result is an increase in greenhouse gas emissions and a marked decline in air quality especially, but not only, in urban areas. The incidence of acute respiratory infections and other diseases increases.

Although international environmental treaties and market forces create economic opportunities for African countries and industries, the region remains vulnerable to imperfections in the provisions of international treaties governing climate and other global changes. African countries act to protect the somewhat limited, but hard-won, gains in industrialization, because these fledgling industrialized sectors are considered essential to further economic and social development. Any abatement strategies which burden them in any way, or which increase their financial and operational risks, are considered unacceptable.

In the Market Forces scenario, the issue of tradeable permits also remains controversial. By investing in projects in developing countries that result either in the transfer of more energy-efficient technologies, or in some other initiative that reduces the level of emissions, developed countries or private sector firms could acquire carbon credits, while the developing countries are assumed to benefit economically from the investments. The underlying assumption is that the marginal cost of a unit reduction in $CO_2$ emissions is lower in developing countries than in developed countries, and that it is, therefore, more cost-effective to seek emission reductions

by appropriate investments in developing countries. The uncertainties surrounding these interventions, particularly their downstream economic and social costs, and the risk that developed countries will only invest in the low-cost, high-gain options—leaving the more costly options to be adopted later by the developing countries alone as their economies mature—causes most countries to hold back from agreeing to these provisions.

Nevertheless, Africa remains vulnerable to the effects of climate change, notably to: the increased incidence of droughts and floods; adverse changes in rainfall patterns; and the spread of diseases, such as malaria, to new zones. Reductions in agricultural production as a result of adverse weather conditions—resulting in increased dependency on food aid, the need for food imports, and reduced revenue from exports—as well as the threats of loss of biodiversity loss, changes in land cover and increased land degradation, all emphasize the region's interest in the issue of global climate change.

The continued depletion of the ozone layer, at least for the first two decades covered by the Market Forces scenario, continues to cause health problems, especially in the southern third of Africa. Although the use of ozone-depleting gases is gradually phased out, in line with the Montreal Protocol, pockets still remain because old and polluting technologies are slow to be replaced. In some cases, efforts to ban the use of certain ozone-depleting substances, such as methyl bromide, which is widely used as a fumigant in agriculture, are resisted because of the lack of cheaper or more effective alternatives, and inadequate assistance to governments to encourage and enforce the change.

• *Central Africa*

In the Market Forces scenario, the countries of this sub-region continue to pollute less, but they suffer from the effects of pollution from the north. The effects of climatic variation become more pronounced, especially in the Sahelian areas and in urban centres. The desert continues to encroach.

Initiatives linked to ozone layer protection and to the reduction of greenhouse effect gas emissions increase at an international level. Monitoring systems of pollution impacts are established, but their results remain mitigated due to non-compliance by some countries in the sub-region.

*In the Market Forces scenario, despite the economic gains from urban and industrial development, most people on the region continue to rely on wood fuels for domestic energy.*

*In some cases, efforts to ban the use of certain ozone-depleting substances, such as methyl bromide, which is widely used as a fumigant in agriculture, are resisted because of the lack of cheaper or more effective alternatives, and inadequate assistance to governments to encourage and enforce the change.*

• *Eastern Africa*

Initially, air quality deteriorates as urban centres expand and as more people, with increased incomes, are able to afford cars. However, measures (standards) are put in place to curb vehicular emissions. Increased tax on fossil fuels forces industries to revert to other alternatives, such as electricity or gas, which have lesser emissions. Climate variability, however, remains an issue, but its impact is reduced because the sector which is most directly impacted by climate variability (that is, agriculture) is adequately funded to offset the impacts through irrigation.

• *Northern Africa*

As the large cities grow larger, air pollution increases. Vehicles and industry remain the main sources of air pollution and emission of greenhouse gases. Nevertheless, as the problem reaches alarming levels, especially in urban and industrialized areas, the major air polluting industries—such as transportation, air conditioning and cement—take appropriate measures to reduce their environmental impacts. Vehicles, especially those used for mass transport, shift to using natural gas instead of fossil fuels. The use of chlorofluorocarbons (CFCs) and other ozone-depleting products in the air-conditioning industry is reduced.

• *Southern Africa*

While virtually all countries of this sub-region have laws to control air pollution, the state of air pollution is not well known. However, it is known that southern Africa's global contribution to air pollution is low, although the situation in South Africa is worrisome. In many countries, air pollution is treated as a health issue and not as an environmental issue, such that monitoring and enforcement are a problem due to different priorities, capacity and resources between health and environment institutions. Nevertheless, global initiatives on the impact of greenhouse gases raise the issue of atmospheric pollution on the sub-regional agenda. The trend is that atmospheric pollution continues to attract attention in the next 30 years, but financial and human resources limit the impact of monitoring and enforcement initiatives.

• *Western Africa*

While the market may tend, initially, to pay no attention to air quality, the nature of the competition forces the

issue. All competitors—as a matter of marketing strategy, public relations and necessity—make 'clean air' an essential part of their activities. The scope of the 'clean air' programmes covers vehicles, agricultural equipment, electricity production generators, sewage and garbage disposals, and other threats to the atmosphere.

• *Western Indian Ocean Islands*

Air pollution increases as forests and woodland are further depleted in order to provide fuel. Continued tolerence of older motor vehicles, with high carbon monoxide (CO) output, exacerbates the impact on urban areas. Port Louis and Antanoravino continue to close down frequently in the summer as a result of urban smog. The burning of baggage as an oil substitute for energy production further pollutes the air, and reduces the cost-effectiveness of solar power systems.

## *Land*

• *Africa*

In the Market Forces scenario, the predominant trend to emerge is one whereby land is used in the most economically efficient way, especially in agriculture. In the drive to develop an export-led market economy, more agricultural land—and particularly the better quality land—is used for producing cash crops and agricultural commodities. Such agricultural production increasingly falls under the control of multinational companies, which have both financial resources and access to overseas markets in order to make these ventures viable. Decisions about agricultural production become increasingly determined by external forces. Agriculture becomes more vulnerable to the vicissitudes of global markets, especially fluctuating commodity prices.

With more of the better quality agricultural land being devoted to export agriculture, the growing numbers of rural poor will be forced to use more marginal lands for subsistence agriculture. Thus, those lands that would normally require substantial technological and other inputs in order to ensure sustainable use end up being used by those who are least able to supply the necessary inputs. The use of these lands is likely to be unsustainable unless efforts are made to provide technological and other support. Without this, soil erosion and nutrient losses are likely to occur, leading to a reduction in soil fertility, and an increase in the siltation of rivers and dams.

The drive to increase the economic returns from land also leads to a diversification of land uses. Many countries in Africa rely increasingly on the development of tourism in order to acquire foreign currency and, therefore, land continues to be set aside and maintained for tourism (or tourism is included in the multiple land use developments). Pressures to offset carbon emissions by industrialized countries through developing carbon sinks in non-industrialized countries, including some in Africa, leads to increased investment in the protection of established forest lands and the reforestation or afforestation of non-forested lands. This displaces some traditional land use practices.

The prospect of employment also attracts increasing numbers of people to areas of tourism development, resulting in concentrated population pressures in attractive areas (that is, coastal resorts). For many immigrants, however, their lack of skills limits opportunities. Consequently, many of them resort to hawking, manufacturing crafts and curios, and other secondary activities. The oversupply of crafts and curios drives down prices, increasing the bargains for tourists, but also increasing their harassment, as hawkers become more relentless in their efforts to sell their goods. This gradually diminishes the attractiveness of some tourist locations.

With the growing economic activity, increasing numbers of people are drawn to urban areas in search of employment. Again, opportunities are limited, because the immigrants are generally unskilled and only qualify as manual labour. Because more people arrive than can immediately find employment, particularly in better-paid jobs, squatter settlements and hastily constructed high-density housing blocks are created. Even in the wealthier suburbs and surrounding areas, the rate of establishment of new houses and housing complexes overwhelms the capacity of town and city planners to coordinate and integrate these developments, leading to unplanned urbanization. The upgrading and expansion of infrastructure (roads, power, water and sanitation) lags behind these developments.

Migration away from rural areas tends to relieve pressure on land resources in some areas, but is more than offset by continued high population growth rates in others. This continues the trend towards unsustainable use of natural resources ,and increases the likelihood and extent of environmental degradation. Changes in lifestyles also affect land use. Increasing demand for meat, particularly among the more wealthy people in urban areas, encourages increased use of land for grazing or, where livestock production is intensified, for fodder and feed grain production.

### • *Central Africa*

Overexploitation leads to the incidence of soil erosion. Access to land is reserved for people who hold the means of production, thus breeding the genesis of land tenure-based conflicts.

### • *Eastern Africa*

The pressure on land is relieved by a thriving industrial sector, which creates employment and takes most people away from the land (farming). Commercial farming becomes more dominant, with measures taken to sustain productivity on the land. Such measures include soil erosion control and fertility enhancement. A functional land market makes it possible for individuals with capital to invest in land or to buy land from those unable to utilize it.

### • *Northern Africa*

The pressure on land, and on other vital resources, mounts. Agricultural land is the most affected, because it gets taken up by urbanization, with increasing numbers of people migrating from rural to urban areas. As the pressure on agricu1tural land increases, in order to provide food for an increasing population, the land becomes overexploited, with all kinds of unsafe and unsustainable farming practices. Soil salinity, water logging, soil erosion and land pollution are some of the types of land degradation which emerge.

### • *Southern Africa*

Soil erosion is probably the most important factor in the decline in agricultural productivity in southern Africa, degrading about 15 per cent of this sub-region's land. This trend is likely to continue over the next 30 years, due to population pressure, skewed land tenure systems and increasing demand for land. A more serious trend, resulting partly from soil erosion, and one that is reflected throughout the world, is the decline in per capita food production. Southern Africa has

*With more of the better quality agricultural land being devoted to export agriculture, the growing numbers of rural poor will be forced to use more marginal lands for subsistence agriculture. Thus, those lands that would normally require substantial technological and other inputs in order to ensure sustainable use end up being used by those who are least able to supply the necessary inputs.*

produced net food surpluses for many years, but in the Market Forces scenario, the drop in per capita food production leads to net food deficits well before 2032. This is despite the promise offered by new seed varieties and by improved agricultural technology, which are unlikely to catch up with population growth.

• *Western Africa*

There is intense competition for land for investment in agriculture and manufacturing. The good rainfall in most of the region makes large-scale farming and the building of agro-industry a profitable investment. Local subsistence farmers are adequately compensated, not only through direct purchases, but by employment options.

• *Western Indian Ocean Islands*

Urban sprawl increases as subsistence agriculture replaces estate management, with a sharp decline in prices for sugar, tea, coffee and fruit. Disused factories from textiles and other small industries are taken over by squatters. Uncontrolled mineral exploitation in Madagascar degrades the surrounding land, and processing pollutes water in rivers, aquifers and coasts. Illegal farming of cannabis and poppy seed overtakes tobacco, as African and Asian influences in drug use permeate the local culture, funded with laundered money from Latin America. Town and country planning is undermined by weakness of government and extensive corruption of officials, promoting urban sprawl and the collapse of protected areas, which are converted into leisure centres and commercial game hunting parks. Remote areas are sold off for dumping imported hazardous waste.

### Biodiversity

• *Africa*

In the Market Forces scenario, the dominant trend affecting terrestrial biodiversity continues to be the fragmentation and loss of habitats, resulting from land transformation. Not only does this reduce the populations of some species to below sustainable levels, but the associated disruption of ecosystem integrity and functioning alters the conditions necessary for the survival of others. Similar changes occur in aquatic ecosystems although, here, the changes are driven by impoundment, sedimentation,

and prolonged isolation of water bodies. Coastal and marine ecosystems, and their biota, are likewise threatened by loss of habitat, pollution and overharvesting of biological resources. In all cases, the trends are driven by the ongoing growth of the human population and its increasing demands for land, water and natural resources.

Concern over the loss of biodiversity continues to stimulate both national and international conservation efforts but, given the magnitude of the countervailing forces, most of these prove to be inadequate. This is exacerbated by the widespread undervaluation of biodiversity, and differing perceptions within society of the economic, environmental and cultural values of biodiversity. Moreover, the assigned values are not just those of people living in Africa but, at a global level, are strongly influenced by outside interests, shaped by tourism and the media. This produces both benefits and drawbacks. People and organizations from outside Africa are willing to invest time and money promoting the conservation of African wildlife, but they also seek to dictate policies on how components of that diversity should be managed and used. For example, trade in wildlife and wildlife products remains a hotly contested issue, even though it can be argued that this would generate more resources for conservation.

Ecotourism continues to grow. Efforts to ensure that local communities benefit financially and in other ways are expanded, but benefits, in many cases, are relatively small. Nevertheless, the sense of empowerment and the prospect of improved benefits at a later date encourage many communities to remain engaged in the management and conservation of their natural resources. This concept of community-based management of natural resources is gradually extended to communities which rely on freshwater, coastal and inshore marine resources. Species are introduced, both deliberately and inadvertently. The spread of alien organisms increases pressure on indigenous species, and contributes to the gradual loss of biodiversity. Extensive cash-crop monocultures reduce on-farm agrobiodiversity. The resulting loss of natural predators, together with other disruptions to ecosystem integrity, create conditions for widespread outbreaks of pests and diseases.

Despite concerns in some quarters over the development and release of GMOs, and the resulting risk of genetic pollution of indigenous species, these are overridden by commercial pressures. The release of GMOs threatens agricultural biodiversity in some areas, especially where farmers depend on maintaining a mix of species and races as a hedge against annual and seasonal variations in farming conditions. This is made worse by the eventual release of GMOs encoded with a terminator gene, which ensures that the seed produced is non-viable, thereby requiring farmers to purchase another batch of seed the following season, instead of retaining some of the seed for planting next year. The strict protection granted to GMO producers by Plant Breeders' Rights, along with the provisions of patents on terminator technology, allow plant breeders to collect revenues every year from the users.

The search for organisms with unique genetic and biochemical properties continues. Again, the shortage of technical knowledge and skills within Africa means that much of this bio-prospecting is carried out by scientists from overseas. Materials stored in gene banks overseas are largely inaccessible to African countries, including those from which the material originated. The benefits of conserving biodiversity are seen by many in Africa to be unfairly shared, with royalties accruing to multinationals rather than to the countries from which the material was derived. This becomes a source of tension and weakness. Africa resolves to continue conserving biodiversity and to cooperate with overseas organizations.

• *Central Africa*

The pressure on biodiversity increases because of the search for raw materials for industries. The pharmaceutical industries develop research on medicinal plants, which are later resold in big markets.

• *Eastern Africa*

Biodiversity takes centre stage as a fulcrum for tourism development. The private sector becomes interested in biodiversity conservation as all possible niches are explored for tourism consumption, for example, bird watching, ecotourism and sub-marine tourism. Government identifies the biodiversity hotspots which should be accorded maximum protection, and gazettes them as national parks or strict nature reserves. Deliberate efforts are directed towards rehabilitating degraded ecosystems and restoring species richness.

Some ecosystems, however, are subjected to so much pressure that their credibility as areas of significant biodiversity importance is diminished, and is eventually overrun by industrial development or urban settlements. Where this involves protected areas, degazetting procedures are undertaken, involving Environmental Impact Assessment (EIA).

• *Northern Africa*

In the Market Forces scenario, biodiversity is under threat from population growth, urbanization, industrialization, and uncontrolled hunting and fishing. The concentration of new developments along the coasts of the Mediterranean Sea and the Red Sea has adverse impacts on fauna and flora. Desert tourism becomes a very popular theme in Northern Africa but, without careful regulation, it impacts the biota of this environment. Due to weak law enforcement, illegal trading in rare birds and animal species continues. Conflicts and wars in some countries are also another threat to biodiversity.

• *Southern Africa*

Biodiversity loss in Southern Africa has been a consequence of human development, as species-rich woodlands and forests have been converted to relatively species-poor farmlands and plantations. However, the number of threatened species could be higher, as the full extent of this sub-region's species diversity is not known. There is a serious lack of species inventories and other baseline data, which are useful for monitoring biodiversity trends. While overexploitation of biological resources is acknowledged in the sub-region, trends show that there is a greater commitment to conserve wildlife, particularly large mammals. The introduction of community-based natural resource management programmes (CBNRMs) in some countries in the sub-region (for example, CAMPFIRE, Peace Parks, Tchuma Tchato and ADMADE) has seen communities playing a role in the management of biological resources. According to the Market Forces scenario, such trends are certain to continue in the next 30 years, and will be strengthened as the CBNRMs expand their activities to other resources, such as forests and fish.

*The release of GMOs threatens agricultural biodiversity in some areas ... This is made worse by the eventual release of GMOs encoded with a terminator gene, which ensures that the seed produced is non-viable, thereby requiring farmers to purchase another batch of seed the following season, instead of retaining some of the seed for planting next year.*

### • Western Africa

Biodiversity remains a major attraction for tourists, and also protects the ecosystem. In an atmosphere where the market is determining progress, competing interests are allowed to promote biodiversity. A number of conservation measures are developed, and are functioning. The government and corporate entities are cooperating, although government has defined its role to one of guidance, whilst the businesses actually implement the measures. In this way, there is little room for destruction of biodiversity as a result of industrial growth.

### • Western Indian Ocean Islands

Progress and maintenance of existing projects are undermined. In the richer countries, some legitimate ecotourism underpins the more commercially appealing projects but, as the quality of tourism declines and as budgets fail, the protected projects become degraded into peep shows. Qualified technical staff migrate, and rare species die off through lack of adequate protection, recorded only on camera and in glass cases as a menagerie alongside the stuffed dodos from Mauritius. Foreign immigration brings with it more alien plant and animal species, which overrun the local delicately balanced ecology. The dying remnants of the rich endemic biodiversity of the region are still to be glimpsed at pre-hunt cocktails in the occasional private parks adjacent to the houses of the rich.

### Forests

#### • Africa

The historical trend of deforestation and degradation of forest areas continues, as a result both of the need for more land for human settlement and agriculture, and of the drive to exploit forest resources (mainly timber) to boost export earnings. The rate of deforestation and land conversion, however, slows down, due to afforestation, decreasing forest exploitation in many parts of Africa and the gradual implementation of existing conventions to use forests in a sustainable manner. The increase in ecotourism provides additional incentives to conserve forests, though this is offset to some extent by the increase in land pressures. In order to secure the future of forested areas, as well as to spread the benefits of their conservation more widely, communities living in the forests and surrounding areas are encouraged to take part in their management, sustained use and conservation. This, too, contributes to slowing down the rate of deforestation.

#### • Central Africa

Trade liberalization, with appropriate technologies, improves the quality of logging. The industrial and artisan exploitation of wood increases the pressure on forest resources, and sparks off the potential of this sub-region for deforestation The strategies to protect the forest, and rational management of forest resources, are difficult to apply, in spite of the number of conventions and initiatives developed to this effect.

#### • Eastern Africa

The forest sector plays a major role, especially in the growing construction industry. Private-sector interest in forestry grows, and most of the plantations are privatized for proper management and profitability. The government focuses more on the protection of forests, where activities are strictly regulated, thus improving the quality of these forests.

#### • Northern Africa

Although the forests in Northern Africa are quite limited, the role they play in stabilizing the sand dunes and providing some quality products, such as gum arabic, cannot be neglected. Unfortunately, the high deforestation rate currently experienced in this sub-region continues, albeit at a slightly lower rate. On the one hand, as the population increases and demand for resources increases alongside it, forests continue to be used as a source of fuelwood and charcoal. On the other hand, natural disasters—such as fires and severe droughts—also take their toll on forests. Despite the acknowledgement of the deforestation problem by governments, the pressures mentioned above outweigh the afforestation and reforestation efforts.

#### • Western Africa

The threat to forests comes with huge commercial farming and massive agro-industrial complexes. The forests have already been experiencing losses, which were not adequately compensated through reforestation. In the Market Forces scenario, the competition of market forces rejuvenates interest in, and compels, reforestation and afforestation.

Environment-sensitive investment policies help competing investors to make concern about the forests part of their plans.

• *Western Indian Ocean Islands*

Forests, including those currently protected, become decimated as the local commercial value of wood for fuel and other uses exceeds the political will to protect it. Another impact is that environmental NGOs get bought out by commercial interests, and dismantled by increasingly right-wing central governments.

## Freshwater

• *Africa*

As a consequence of the growing numbers of people in Africa, the availability of water, both absolutely and per capita, declines overall (although accessibility in urban areas increases whilst, in rural areas, it generally decreases). This population increase is coupled with greater urbanization (people in urban areas use more water per capita than those in rural areas), and greater demand for water for agricultural and industrial development. Demand for water increases across all sectors of the economy, leading to further water scarcity. There is more competition and conflict over available freshwater, both between economic sectors and, in some cases, between countries which share a common water source.

In an attempt to meet the rising demand for water, more dams are built and more boreholes are established. There is a marginal increase in the proportion of water that is recycled. The increased extraction of groundwater results in a decline in groundwater levels. Contamination of groundwater—caused by leaching of agricultural and industrial chemicals, and human and animal waste—causes deterioration in groundwater quality. The area of land under irrigation increases, supplied from both stored surface water and groundwater sources but, in some cases, inappropriate irrigation techniques result in salinization of the soil, eventually causing the land to be abandoned.

In the absence of sound environmental management, water pollution—generated through increased industrial and agricultural activity—leads to higher incidences of eutrophication and the spread of water weeds, such as water hyacinth (Eichornia crassipes) and water cabbage (Pistea stratiotes). Microbial contamination also increases, as does the incidence of waterborne diseases, particularly as more still and stagnant water bodies are created. Reductions in land cover result in more run-off, erosion and, eventually, sedimentation of rivers.

• *Central Africa*

The pressure on water resources increases with population growth. There is a rise in water demand for agricultural and industrial production, and for the urban population. The Sahelian areas are abandoned to desert encroachment. There are many conflicts, linked to water issues, between breeders and agriculturists.

• *Eastern Africa*

Freshwater supply remains the main focus of the governments of this sub-region. Measures are undertaken and investment is targeted to reduce the problems in the deficit areas of Ethiopia, Kenya, Somalia, Djibouti and Eritrea. Increased incomes from the industrial sector makes these investments possible. The concept of rainfall harvesting is actively promoted in the countries of Uganda, Rwanda, Burundi, Kenya and Ethiopia, where precipitation is adequate. This increases freshwater access in rural areas.

• *Northern Africa*

If a large increase in population is translated into higher demands for resources, water is one of the most directly affected resources. In the Market Forces scenario, competition for water between various sectors, especially the agricultural and industrial sectors, increases, as each sector tries to manage the very limited resources available. The problem of water scarcity in this sub-region only gets more acute. Unsafe exploitation of groundwater resources, due to over-abstraction and pollution, reaches its worst levels in 2032, especially in Egypt and Libya. Sea water intrusion along the coast of the Mediterranean Sea becomes a major water-quality problem, and a threat to the sustainability of shallow groundwater wells. Water reuse is on the increase, but the quality of the reused water poses serious risks to the environment. Water recycling is not yet widely implemented, due to the high cost of treating the water. The number of desalination plants along the long coasts increases, but the technology is still only feasible for wealthy industries, such as tourism. Water, the scarcest resource in the sub-region, becomes one of the factors which limits further development. A serious review of the integrated water resources management programmes of the Northern African countries is called for, in order to rescue the badly hit vital resource.

#### • Southern Africa

The demand for water continues to rise in response to population growth and industrial development. Groundwater continues to be the dominant source of water for rural people. Due to the high demand for water, wetlands continue to be threatened, not only by human factors, but also by natural factors such as drought. The beginnings of a sub-regional water strategy is a positive trend, which sees growth in water-sharing arrangements between water-rich and water-poor countries. However, the water transfers and the increased use of water storage could have a negative impact on the environment. As more dams are built, there are fewer natural rivers, and a substantial loss of habitat. As such, the next 30 years sees: a drop in both river and dam water levels; damage to floodplains, due to loss of annual flooding; and disruption to estuaries, when the mix of fresh and salt water changes.

#### • Western Africa

Most of the countries of Western Africa have a fairly good supply of freshwater. Existing bodies of water are regularly supported by good annual rainfall. More sophisticated management techniques and measures ensure the all-year availability of freshwater in most of this sub-region. Of course, the countries that are being encroached upon by the Sahara Desert apply more resources, which are assessed from investments in the integrated economies of the sub-region.

#### • Western Indian Ocean Islands

Freshwater quality declines in the middle-income countries, as investment in public sector services declines, and as the richer groups concentrate in safe water areas or have their own purification systems. Water in the poorer countries becomes a scarce commodity. Maintenance of water pumps is reduced, and the growing population creates more pollution of existing water sources.

### Coastal and marine environments
#### • Africa

Coastal and marine environments are relatively rich, in terms of resources and economic opportunities. Consequently, in the Market Forces scenario, they come under greater pressure, as more people seek to take advantage of those opportunities. Environmental degradation—characterized by the loss of coastal barrier communities, such as: coral reefs, mangroves and other plants; coastal erosion; pollution; and the depletion of fish stocks—is widespread in areas where development is unregulated. In other areas, where careful coastal-zone planning is implemented and maintained, the high environmental qualities of these ecosystems are maintained, and sustainable benefits accrue. In particular, opportunities for properly managed tourism development create both the incentives and the money needed for managing these coastal and marine resources.

#### • Central Africa

The coastal and marine resources exploitation by the private sector increases erosion and the destruction of mangrove swamps.

#### • Eastern Africa

A rapidly growing tourism industry and affluence lead to extensive development of resorts on the beaches. Governments recognize the importance of coastal resources, and move in to strictly regulate the developments at the coast. Sensitive ecosystems at the coast are gazetted for protection. Discharge of effluents into these ecosystems is strictly controlled. There is the capacity and the institutional framework to monitor coastal development. Marine resources harvesting is regulated through licensing and policing procedures.

#### • Northern Africa

The growth in population numbers and densities along the coasts of the Mediterranean Sea increases marine pollution. Untreated wastewater is the main source of pollution, causing eutrophication in coastal waters. In addition to high pollution densities, other risks to the coastal zone and to marine life are: the oil mining industry; marine tourism; and the large increases in the numbers of ships visiting the ports, as the sub-region opens up to free trading with the global market. The increased population puts more demand on fish, which is sometimes caught using illegal methods, such as using poisons.

• *Western Africa*

Some 13 of the 16 countries of this sub-region have access to the Atlantic Ocean. This may be both a blessing and a curse. The curse could be the temptation by unscrupulous elements to dump all manner of hazardous wastes at sea. However, in the Market Forces scenario, with vigilant governments of the sub-region working together and with the ensured participation of corporate bodies, the seaways are properly patrolled, and measures for cleaning up the coastlines are brought into operation. Fishing is controlled, whilst ensuring that the fishing business remains profitable for both the small fishermen and for corporate entities.

• *Western Indian Ocean Islands*

Coastal waters become increasingly polluted and overfished. The coral lagoons become salt-water, dead-sand seas, devoid of aquatic life. Uncontrolled high-speed motor watersports dominate the scene, and add to the oil and noise pollution of the areas. Deep-sea fishing is industrialized and internationalized. The region fails to establish territorial command of the oceans to which it has legitimate claims, and the poorer countries sell their rights, thus undermining the potential for regional negotiations. Lack of international policing gives rise to commercial security protection and to violent piracy, with continual armed disputes, disrupting other sea trade.

*Urban areas*
• *Africa*

In the Market Forces scenario, both the proportion of people living in urban areas and the rate of growth of urban populations increase, as does the number of cities in Africa with populations greater than one million people. The growth in urban populations is driven mainly by the migration of people from rural areas, who come to the towns in order to seek employment and opportunities for a life outside agriculture. A disproportionate number of immigrants are men—mostly young adults. The rate of urban population growth initially outstrips the capacity of the municipal and central government authorities to provide the necessary services and infrastructure, so that the number of slums and unplanned peri-urban settlements increases.

These changes have generally negative effects on the environment in the vicinity of towns, which provides fuelwood for the increasing numbers of urban poor. In terms of air, water and waste pollution, many urban environments become degraded, although environmental quality does improve in a few of the wealthier cities. The generally negative effects of urbanization are offset, to some extent, by some positive changes. As the cities grow, the proportion of the population with access to electricity, piped water and sanitation also increases, and water management services become more effective. Over time, these positive effects on the environment are further enhanced by a marginal decline in the proportion of the population living in poverty in slums and other unplanned settlements.

• *Central Africa*

The development of cities continues unabated as farmers continue to migrate from rural areas. The private sector occupies a vital place in the urban management process.

• *Eastern Africa*

There is steady urban growth, but this is planned and the quality of services for the urban dwellers is improved. Industrial emissions and discharges are controlled through regulations, and by industries being located away from major settlements and sensitive ecosystems. Urban authorities are given adequate capital through streamlined revenue collection systems, and are able to deliver improved services including, refuse collection and street lighting. There is more private sector involvement in the provision of services in urban areas.

• *Northern Africa*

In the Market Forces scenario, some of the largest cities in Northern Africa are expected to grow larger. The consequence of this is that many resources and services become very stretched to satisfy increasing demands. Large numbers of people are deprived of basic services, such as safe water, sanitation and power supply. Air pollution from the large numbers of vehicles in congested cities reaches alarming levels. However, appropriate attention and good measures are taken to reduce the problem. The large volumes of solid waste generated in urban areas overburden the municipalities, which are responsible for managing such waste. Improper incineration of solid waste adds to the problem of air pollution.

### • Western Africa

In the Market Forces scenario, the explosive rural-urban migration—which is associated with locating industries, educational and health facilities in the cities—is controlled. With huge commercial farms, cooperatives and agro-industrial plants situated in rural areas, pressures in the urban centres are reduced. Some industrial plants are actually relocated from major urban centres to locations formally considered to be 'countryside'. Urban authorities continue to encourage the cooperation of business and communities to clean up, repair or construct facilities to create a safer urban environment. Competition begins to show signs of renewal.

### • Western Indian Ocean Islands

Urban areas become the living quarters of the poor: overcrowded, polluted, and with ever fewer jobs and services—the seat of misery and disease. The richer classes move to the deserted countryside, living off foreign earnings and the sale of land to wealthy immigrants. Town planning and maintenance are restricted to the few multinational commercial areas, leaving the rest with poor water, sanitation, pollution, overcrowding, potholed roads, crime and destitution. Coastal areas and the countryside become the sought-after areas for commercial immigrants and retired people with fixed incomes from abroad. This forces up land prices and drives the indigenous population into the squatter areas of towns, whose infrastructure fails to keep pace with the growth of the population.

## THE POLICY REFORM SCENARIO

### Introduction

The beginning of the millennium saw a renewed commitment to address issues of sustainability and the environment. A consensus emerged on the urgent need to temper what had come to be called the Market Forces scenario, with policies to secure environmental resilience and to sharply reduce poverty. The Policy Reform scenario is not a radical deviation from the Conventional Development scenario. The emphasis on economic growth, trade liberalization, privatization and modernization endures. The integration of the global economy proceeds apace, as poorer regions converge very gradually toward the model of development of the rich countries. The values of individualism and consumerism persist, transnational corporations continue to dominate the global economy, and governments modernize their economies and social welfare structures. The defining feature of the Policy Reform scenario is the emergence of the political will to constrain market-driven growth with a comprehensive set of sustainability policies.

The Policy Reform scenario is based on a set of social and environmental goals adopted by the international community. These guidelines are adjusted periodically in light of new information. Social and environmental targets are set at global, regional and national levels, and include a mix of economic reform, regulatory instruments, voluntary actions, social programmes and technology development.

Unlike the Market Forces scenario, the Policy Reform scenario tempers market-driven prescriptions with strong social and environmental policies. It thrives on the harmony between different stakeholders and otherwise divergent policies. It is consistent with development which the Managing Director of the IMF, Michel Camdessus, said not only promotes liberalization of trade and capital movements, but also emphasizes transparency, accountability, democratic governance, fighting corruption, alleviating poverty, gender equality, increasing aid, debt relief and market access to developing countries (Raghavan 2000).

The assumptions of the Policy Reform scenario may be summarized as follows:

- It is similar in many ways to the Market Forces scenario.
- It is based on a set of social and environmental goals adopted by the international community, and set at global, regional and national levels.
- There is an emergence of the political will to constrain and to guide market-driven growth with a comprehensive set of sustainability policies.
- Policy initiatives for achieving goals are regionally differentiated, but include a mix of economic

reform, regulatory instruments, voluntary actions, social programmes and technological development.

- The 'western' model still prevails, and 'western' values still spread.
- There is less trust in automatic positive results from markets, and more emphasis on targeted policies.

### The narratives

The Policy Reform scenario is the Market Forces scenario with a human face. It not only embraces the market-driven prescriptions of the Breton Woods institutions—the IMF and the WB—but it is also strong on social and environmental policies. The Market Forces scenario is premised on the so-called Washington Consensus, which says that good economic performance requires liberalized trade, macroeconomic stability and getting prices right (Stiglitz 1998). The Washington Consensus supporters 'unreservedly promote free trade, financial liberalization and foreign investment incentives, business deregulation low taxes, fiscal austerity and privatization, and flexible labour markets' (Bond 2000; see also Box 4.1).

For Africa, the Policy Reform scenario offers an opportunity for the region to break with more than four decades of 'unfulfilled promises of global development strategies' (OAU 1980). For instance, in the years between decolonization, which began with Ghana in 1957, and the democratization of South Africa in 1994, the region has been 'unable to point to any significant growth rate, or satisfactory index of general well-being' (OAU 1980). Some of the important challenges facing Africa at the beginning of the 21st century, therefore, include: a population which is growing rapidly, at a rate which is faster than food production and which is beyond the capacity of some resources to satisfy such demand; growing poverty in both rural and urban areas; millions of refugees, due to wars in different sub-regions; growing urbanization, introducing new environmental issues; HIV/AIDS; land degradation, particularly desertification; deforestation; recurrent droughts; increasing demand on the finite water resources; water pollution; and biodiversity loss.

Since the Stockholm Conference on the Human Environment in 1972, Africa has participated in many conferences, such as: the 1990 World Conference on Education for All; the 1990 World Summit for Children; the 1992 United Nations Conference on Environment and Development; the 1993 World Conference on Human Rights; the 1994 International Conference on Population and Development; the 1995 United Nations Fourth Conference on Women; and the 2000 Millennium Summit. In addition to these international initiatives, African countries have also convened their own important meetings, which have set targets for economic and social development, and environmental management. Meetings under the auspices of the Organization of African Unity (OAU), which led to the adoption of the Lagos Plan of Action—the region's blueprint for economic development—in 1980 have helped to highlight the challenges facing the region. Under the Lagos Plan of Action, African leaders emphasized that 'Africa's huge resources must be applied principally to meet the needs and purposes of its people'. They also emphasized the need for Africa's virtually 'total reliance on the export of raw materials' to change, and for the need to mobilize the region's entire human and material resources for the development of Africa (OAU 1980).

---

### Box 4.1 Camdessus speaks at UNCTAD-X Interactive Sessions

'Now we know, it is not enough to increase the size of the cake. How the cake is shared is equally relevant to the dynamics of development… it is recognized that the market can have major failures, that growth alone is not enough or can even be destructive of the natural environment or precious social goods and cultural values.

'Only the pursuit of high quality growth is worth the effort – growth that can be sus-tained over time… growth that has the human person at its centre… growth based on continuous effort for more equity, poverty alleviation, and empowerment of poor people, and growth that promotes protection of the environment and respect by national cultural values… a striking and promising recognition of a convergence between a respect for fundamental ethical values and the search for efficiency…

'… the new emerging paradigm, rooted in fundamental human values, taken together with a better ability to prevent and manage the crises, is a distinct and positive chance of our times… a new perception of globalization is emerging… a call for common action to trans-form globalization into an effective instrument for development. Globalization can be seen in a positive light, not what some have portrayed it to be, a blind, potentially malevolent force that needs to be tamed… a logical exten-sion of the same basic principles of economic and human relations that have already brought prosperity to many countries…'

Source: Raghavan 2000

The Lagos Plan of Action is one of many measures adopted by the region which set qualitative and/or quantitative targets that should have been met by the new millennium. Unfortunately, many of these targets remain unmet, largely because of errors of judgement, of both omission and commission. Nonetheless, the 21st century marks the beginning of a new dawn. The symphony created by an informed populace, fully familiar with its rights; the commitment by political leaders to serve their people rather than their egos; the development of strong legal and institutional frameworks; the willingness by all stakeholders to constantly keep their development plans under review; the development of a strong entrepreneurial base; and breakthroughs in science and technology see Africa claim its place as one of the leading regions in the world. Africa takes policy reform seriously, looking within for any shortcomings and enhancing its strengths.

## Environmental consequences
### Atmosphere
#### • Africa

In the Policy Reform scenario, the increased economic activity across Africa—spearheaded by Côte de Ivoire, Egypt, Nigeria and South Africa—sees the number of least developed countries in the region go down to just five, compared to 33 at the beginning of the century. Increased manufacturing leads to serious problems of air and water pollution. Governments introduce stringent measures to curb such pollution, including the polluter pays principle and trade-in pollution permits. The polluter pays principle is strictly enforced, and companies start to introduce self-policing measures. These measures subsequently see pollution gradually decrease as the permits increasingly become more expensive, thus making the products of polluters more costly and less competitive.

#### • Central Africa

The harmonization of regulations allows for pollution control, but the sub-region continues to suffer the consequences resulting from pollution from the north. The development of an air pollution observatory helps to intervene with regards to levels of pollution.

#### • Eastern Africa

The issue of atmospheric pollution takes centre stage as public health policies become strict and assertive.

Vehicle emissions are strictly regulated and alternative sources of energy, such as solar power, is promoted. The Montreal Protocol on ozone-depleting substances (ODS) is enforced throughout the sub-region, with the customs departments strictly monitoring the illegal movement of such substances. The use of methyl bromide is phased out in the flower industry.

#### • Northern Africa

In the Policy Reform scenario, governments take good steps towards improving and preserving air quality in this sub-region. More people switch from using private cars to using better and reliable public transport services. Vehicles which run on natural gas are offered lower licensing fees. Solar energy is used in tourism and housing developments, because the units are manufactured locally and, hence, become affordable. The emission of greenhouse gases from industry, power generation and some agricultural activities is controlled through government-set regulations. The cooling and air-conditioning industries are banned from using CFCs and other environmentally harmful products. The overall improvement in air quality has positive effects on public health.

#### • Southern Africa

Improvements in early warning systems result in the sub-region being better prepared to handle climate variability. The impacts of climate variability become very much reduced, as a result of new technologies producing seed varieties which can withstand variable climatic conditions, and also as a result of better housing.

#### • Western Africa

In the Policy Reform scenario, as a result of years of scientific research and successes elsewhere, a 'clean air' policy is institutionalized—at the level of the home, workplace and community, and across national borders—through education, broad consultation and encouragement, led by government. All major environmentally harmful products which cause air pollution are banned or controlled. Vehicle, aircraft and air conditioning gases, gases for home use and other gases are controlled. Industrial pollution remains at current levels of manageability. The major oil producing country, Nigeria, leads the 'clean air' campaign by putting adequate measures in place. Major petroleum refineries in the sub-region are modernized, and use processing methods which have very few pollutants.

• *Western Indian Ocean Islands*

Pollution levels are reduced in intensity, but careful monitoring shows that they are now extending into more urban areas as the use of motor vehicles increases.

## Land

• *Central Africa*

In the Policy Reform scenario, modern law and traditional law, in terms of land use management, are harmonized. This harmonization entails the reduction of conflicts linked to land use. The procedures of acquiring land occupancy papers are simplified, and the costs reduced. The ongoing population pressure remains controllable, and is managed according to the areas with a greater population density.

• *Eastern Africa*

Land reform programmes are supplemented by comprehensive land use policies. All major land use programmes undergo EIAs to identify mainstream environmental considerations. Universal primary education policies increase literacy, and have a ripple effect in contraceptive prevalence and family planning. There is a subsequent reduction in the population growth rate, thus relieving pressure on land. Urban migration is reduced, because there are deliberate policies to increase land-based employment opportunities in rural areas.

• *Northern Africa*

The relative slowing down in the population growth rates eases demand on land. The pressure on resources is further relieved as governments give more emphasis on the planning and implementation of land use plans. The reduction in the numbers of migrants from rural to urban areas helps to reduce the rapid unplanned expansion of urban areas. Due care is given to the problem of the loss of good agricultural land due to soil salinization in agriculture, and organic fertilizers are employed.

• *Southern Africa*

Land reforms in Southern Africa are guided by principles of good governance and a taxation system in order to maximize productivity and profitability, leading to equitable land distribution. Cash farming becomes a predominant activity, with the sub-region engaging itself in farming activities in which it has comparative advantages. Food security is achieved through trade.

• *Western Africa*

In the Policy Reform scenario, there are successful land reform policies, which depart from both colonial land holding laws and traditional holdings laws. The new policies result from an evolutionary strategy under which landed city dwellers and traditional chiefs in rural areas allow the government to mediate arrangements for both industries and communities to become beneficiaries. Ferocious competition becomes a thing of the past, and conflicts are, therefore, minimized.. There is an acceptance that the reforms are benefitting the greater society. The resulting development, though slow, is planned and peaceful.

• *Western Indian Ocean Islands*

In the Policy Reform scenario, urbanization continues, whilst forest and woodland areas remain protected and extended. Agriculture in Madagascar gives way to new exploitation of mineral resources in controlled and well-managed developments, respecting environmental standards. Ecotourism promotes protection for the remoter parts of the region, serving as a model for the re-examination of western practices in countryside planning. Sugar kibbutz estates are developed by small planters, using manual methods and traditional animals for power. Traditional sailing boat flotilla holidays become well established. These two developments reinforce traditional land use under model conditions, and revive slackening parts of the economy and culture. Populations of giant tortoises are reestablished as new tourist attractions in Rodrigues and Bird Island.

## Biodiversity

• *Africa*

A technology levy on all tourism receipts generates revenue to undertake comprehensive research on Africa's biodiversity, leading to the identification of numerous species—both fauna and flora—which had not been recorded before. This research adds tremendously to the global understanding of the region's biodiversity, and also generates further tourism interest in Africa.

• *Central Africa*

The harmonization of legislation relating to conservation helps to exploit natural resources better. Ongoing processes of decentralization continue, and lead to the involvement of various stakeholders in the management of biodiversity. Practical and realistic laws define the frameworks for the conservation of natural resources, by defining the right to natural resources use by the sub-region's populations. The populations see their recognition and participate in the conservation of natural resources. The laws define conditions necessary for further laboratory research in, and commercial exploitation of, non-wood forestry products.

• *Eastern Africa*

Biodiversity is the most important asset of this sub-region, because it forms the basis of the growing tourism industry. In the Policy Reform scenario, tourism continues to make a significant contribution to the sub-region's GDP and foreign exchange earnings. Policies to protect sites with unique biodiversity are established and enforced. Deliberate efforts are directed towards curbing illegal activities, such as poaching and insecurity, in the protected areas. International and regional conventions and agreements on biodiversity resources are actively implemented, and external donor assistance is sought to enhance institutional capacity building for those institutions charged with the responsibility of conserving biodiversity resources.

• *Northern Africa*

Despite the relative increase in population numbers, planned urbanization and industrialization ensure that measures are taken to protect the natural environment. Industrial activities near protected areas are totally banned. Industries that seriously affect biodiversity through harmful waste products are forced to comply with environmental laws or pay heavy fines. The limited areas of wetlands, where a wide variety of fauna and flora live, are all declared protected areas. International technical and financial assistance is sought in order to restore the unique ecosystems of those areas.

• *Western Africa*

In the Policy Reform scenario, a large number of areas throughout this sub-region have been declared protected areas, in the interest of biodiversity. As a result, a great deal and variety of fauna and flora are alive. Harmful activities, including the dumping of hazardous substances, are banned, and public education is paying off. Self-regulatory bodies and communities are functioning properly in order to protect the environment.

• *Western Indian Ocean Islands*

The region builds strongly on the lead already established in species protection. However, it fails to maintain its momentum because of the inability to attract sufficient local scientists, brain drain and the failure in the reform of civil service to offer market rates for this scarce international resource. This means that much of the world leadership is lost. Many species fail to be saved from extinction, including various species of molluscs in Madagascar, turtles from the Comoros and the pink pigeon from Mauritius, after Round Island becomes the regional headquarters for ecotourism, upsetting the birds' habitat.

### Forests

• *Africa*

In the Policy Reform scenario, dependence on biomass, the traditional fuel in Africa—which supplied 52 per cent of all energy requirements in the region in the 20th century (WRI 1994)—is reduced, because people have more energy choices. Both public and private power utilities compete to provide electricity to both urban and rural areas, making such services more reliable. The result is that the rate of deforestation, due to fuelwood demand and charcoal production, is reduced considerably.

• *Central Africa*

The legal framework clarifies the involvement of the local population and defines, with their participation, the conditions of their involvement. Local populations participate in the conservation of forest resources. The industries exploit forests within a revamped legal framework. New technologies are developed and are introduced in logging. The research works continue, with the involvement of research programmes in forest planning through the use of modern technologies.

• *Eastern Africa*

This sub-region has 11.4 per cent of the total forest cover in Africa, making forests one of the key resources in the region. Realizing the great economic and social value of the resource, and the sub-region's comparative advantage in the Africa region, measures—including policy reviews—are undertaken to increase the level of conservation and to increase forest cover. Private investment is attracted to the sector, initially in the area of forest exploitation, but gradually moving into private plantations. The sub-region becomes a major supplier of timber products to the neighbouring sub-regions of Southern and Northern Africa. The role played by forests in catchment protection for the Nile waters is acknowledged, and bilateral agreements are reached to increase investment in catchment forestry, with significant contribution from the lower riparian states. Carbon trading becomes an active commercial transaction for the countries of Uganda, Rwanda, Ethiopia and Kenya, all of which have significant forest coverage and are willing to maintain the forests for carbon sequestration. Forests also continue to be the major source of energy, especially for the poor segments of the population, and governments take deliberate steps to promote agroforestry in their poverty eradication action plans.

• *Northern Africa*

The governments realize that, despite the fact that forests only occupy very small areas of this sub-region, their environmental and economic values are much bigger. The long issued legislations regarding forests are reviewed and strengthened. More importantly, they are strictly enforced in order to achieve their goals. The impact of the forest protection and restoration efforts does not show up quickly, but certainly became identifiable by 2032. Nevertheless, nature still plays its role, and accounts for losing some forest areas. As the sub-region understands how to deal with such disasters, their impacts are reduced. Forests remain the primary source of cheap energy for poor people, but their use is on such a limited scale that it does not adversely affect the resources.

• *Southern Africa*

New technologies are found through the channelling of more funds into research and development in microenergy sources, resulting in lower levels of deforestation. However, use of the abundant fossil fuels (coal) grows, but using technologies that are less polluting.

• *Western Africa*

In the Policy Reform scenario, the priority in dealing with forests in this sub-region is forestation in the countries that are being encroached upon by the Sahara Desert, and reforestation in previously forested countries. Evidence abounds about the successes of these policies. Because of the successes, a good balance is struck between reconstruction and other activities which require forest products, and maintaining a healthy environment.

• *Western Indian Ocean Islands*

Reforestation gathers momentum and exotic tree varieties are reduced, remaining largely in private gardens, with minimal influence on ecosystems. Mango swamps, however, suffer continual degradation as beach areas are opened up for tourism.

*Freshwater*

• *Africa*

In the Policy Reform scenario, although an increasing population exerts pressure, particularly on water and land resources, the introduction of integrated water resources management ensures that the needs of the people are adequately met, even though the resources available are much reduced compared to the beginning of this century.

Water distribution networks are upgraded in order to minimize water losses. Appropriate technology enables local authorities and the private sector to monitor water distribution 24 hours day, ensuring that burst pipes are repaired as soon as leakages occur. The technology also helps to reduce water piracy along distribution lines. A complete mapping of urban groundwater supplies, using the latest technology in geographic information systems, enables authorities to monitor excessive abstraction of water and to enforce punitive tariffs.

Irrigation equipment is improved in order to reduce water losses due to seepage and evaporation. The revolutionary technology enables Africa to intensify agricultural production throughout the year. Dependency on rainfed agriculture, particularly commercial farming, is reduced, lowering food insecurity at different levels, including at the family level.

- *Central Africa*

The optimal management of financial resources helps to ensure the provision of potable water in cities and rural areas, through specific programmes. For the Sahelian areas, the problem of desertification continues, but the development of water management policies and 'Green Sahel' programmes helps to ensure the replanting of wood in arid areas. The practice of irrigated agriculture develops, and ensures an export-oriented agricultural production, which earns foreign currency.

- *Eastern Africa*

Government policies to promote industrial development and to increase access to safe drinking water to majority of the population make freshwater a focal issue in government strategies. The private sector becomes a key player in the water sector, as realistic values are attached to freshwater. As its value increases, appropriate measures are undertaken to protect and tap this valuable resource. Ambitious projects are initiated to take water to deficit areas where the demand and price are attractive, for example, the lowlands of Ethiopia or urban centres such as Nairobi, which suffer from periodic shortages. Appropriate water pricing reduces wastage and promotes the conservative use of water in the sub-region. Social development programmes are designed and implemented by governments, focusing on the supply of freshwater to the poorer segment of the population.

- *Northern Africa*

In the Policy Reform scenario, water resources remain the main focus of governments, because of their extreme importance to development. Many countries still suffer from water scarcity in 2032, due to limited renewable resources and an arid climate, in spite of the reductions in the rate of population growth. Water master plans are drawn, and integrated water resources management frameworks are formulated and implemented, under appropriate institutional frameworks.

Egypt and the Sudan seek to increase their supplies of water from the River Nile by integrating their development projects and becoming closely involved with the rest of the riparian countries when formulating their water policies. Egypt and Tunisia adopt groundwater-recharging schemes while, in Morocco, a drought management plan is drawn up. New laws are passed through the legislative system, which sets strong penalties for quantitative and qualitative water misuse. Extensive public campaigns are launched to educate the people about the water resources problem.

- *Southern Africa*

Proper accounting and economic valuation of water results in its efficient usage. The effects of water scarcity are less prominent as consumers, especially large ones such as irrigation agriculture, can no longer afford to be wasteful of the resource, which is charged in economic terms.

- *Western Africa*

Every country in Western Africa gets a degree of rainfall. It ranges from as little as 0.1 millimetres in some parts of Mauritania, Mali and Niger to an annual average of more than 4000 millimetres in some parts of Liberia, Sierra Leone and Guinea. The heavy rainfall in most of the sixteen Western African countries closest to the coast assures these countries many bodies of water. In the Policy reform scenario, management of these bodies of water is transformed, thereby guaranteeing supplies of freshwater all year round. The countries that are less endowed within the context of the successful economic integration policies of ECOWAS also derive benefits.

- *Western Indian Ocean Islands*

In the Policy reform scenario, some profound damage to aquifers proves irreversible in the medium term. Continual problems arise as a result of the heavy use of water for irrigation by agriculture and by tourism, which is reluctant to opt for recycled water and is able to pay higher tariffs for metered use.

### Marine and coastal environment

- *Central Africa*

In the Policy Reform scenario, flexible laws allow for the involvement of new economic operators in coastal areas. The pressure on coastal and marine resources increases, although their exploitation has real economic impacts.

- *Eastern Africa*

The Policy Reform scenario redirects the focus in this sub-region regarding the issue of marine pollution. The countries of Kenya, Somalia, Djibouti and Eritrea develop stringent pollution control regulations in order to deal with effluent discharge and agricultural run-offs. The coastal towns are obliged to put in place pre-treatment facilities for industrial and domestic sewers.

International laws concerning territorial rights in the sea are enforced, in order to eliminate piracy and illegal fishing. Marine productivity gradually increases, making it the top contributor to the GDP of the four countries of this sub-region.

• *Northern Africa*

The success of the Policy Reform scenario in redistributing the large populations of Northern Africa, such that population densities do not increase or are sometimes lowered, shows its benefits on coastal areas. The countries of Northern Africa take necessary measures for reducing marine pollution, by developing infrastructure for treating wastewater before it is disposed of in the sea. Laws and regulation are issued or revised in order to protect the coastal areas and water bodies from unplanned development and the associated environmental impacts. EIAs are mandatory for projects with potential impact on coastal and marine resources. Field inspection by specialized government agencies ensures that such projects follow the mitigation measures proposed in their EIAs. Many new coastal areas and inland water bodies are declared as protectorates, with emphasis on the ecosystem in the Red Sea.

• *Western Africa*

In the Policy Reform scenario, the notoriety about Western African beaches being used for dumping wastes is now a thing of the past. With increased sanitation, and proper rubbish disposal policies and programmes, the beaches have become places for pleasure. Massive tourist enterprises, interspersed with coastal and marine industries, can be seen along the coast. International conventions opposed to the dumping of hazardous wastes are respected.

• *Western Indian Ocean Islands*

Erosion protection schemes reduce overall levels of loss of coastline from natural processes, but littoral urban development and ever-expanding tourism affect the natural terrain and its ecology. Reestablishing in-shore fishing proves more protracted than envisaged, because

of continual infringement of closed season rules, and the use of fine nets for coastal fishing in small boats. Deep-sea protection arrangements prove satisfactory, but the sub-region is slow to respond to opportunities, with continual haggling over the internal division of territory, and the sharing of protection and development costs.

### Urban areas
• *Africa*

In the Policy Reform scenario, clean water and sanitation has become virtually equally available in rural and urban areas, as a result of improvements in infrastructure. Most rural residents have access to piped water. Tax and other incentives encourage entrepreneurs to invest in both urban and rural areas, thus reducing rural-urban migration, and lowering pressures on services and the environment in urban areas. The investment in rural areas also helps to diversify the agrarian economy, which is dominant in many parts of the region. The diversification of the rural economy helps to eliminate the conversion of fragile ecosystems into agricultural land, a problem which had begun to manifest itself increasingly in the last decade of the 20th century.

• *Central Africa*

The Policy Reform scenario fosters better management of the cities, with the involvement of various parties in defining blueprints for urban planning and land occupancy plans. The parties include, among others, territorial and local authorities, and civil society organizations. The issue of satellite urbanization is better controlled, with private companies, regional and local authorities, and civil society organizations in charge of urban regional development all playing a role. Social welfare departments are mostly privatized, thus increasing the cost of access to quality services. The poverty gap widens between the richest and the poorest, but a good standard of living remains within the reach of middle-class citizens. The problem of sanitation (waste management) remains an issue.

• *Eastern Africa*

More planned and less congested urban centres emerge as land use planning policies are implemented. The majority of urban dwellers have access to clean water and sanitation. Solid waste management problems are tackled with the involvement of the private sector. Crime rates are reduced as better investment policies result in a vibrant private sector, which is able to create youth employment.

- *Northern Africa*

In the Policy Reform scenario, efforts to reduce population densities in large cities, and well-planned urban development, improve the urban environment. Safe drinking water, sanitation, transportation and power supply are all available to old, as well as to new, cities. As public transport improves, more people use it in place of using private vehicles, thus reducing the air quality problem. A proper system for solid waste management is adopted, and better education sees people care more about their sensitive urban environment. Industries located within urban areas are forced to control their gas emissions and waste, in order to minimize their impact on the environment.

- *Southern Africa*

Provision of social services is totally privatized, resulting in proper waste disposal in urban areas and the reuse of all recyclable waste. However, the pace of urbanization remains high, and the problem of waste disposal in informal settlements remains an issue.

- *Western Africa*

In the Policy Reform scenario, the success of rural development programmes stem rural-urban migration. Self-managing communities are in charge of all social services. Very small management units cooperate where necessary to provide services. Urban planning has become far more advanced, with equal access to services being central.

- *Western Indian Ocean Islands*

Urban areas continue to expand in all parts of the region, with insufficient attention to urban planning. Mismanagement, and political corruption in granting unwarranted permissions, also greatly affect development and lead to abuse of the urban environment.

## THE FORTRESS WORLD SCENARIO
### Introduction

In the Fortress World scenario, the failure of the world to heed the need for strong policy reforms on the environment leads to a state of complacency, with governments retreating from social concerns and responsibilities. In such a situation, development declines as poverty rises. Environmental conditions deteriorate as pollution, climate change, land change and ecosystem degradation interact to amplify the crisis. Environmental degradation, food insecurity and emergent diseases foster a vast health crisis. Free market values are unable to constrain environmental externalities. The affluent minority is alarmed by rampant migration, terrorism and disease, and reacts with sufficient cohesion and strength to impose an authoritarian 'Fortress World', where they flourish in protected enclaves in rich nations and also in strongholds in poor nations. The fortresses are bubbles of privilege amidst oceans of misery. The élite halts barbarism at its gates, and enforces a kind of environmental sustainability.

At the African regional level, at the turn of the 21st century, there was optimism about the future of Africa. The region's countries have made significant economic, social and political progress over the past decade and, particularly, over the past four years. However, such optimism, and the improvements made, are fragile. Great challenges haunt the African future, and the threat of reversal of any progress made is, in many cases, real. Among the challenges threatening the sustainability of progress in Africa are:

- fast population growth, and rampant migration and urbanization;
- the spread of poverty (about 50 per cent of the population in sub-Saharan Africa, and 20 per cent in North Africa, reside in absolute poverty);
- vulnerability to adverse external shocks, such as the 1997–98 Asian crisis;
- failure to recapitalize Africa;
- policy reversal;
- weak overall governance, and conflicts;
- the spread of diseases, including HIV/AIDS and malaria; and
- foreign and domestic debt, and debt servicing.

With these challenges, African countries are led to the emergence of a Fortress World scenario, in which

African societies are split into two groups: a small group of élite and public officials, who live in a relatively prosperous conditions, but in a highly protected world; and a poor majority, deprived of basic services and rights. The irony of a fortress world crisis is the suffering, hardship and impoverishment incurred by the vast majority of people at a time when a minority of élites live modern, prosperous lives. The fortress world is a grim outlook for the future, in which social and environmental problems lead increasingly to the authoritarian 'solutions' of a minority of affluent people. Under such circumstances, members of the élite organize themselves to live in protected enclaves, while the poor majority outside of this fortress have few options and resources. A fortress world in Africa could eventually lead to the complete breakdown of society, and also to the emergence of new paradigms for a brighter future.

The assumptions of the Fortress World scenario may be summarized as follows:

- Increasing social and environmental problems lead to authoritarian 'solutions'.
- Members of the élite live in protected enclaves. These may or may not involve a physical wall, and they may be within a country or between countries.
- Those in the fortresses reap the benefits of globalization. Those outside the fortresses have few options and few resources, and are excluded from the privileges of the élite.
- Components of the environment may actually improve under this scenario, because valuable environmental resources are controlled by the élite.
- This improvement is not necessary unsustainable, but it may not be feasible to maintain it for an indefinite time.

## The narratives

The fortress world is not an invention of today, but a historic norm. Looking into human history, there are many examples of the élite living in prosperity in protected fortresses, while the majority of the public are poor, working mainly for the benefit of that élite. In some cases, the élite imposed taxes and fines—a reflection of the breakdown and failure of policy reforms undertaken by governments, and of the loss of coping capacity to meet the challenges and evolving world trends of globalization and trade liberalization.

In the Fortress World scenario, before the complete breakdown of the whole society, the élite perceive and comprehend the dangers of falling into complete anarchy and chaos, and organize themselves into enclaves or strongholds to protect their interests, families, businesses and assets. They create strong alliances amongst themselves at national and regional levels, through networks across continents. Furthermore, these alliances are well connected to global systems, and are driven by the interests and mechanisms of such systems, especially through the multinational companies which operate within the élite strongholds or fortresses. With the increase and dominance of cynical attitudes amongst these alliances, economic and social welfare in the region are not directed at improving the general well-being of the public majority, but at protecting the privileges of the rich and powerful élite. This situation paves the way to increasing tensions and disputes over issues of wealth and power between individuals, institutions, governments, factions and ethnic groups. Increasing tension outside the fortress produces a siege mentality within the élite, who feel their security threatened. This leads to high investment in security.

A combined effect of interacting driving forces has led to the rapid impoverishment of the African region: SAPs have failed to realize the economic reform they were set to achieve; unfavourable international trade terms have marginalized Africa in the global economy; and servicing and repayment of foreign debt have paralysed economic advancement in Africa. A number of external shock waves have led to a major crisis in the region. These include: the collapse of international commodity prices; world economic recession, for example, the Asian crisis of 1997–98; deterioration in governance; several regional conflicts; and the spread of epidemic diseases, such as AIDS.

In the Fortress World scenario, there is a general decline in the capacity of the state to perform adequate developmental tasks, including even routine administration functions, outside of the fortress. The state thus fails to meet the basic needs of the people, leading to erosion of legitimacy, and the breakdown of peace and security. The accountability and transparency of the government and public officials erode or vanish. The security and protection of the élite and government officials become a top priority issue. To safeguard their interests, they mobilize all possible security resources, from the public to the private, including police, armed forces, militia and even mercenaries. The regional system, in turn, is well connected to an updated system of global apartheid, dominated by a minority of rich-country élite. Therefore, the enclaves of urban-based élite get control of wealth and power, while the poor majority have no privilege to exercise any leverage on public policy or budgetary choices.

The economy and the structural systems of the élite are strongly influenced, if not completely driven or controlled, by multinational corporations. Their systems and laws override and impose on the local, weakened state systems. Nevertheless, the informal sector starts to play a key role in the local economy of the poor groups, but primarily caters to its own needs and demands (food, housing and transport).

Corruption in all forms spreads at all levels, within and between countries, and at the interface between the public and private sectors. This is not only in the form of money or in-kind payment but, more widely in the African context, in the form of favouritism (nepotism) towards relatives and friends. The spread of corruption has its toll on society: losses in economic efficiency; disintegration of the work ethic; damage to the moral fabric of society; the distortion of incentives and distribution; and loss of political legitimacy.

Two variants of the Fortress World scenario could unfold in Africa. One variant is a self-generated fortress world, driven mainly by African issues—such as those that have been haunting the region for the past 50 years—while most of the rest of the world remain unaffected and possibly prosperous. The other variant is a global fortress world, in which the world economic, social and political systems collapse, with severe manifestations reflected in Africa, which is the most vulnerable region to such global failure.

The tides of geopolitics surge across Africa, with regional alliances influenced by external forces. The western world, in general, prefers to deal with Africa as sub-Saharan Africa. Europe, specifically since the colonial era, has dealt with Northern Africa in the Mediterranean context. More recently, there is even greater interest in this Mediterranean context, with the

European perception of Northern Africa and the Middle East as an intimate geographic extension that has economical, political and cultural dimensions which have a direct impact on Europe. The 'Euro-med' partnership is a manifestation of such interest. In the Fortress World scenario, this background has implications for regional alliances between the élite of Northern Africa and Europe. Such external views of a divided Africa go against the true African view—originating from within the region—of one united Africa. This African view was the driving force for the creation of the OAU, which works to unite and strengthen the integration of all the African countries. The alliances in a fortress world, driven by individuals and special groups interests, are likely to weaken the trend of African unity and integration.

At the turn of 21st century, it is no surprise to see many aspects and elements of the fortress world manifesting themselves in all societies, in all countries. Examples of fenced, highly secured residential complexes, and segregated private schools and clubs for the élite, are common in most world societies. In many countries, businesses are concentrated in the hands of a small group of élite, with strong barriers to prevent outsiders breaking into their systems and obtaining any business opportunities. With the current economic, social, political and environmental conditions in Africa, some people argue—with mounting evidence—that the fortress world is already dominating many African countries and even sub-regions, from Western to Central to Eastern Africa.

## Environmental consequences
### Atmosphere
• *Africa*

In the Fortress World scenario, the African environment has been negatively impacted, except in some protected or isolated areas away from human pressure, where limited improvement can be seen. Poor economic performance has driven Africa into fierce competition over ever-dwindling natural resources. Atmospheric emissions remain modest and the region remains more vulnerable to the imperfections of the provisions of international treaties related to climate change. For example, emissions trading could result in Africa losing cheap credits and having to pay more later per unit of emission reduction. Africa is also more vulnerable to the effects of climate change; for example, malaria extends to new zones. Urban air quality declines in most urban areas, and effects are felt in other areas, because of the failure of mitigation measures, and an increase in urbanization and polluting industries. This leads to incidents of respiratory diseases, such as asthma. In some areas, the quality of air remains the same. Africa is pressured to adapt afforestation and reafforestation programmes, for the good of the polluting west. However, access to large forest areas becomes controlled by the élite. Under continued international pressure, Africa bans totally the use of ozone-depleting substances, such as methyl bromide, with inadequate assistance in the changeover to alternatives, which may be more expensive.

• *Central Africa*

In the Fortress World scenario, this sub-region continues to pollute less, but it suffers the consequences of pollution brought about by industries in the north. The consequences are observed through variations and changes in climate, producing some effects on seasonal patterns and agricultural production. International protection conventions, such as the 1971 Ramsar Convention on Wetlands and the 1992 Kyoto Protocol, force countries not to continue exploiting forest resources any longer, in order to preserve the natural resources of the Congo Basin for carbon sequestration.

• *Eastern Africa*

The impact of climate variability is more devastating, especially for the vulnerable poor, who are eking out a living outside the fortress. Measures to control emissions are not implemented, because people, including those in the fortress, are more concerned with personal security and survival. Air quality deteriorates both within and outside the fortress, making the environment unattractive for investment.

• *Northern Africa*

The air quality in many parts of Northern Africa sees some improvement, due to the interest of the élite in improving it. However, there are differences in the air pollution levels in the rich and poor areas, because of the concentration of industries in poor areas. The élite

protect their fortresses against potential climatic changes such as floods, while the poor are left vulnerable to such elements.

• *Southern Africa*

Southern Africa continues to suffer the effects of climate change. At the same time, the sub-region is forced to exploit its forest resources due to demands for firewood. Unfortunately, this activity decimates an important carbon sink. On the other hand, developed countries, particularly the USA, take their time in ratifying important multilateral environmental agreements, such as the 1992 Kyoto Protocol, because they want to protect their industries, as well as to maintain their lifestyles.

• *Western Africa*

In the Fortress World scenario, élites consider their farms to be environmentally friendly places, free from the pollution of the cities. They protect large stretches of the countryside from environmental abuse. Because some of these farmers may themselves be in government, or have influence over those who occupy key positions in government, they ensure legislation to protect their farmlands and, thus, the environment. However, the landed aristocrats have no interest in the barren lands and, therefore, condemn them for dumping, irrespective of who lives there.

• *Western Indian Ocean Islands*

The atmosphere in this sub-region is surprisingly little affected by events, except in Madagascar, where the absence of the advantages of sea winds, enjoyed by the smaller islands, results in worsening air quality. Nevertheless, Port Louis, Antananarivo and Victoria have periods of closure due to urban smog, and seek technical advice.

### Land

• *Africa*

With stagnant to declining economic growth, Africa remains essentially a subsistence economy. Land resources, however, have been subjected to increasing pressure, as a result of population increase and climate change. There is declining productivity in grazing and agricultural lands, due to a combination of: inequitable land distribution; poor farming methods and unfavourable land tenure; ownership systems; and inefficient irrigation systems. The rapidly increasing populations of humans and animals translate into overexploitation of water, land, forest and pasture resources, through overcultivation, overgrazing, deforestation and poor irrigation practices.

In the Fortress World scenario, indigenous African farming systems and cultivation techniques –adapted to local ecological conditions and sensitive to the preservation of fragile natural resources—collapse, except in pockets here and there. Meanwhile, agriculture practices that are not adapted to the fragile African soils significantly degrade the land and reduce its productivity. The rapid impoverishment of rural traditional systems results in the abandonment of relatively benign methods of exploitation of nature, and in their replacement with aggressive methods, which assume that natural resources are limitless. With stringent economic conditions, and denied access to land, poor peasants cultivate marginal land, leading to declining productivity. Resilience to environmental and social changes deteriorates outside of the fortresses, contrary to the situation inside the enclaves.

The pressure on land, vegetation and water supplies has made Africa increasingly prone to food security crises. Dramatic increases in populations, combined with increased land degradation, has resulted in the decline of per capita food production. Annual population growth has exceeded increases in food production, creating a chronic food crisis for the poor, and widening the food gap. In the bubbles of élite, the situation is the reverse. Food production exceeds the needs of the small élite population. In the Fortress World scenario, a positive survival practice of urban poor is the increasing practice of urban agriculture—the production of food and non-food crops, and animal husbandry, in built-up areas. However, most urban agriculture remains largely unrecognized and unassisted, if not outlawed or harassed.

In the Fortress World scenario, environmental refugees and conflict-displaced people migrate to areas of greater food availability—mainly where they can receive food aid. However, with emergency aid decreasing, it becomes inevitable for rural people to migrate to towns or cities, considerably swelling the numbers of the urban poor. In general, the élite become the class of resource-extractors, driven by the global

market economy, and impoverishing both the environment and subsistence resources for rural people.

The chaos and weakness of systems of governance precipitate insecure land tenure as the norm for those outside the fortress, thus providing no incentive for sustainable land management. In contrast, the élite secure land tenure for their land. Thus, their land is preserved and properly managed, using modern and enhanced agriculture systems.

• *Central Africa*

Dualism (modern law and traditional law) in the area of land management increases the pressure on land, and leads to land use conflicts in the cities as well as in the countryside. The long and expensive procedures to obtain certificates of occupancy for land induce real estate to drift into illegality. In high-density localities, there is mass movement of people to settle elsewhere, thus sparking off conflicts with the other ethnic groups. The élites increasingly invest in the development of farms, in order to combat food insufficiency resulting from these movements.

• *Eastern Africa*

In the Fortress World scenario, land degradation outside the fortress is rampant, due to overcrowding and insecure land tenure. Investment in land improvement, especially outside the fortress, is very limited. Soil erosion accounts for more than 90 per cent of environmental degradation in this sub-region by 2020 (in 2002, it accounts for about 80 per cent).

The land reform programmes which five countries in sub-region—namely, Uganda, Kenya, Rwanda, Ethiopia and Eritrea—were pursuing are abandoned. The poor are driven off the prime land into more fragile areas, such as wetlands, steep slopes and semi-arid zones. Conflicts over resource use and access increases, particularly in the Horn of Africa (Ethiopia, Eritrea, Somalia and Northern Kenya).

• *Northern Africa*

Land is overexploited and, eventually, degraded by the poor sectors of society. The élite-controlled governments focus on those resources that are vital to their success and well-being, with land being left mainly to the poor. With the exception of relatively small areas that are well preserved by and for the élite, most of the land resources are degraded, due to unsustainable practices. Land use planning has become almost non-existent, and ad hoc development spreads. Allocation and access to land becomes grossly inequitable.

• *Southern Africa*

In the Fortress World scenario, in southern Africa, where land reforms are either being implemented or contemplated, radical change in land ownership occurs. Previously disadvantaged members of society are given the opportunity to own land, but they have limited knowledge of farm operations and a lack of resources to acquire inputs. As a result, land is degraded further, as a result of inappropriate farm practices, while food production declines because of lack of necessary inputs. On the other hand, the politically élite are better resourced, but their focus is on cash farming, and this has serious implications for food security.

• *Western Africa*

The landed class acquires more and more land for commercial farming. In the process, the peasant or subsistent farmers are displaced, and are forced to become farm workers. The big farm owners and their families barricade themselves in their luxury farm homes, protected by security guards and dogs.

• *Western Indian Ocean Islands*

Land is devastated by warfare, civil strife, and reversion to subsistence agriculture. Soil erosion continues to be a major problem, as a result of massive deforestation in all areas. Some 80 per cent of the urban settlement areas of the Seychelles are lost through sea level rises of two metres. Millions of hectares in Madagascar are flooded by a devastating cyclone (*Indira*), and much of the land falls out of use through civil war. Vast unplotted areas of land are thought to have been laid with mines.

**Biodiversity**

• *Africa*

In the Fortress World scenario, biodiversity and its ecosystems come under severe pressure, with natural ecosystems reduced to small pockets of protected areas with limited access, mainly to the élite. National and international conservation and protection efforts decline, as a result of: lack of biodiversity conservation frameworks; strategic and financial resources; and

unfair activities which cause significant habitat destruction, resulting in a greater number of extinctions and species under threat. Furthermore, as ethics and cultural values degrade, trade regulatory mechanisms completely break down, legitimizing trade in endangered species. Loss of biodiversity and invasion by alien/exotic species are increasingly widespread, causing increased outbreaks of pests and disease, resulting from a lack of natural predators and ecosystem destabilization.

Unfair sharing of the benefits of biodiversity continues, with royalties accruing to multinationals rather than African source countries. Gene banks are mostly located in the west, and are inaccessible to poor African countries, which are among the main source of the gene materials. The long undervalued biodiversity becomes valued, but is overshadowed by inequity and market forces. Biodiversity comes under use in ecotourism, which is controlled and managed by the élite, who also reap much of the benefits.

Patenting of GMOs continues to discriminate against African countries, and to threaten agricultural biodiversity, especially of wild species. For example, there is narrowing of gene biodiversity through the introduction of terminator genes, and genetic pollution of indigenous species. In addition, Africa faces new weed and pest problems, arising from GMOs.

• *Central Africa*

Lack of technological development does not allow for developing and enhancing local knowledge, especially as far as medicinal plants are concerned. Natural resources continue to be exploited by the laboratories from the north, who develop research projects in this sub-region. The risk of specific species disappearing is on the increase. The practice of (traditional, commercial and sporting) hunting and fishing increases the pressure on the species which, in turn, are subject to disappearance, in spite of conservation programmes. The development of ecotourism is spearheaded by private companies, which do not bring about notable impacts on local economies.

• *Eastern Africa*

Critical ecosystems and forests are more strictly protected, and the governments in this sub-region strengthen the policing functions of the relevant agencies. However, as the Fortress World scenario persists, the cost of protecting large numbers of sites increases, and the focus shifts to prime sites—including national parks, such as Bwindi, Mgahinga, Tsavo, Masa, Maara and Budongo, and so on—which have a high potential of attracting income from tourism. Areas with very high protection costs—such as Lake Mburu in Uganda, and Lakes Nakuru and Naivasha in Kenya, and so on—are degazetted.

### • Northern Africa

The impact of the Fortress World scenario on biodiversity is similar to the impact on coastal and marine areas. The élite minority put some areas under their control, where industrial, recreational and tourism developments take place. Only in those developments where biodiversity has a significant value is it well protected. In industrial developments, preserving the natural environment is very low on the list of priorities. On the other hand, the rapid urbanization of the poor minority causes serious damage to biodiversity. Some people seek to make a quick profit by illegally selling rare species, either to the élite or abroad.

### • Southern Africa

Due to lack of knowledge of the whole range of their biological resources, most Southern African countries continue to have important genetic resources pirated from the region. At the same time, high costs for medical treatment and increasing rates of AIDS infection turn many to traditional medicine, and this results in the overharvesting of certain species.

### • Western Africa

In the Fortress World scenario, the élite are consumed by the insatiable desire for making money. They realize that, by protecting the fauna and flora, they attract ecotourists, and earn huge profits on investment in the tourist industry. Therefore, there are numerous sites which promote biodiversity, although such sites are always almost linked to some investment in tourism. Traditional forms of promoting biodiversity are undermined as people are displaced. Survival becomes the paramount concern for the displaced who, quite often, prey on rare species, either by selling to visitors or by using some as food, where applicable. The landed class in the fortress world realizes that huge commercial

farming affects forest resources and, therefore, engages in some reforestation. They feel the effect of climate change, which also has an impact on agricultural output. Efforts to diversify into the construction industry are hindered by low forest resources and increased demand for wood, and the élites push seriously for conservation of the remaining forests.

### • Western Indian Ocean Islands

Biodiversity declines steeply during this period, with serious loss of marine and bird life, and with no large mammal survivals, apart from humans. Invasions of vultures and other birds of prey are reported in Madagascar, and sightings made in the Comoros.

## Forests

### • Africa

In the Fortress World scenario, deforestation and degradation of forest areas continue at higher rates, except in some areas, where they might be recovered. The élite, tempted by the high demand for forest products in the global market, act as resource extractors, and overexploit the forest resources. Ironically, they safeguard some forest areas under international pressure. Some of the remote forest areas, away from population pressure, are also saved, such as parts of the rainforests of the Congo Basin. Poor people fall back on extensive use of the forest resources which they have access to, as a source of energy, food and shelter. Commercial exploitation of medicinal plants contributes to accelerated deforestation. Forest wood is also used commercially, for the production of crafts for trade. The introduction of alien species and forest plantation play a significant role in modifying the structural composition of forests.

### • Central Africa

The increase in the poverty of peasant farmers, and the fall in prices of agricultural produce, lead to more pressure on the forests, which are the primary sources of revenue. 'Unbridled' logging is witnessed. The industries which are involved in logging do not take the regulations into account and, thus, engage in illegal logging activities. The export of logs limits the economic impact of logging in the local and national economies. Non-wood forestry products experience a boost in their development, with greater pressure brought to bear on them. This is also true of medicinal plants and forestry products for domestic use. The consequences of deforestation and desert encroachment are disastrous for the Sahelian areas (Northern Cameroon, Central African Republic and Chad).

• *Northern Africa*

In the Fortress World scenario, the future of forests in Northern Africa is largely determined by the way they are seen by the élite. Because forests in this sub-region do not have much significant value as a source of raw material for industry—apart from the production of gum arabic—the élite do not give them much attention. Forests are left to be overexploited by the poor majority as a cheap source of fuelwood, charcoal and wood for small shelters. Unfortunately, the poor, who struggle to satisfy their basic needs before they start to worry about the environment, do not see the environmental values of the forests.

• *Southern Africa*

Deforestation worsens, especially in places surrounding urban areas, due to increased poverty. High population growth rates imply that demand for forest resources, especially firewood, grows. Tree types that produce a lot of heat and less smoke are selectively felled, resulting in changes in the structure and composition of forests.

• *Western Indian Ocean Islands*

Forest and woodland areas are reduced by 75 per cent throughout the region, as a result of warfare, subsistence farming and loss of other sources of energy, and through neglect.

### Freshwater

• *Africa*

Water resources have become inadequate in quantity and quality in Africa, mainly because of poor management resulting from chaotic circumstances dominating the region. Although there is abundant water supply in Africa, there are sub-regions and countries which have water scarcity. Water inadequacy results in poor health, low productivity and food insecurity, and constrains economic growth. In the Fortress World scenario, the vast majority of African countries fail to conserve and protect water as a valuable, but vulnerable, resource.

As a result of increased poverty, access to water and sanitation has been in increasing decline in Africa. This, in turn, causes high incidents of communicable diseases, which diminish and hinder economic development. The result is less investment in developing water resources which, in turn, leads to reduced water availability—a vicious spiral between cause and effect. With the crises of endemic poverty and pervasive underdevelopment, water in Africa has been subject to inefficient, inequitable and unsustainable use. The élite ensured abundant share of available water resources to meet their specific needs, while the poor majority continued to suffer even while playing a central role in managing and safeguarding these water resources. Water continues to be underused in energy generation as the hydropower generation infrastructure collapses, due to poor maintenance and lack of resources for modernization.

In the Fortress World scenario, the infrastructure for the management of water resources deteriorates, with inadequate institutional and financial arrangements, lack of data and weak human capacity. Problems with water and land further aggravate food problems. Groundwater, especially in shallow aquifers, is highly degraded and depleted, a situation which becomes severe in Northern Africa, where there are very few renewable water resources. Water interdependency is high in Africa. Nevertheless, regional cooperation on transboundary water issues is not only weakened further, but is strained by escalating tensions and conflicts, as openness and transparency are eroded.

• *Central Africa*

Certain rural areas (northern Cameroon, northern Central Africa, Chad) are already suffering from lack of access to potable water and water supply. In the Fortress World scenario, this phenomenon becomes more pronounced, with the advent of desert encroachment due to deforestation. As far as agriculture is concerned, production is reduced and conflicts worsen due to water resource issues, with conflicts between breeders and agriculturists, and between élites and indigenous populations (who either own the resources or who do not have access to adequate means of production). In view of the lack of sewage water recycling, high population growth and an obsolete hydraulic (sewage water) system, urban areas experience shortages in potable water, and there is an upsurge in water-borne diseases.

• *Northern Africa*

If the unequitable allocation of natural resources per capita is the symptom of the Fortress World scenario, it is specifically apparent in the case of valuable resources, such as land and water. The industry-driven élite control most of the freshwater resources, whilst leaving marginal water resources to the poor majority. Being industry-driven, the élite allocate more water to the industrial sector, depriving irrigated agriculture of a much-needed vital input. However, the improper management of water with different qualities has an impact, with increased pollution and degradation of many sources of surface water and groundwater. The élite are not concerned with the problem, because they use desalinated water for supplying their domestic needs.

• *Southern Africa*

In the Fortress World scenario, the water supply situation continues to be precarious for most countries in this sub-region. The situation is particularly bad for South Africa, Zimbabwe and Malawi, resulting in drastically reduced agricultural outputs. Even with transboundary initiatives, water demand management and water recycling, supply does not meet demand. Conflicts between water users emerge, with disadvantaged groups, such as women and the elderly, being the losers in the competition for the resource. Water-borne diseases, which had hitherto been under control, emerge once more.

• *Western Africa*

Concerned only about their individual and family comfort, the élites in the fortress world appear to be satisfied with meeting their direct freshwater needs. Once these needs are met, they pay little heed to problems affecting the rest of society. Only when an epidemic breaks out, which denies them the labour of their farmworkers, do they develop ways and means of managing freshwater resources.

• *Western Indian Ocean Islands*

In the Fortress World scenario, pollution, and the destruction of dams and smaller water storage systems, pipes and pumps in civil strife, create continual water shortages and drought throughout this sub-region, with a devastating impact on animal and human life.

### Coastal and marine resources

• *Africa*

In the Fortress World scenario, urbanization and population movement to coastal areas continue at higher rates. The demand for marine resources for food and shelter increases. The élite utilize Africa's unique coastal and marine resources for tourism, and aggressively market them globally, through multinational companies. An increase in eco-tourism opportunities contributes to sustainable management of resources, such as forests, coastal and marine resources. However, fulfilling the needs of locals and tourists results in overfishing.

Marine pollution increases locally as water treatment of sewage declines. In some poorer areas, there are no sanitation systems and no sewage treatment. Valuable coral reefs and mangrove forests become increasingly vulnerable, and threatened with human activities. However, some areas improve as a result of stagnant development activities and protection by the élite.

• *Central Africa*

In spite of laws in force in this sub-region, the exploitation of coastal areas continues, leading to an increase in coastal erosion. Coastal and marine pollution continues unabated, because of population pressure resulting from housing development The insular countries of the sub-region run the risk of disappearing as polar ice cools, creating an upsurge of ocean water levels.

• *Northern Africa*

In the Fortress World scenario, the environment along the coastal and marine areas remains quite diversified. The population and densities of people in old developments along the Mediterranean coast suffer from further congestion. The rate of infrastructure development is much slower than the rate of population increase, leading to severe pollution of the coasts and seawater. On the other hand, the élite develop new areas along the coasts of the Mediterranean Sea and the Red Sea for recreation and water tourism. Every care is taken to minimize the environmental impacts of those new developments. Nevertheless, the élite overexploit the soil along the coasts, neglecting the serious damage this causes to coastal waters and marine life. Many rare marine species are threatened or totally destroyed. Unsafe abstraction of groundwater within the coastal zone of the Mediterranean Sea causes seawater intrusion and, hence, renders many wells unusable.

• *Southern Africa*

Uncontrolled development in coastal areas accelerates coastal erosion, at the same time polluting the seas, due to lack of adequate facilities to collect and properly dispose of rubbish. Growth in population in coastal areas, such as Maputo, Dar es Salaam and Durban, exacerbates the overexploitation of marine resources, including mangroves, fish and prawns.

• *Western Africa*

In the Fortress World scenario, while the élite begin spending most of their time on the large, environmentally friendly farms, they maintain their showcase or dream houses in the cities—quite often in close proximity to the Atlantic Ocean or the big rivers. Although, initially, each family was concerned only about its immediate environs, or the piece of land adjoining the beaches and waterfronts, they realize that they are unable to enjoy their prized places if the waters are polluted. Thus, they participate in the exercise of cleaning the beaches and waterways. Their representatives in government put in place appropriate policies and laws to protect coastal and marine areas. The élite are, however, very uncooperative when policies tend to restrict the exploitation of coastal and marine resources in a way that affects their investments in the fishing industry.

• *Western Indian Ocean Islands*

Coastal and marine areas are seriously affected by military operations, piracy and overfishing, using fine nets, gelignite and seabed trawling. Oil pollution from major tanker spills, following naval action in pirate wars in the 2020s, scars coastlines and causes irretrievable damage to marine stocks. Sea mines laid in the same period remain a hazard to all regional marine traffic, causing continual loss of shipping and life, and diverting away most trade from other regions.

## Urban areas

• *Africa*

In the Fortress World scenario, rapid urban growth exerts pressure for housing and infrastructure investments, in order to accommodate the rapidly growing population. However, the system of urban governance is responsive to the needs of the élite, and not to the needs of the poor majority. This situation leads to the mushrooming of informal housing and slum areas, which account for the majority of African urban dwellers. These informal settlements are not provided with adequate transport, water, sanitation, electricity and health services. People living in these areas are exposed to various health hazardous. Housing finance systems are non-existent, or are limited to élite bubbles.

The deterioration of the urban economy and environment leads to widespread urban poverty, beyond known historical norms. As a consequence, crime becomes a major problem in most African cities, and worsens with increased poverty and deteriorating living conditions. The streets of most cities become completely unsafe, even in daylight.

The élite, who live in enclaves, produce much higher municipal waste, per capita, than those outside of the enclaves. This waste is dumped, without regard for costly hygiene measure, in sites outside of the gated cities inhabited by the poor. This exacerbates land and water degradation, leading to the spread of disease.

The African region is the cradle of many of the world's oldest civilizations and cities, and it traditionally attracted a significant share of worldwide tourism. In the Fortress World scenario, instead of revitalizing and preserving these assets to empower communities and to strengthen their economic base, these assets become threatened by neglect and environmental degradation. The élite, using authoritarian solutions, succeed in protecting some of the cultural heritage, but their incentive is to reap the benefits and to project a civilized image to the international community.

• *Central Africa*

This sub-region continues to witness an increased concentration of populations in urban areas. The Fortress World scenario is characterized by uncontrolled urban growth, outdated urban settlement planning and increased poverty in cities. The high population pressure affects the consumption of natural resources, and is linked to the poverty of city-dwellers. Uncontrolled urbanization leads to the emergence of places in which no provision is made for green spaces and, where such provision is made, it is not respected. Because of lack of solid and liquid waste planning and management programmes, the issue of sanitation becomes more acute, and impacts the health of the population through frequent outbreaks of new diseases.

• *Eastern Africa*

In the Fortress World scenario, proliferation of slums is rampant, as more people are displaced from prime agricultural land by landlords and forced to move into the cities to look for other opportunities. The infrastructure in the low-income, urban areas deteriorates rapidly as the city authorities become unable to cope with the rate of influx of people. While, in 2002, only 3 per cent of Addis Ababa is planned, and 28 per cent and 40 per cent of Kampala and Nairobi respectively, these percentages decline as the planning fails to keep pace with the rate of development and expansion.

Crime rates increase, especially in the informal settlements, and this spills over to the high-class residential areas, forcing those living in these areas to increase expenditure on security measures, such as high walls, electric fences and armed security guards. As the situation persists, middle-class and high-class residents move out of the cities, and create their own secure settlements in peri-urban areas, such as Mukono in Kampala, Ngong in Nairobi and Debra Berhane in Addis Ababa.

• *Northern Africa*

There are large differences between the environments in the élite fortresses and in poor areas. The élites being much fewer in numbers, richer and better educated, the environments of the fortresses are well protected. Basic services are not only available, but also very much more reliable than they are in the poor areas. The environment is still rich, and is sustained by measures for environmental protection. Careful planning of the fortresses ensures that air quality remains high, because of low population densities and sufficient green areas. No significant industrial waste is produced inside the fortresses, because industries are concentrated outside the fortresses, where the poor labourers live.

Nevertheless, sewage water is not treated before disposal, because it is disposed of in the areas of the poor. The picture in poor areas is totally different, with air pollution, unsafe drinking water, poor sanitation and solid waste accumulation being the main elements.

• *Southern Africa*

The concentration of populations in well-resourced areas implies that, in the Fortress World sceanrio, the Southern African countries continue to witness phenomenal urban growth, with inadequate provision of basic services, such as water and sanitation. Informal peri-urban settlements proliferate, resulting in disease, destitution and moral decay, which further worsens the spread of HIV/AIDS. The areas with high population concentrations witness further land degradation, water pollution, habitat destruction and deforestation.

• *Western Africa*

There is nowhere more expressive than this sub-region of the dichotomy between the élites in the fortresses and the poor in the shanties. Well laid-out and year-round green areas in urban centres are occupied by the rich and powerful. There are first-class, efficient rubbish disposal systems, proper sanitation, effective sewage systems, and good roads and waterways. The poor work to maintain the comforts of the élite. At the end of the day, the workers return to a totally different life—one of filth, unsafe water, non-existent rubbish disposal systems, poor sanitation and non-functioning sewage systems.

• *Western Indian Ocean Islands*

In the Fortress World scenario, urban areas expand as populations flee from the devastated countryside, making existing settlements unsustainable, deprived of adequate water, sanitation, power and services. Many are scenes of continual pillage, violent crime and misery. A few have been converted to religious retreats by fundamentalist groups of all persuasions, prepared to fight to death for the preservation of their rights.

## THE GREAT TRANSITIONS SCENARIO

### Introduction

The vision of the Great Transitions scenario stems from developments at the start of the new millennium. These include a conviction regarding the need to embrace a new sustainability paradigm—one which transcends the dictates of both the Market Forces scenario and the Policy Reform scenario and one which, at the same time, prevents the occurrence of the ills associated with the Fortress World scenario. Associated with these is a philosophical dimension, at both personal and group levels, which holds that an end must be put to consumerism as way of life, and that a search must be made for issues that can provide a renewed sense of meaning and purpose to life. Consequently, the values of simplicity, tranquillity and community begin to displace the values of consumerism, competition and individualism. Voluntary reduction in work hours frees time for study, art and hobbies.

In the Great Transitions scenario, lifestyles become simpler, in a material sense, and richer, in a qualitative sense, as the old obsession with possessions gives way to intellectual and artistic pursuits. In the new sustainability paradigm, markets remain critical, in terms of achieving efficiency in the production and allocation of goods, but well-designed policies constrain the level and structure of economic activity, so it remains compatible with social, cultural and environmental goals. A variety of mechanisms enforce these principles, including regulation, international negotiation and market signals, such as revised tax systems which discourage the production of environmental 'bads', and which reward restorative practices. Environmental, economic and social indicators track real progress at all scales—business, regional, national and global—giving the public an informed basis for seeking change.

The assumptions of the Great Transitions Scenario may be summarized as follows:

- Neither the Market Forces scenario nor the Policy Reform scenario possesses strategies that are adequate for addressing the ills of the assault on the environment.
- Furthermore, given current trends in the adoption and effectuation of treaties on environmental issues, policies alone cannot be sufficiently effective against social inequities and environmental uncertainty.

---

**Box 4.2  The twin challenge: the challenge of Africa and the challenge of the future**

*'Thus, Africa today needs both new questions and smaller errors. The project is primarily concerned with the former. The assumption being that they are a prerequisite for adequately tackling the latter. There is sufficient scientific competence available in Africa today for new knowledge to be increasingly generated from within, as it were, through the realignment of research agenda away from the 'short term' and 'applied' towards the longer term and more basic questions affecting the continent's future …*

*'A first step in this direction is to examine critically the conventional wisdom as expressed in dominant policy documents related to Africa's present and future … The second step is the development of alternative future scenarios for Africa that challenge the 'surprise free' projections of the current perspectives.'*

Source: Achebe and others 1990

---

- While market forces are not abandoned as a policy tool, social, cultural and environmental goals take precedence in thinking about development.
- For Africa (and perhaps for the whole world), notions of sustainability fundamentally change the values and lifestyles of peoples (an African Renaissance).
- In general, there is a cultural renaissance, which is not only critical of past behaviour and effects on the environment, but which also outlines new ways of thinking, and which fosters environmental goals.
- The affluent, having become disillusioned with consumerism, other ills of society and the negative impacts of development on the environment, undertake steps to develop new values and value systems. These are gradually introduced, and promote a new set of ethics in society.
- A new generation of thinkers, leaders and activists join and shape national and global dialogue towards environmental sustainability.

The Great Transitions scenario represents a very optimistic view of the development of the environment in Africa, as well as all over the world. Nevertheless, it is not as utopian as it looks at first examination, because its tenets are perfectly achievable, given the right atmosphere. As mentioned in Box 4.2, and discussed extensively in Beyond Hunger, Africa needs a resurgence at many dimensions. The beginning of the millennium is a good time to start such an exercise. Africa must ask new questions, and must challenge the conventional wisdom that has tied the region down for too long. Africa must be ready for a surprise, rich future.

### The narratives

The major strategies through which the Great Transitions scenario will evolve are not difficult to imagine. Achebe and others (1990) have argued that these are new sets of strategies, which differ from current approaches to thinking about development at conceptual, methodological, institutional, operational and financial levels. For example, while this African Renaissance vision of development is conceptually dialectic and beyond crisis, it is unlike conventional wisdom regarding development, which is unilinear and crisis-oriented. Furthermore, the African Renaissance vision is methodologically 'surprise-rich, inductive and retroductive', as against conventional wisdom, which is 'surprise-free, deductive and predictive'. Operationally, the strategy: is locally owned and initiated; is supportive, nurturing and people-intensive; has views which depart from the donor-fed and controlled, directive and preemptive; and has capital-intensive visions of conventional wisdom. The institutional set-up was state-centred, concentrated and monopolistic while, in the African Renaissance vision, it is 'grassroots oriented, multiple, dispersed and pluralizing' (Achebe and others 1990).

The attributes of an African Renaissance are based on visions of a desirable and environmentally sustainable future. These, indeed, are similar to the attributes of the Great Transitions scenario. The beauty of a Great Transitions scenario for Africa is that there are already bodies of ideas among great thinkers in the region, as well as within government circles, regarding the processes that lead to the so-called Big Lift scenario (see below). Current moves by the leaders of African countries to create the Africa Union to replace the OAU, and in the development of the 2001 Omega Plan for Africa and the 2001 Millennium Africa Recover Plan (MAP) (see Chapter 1), are steps in the right direction. So, too, is the evolution of the New Partnership for Africa's Development (NEPAD) (see Chapters 1 and 5). These ideas will continue to crystallize, and will become major issues, as other regions of the world begin to see

the wisdom of the Great Transitions scenario.

In the monumental work produced by Achebe and others in 1990, attempts were made to compare the current situation with the expectations of the African Renaissance vision. The scenario described as the Big Lift produced comforting levels of development. For instance, where current projections could put the population of Africa at some 2 200 million in 2057, with growth rates as high as 2.5 per cent per year, the Big Lift scenario would put it at 2 500 million, but with only a 1.5 per cent per year growth rate. In the Big Lift scenario, with a projected time-frame of 2057, the literacy level would have risen to 95 per cent. More important, however, is that life expectancy would have increased to 80 years, while GDP per capita would have risen to US$7 800. Food production and capital goods production would have risen considerably, and the environment would have started to recover. Forest areas would have reverted to sizes that were in existence in 1957. Arable land would have increased four times over what it was in 1957, while electricity use would have multiplied by a factor of more than 200 (see Table 4.2).

(The arguments of the Big Lift scenario, which are similar to those of the Great Transitions scenario, are that we need new paradigms to deal with development and with the creation of a sustainable environment. We have hitherto used 'the evolutionary paradigm, the gradual incremental unfolding of the world system in a manner that can be described by surprise-free models with parameters derived from a combination of time-series and cross-sectional analyses of the existing system' (Achebe and others, 1990).

Of course, the Great Transitions scenario is expected to usher in better educational facilities, greater empowerment of all people, and especially women, and absolute reductions in poverty levels, through enlightened policy reforms. It is also expected to engender greater political consciousness and commitments at local, national, regional and international levels, through visionary leadership, the eradication of corruption and improved economic performance.

The Great Transitions scenario can, therefore, be seen as involving situations where a new emphasis would be placed on issues including: the content and structure of education and training; culture; governance; and the creation of effective organs and

## Table 4.2 The African Renaissance and the Big Lift scenario

| | Actual 1957 | Actual 1987 | Projection 2057 | Big Lift 2057 |
|---|---|---|---|---|
| **Demography** | | | | |
| Total population (million) | 277 | 599 | 2 200 | 2 500 |
| Population growth rate (per cent/year) | 2.3 | 3.1 | 2.5 | 1.5 |
| Infant mortality (per 1000 births) | 182 | 181 | 10 | 8 |
| Life expectancy at birth (years) | 40 | 53 | 77 | 80 |
| **Economy and agriculture** | | | | |
| GDP per capita (1980 US$) | 450 | 815 | 3 800 | 7 600 |
| Capital goods production (million 1975 US$) | 127 | 1 273 | 75 800 | 115 000 |
| Agricultural production (FAO in-dex) | 63 | 115 | 1 000 | 2 000 |
| Food supply per capita (calories) | 2 060 | 2 094 | 3 200 | 6 000 |
| **Human resources** | | | | |
| Literacy rate (per cent) | 16 | 53 | 80 | 95 |
| Scientists and technologists (per million inhabitants) | 15 | 103 | 270 | 1,000 |
| **Natural resources and environment** | | | | |
| Arable land (million ha) | 177 | 221 | 365 | 500 |
| Energy consumption per capita (kg coal equivalent) | 180 | 451 | 2 000 | 3 600 |
| Forested areas (million ha) | 1 580 | 1 315 | 920 | 1 500 |

*Source: Achebe and others, 1990*

institutions, working in harmony to create the desired future. The Great Transitions scenario also involves increased regional cooperation on environmental issues, such as water and food availability, mineral resources exploitation, and wildlife management. The goals of a desired and sustainable future require much more imagination than is available in the Policy Reform scenario but, as in this scenario, 'backcasting' is a major tool of analysis.

## Environmental consequences
### Atmosphere
#### • Africa

In the Great Transitions scenario, the atmosphere has been eliminated of most unwanted substances, as technology to clean up the atmosphere has been developed and used in Africa. Some gases still persist in the atmosphere, but technology to clean them is available. Furthermore, emission levels have been reduced, as a result of cleaner production and transport technologies. Urban air quality and energy use efficiency improves tremendously, due to improved awareness and technology, and changes in attitudes. African countries phase out the use of ozone-depleting substances within the context of a co-partnership in industrial development as opposed to recipient of industrial finished goods.

#### • Central Africa

The next 30 years witness a slight change in climate. The treaties on reducing the production of greenhouse-effect gases are applied by all countries in this sub-region, in order to reduce air pollution.

- *Northern Africa*

In the Great Transitions scenario, the redistribution of the population between urban and rural areas helps to alleviate congestion and its associated air pollution. Air quality, particularly in large urban areas, significantly improves, due to the wide replacement of fossil fuels with natural gas and solar energy. Strict environmental laws stop the depletion of the environment and improve air quality. The emphasis on increasing green areas and on maintaining forests helps the other measures adopted for improving the air quality.

- *Southern Africa*

The next 30 years see little change in climate variability, because it is a natural phenomenon that takes a long time to evolve. However, non-polluting industrial and domestic technologies result in greatly reduced atmospheric and indoor pollution.

- *Western Africa*

The need for clean air, and clearer and pollution-free skies, is central to policies at all levels. Appropriate legislation is respected, and has widespread public support.

- *Western Indian Ocean Islands*

Following the priority given to environmental rejuvenation, prospects improve for clearer skies and a cleaner environment within the first decades of the new century.

## Land

- *Africa*

In the Great Transitions scenario, there is increased equitable access to land. Furthermore, conflicts over land-based resources are reduced, and there is amicable resolution of any conflicts that do arise. Major rehabilitation of marginal and degraded lands have taken place, and there is planned development and rational land use—within countries and within the region. Land markets open up (including purchasing land in any country), rural-urban disparity decreases and there is greater incentive for sustainable land use. Opportunity costs for environmentally benign use of the land are recompensed. Environmental management is applied to ecosystems (that is, watersheds) within the regions, and is not restricted by national political considerations. National boundaries become less

significant —nomadic and pastoralist communities are allowed greater movement across borders (returning to traditional practices) without any hindrances whatever.

- *Central Africa*

Land tenures are improved and facilitated by simple and cheaper procedures. Regional development plans are drawn up, updated and applied. Agricultural production is better improved, both qualitatively and quantitatively, for domestic and foreign markets.

- *Eastern Africa*

Implementation of land reform policies and the emergence of a land market see more value being attached to land and a significant decrease in land degradation, as individuals take deliberate measures to maintain productivity of their land parcels. The emergence of an active land market, however, means that the poor segment of society is pushed off the prime land into marginal areas, or to seek employment on land owned by the rich. Increased productivity on the sustainably utilized land, however, ensures food security in the sub-region, and an increase in land-based income means that more people—even those without land, but with employment—are able to afford food.

- *Northern Africa*

In the Great Transitions scenario, the diversification of the economies lowers the high demand on specific natural resources, including land. As the role which industry plays in the economy increases, the demand on agricultural land is significantly reduced. In addition, the traditional role that agriculture used to play changes, as farmers switch to growing cash crops. The adoption of modern irrigation techniques, replacing chemical fertilizers and pesticides with organic types, economizing on water use and using better-quality water all help to maintain the sustainability of agricultural land. Due attention is given to the problems of drought and soil erosion, and new land stabilization techniques are introduced.

- *Southern Africa*

Issues of inappropriate land tenure, inequitable access to land, and land degradation remain dominant in Southern Africa in the Great Transitions scenario. However, growth in the adoption of democratic systems of governance may result in a more secure system of land tenure, resulting in the creation of a sense of ownership, and in a decrease in levels of land degradation. On the other hand, improvements in agricultural technologies, and the provision of adequate extension services, result in higher agricultural yields—and a step towards achieving food security. Emphasis is also given to agricultural practices which have the greatest possible potential, implying that total sub-regional food security is achieved through trade.

• *Western Africa*

In the Great Transitions scenario, after years of 'trial and error' practices based on traditional and colonial legacies, a major transition towards modern land use policy has begun. Land belongs to the people. Appropriate governing structures are put into place, in which governments, investors and the people consult and reach agreements on the uses to which a given piece of land are put. All sides benefit and, therefore, do everything to protect the land as partners. This reduces conflicts in the sub-region.

• *Western Indian Ocean Islands*

The new wave of government policies on land are warmly greeted by the international community, and the sub-region is duly rewarded for its about turn on conservation.

## Biodiversity

• *Africa*

Biodiversity, ecosystems and habitats receive adequate national and international protection, as result of policies and practices. The value of biodiversity, ecosystems and habitats is recogniszed globally and nationally, with a fair share of benefits from sustainable use accruing to local communities and national governments. Trade in rare species is fairly regulated, and is driven by ethics and moral values. Incremental loss of biodiversity due to human activity is reduced to zero, and there are no endangered species.

• *Central Africa*

In the Great Transitions scenario, there is active collaboration between modern conservation powers and methods and traditional conservation powers and methods, in order to end biodiversity loss. Greater sub-regional cooperation is developed, in terms of biodiversity conservation, with regard to transborder reserves (present-day situation in Cameroon, Gabon, Congo and Central Africa). There is also development and strengthening of sub-regional programmes, such as the Programme Régional de Gestion de l'Information Environnementale (PRGIE), which put data on the state of the environment at the disposal of decision makers, and which ensure the conservation of Protected Areas Resources. The biodiversity protection of class A species of the Convention on International Trade in Endangered Species (CITES) and other endangered species is ensured by all countries in the sub-region. The conservation programmes, and access to environmental information developed within the sub-region, are sustained by international financial backers and cooperation funds.

• *Eastern Africa*

Pressure from global stakeholders focuses the attention of national governments on biodiversity management. More national resources are channelled into the implementation of relevant conventions. National laws are enacted or reviewed, to bring them in tandem with these conventions. The countries of the sub-region promote new products and processes which enhance the potential biodiversity resource, including medicinal plants. This elevates the profile of biodiversity resources and the need to protect them.

• *Northern Africa*

In the Great Transitions scenario, biodiversity in this sub-region benefits from the overall improvement in environmental conditions. One key factor causing serious damage to biodiversity and to the ecosystems of water bodies was the disposal of very low-quality wastewater. As the quality of wastewater greatly improves, so does the health of biodiversity. The whole ecosystem shows symptoms of recovery, after long deterioration. Public awareness and better education help the governments to preserve biodiversity, as people take an active role in the process.

• *Southern Africa*

Collaborative conservation efforts between traditional practices and modern research methods result in a slowdown in the loss of biological diversity, while widespread EIAs reduce loss of habitat. Cooperation in the management of migratory species, and the use of international instruments, such CITES, also result in the conservation of biodiversity.

- *Western Africa*

In the Great Transitions scenario, comprehensive policies and appropriate legislation are in force to protect special species of flora and fauna. Large tracts of lands, including wetlands, are under protection, to ensure that biodiversity is not destroyed by development activities. These successes are promoting ecotourism.

- *Western Indian Ocean Islands*

No large mammals in this sub-region survived the previous era, apart from humans, and the region suffered serious loss of marine and bird life. However, in the Great Transitions scenario, institutions gather mating pairs of marine, land species and birds and—either by cloning from deep-frozen DNA or through natural breading programmes—achieve remarkable species survival. Following the establishment of a new regional government, there follows a re-transfer of pools of regenerated species. The pink pigeon becomes a regional icon, alongside the dodo. The invasions of vultures and other birds of prey, which created problems in Madagascar and the Comoros in the past, are dispersed, with the introduction of sparrow hawks and eagles. These also serve to provide natural control of the growing problem of urban pigeons, which are a hazard to humans and aircraft alike.

### Forests

- *Africa*

In the Great Transitions scenario, real recognition is given to sustainable uses of forest resources for medicinal and other purposes. Consequently, sensitive and important habitats are protected. Communities are environmentally aware, and are empowered to care for the earth. Areas of forests increase and forest quality improves, as a result of the realization of the true value of forest resources, and improved forest management. Integrated and sustainable development management ensures minimal degradation of the human-environment system. Human and environmental vulnerability are minimized. The capacity of NGOs and

civil societies is ennhaced, and they are empowered to play a more significant role in environmental management. Higher protection of the environment and of fragile ecosystems is the order of the day.

- *Central Africa*

This sub-region continues with the development and use of light technology by trained people, in order to make rational use of forest resources. The existence of centres and research institutes, and training in forestry (wood-based occupations), makes Central Africa a specialist in forestry training and research.

- *Eastern Africa*

The pressure on forestry resources in this sub-region is sustained and, in some areas, increased by events triggered by the Great Transitions scenario. Increased liberalization opens up the forest sector to both national and foreign direct investment for high quality logs. With the existing net deficit, increased investment in wood conversion tips the deforestation scale. However, the situation ameliorates in the latter parts of the second decade of the scenario, as private sector investment in the afforestation programme starts to yield some harvests. The emergence of regional blocks, such as the East African Community (EAC) and the Inter-Governmental Authority on Development (IGAD), and associated free trade policies, also increase the movement of forest products from resource-rich member states to those with deficits, ranging from Uganda to Kenya, and Ethiopia to Eritrea. A lucrative trade with favourable tax incentives is the most likely to lead to mining of the resource. This is further exacerbated by the improvement in transport technology and systems, enabling faster movement of goods in the sub-region.

Technology development, however, negates some adverse impacts. Improved technology expands the scope of species utilized, and improves wood processing through recycling and the use of small-dimension timber and so on. Tree improvement biotechnology leads to high productivity plantations.

- *Northern Africa*

In the Great Transitions scenario, the rate of deforestation is greatly reduced, as governments of this sub-region strengthen and enforce legislation governing the exploitation of forests. Better education and public awareness programmes—albeit relatively limited—teach people the importance of forest resources. Droughts and shortage of water resources seem to be the only factors hindering afforestation and reforestation efforts. Because the Northern African countries are able effectively to apply integrated water resources management techniques, they manage to restore the natural forests, or to grow human-made ones. In the Sudan, the value of gum arabic is not overlooked by the forest protection policy, as careful planning ensures the double benefit of producing the gum and preserving the forests.

• *Southern Africa*

Cheap micropower technologies gain prominence, reducing the pressure on forests for energy. There is significant growth in technologies, such as solar power, which do not need wood, and efficient wood stoves, such the 'tso-tso' stove, which requires very little fuelwood. The use of better fuels, and the adoption of better heating and cooking technologies, reduce both indoor and outdoor pollution, while concerted efforts by the developed world to reduce greenhouse gas emissions ameliorate the effects of climate change.

• *Western Africa*

As a matter of policy, every tree which is destroyed is replaced by two new trees. Not only is there reforestation, there is active afforestation. The benefits are felt throughout the sub-region, especially in the Sahel parts of Western Africa, where Burkina Faso, Mali, Niger, Mauritania and Senegal lead the way.

• *Western Indian Ocean Islands*

Massive reforestation is in hand, with the target of planting ten trees per person per year in the region over the next decade (2025–35). This ends the devastating decade (2015–25) of barbaric depletion of forests and woodlands.

### Freshwater

• *Africa*

In the Great Transitions scenario, water remains vital to both the agricultural and the industrial development process. Unfortunately, a large proportion of the region is semi-arid, with very low annual rainfall. Droughts have, therefore, become a major feature of weather and climate

all over Africa, and the effects have been difficult to contain. In many countries, there is uneven distribution of water and land suitable for agriculture, and this affects the potential of agricultural production. Serious stress on water occurs in countries such as Djibouti, Mauritania, Somalia, Algeria and Sudan, and in many countries in Western Africa, Southern Africa and Eastern Africa (Chenje and Johnson 1996). The problem of water shortages are carefully addressed and eliminated.

• *Eastern Africa*

Localized freshwater availability continues to be a problem, especially in the semi-arid parts of the sub-region (Somalia, Eritrea, Djibouti and north-eastern Kenya). In the Great Transitions scenario, investment in large-scale water redistribution schemes is likely to take place in the next 30 years, particularly focusing on the water of the Blue Nile and its tributaries. Large-scale irrigation schemes may be developed in the Ethiopian lowlands. Concerted diplomatic moves are witnessed in the direction of shared water resources, in order to increase equitable distribution in the sub-region. Sub-regional groupings around the shared water resources, such as the Nile Basin Organization (NBO), become stronger and more influential in driving economic development in the sub-region.

• *Northern Africa*

In the Great Transitions scenario, the existing water scarcity problem absorbs the full attention and consideration of governments in the sub-region. The sub-region takes sufficient measures to ease the problem, through careful and sustainable management of conventional water resources. Furthermore, the management of non-conventional resources—such as rainfall harvesting, desalination and water recycling—are implemented and adapted to the local conditions. The industrial revolution experienced in the sub-region helps to lower the cost of these new techniques, to the extent that they become affordable to many sectors. In addition, all the countries adopt integrated water resources management programmes to effectively manage all available water resources. In Egypt and the Sudan, the concept of integrated water resources management is extended across the boundaries of each country, to include the rest of the Nile riparian countries. The experience gained whilst implementing integrated water resources management procedures is exchanged by experts during regular meetings organized at the sub-regional level.

• *Southern Africa*

Water scarcity is a big issue in Southern Africa, and may continue to be dominant during the next 30 years. In the Great Transitions scenario, for the sake of human sustainability, small-scale water-harvesting technologies may become prominent in the sub-region. Wasteful water consumption patterns, such as inappropriate irrigation practices, may slowly be replaced with more efficient systems, such as drip or micro-jet irrigation, as well as proper economic accounting and costing of water.

• *Western Africa*

With advanced water management and protection skills, there are freshwater supplies all year round, and they are available in all parts of the region. The net benefits are shared throughout the sub-region under the ECOWAS agreements and protocols.

• *Western Indian Ocean Islands*

In the Great Transitions scenario, the hitherto lack of enforcement of public health legislation on water protection becomes a thing of the past. Governments call for community service orders to enlist communities in environmentally clean projects, as an alternative to custodial care.

### Coastal and marine resources

• *Africa*

In the Great Transitions scenario, there is less pressure on coastal zones and lagoons. Effective integrated management leads to the sustainable use of coastal and marine resources, and the sustainable use and management of marine resources.

• *Central Africa*

The issue of coastal and marine erosion is reduced, as a result of to national and regional policies. However, the reconstruction of damaged areas requires some time.

• *Northern Africa*

The coastal and marine environments are very valuable resources to Northern Africa, and are well protected and managed. In the Great Transitions scenario, laws regulating coastal areas are issued, and are strictly enforced. New developments in sensitive areas are totally banned, and tight control is imposed on developments in other areas. Protection measures include: the prohibition of any direct disposal of all types of waste in the sea; rejecting licenses for projects which cause sedimentation or erosion of shorelines; and issuing guidelines to boats, divers and fishermen regarding safe practices in territorial waters and the protection of marine life. Despite the fact that water tourism is encouraged, strict guidelines for the development and management of tourism establishments in coastal areas, issued in order to minimize their impacts on the coastal environment, are willingly enforced.

• *Southern Africa*

In the Great Transitions scenario, the rates of coastal erosion are greatly reduced, thanks to effective national, regional and global policies. Greater environmental awareness results in the sustainable harvesting of marine resources, although more time is required to replenish stocks.

• *Western Africa*

A sub-regional policy is enforced, which provides that landlocked countries are as entitled to the Atlantic coast as are those actually adjoining the coast. Under this policy, there is a collective responsibility: to protect the ocean and waterways; to police the high sea for vessels which carry and dump hazardous waste; to control fishing; to build environmentally friendly harbors; and to keep beaches clean for tourism.

• *Western Indian Ocean Islands*

In the Great Transitions scenario, a regional integration movement is revived to save the marine and coastal livelihoods of in-shore fisherman. There is a reawakening of interest in the development of deep-sea fishing, with the inauguration of the regional shoal world integrated satellite monitoring system (SWIMS), promoting fisheries and protecting endangered fish species. Equipment for detecting shoal size, species mix, age and edibility has the capacity to filter metal and plastic from the sea floor. This transforms coastal areas which were unregulated dumping grounds for domestic and industrial waste in the sub-region.

## Urban areas

### • Africa

In the Great Transitions scenario, the proportion of people living in urban environments increases. However, there are also marked increases in access to water and sanitation. There is a marked decrease in people living below the poverty line, and in people living in urban slums and unplanned settlements. There is also a greater control of waste management.

### • Central Africa

Urban areas continue to grow, albeit under much greater control. The involvement of non-governmental actors is important, and improves urban management (access to land use, housing and social infrastructure dispositions). The system of revamping this sector functions, and reduces its effects on health.

### • Eastern Africa

Nearly 50 per cent of the population in this sub-region are living in urban areas, attracted by increased employment opportunities and availability of services. The social infrastructure is overstretched for the next 30 years, as investment in infrastructure development continues to lag behind the population increase. The housing shortfall currently stands at about 35 per cent but, in the Great Transitions scenario, this certainly reduces, with more involvement of the private sector in the housing industry, despite the influx of people to urban centres.

### • Northern Africa

In the Great Transitions scenario, the equitable distribution of public expenditure amongst urban areas, and the emphasis on underdeveloped areas, brings them to the same status. As the economic situation improves and many services become privatized, more people have access to and can afford the services. Unleaded petrol and natural gas widely replace other fossil fuels, as polluting industries are moved out of urban areas to newly and specially developed areas, where planning has taken environmental issues into consideration. Advancement in technology means that environmentally harmful equipment, such as air conditioning units, are replaced with new, safer technologies.

### • Southern Africa

Urban growth continues, but in a more planned fashion, as more people become aware of the human risks and environmental dangers of shanty settlements. This awareness is supported by growth in incomes, which enable many to afford decent housing.

### • Western Africa

Balanced and planned development activities and programmes in both urban and rural areas stem urban migration. Pressure on urban services is reduced. Rural areas experience a flow of former urbanites, especially retirees and young professionals, who are gainfully employed or re-employed in their own localities and communities.

### • Western Indian Ocean Islands

Urban regeneration programmes enrol groups of people in public works programmes. These lead to social education which transforms destructive tendencies to productive avenues for urban governance and the provision of services. Religious retreats—established in earlier years by fundamentalist groups of all persuasions, who were prepared to fight to the death for the preservation of rights—are transformed into business, educational and development retreats for reflection and social development projects.

## DISCUSSIONS AND SYNTHESIS

### INTRODUCTION

Humankind has, through the ages, employed various methods and devices to seek to know what the future holds. Indeed, there is a school of thought that humanity's concern with knowing the future, and the impulse to propitiate the future in order to avoid catastrophes and to produce blessings, may well have been part of the drive for knowledge. Be that as it may, any explorations about the future remain a major undertaking, and one which is designed to elucidate and educate. It is even more challenging when the subject of study concerns humanity in its domain, and humanity's ability to ensure self-preservation.

In our endeavour to chart a new course for the development of the African environment, we have made recourse to all the tools available at our disposal. For instance, as Achebe and others (1990) wrote in Beyond Hunger, 'thinking about the future requires faith and vision, mixed with philosophical detachment, a rich emotional life and creative fantasy as well as the rigorous and orderly tools of science'. Consequently, in looking at the future of Africa from 2002 to 2032, we have combined all these efforts to produce quantitative descriptions and rich narratives of the various scenarios, which we believe will assist in the major steps that must be taken in the new millennium.

AMCEN's concern with developing an AEO represents a significant landmark in the region's development process. It should also constitute a threshold in the way in which we relate to the environment. It is this understanding that constitutes the framework for the analysis described in this chapter. The four scenarios—the Market Forces scenario, the Policy Reform scenario, the Fortress World scenario and the Great Transitions scenario—were developed to facilitate the discussions on the future of our environment. The scenarios are simple conceptualizations of how the environment might develop in the next 30 years. The truth is that this categorization is convenient, and is not meant to indicate some compartmentalization of the processes of change. As we have shown in the narratives, it is possible, even within countries, to find aspects of each scenario occurring. The Fortress World scenario— though a doomsday scenario, which is heralded by apathy and the neglect of issues of sustainability—has many features that are present today in the socioeconomic systems of many African countries.

Nevertheless, writing about the future can be one of the most dangerous undertakings, because the future is essentially unknown and uncertain. It can also be a very challenging task, because it can be full of suspense and surprises. This is our understanding. What we have provided in this work, therefore, is both the quantitative and qualitative evaluation of the future of our region, in the hope that the minds of people in positions of authority, who can make and execute policies, are drawn to the crucial issues raised regarding desirable and undesirable trends. In this final section, we shall attempt a recapitulation of the findings.

## THE SCENARIOS

We can draw some general conclusions from the scenarios about the state of the environment not only in Africa as a region, but also in its sub-regions. Some of these are:

- While some climatic change may be inevitable in Africa, the scenarios will differ mainly in the ability of different sub-regions to put into operation adequate strategies for coping with these consequences. African countries are very vulnerable to all forms of hazards and disasters.

- Tourism of all kinds may become a major force, promoting environmental change. However, the effects of tourism could be either positive or negative, depending on how its development is handled, as well as how equitably the benefits of tourism are shared.

- The land issue remains critical, as the need to ensure effective and acceptable land reform becomes crucial not only for rural areas, but also for urban land development.

- Water pollution is likely to emerge as a serious problem, in addition to water inadequacies in some regions. Consequently, controlling water pollution could have numerous advantages and benefits. For instance, limiting water pollution could offer opportunities for cost-sharing, as different groups might benefit.

- Associated with the land issue are the issues of deforestation and desertification, which have already become major problems in many parts of the region. Compounded by sourcing fuels through tree cutting and wood burning, the state of the environment remains one to be watched most carefully.

Nevertheless, the scenarios described in this work all have one thing in common. They are forms of peeping into the future from a current situation: the state of the environment. In other words, they are humanity's attempts to conceive the future for the purpose of evolving a society which benefits humankind, not only in our present circumstances, but also in all future endeavours. For this reason, the scenarios are greatly influenced by our current situation, and by the operating driving forces of the world economic system. In the context of Africa, the major driving forces are globalization and the unsustainable level of the world environment system. While these represent major uncertainties, as we progress through the new millennium, they remain important considerations in the way in which they impact the environment at global, regional, sub-regional, national and local levels.

The Market Forces scenario is predicated on the assumption that existing strong linkages with international financial institutions, multinational corporations and global markets will continue to drive the world economic system which, in turn, will impact the African regional environmental system with its catalogue of goods and bads. Furthermore, there will be effective mechanisms through which environmental bads are processed and converted into forms that ensure the sustainability of the African environment. Even at the beginning of the 21st century, this assumption has become more theoretical than practical, both in its interpretation of the development process and in its attempts to order the impacts on the environment. In the African situation, market forces have brought more socioeconomic problems than were experienced in earlier decades, and people have neglected to consider the negative impacts of their activities on the environment. The peculiar position of Africa as the backwater of the advanced countries, coupled with the almost obliterating levels of poverty in the region, make consideration of environmental sustainability secondary. Market forces, consisting of economic principles of development, have always played themselves out to the detriment of African people. Africa and Africans continue to remain vulnerable to unrestrained economic exploitation of resources, and this trend leads not only to unsustainable patterns of living, but also to unsustainable assaults on the environment (see Figure 4.13).

The Policy Reform scenario is a natural reaction towards seeking a balance between socioeconomic development and environmental sustainability. Current threats to the environment—as seen in rising levels of $CO_2$ in the air; the inability to manage all forms of solid and liquid waste, especially in the cities; the continued destruction of forests and biodiversity; and the decreasing levels of environmental health—are indications of the inability of market forces to order reactions to environmental sustainability. The Policy Reform scenario, therefore, proposes to achieve a balance in sustainable economic and social development while, at the same time, respecting many environmental issues. Under these conditions, targets are set by regional organizations, such as the OAU and AMCEN, and serious efforts are made to adopt Multilateral Environmental Agreements (MEAs) and protocols, and to make these operational. We are all witness to the fact that the ratification of these treaties is a far cry form their actualization in individual countries. Consequently, the problem remains of managing the environment in a sustainable manner through the mechanism of MEAs and policies (see Figure 4.13).

A major problem of the Policy Reform scenario lies in its inability to go all the way in outlining requirements for sustainable social and environmental development, partly because of the strong interplay of market forces in the scenario. There are also other political and national considerations in the implementation of MEAs. Furthermore, the danger also remains that inadequate and inappropriate attention to environmental issues, in the short term and in the long term, may lead to the breakdown of law and order. Many of the current tensions and disturbances in Africa may be traced to one form or other of the numerous environmental pressures in the region.

Of course, the movement to a Fortress World scenario is much closer than is usually envisaged. Wealth is distributed in a lopsided manner—between people, between urban and rural areas, and between regions—and it is difficult to introduce policies that can reasonable alter these distributions of wealth. Within any of these units, there are enough tensions to lead to a breakdown of law and order. And so the wealthy have already learned how to protect themselves. The crime levels in cities, and the fact that these are perpetuated openly, are indicative of a gradual breakdown of law and order, which could be a reaction against the society and those who manage it. It is, as yet, an unorganized reaction, but it is already a dangerous trend. Coupled with other forms of tensions—such as those which could arise from religion, and from political alliances which the wealthy have managed to exploit for selfish reasons—a Fortress World scenario is very much just around the corner.

The Fortress World scenario has the potential to destroy the environment and, possibly, the whole of humankind (see Figure 4.13). Since there will be no

**Figure 4.13 The scenarios compared**

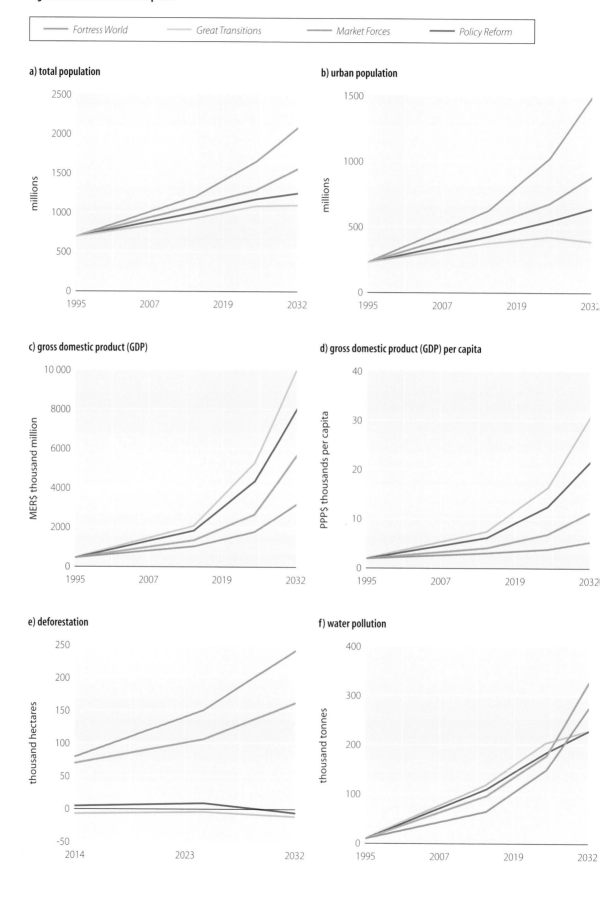

coordinated efforts to ensure environmental sustainability, the situation regarding most resources is one of overexploitation, either to meet the needs of the élite or to ensure the basic survival of the poor. Environmental conditions, therefore, deteriorate, as pollution, climate change, land change and ecosystem degradation interact to amplify the crisis. Environmental degradation, food insecurity and emergent diseases foster a vast health crisis, as both free market values and reformist tendencies become incapable of handling environmental externalities. The affluent minority, alarmed by rampant migration, terrorism and disease, reacts with sufficient cohesion and strength to impose an authoritarian Fortress World, where they flourish in protected enclaves. The fortresses are bubbles of privilege amidst oceans of misery.

The Great Transitions scenario, on the other hand, may arise from two totally unrelated considerations. Given that a Fortress World scenario is destructive to the environment and to humanity, it could be a harbinger of the urgent need for all humankind, including Africans, to seek alternative ways of managing the environment. Depending on the level of destruction that people's activities would have wrought on the environment by the time this need arises, a Great Transitions scenario may or may not be sustainable (see Figure 4.13).

On the other hand, the seeds of change are already being germinated all over Africa, and governments and NGOs are already aware of the need to evolve a new course and a new perspective on environmental and development issues. The intellectual community, too, as shown in various aspects of the description of the Great Transition scenario, are already clamouring for a new dictum on development. The honest ones among the developed countries already accept the fact that they have misled the developing world for too long (Gotlieb 1996). Africa's democratically elected leaders have also become more responsible and humane, and are anxious to subscribe to the issue of environmental sustainability. They, however, need to be convinced of the inability of the Market Forces scenario and the Policy Reform scenario to lead Africa to the promised land (see Box 4.3). Everybody detests the idea of a Fortress World scenario evolving and playing itself out in Africa.

Given the current trend in Africa and all over the world, it may be surmised that a Market Forces scenario may only be plausible within a short time—say, ten years—at about which point probabilities are high that a number of branching points will start to emerge. Possible branching points may include:

- Increasing social pressures for government reform, and more concerted attention to environmental and social issues. It is this that may lead to a full blown Policy Reform scenario.
- Increasing concentration of wealth and power. When posited with existing levels of corruption, there will be a breakdown in civil society and a rise in political anarchy. It is this that may lead to lead to the Fortress World scenario.
- Of course, widespread rejection of the 'IMF Dream', coupled with the development and adoption of new technologies, approaches and visions. This is what may lead to the Great Transition Scenario.

Nonetheless, the point being made is that the environment is so valuable and the inhabitants so precious that the future need not be left to chance or to some curious form of evolution. The scenarios described in this report have shown the unacceptability of a business-as-usual approach to environmental issues. It has also shown the inadequacy of an acceptance of a Policy Reform scenario. Something that has not worked in the past is not likely to work now or in the future unless the constraints that made it non-functional in the first instance are removed. Certain forces that militate against the removal of these constraints dominate the world political and economic system.

Africa has been very vulnerable to many events happening all over the world or within the region itself. For instance, the consequences of environmental disasters, such as floods and drought, continue to haunt the inhabitants of many countries, and remain one of the great challenges to governments. Of course, the problem of hunger is mostly felt in Africa, where more than 75 per cent of the population live below the poverty line. Many of these aspects of vulnerability have been discussed and described in Chapter 4, and need not be repeated here. But the lesson is that, unless concrete steps are taken in the way we use the environment, the sufferings and problems of the 20th century will be child's play compared to what lies in store. Wherein, then, lies our future in Africa?

For some time, the Great Transitions scenario may remain an enigma to both policy and practice. Yet

### Box 4.3 Disenchantment

*'Development theory has undergone constant revision and has been informed by a broad spectrum of approaches. One can conclude that the problems of development are larger than the highly heterogeneous policies, plans and programmes undertaken by the full spectrum of economic approaches. It is increasingly evident that our understanding of the reasons why one-fifth of humanity enjoys unprecedented wealth while the other four-fifths live in various states of poverty and privation is incomplete. Clearly, conventional analyses of socio-economic, cultural, ecological and political phenomena are limited in their ability to elucidate the multiple problems we now encounter globally.*

*'The thesis presented here is that development itself, or at least the concepts we use to define it, are deficient. We can speak of a global problematic embodying three broad sets of phenomena; trenchant poverty, environmental degradation and socio-political unrest. Related to this problematic is existential malaise, particularly in those societies considered the most "developed". This global problematic is not transient. It speaks to profound issues about who we are that have become repressed in our individual and social consciousness.'*

Source: Gotlieb 1996

therein lies the hope for Africa and the African environment. It will remain an enigma for many reasons. First, it is going to be difficult to convince the peoples of Africa that the future of humankind lies in the Great Transitions scenario. Secondly, African countries are not all at the same level of awareness and socioeconomic development, so the idea of a Great Transitions scenario could still look like a dream to some. Furthermore, the expectations in the Great Transitions are things that take time to mature. Take, for instance, the issue of good governance, which democracy represents, and its almost universal acceptance as the best form of government to promote development. In recent times and in many countries, including some Indian Ocean Islands, there have been reports of military takeover of governments. It is, therefore, possible to expect that, even when there could be a belief in the principles of the Great Transitions scenario, the playing field in Africa is far from level. There is a need to emphasize the need for this awareness. Wherein lies the future of Africa?

In order to answer this question meaningfully, we

like to draw on the relationship between the Policy Reform scenario and the Great Transitions scenario. Both scenarios are forms of 'backcasting' (see Figure 4.4), whereby desirable futures, and device mechanisms for manipulating the system to meet targets, are defined. The future of the African environment lies in the ability of governments and ministers of the environment to appreciate that current policies and practices remain grossly inadequate when it comes to meeting the demands of a sustainable environment. There are many MEAs currently in operation. These agreements and protocols were designed to assist Africans to cultivate some respect for the environment. We have shown that current rates of population growth, and the demand on resources, render these efforts inadequate. Of course, many of them are a distant cry from what is required to move towards a Great Transitions pattern of environmental sustainability. As a first step, governments should review these MEAs, and create mechanisms in order to ensure their compliance by differ-ent countries in the region. Governments should:

- devise systems for identifying requirements for a sustainable environment;
- set new targets based on the requirements of a Great Transitions pattern of environmental growth (see narratives);
- design mechanisms to steer national, sub-regional and regional poli-cies towards meeting these new targets (this could imply the institution of new memorandums of understanding, based on the new environmental standards); and
- encourage governments in different countries to manage their environments conscientiously, in line with these new standards and targets.

## CONCLUSION

The goals of sustainable development are summarized in *Our Common Future*, or the Brundtland Report, which states that we must meet 'the needs of the present generation without compromising the ability of future generations to meet their own needs'. These needs could be economic, political, socio-cultural and ecological. Thus, the management of the environment has always presented people with many problems and challenges, largely because of its complex nature, and of the com-plex interactions and relation-ships within it. Furthermore, since environmental management value-laden, there is usually more than one way of conceptualizing and expressing the value of the environment and its resources. The result is that environmental management normally requires that we deal with several elements of the environment, and with several perspectives reflecting the different concepts and perceptions held by different societies, belief systems and interests.

We have constructed scenarios from: our understanding of current conditions and driving forces; a vision of the future; and a coherent story of a process of change, leading to that future. We have used both imagination and science as ingredients for generating effective scenarios, and we have made quantitative assumptions across a range of dimensions: economic growth and structure, population, technology, resources and the environment. We have used the scenarios to take the inherent uncertainty in future development as the point of departure, and to seek to formulate plausible stories about alternative possibilities that can emerge from current conditions and driving forces.

As a first step in scenario development, we need to be able to calibrate relationships between the more discerning variables or driving forces with what occurs in real-life situations. While data currently preclude a full investigation along these lines, they are undoubtedly necessary for the future development of scenarios. However, for studies which link vulnerability to scenario development, we need to be able to define the stress variables, such as people under stress for water or people living within a certain distance of a potentially dangerous hazard.

Nevertheless, there seems to be no time more opportune than now to discuss issues that affect the sustainability of the African environment. The reasons are many, but perhaps most important is that there is a unity of purpose in the minds of African leaders on the urgent need to eradicate poverty from the region and to give development a much-needed direction. For instance, NEPAD seeks 'to build on and celebrate the achievements of the past, as well as reflect on the lessons learned through painful experience, so as to establish a partnership that is both credible and capable of implementation. Africa must not be the wards of benevolent guardians; *rather they must be the architects of their own sustained upliftment*' (authors' emphasis). Furthermore, NEPAD centres around African ownership and management, issues that were considered germane to development of a Great Transitions scenario.

## REFERENCES

Achebe C. and others (1990). *Beyond Hunger in Africa 2057: An African Vision*. Heinemann Books, New Hampshire

Bond, P. (2000). *Washington Conflict—Not Consensus—Over Global Financial Management*. Available on: http://www.aidc.org.za/archives/pbond_washington_conflict.html

Chazan N. and others (1992). Regimes in Independent Africa. In *Politics and Society in Contemporary Africa* 2nd edition. Lynne Rienner Publishers, Boulder

Chenje, M. and Johnson, P. (eds) (1996). *Water in Southern Africa*. SADC/IUCN/SARDC, Maseru, Lesotho/Harare

ECA (1999) African Development Forum—Strengthening Africa's Information Infrastructure. Addis Ababa

ECA (2000). *The ECA and Africa: Accelerating A Continent's Development*. Addis Ababa

Emeagwali, G. (2000). *Colonialism and Africa's Technology*. Available on: http://members.aol.com/afriforum/colonial.htm

Gallopin, G., Hammond, A., Raskin, P. and Swart, R. (1997). *Branch Points: Global Scenarios and Human Choice*. Stockholm Environment Institute, Stockholm. Available on: http://www.gsg.org.

Gotlieb, Y. (1996) *Development, Environment and Global Dysfunction*. St Lucia Press, Delray Beach, Florida

IIED (1997). *Southern Africa Beyond the Millennium: Environmental Trends and Scenarios to 2015*. Prepared by Dalal-Clayton, B. International Institute for Environment and Development. London, United Kingdom

Kemp-Benedict (2001) *Training Workshop on GEO-3: Scenario Development, African Region, 18-20 June 2001; Follow-Up Report*. UNEP, Nairobi

Makinwa-Adebusoye, P. (2000). Population and Development. The APIC/ECA Electronic Roundtable International Policies, African Realities. Also available at http://www.africapolicy.org/rtable

OAU (1980). *Lagos Plan of Action for the Economic Development of Africa: 1980–2000*. OAU, Addis Ababa

OECD (1995). *Preparing for the Future: A Vision of West Africa in the Year 2020*. Organisation of Economic Co-ordination and Development, Paris, France

Raghavan, C. (2000). What Washington Consensus? *I never signed Any— Camdessus*. South-North Development Monitor. Available on: http://www.twnside.org.sg/title/signed.htm

Raskin P. D. (2000a) *Regional Scenarios for Environmental Sustainability: A Review of the Literature*. Stockholm Environment Institute, Boston Center, Boston

SARDC, 1994, The State of the Environment in Southern Africa, SARDC/SADC ELMS/IUCN, Johannesburg

Stiglitz, Joseph (1998). More Instruments and Broader Goals: Moving Toward the Post-Washington Consensus. Paper presented at the 1998 WIDER Annual Lecture, 7 January 1998, Helsinki

UNDP (2000) *Human Development Report*. Oxford University Press. New York

UNEP (2000) *Global Environmental Outlook 2000*. UNEP, Oxford University Press, New York

UNEP (1989). *Report of the African Ministerial Conference on the Environment on the Work of its 3rd Session*. 10–12 May 1989, Nairobi

UNPD (1999). *World Urbanization Prospects*: Population Division, Department of Economic and Social Affairs, United Nations Secretariat.

Vavi, Z. (1999). *Africa: Statements on Globalization*. Meeting on Globalization and Social Justice, Trade Union View

World Bank (2000). *World Development Indicators Database*, July 2000.

World Commission on Water (2000). *The Africa Water Vision for 2025: Equitable and Sustainable Use of Water for Socio-economic Development*. World Water Commission on Water for the 21st Century

WRI (1994). *World Resources Report 1994–95*—People and the Environment. WRI/World Bank/UNDP/UNEP, Washington D.C.

# CHAPTER 5

## POLICY RESPONSES, ANALYSIS AND ACTION

# CHAPTER 5

## POLICY RESPONSES, ANALYSIS AND ACTION

### INTRODUCTION

This chapter provides a synthesis of the issues covered in the previous chapters of AEO, followed by an analysis of policy responses for the implementation of a sustainable environment and development agenda for Africa. The chapter closes with a key output of the AEO process: 31 recommendations for specific actions by policy makers.

### OVERVIEW

The wealth of a nation is measured by its total national capital, that is to say, the sum of its human-made capital, natural capital, human skills capital and social capital (Serageldin 1994). This is illustrated in easily memorizable form in Box 5.1. For development to be sustainable, the stock of national capital at any given time in the future must be greater than the current amount.

Africa has increased its total capital over the past 30 years. In spite of the challenges the region has faced from colonial times to the present, its overall stock of total capital—both absolute and per capita—has

*Compared to 30 years ago, African countries are increasingly democratizing, are devolving power from the centre to lower levels, and are empowering communities and civil society organizations to participate meaningfully and effectively in decision making.*

---

**Box 5.1  The stock of national capital**

TNC = HMC + NC + HSC + SC

Where:
    TNC = total national capital
    HMC = human-made capital
    NC = natural capital
    HSC = human-skills capital
    SC = social capital

---

increased when measured in terms of GDP (UNDP 2000). Table 5.1 shows this trend for selected African countries, for 1975, 1985 and 1998. The individual components of total capital are considered in a little more detail, below:

● In the past three decades, Africa's stock of **human-made capital** (for example, buildings, highways and factories) has increased, as indicated by the proliferation of urban areas in the region.

● **Natural capital** (soils, forests, minerals, oil and gas, fisheries, wildlife) has declined, as evidenced by the degree of environmental degradation already described in this report.

● Levels of investment in education indicate that, in absolute terms, Africa's **human-skills capital** has increased over the past 30 years (UNDP 2000), although this increase is currently threatened by the poor working conditions and inadequate motivation of the region's skilled labour force. This is evidenced by Africa's 'brain drain' problem.

● Africa's **social capital**—the sum of democratic governance systems, social services, institutional capacity and the empowerment of women and other marginalized groups—can also be surmised to have increased. Compared to 30 years ago, African countries are increasingly democratizing, are devolving power from the centre to lower levels, and are empowering communities and civil society organizations to participate meaningfully and effectively in decision making.

However, this apparently positive picture is somewhat deceptive. Four reasons for this are outlined below:

● While absolute and per capita amounts of total capital have increased on a region-wide basis, there are wide disparities between countries, with increases in some and declines in others. For example, over the past 30 years, per capita GDP has tripled in Mauritius, almost doubled in Seychelles and increased to a degree in Tunisia. On the other hand, it has declined in Gabon, Ghana, South Africa and Sierra Leone, and has more or less stagnated in Kenya. In Ghana, Sierra Leone and South Africa, per capita GDP was higher in 1975 than in 1998.

- The rate of growth of Africa's total capital is lower than that of all developing countries combined, which means the increase is not sufficient to bring about significant change.

- The distribution of total capital is not equitable. For example, in 1987–98, the poorest 20 per cent of Sierra Leone's population benefited from a paltry 1.1 per cent of income consumption (that is to say, only 1.1 per cent of the 'national cake'). The picture is little better for the rest of Africa, where the proportion was less than 10 per cent (UNDP 2000).

- Lastly, and perhaps most importantly, the increase in Africa's total capital has been achieved through the transformation of natural capital into other forms of capital. In principle, such transformation is not a reason for concern so long as the total sum of capital is increasing. In Africa, however, there is reason for concern, because the process of transformation has been unsustainable and continues to be so. For example, in many parts of Africa the rates of harvesting from forests, fisheries and wildlife resources exceed their sustainable levels, and utilization of resources has been wasteful.

Overall, while some African countries, such as Seychelles and Mauritius, have significantly improved the quality of life of their people, the majority of African countries are in the low human development index (HDI) category (UNDP 2000). As of the year 2000, Cape Verde Islands, Ghana and Kenya were the only three countries in the western and eastern African sub-regions to have medium HDIs; the rest of the countries were in the low HDI group (UNDP 2000).

The link between environment and development is particularly strong in Africa because national economies are dependent on agriculture and natural resources at the primary production and processing stages. The combined contribution of agriculture and industry (largely natural-resource based) to GDP is significant, especially in those countries with lower HDIs. For example, in Sierra Leone—the country with the lowest HDI in the world—agriculture and industry represent 68 per cent of GDP (UNDP 2000). Yet many African countries, especially those in the eastern and western sub-regions, belong to this category. Sound management of the environment, therefore, has important implications for rural livelihoods, overall economic growth and better quality of life.

Although African countries have made some improvements in environmental management, many challenges still remain and should be addressed. For example, the current levels of land degradation; deforestation; loss of biodiversity; overharvesting of natural resources; atmospheric pollution; lack of access to clean and safe water and sanitation services; and poor urban conditions are manifestations of remaining unfavourable conditions. If nothing is done, these factors will combine to undermine Africa's prospects for sustainable development.

Furthermore, the inadequacy of economic opportunities in Africa, the existence of trade barriers and farming subsidies in the developed countries, and the declining state of the region's environment mean that its people are becoming increasingly vulnerable to adverse changes in the environment. Many African countries are not adequately equipped to deal with

*The link between environment and development is particularly strong in Africa because national economies are dependent on agriculture and natural resources at the primary production and processing stages.*

| Table 5.1  Trends  in per capita GDP in selected African countries (in US$) | | | |
|---|---|---|---|
|  | **1975** | **1985** | **1998** |
| Seychelles | 3 600 | 4 957 | 7 192 |
| Mauritius | 1 531 | 2 151 | 4 034 |
| Tunisia | 1 373 | 1 771 | 2 283 |
| South Africa | 4 574 | 4 229 | 3 918 |
| Gabon | 6 480 | 4 941 | 4 630 |
| Ghana | 411 | 328 | 399 |
| Kenya | 301 | 320 | 334 |
| Sierra Leone | 316 | 279 | 150 |
| Sub-Saharan Africa | 780 | 1 170 | 1 520 |
| All developing countries | 720 | 1 520 | 3 260 |

Source: UNDP (2000)

Sound environmental management has important implications for rural quality of life and livelihood.

*Mark Edwards / Still Pictures*

natural disasters, such as floods, droughts and earthquakes, and emerging health problems such as the HIV/AIDS pandemic (discussed in Chapter 3).

Africa's challenge is captured in a statement attributed to H.E. President Olusegun Obasanjo, of the Federal Republic of Nigeria, and reproduced in Box 5.2. In the face of such a challenge, defining paths to sustainable development is a necessity if Africans—especially those in the eastern and western sub-regions where HDIs are low—are to achieve the better

quality of life they deserve and are to be able to improve their environment.

## SCOPE OF ACTIONS AND DIFFICULTIES OF ASSESSMENT

African states have put in place a number of policy responses to address environmental issues. Many new national environmental policies, laws and regulations have been introduced and African countries are parties to a number of multilateral environmental agreements (MEAs). The catalogue of documents relating to environmental management clearly shows that—on paper at least—problems have been addressed extensively (NEMA 2001). However, these remain mere intentions unless implemented. Moreover, even after implementation, there is a need to verify that effects on the environment are positive and adequate (NEMA 2001).

Quantitative assessment of the success or failure of policy initiatives and developments is not an easy task. African states face the same problems as the rest of the global community where analysis of policy responses is concerned. Global experience indicates that assessment of the effects of implementation and of efficiency is made particularly difficult by uneven monitoring, poor and missing data, a lack of indicators and continuous reporting, and paucity of data on the environmental situation before and after implementation (UNEP 1999). Furthermore, there are

---

**Box 5.2 The African challenge**

'We are all aware of the problems and challenges facing our continent today. Almost 15 years after the establishment of AMCEN and, indeed, eight years after the Earth Summit in Rio de Janeiro, Brazil, our region is still bedevilled by many problems. We are still contending with land degradation and natural, as well as man-made, disasters. Our forests and forest resources are being indiscriminately exploited and depleted, our coastal and marine resources are being degraded, and we still have enormous problems with water supply and availability, quantitatively and qualitatively. Many of those problems result from the unplanned and unsustainable manner in which the region's natural resources, including its diverse ecosystems, are being exploited.

'These difficulties are further aggravated by broader

environmental problems of planet Earth, such as ozone layer depletion and climate change, which continue to threaten the survival of mankind. In addition, Africa has unfortunately been an easy dumping ground for toxic and hazardous wastes and obsolete chemicals and technologies. Add to these the intractable difficulties of crippling debt burden, a population growth rate that is spiraling out of control, pervasive and frequent violent conflicts on the continent … and one gets a picture of an Africa that awesomely challenges each and every one of us to find immediate solutions.'

**H.E. Olusegun Obasanjo,**
**President of the Federal Republic of Nigeria**

UNEP (2000)

no suitable mechanisms, methodologies or criteria to determine which policy contributes to which change in the state of environment. Such problems often prevent valid comparisons between the current situation and what would have happened if no policy action had been taken. A more complete and precise analysis will require the development of better mechanisms to monitor and assess the effects of environmental policies on the quality of the environment (UNEP 1999).

It is clear from this overview that there is a need to review and recommend achievable actions at national, sub-regional and regional levels, and to consider their implications for implementation at these levels and for the global environmental agenda. The difficulties outlined above notwithstanding, the remaining sections of this chapter identify appropriate policy responses, analyse their implications and provide recommendations for action.

## POLICY RESPONSES AND ANALYSIS

The history of social, economic and environmental development presented in Chapter 1 demonstrates that African countries have risen to the challenge of environmental degradation. They have developed a collective will to address environmental and related issues, and have created institutions to translate that will into concrete results. Some milestones in this process are recapitulated below.

- In 1968, African governments signed the Algiers Convention on the conservation of nature and natural resources.
- Efforts to use and manage natural resources in a sustainable manner doubled after the 1972 United Nations Stockholm Conference on the Human Environment.
- In 1980, under the auspices of the Organization of African Unity (OAU), an extraordinary summit of African heads of state and governments led to the adoption of the *Lagos Plan of Action*—Africa's blueprint for economic development, which helped to highlight the challenges facing the region.
- In 1985, African countries established the AMCEN which, over the past 15 years, has made concrete achievements in providing region-wide leadership, awareness raising and consensus building on global

and regional environmental issues, and in enhancing the skills necessary for African governments to manage their environment and participate in global environmental negotiations (UNEP 2000). In spite of these achievements, AMCEN's leaders realize that the environmental challenges facing Africa are immense and are becoming increasingly complex. Meeting these challenges will require more human and financial resources, increased global, regional and sub-regional cooperation and efforts by individual African states, which must be combined with a strong political will, commitment and good governance (UNEP 2000).

- In 2001, African heads of state agreed to transform the OAU into the African Union. They also agreed on the *New African Initiative*, a plan of recovery forming part of the *New Partnership for Africa's Development* (NEPAD). This was a milestone in the quest for a new path to sustainable development (AMCEN 2001).
- During the 55th Session of the United Nations General Assembly (September 2000), African governments endorsed the six fundamental values which should underpin international relations in the 21st century, namely: freedom, the equality of nations, solidarity, tolerance, respect for nature, and shared responsibility (AMCEN 2001).

In addition to the regional and sub-regional initiatives mentioned above, there have also been country level efforts, a summary of which is presented in Annex 1. In their efforts to address environmental degradation, African countries have focused on a range of policy responses. These are examined individually below. Some failures and weaknesses in their implementation are analysed in Annex 2.

### INTRODUCTION OF MACROECONOMIC AND SOCIAL POLICIES

Governments use policies to influence the structure and operation of economies, with the aim of attaining goals and targets for development and economic growth. To do this, they use economic, financial, legal and institutional instruments to encourage or to discourage particular types of economic activities at macroeconomic or sectoral levels (Mogaka and others

2001). For example, macroeconomic policies are used throughout Eastern and Southern Africa to manipulate exchange rates, money supply and interest rates in order to achieve economic growth, to stimulate employment and investment, and to generate foreign exchange. Agricultural policies in the sub-regions, on the other hand, have long made use of combinations of subsidies, taxes and credit arrangements to promote the goals of food security, increased export earnings and rural income generation (Mogaka and others 2001).

African states have endorsed poverty reduction as a priority goal, the foundation of which is sound macroeconomic policies and strategies, ensuring both sustainable broad-based economic growth and macroeconomic stability. They have also become more outward-looking and have put in place trade liberalization policies, structural reforms in agriculture, and monetary policies that aim at maintaining low inflation, a stable exchange rate, lower interest rates and fully convertible currencies. Collectively, these responses are aimed at making Africa a part of the global village. The perceived benefit of greater globalization is the reduction of poverty in Africa.

Food insecurity is a measure of poverty. Table 5.2 shows that many countries have improved their daily per capita calorie intake, although some sub-Saharan countries have done the opposite. The situation is similar for per capita protein and fat supply, further strengthening the argument for greater attention to poverty reduction.

Some macroeconomic policies have had negative effects, exacerbating poverty instead of reducing it. For example, during the 1980s, most of the countries of Eastern and Southern Africa faced economic stagnation, declining growth, and increasing public sector and trade deficits. These worsening economic conditions, and the economic stabilization and structural adjustment measures introduced in the 1990s to overcome them, resulted in considerable contraction of the economy, a decline in rural living standards, and a fall in income and employment (Mogaka and others 2001). As is often the case, increased poverty made people more reliant on natural resources to meet their daily needs or to generate income, and this increased pressure led to overexploitation of resources. In other words, macroeconomic policies exacerbated poverty in the two sub-regions and contributed to further environmental degradation. This was certainly the case in Tanzania,

## Table 5.2  Food security and nutrition

| Country/region | Daily per capita calorie supply | | Food production index | Daily per capita protein supply | | Daily per capita fat supply | |
| | 1970 | 1997 | (1989–91 = 100) | 1970 | % Δ 1970–97 | 1970 | % Δ 1970–97 |
|---|---|---|---|---|---|---|---|
| Seychelles | 1 930 | 2 487 | 143 | 79 | +52.2 | 72 | +112.7 |
| Mauritius | 2 355 | 2 917 | 109 | 72 | +43.2 | 87 | +72.0 |
| Tunisia | 2 255 | 3 283 | 122 | 88 | +55.0 | 93 | +45.6 |
| South Africa | 2 831 | 2 990 | 97 | 77 | +2.9 | 77 | +12.8 |
| Gabon | 2 183 | 2 556 | 111 | 73 | +18.7 | 55 | +44.4 |
| Ghana | 2 242 | 2 611 | 144 | 49 | -0.4 | 32 | -20.5 |
| Kenya | 2 187 | 1 976 | 105 | 52 | -19.0 | 47 | +40.2 |
| Sierra Leone | 2 449 | 2 035 | 101 | 44 | -11.3 | 58 | -13.6 |
| Sub-Saharan Africa | 2 271 | 2 231 | | 53 | -4.1 | 46 | +2.8 |
| All developing dountries | 2 145 | 2 663 | | 67 | +27.5 | 59 | +79.6 |
| World | 2 358 | 2 791 | | 74 | +19.7 | 72 | +42.2 |

**Box 5.3: Macroeconomic policy impacts on the forest sector in Tanzania—the case of structural adjustment and the agricultural sector**

About 40 per cent of Tanzania's land area is covered by forests and woodlands. Macroeconomic reforms introduced over the past two decades have impacted on these forest and woodland resources. Research on this subject has led to the conclusion that deforestation in Tanzania is not linked to issues of forestry alone; it is intimately related to questions of public policies, and economic and social forces.

The effects of structural adjustment on Tanzania's forests and woodlands have been particularly intense. In the late 1980s and 1990s, a series of economic reforms was introduced in response to a series of economic crises. These reforms aimed at restoring balances in the economy and at creating a basis for sustainable growth by liberalizing key markets from excessive state control. The agricultural sector was a major focus of these reforms. The role of the state in the marketing of outputs and inputs was diminished considerably with the private sector assuming an increased role. At the same time, an increasingly liberalized economic environment was accompanied by a devaluation of the local currency, and a considerable increase in inflation rates. These and other conditions led to a decline in per capita income and a rise in the cost of living, making it increasingly difficult

for both urban and rural dwellers to make ends meet. They also had major impacts on the way in which land and other natural resources were used. Examples are given below.

- Devaluation increased the price of imported inputs, agro-chemicals and machinery. As these became more expensive, farmers reduced or abandoned their use, accelerating extensive agriculture which required the clearing of woodland and forest to increase production.
- Removal of price controls and parastatal subsidies created more space for trading in crops, which translated into a greater market demand for crops, and greater production. Since the private sector has failed to assume many of the more extension-based roles of government, many farmers remain uninformed about sustainable farming practices and agricultural expansion has often occurred at the expense of the environment.
- Falling yields, linked to poor extension and farming practices and to relatively higher costs of inputs, have encouraged farmers to expand production through extension, often into forests and woodlands.

*Source: Shechambo, 1999*

where structural adjustments had a negative impact on the forest sector (see Box 5.3).

In addition to macroeconomic policies, African states have introduced sectoral policies aimed at stimulating output, employment and income and, hence, poverty reduction. However, using economic instruments to stimulate sectors can be accompanied by risks to the environment. There are incidences of this in Africa, including promotion of the energy sector and urban development, with a risk of overexploitation of forest resources; promotion of the agriculture, mining and infrastructure sectors, leading to woodland and forest clearance; and promotion of industrial and manufacturing sectors, which generate wastes and pollutants that undermine environmental quality. Table 5.3 illustrates sectoral economic policy incentives and disincentives for sustainable forest utilization and management in Zambia.

Industrial sector activity is often accompanied by risks to the environment.

*UNEP*

**Table 5.3  Sectoral economic policy incentives and disincentives for sustainable forest utilization and management in Zambia**

| Policy | Economic incentives | Economic disincentives | Gaps and omissions |
|---|---|---|---|
| Macro | Incorporation of sustainable development concerns<br><br>Liberalization of forest prices and markets<br><br>Empowerment of private sector and communities | Continuing promotion and protection of sectors reliant on forest land and resources | Poor recognition of the role of forests and trees in national income, employment and economic growth |
| Agriculture | Land and environmental conservation and restoration<br><br>Promotion of sustainable farming practices | Punitive and restrictive approach to natural resource conservation<br><br>Main focus on optimizing agricultural production | Lack of consideration of role of trees in agricultural systems<br><br>Lack of recognition of dangers of agricultural conversion of forest land |
| Land | Definition of land tenure and ownership<br><br>Provisions for land management | Punitive and restrictive approach to natural resource conservation<br><br>Main focus on optimizing agricultural production<br><br>Unclear rights and tenure over trees and forests<br><br>Unclear role of traditional authorities in natural resource management<br><br>Lack of land use policy and guidelines | Little mention of forests or trees, their tenure or management |
| Water | | Focus on increasing water abstraction and use<br><br>Under-priced water | Lack of consideration of upstream catchments |
| Energy | Improvement in woodfuel supply, production and marketing | | Lack of consideration of role of forests in hydropower |
| Authority and decision making | Enforcement of controls on forest use and conversion | Failure to empower communities and minimize group's and individual's rights over trees and forests<br><br>Allocation of land and resources based on goals other than sustainable forestry | Insufficient emphasis on role of forests in livelihoods and development |

Source: PFAP (1998)

Economic instruments in support of agricultural policy goals are, arguably, those that have had the most detrimental effect on the environment. Agriculture has long been promoted as a key sector for development and growth in Africa, and for pursuit of national goals of food security, rural income generation and export earnings. The range of economic instruments used in support of the sector is well-documented, and has mostly involved manipulating of fiscal, financial, price and market mechanisms. Examples are: imposition of relatively lower tax rates on agricultural land uses (Barnes and de Jager 1995); subsidies to inputs; government intervention in marketing; preferential credit arrangements; relief on taxes and duties; and high spending on research, extension, development and marketing (Mogaka and others 2001). Use of such mechanisms has led to an artificial inflation of the profitability of agriculture and has often encouraged the spread of farming activities at the expense of the environment (Mogaka and others 2001).

Therefore, when African states introduce macroeconomic and sectoral policies aimed at reducing poverty, they must take care to ensure that none of the planned improvements in economic growth are at the expense of the environment. This is particularly relevant as African governments embark on the modernization of agriculture as a means of reducing poverty and for overall modernization of their economies.

Apart from macroeconomic and sectoral economic policies, social policies can also have significant consequences for environmental management. For example, the underlying cause of much of Africa's widespread poverty is the high rate of population growth, and it is poverty that forces people to overexploit their natural resources and thus degrade their environment. Social policies that tackle the underlying cause of poverty therefore indirectly support environmental improvement. Family planning and other population growth control strategies have been introduced in some African countries and are beginning to yield positive results (UNDP 2000). Better education can also provide a way out of the poverty trap, and there are efforts in African countries to ensure 'free education' for the young to attain the goal of universal literacy.

In some countries, the health policies introduced have emphasized preventive rather than curative strategies. Figure 5.1 and Figure 5.2 show the impacts of policy responses that improve accessibility to safe water and sanitation services. In both cases, there is a direct inverse relationship between infant mortality and the increasing percentage of the population that has access to safe water and sanitation. Removing unhealthy environmental conditions can therefore contribute to reducing vulnerability to disease (see Chapter 3).

## RATIFICATION OF MULTILATERAL ENVIRONMENTAL AGREEMENTS (MEAs)

The majority of African states have ratified the MEAs that are of relevance to the region, at both global and regional levels. MEAs are recognized as the primary instruments for state commitment to the pursuit of sustainable development (UNEP/ SIDA 1996). The main MEAs of the past two decades have covered areas of critical importance for the management of environmental

•

*Use of such mechanisms— subsidies to inputs, perferential credit arrangements, etc.— has led to an artificial inflation of the profitability of agriculture and has often encouraged the spread of farming activities at the expense of the environment.*

•

**Figure 5.1 Safe water and child health in African countries**

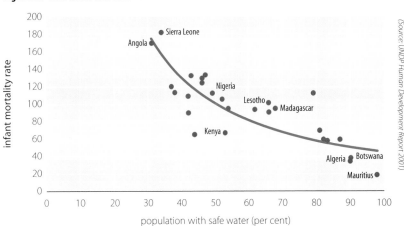

(Source: UNDP Human Development Report 2001)

**Figure 5.2 Safe sanitation and child health in African countries**

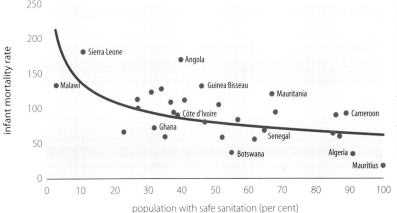

(Source: UNDP Human Development Report 2001)

resources. They include: new and additional resources for environmental programmes; technology transfer; mechanisms for addressing vital matters such as the loss of biological diversity and poverty alleviation; and institutional frameworks for dealing with environment and development concerns (UNEP/SIDA 1996). Although the various global agreements clearly give grounds for hope where management of the environment is concerned, actual achievements have been very limited (UNEP/SIDA 1996). The agreements signify a collective will to address environmental problems, but many African countries have been unable to benefit from the full potential offered by the global MEAs, and have even found themselves unable to effectively implement the necessary provisions of the MEAs they have ratified (UNEP/SIDA 1996). Furthermore, even regional and sub-regional environmental agreements have been difficult to operate, largely due to lack of adequate and sustainable financial and human resources. Examples are the Abidjan and Nairobi conventions. Both of these were developed in the 1980s under the auspices of UNEP's Regional Seas Programme. However, the Nairobi Convention took 11 years to come into force and neither convention succeeded in establishing a fully operational Regional Coordinating Unit (RCU). Under impetus from African governments, UNEP is now taking steps to compensate for these delays and shortcomings and a Joint Secretariat for the conventions has been set up to coordinate and build synergies between ongoing projects and programmes in Central, Western and Eastern Africa. But success has also been achieved, when financial assistance has been made available. An example is the Nile Basin Initiative (NBI). Launched in 1999, the NBI is an initiative on the part of the riparian countries to establish a basin-wide framework to fight poverty and promote economic development in the Nile Basin area. The Nile Basin is home to around 160 million people and, although it has a rich natural endowment of high mountains, tropical forests, woodlands, lakes, savannas, wetlands, arid lands and deserts, it is characterized by poverty, instability and environmental degradation. Furthermore, its population is expected to double in the next 25 years placing, increasing the stress on water and other natural resources. The NBI is based on a shared underlying vision, 'to achieve sustainable socio-economic development through the equitable utilization of, and benefit from, the common Nile Basin water resources'.

*... there are more than 500 multilateral agreements in existence and, while African countries are not signatories to them all, the sheer number of agreements is too great to be managed by African countries with small economic bases.*

Multilateral Environment Agreements (MEAs) aim at protecting Africa's unique biodiversity.

*UNEP*

And, finally, there are more than 500 multilateral agreements in existence and, while African countries are not signatories to them all, the sheer number of agreements is too great to be managed by African countries with small economic bases.

## PROMOTION OF REGIONAL AND SUB-REGIONAL COOPERATION

African states are participating actively in various international fora aimed at developing collective responsibility for the environment. This is the case of the MEAs, which the majority of African countries have ratified under impetus from AMCEN and with technical support from UNEP. The decision of African states to establish AMCEN was a key enabling factor for improvement of environmental management in Africa and for successful policy response. AMCEN's efforts have been further strengthened by sub-regional organizations devoted to economic cooperation and environmental management.

Considering the number of regional and sub-regional groupings, Africans appreciate the contributions of these organizations to economic development and environmental management. Unfortunately, many of them lack financial sustainability. It is, therefore, critical that suitable institutional capacity-building and financial mechanisms be developed for these organizations.

## INTRODUCTION OF ENVIRONMENTAL POLICIES, LAWS AND INSTITUTIONS

Perhaps the greatest effort in policy responses to combat environmental degradation in Africa has been in the area of environmental policy and legal reform. Not long ago, most African countries had limited institutional instruments for environmental management, or had instruments that were outdated or sectoral and, therefore, narrowly focused. The National Environmental Action Plan (NEAP) processes adopted by some African countries have allowed them to formulate relevant environmental policies and to enact new laws. New environmental policies have also provided guidance for the formulation or review of sectoral policies and, subsequently, of laws.

There are a great many policies, laws and regulations in place in most African countries which, at first sight, should provide a sufficient basis for sound environmental management. More could be formulated or drafted if need be, although more policies does not necessarily mean better environmental management. However, the fact that Africa's environment continues to deteriorate in spite of such a substantial body of policies, laws and regulations may be an indicator of a low level of implementation and, particularly, of enforcement. The sectoral approach to environmental management often results in contradictory laws. However, other problems, such as inadequate finances and human resources capacity, hinder effective implementation.

Although African states have improved the policy framework for more effective management of the environment, there is need for more. For example, there is a need to develop policies governing the management of transboundary resources, and a need to ensure that the policies of neighbouring countries are in harmony with one another. Other policy gaps include frameworks for access to genetic resources and the management of indigenous knowledge.

New environmental laws have also facilitated the creation of institutions responsible for coordinating, supervising and monitoring environmental management in African countries. Horizontally, these relate to the various sectoral agencies of government; vertically, they relate to lower levels of government and civil society. African countries are investing heavily in building the institutional capacity for better environmental management. New institutions have been created and, sometimes, old ones rehabilitated. However, many of these countries are experiencing significant shortages of skilled personnel, partly as a result of the 'brain drain'. Some of the poorer countries, such as Uganda, have embarked on Universal Primary Education (NEMA 2001), but such commendable efforts will take time to yield results. There is, therefore, a need to provide training opportunities for those in the relevant institutions to fill the skills gap in the short and medium terms. African countries will also need to address the 'brain drain' issue, by offering their trained personnel meaningful employment opportunities and better working conditions.

While the new national institutions for environmental management represent significant improvements over the previous ones, the viability of some of them is questionable because of heavy reliance on external financing. A number are currently being financed through Official Development Assistance (ODA) from bodies such as the World Bank. Mechanisms are therefore needed to make these institutions financially self-sustaining. Also, by their very nature as national institutions, the new agencies have a limited capacity to address sub-regional and cross-border environmental issues. While sub-regional environment and development organizations such as the Inter-Governmental Authority on Drought and Development (IGAD) exist, they too are, to some extent, limited by their mandates. Africa needs a strong institution that can negotiate, lobby and monitor and, at the same time, encourage harmonization of environmental management approaches. Such an institution, which could possibly be an arm of AMCEN, currently does not exist.

## DECENTRALIZATION OF ENVIRONMENTAL MANAGEMENT

Policy responses relating to environmental governance include the decentralization of management responsibilities from central to lower levels of government. They also include involvement of communities in the planning and management of environmental resources. In theory, and at least in the long term, decentralized environmental management should be seen as the right thing to do. However, in the

*... the fact that Africa's environment continues to deteriorate in spite of such a substantial body of policies, laws and regulations may be an indicator of a low level of implementation and, particularly, of enforcement.*

short term—since many African governments have inadequate capacity for environmental management at the centre—it is likely that this constraint will be even more pronounced at lower levels of governance. There is, therefore, a need to identify mechanisms for the meaningful decentralization of environmental management responsibilities, including building capacity at the lower levels of administration.

## IMPROVING ENVIRONMENTAL COMPLIANCE

The history of the management of environmental goods and services indicates that the 'command and control' approach (i.e. regulations and controls introduced and operated by central authorities) has not been very effective unless accompanied by a strong level of compliance enforcement. African countries, in search of alternative approaches for better environmental management, are increasingly beginning to consider economic instruments (incentives and disincentives) as a tool for promoting appropriate behaviour and attitudes towards the environment. However, if economic instruments are to be applied appropriately and effectively, the value of environmental goods and services (among other things) needs to be established, even if not very precisely, through economic evaluation of natural resources. Knowledge of these values would assist political leaders in the making of informed decisions as to the mechanisms needed for environmental protection and for conservation of natural resources. The use of economic instruments also requires that appropriate management institutions be put in place and that attitudes change, something which will very probably take longer to achieve. A practical approach would, therefore, be to use mixes of economic instruments and of the command and control approach, adapted to individual African countries.

## PREPARATION OF STRATEGIES AND ACTION PLANS

As part of the NEAP process, African governments have developed strategic action plans to facilitate the implementation of environmental policies. Sectoral and cross-sectoral action plans have also been prepared (for example, National Biodiversity Strategies and Action Plans). African countries have devoted a lot of effort to the preparation of strategies and action plans. Unfortunately, most of the activities specified in plans remain on paper, largely because of low levels of implementation funds. There is a clear need for governments to demonstrate their commitment to the environment through better allocation of budgetary resources. Moreover, there is often a feeling that some strategies and action plans are prepared as a direct response to the need to fulfil global, regional and sub-regional agreements, and not as a result of careful consideration of national priorities or through consensus reached with national stakeholders.

## PREPARATION OF RESOURCE MANAGEMENT PLANS

Regional, sub-regional, national and sub-national management plans have been developed for natural resource conservation and environmental protection as part of the response to combat environmental degradation. Whether for forestry, biodiversity, wetlands or wildlife, governments and NGOs have prepared resource management plans, sometimes with the full support and participation of local communities. However, both their levels of implementation and their impacts have been lower than anticipated. This begs the question as to whether such plans are too ambitious, inappropriate, or have overlooked some critical assumptions. In most cases, implementation of resource management plans is assumed to be the responsibility of government and, given that the resource envelopes available to governments are meagre and uncertain, it is not surprising that implementation of such plans has suffered in the past. There is, therefore, a need to identify innovative ways of relieving African governments of some of the responsibilities of financing the implementation of resource management plans. For example, a national NGO or community-based organization could be given a lease to manage a given resource area (such as a wildlife protected area) on behalf of the government, in a win-win partnership. African states could also pool resources and cooperate in the preparation and implementation of natural resource management plans, especially transboundary ones, where there are mutual benefits to be derived.

●
*... there is often a feeling that some strategies and action plans are prepared as a direct response to the need to fulfil global, regional and sub-regional agreements, and not as a result of careful consideration of national priorities or through consensus reached with national stakeholders.*
●

Affordable appropriate technology would enhance the quality of life of Africa's rural people.

*Hartmut Schwarzbach /Still Pictures*

## IMPROVING THE KNOWLEDGE BASE FOR INFORMED DECISION MAKING

Although a number of regional, sub-regional and national organizations are involved in various aspects of environmental management, research capacity in this area is relatively weak, especially at sub-regional and national levels. Research organizations lack sustainable sources of funding and there are, in many cases, inadequate incentives for African researchers to engage in more meaningful research, including research into the state of the region's environment, and preparation of various scenarios for sustainable development paths and modelling of vulnerability parameters. If such incentives are not provided, Africa will continue to experience a 'brain drain' and will remain a technological backwater condemned to 're-invent the wheel'. There is also room for the private sector and civil society to participate in enhancing the knowledge base for sound environmental management in African countries.

African states have a wealth of indigenous knowledge, some of which has been used in the past to cope with natural disasters, and which could still be used to address present and future environmental challenges. However, value must first be attached to this heritage, and the issues of its ownership and of protection of the associated intellectual property rights must also be addressed.

On the other hand, some of the technologies that African countries require to promote sustainable development in general, and to reduce poverty and enhance environmental management in particular, are available globally. However, the terms and conditions of access to these technologies do not appear to be favourable to African countries, despite the provisions of Agenda 21.

It was agreed at UNCED (the 1992 'Earth Summit') that, in order to facilitate the implementation of Agenda 21, the developed world would assist African states with appropriate technologies on affordable terms. Only limited success has been achieved in this area so far, because many development partners have reneged on this agreement.

African countries, for their part, should cease to 're-invent the technological wheel', and should invest in the development or adaptation of suitable technologies and techniques. They should also address the subject of indigenous technologies. That is to say, African countries should inventory, document and disseminate information on available indigenous technologies that are appropriate to and, by their very nature, affordable for environmental management.

## BETTER VALUATION OF ENVIRONMENTAL RESOURCES

Africa can provide very diverse and significant environmental goods and services, such as carbon sequestration by its forests and a wide range of other options arising from its rich biodiversity. However, exploitation of the region's wealth has so far served to meet global needs, with the benefits accruing to the global community while Africa's people remain in poverty. This is a clear indication that Africa's environmental goods and services are undervalued and that the people of Africa, especially the rural poor, are bearing disproportionately high opportunity costs in conserving the region's environmental resources. It is, therefore, imperative that African countries begin to price their environmental goods and services appropriately, and to ensure that fair compensation is derived. They also need to add value to environmental goods and services, and to export them in processed and enhanced forms, in order to obtain better returns. Kenya, Mauritius and Seychelles have demonstrated that, with proper pricing and service delivery, the tourism potential of environmental goods and services can contribute to enhanced quality of life, as

*African states have a wealth of indigenous knowledge, some of which has been used in the past to cope with natural disasters, and which could still be used to address present and future environmental challenges.*

*It was agreed at UNCED (the 1992 'Earth Summit') that, in order to facilitate the implementation of Agenda 21, the developed world would assist African states with appropriate technologies on affordable terms. Only limited success has been achieved in this area so far, because many development partners have reneged on this agreement.*

Activities such as ecotourism have the potential to add value to environmental goods and services, bringing much-needed resources to local and national economies.

*Nigel Dickinson / Still Pictures*

shown by their relatively high HDIs (UNDP 2000). Similarly, Gabon, South Africa and Tunisia have been able to utilize their natural resources to achieve better levels of human development (UNDP 2000). It is fair to say that the global community stands to lose the region's resources if Africa's rural communities do not share equitably in the benefits of biodiversity conservation.

## ENVIRONMENTAL MANAGEMENT TOOLS

Various information systems have been put in place for informed decision making in environmental management. Early warning systems are being used for better management of natural disasters, and tools such as remote sensing and geographical information systems (GIS) are in use. The information generated is managed through physical databases.

Many of the interventions required to arrest environmental degradation in African countries call for relatively elaborate organizational and management systems. Key among these are databases and information management systems, particularly for monitoring. While some such systems are in place, more needs-based systems should be created. African countries need better early warning and detection systems to help them, for example, in managing climate variability or in addressing illegal trafficking of toxic, hazardous and radioactive wastes. Creation of these systems is expensive, a key constraint for many African countries.

*Whereas, previously, NGOs had to gain a seat at the policy-making table, today they are necessary participants in all aspects of development programming from the donor and, increasingly, from the national government perspective.*

African states have also introduced a number of tools that facilitate better management of the region's environment. These include environmental assessments (EIAs, reviews and audits), regulations, standards and environmental information systems. African countries also now produce regular national state-of-environment reports. Better environmental management information systems have also been set up to facilitate the collection, storage, analysis and dissemination of environmental information, as a key component of regular monitoring. National, sub-regional and regional environment information networks also exist.

## CIVIL SOCIETY PARTICIPATION IN ENVIRONMENTAL MANAGEMENT

The increasing role of NGOs in environmental management became evident during the UNCED (1992). This event was attended by some 8 000 NGOs from more than 160 countries; the Habitat II Conference, in 1996, was attended by representatives of more than 500 local authorities. The role played by NGOs in environmental management and other areas of development work in Africa has changed considerably over the past two decades. NGOs—once perceived by some governments as subversive elements in the development process—have become, in many cases, centre stage performers. Expectations as to what NGOs can and must contribute to development have thus changed dramatically. Whereas, previously, NGOs had to gain a seat at the policy-making table, today they are necessary participants in all aspects of development programming from the donor and, increasingly, from the national government perspective. The question of the capacities of the NGO sector in Africa is sensitive and sometimes controversial, but the consensus seems to be that NGOs are strong at promoting local participation, and that they fill a niche in certain aspects of sectoral work. They are, on the other hand, seen as relatively weak in complex, multi-component projects. Some NGOs are effective at lobbying and advocacy while others, the majority in Africa, are fairly weak in both technical and institutional aspects of project or programme planning and implementation (Brown and McGann 1996). Although the quality and capacity of NGO work in Africa is improving, their capacities need to be strengthened.

## PROMOTING PUBLIC AWARENESS AND PARTICIPATION IN ENVIRONMENTAL MANAGEMENT

Broad public participation in decision-making is an important element of Agenda 21 because, combined with greater accountability, it is basic to the concept of sustainable development (UNEP 1999). However, if people are to participate effectively, they must first be aware of the problems. Agenda 21 also recognizes that there is a considerable lack of awareness of the interrelated nature of human activities and the environment, due to inaccurate or insufficient information. Increasing public awareness is, therefore, a prerequisite for action, and is an essential element of any educational effort to stimulate or strengthen attitudes, values and actions which are compatible with sustainable development.

Agenda 21 devotes separate chapters to involving many different groups including women, children and youth, indigenous people, NGOs, local authorities, workers and trade unions, business and industry, scientists, technologists and farmers (UN 1993). The belief is that both individuals and members of these groups are the best source of knowledge about the causes of, and remedies for, many environmental problems (UNEP 1999, NEMA 2001). Public participation enables such knowledge, skills and resources to be mobilized and fully employed, and the effectiveness of government initiatives to be increased (UNEP 1999).

African countries by themselves, and through the support of such organizations as UNEP and AMCEN, have increased public awareness about the environment, and have encouraged participation. There is public participation in the formulation of policies and strategic plans and, in major projects, through the environmental impact assessment process. In Uganda, the right of the public to participate in environmental matters is enshrined in the country's constitution, adopted in 1995 (GoU 1995). Formulation of Uganda's long-term development perspective, Vision 2025, involved extensive consultations. It is, as a result, a product of consensus (MoFPED 1999).

Apart from statutory requirements or government policy directives, public participation in environmental decision making is also being promoted from within civil society. Various projects, both international and from

Youth awareness of environmental issues is an important building block for sustainable environmental management at the level of the individual.

*Gilles Nicolet / Still Pictures*

local non-governmental and community-based organizations, have promoted public participation. Public participation is also a strong aspect of decentralized environmental management. In encouraging transparency, accountability and ownership, African countries have welcomed the contribution of public participation in decision making, including in the area of environmental management.

### GENERAL GOVERNANCE

Democratic governance is beginning to take root in Africa. There are now far fewer military dictatorships, and the military *coups d'état* common in the 1960s and up to the 1980s are much less frequent. This shift has facilitated a better focus on sustainable development. Through the efforts of AMCEN and sub-regional organizations, and at national level, there is strong political support for better management of Africa's environment. The environment now features prominently in political pronouncements across Africa. In those African countries where efforts have been made to devolve political power to lower levels of government, this augers well for the principles of decentralized environmental management. However, the capacities of lower levels of government and community-based organizations will have to be built to equip them for their new roles as environmental planners and managers.

## DEFINING A DEVELOPMENT PATH

Having been used as a laboratory for a variety of economic and social development paradigms in the past, African governments have had the opportunity to learn important lessons useful for defining appropriate paths for the 21st century. Clearly, the Great Transitions scenario—requiring a development paradigm in which responses to the challenge of sustainability are based on new values, and on more humanistic forms of social and economic organization—is the most appealing (see Chapter 4). For the present, however, this must be seen as utopian, given the present situation of African countries. The challenge, therefore, is to select paths that will lead these countries from their present situations through the other scenarios and, ultimately, to the Great Transitions scenario, to be reached at some pre-determined point in the future. How this should be done and the time it will take will depend on the capacities and capabilities of individual African countries. As described earlier (Chapter 4), both the Fortress World scenario (in which the wealthy protect themselves in enclaves, while environmental stress elsewhere leads to fragmentation, extreme inequality and widespread conflict) and the Market Forces scenario (where market-driven global development leads to converging values and patterns of development) are undesirable, and are unlikely to lead Africa towards sustainable improvements in quality of life in the long run. The Policy Reform scenario (incremental policy adjustments steering conventional development toward environmental and policy reduction goals) is a good start, given the current situations of most African countries, but it, too, must ultimately give way to the Great Transitions scenario as a sustainable development option. The newly established African Union will have a significant role to play in mapping an overall development framework for the region. It will then be up to AMCEN to ensure that the African Union incorporates considerations of environmental sustainability when selecting a broad development framework.

## RESOURCE MOBILIZATION

Africa needs external support if it is to succeed in reversing the current trend in environmental degradation. As discussed in earlier chapters, the extreme poverty suffered by many Africans is a major factor contributing to the degradation of Africa's environment. For example, the fact that in Sierra Leone the poorest 20 per cent of the population benefits from only 1.1 per cent of income consumption indicates that the majority of its people are deriving livelihoods from subsistence activities largely based on environmental goods and services. Should this trend persist, environmental degradation will continue unabated, and at tremendous cost to the country. Uganda is another example—a conservative estimate puts the annual cost of environmental degradation at 4–12 per cent of the country's GNP (NEMA 2001, Slade and Weitz 1991). If corrective actions are not financed and put in place, the cost of this degradation is likely to increase.

It was agreed at the UNCED (1992), where most African states were represented, that implementing Agenda 21 would require new and additional financial resources. African countries have received external assistance for environmental management, but the resources received have been inadequate, partly because the new and incremental funds being made available to them are less than those agreed at UNCED, and partly because procedural difficulties are hindering access by many African countries to the little funding that is available.

External aid nevertheless represents a significant share of national budgets in Africa—especially in sub-Saharan Africa—and dependence on aid is even more pronounced when it comes to investment in environmental management. Dependence on external aid raises concerns in Africa as to adequacy of funds, sustainability of interventions, and freedom to reflect national priorities rather than priorities perceived by donors. However, in spite of such concerns, incremental funding will still be required if African countries are to continue with implementation of Agenda 21. Moreover, access to funding will have to be easier than at present, and it must be largely in the form of grants rather than loans, however soft.

While it is clear that Africa needs external support in its efforts to reverse environmental degradation, African

## Box 5.4 Rationale for and benefits of investing in environmental management

### Rationale

- Halting or reversing of environmental degradation so as to guarantee improved productivity of the environment, with a view to accelerating sustainable economic growth and improving human welfare.

- Building and strengthening of human institutions and capital in environmental management, to allow continual response to new demands and challenges.

- Holding open future options for resource conservation and development, by formulating good policies so that irreversible losses are avoided, and positive conservation culture and attitudes are inculcated.

### Expected benefits

- Increased earnings at macro and micro levels, due to improved productivity of biotic elements.

- Avoiding losses of future income (for example, by controlling the impact of soil erosion on agricultural productivity).

- Avoiding future costs (for example, replacing lost soil nutrients, extensive curative medical care, etc.).

- A healthy and productive labour force.

*Source: adapted from NEMA (2001*

governments also need to recognize that environment is a priority area for investment. They could, for example, include environment as a priority area for intervention in national poverty reduction strategies.

Furthermore, African states often put too much emphasis on accessing external sources of funds, almost to the exclusion of domestic resources. They should—if they are to become more self-reliant—be more proactive in identifying and developing creative mechanisms to generate funds from the region's significant environmental resources. They could improve the generation of non-tax revenues from environmental resources and services by moving towards charging economic rates for them. Valuation of their environmental resources on the basis of total economic value would allow African governments to introduce more appropriate taxes and to develop non-tax sources of revenue, such as user fees. The rationale for, and the benefits of, investing in environmental management are presented in Box 5.4.

The collective will of African states to arrest and to reverse environmental degradation exists, as evidenced by the wide range of responses presented in Annex 1. African states, and regional and sub-regional organizations, also want to do more, as illustrated by recent resolutions (AMCEN 2001). The main limitations are financial—mobilization of additional resources is, therefore, a priority.

## THE PROBLEM OF DEBT

African countries are generally poor and heavily indebted, although the level of indebtedness is declining (see Figure 5.3). Nonetheless, however small the debts may be in absolute amounts, when indexed to debt servicing abilities, they are still a constraint. Despite these odds, African countries have been able to leverage additional funds from external sources for investments in environmental management and, by and large, they have enjoyed goodwill from development partners. Key sources of external financial inflows include loans from the multilateral and regional development banks, and grants from bilateral donors and other agencies.

**Figure 5.3 Level of indebtedness to African countries**

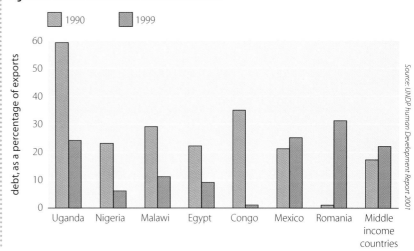

*Source: UNDP Human Development Report 2001*

# PROPOSALS FOR ACTION

A key aim of the AEO process has been to identify 'achievable action items' for recommendation to policy officials, and to AMCEN as Africa's environmental body. Urgent actions necessary to reverse the current trends in environmental degradation in Africa have been derived from the recommendations and analyses presented in the preceding chapters of this report. These actions are detailed below in the form of a 31-point list. They are summarized in a matrix presented in Annex 3. Actions are grouped in the following categories:

- reducing poverty;
- arresting environmental degradation directly; and
- promoting cross-cutting actions.

Implementation of the recommended actions is principally the responsibility of African governments, with technical assistance from AMCEN and sub-regional organizations. In turn, African governments, AMCEN and the sub-regional organizations may enter into partnerships with sub-regional, national and international organizations, to further facilitate implementation.

## REDUCING POVERTY

Poverty is a complex multidimensional problem. In Africa, poverty is one of the drivers of environmental degradation, largely because the poor have limited choices and depend heavily on the natural resource base. There is no uniform solution to the problem of poverty. Country-specific programmes to tackle poverty, and sub-regional, regional and international efforts supporting national efforts, are needed. At national level, a specific anti-poverty strategy is, therefore, one of the basic conditions for ensuring sustainable development. Many African countries have prepared and are implementing poverty reduction strategies and plans.

Actions which are directly relevant to the environment, and which are imperative if Africa is to reduce poverty, are as follows:

- **Endorsement and promotion of the principles of sustainable development**
  The African Union in general and, where the environment is concerned, AMCEN in particular, need to persuade the global community to adopt

the *New African Initiative*—a recovery plan in the *New Partnership for Africa's Development* (NEPAD)—as the framework for sustainable development in Africa, and to accelerate setting up of the necessary mechanisms of the World Solidarity Fund (WSF). National governments also need to increase efforts towards attaining the poverty reduction goals of the United Nations Millennium Declaration, December 2000. In the Declaration, world leaders agree among other things, to set specific targets to halve the proportion of people who live in extreme poverty, and to reach development goals.

- **Acceleration of industrial development**
  Acceleration of industrial development is necessary to provide employment and to raise the financial resources needed to stimulate economic growth. In this respect, regional cooperation is required in order to raise the industrial productivity and competitiveness of African states to international levels. National efforts should also be devoted to promoting the development of micro-, small- and medium-sized enterprises. The focus should be on agricultural commodities and natural resources, in order to add value to Africa's traditional exports. However, any national industrialization strategy must be environmentally sustainable and must not be a contributor to further environmental degradation.

- **Increase of sustainable agricultural production**
  National governments must increase financing for the agricultural sector. Regional and international support is needed for implementation of the UNCCD. Similarly, regional and international support is needed to persuade the developed countries to remove agricultural subsidies which are currently blocking entry of African agricultural products into their markets, and which are encouraging the dumping of products onto the African market. Regional support is needed to convince the developed countries to apply the precautionary principle to genetically modified organisms (GMOs) which have unknown, but potentially dangerous consequences for agricultural production in Africa. African

Sustainable agricultural production techniques reduce negative impacts on the environment.

*Ron Giling /Still Pictures*

governments should also promote sustainable agricultural production techniques to avoid the adverse impacts of the sector on the environment.

● **Promotion of human health, well-being and development**

African governments must ensure greater access to affordable primary and secondary health care and medical technology. They also need to improve environmental and social conditions, which are responsible for spreading diseases, and to build the capacity of local communities to improve their living conditions. International partnership is required to make both preventive and curative health care available. Regional and sub-regional technical assistance and national efforts are needed to provide access to medicine at affordable prices, while promoting public health and nutrition. National governments need to empower African women in social and economic development, and to strengthen the skills of the region's youth. Regional and sub-regional technical assistance is also required to complement national efforts in promoting human-resource development and capacity building, including universal primary and secondary education.

● **Advocacy for better terms of trade**

Regional lobbying is needed to support the efforts of African governments in persuading developed countries to open up their markets, and to eliminate subsidies on agriculture, textiles and other export products competing with those of the region.

● **Generation of increased domestic financing for sustainable development**

Efforts are required at all levels—national, sub-regional, regional and international—to promote foreign direct investment in Africa. The developed countries and the Bretton Woods institutions should cancel Africa's external debt. Efforts should also be intensified to persuade developed countries to adhere to the United Nations target of 0.7 per cent of GNP for ODA, and to ensure that Africa gets its fair share. New partnerships with UNDP, UNEP and the World Bank are needed to increase the resources available, and to improve on operational and project implementation procedures of the Global Environment Facility (GEF).

● **Improvement of infrastructure and sustainable human settlement patterns in Africa**

Improvements are needed in infrastructure and sustainable human settlement patterns in Africa, in order to reduce congestion and pollution. National governments need to improve access to, and the affordability and reliability of, infrastructural services. African governments need the support of AMCEN in mobilizing external resources for implementation of the Habitat Agenda and the declaration of the 25th United Nations Special Session, to achieve sustainable human settlements in Africa.

● **Improvement of the scientific and technological base in Africa**

AMCEN needs to identify scientific and technological gaps, particularly relating to environmental management, and should guide African governments in accessing appropriate indigenous and external technologies in order to enhance environmental management and economic development.

## ADDRESSING ENVIRONMENTAL DEGRADATION DIRECTLY

The environment is the basis of human health, wealth, well-being and security. The majority of Africans derive livelihoods directly from the natural resources of the environment. Humankind in general has also benefited significantly from Africa's natural resources, and from environmental services, such as carbon sequestration. However, deterioration of the environment in many parts of Africa over the past three decades has left millions of people more vulnerable to adverse environmental change than before. The African countries affected need to halt and, where possible, even reverse the current trend in environmental degradation.

If environmental degradation in Africa is to be halted and reversed, the following actions are imperative for the region.

● **Reduction and halting of activities that lead to land degradation**

Efforts at national and sub-regional levels should be focused on promoting campaigns of environmental information, education and communication. African governments, for their part, need to ensure that they are in a position to implement the UNCCD in a timely and effective manner and, together with AMCEN, they need to ensure that the UNCCD is acknowledged as a sustainable development convention. National governments (with technical help from AMCEN if necessary) and sub-regional organizations should prepare sound land use policies and plans, where these do not exist. National governments also need to improve

systems that address gender considerations, where these, too, are lacking. They also need to put in place mechanisms to protect Africa's cultural and historical heritage.

● **Conservation and sustainable management of Africa's rich biodiversity**

The current 'ecosystem approach' to biodiversity conservation is too narrow. More comprehensive regional, sub-regional and national efforts are needed to promote both an ecosystems and landscape approach, with an emphasis on sustainable development in a wider context Similarly, African governments—with the technical assistance of AMCEN and sub-regional organizations, where necessary—need to develop and implement national legislation for the protection of the rights of local communities, farmers and breeders, and for the regulation of access to biological resources and biosafety, in line with the OAU Model Law. Sub-regional groupings and national governments must also endeavour to rehabilitate degraded wetland areas. AMCEN's technical assistance will also be required to strengthen the Lusaka Agreement, in order to facilitate implementation of the provisions of the Convention on International Trade in Endangered Species (CITES), and to safeguard the conservation of eastern and southern African wildlife. Where there has been extensive poaching in wildlife protected areas, the relevant sub-regional organizations and national governments should carry out rehabilitation work through species re-introduction and habitat restoration. National governments, with the technical assistance of AMCEN, should document and disseminate indigenous knowledge and practices that are applicable to conservation. Given that Africa's biodiversity is not fully understood, regional groupings and national governments should, with international assistance, make biodiversity inventories and document significant landraces (types of seeds used in traditional agriculture). National governments should promote the establishment of *ex-situ* conservation facilities for rare, vulnerable and endangered species. They should also, through sub-regional, regional and international partnerships, promote the conservation of agricultural biodiversity, the backbone of rural

Land tenure and land use policies and plans need to be addressed to reverse degradation, and to promote equity and food security.

*USGS – EROS Data Center*

Conservation practices play a important role in protecting Africa's threatened species.

*UNEP*

livelihoods and the engine of economic growth in many parts of the Africa region. Such efforts should include the promotion of *in-situ* conservation of landraces of important agricultural crops.

● **Reduction of high rates of deforestation in parts of Africa**

National governments need to devote and access additional financial resources and technology to implement the provisions of the International Panel on Forests (IPF)/International Forest Forum (IFF) of the United Nations Forum on Forests (UNFF). Similarly, African governments need to promote access to affordable energy for sustainable development, especially in rural areas. They should also promote research on and development of clean energy technologies, efficiency of energy supply and usage, and efficient uptake of renewable resources. The West Indian Ocean Island States of Africa need to revitalize the Barbados Programme of Action for the Sustainable Development of Small Island Developing States. AMCEN should, through technical assistance, contribute to strengthening of the operations of the African Timber Organization (ATO), and should support the Yaounde Declaration. National governments should promote collaboration in forest management between forestry authorities and local communities. They should also review the

pricing of forest products to reflect their true economic value, in order to provide better earnings, discourage wasteful exploitation and promote more efficient utilization of non-wood forest products. African governments need to rehabilitate degraded forest areas. They also need to attract private sector investment to forestry. AMCEN needs to guide African states in researching and documenting the medicinal values of their forests.

● **Mitigation of the adverse impacts of climate change and other atmospheric conditions**

AMCEN should be an advocate for finalization of the Kyoto Protocol. African governments need to lobby the global community to operationalize the Climate Change Fund (CCF) for developing countries, as well as the Special Fund for least developed countries. Similar assistance is required to facilitate African countries' access to the adoption of cleaner technologies, in order to reduce industrial emissions. Africa needs combined efforts at regional, sub-regional and national levels to establish climate modelling programmes and early warning of rainfall variations. African governments also need to: ratify the United Nations Motor Vehicle Emissions Agreement; improve enforcement of emission standards and regulations; and promote the use of unleaded gasoline. The latter would be an expensive option for Africa, but it has environmental benefits and, therefore, a positive cost-benefit ratio. AMCEN should facilitate the north-south exchange of experience and knowledge between experts and provide for the transfer of know-how between African countries. IGAD countries, with technical assistance from AMCEN, should support and make operational the Strategy for the Elimination of Hunger in the Horn of Africa, an area experiencing extreme levels of rainfall variation, in terms of both amount and frequency. Through partnerships at international, regional and sub-regional levels, African states need to improve their understanding of the likely social impacts of atmospheric pollution and to quantify them.

● **Improvement of waste management practices**

Partnerships are required to put in place the funding and capacity required for effective management of

Enforcement of waste management and disposal regulations is necessary to improve urban pollution.

*UNEP*

non-hazardous wastes. AMCEN should make recommendations to counter the export of obsolete capital goods and equipment to Africa. Through partnerships at international, regional and sub-regional levels, African governments should implement the various conventions governing the generation, storage, transport, and transboundary movement and disposal of hazardous wastes, including radioactive wastes.

● **Promotion of environmentally sound management of chemical products**

African states need to establish partnerships to support the management of chemical products, in accordance with Chapter 19 of Agenda 21, and the Rotterdam and Stockholm conventions.

● **Improvement of access to and quality of freshwater resources**

Recognizing the limited financial resources at their disposal, African governments need to promote public-private partnerships in water resource management. They also need to: develop appropriate standards for water quality; promote the use of economic instruments (incentives and disincentives) for water resource management, including the application of the polluter pays principle; and introduce appropriate water pricing policies and mechanisms. International partnerships, such as the Nile Basin Initiative, are required to support regional and sub-regional water resource management bodies. National governments, with the

assistance of sub-regional organizations, should inventory important water catchments and develop guidelines for their sustainable utilization. They should also promote integrated water resource management and development as a standard practice. African governments, through sub-regional and international support, should improve general access to freshwater resources. AMCEN should promote the popularization of environmentally-sound, low-cost technologies for water harvesting. African governments should also address water quality issues. In particular, where necessary, they need to establish, and subsequently enforce, national effluent discharge regulations and standards. They should also increase investments in sewerage management, in order to improve freshwater quality.

● **Improvement of living conditions in urban areas**

African governments need to plan urban development appropriately for sustainable development. Furthermore, they need to formulate appropriate human settlement and waste

Improved water quality and access is possible through effective public-private sector partnerships.

*UNEP*

management policies, laws and regulations, and to promote private sector participation in improving urban infrastructure and the provision of municipal services. African governments should also fulfil their national obligations under the Habitat Agenda, and should prepare integrated water and waste management strategies and action plans.

## PROMOTING CROSS-CUTTING ACTIONS

There are a number of cross-cutting actions which, if carried out, would help to halt and even reverse environmental degradation and reduce human vulnerability. Such actions target: improving coping capacities; promoting increased regional and sub-regional cooperation; mobilizing domestic financial resources; enhancing institutional capacity; promoting greater involvement of non-governmental organizations; addressing policy failures; defining sustainable development paths; promoting good governance; enforcing compliance; and setting targets and monitoring performance.

Cross-cutting actions that are strongly recommended are outlined below:

● **Enhancement of the coping capacities of Africa's population, with regard to adverse environmental change and reduction of environmental insecurity**

African governments need to enhance their capacity to anticipate natural disasters and to be able to cope with the ensuing impacts. They also need to invest in early warning mechanisms and disaster preparedness planning, including the formulation of appropriate policies, laws and regulations. They should take steps to promote healthy living that is respectful of the environment, in order to reduce the incidence of diseases associated with environmental degradation. They should also endorse measurement of vulnerability as an important indicator of the state of the environment and, therefore, establish vulnerability assessments and early warning systems.

● **Promotion of human resources development**

African governments should assess their human resources needs for improved environmental planning and management. They should then put in place training programmes to fill identified gaps. AMCEN, in partnership with sub-regional and international organizations, should provide technical assistance to complement the efforts of national governments, particularly in the areas of identifying and strengthening the capacity of African centres of excellence in environmental planning and management. It should also promote intra-continental exchanges of expertise, collaboration and networking.

● **The promotion and enhancement of multi-level cooperation**

African governments should seek partnerships with the international community, in order to support the operations of the newly formed African Union, and the institutions for sub-regional cooperation and economic integration, such as: the Indian Ocean Commission (IOC); Inter-Governmental Authority on Drought and Development (IGAD); East African Community (EAC); Inter-State Commission to Combat Drought in the Sahel (CILSS); Economic Community of Central African States (ECCAS); the Central African Economic and Monetary Union (CEMAC); the Arab Maghreb Union (AMU); Economic Community of West African States (ECOWAS); and Southern Africa Development Community (SADC). AMCEN and sub-regional organizations should facilitate the provision of technical assistance to African governments in formulating programmes of action to support the management of shared water and other transboundary environmental resources.

● **Mobilization of domestic financial resources for environmental management**

African governments should make far greater efforts to mobilize domestic resources for investment in environmental management. As an initial step, they need to recognize that environment is a priority area for investment when allocating budgetary resources. Attracting private sector investment for environmental management is also important, and AMCEN and sub-regional organizations should facilitate provision of the technical assistance African governments need to develop win-win strategies. African governments also need to be able to evaluate their natural resources accurately, if they are to develop revenues from sources other than tax. AMCEN should provide them with the technical assistance and accounting methods they require to make such evaluations.

- **Enhancement of institutional capacity to coordinate, monitor and supervise environmental management in Africa**

  African governments should support AMCEN in defining an appropriate institutional structure which can coordinate, monitor and supervise environmental management on a region-wide basis, and which can provide a stronger voice for Africa in international negotiations and deliberations.

- **Promotion of greater involvement of NGOs in environmental management**

  African governments should encourage stronger national NGO partnerships in environmental management. Likewise, AMCEN should promote the greater involvement of regional and sub-regional NGOs in environmental management.

- **Greening development plans and strategies at all levels**

  AMCEN should provide technical assistance to African governments to build capacity for greening of development plans and strategies at national and sub-national levels. In turn, African governments should make the greening of such plans mandatory, and a precondition for allocation of financial resources to the various sectors of the economy and lower levels of government.

- **Focus on policy failures**

  African governments have introduced various macroeconomic, social, environmental and sectoral policies to promote sustainable development, and have improved environmental management. Some of these policies have, however, not been very effective and, therefore, need to be reviewed to remove contradictions and other underlying causes of failure. African governments, in partnership with sub-regional, regional and international organizations, need to create capacity for policy

Public awareness of environmental issues builds commitment to better practice.

*Mark Edwards/Still Pictures*

analysis, where this is necessary, to ensure timely detection of failures in implementation. They also need to demonstrate greater commitment to the implementation of policies that are adopted.

- **Promotion of greater public awareness**

  African governments and AMCEN, with support from UNEP and other international organizations, have invested significantly in creating greater public awareness of environmental matters and sustainable development. These efforts need to be sustained, particularly at community level. Given the increasing levels of urbanization and industrialization in the region, African governments, in partnership with sub-regional, regional and international organizations, need to increase public awareness of 'brown issues', such as atmospheric pollution.

- **Promotion of environmental education**

  African governments, in partnership with sub-regional, regional and international organizations, need to invest in the formulation and implementation of formal and informal environmental education strategies, where these do not exist.

● **Improvement of environmental information systems**

African governments need to improve environmental information systems as a basis for sound decision making. AMCEN, sub-regional and international organizations should, in partnership with African governments, promote the creation of physical databases at sub-regional and regional levels, and should enhance networking and collaboration between African states.

● **Definition of sustainable development paths**

African countries are at different points along the continuum of development paths which extends from the Fortress World scenario to the Great Transitions scenario discussed in Chapter 4. Where necessary, AMCEN should facilitate the provision of technical assistance to member countries, in order to design feasible and sustainable development paths to lead them from their present positions towards the Great Transitions scenario, within the overall framework of the new African Union and the New Partnership for Africa's Development.

● **Promotion of good governance**

Although there have been improvements in governance in the Africa region, African governments still need to show real political will and commitment in several areas which, ultimately, aggravate environmental degradation. First, a lack of democratic institutions and the persistence of corruption in some places contribute to inefficient use of resources. AMCEN, and sub-regional and international organizations, should encourage African governments to address the problem of corruption and to develop effective governance regimes that are favourable to sustainable development, where these are not in place. Conflicts and their aftermath also impact on the environment when, for example, refugees and displaced persons, through no fault of their own, increase the burden on already over-stretched resources in a host country. To address such problems, African

governments should use conflict-minimization strategies to help to promote peaceful coexistence, and to avoid situations that lead to displacement of people. Resource scarcities have been, and continue to be, major causes of conflict among states. AMCEN and sub-regional groupings should, therefore, assist African countries in effectively managing transboundary resources, so as to avoid conflict.

● **Enforcement of compliance**

Although there are various laws and regulations governing environmental management in Africa, their enforcement is generally weak. Greater and more effective capacity is needed for the enforcement of existing laws and regulations. AMCEN, in partnership with international and sub-regional organizations, should identify strengths and weaknesses in compliance and enforcement, and recommended areas, to build the capacity of law enforcement agencies and their judiciaries, in order to allow them to better appreciate their roles in environmental management.

● **Setting targets and monitoring performance**

African governments need to demonstrate to the rest of the world, and to their own people, real improvements and success stories in environmental management. This calls for the measurement of performance over time which, in turn, requires setting targets and monitoring of programmes. AMCEN should persuade African governments to agree on regional targets for environmental management. Furthermore, through partnerships with national, sub-regional and international organizations, AMCEN should monitor progress towards the attainment of the agreed targets for environmental management, at least in the medium term. AMCEN should also seek support for continued production of the AEO as part of the monitoring process. African governments should support the production of sub-regional state-of-environment reports that would subsequently feed into the AEO process, where such reports do not currently exist.

## CONCLUSION

Africa's environment has deteriorated steadily, with poverty being the main cause of that degradation, and with the poor being its direct victims. High levels of poverty—in combination with increasing instances of climate variability and natural disasters, internal institutional weaknesses in Africa, and unfair trading practices in developed countries—have made Africans more vulnerable physically, psychologically and economically. The collective African capacity to cope with increasing vulnerability is also generally low.

Given the magnitude of these problems, it may appear as if African governments, and sub-regional and regional organizations, are doing nothing to solve them. This is not so. They have initiated steps to halt or even reverse environmental degradation, although initiatives are now required for more effective implementation of policies and strategies that have been adopted.

It is also significant that Africa has a vision for sustainable development embodied in the newly formed African Union and the New Partnership for Africa's Development. There is great optimism that Africa can catch up with the rest of the world and even surpass it, using the Great Transitions scenario. However, if this is to be achieved, African countries need, amongst other things, to:

- reduce poverty;
- improve the state of the environment;
- improve management systems;
- reduce vulnerability to adverse environmental changes;
- promote regional and sub-regional cooperation;
- mobilize additional financial resources; and
- create an effective institutional structure to holistically manage the environment on a region-wide basis.

African governments must show greater political will and commitment to solving environmental problems, and must be prepared to devote their own financial and human resources to practical environmental action. They must also address the issue of corruption if they are to improve efficiency in utilization of resources, and they must embrace the democratic process for better governance.

African national governments, sub-regional organizations, AMCEN and the international community are encouraged to contribute to implementation of the specific activities proposed in the 31 areas for action identified above. A key responsibility lies with AMCEN, and with sub-regional groupings and national governments, to mobilize adequate technical, human and financial resources for the implementation of these activities. The international community is also urged to support the efforts of the national governments, sub-regional organizations and AMCEN, in the spirit of the New Partnership for Africa's Development.

## REFERENCES

AMCEN (2001). *Report of the Special Session held on 16th October, 2001*. African Ministerial Conference on Environment. UNEP, Nairobi

Barnes, J. I and de Jager, J.L.V. (1996). *Economic and Financial Incentives for Wildlife Use on Private Land in Namibia and the Implications for Policy*. Windhoek, Namibia

Brown, M. and McGann, J. (1996). *A Guide to Strengthening Non-governmental Organization Effectiveness in Natural Resources Management*. The PVO-NGO/NRMS Project: Funded by the U.S. Agency for International Development. Washington D.C.

GoU (1995). *The Constitution of the Republic of Uganda 1995*. Government of Uganda, Kampala

MoFPED (1999). *Vision 2025. A strategic Framework for National Development*. National Long Term Perspective Studies Project. Ministry of Finance, Planning and Economic Development, Kampala

Mogaka, H. and others (2001). *Economic Aspects of Community Involvement in Sustainable Forest Management in Eastern and Southern Africa*. Forest and Social Perspectives in Conservation, No. 8. IUCN Eastern Africa Programme. IUCN—The World Conservation Union. Gland

NEMA (2001). *The State of Environment for Uganda 2000*. National Environment Management Authority, Kampala

PFAP (1998). *The Policy and Legal Framework to Forest Management in Central Copperbelt and Luapula Provinces, Zambia*. Provincial Forestry Action Programme, Forestry Department, Ministry of Environment and Natural Resources, Ndola

Serageldin, I. (1994). Making development sustainable. In Ismail Serageldin and Andrew Steer (eds). Making Development Sustainable—From Concepts to Action. Environmentally Sustainable Development Occasional Paper Series No. 2. The World Bank. Washington D.C.

Shechambo, F. (1999). Macroeconomic policy incentives and disincentives for biodiversity conservation: the case of structural adjustment in Tanzania. Paper presented at workshop on *Economic Incentives for Biodiversity Conservation in Eastern Africa*. IUCN Eastern Africa Programme, Nairobi

Slade, G. and K. Weitz. (1991). *Uganda Environmental Issues and Options*. A Masters Dissertation. Unpublished. Duke University, North Carolina

UNDP (2000). *World Development Report*. United Nations Development Programme. Oxford University Press, New York

UNEP (2000). *AMCEN: Mapping the Future*. The United Nations Environment Programme. Nairobi

UNEP (1999). *Global Environment Outlook 2000: UNEP's Millennium Report on the Environment*. United Nations Environment Programme. Nairobi

UNEP/SIDA (1996). *Multilateral Environmental Agreements: Relevance, Implications and Benefits to African States*. United Nations Environment Programme, Regional Office for Africa (ROA), Water Branch/Swedish International Development Cooperation Agency. Nairobi

United Nations (1993). *Agenda 21*. New York

## ANNEX 1: SUMMARY OF KEY POLICY RESPONSES ACROSS AFRICA BY THEMATIC AREA

| ISSUES | KEY POLICY RESPONSES |
|---|---|
| Environment and development | • Development of national strategies for sustainable development (NSSDs), and in some cases, National Conservation Strategies (NCSs). <br> • Translation of the global Agenda 21 into *National Agenda 21s* and *Local Agenda 21s*. <br> • Establishment of fully fledged ministries of environment and environmental protection authorities or agencies. <br> • Improvements in sub-regional and regional coordination of environmental management. |
| Poverty | • Preparation of poverty reduction strategy papers and poverty eradication action plans. <br> • Formulation of sustainable livelihood strategies. <br> • Promotion of south-south and intra-Africa trade. <br> • Lobbying for greater access to developed country markets and, in general, removal of trade barriers. <br> • Modernization of agriculture. |
| Climate variability | • The majority of states are parties to the United Nations Framework Convention on Climate Change (UNFCCC) and the Convention to Combat Desertification (UNCCD). <br> • Several countries have produced National Action Plans in accordance with UNFCCC. <br> • Establishment of Early Warning Systems. <br> • Establishment of food reserve programmes. <br> • Crop research to identify drought resistant varieties. <br> • Improving housing design and construction. <br> • Urban planning to reduce vulnerability of human populations. |
| Climate change | • Ratification of UNFCCC and the Kyoto Protocol. <br> • Undertaking of Activities Implemented Jointly (AIJs) projects through joint ventures with the private sector of developed countries. <br> • Development of National Communication Strategies to provide detailed inventories of emissions and sinks, and programmes to mitigate the impacts of climate change. <br> • Exploration of options for further exploitation of alternative sources of energy (for example, solar, wind, micro-hydro and biomass), particularly by countries of Northern and Southern Africa. |
| Air pollution | • Establishment of air quality standards and guidelines. <br> • Monitoring of ambient air quality. <br> • Operations of the Air Pollution Impact Network for Africa (ALPINA), a network of scientists, policy makers and NGOs established to provide information on air pollution, methodologies and databases, and to bridge the gap between information and policy making. <br> • Upgrading public transport systems, imposition of age limits for private and commercial vehicles, and the provision of subsidies for switching to unleaded fuels. <br> • Preparations of plans for adoption of cleaner technologies to reduce industrial emissions. |

| ISSUES | KEY POLICY RESPONSES |
|---|---|
| Land degradation | • Formulation of land use policies and plans, including zoning. |
| | • Land reform (for example, land redistribution and resettlement). |
| | • Capacity building. |
| | • Development of environmental management programmes (district, national environment action plans). |
| | • Promotion of community-based natural resources management (CBNRM) projects. |
| | • Development of erosion hazard mapping. |
| | • Regional initiatives for the conservation and utilization of soils, for example, Southern African Regional Commission for the Conservation and Utilization of the Soil (SARCCUS). |
| | • Regional initiatives to combat desertification (SADC, Sub-Regional Action Programme). |
| | • Reducing the rate of growth of the human population. |
| | • Environmental education programmes. |
| | • Promotion of private sector involvement in land management issues. |
| | • Universal primary education. |
| | • Plans for modernization of agriculture. |
| | • Ratification of a large number of international conventions. |
| | • Irrigation. |
| Habitat loss | • Increase in the number and extent of protected areas. |
| | • Ratification of conventions related to biodiversity, CBD in particular, and RAMSAR and CITES. |
| | • Promoting community based natural resource management programmes (CBNRM). |
| | • Development of national environmental action plans and conservation strategies. |
| | • Promotion of sub-regional cooperation in conservation. |
| | • Formulation of natural biodiversity strategy and action plan (NBSAP). |
| Species loss | • Ratification of conventions related to biodiversity, CBD in particular, but also RAMSAR and CITES. |
| | • Species reintroduction. |
| | • *Ex-situ* plant propagation in nurseries. |
| Alien invasive species | • Tightening controls on imports and spraying of aircraft (and in some cases disinfecting of passengers too). |
| Inadequate attention to indigenous knowledge and intellectual property rights | • Gene banking (Southern Africa). |
| | • Reform of policies to assign intellectual property rights to certain countries, communities or individuals. |
| | • Establishment of resource centres across Africa that focus on identification and dissemination of indigenous or traditional knowledge and practices. |
| | • Using indigenous knowledge in the treatment of HIV/AIDS (Tanzania). |
| Deforestation | • Improving forest harvesting sustainability through removal of subsidies for commercial logging and privatization of state-owned forests. |
| | • Ensuring greater stakeholder participation in forest management through, amongst other things, partnerships between state or private and local communities. |

| **ISSUES** | **KEY POLICY RESPONSES** |
|---|---|
| | ● Use of technologies, such as remote sensing and geographic information systems, to provide more accurate information. |
| | ● Formation of the ATO, whose member states collectively control over 80 per cent of Africa's natural forests. |
| | ● Development of implementing and indicator programmes through the Dry Zone Africa Process (Southern African States). |
| | ● Having some forest areas certified by the Forest Stewardship Council (Southern African States). |
| | ● Including forests in wildlife protected areas such as national parks to accord them greater conservation status. |
| | ● Proposal for a consortium approach to ease access to funding (African Development Bank). |
| Limited access to water resources | ● The United Nations International Drinking Water Supply and Sanitation Decade (1980–90). |
| | ● Africa 2000 Initiative of 1994, by the World Health Organization (WHO)'s Africa Regional Office. |
| | ● Construction of dams on almost all major rivers in Africa to provide water storage capacities, hydro-electric power, and to supply domestic, industrial and agricultural users. There are more than 1 200 dams in Africa. |
| | ● Revision of water policies and pricing mechanisms, as measures to manage demand and encourage more conservative water use. |
| | ● Recycling of wastewater as irrigation water, and upgrading of reticulation networks. |
| | ● Increasing favour for integrated water resource management (IWRM) in several countries. |
| | ● Public-private partnerships in water resource management and water supply programmes. |
| | ● Establishment of international agreements and protocols, either as proactive measures or in response to escalating conflict over shared water courses (for example, The Nile Basin Initiative, the Regional Programme for the Sustainable Development of the Nubian Sandstone Aquifer, and the SADC Protocol on Shared Water Courses). |
| Poor water quality | ● Development of wetlands policies and/or conservation strategies (for example, Ghana, South Africa and Uganda). |
| | ● Establishment and enforcement of effluent water standards. |
| | ● Rehabilitation of existing wastewater treatment facilities as measures to control water quality. |
| | ● Incorporation of the polluter pays principle in many policies and legislation. |
| | ● Schemes for improving drainage, purification and decontamination of freshwater systems, and public awareness campaigns. |
| Coastal erosion | ● Declaration of marine protected areas (MPAs). |
| | ● Integrated environmental management, particularly integrated coastal zone management (ICZM). |
| | ● Promulgation of laws and regulations requiring environmental impact studies to be carried out before development proceeds in the coastal zone or hinterland. |
| | ● Sub-regional and regional agreements. |
| | ● Ratification of several international conventions aimed at enhancing conservation of natural resources. |
| | ● Support for capacity building and access to financial resources. |

| ISSUES | KEY POLICY RESPONSES |
|---|---|
| Marine and coastal pollution | ● Ratification of international agreements, such as the Convention for the Prevention of Pollution from Ships (MARPOL), the Convention for the Protection, Management and Development of the Marine and Coastal Environment of the Eastern African Region (Nairobi Convention), and the Convention for Cooperation in the Protection and Development of the Marine and Coastal Environment of the West and Central African Region (Abidjan Convention).<br>● Participation in UNEP's Regional Seas Programme.<br>● Public health legislation.<br>● Clearing of coastal areas. |
| Overharvesting | ● Various management measures, including minimum net size limits, bag limits, use of appropriate fishing gear and closed seasons.<br>● International agreements between African countries, and between African and European or other international fisheries (the United Nations Law of the Sea). |
| Sea level rise | ● Construction of groynes, sea walls and other physical barriers.<br>● Signing of the Convention for Cooperation in Protection and Development of the Marine and Coastal Environment of the West and Central African Region (Abidjan Convention). |
| Poor urban conditions | ● Increased production of low-cost housing stocks, and introducing housing subsidies for low-income groups.<br>● Creation of the United Nations Commission for Human Settlements (Habitat) and Local Agenda 21.<br>● Revision or formation of constitutions and national legislation to promote the right to adequate shelter.<br>● Revision of policies to recognize women's rights to own property.<br>● Land reform.<br>● Formulation of environmental policies.<br>● Development of integrated water policies and waste management strategies.<br>● Privatization of municipal services in an effort to improve coverage and maintenance.<br>● Development of effluent standards and tighter controls on waste management.<br>● Housing programmes, subsidies for low income families, poverty alleviation programmes, and decentralization strategies.<br>● Attainment of international best practices and awards (Angola and Sudan). |
| Vulnerability | ● Establishment of ministries (departments) responsible for disaster preparedness, prevention and management.<br>● Formulation of policies and action plans for disaster prevention and management.<br>● Formulation and implementation of poverty reduction strategies.<br>● Establishment of early warning systems.<br>● Land use planning. |
| Future outlook | ● Preparation of long-term perspectives, strategic framework for national development (National Vision 2025).<br>● Preparation of poverty reduction strategies, based on national Vision 2025s.<br>● In some cases (for example, Uganda), preparation of development plans for lower levels government in conformity with the national poverty eradication action plans. |

## ANNEX 2: SELECTED EXAMPLES OF FAILURES, WEAKNESSES AND GAPS IN ENVIRONMENTAL MANAGEMENT BY THEMATIC AREAS

| AREA | FAILURES, WEAKNESSES AND BARRIERS FOR IMPLEMENTATION |
|---|---|
| **A. Protection of the environment** | |
| A1. Atmosphere | ● Weak early warning system and low capacity for prediction of climate variability.<br>● In some African states, inadequate integration of transport systems with urban and regional settlement strategies, due to absence of land and land use policies. |
| A2. Toxic chemicals | ● Lack of risk assessment and of dissemination of information.<br>● Labelling of chemicals not sufficiently understood by the majority of Africans.<br>● Inadequate industry response to risk reduction programmes.<br>● Inadequate national coordinating mechanism for liaison between all parties involved in chemical safety activities.<br>● Weak national enforcement programmes for prevention of illegal international traffic in toxic and dangerous products. |
| A3. Hazardous wastes | ● Inadequate industry response to treat, recycle, re-use and dispose of wastes at source.<br>● Inadequate information network and alert systems to assist with detection of illegal traffic in hazardous wastes. |
| A4. Solid wastes and sewerage related issues | ● Commitments to achieving certain benchmarks by year 2000 have not been met. African states agreed that, by 2000, they would:<br>• ensure sufficient national capacity for waste management;<br>• promote sufficient financial and technological capacities at national and local levels;<br>• establish waste treatment and disposal quality criteria; and<br>• ensure that 75 per cent of solid waste generated in urban areas is collected, recycled or disposed of in an environmentally safe manner. |
| A5. Radioactive wastes | ● No significant activity in this area in most African states. |
| **B. Natural resources** | |
| B1. Land resources | ● Absence of planning and management systems.<br>● Little community involvement in information gathering. |
| B2. Combating deforestation | ● Inadequate information base on status of resources and rates of deforestation.<br>● Inadequate valuation of forest resources. |
| B3. Combating desertification and drought | ● Low capacity for drought preparedness and drought relief schemes.<br>● Absence of comprehensive anti-desertification programmes integrated into national development plans and national environmental planning.<br>● Inadequate popular participation and environmental education focusing on desertification control and management of effects of drought. |
| B4. Sustainable development for mountainous areas | ● Lack of database or information systems to facilitate integrated management and environmental assessment of mountain ecosystems. |
| B5. Sustainable agriculture and rural development | ● Africa's agriculture still remains low-input/low-yield, and is therefore unsustainable.<br>● Limited opportunities for non-farm employment.<br>● Limited incentives to promote land conservation.<br>● Inadequate attention given to indigenous knowledge in agriculture.<br>● Both plant and animal genetic resources are poorly inventoried or documented. |

| AREA | FAILURES, WEAKNESSES AND BARRIERS FOR IMPLEMENTATION |
|---|---|
| B6. Conservation and biodiversity | • Biodiversity resources poorly inventoried. |
| B7. Environmentally-sound management of biotechnology | • The potential contribution of biotechnology to sustainable development in Africa is unknown or, at best, under-estimated. |
| B8. Water bodies, shoreline and aquatic resources | • Fisheries research focused largely on a few selected species, neglecting the remaining water bodies.<br>• Absence of land use zones for shoreline areas.<br>• Inadequate resource inventory and management planning. |
| B9. Protection of quality and supply of freshwater | • African states agreed that, by 2000:<br>  • all urban residents would have access to at least 40 litres per capita per day of safe water; and<br>  • 75 per cent of the urban population would be provided with on-site or community facilities for sanitation. |

# ANNEX 3: PROPOSED ACTION AREAS, ACTIVITIES AND RESPONSIBILITIES

| Category | Specific Actions | International | AMCEN | Sub-regional | National |
|---|---|---|---|---|---|
| **A. Reducing poverty** | | | | | |
| A1. Sustainable development | Adopt New African Initiative (NAI) as framework for sustainable development in Africa. | ● | | | |
| | Promote attainment of the poverty reduction goals of the Millennium Declaration. | ● | ● | | ● |
| | Accelerate setup of necessary mechanisms of the World Solidarity Fund (WSF). | ● | | | |
| A2. Accelerated industrial development | Assist with industrial productivity and competitiveness of African industries. | ● | ● | | |
| | Promote development of micro, small- and medium-size enterprises, with focus on agro industry. | ● | ● | | ● |
| | Promote use of environmentally benign industrial technologies and techniques. | | | ● | ● |
| A3. Increasing agricultural production | Increase national financing for the agricultural sector. | | | | ● |
| | Support implementation of the United Nations Convention to Combat Desertification (UNCCD). | ● | ● | | |
| | Urge developed countries to remove agricultural subsidies and apply the precautionary principle to genetically modified organisms. | ● | ● | | |
| | Promote sustainable agricultural production techniques. | | | | ● |
| A4. Promote human development | Ensure greater access to affordable primary and secondary health care, and medical technology. | | | | ● |
| | Improve environmental and social conditions that are responsible for spread of diseases, and build the capacity of local communities. | | | | ● |
| | Assist Africa in making both preventive and curative health care available. | ● | | | |
| | Take all necessary measures to provide access to medicine at affordable prices, and promote public health and nutrition. | | ● | ● | ● |
| | Empower women in social and economic development. | | | | ● |
| | Promote human resources development and capacity building, including universal primary and secondary education. | ● | ● | ● | ● |
| | Strengthen skills of youth. | | | | ● |

| Category | Specific Actions | International | RESPONSIBILITIES AMCEN | Sub-regional | National |
|---|---|---|---|---|---|
| A5. Trade and market access | Open markets and eliminate subsidies on agriculture, textiles and other export products of interest to Africa. | ● | ● | | |
| A6. Increased financing for sustainable development | Promote foreign direct investment in Africa. | ● | ● | ● | ● |
| | Cancel debt of African states. | ● | | | |
| | Adhere to the United Nations target of 0.7 per cent of GNP for official development assistance (ODA). | ● | | | |
| | Increase the resources of, and improve upon operational procedures and project implementation of, the Global Environment Facility (GEF). | ● | | | |
| A7. Improving infrastructure and sustainable human settlement | Improve access to, and the affordability and reliability of, infrastructure services. | | | | ● |
| | Mobilize resources for the implementation of Habitat Agenda and the declaration of the 25th United Nations Special Session to achieve sustainable human settlements in Africa. | ● | | | |
| A8. Promoting science and technology | Assist African countries in their efforts to gain access to new technologies, particularly information and communication technologies and create conditions for the development of indigenous technologies to enhance economic development. | ● | | | |

## B. Improving the state of environment

| Category | Specific Actions | International | RESPONSIBILITIES AMCEN | Sub-regional | National |
|---|---|---|---|---|---|
| B1 Reducing land degradation | Promote campaigns of environmental information, education and communication. | | | ● | ● |
| | Ensure timely and effective implementation of UNCCD and acknowledge it as a sustainable development convention. | ● | | | |
| | Encourage the production of land, and land-use policies and plans. | | ● | ● | ● |
| | Improve land tenure and land ownership systems that also address gender considerations. | | | | ● |
| | Protect cultural and historical heritage. | | | | ● |
| B2. Conserving biodiversity | Promote landscape approaches to biodiversity conservation. | | ● | ● | ● |
| | Develop and implement national legislation for the protection of the rights of local communities, farmers and breeders, and for the regulation of access to biological resources and biosafety, in line with OAU Model Law. | | ● | ● | ● |
| | Rehabilitate degraded wetland areas. | | | ● | ● |

| CATEGORY | SPECIFIC ACTIONS | RESPONSIBILITIES | | | |
|---|---|---|---|---|---|
| | | INTERNATIONAL | AMCEN | SUB-REGIONAL | NATIONAL |
| B2. Conserving biodiversity (continued) | Strengthen the Lusaka Agreement. | ● | ● | | |
| | Rehabilitate degraded conservation areas through species reintroduction and habitat restoration. | | | ● | ● |
| | Document and disseminate indigenous knowledge and practices applicable to conservation. | | ● | | ● |
| | Improve upon biodiversity inventory, and document landraces. | ● | | ● | ● |
| | Promote *ex-situ* conservation facilities for rare, vulnerable and endangered species. | | | | ● |
| | Promote the conservation of agricultural biodiversity. | ● | ● | ● | ● |
| B3 Reducing deforestation | Procure financial resources and technology transfer to Africa for implementing the provisions of the International Panel on Forests (IPF)/International Forest Forum (IFF) of the United Nations Forum on Forests (UNFF). | ● | | | |
| | Promote access to affordable energy, especially in rural areas, for sustainable development. | ● | | | ● |
| | Promote research on and development of clean energy technologies, efficiency of energy supply and usage, and efficient uptake of renewables. | ● | | | |
| | Revitalize the Barbados Programme of Action for the Sustainable Development of Small Island Developing States. | ● | | | |
| | Strengthen the African Timber Organization. | ● | ● | | |
| | Support the Yaounde Declaration. | ● | ● | | |
| | Promote collaborative forest management between forestry authorities and surrounding communities. | | | | ● |
| | Support rehabilitation of degraded forests. | ● | | | ● |
| | Revise the pricing of forest products to reflect true economic values. | | | | ● |
| | Support documentation of the medicinal value of forests. | ● | ● | ● | ● |
| | Encourage private sector participation in forest establishment and management. | ● | | | ● |
| | Promote the greater utilization of non-wood forest products. | | | | ● |

| CATEGORY | SPECIFIC ACTIONS | INTERNATIONAL | RESPONSIBILITIES AMCEN | SUB-REGIONAL | NATIONAL |
|---|---|:---:|:---:|:---:|:---:|
| B4. Protecting coastal and marine environment | Harmonize, coordinate and ensure compliance of regional and international laws and agreements related to seas. | ● | ● | ● | ● |
| | Revitalize the Nairobi and Abidjan agreements. | ● | ● | | |
| | Promote the establishment and sound management of marine protected areas in the freshwater lakes of Africa. | ● | | ● | ● |
| | Improve on fishery stock inventory and monitoring. | ● | ● | ● | ● |
| | Support formulation of marine resources management plans, especially integrated coastal zone management plans (ICZMPs). | | | ● | ● |
| | Promote aquaculture. | | | | ● |
| B5. Mitigating the adverse impacts of climate change and other atmospheric conditions | Finalize agreement on the Kyoto Protocol. | ● | ● | | |
| | Operationalize the Climate Change Fund (CCF) for developing countries, as well as the Special Fund for least developed countries. | ● | | | |
| | Facilitate access to and adoption of cleaner technologies to reduce industrial emissions. | ● | | | |
| | Establish climate modeling programmes and early warning of rainfall variations. | | ● | ● | ● |
| | Ratify the United Nations Motor Vehicle Emissions Agreement. | | | | ● |
| | Facilitate north-south exchanges of experience and knowledge between experts, and provide for transfer of know-how between the various areas of Africa. | ● | ● | | |
| | Improve enforcement of emission standards and regulations. | | | | ● |
| | Support operationalization of the Strategy for the Elimination of Hunger in the Horn of Africa. | ● | | | |
| | Promote the use of unleaded gasoline. | | | | ● |
| | Study the social impacts of atmospheric pollution. | ● | ● | ● | ● |
| B6. Waste management | Put in place the required funding and capacity for effective management of non-hazardous waste. | ● | | | |
| | Implement regional and international conventions on the generation, storage, transport, and transboundary movement and disposal of hazardous waste, including radioactive waste. | ● | ● | ● | ● |
| | Take steps to counter the exports of obsolete capital goods and equipment to Africa. | ● | ● | | |

| CATEGORY | SPECIFIC ACTIONS | INTERNATIONAL | RESPONSIBILITIES AMCEN | SUB-REGIONAL | NATIONAL |
|---|---|:---:|:---:|:---:|:---:|
| B7. Environmentally sound management of chemical products | Assist and support African states with the management of chemical products, in accordance with Chapter 19 of Agenda 21, and the Rotterdam and Stockholm conventions. | ● | | | |
| B8. Improving access to freshwater resources | Promote public-private partnerships in water resource management. | | | | ● |
| | Support regional and sub-regional water resource management bodies, such as the Nile Basin Initiative. | ● | | | |
| | Inventory watersheds and develop guidelines for sustainable utilization. | | | ● | ● |
| | Develop standards for water quality. | | | | ● |
| | Improve access to freshwater resources. | ● | | ● | ● |
| | Promote integrated water resource management and development. | | | ● | ● |
| | Promote the use of economic instruments for water resource management. | | | | ● |
| | Popularize environmentally sound low-cost technologies for water harvesting. | ● | ● | | ● |
| B9. Improving urban areas | Provide support for urban development planning. | ● | | | ● |
| | Promote construction of environmentally sound low-cost houses to overcome congestion in poorer segments of urban areas. | | | | ● |
| | Support African countries to fulfil national obligations under *Habitat Agenda*. | ● | | | |
| | Assist African countries to prepare integrated water and waste management strategies. | ● | | | |
| | Formulate human settlement policies, laws and regulations. | | | | ● |
| | Formulate solid waste management policies, laws and regulations. | | | | ● |
| | Promote private sector participation in improving urban infrastructure and provision of municipal services. | | | | ● |

| CATEGORY | SPECIFIC ACTIONS | INTERNATIONAL | RESPONSIBILITIES AMCEN | SUB-REGIONAL | NATIONAL |
|---|---|---|---|---|---|
| **C. Promoting cross-cutting actions** | | | | | |
| C1. Reducing the vulnerability of Africans | Reach agreement on the need for an effective governance regime for sustainable development. | | ● | | |
| | Access resources and support for mechanisms to prevent, manage and resolve conflicts, and to satisfy the needs of refugees and internally displaced persons and their host countries. | ● | | | |
| | Use conflict-resolution mechanisms to discourage situations leading to internal displacement of peoples. | | | | ● |
| | Encourage the full participation and consideration of the views of all major groups in matters concerning sustainable development and environmental management. | | | ● | |
| | Prepare disaster preparedness, prevention and management plans. | | | | ● |
| | Show greater political commitment to addressing environmental degradation. | | | | ● |
| | Address the problem of corruption to increase resource-use efficiency. | | | | ● |
| C2. Enhancement of regional and sub-regional cooperation | Provide support to the newly created African Union. | ● | | | |
| | Agree on a programme of action to support regional shared water initiatives, and other transboundary environmental resources. | | ● | ● | |
| | Support institutions of sub-regional cooperation and economic integration, such as IOC, EAC, ECOWAS, SADC | ● | | | |
| | Reach agreement on the need for an effective governance regime for sustainable development. | | ● | | |
| C3. Mobilizing financial resources | Provide new and additional financial resources to African states for environmental management and reduction of the vulnerability of Africans to adverse environmental changes. | ● | ● | ● | |
| | Increase efforts in mobilizing domestic resources through taxes, fines and resource user fees. | | | | ● |
| | Ensure environment has higher profile in the budget allocation process, including disbursement of Poverty Action Funds. | | | | ● |
| | Develop and promote win-win strategies for encouraging private sector investment in environmental management. | | | | ● |

| CATEGORY | SPECIFIC ACTIONS | RESPONSIBILITIES | | | |
|---|---|---|---|---|---|
| | | INTERNATIONAL | AMCEN | SUB-REGIONAL | NATIONAL |
| C3. Mobilizing financial resources (continued) | Develop capacity for natural resource accounting and valuation to facilitate proper pricing of environmental goods and services. | | ● | ● | |
| C4 Improving institutional capacity | Improve the institutional capacity to monitor, coordinate and supervise environmental management, holistically and on a continent-wide basis. | ● | ● | | |
| C5. Enhancing the participation of civil society | Build capacity of national NGOs to become effective partners in environmental management. | | | | ● |
| | Promote the establishment and operations of regional and sub-regional NGO fora. | | ● | ● | |
| C6. Emphasizing sustainable development | Build capacity for greening development plans at national and sub-national levels. | ● | ● | | |
| | Make greening of accounts mandatory and a precondition for allocation of financial resources. | | | | ● |

# ACRONYMS AND ABBREVIATIONS

| | |
|---|---|
| ACOPS | Advisory Committee on the Protection of the Sea |
| ADB | African Development Bank |
| ADMADE | Administrative Management Design for Game Areas (Zambia) |
| AEC | African Economic Community |
| AEO | Africa Environment Outlook |
| AIDS | acquired immunodeficiency syndrome |
| AMCEN | African Ministerial Conference on the Environment |
| AMU | Arab Maghreb Union |
| APINA | Air Pollution Impact Network for Africa |
| ArabMAB | Arab Man and Biosphere (network) |
| ARI | acute respiratory infection |
| ASCN | African Sustainable Cities Network |
| ATO | African Timber Organization |
| AU | African Union |
| CAIP | Cairo Air Improvement Project |
| CAMRE | Council of Arab Ministers Responsible for the Environment |
| CARPE | Central Africa Regional Programme for the Environment |
| CBD | Convention on Biological Diversity |
| CBNRM | Community-Based Natural Resource Management |
| CCF | Climate Change Fund |
| CEDARE | Centre for Environment and Development for the Arab Region and Europe |
| CEFDHAC | Conference on Ecosystems of Dense Humid Forests in Central Africa |
| CEMAC | Economic and Monetary Community of Central Africa |
| CESP | Country Environmental Strategy Papers |
| CFC | chlorofluorocarbon |
| CILSS | Inter-State Committee to Combat Drought in the Sahel |
| CITES | Convention on International Trade in Endangered Species of Wild Fauna and Flora |
| CMS | Convention on the Conservation of Migratory Species of Wild Animals |
| CO | carbon monoxide |
| $CO_2$ | carbon dioxide |
| COMESA | Common Market for Eastern and Southern Africa |
| DEAW | Division of Environment Assessment and Early Warning |
| DRBC | Drill Rehabilitation and Breeding Centre (Nigeria) |
| DRC | Democratic Republic of Congo |
| EAC | East African Community |
| EAEC | East Africa Economic Community |

| | |
|---|---|
| ECCAS | Economic Community of Central African States |
| ECOFAC | Ecosystèmes forestiers d'Afrique Central |
| ECOMOG | Monitoring Group (ECOWAS) |
| ECOWAS | Economic Community of West African States |
| EEZ | Exclusive Economic Zone |
| EIA | Environmental Impact Assessment |
| ENSO | El Niño Southern Oscillation |
| EU | European Union |
| EWS | Early Warning Systems |
| FAO | Food and Agriculture Organization of the United Nations |
| GDP | gross domestic product |
| GEF | Global Environment Facility |
| GEO | Global Environment Outlook |
| GHG | greenhouse gas |
| GIS | Geographic Information Systems |
| GM | genetically modified |
| GMO | genetically modified organism |
| GNP | gross national product |
| GOOS | Global Ocean Observing System |
| GoU | Government of Uganda |
| GPA | Global Programme of Action for the Protection of the Marine Environment from Land-Based Activities |
| HDI | Human Development Index |
| HIV | human immunodeficiency virus |
| HYCOS | Hydrological Cycle Observing Systems |
| IADD | Inter-governmental Authority on Drought and Development |
| ICCON | International Consortium for Cooperation on the Nile |
| ICZM | Integrated Coastal Zone Management |
| IFF | International Forest Forum |
| IGADD | Inter-Governmental Authority on Drought and Development |
| IGAD | Inter-governmental Authority on Development |
| IIED | International Institute for Environment and Development |
| IMF | International Monetary Fund |
| IOC | Indian Ocean Commission, Mauritius |
| IOC | Intergovernmental Oceanographic Commission of UNESCO |
| IPCC | Intergovernmental Panel on Climate Change |
| IPF | International Panel on Forests |
| IT | information technology |

| | |
|---|---|
| ITCZ | Inter-Tropical Convergence Zone |
| ITTO | International Tropical Timber Organization |
| IUCN | World Conservation Union |
| IWRM | Integrated Water Resource Management |
| KICK | Kisumu Innovation Centre-Kenya |
| KWS | Kenya Wildlife Service |
| LCBC | Lake Chad Basin Commission |
| LIFE | Living in a Finite Environment (Namibia) |
| LVEMP | Lake Victoria Environment Management Programme |
| MAP | Mediterranean Action Plan |
| MAP | Millennium Africa Recover Plan |
| MAP | Millennium Partnership for the African Recovery Programme |
| MARPOL | Convention on the Prevention of Pollution from Ships |
| MBIFCT | Mgahinga Bwindi Impenetrable Forest Conservation Trust |
| MEA | Multilateral Environmental Agreement |
| MoFPED | Ministry of Finance Planning and Economic Development |
| MPA | Marine Protected Area |
| NBI | Nile Basin Initiative |
| NBO | Nile Basin Organization |
| NBSAP | National Biodiversity Strategy and Action Plan |
| NCC | National Climate Committee (Mauritius) |
| NCS | National Conservation Strategy |
| NEAP | National Environment Action Plan |
| NEMA | National Environment Management Authority |
| NEPAD | New Partnership for Africa's Development |
| NGO | Non-Governmental Organization |
| NPCAD | National Plan of Action to Combat Desertification |
| NPP | Net Primary Productivity |
| NSA | Nubian Sandstone Aquifer |
| NTFPA | National Tropical Forestry Action Plans |
| OAU | Organization of African Unity |
| ODA | Official Development Assistance |
| ODS | ozone-depleting substance |
| OECD | Organization for Economic Cooperation and Development |
| PACSICOM | Pan-African Conference on Sustainable Integrated Coastal Management |
| PAH | polycyclic aromatic hydrocarbon |
| PCB | polychlorinated biphenyl |
| PERGSA | Protection of the Environment of the Red Sea and the Gulf of Aden (Organization for) |
| POM | polycyclic organic matter |

| | |
|---|---|
| POP | persistent organic pollutant |
| PRGIE | Programme Régional de Gestion de l'Information Environnementale |
| PTA | Preferential Trade Area |
| SADC | Southern Africa Development Community |
| SAP | Structural Adjustment Programme |
| SARCCUS | Southern African Regional Commission for the Conservation and Utilization of the Soil |
| SARDC | Southern African Research and Documentation Centre |
| SIDA | Swedish International Development Agency |
| SWIMS | Shoal World Integrated Satellite Monitoring System |
| TLA | Tree Lovers Association (Cairo, Egypt) |
| UN | United Nations |
| UNGA | United Nations General Assembly |
| UNCBD | United Nations Convention on Biological Diversity |
| UNCCD | UN Convention to Combat Desertification in Countries Experiencing Serious Drought and/or Desertification, Particularly in Africa |
| UNCED | United Nations Conference on Environment and Development |
| UNCHS | United Nations Commission for Human Settlements (Habitat) |
| UNCLOS | United Nations Convention on the Law of the Sea |
| UNDP | United Nations Development Programme |
| UNECA | UN Economic Commission for Africa |
| UNECE | UN Economic Commission for Europe |
| UNEP | United Nations Environment Programme |
| UNESCO | United Nations Educational, Scientific and Cultural Organization |
| UNFCCC | UN Framework Convention on Climate Change |
| UNFF | United Nations Forest Forum |
| USAID | US Agency for International Development |
| WB | World Bank |
| WCED | World Commission on Environment and Development |
| WCS | World Conservation Strategy |
| WFP | World Food Programme |
| WHO | World Health Organization |
| WMO | World Meteorological Organization |
| WSF | World Solidarity Fund |
| WSSD | World Summit on Sustainable Development |
| WTO | World Trade Organization |
| WWF | World Wildlife Fund |

# CONTRIBUTORS

Those listed below have contributed to AEO in a variety of ways, as authors, reviewers, participants in AEO consultations and survey respondents.

**Abdel Farid Abdel-Kader**
Centre for Environment and Development for the Arab Region & Europe (CEDARE), Egypt

**Ahmed Abdel-Rehim**
Centre for Environment and Development for the Arab Region & Europe (CEDARE), Egypt

**Hamid Abdoolakhan**
Programme Régional Environnement COI, Mauritius

**Sherif Abdou**
Centre for Environment and Development for the Arab Region & Europe (CEDARE), Egypt

**Mohamed A. Abdrabo**
Institute of Graduate Studies and Research, Alexandria University, Egypt

**Wilna Accouche**
Ministère de l'environnement et du Transport, Seychelles

**Maha Akrouk**
Centre for Environment and Development for the Arab Region & Europe (CEDARE), Egypt

**Emmanuel K. Alieu**
Ministry of Agriculture, Forestry & Marine Resources, Sierra Leone

**Hossam Allam**
Centre for Environment and Development for the Arab Region & Europe (CEDARE), Egypt

**Gaston Andoka**
Ministère de l'Environnement, Congo
Coordonnateur National ADIE-PRGIE Congo

**Emile Amougou**
Ministère de l'Environnement et des Forêts, Cameroon

**Willy Andre**
Ministère de l'environnement et du Transport, Seychelles

**Linda Arendse**
CSIR-Environmentek, South Africa

**Franck Attere**
WWF CARPO, Gabon

**Rajen Awotar**
Council for Development Environmental Studies & Conservation (MAUDESCO), Mauritius

**Bola Ayeni**
Department of Geography, University of Ibadan, Nigeria

**Salim Bachou**
Consulting Economist, Uganda

**Marcel Baglo**
Agence Béninoise pour l'Environnement, Benin

**Anna Ballance**
UNEP/GRID-Arendal, Norway

**Abou Bamba**
Network for Environment and Sustainable Development in Africa (NESDA), Côte d'Ivoire

**Patricia Baquero**
Ministry of Land Use and Habitat, Seychelles

**Louis Guyto Barbe**
Division de l'Environnement, Seychelles

**H. Beekhee**
Economist, Mauritius

**John F. Benson**
Centre for Research in Environmental Appraisal and Management, School of Architecture, Planning and Landscape, University of Newcastle, UK

**Sitotaw Berhanu**
Environment Protection Authority, Ethiopia

**Wilfrid Bertile**
Indian Ocean Commission, Mauritius

**Didier Biau**
Direction Régionale de l'Environnement, Réunion

**Henriette Bikie**
Global Forest Watch, Cameroon

**Edelmiro Castano Bizantino**
Association pour le Développement de l'Information Environnementale (ADIE), Programme Régional de Gestion de l'Information Environnementale (PRGIE), Central Africa, Equatorial Guinea

**Innocent Bizimana**
Ministry of Agriculture, Livestock, Environment and Rural Development, Rwanda

**Foday Bojang**
Organisation of African Unity (OAU), Ethiopia

**Q. Chakela**
Consultant, Lesotho

**Geofrey Chavula**
Department of Environment Affairs and Tourism, Malawi

**Thomas Chiramba**
Southern African Development Community (SADC), Water Sector Coordinating Unit, Lesotho

**Hennie Coetzee**
FOSA Expert Advisory Group, South Africa

**Judie Combrink**
Dept. of Environmental Affairs and Tourism, South Africa

**Athanase Compaoré**
Global Water Partnership, Burkina Faso

**Berhe Debalkew**
Intergovernmental Authority on Development (IGAD), Djibouti

**Charl De Villiers**
Consultant, South Africa

**Anne-France Didier**
Direction Régionale de l'Environnement, Réunion

**Amadou Moctar Dieye**
Centre de Suivi Ecologique, Sénégal

**Koulthoum Djamadar**
Direction Générale du Plan, Comoros

**Bougonou Djeri-Alassani**
Togo

**G. Domingue**
Ministry of Land Use and Habitat, Seychelles

**Clement Dorm-Adzobu**
Water Resources Commission, Ghana

**Charles-Elie Doumambila**
Association pour le Développement de l'Information Environnementale (ADIE),
Programme Régional de Gestion de l'Information Environnementale (PRGIE),
Central Africa, Gabon

**Tewolde Berhan Gebre Egziabher**
Environment Protection Authority (EPA), Ethiopia

**Mohammed El-Anbaawy**
Faculty of Science, Department of Geology, Cairo University, Egypt

**Khaled El-Askari**
Water Management Research Institute, Egypt

**Aly El-Beltagy**
National Institute for Marine and Oceanography, Egypt

**Jean Christophe Elembo**
Association pour le Développement de l'Information Environnementale (ADIE),
Programme Régional de Gestion de l'Information Environnementale (PRGIE),
Central Africa, Democratic Republic of Congo

**Dr Hisham Elkadi**
School of Architecture, Planning and Landscape, University of Newcastle, UK

**Dina El-Naggar**
Egyptian Environmental Affairs Agency, Ministry of Environment, Egypt

**Mahmed Eltawil**
Tawil Consultants, Architects, Planners and Environmental Engineers, Egypt

**Rachid Firadi**
Ministère de l'Amenagement du Territoire, de l'Urbanisme, de l'Habitat et de
l'Environnement, Morocco

**Peter G. H. Frost**
University of Zimbabwe, Zimbabwe

**Richard Fuggle**
University of Cape Town, South Africa

**Prudence Galega**
Network for Environment and Sustainable Development in Africa
(NESDA), Cameroon

**Troy Govender**
External Specialist Advisor to UNEP, Children, Youth & Sport Programmes,
South Africa

**Betty Gowa**
National Environment Management Authority (NEMA), Uganda

**Jacob Gyamfi-Aidoo**
EIS-Africa, South Africa

**Craig Haskins**
Cape Metropolitan Council Administration, City of Cape Town, South Africa

**Ahmed Hegazy**
Botany Department, Faculty of Science, Cairo University, Egypt

**Hamadi Idaroussi**
Ministère de la Production et de l'Environnement, Comoros

**S. K. Imbamba**
Consultant, Kenya

**Phoebe Ayugi Josiah**
Health Environment and Population Consultancy, Kenya

**Godfrey Kamukala**
Consultant, Tanzania

**Yemi Katerere**
IUCN Regional Office for Southern Africa, Zimbabwe

**Eric Kemp-Benedict**
Stockholm Environment Institute (SEI), USA

**Bowdin King**
ICLEI, Zimbabwe

**Consolata Kiragu**
National Environment Secretariat, Kenya

**Evans Kituyi**
African Centre for Technology Studies (ACTS), Kenya

**Etienne Kayengeyenge**
Ministère de l'Aménagement du Territoire et de l'Environnement, Burundi

**Michael Koech**
Kenya Ambassador to UNEP, Kenya

**Tiékoura Koné**
WWF International, Western Africa Regional Programme Office, Côte d'Ivoire

**Koffi Kouakou**
Timbuktu Ventures, South Africa

**Andries Kruger**
South African Weather Service, South Africa

**Tendai Kureya**
SAFAIDS, Zimbabwe

**Elton Laisi**
CEDRISA, Malawi

**Christian Leger**
Direction Régionale de l'Environnement, Réunion

**Francis Coeur de Lion**
GIS Centre, Seychelles

**Festus Luboyera**
Deptartment of Environmental Affairs and Tourism, South Africa

**Clever Mafuta**
SARDC-IMERCSA, Zimbabwe

**Jean Roger Mamiah**
Association pour le Développement de l'Information Environnementale (ADIE),
Programme Régional de Gestion de l'Information Environnementale (PRGIE),
Central Africa, Cameroon

**Pierre Mangala**
Ministère de l'Environnement, République Centrafricaine Coordonnateur National
ADIE-PRGIE Centrafrique

**Anna Mampye**
Department of Environmental Affairs & Tourism, South Africa

**Denis Eddy Matatiken**
Division de l'Environnement, Seychelles

**Simon Mbarire**
National Environment Secretariat, Kenya

**Mireille Mbombo**
C.R.E.F. Central Africa, Cameroon

**Michel Mbomoh-Upiangu**
Association pour le Développement de l'Information Environnementale (ADIE),
Programme Régional de Gestion de l'Information Environnementale (PRGIE),
Central Africa, Gabon

**Jean Boniface Memvie**
Ministère de l'Environnement, Gabon

**Beyene Zigta Mesghenna**
Department of Environment, Eritrea

**Watipaso Mkandawire**
Common Market for Eastern and Southern Africa, Zambia

**Rajendranath Mohabeer**
Commission de l'Ocean Indien, Mauritius

**Yagoub Mohamed**
Ministry of Environment and Physical Development, Sudan

**Jobo Molapo**
Southern African Development Community (SADC), Environment and Land
Management Sector, Lesotho

**Santaram Mooloo**
Ministry of Environment, Mauritius

**Yakobo Moyini**
Environmental Management Associates (EMA), Kampala

**Sam Moyo**
Independent Policy Analyst, Zimbabwe

**Lucy Mulenkei**
Indigenous Information Network, Kenya

**Maria Mutama**
SARDC-IMERCSA, Zimbabwe

**Catherine Mutambirwa**
SARDC-IMERCSA, Zimbabwe

**Jocselyne Mutegeki**
Environmental Management Associates (EMA), Uganda

**Leonard Ntonga Mvondo**
CREF Central Africa Cameroon. GEO/AEO Focal Point in Central Africa. Association pour le Développement de l'Information Environnementale (ADIE), Programme Régional de Gestion de l'Information Environnementale (PRGIE), Central Africa, Cameroon

**Benjamin Nami**
Cameroon, Consultant

**Fatou Ndoye**
Network for Environment and Sustainable Development in Africa (NESDA), Côte d'Ivoire

**John Nevill**
Ministère de l'environnement et du Transport, Seychelles

**Protasius Nghileendele**
Ministry of Environment and Tourism, Namibia

**Deborah Nightingale**
Environmental Management Advisors, Kenya

**Steven Njinyam**
CEMAC, Central African Republic

**Marie Tamoifo Nkom**
Association Jeunesse Verte du Cameroun, Cameroon

**Etienne Ntsama**
Etablissement Ntsama et Fils, Cameroon

**Zacharie Nzooh**
ECOFAC, Cameroon

**Charles Obol**
Southern African Development Community (SADC), Environment and Land Management Sector, Lesotho

**Agnes F. Odejide**
FOSA Expert Advisory Group, Nigeria

**Peter Ondiege**
KEIPET Consultants Ltd., Kenya

**Mohamed Youssouf Oumouri**
Ministère de l'Environnement, Comoros

**Rajesh Parboteeah**
Consultant, Mauritius

**Rolph Payet**
Ministère de l'environnement et du Transport, Seychelles

**Joyce Phoshoko**
Department of Environmental Affairs & Tourism, South Africa

**Fatou Planchon**
Centre de Suivi Ecologique, Senegal

**Danny Poiret**
Ministère de la Santé, Seychelles

**Johannes Rudolph Pretorius**
Department of Environmental Affairs & Tourism, South Africa

**Deepnarain Prithipaul**
Ministère de l'Environnement et du Développement Rural et Urbain, Mauritius

**Fouad Abdou Rabi**
ONG AIDE, Comoros

**Georges Rafomanana**
Ministère de l'Environnement, Madagascar

**Côme Ramakararo**
Ministère de l'Environnement, Madagascar

**A. Ramsewak**
Ministère des Affaires Etrangéres et de la Cooperation Regionale, Mauritius

**Pierre Randah**
CEMAC, Central African Republic

**Jean de Dieu Ratefinanahary**
Ministère des Affaires Etrangères, Madagascar

**Herisoa Razafinjato**
Office National pour l'Environnement, Madagascar

**J. L. Roberts**
Ministry of Health and Quality of Life, Mauritius

**Soonil Dutt Rughooputh**
Université de Maurice, Mauritius

**Renison K. Ruwa**
Kenya Marine and Fisheries Research Institute, Kenya

**Osama Salem**
Consultant, Egypt

**Munyaradzi Saruchera**
Programme for Land and Agrarian Studies, South Africa

**Craig Schwabe**
GIS Centre, Human Sciences Research Council (HRSC), South Africa

**Charles Sebukeera**
National Environment Management Authority (NEMA), Uganda

**Daniel Sibongo**
Consultant Comms, Zimbabwe

**Jay Singh**
Department of Environmental Affairs & Tourism, South Africa

**Didier Slachmuylder**
Indian Ocean Commission, Mauritius

**Soondaree Devi Soborun**
Ministry of Land Transport, Shipping and Port Development, Mauritius

**Lovemore Sola**
GEOFLUX, Botswana

**Nouri Soussi**
Ministry for Environment and Land Management, Tunisia

**Blondeau Talatala**
Ministère de l'Environnement et des Forêts, Cameroon
Coordonnateur National ADIE-PRGIE, Cameroon

**Marie Tamoifo**
Jeunes et Environnement, Cameroon

**Nicodème Tchamou**
Central Africa Regional Programme for the Environment (CARPE), Central Africa, Gabon

**Jonathan Timberlake**
Biodiversity Foundation for Africa, Zimbabwe

**Gabolekwe Lesole Tlogelang**
Office of the President, Botswana

**Tsala Abina**
Secrétariat Permanent à l'Environnement, Cameroon

**Jean Pierre Vandeweghe**
Association pour le Développement de l'Information Environnementale (ADIE), Programme Régional de Gestion de l'Information Environnementale (PRGIE), (Central Africa), Gabon

**Michel Vieille**
Ministère de l'environnement et du Transport, Seychelles

**Ahmed Wagdy**
Centre for Environment and Development for the Arab Region & Europe (CEDARE), Egypt

**Enock Wakwabi**
Kenya Marine and Fisheries Research Institute, Kenya

**Conmany B. Wesseh**
Center for Democratic Empowerment, Cote d'Ivoire

**Jessica Wilson**
Environmental Monitoring Group, South Africa

**Tesfaye Woldeyes**
Environment Protection Authority (EPA), Ethiopia

**Cletus Wotorson**
Mineral Resources Policy Development Expert, USA

**Alaphia Wright**
University of Zimbabwe, Zimbabwe

**Joseph Ipalaka Yobwa**
Ministère de l'Environnement, République Démocratique du Congo Coordonnateur National ADIE-PRGIE, RDC

**Rose Don Zoa**
Ministère de l'Environnement et des Forêts, Cameroon

## UNEP

Subramonia Ananthakrishnan

Jacquie Chenje

Munyaradzi Chenje

Salif Diop

Sheila Edwards

Beth Ingraham

Kagumaho Kakuyo

Rungano Karimanzira

Jesper Kofoed

Dave MacDevette

Strike Mkandla

Naomi Poulton

Megumi Seki

David Smith

Anna Stabrawa

Thomas Tata

Sekou Toure

Laura Williamson

## Other UN bodies

**Mr Kwame Awere-Gyekye**
ECA, Ethiopia

**Mr Jean Louis Blanchez**
FAO, Italy

**Mr Seraphin Dondyas**
FAO Consultant, Gabon

**Mr Michel Laverdiere**
FAO, SAFR, Zimbabwe

**Dr Ousmane Laye**
ECA, Ethiopia

**Dr C. T. S. Nair**
FAO, Italy

**Ms Ada Ndeso-Atanga**
Consultant, FAO Regional Office for Africa, Ghana

**Mr Ojijo Odhiambo**
UNDP, Kenya

**Dr Kwadwo Tutu**
ECA, Ethiopia

**Mr Hassan Musa Yousif**
UNOPS/NLTPS/African Futures, Côte d'Ivoire

## Institutions

Association pour le Développement de l'Information Environnementale, Programme Régional de Gestion de l'Information Environnementale (ADIE-PRGIE)

Central Africa Regional Programme for the Environment (CARPE), Cameroon

Centre for International Forestry Research (CIFOR)

Communauté Economique et Monétaire de l'Afrique Centrale (CEMAC)

ECOFAC Cameroon

Global Forest Watch Cameroon

Institut National de Cartographie du Gabon (INC Gabon)

IUCN, Regional Office for Central Africa

Ministry of Environment and Forestry, Yaounde, Cameroon

Network for Environment and Sustainable Development in Africa (NESDA), Cameroon

Programme de Développement Participatif Urbain

Shoals of Capricorn, Seychelles

WWF Cameroon

WWF Gabon

# INDEX

Index prepared by Indexing Specialists (UK) Ltd.,
Hove, East Sussex, UK

The Swedish oceanographic research ship, the *Albatross,* made a world-wide voyage in 1947–48. This expedition, led by Professor Hans Pettersson, made many geological and physical observations in the Atlantic, Indian, and Pacific oceans.

The *Calypso* is operated by Commandant Jacques-Yves Cousteau, a naval officer turned oceanographer. Cousteau's most important contribution to oceanography has been his development of the aqualung and its use for research at depths down to 65 fathoms. He has also much improved techniques of underwater photography.

The *Discovery*, completed in 1962, is a research ship built for Britain's National Institute of Oceanography. Displacing 3000 tons and 260 feet long, it has a range of 15,000 miles. The ship will accommodate 20 scientists and a crew of 43, and contains over 3500 square feet of laboratory space. It is powered by a diesel-electric engine. The neutral buoyancy float below is one of the newest instruments developed to measure current speeds.

Dr. William Beebe's steel bathysphere, built in 1929, was the first effective craft for manned dives into deep water. The steel chamber, with walls over an inch thick, was five feet in diameter. Beebe's deepest dive, over half a mile down, was made in 1934 in the Atlantic Ocean near Bermuda.

The modern bathyscaphe is superior to the bathysphere in that it does not have to be attached by cable to a ship at the surface. Pioneered by Professor Auguste Piccard, the bathyscaphe can descend to the deepest parts of the ocean, then rise to the surface by jettisoning weights of lead shot stored beneath its float.

# World Beneath the Oceans

T. F. Gaskell

with paintings by Barry Evans

# World Beneath the Oceans

Nature and Science Library: The Earth
*published for*
The American Museum of Natural History
*by*

The Natural History Press / Garden City, New York

The Natural History Press, publisher for the
American Museum of Natural History, is a division
of Doubleday & Company, Inc. The Press is
directed by an editorial board made up of
members of the staff of both the Museum and
Doubleday. The Natural History Press has its
editorial offices at The American Museum of
Natural History, Central Park West at 79th Street,
New York 24, New York, and its business offices
at 501 Franklin Avenue, Garden City, New York.

Editorial Adviser        Roy A. Gallant
Editor                   Kit Coppard
Designer                 Judy Hannington
Design Consultant        Edwin Taylor
Research                 Nan Russell-Cobb

First published in the United States of America in 1964 by
The Natural History Press, Garden City, New York
in association with Aldus Books Limited

Library of Congress Catalog Card No. 64–10005
© Aldus Books Limited, London, 1964

Printed in Italy by Arnoldo Mondadori, Verona

# Contents

# 1 Explorers of the Deep

When we use the expression "exploration of the seas" it brings to mind the famous navigators of centuries past: men like Captain James Cook, who explored and charted vast stretches of the Pacific Ocean in the 18th century; and Ferdinand Magellan, whose skill and courage led to the first circumnavigation of the globe in the 16th century.

The sea explorers of the 20th century no longer look for undiscovered continents. Instead they probe downward into the oceans, into the sediment carpet covering the sea floor, and into the rocks beneath the sediments. It is this sort of exploration that will provide answers to the great questions of our century: How were the seas formed? What is the Earth made of, and how did it come to be the way it is today? To answer these questions about our planet it is no use probing downward on the land only. Over two thirds of the Earth's surface that lies hidden beneath the oceans must also be explored.

Why is it that we know so little about the sea, and about the land beneath it? You have only to try looking underwater to discover one good reason. In clear water you may be able to see things up to several yards away, and in some places to take remarkably clear undersea photographs, as did Commandant Jacques-Yves Cousteau and his aqualung divers. But you cannot look down over the side of a ship and see the sea bed because the water is not transparent. It is therefore impossible to take aerial photographs of the sea bed, as we can of the land. To make things worse, the surface of the sea is always in motion, so that carrying out experiments at sea is nearly always a tricky business.

It has been only during the last 100 years or more that proper scientific investigations could be made of the sea. Until reliable navigational and other technological aids had been developed, scientists could not hope to get detailed and sound information about the oceans. Obviously it does little good to measure a current in the ocean if your navigator cannot tell you exactly where you are! It is the new inventions that have really allowed oceanographers to forge ahead. Electronic devices now make it possible to record what goes on in the dark ocean depths.

With the echo-sounder we can get an accurate idea of the shape of the sea floor; with other instruments we can learn about the rocks underlying the 1500-foot-thick sediment carpet. Modern drilling methods developed by oil companies to find underwater oil could possibly be employed for the well-known Mohole Project. In this project oceanographers hope to drill a hole through the crust of the Earth in order to obtain a sample of the *mantle*—the material making up more than three quarters of our planet.

Apart from scientific probes designed to tell us more about our planet and its oceans, there are other reasons why we must learn more about the sea. As our planet becomes more and more crowded, we will have to find improved ways of feeding the increasing

Man has had to build special vessels such as the bathyscaphe to protect himself from the hostile environment found at the bottom of the sea. Conventional submarines would be crushed like egg shells if they attempted to explore the deepest trenches, where the pressure is nine tons per square inch and the temperature is near 0° C.

world population. In just the same way that farming of the land has been improved by careful observation and experiment, so the harvest of fish and other products of the sea can be increased. Also, we know far too little about the movements of deep currents to be sure of what is happening to the radioactive wastes and other industrial by-products that are being dumped into the seas. To what extent they are poisoning plant and animal life we cannot say. We also want to know more about waves so that we can design more efficient ships and harbors. The oceanographer's list of inquiries about the seas is endless.

### Early Beliefs about the Sea

For many centuries men were frightened of the sea. As late as the 1800s many intelligent people still believed that the sea was bottomless, except for the fringe of shallow water around the coasts. There was also the peculiar idea that when ships sank they stopped in midwater a mile or so down, and there they remained in a sort of nether region. The explanation was that the ocean became denser and denser with depth because of the great weight of water above. It is quite true that the water does become more dense with depth,

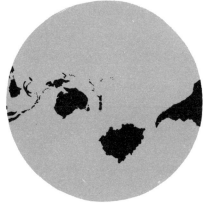

Land masses (above) lie mainly in the northern hemisphere. The oceans (below) lie mainly in the southern hemisphere. In the world of the almost landlocked Babylonians the sea was unimportant; nor did the Greeks adventure much beyond the Mediterranean (see chart); they preferred coastal navigation.

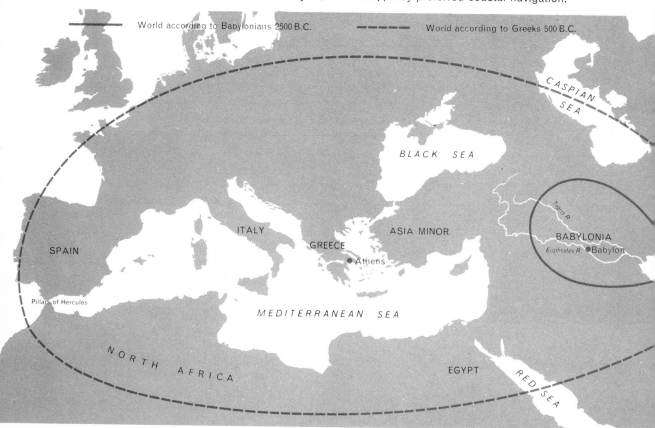

——— World according to Babylonians 2500 B.C.          – – – World according to Greeks 500 B.C.

CASPIAN SEA

BLACK SEA

SPAIN

ITALY

GREECE
• Athens

ASIA MINOR

BABYLONIA
Tigris R.
Euphrates R. • Babylon

Pillars of Hercules

MEDITERRANEAN SEA

NORTH AFRICA

EGYPT

RED SEA

but only by a very small amount. To support an iron ship the density would have to increase by many times its normal value. You can be certain that if you drop anything heavier than water overboard from an ocean liner it will sink to the bottom—as do the skeletons and shells of the animals of the sea, and even the small particles of cosmic dust that rain down from outer space.

The earliest writings about the Earth and its seas came from the eastern part of the Mediterranean a few thousand years before Christ. At that time the shape of the land and the sea was very much the same as it is today. But in those days the known world was only a small part of what we know today. The Babylonians, in about 2500 B.C., saw the world as centered on the great Tigris and Euphrates rivers. The floods that periodically swept down from these rivers, ruining crops and villages, taught these ancient people something about the forces of nature. The Babylonians believed that the universe was like a box, with the Earth as the flat floor on the inside. There was a high snowy region in

This "map" of about 600 B.C. shows Babylon encircled by water—perhaps representing the Mediterranean, Black, and Red seas, and Indian Ocean, about which the Babylonians knew little.

the center of the floor, and it was here that the Euphrates was believed to have its source. All around the land, like a moat encircling a castle, was water; beyond the moat were the celestial mountains that held up the sky.

The Greeks left many maps and legends of what they thought the Earth and its oceans were like. Not until men began to set out and "see for themselves" were some of the stories and legends proved false. One Greek explorer, Pytheas of Massilia, set out around 325 B.C. on a voyage to Britain. There he heard of a peculiar land called "Thule," where the water, the Earth, and the air "all became as one." Thule was actually the icy coast of Norway or Iceland, where mist is often present. To Mediterranean people—accustomed to a sunny climate—it must have appeared that the "air" blended with the Earth and sea.

These early voyages outside the Mediterranean called for great bravery and skill in seamanship. The Greeks, and the Phoenicians before them, did not know if they would ever be able to get back home once they had sailed through the Pillars of Hercules (which we today call the Strait of Gibraltar)—the 20-mile channel between the Mediterranean and the Atlantic—into the unknown sea beyond.

The Greeks believed the world to be encircled by a river controlled by the god Oceanus. He was said to live beyond the Strait and to have power over the waters outside the Mediterranean. He controlled the currents in the sea, which aided ships but which could also make it difficult for them to go forward. The current that flows into the Mediterranean from the Atlantic must have discouraged or prevented many early explorers from entering the Atlantic. Today we know of a second current—a deep one that flows in the opposite direction through the Strait. Submarines find it a nuisance when they are trying to enter the Mediterranean. The ancient Greeks and Phoenicians must also have known about this deep current. From time to time they most likely found wreckage or corpses that had been carried by the current out into the Atlantic.

Poseidon was another of the Greek sea gods. He could beat the sea with his trident to

When plying the Mediterranean trade routes in ships like these (above), Greek sailors invoked protection against storm and shipwreck by making sacrifices to Oceanus (below), the god who controlled the waves and currents.

raise storms and hurricanes, and he could split rocks with such force that he was responsible for earthquakes. To appease him, with the hope of being favored with gentle winds, the Greeks built temples in his honor on dangerous headlands and cliffs. These are the places where today we build lighthouses to guide ships' navigators.

Stories of storms and perils of the sea are found in many ancient writings, and many historians would be pleased if oceanographers could produce some evidence that the stories have some basis in fact. Plato, for instance, described a great empire called Atlantis, which around 9000 B.C. was supposed to have conquered all the known world. The Greeks held out against Atlantis and were saved at the last moment, when Atlantis was supposed to have been engulfed overnight, disappearing without a trace. Legends of other mysterious islands are met with in the folklore of Europe: the Portuguese Antilia (perhaps the equivalent of Atlantis) and the Greek Isles of the Blest.

It is always difficult to know how much

faith to put in old legends of this sort. The catastrophes that we see today, when earthquakes and tidal waves destroy whole towns, could easily have been exaggerated by imaginative storytellers. The biggest exaggerations are usually made about things that are not understood, and the great oceans have always been a source of wonder and awe.

Some of the best tales of the sea are those about sea "monsters." In the *Just So Stories* by Rudyard Kipling, King Solomon wanted to feed all the animals in the world in a single day. He had all the food ready when one animal appeared from the sea and ate the lot. The animal explained ". . . I am the smallest of thirty thousand brothers, and our home is at the bottom of the sea. . . . Where I come from we eat twice as much as that between meals."

Surely the whale must have appeared to be a "monster" to men who saw one for the first time. And what evidence do we have that there are not other large animals that live permanently at great depths? There are many stories about giant "sea serpents." Norsemen voyaging to Iceland and Greenland told of sea serpents rising out of the ocean to snap off mouthfuls of men on the deck. It may be that these "sea serpents" were really giant squid. Although their bodies are only 12 to 14 feet long, their paired tentacles can exceed 40 feet in length and may make snakelike movements. Some evidence also exists that giant eels may be found at great depths, but proof must await the invention of gear that will catch large creatures in deep water.

### How the Oceans were Formed

To understand how the ocean waters were formed, and how the sea floor has changed and shifted through the ages, we must go back 4500 million years to the very beginnings of our planet. There are many theories that try to explain how the Earth was formed, and one day we hope to be able to say for certain which one is correct. Most of the theories, however, agree that our planet began its life as a great globe of gas and cosmic dust that eventually condensed into a solid mass. As

This section of the Carta Marina published in 1539 by the Swedish priest Olaus Magnus shows the Norwegian coast and part of the northern Atlantic. Sea monsters that decorate many maps of this period were probably based on a mixture of legends, old sailors' tales, and the mapmakers' imaginations.

15

the hot gases cooled, they first became a molten liquid; next a separation of the materials making up the liquid took place. The heaviest material, which we think is mostly iron, sank toward the center of the planet and formed the core, which has two parts: the inner solid core and the outer liquid core. The radius of the whole core is about 2150 miles. Around the core is a wrapping of rock called the *mantle*. This is approximately 1800 miles deep, and it forms the bulk of the Earth. The third and last layer, called the *crust*, forms the surface of our planet and is a very thin skin with an average depth of about 20 miles.

The crust was the first solid layer of the Earth to be formed. It floated on top of the molten mantle rock, like the skin on a pan of hot sauce. Then, as the underlying mantle cooled, great masses of steam were expelled from the solidifying crustal rock. This steam, together with other gases, formed the atmosphere. As soon as the Earth's surface cooled enough to support water without boiling it back into the atmosphere, rain fell and began filling the great basins that eventually became the oceans.

If the Earth's cooling process had been as simple as we have described, there would have been a smooth outer skin of crustal rock with a uniform covering of sea above it. For some reason the continental (granitic) rocks clustered in groups, leaving deep gulfs in between. One possibility that has been suggested to explain this clustering is the following: In the early stages of cooling, after the crust had solidified, a large piece of the Earth broke off and became the Moon. This could have been caused by the attraction of the Sun, which at that time must have raised great tides in the liquid inside the Earth.

The Moon, in this theory, was thought to have been torn out of what is today the Pacific Ocean basin. Since the crust by this time was solid, and was left with a great piece torn out of it, it could not again flow evenly to cover the newly exposed mantle. Once a large area of the mantle was laid bare it, too, would solidify. Thus the Earth would be divided into areas of continental rock with large spaces between, the spaces serving as

containers to receive the water from the atmosphere when it condensed and formed the oceans.

During the millions of years of the Earth's history, vast changes have taken place both on the land and on the sea floor. It may be that the continents—originally grouped as one great land mass, or in an arrangement different from the one we know today—are "wanderers." Evidence today suggests that they are. The skin of crustal rock has wrinkled into mountain ranges and deep valleys. As we shall see in the following chapter, an impressive array of mountains and valleys also stretches over the sea floor.

The reason why the world is not completely covered with mountain ranges after thousands of millions of years of wrinkling is that only the newer mountains remain today. The old ones have been worn down to their roots, like the teeth of an old horse. Frost splits the rock, glaciers grind it up, and rain sweeps the debris down to the rivers, which carry it out to sea. There the ground-up rock is deposited, forming new layers of rock—called sedimentary rocks. Sometimes the sediments are soft, like clay, at other times they set like cement and form limestone and sandstone. Whether they are hard or soft, they are all called *sedimentary* rocks, and generally they are spread out in flat layers. In the course of time great geological changes twist, bend, and fold these layers, producing the complex rock strata that can be seen in the mountains on land. Because this process goes on continually, the Earth's crust is reworked over and over again on land and in the shallow parts of the sea alike.

## Diving to the Bottom

Man's first glimpses of the ocean floor were modest ones compared with what we are able to "see" today. The early divers of centuries past searched the shallow coastal sea floors for food, for items that could be traded, and sometimes just out of curiosity.

When archaeologists dig down layer by layer through the mound-remains of settlements where people lived many thousands of

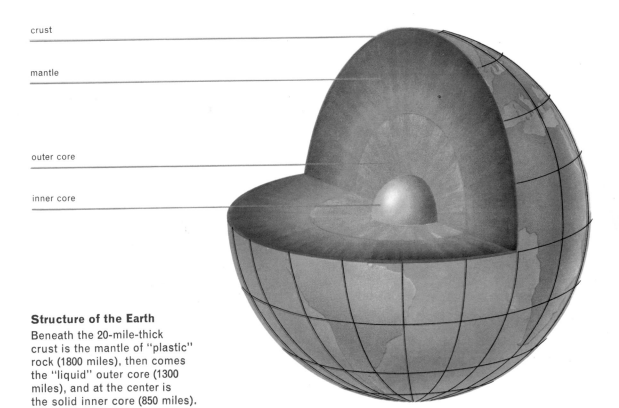

crust

mantle

outer core

inner core

## Structure of the Earth

Beneath the 20-mile-thick
crust is the mantle of "plastic"
rock (1800 miles), then comes
the "liquid" outer core (1300
miles), and at the center is
the solid inner core (850 miles).

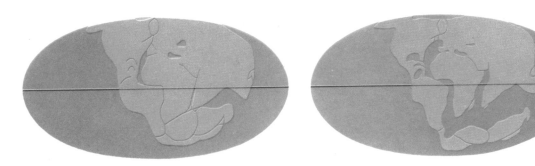

Many geologists believe that the Earth's
continents were originally grouped as one huge
land mass (above) and that they gradually
"wandered" to their present positions.

This greatly simplified profile (below) of the
Earth's surface along the equator shows that
the sea floor, like the land, consists of
mountains, valleys, and plains.

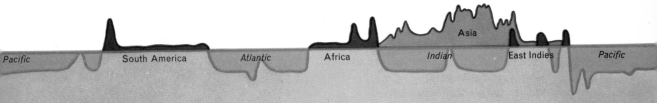

Pacific     South America     Atlantic     Africa     Indian     Asia     East Indies     Pacific

years ago, they often find great piles of oyster shells—proof that early men depended on the sea. And we know that many thousands of years later the Phoenicians discovered that a certain shellfish (*Murex*) produced a secretion that could be used as a deep purple dye. Sponges and pearls were also regarded as valuable trade products, as they are today; and in some parts of the world diving for them has changed little over the years. In the Persian Gulf, for instance, men still collect pearls by diving without any special equipment, just as they did in the Mediterranean in 1000 B.C. They are carried to the bottom by a stone attached to a rope, and when they have filled their basket with oysters they signal to be hauled up with their load.

Because a man cannot hold his breath and swim around for much more than two minutes, some artificial means of feeding air to the diver had to be found if his work on the bottom required that he stayed down for any length of time. One of the earliest devices was the diving bell. You can demonstrate how the diving bell works by holding a glass upside down and pushing it into a bucket of water. Although the water is forced a little way up into the glass, a good supply of air is left above it.

Alexander the Great, who developed an interest in marine biology while he was a pupil of Aristotle, is said to have explored the sea floor in a diving bell that permitted him to gaze out on the underwater world. Since that time large, steel diving bells have been designed. Capable of holding several men, they have been used for salvage and construction work. Because the men inside gradually use up the oxygen in the original supply of air trapped within the bell, a pipe has to be connected to compressors that provide a constant supply of fresh air. At best, diving bells are clumsy and heavy affairs and are not used much for underwater work now. They have been replaced by a variety of equipment that allows divers to move freely over the sea bed—whether they are working on a bridge foundation, salvaging an old wreck, or surveying a submerged archaeological site.

Man has experimented with diving equipment for over 2000 years. One of the first to explore the sea floor was Alexander the Great (above), who used a simple diving bell. An 18th-century improvement (below), used for salvaging, took down its own supply of air in a barrel.

In one of these 17th-century inventions pressure of water on the diver's body would prevent him from breathing, even though the suit was connected to the surface by a tube. In the forerunner of the aqualung (below) the diver's head was inside an air-bag; within limits it is a workable device.

Like the diving bell, diving suits have their roots in ancient history. We now believe that divers were used during the siege of the Phoenician city of Tyre by Alexander the Great to destroy underwater defenses built to stop attacking ships. These divers were early versions of the frogmen whose mission it was to destroy underwater obstacles during World War II. It is possible that some of the divers engaged in warfare during Greek times had a simplified form of diving suit, for there are reports of their using long breathing tubes held at the surface by floats. Leonardo da Vinci, the inventive genius of the Renaissance, provided his frogmen with a skin bladder filled with air to breathe while they sabotaged an enemy ship from below.

The principles of these two devices—feeding a diver air through a pipe from the surface, or giving him a container of air that he can take down with him—form the basis of all our diving equipment today, whether aqualung, snorkel, or modern diving suit.

The modern diving suit was invented by an Englishman named Augustus Siebe in 1819. It consists of a watertight suit and the familiar metal helmet with strong glass windows. Air—a mixture of oxygen and nitrogen—is fed to the diver through a hose connected to a pump located in a boat, and can be regulated by valves so that the diver may breathe comfortably as he works. Heavy boots keep him upright while heavy weights on his chest and back prevent him from floating up to the surface. With the Siebe diving suit men can go down to about 600 feet.

Although divers have gone much deeper than this, deep diving is very difficult; and even the most experienced diver must take great care when he is brought up to the surface. The reasons for this caution are not hard to understand. At a depth of 600 feet, say, the air pumped to the diver is under great pressure—about 16 times the pressure of the air we breathe normally. This high pressure causes some of the nitrogen in the diver's air supply to dissolve in his blood, and this is where the trouble begins. If the diver comes up to the surface too quickly, the nitrogen that has entered his bloodstream bubbles out in

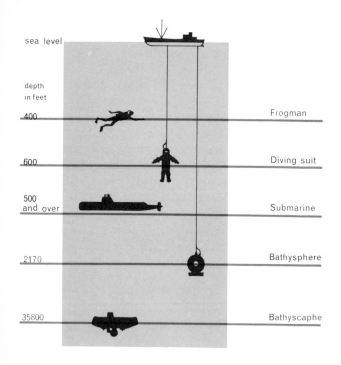

| | |
|---|---|
| sea level | |
| depth in feet | |
| 400 | Frogman |
| 600 | Diving suit |
| 500 and over | Submarine |
| 2170 | Bathysphere |
| 35800 | Bathyscaphe |

Above are maximum depths
at which various kinds of
diving apparatus have been
used. Aqualungs (below) are
suitable only at 300 feet or
less but allow the diver
to move about freely.

much the same way that the carbon dioxide gas in a bottle of fizzy lemonade bubbles out of the liquid when you lower the pressure by removing the cap. Instead of diffusing harmlessly through the lungs, the dissolved nitrogen bubbles out into the diver's joints and bone marrow, causing agonizing pain known as the "bends." In extreme cases bubbles form in the heart and brain; these are usually fatal. After a long dive, a diver must be brought back to normal pressure slowly, to enable the blood to give up nitrogen through the lungs. Nitrogen under pressure also causes a form of drunkenness. In very deep dives a mixture of helium and oxygen avoids this danger, while the low density helium is easier to breathe when highly compressed. Even oxygen is poisonous at high pressures, but we can avoid disaster by reducing the percentage of oxygen in the gas mixture.

## The Birth of Modern Oceanography

There are scattered around the world today more than 500 marine research stations, many of which have their own ships that put to sea for months at a time to study one or more aspects of the oceans. While much of our knowledge of the oceans has come about since World War II, when a variety of new equipment made it possible to probe the oceans with new tools, a lot of valuable work had been done before the age of steam.

As early as 1673 the English chemist Robert Boyle had taken an interest in the saltness of the seas. He realized then that salinity in any part of the ocean depends on the amount of evaporation and rainfall—the more evaporation, the greater the salinity, as in the Red Sea. Temperature measurements were more difficult to obtain in the early days. The system used by George Forster on Captain Cook's voyage in the late 1700s was to lower a wooden bucket to the desired depth, close a set of valves in order to trap the deep-water sample inside the bucket, then raise it to the surface and immediately record the temperature of the water sample. Because the temperature of the trapped sample was bound to change by at least a small amount on the way

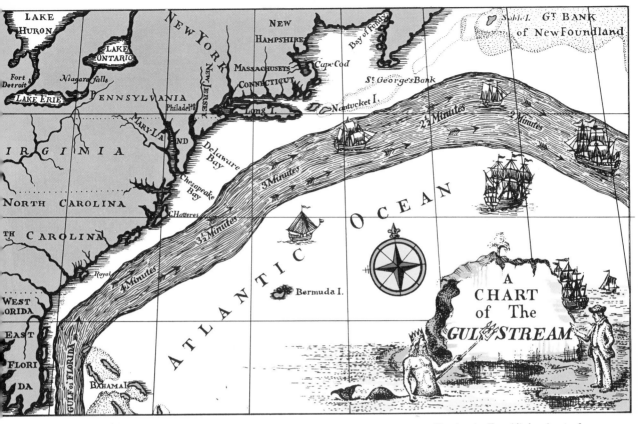

Benjamin Franklin's chart of
the Gulf Stream, published in
1770, helped mariners avoid
the eastward-flowing current
on the Atlantic run from
Europe to North America.

up, it was impossible to get an accurate reading this way. Later expeditions by the British ships H.M.S. *Lightning* (1858) and H.M.S. *Porcupine* (1859), using improved methods, were able to report temperature changes that suggested that the deep oceans were not quiet and still, but that they circulated as did the surface waters.

In 1770 Benjamin Franklin published the first chart of the Gulf Stream, and during the following century the American naval officer Matthew Maury made a detailed study of winds and currents. It was these studies that led to our present-day meteorological offices. Maury also played an active part in obtaining information for the laying of an underwater telegraph cable linking Europe and America.

There were many people who said such an idea was absurd. Some of them believed that the cable would be destroyed by sea monsters; others claimed that it would disappear into bottomless chasms. Maury encouraged systematic soundings of the North Atlantic, and eventually a chart of the ocean bed showing where the cable could be laid was prepared. Although the first attempt to lay the cable failed (when the cable broke at a depth of 2000 fathoms and could not be recovered), the second attempt in 1866 was successful.

The laying of this transatlantic cable stimulated new interest in oceanography. Six years after it was completed the most famous oceanographic expedition of them all set out —the three-year voyage around the world of

the British ship H.M.S. *Challenger*. The *Challenger* expedition made over two hundred soundings, including a record sounding of 4500 fathoms (which was made only 50 miles from what is now known to be the deepest part of the ocean). They also collected an enormous number of samples of material from the sea bed, and of animals both from the sea floor and from various levels of the oceans. During her voyage the *Challenger* probably amassed more new data on all the aspects of the sea than had been collected up to the time of the voyage.

Today there are many oceanographic research ships from many countries exploring the deep oceans. One of the most useful special experiments (now being carried out) is the drilling of holes through to the bottom of the soft sediments coating the floors of all the oceans. These sediments have been raining down to the sea floor ever since the oceans were formed. Core samples of the complete thickness of the sediment carpets of all the oceans will reveal the history of the oceans in a more direct way than can any other of our oceanographic tools.

The oceanographer's interests are far-reaching: He wants to know more about the motion of currents found at the surface and at great depths; about the shape of the sea floor; about the temperature and salinity of the water; about plant and animal life in the seas; about tides, waves, and the forces that generate them. Let us now turn to these subjects and find out how much we know about the world beneath the oceans.

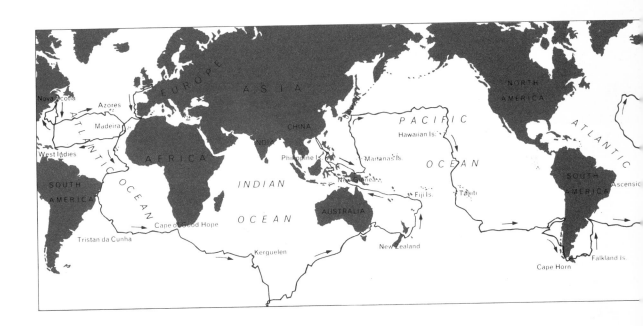

Left: A dredge is landed on the deck of H.M.S. *Challenger*. In the deep parts of the oceans, lowering and raising the dredge took many hours. Above the dredge is an "accumulator"—a spring made of rubber bands—to prevent the ship's movement from straining the dredge line.

The voyage of the *Challenger*, whose route is shown on this map, began in 1872 and continued for over three years. The crew made 250 depth soundings, recorded water temperatures at different depths, and collected animal, plant, and mineral samples from the sea bed.

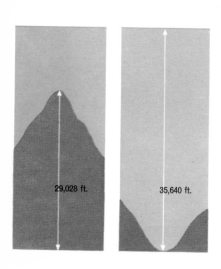

29,028 ft.

35,640 ft.

The ocean floor (above) has scenery just as dramatic as any found on land. The abyssal plains that comprise the true bottom are interspersed with towering volcanic ranges, flat-topped mountains, steep-sided canyons, and trenches that may plunge to depths of several miles. Left: Mount Everest would fit comfortably into the Mariana Trench without breaking the surface of the Pacific Ocean. (Vertical scale of trench exaggerated.)

# 2 Journey to the Bottom

What can we expect to find on the floor of the oceans? For many years people thought the sea floor was flat and featureless, except where the peaks of underwater mountains rose above the surface to form islands. Today we know that most of the ocean floor is between 2800 and 3000 fathoms deep. Since six feet equal one fathom, the *average* depth works out to about 17,500 feet, or more than three miles.

But sometimes, as we will see later, great undersea trenches are cut into the ocean floor. The greatest known depth yet discovered is in the Pacific Ocean, and it is called the Challenger Deep in the Mariana Trench. Mount Everest, which is more than five miles high, would fit into this trench without its peak breaking the surface of the ocean.

Imagine that we are standing on the floor of the Pacific Ocean with all of the water drained away. Around us we would see beautiful mountains rising sharply from the ocean bed, like the tidy volcanic peaks seen in

Japanese prints. Some of the mountain tops would be like crowns, with coral rock taking the place of the jewels, and most of them would be at exactly the same height, for they are atolls that rise only a few feet above sea level. There would be many other flat-topped peaks of lesser height, just as if a great knife had been used to slice off the tops. Among the flat-topped mountains would be volcanic cones of all sizes, some smoking, showing that they were still growing by means of volcanic eruptions.

The world map on page 28 shows how the ocean floor would look if all the water were taken away. Many of the undersea volcanoes lie along the top of a very gentle rise in the sea bed. Some of the mountains on the ocean floor are very large and form great extended ranges like the Mid-Atlantic Ridge, which runs down the middle of the Atlantic Ocean—from Iceland in the north to below Cape Town in the Southern Hemisphere. It is as if someone had attempted to build a wall to separate Europe and Africa on one side from North and South America on the other. This wall continues round into the Indian Ocean and extends below Australia to the Pacific Ocean, where it is called the East Pacific Rise. The East Pacific Rise finally ends near the great mountain chain that runs down the West Coast of America.

In the same way that the Earth's crust has been wrinkled to form mountains on land, so it has been bent and buckled to break up the flat oceanic floor. In the Pacific there are several great tears in the ocean floor, where the Earth's crust has given way to forces greater than it could stand. These tears are like underwater cliffs, or escarpments, and they stretch out for hundreds of miles from the California coast. So we see that although

a large part of the ocean floor is roughly flat at an average depth of about 17,500 feet, it is by no means like a dull, flat desert all over. The sea floor is really the base on which the Earth has been built up—a base that has been split, bent, and pushed up and down. Looking up from this base, we would see the continents as great towering plateaus that take up only about a quarter of the surface of the Earth. This is the best way to think of the Earth today. The ocean floors form the real surface of the Earth, with islands and continents rising up from them. This view is a change from bygone days when everyone regarded the land as being the most "important" part of the Earth and the oceans as merely the uninteresting and unknown gaps in between.

## Mapping the True Bottom

We cannot see very far into the oceans, and obviously we cannot drain all the water away for a closer look. How, then, do we know so much about what is hidden by two to three miles of water? Because the surface of the sea is fairly level, if we continuously measure the depth of the ocean beneath the surface we should be able to draw a profile of the sea floor.

In days past the method of sounding the depth was simply to lower a lead weight on a line and feel the slackening of the pull on the line when the weight hit bottom. This heaving-the-lead method was used in times of old when ships were nearing dangerous shoals, and two hundred fathoms was the longest sounding line that was normally carried, but this would not reach anywhere near the bottom of the oceans. During the 1872 *Challenger* expedition a special grass-

granitic rock      ocean      basaltic rock      granitic rock

This sectional diagram shows that the floor of the ocean, consisting mainly of basaltic material, forms the real surface of the Earth. The continental land masses (mainly granitic) rest upon the basalt and occupy only about a quarter of the Earth's total surface.

rope line was constructed in order to sound several miles. But the weight of several miles of rope is so great that the material chosen for the rope had to be both strong and light. Even so, the rope was as thick as your finger —one inch in circumference—and the taking of a sounding was a major operation. This is how it was described by the engineering officer of H.M.S. *Challenger*:

"On commencing the operation of sounding, the weighted sounding rod, the water-bottle, and the thermometers are suspended from the line, and lowered from the sounding-bridge by reversing the engine for 500 fathoms; the line is then let go and allowed to run out freely. As it runs out the exact time of each 100-fathom mark entering the water is registered, and set down in its appropriate column in a book provided for that purpose. These intervals gradually increase in duration as more line is paid out, the weight having to overcome the friction of the line in the water, which becomes greater with the amount run out. They will, however, be found to increase in regular proportion, so that when four minutes are taken up by one interval, the weights have reached the bottom, or a depth of between 2000 or 3000 fathoms has been obtained."

No wonder they made only 250 deep soundings during the whole three-year voyage! The weight and the sample bottle and the thermometers all had to he hauled laboriously inboard when bottom had been reached, and this took several more hours. Today the echo-sounder can take a sounding in a few seconds—the time it takes a sound wave to travel down from the ship to the sea bed and bounce back again. Sound waves travel in water at about a mile a second, so that in a three-mile depth of ocean the sound

Soundings with a weighted rope or wire have been used to determine depth of water since classical times. In mapping the floor of the oceans this laborious method has been replaced by the quicker and more accurate echo-sounder that measures water depth by bouncing sound signals off the sea bottom and automatically records its findings on a tracing.

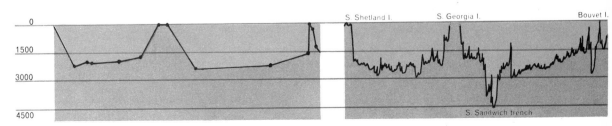

Comparison of wire and echo soundings of the same area of the South Atlantic Ocean floor. Above: A profile made from 13 wire soundings. Right: A much more detailed profile based on 1300 echo soundings. In both profiles the vertical scale is greatly exaggerated.

World map of the ocean floor. From the coast of each land mass the continental shelf (light green) slopes gradually to a depth of 50 to 100 fathoms. In some regions, such as the California coast, the shelf is narrow; in others, such as many coasts in Southeast Asia and the eastern coast of America, it stretches out for more than 100 miles. At the edge of the shelf is the continental slope where the sea floor is steeper, gaining depth at the rate of about one foot to every 15 feet. The base of this slope—where it meets the floor of the ocean—is the true edge of the continent. The Atlantic Ridge, a vast undersea mountain range, runs down the center of the Atlantic Ocean, turns eastward into the Indian Ocean, continues south of Australia and across the southern and eastern Pacific Ocean.

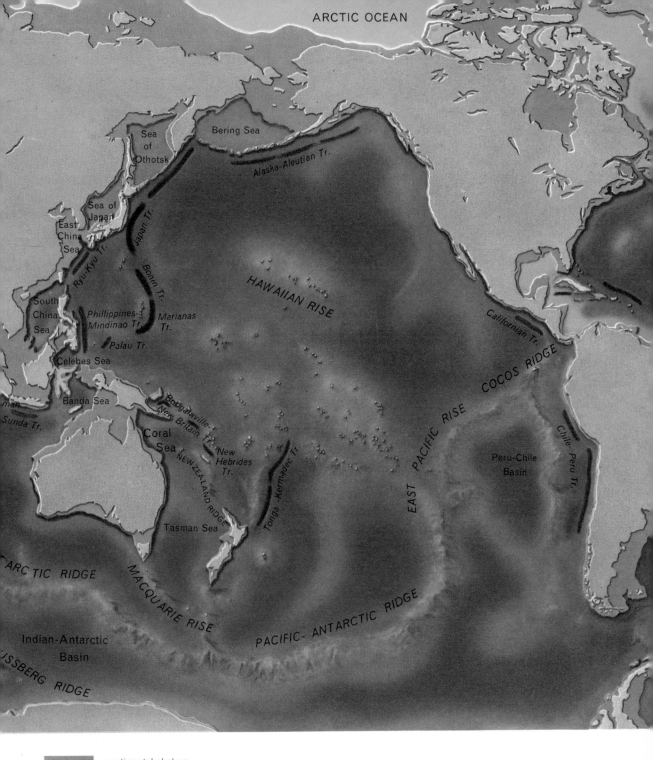

ARCTIC OCEAN

Sea of Othotsk

Bering Sea

Alaska-Aleutian Tr.

Sea of Japan

East China Sea

Japan Tr.

HAWAIIAN RISE

Ryu-Kyu Tr.

Bonin Tr.

South China Sea

Phillippines-Mindinao Tr.

Marianas Tr.

Palau Tr.

Celebes Sea

Californian Tr.

COCOS RIDGE

man

Banda Sea

Bougainville Tr.

New Britain Tr.

EAST PACIFIC RISE

Chile-Peru Tr.

Sunda Tr.

Coral Sea

New Hebrides Tr.

Tonga-Kermadec Tr.

Peru-Chile Basin

NEW ZEALAND RIDGE

Tasman Sea

ARCTIC RIDGE

MACQUARIE RISE

PACIFIC- ANTARCTIC RIDGE

Indian-Antarctic Basin

ISSBERG RIDGE

continental shelves

mid-oceanic ridges

shallower basins

abyssal plains

trenches

and echo take only about six seconds. It is not necessary to wait for one sound wave to arrive back at the ship before sending out another pulse of sound. Pulses can be sent out every second and usually at any one time several pulses are traveling through the water, some on their way down to the sea bed, some on their way back. So the echo-sounder can provide a continuous record of the depth of water, and can draw out profiles of the sea bed showing the exact shape of underwater volcanoes and undulating hills and valleys.

The echo-sounder consists essentially of a source of sound signals originating from the ship's hull. Some early models had a hammer that struck the bottom of the ship and sent out a noise. But modern echo-sounders transmit pulses of ultrasonic waves by means of an electrical oscillator. The instant a sound pulse is sent out, a zero mark is made by a pen on a moving paper chart. When the echo comes back another mark is made. As it writes on the chart, the pen sweeps slowly across the paper. If it is a deep echo, which takes a long time to return to the ship, the pen moves a long way over the chart; if it is a shallow echo, the pen makes its mark early on the chart. Thus the pen sweeping across the paper records a shape representing the sea bed beneath the ship. In the past, when lead soundings were taken only occasionally in deep water, it was impossible to draw a proper map of the sea bed, as it would be impossible to make a map of Switzerland if we knew only the heights of Geneva, Mont Blanc, and Grindelwald.

Mapping the sea bed is the work of the hydrographic departments of the navies of the world. The hydrographic departments are always at work, checking old charts, mapping shifting sandbanks (like the famous Goodwin Sands in the Strait of Dover), and making

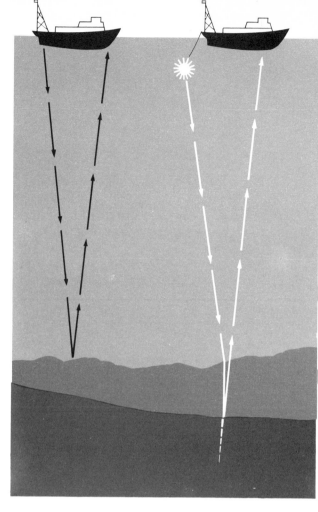

The echo-sounder signal (left, above) provides an accurate record of the shape of the sea bed. The sparker (right) emits an underwater bang that is loud enough to penetrate the soft mud on the bottom and reach the underlying rock.

The cut-away diagram below (much exaggerated in the vertical scale) shows some typical features of the ocean floor. Many underwater volcanoes never reach the surface of the sea. Others, like the guyot, sink below the surface, their tops worn flat by waves.

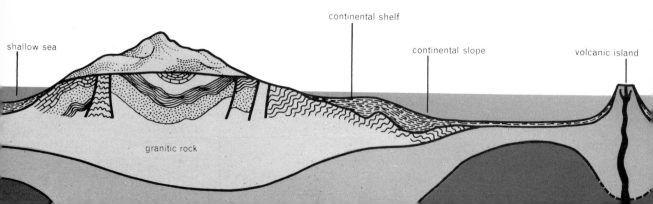

shallow sea

continental shelf

continental slope

volcanic island

granitic rock

lists of all the lights, beacons, and other marks that help the navigator. In order to draw an accurate chart of the oceans it is necessary to know as nearly as possible the exact position where soundings have been taken. At sea, however, the only fixed points to measure from are the stars, so the navigator must keep a continuous record of the ship's changing position by taking star sights and by measuring the ship's speed and direction. Whenever the ocean surveyor wishes to make a careful study of a particularly interesting section of the sea bed, he puts out a marker buoy. The buoy is anchored to the bottom of the sea by a fine line, and it will remain in position for several days—provided a storm does not blow up. The ship can then steam around the buoy making soundings and plotting its position by watching the buoy on radar.

Although there are many survey and oceanographic ships steaming over the oceans, there is an enormous amount of deep sounding that still remains to be done. In fact only about one tenth of the total area of the oceans has thus far been sounded. As you might expect, the relatively shallow continental shelf around the land is better known than most other parts of the ocean. The continental shelf is the gently sloping submerged border of the land shown colored green on the chart on page 28. These shelves usually extend out to an average of about 40 miles. In some parts of the world, such as along the California coast, the shelf is narrow. But along the Atlantic coast of America, for instance, the shelf stretches out to more than a hundred miles. The slope of the shelf is very gradual, dropping about one foot in every 500 feet out. At the edge of the shelf the sea bed is by no means smooth or flat. Soundings tell us that great valleys and steep-sided canyons are cut into the shelves.

We shall see how these canyons have been formed when we consider the sediments of the sea bed in the next chapter.

The change from shelf to deep ocean—that is, from a depth of 50 or 100 fathoms to 2000 or 3000 fathoms—takes place in a short distance. The average steepness of this continental slope is a one foot drop in every 15 feet out. The continental slope seems to be the true edge of a continent, and its base is where the continent joins the real floor of the ocean. Mud and silt, which is endlessly carried down to the sea by rivers, flows out over the shelf and continues down the slope and out onto the sea floor. Water movement on the shallow shelves is enough to keep the mud stirred up. It is only in the more peaceful deep ocean that the particles of mud can sometimes come to rest. But even in the deeper water the mud sometimes flows as great rivers and spreads into enormous pools that cover many parts of the ocean floor.

## Echoing through the Mud

Dr. J. B. Hersey, a physicist of the Woods Hole Oceanographic Institution in Massachusetts, is one of the oceanographers who study these flat pools of sediment on the sea bed. They are called *abyssal plains* because on a chart they appear as large, flat areas hundreds of miles across. At one time it was supposed that they formed the true floor of the ocean, but echo sounding has given us quite a different picture. Hersey, using an echo-sounder that can record changes in depth of one fathom, even where the water is 3000 fathoms deep, verified that abyssal plains are not quite flat. Sometimes they slope very gently in a direction away from the land, and sometimes they slope toward the center of the plain. This is just what we would expect if the plains had been formed

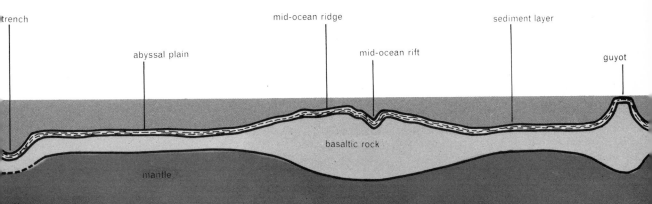

trench

abyssal plain

mid-ocean ridge

mid-ocean rift

sediment layer

guyot

basaltic rock

mantle

An island is born as an erupting volcano breaks surface 150 miles off the Pacific coast of Japan.

by mud pouring down the continental slope and filling up hollows in the sea bed.

The soundings made by Hersey also showed that abyssal plains are not all at the same depth. In some cases the mud from one abyssal plain has flowed downhill and onto another plain at a lower level. A detailed chart of such an area resembles a map of a lake with a river running out of it and flowing into a second lake.

Dr. Hersey has made another kind of echo-sounder to help him in his deep-sea work. This one is called the *sparker* because it makes a loud bang underwater by means of an electric spark. It is just as if a sparking plug from a car were worked underwater; in fact the very first sparker used was an ordinary car spark plug. The sound from the underwater spark is strong enough to go through some of the soft mud covering the sea bed. When the sparker is used to examine the edges of abyssal plains, the sound waves reveal the hard rock lying beneath the sediment-mud. In some cases the mud itself shows up in layers, caused by separate flows of mud rivers on the ocean bed.

There are some flat plains in the Pacific that are different from abyssal plains. These often occur around volcanic islands that erupt and spill lava-flows into the surrounding sea. Sometimes, however, lava pours out from great cracks in the sea bed itself, so that a flat plain is formed where there is no island nearby. These plains are being studied by Dr. W. Menard and other geologists from the Scripps Institution of Oceanography at La Jolla, California, in order to learn more about the shape of the Pacific sea bed. In the Indian Ocean the sea bed is flat for 600 miles to the southeast of Ceylon. During the next few years oceanographers will be making a determined effort to learn more about the Indian Ocean; only a few expeditions have been there so far. They may find that the long, flat plain making up the ocean floor has something to do with the great lava flows of southern India.

We now know that volcanoes are responsible for much of the underwater scenery in the oceans. Some volcanoes keep on erupting and building themselves higher year after year until finally they break the surface as islands. Because so much lava must be erupted before an island grows three miles upward from the ocean floor, many underwater volcanoes never manage to reach the surface. But these peaks can be plotted by the echo-sounder. Frequently a new underwater peak is found by an oceanographic expedition in some part of the world. Gradually we are collecting enough information to allow us to draw complete charts of the oceans.

One of the geologists who has learned a great deal about the ocean with the help of the echo-sounder is Professor H. Hess of Princeton, New Jersey. In the Second World War, Hess was a naval officer and made many journeys across the Pacific. During long hours on watch he kept an eye on the echo-sounder, even though the ship was in deep water and in no danger of running onto a reef or a sandbank. Every now and then the echo-sounder pen would draw a neat little underwater volcano—with its peak maybe a mile below the ship. Sometimes a different kind of underwater island appeared on the record—an island with a flat top. These flat-topped sea mountains are called *guyots*. Hess named them after Arnold Guyot, a famous 19th-century geologist.

Guyots look like volcanoes that have had their tops chopped off—and this is just what has happened. At one stage of their growth these odd-looking peaks must have reached up to the sea surface as proper islands. But for some reason they sank, and since they sank very slowly—over thousands of years—the sea gradually wore the tops away. The sinking continued; so today we find these flat tops deep down below the surface. The reason for the sinking is one of the geological puzzles that oceanographers are trying to solve today—as we shall see later on, when we describe what is hidden beneath the sea floor.

Most of the underwater volcanoes and guyots that we discover with the echo-sounder grew millions of years ago. We know how they grew because sometimes a new island is born out of the sea even today. A few years ago smoke was reported in the Philippine Sea to the south of Japan. After days of violent eruption an island appeared above the sea, remaining only for a short time because the soft volcanic ash was soon washed away by waves. During the eruption a tragic accident occurred: A Japanese oceanographic ship sent to explore the new island disappeared. The volcano must have erupted and swamped the ship when the oceanographers were almost on top of the island.

While mountain ranges and chains of volcanic islands rise up from the sea bed in some places, in others great winding valleys, called *trenches*, have been cut into the ocean floor, particularly around the eastern edge of the Pacific Ocean. These trenches are about 50 miles wide, and some of them go down to a

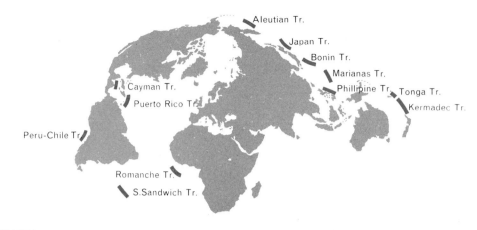

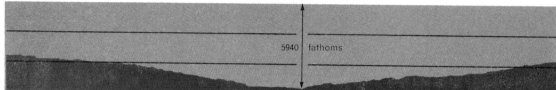

5940 | fathoms

Positions of the main oceanic trenches are shown on the map above. The profile of the Challenger Deep (below map) shows how even the largest trenches slope quite gradually to their maximum depths.

depth of nearly six miles—about twice the average depth of the ocean floor. The deepest known trenches run southward along the east of Japan and the Mariana Islands, then running west toward the Philippine Islands. There are others south of the equator next to the Kermadec and Tonga islands. The relatively narrow width of these trenches, if considered in relation to the Earth's circumference of 25,000 miles, shows that they are merely folds in the Earth's crust. As you sail over one of these deep trenches, with the echo-sounder recorder going all the time, you find that the slope of the sea floor is fairly gentle at first. Then quite gradually the side of the underwater valley becomes steeper, sloping like a mountainous road. In some parts there are steep cliffs, and in other places it is almost flat. But the average slope continues down to nearly 5000 fathoms, where the valley rounds off at the bottom.

The Challenger Deep in the Mariana Trench is not called after the famous ship of that name. In 1951 a new survey vessel, H.M.S. *Challenger*, working at about 11° north of the equator, was the first to record the great depth of 5940 fms. Since then, this

depth has been confirmed by the Russian research ship *Vityaz* and by United States expeditions. On board an oceanographic ship you would realize the great depth if you watched the echo-sounder: It takes $14\frac{1}{2}$ seconds for the signal to travel to the bottom of the valley and back again. To verify the echo-sounder reading, oceanographers on board the *Challenger* lowered a piece of iron weighing 140 pounds on a steel wire until it touched bottom. It took an hour and a half!

What forces within the Earth cause the trenches? They are probably formed because the crust of the Earth is being squeezed constantly—just as folds appear in the carpet when you push it from one end with your foot. But another kind of deep-ocean valley has been discovered recently, and it seems to have been formed in just the opposite way —by a tearing-apart of the earth's crust. Thus sometimes the Earth's crust is pushed and folded, and at other times it is split apart.

All down the middle of the Mid-Atlantic Ridge—that great underwater mountain range that forms a wall in the center of the Atlantic Ocean—there is a steep-sided valley. This valley has been tracked by continually

A photograph taken during underwater **location** work on the film *20,000 Leagues under the Sea.* The use of aqualungs is rarely **possible** at depths greater than 300 feet.

sounding over it with an echo-machine. In some places the valley appears to be blocked. This is what we should expect because we know there are active volcanoes in the Mid-Atlantic Ridge, and these will sometimes fill the valley with lava. Then again, we know that earthquakes will cause landslides that may block the valley. However, there are signs of splitting all the way down the Atlantic Ocean. Sometimes the splits stop and new ones appear a little to one side, just like cracks and splits in a piece of wood. But they all follow the grain of the mountains—which is a winding path down the ocean bed. Why is this great crack occurring in the crust? It seems to be a sign that the continents on either side of the Atlantic Ocean are moving apart. We do not know whether this is true, but as more facts are collected we gradually begin to understand what is happening. Then we try to think of convincing reasons why it should happen. Although man has contemplated the oceans for a long time, it is only recently that new tools—like the echo-sounder and sparker—have begun to tell us what actually goes on three miles under the sea surface.

## Filming the Depths

One way of looking at the sea bed is to take photographs of it; but this has not been done until recently, simply because it is very difficult to use a camera under three miles of water. Dr. A. S. Laughton, of the British National Institute of Oceanography, is among those oceanographers who have made a camera that works in deep water. The pictures he has taken are not only very useful in telling us what goes on at the sea floor, but some of them are also strikingly beautiful.

The first thing that must be done to take deep-sea photographs is to make a strong metal box with a thick glass window. The camera, with its delicate lenses and sensitive film, must be kept dry and protected from the very great water pressure. It is also necessary to have a flash unit because there is total darkness three miles below the surface. And to obtain clear pictures, the camera must be in focus. Laughton arranges his camera so that it always takes its pictures when it is at a fixed distance above the sea bed. The camera with its flash unit is lowered on a strong wire line. Hanging about 20 feet beneath the camera is a heavy weight. The wire is payed out, and when the camera is exactly the right distance from the bottom to be in focus, the weight hits the bottom and triggers a switch that fires the flash, and the camera takes a photograph. The film in the camera automatically winds on and is ready for the next picture, which is taken simply by raising the camera and weight off the bottom and then lowering it again.

Sometimes many interesting things appear in undersea photographs. They show us that the deep ocean floor is not all dull and flat, and that it is by no means only a silent grave-yard of things that have sunk to the bottom. The earliest photographs showed ripple marks in the sand of the sea bed, which meant that there must be movement of the water along the bottom. Other pictures showed tracks of animals that had been crawling about on the sea floor. Occasionally photographs show plant life growing, and once or twice a fish has been caught by the camera, swimming comfortably at a depth of more than a mile, where the pressure is over 2000 pounds per square inch.

It is usually possible to determine from the photographs whether the sea bed is made of mud or rock. The camera thus helps translate some of the changes that appear on the echo-sounder recorder. In some cases, however, even the photographs cannot be read proper-ly, but they do tell us when it seems to be worth the effort to lower a dredge and bring a sample up from the sea bed. So all the different instruments at the disposal of the oceanographer are used in concert. In some canyons, for example, the echo-sounder does not record an echo if the side of the canyon is too steep. In fact, even a small slope gives a confusing picture on the recorder, because the sound waves tend to bounce back at right angles to the slope. If a steep cliff is suspected from the results of the echo-sounder, then a check can be made by taking photographs.

Commandant Jacques-Yves Cousteau has mounted a camera on a skid that he tows along the sea bed, taking photographs every few seconds. He has been able to learn a great deal about the rifts in the Mid-Atlantic Ridge

in this way from the bathyscaphe. The next development will no doubt be a deep-water television camera. Television is being used in shallow and deep water to inspect wrecks and watch the work of divers.

The chief difficulty with underwater cameras and television lies in the fact that sea water is not very clear. Thus it is almost im-possible to take pictures of large areas of the ocean in the same way that we take aerial photographs of the land. It seems unlikely that we shall ever be able to take sea-bed photographs from thousands of feet up. But it may be possible to improve the echo-sounder, so that we can scan the sea bed with moving beams of sound waves and in this way obtain a sort of television view of the sea floor. In the meantime, the close-ups that we obtain with underwater cameras are teaching us a great deal.

**Exploring the Deep**

It is not a long step from photography to direct exploration of the bottom. The only trouble is that human beings are about as

fragile as the lenses of our cameras, and we have to be protected by pressure-tight enclosures if we wish to go to the bottom of the oceans. In 1930 a pressure-tight bathysphere with windows was designed and constructed by Otis Barton, who invited Professor William Beebe to participate in the dives, since he, Barton, knew nothing about marine life. In 1934 the bathysphere descended more than half a mile below the surface, and this achievement showed the practical possibilities of underwater investigation by means of this sort.

The steel sphere was lowered on a cable from a ship at the surface, which meant that it could not be maneuvered. This meant also that it was impossible to follow interesting-looking animals or even to be certain of looking in a desired direction. This is why free-moving underwater craft have been attracting the attention of oceanographers. Although it may be difficult to make a deep-ocean submarine, there is so much to be gained from having one that the effort is worthwhile.

Professor Auguste Piccard was another pioneer of the deep water. In 20 years' time the bathyscaphe (from the Greek, meaning "deep boat") that Piccard designed will probably be regarded as a primitive machine, but it has already shown the way to the deepest part of the ocean. Three bathyscaphes have been built, and it is the third one, constructed in 1953 by Piccard himself, that dived to the deepest point on earth, seven miles down in the Pacific Ocean. On January 23, 1960, this bathyscaphe, named *Trieste*, and owned by the United States Navy, descended to a record depth of 35,800 feet, to the bottom of the deep valley in the Mariana Trench off Guam—the famous Challenger Deep. Piloted by Piccard's son, Jacques, and Lieutenant Donald Walsh of the United States Navy, the *Trieste* took four hours and 48 minutes to

Left: Dr. A. S. Laughton using the camera he designed for photographing the deep-sea bed.
Right: A cut-away diagram of the bathysphere.

main cable

communications hose

switch box

barometer

humidity recorder

air blower

hatch door

oxygen tank

oxygen tank

air purifier

central observation window

telephone and battery

searchlight

descend and three hours and 17 minutes to return to the surface.

The bathyscaphe consists of two main parts. One is the sphere carrying the crew. This has to resist the full pressure of the depths, while the crew inside breathe air at atmospheric pressure. The $6\frac{1}{2}$-foot-wide sphere is made of $3\frac{1}{2}$-inch cast steel and is fitted with conical windows of lucite, held tightly in position by external water pressure. The rest of the bathyscaphe—the hull made of thin plating 0.16 inch thick—carries batteries, ballast, motors, and searchlights, but its main function is to provide buoyancy like a balloon. Gasoline, a liquid lighter than sea water, is contained in the hull tanks. Gasoline has the advantage over air in being only slightly compressible. Air buoyancy tanks would collapse under pressure if they were not open to the sea on their underside, and if they *were* open the pressure would reduce the air's volume progressively so that the deeper the craft dived the less buoyancy it would have. Of course a much greater volume of gasoline than air is needed to provide the necessary buoyancy, in fact 500 times as much. (That is why the hull is so large in proportion to the cramped sphere with barely room for two men plus instruments.) To prevent straining the thin hull plates, these tanks are open at the bottom to the sea, so that the pressure inside and out is identical.

The operation of the bathyscaphe is ingenious but simple. It starts by being lighter than water so that the crew can embark and be sealed into the sphere. At this stage there is air in the air-lock shaft leading to the entrance hatch. Lead-shot ballast has been placed in position, the air lock is flooded with water and the craft sinks. When the crew want to surface, the ballast is dropped and the "lift" of the gasoline brings the bathyscaphe up.

There is also the possibility of making the equivalent of an underwater airplane. This waterplane would be constructed to drive itself down through the water by means of propellers and wings just as an airplane raises itself through the atmosphere. This would be as substantial a development as that of airplanes from balloons.

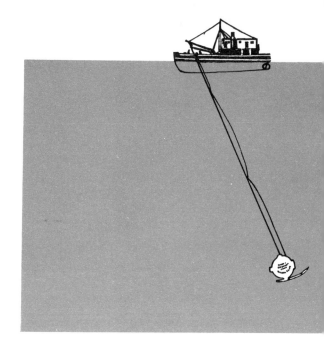

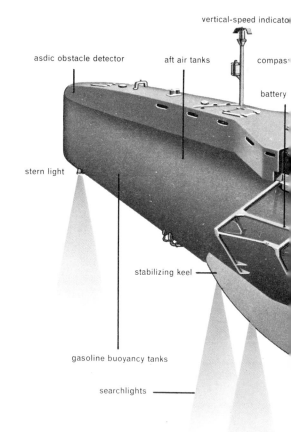

vertical-speed indicator

asdic obstacle detector

aft air tanks

compass

battery

stern light

stabilizing keel

gasoline buoyancy tanks

searchlights

Even the most efficient submarine, such as the *Aluminaut* now being built of aluminum in the United States, especially for deep diving, needs powerful searchlights. In clear water, the human eye can just detect faint light at about 1500 feet down; below this the depths of the sea are in total darkness. Vision is therefore limited by the length and strength of the beam thrown by the searchlights. Turbulence may also obscure the view of the rocks, animals, and the sea bed.

Left: The bathysphere hung from a cable attached to a surface vessel and could not be maneuvered about.

The free-moving bathyscaphe invented by Prof. Auguste Piccard in the 1930s overcame the limitations of the bathysphere. The French *FRNS3* (below) dived to 13,125 feet in 1953; the *Trieste* piloted by Piccard's son reached the bottom of the Challenger Deep in 1960; it is very similar to the *FRNS3*. The forward and aft air tanks give extra buoyancy when the craft is being towed; they flood with sea water when the bathyscaphe submerges.

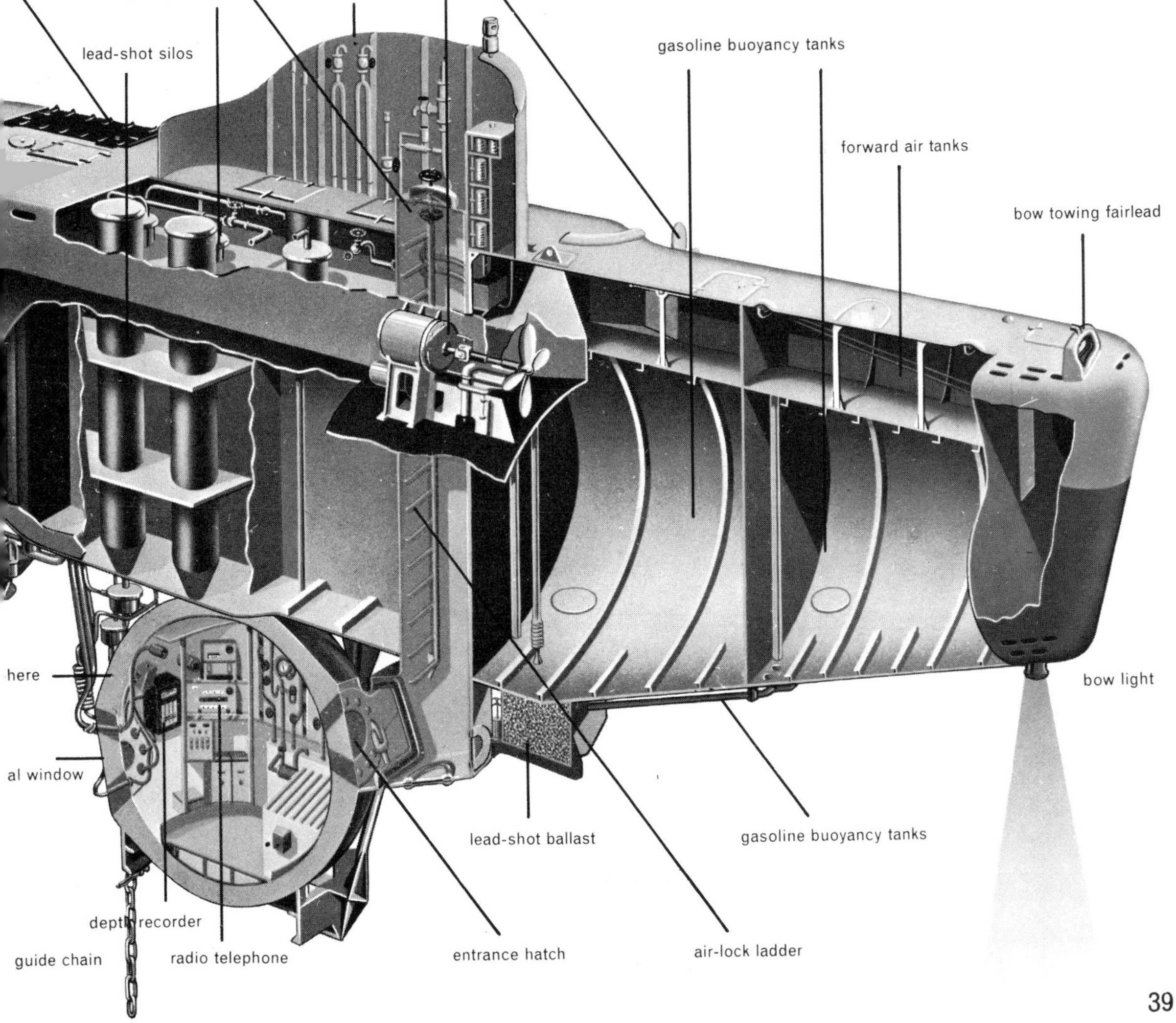

air-lock entrance

electric motors

detachable gasoline tanks

conning tower

lead-shot silos

gasoline buoyancy tanks

forward air tanks

bow towing fairlead

here

al window

lead-shot ballast

gasoline buoyancy tanks

bow light

depth recorder

guide chain

radio telephone

entrance hatch

air-lock ladder

# 3   The Carpet of Sediments

If you keep a tank of tropical fish, you probably find it necessary to clean out the carpet of fine rubbish that collects on the gravel at the bottom every few weeks. The rubbish is sediment. On a much larger scale sediment rains down from the sea onto the sea bed, providing a layer of soft mud that contains many things of interest to the oceanographer. If we scrape up some of the deep-sea sediment we may find old sharks' teeth, the skeletons of microscopic animals, roundish nodules of manganese, or ash from volcanic eruptions that took place many centuries ago.

The oceans have been collecting their rubbish for hundreds of millions of years, until there is now a layer of mud and other materials averaging 1000 feet thick on the sea floor. This layer contains a wide variety of relics from the past, and from them we can read back page by page into our planet's geological history.

## The Rain of Sediment

Where has all the sediment come from? To answer this question we must first under-

stand what is constantly going on around us on land. You must have noticed how muddy a river becomes after heavy rain. The mud has been washed into the river by small streams that are gradually eating away the soil. The soil in its turn was formed by the action of the weather on rocks, helped by the acids produced by decaying plants and by the strong, burrowing roots of plants and trees. Part of the weathering is caused by frost. When water contained in the rocks freezes and expands, it acts as a wedge that breaks the rocks apart. In the mountains,

The floor of the oceans is overlaid with a carpet of sediment that averages over 1000 feet in thickness. Much of the sediment—sand, mud, and clay—comes from the land. The painting above shows how sedimentary material, carried off the face of the land by rivers, flows down the continental shelf, spills over the steeper continental slope and is deposited on the ocean floor. Here the slope is gentler but may be just enough to carry the sediment from one abyssal plain to another deeper one farther out. The perspective of the picture is deliberately distorted; a line drawn from the coast to the abyssal plain toward the bottom of the picture would represent a true distance of perhaps 250 miles.

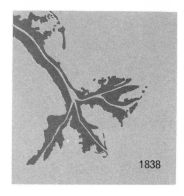

1838

Mississippi Delta

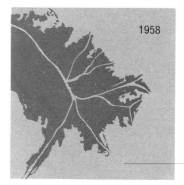

1958

Some land-derived sediments never reach the ocean floor but build deltas at the mouths of rivers. The two diagrams above show (top) the delta of the Mississippi in 1838 and (below) its growth since that time. The photograph at left is a recent view of the same area.

where temperatures are below zero for much of the time, snow packs down to form great glacial rivers of ice that grind their way down valleys, scouring off soil and rock along the way. Although we cannot see these glaciers moving, we can prove that they are sliding downhill all the time. In the Antarctic and Greenland great icebergs are split off from the glaciers during the summer months and fall into the sea.

All this wearing down of the rocks is called *erosion*. Although it takes place very slowly —so that the shape of the land is not noticeably changed from one year to the next—it has been going on for millions of years. Sometimes, however, we can see quite sudden changes. After a storm, when the waves have been beating against a soft cliff, large sections of the cliff may suddenly slip into the sea, sometimes carrying houses with them; or

large pieces of a river bank may crack off and slide into the water.

The Grand Canyon in the United States offers the most spectacular display of erosion visible to us. At the bottom of the canyon is a raging river that is eating the rock away at the rate of about half a million tons a day. In the course of millions of years the river has carved out the canyon to a depth of a mile and to a width of about 10 miles at the top.

Rivers flowing rapidly down the mountains often roll great boulders along. When their flow becomes slower—when they reach a flat plain or by the time they flow into the sea —most of the pebbles and large sand grains sink to the bottom. For the most part only fine sand and mud are left suspended in the water, and even much of this material has settled out by the time the river water mixes with the water of the ocean.

If you look at any detailed map of a coast you can see that many of the large rivers build *deltas* at their mouths. These deltas are really new land being built up. So the sediments that have been eroded away from the land and washed down to the sea are not "lost." Had this been so, all the continents would long ago have been worn down flat so that the whole Earth would be one vast sea. The sediments that are deposited at river mouths and offshore gradually form new land. Sedimentary rocks have been formed in this way—first by erosion and then by the pressure exerted by further layers of sediment. Sand grains form sandstone, particles of soft mud form clay, and where large numbers of dead animal shells are found chalk and limestone are formed on the floor of the sea.

After millions of years of erosion the layers of clay, sandstone, and chalk that have been formed in shallow water by the material brought down by rivers may be heaved up by great forces inside the Earth so that they become dry land, sometimes in the form of mountains. In the Persian Gulf today the rivers are gradually filling up the northern end of the Gulf. In the last 5000 years the ancient port of Ur of the Chaldees has been left more than 100 miles inland. The rocks that contain the great oil deposits of Persia were formed when an ancient gulf was filled in from 10 to 100 million years ago. This old gulf went on sinking until it collected a thickness of nearly seven miles of limestone, sandstone, salt, and clay. Then forces within the Earth—which we do not fully understand—squeezed these rocks up into mountains, and in the folds of some of them oil and gas were trapped.

This process of change from one kind of rock to another has been going on throughout the 4500 million years of the Earth's life. Most of the land that we see above water today has at one time or another been under water, where it has gathered new layers of rock, but not often has it been part of the *deep* ocean floor. It has merely been under shallow water a few hundred feet deep, like the present floor of the North Sea and the

shallow bottom linking Alaska and Siberia. These layers of new rock extend to the edge of the continental shelves, so it is not surprising that our echo-sounder records show deep underwater canyons that were cut by rivers at a time when the edge of the shelf was in one of its above-water periods.

The deep oceans, where the water extends three miles to the bottom, have never been filled up by sediments. Over the years they have been collecting only a few sedimentary crumbs from the continental tables while the process of erosion and deposition has been going on. In the deep oceans there is no rain or frost to break up the bottom rocks, so we can expect to find a different account of the Earth's history in the deep ocean sediments than on or near the land. However, as we shall see later, there are disturbances even three miles under the sea, so the evidence that nature has left must be sorted carefully.

The sand and mud that the rivers bring down to the sea may build out from the land and form a delta that will one day become new land. Sometimes the tides and waves agitate the delta mud and carry it seaward as far as the edge of the continental shelf. Over the years it accumulates on the shelf's edge until one day it breaks away and tumbles down the steep slope as an underwater landslide. Although these *turbidity currents*, as they are called, deposit some sediment on the deep ocean floor near the edge of the continents, most of the continental rubbish remains near the land. Some sediment, however, is carried far out into the deep oceans by currents. It consists of very small particles of red clay about one ten-thousandth of an inch across. The clay has been worn from the continents by the process of erosion and is stained red by small traces of iron that it has brought with it. Once in the grip of the currents, the clay sediment can be carried to all parts of the oceans. And because the particles are so fine, it may take many years for them to fall down through the three miles of water before they reach the bottom.

Red clay was first discovered by scientists of the *Challenger* expedition of 1872, when they dredged up samples from the deep sea

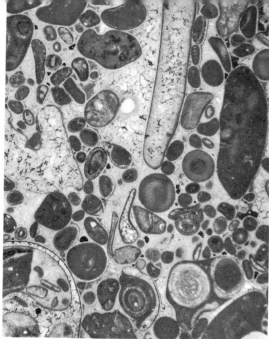

In many parts of the ocean the skeletons of dead animals in the plankton sink to the bottom and form an important part of the sediment layers. The engraving (left) is of Globigerina, a common planktonic animal. Embedded in the section of sediment above are a number of fossilized Globigerina skeletons. Made of calcium carbonate, the skeletons will eventually help to form a chalk or limestone layer in the sediment.

bottom. This red clay collects very slowly, less than an inch in thousands of years. It underlies the areas of the oceans where the great eddies occur and these areas seem to be less productive of plant life. Red clay contains less organic matter and seems to support less animal life than do other sediments.

The continents, then, provide the sea bed with at least a portion of its sediment carpet. But the fine clay particles are not the only ones that rain slowly down to cover the sea floor. Animal and plant remains also contribute to form sediments and are called "oozes." Associated with red clay, mainly in the equatorial region of the Pacific Ocean is a wide belt of Radiolarian ooze. This is formed by the skeletons of dead unicellular animals, the Radiolaria, which are part of the floating population of the sea, the *plankton*. Their microscopic skeletons of hard silica sink into the depths, when the animals die, to

form one of the sediments derived from the plankton. Another sediment formed in the same way is known as Globigerina ooze. This covers about 45 per cent of the ocean floor and consists of the calcium carbonate skeletons of another class of unicellular animals, the Foraminifera. The third kind of sediment, the Diatom ooze, comes from the plant population of the plankton. Diatoms are microscopic plants found in great abundance in the sea. They have (like the Radiolaria) skeletons of silica.

Of course the hard shells, skeletons, and bodies of the larger animals also contribute to the contents of the sediments and all this dead material is gradually broken up by the action of bacteria. In good garden soil, the bacteria are of great importance because they decompose plant and animal remains. In the sea, and on the sea floor and in the oozes, they act in just the same way.

The photograph above was taken with a deep-sea camera at a depth of about 1700 fathoms. It shows a small area of sediment on the floor of the Atlantic Ocean about 350 miles off the west coast of Spain. The well-defined and symmetrical ripples formed on the surface of the sediment, together with the sand drifts that are piled against the sides of the small boulders, show that regular movement of the water occurs even at this depth.

In addition to the clay sediments washed away from the land, and the animal and plant sediments originating in the sea, there is a third source of supply. The atmosphere plays a part in the accumulation of material on the sea floor. Any fine dust floating in the air will eventually fall to Earth. If it falls on the sea it will gradually sink until it joins with the other ocean sediments. Part of the dust starts from the land, blown to sea by strong winds, and sometimes sucked up by atomic explosions. Occasionally nature provides volcanic explosions that throw fine ash high into the atmosphere. In 1883, when the Krakatoa volcano blew a crater three and a half miles across in the sea near Java, so much ash was thrown into the atmosphere that especially colorful sunsets were seen for months afterward. In the end, however, the volcanic ash finds its way to the sea bed with all the other material.

There is yet another source of sedimentary material—one that has nothing to do with our planet. This is a gift from outer space and reaches us in the form of meteorites. Some meteorites are large enough to produce a streak of light as they heat up on entering the atmosphere. Others are still larger and make gigantic craters when they strike the Earth with explosive force. But apart from these large meteorites there is a constant rain of tiny meteorites that arrive unseen at the Earth's surface. Because three quarters of our planet is covered by sea, the greater part of this meteorite dust ends up in the oceans. Each year several million tons of this material finds its way to the sea floor. The oceanographer has ways of measuring how old it is, so that the dust from outer space can become a clock that registers the age of the ocean sediments. It is especially useful for fixing the age of the deeper sedimentary layers.

## Sampling the Bottom

Since the material making up the carpet covering the sea floor originates from so many different places, the deep-sea sediments contain many clues that can help us in our study of the Earth's history. When the day comes when we are able to examine the whole thickness of the sediment carpet, we will learn many things. From the mud deposits we will come to understand more about the erosion of the continents in the past. The animal and plant sediments will help us piece together the story of the growth of life in the sea. The volcanic ash and meteorite dust will tell us when the Earth had periods of particularly violent volcanic activity, and, quite possibly, some surprising things about outer space as well. But first we have to collect the samples and then find a way to sort out the different kinds of sediment so that each one can tell its own story. Before we see how this is done there are, unfortunately, two things that must be considered because they tend to disturb the evidence provided by the carpet of sediment. If we understand how these disturbances work, however, we may turn them to our advantage.

Canyons like this (above) in the continental slope may be cut by turbidity currents carrying river-borne sediments from the continental shelf to the ocean floor. Below: This deep-sea dredge is used by oceanographers to collect sediment samples.

When ships drag a dredge along the floor of the northern Atlantic Ocean they often bring to the surface a haul of some rock as well as soft sediment. You might regard this as a sign that the top layer of sea bed is made of hard rock. The rocks certainly could not have been carried hundreds of miles from land by sluggish ocean currents. At first sight, then, they seem to have been formed where they are found. But careful examination of the rocks shows that they have scratches on them. To the geologist this means that they have been dragged along by a glacier.

Immediately this supplies the clue to the whole history of *erratic* boulders. These rocks did not originate on the sea bed, but have been carried out from land into deep water by icebergs. As the ice gradually melted, the stones were released and dropped to the bottom of the sea. Once we realize this method of rafting land material by floating ice, we can follow the extent of the movement of icebergs in past ages. Although it is a nuisance to have the sediment layer confused by *erratics*, as they are called, we can use the observations if we are careful.

The technique of dredging that oceanographers use today is similar to that used by the *Challenger* in the last century. A heavy bag made of chain mail, fitted with a strong steel mouth, is dragged very slowly behind the ship by a strong cable. We can tell when the dredge is sliding along the sea bed by feeling the taut wire line, because the line shudders and vibrates as the dredge jumps along the bottom. But it is not easy to do this; often the dredge planes along in the water just when we want to collect a sample.

Hersey, of Woods Hole, has fitted an echo-sounder to the wire a little way ahead of his dredge. This means that from the ship's deck he is given a record of when the wire is riding at just the right depth for the dredge to work properly.

Another modern improvement over the old method is to take photographs with the underwater camera to find out what a particular section of the sea bed is like. If it looks promising, then the dredge can be lowered. In the Puerto Rico Trench photographs showed that hard rock cliffs formed the slopes of the underwater valley. By careful dredging it was possible to break off a piece of rock; this was probably the first sample of the Earth's crust to be gathered from beneath the ocean.

Although we can take precautions and avoid the mix-up of rocks that has been caused by icebergs, there is another form of mixing that is more difficult to unravel. Since it plays a major part in shaping the sea floor it is the subject of a great deal of oceanographic investigation.

We once regarded the floor of the deep oceans as a quiet place, with a snowfall of dust and shells floating gently down from above, and perhaps with a few unusual fishes and burrowing animals silently going about their business. The picture was not marred by any great disturbance of the sediment. In the sediments we expected to be able to read the Earth's history page after page in an orderly way, starting from the time when the oceans were formed. We know that many areas of the sea floor are far from being the quiet places we once imagined. Rushing torrents of mud mixed with water sweep down over parts of

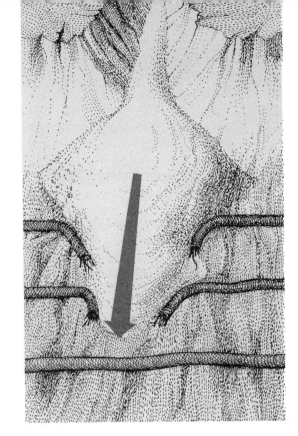

In 1929 turbidity currents set off by earthquakes snapped a series of transatlantic telegraph cables on the continental slope south of Newfoundland. From the time interval between the break in each cable it was shown that the currents were traveling at about 50 miles an hour on the steepest gradients.

the sea floor, fluttering the pages of evidence that are written in the sediments. At the same time these torrents interleave the sediments with coarse material from the continents.

The turbidity currents do more to mix up the sediments than any other process we know of. They are formed by mud, which sinks because it is heavier than water. Where the sea bed slopes, mud that has accumulated near the edge may start to flow downhill, just as a river flows. If the slopes are steep and long, the mud gradually gathers speed until it moves at 50 miles an hour or more. Once it has started, the force of the current sweeps up older bottom sediment, and the torrent continues to rush downward. Gradually it slows down, cutting a river bed along the way, and when its energy is waning it ends up by flowing onto one of the abyssal plains.

It may seem strange that a stream of mud can move through water at such high speeds. But think for a moment of what happens when avalanches occur on land. Those that rush down a mountainside at more than a 100 miles an hour are caused by snow mixing with air; the whole mass of material can be thought of as "floating" down the mountain slope. In a similar way the mud mixes with water in the turbidity currents. Although we have not been able to sit at the bottom of the sea and watch a mud torrent rush down a continental slope, we have similar things on land that make us believe that turbidity currents can be regarded as undersea avalanches.

One very good piece of evidence in favor of turbidity currents has come to light through damage done to underwater telegraph cables. These cables run along the floor of the oceans and are in constant use. If one of them is broken the operator knows the exact time when the damage occurred. Every now and then a whole group of cables breaks on the same day. Dr. Bruce Heezen of the Lamont Geological Observatory, Columbia University, New York, studied a group of these broken cables. He discovered that they had been snapped in a definite order—those in shallower water first, then those in deeper water later—as if someone had drawn a giant knife across the sea bed. There were several hours between the first break and the last. When the distances between breaks and the time of each break were known, the speed of the turbidity current along the cable path could be worked out. It had traveled at about 50 miles an hour on the steep slopes, then had slowed down to 12 miles an hour on the flatter sea bed. Each cable that lay in its path had been snapped as if it had been a piece of string.

If turbidity currents begin at the top edge of the continental slope, they should take with them a large amount of sediment that belonged to the continents. When Heezen sampled some parts of the Atlantic, where turbidity currents were suspected, he found

This corer is used to obtain sediment samples. The corer, weighted down by rings (by the man's hand), is forced into the sea floor by gravity or the use of an explosive, and the sediment enters the hollow tube at the bottom.

that the sediment was not just plain red clay or Globigerina ooze. There were alternating bands of sand and silt, together with fossils formed from shallow-water animals.

If you did not believe in turbidity currents you could say that great lumps of sediment simply break off the edge of the continental shelf and tumble in boulder fashion down the continental slope to the ocean floor. No doubt this does happen sometimes, but it does not explain the long distance that the sediment travels along the bed of the ocean—from one

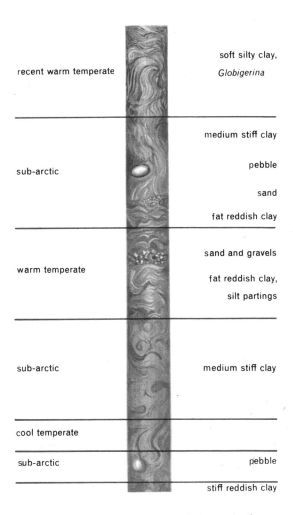

| | |
|---|---|
| recent warm temperate | soft silty clay, *Globigerina* |
| sub-arctic | medium stiff clay |
| | pebble |
| | sand |
| | fat reddish clay |
| warm temperate | sand and gravels |
| | fat reddish clay, silt partings |
| sub-arctic | medium stiff clay |
| cool temperate | |
| sub-arctic | pebble |
| | stiff reddish clay |

Diagram of a core taken at 2250 fathoms in the Atlantic. On the right are details of the sediment layers; on the left are deductions made about the climate in this region at the times the layers were deposited.

flat abyssal plain, through underwater river beds to a lower abyssal plain. When we examine these abyssal plains with the sparker, we find that there are several layers of different material—just as we would expect if every now and then a new load of sediment had been spread over the sea bed by our fast underwater rivers. The turbidity currents often follow the canyons that are cut in the sides of the continental shelf and they clean these canyons out. It is possible that the turbidity currents also cut valleys of their own in the softer parts of the shelf.

Turbidity currents are the second of the disturbances we mentioned earlier, simply because they mix the continental sediments with the rain of small particles originating in the oceans themselves. Furthermore, those sediments that form in layers on the sides of the underwater mountain ranges and volcanoes also slide down as turbidity currents. So even when we go well away from the continents we may find a jumbled-up mass of sediments instead of an orderly series of layers.

Dr. J. D. H. Wiseman, of the British Museum of Natural History, is one of the geologists who is tracing the past story of the Earth by examining its sediments. He has studied cores from the bed of the ocean to determine the exact dates of the ice ages, and to follow the changes that have taken place in the temperature of the sea. He believes that there are only a few places where really undisturbed sediments will be found. One of these is the flat tops of those seamounts out of reach of turbidity currents. The trouble is that many of the smaller flat-topped hills or mountains are not nearly as old as the oceans. They are volcanoes that have erupted from the sea bed fairly recently in geological time. They are not, then, covered by the entire thickness of sediments that would tell the complete life story of the seas.

How do we discover that the sea-bed sediments are mixed with sands brought down by turbidity currents? To learn this we must go deeper than the few inches of scraping that the dredge can manage. We must take *cores*. The simplest way to take a core is to push a hollow tube as deeply into the soft sea bed as we can. Some of the earliest cores from the deep oceans were obtained by Dr. Charles Piggot, who made an underwater gun. The gun was lowered to the sea bed on a wire, with its muzzle pointing downward. When the muzzle touched bottom a charge at the top of the gun fired automatically and the whole gun barrel was forced into the sediment. It was then pulled back to the surface with a barrel full of core, about 10 feet long. The core was pushed out of the gun by means of a ramrod, so that it could be examined in the laboratory.

It did not take long before oceanographers realized that a heavy lead weight was as good as an explosion to drive the tube into the soft sediments. Today many cores are collected with the "free-fall" corer, a long steel tube about two inches in diameter with a 500-pound lead weight attached to the top. The corer is lowered by a wire from a winch on board the ship. To make the corer fall quickly so it will go deep into the sediment a large weight is hung about 100 feet below the device. When the weight touches bottom it trips a switch that lets the corer fall freely. The corer is still connected by a spare loop of wire, so it can be pulled back to the surface with its sediment sample.

If you have ever tried to push a tube deeply into the sand at a beach you know that it goes in easily to start with, but after a short distance it becomes stubborn because the sand becomes packed ahead of the tube. To force it in deeper you have to twist the tube into the ground. One way of getting deep cores was invented by Dr. Börje Kullenberg of Sweden, who collected cores from all over the world during the famous *Albatross* expedition in 1947. Kullenberg fitted his corer with a piston that sucked the sediment core up into the tube. This improvement has enabled oceanographers to bring up cores 50 and 60 feet long from the deep oceans. Once the cores are safe on deck, they are sealed at both ends and carried back to the laboratory where they are sliced down the middle for examination and testing.

## Reading the Sediments

Sometimes the cores show thin layers of ash, which could mean that a volcano had erupted nearby. Sometimes they show layers of sand brought down by turbidity currents. Even when the core looks much the same all the way along, careful analysis will tell a story. We can, for instance, use a special instrument to measure the ratio of light oxygen atoms to heavy oxygen atoms in a core sample. This tells us whether the sea was hot or cold at the time that a particular sediment layer was formed. Water contains atoms of both heavy and light oxygen. When it evaporates it loses more light atoms than heavy ones. So when we find a large amount of heavy oxygen in a particular layer of the sediment core, it means that the sea was warm at the time the sediment was settling to the bottom.

Another of the elements, carbon, provides oceanographers with a natural clock for telling how old the sediments are. Throughout the years there is a constant rain of radioactive carbon from the upper atmosphere. A definite amount of this isotope—known as Carbon-14—is present in the carbon dioxide of the air. When carbon dioxide is dissolved in the oceans it eventually becomes part of the shell of a sea animal. The absorption of Carbon-14 stops when the animal dies and that contained in the shell at once begins to decay into ordinary carbon. Therefore, the older a fragment of shell is, the smaller will be the amount of Carbon-14 it contains. If we carefully measure the amount of Carbon-14 locked up in a shell fragment brought up in a core sample, we can calculate the age of the sediments as far back as 30,000 years.

But this is only a small fraction of the age of the oceans. Even so, it is a useful piece of information because it means that we can see how fast the sediments are collecting. There are other natural clocks that tell us the ages of the sediments over millions of years. They are based on the atoms of lead and uranium or of argon and potassium. Once we can be sure of the real ages of the sediments we can then work back and trace the history of the oceans and try to answer some of our questions. Have the oceans always been where they are now? Was the climate enormously different during certain periods of the Earth's life? When did life begin on the Earth?

We are still in the early stages of learning to tell time with our isotope clocks, but gradually we are making progress. In the meantime the geologist looks at the sediment cores under a microscope. Small shells that appear are rather like those we collect at the shore. They are *fossils*, ancient remains of animals that are valuable time markers for the geologist. A staggering variety of animals has developed since the first life appeared on our planet hundreds of millions of years ago. The animals we see around us today look much like their ancestors; if we look back over many generations we find small differences. Sometimes there is an extra twist in a curly shell. It is not difficult to make a collection of all sorts of animals that have changed their shapes in the course of millions of years. Once we have a record of the changes, we can then place our core-sample fossils in time.

On page 44 we have already mentioned the Foraminifera which form the Globigerina ooze. "Forams" serve as labels to our sediments. For example, the very ancient forams are made of fragments of shell "cemented" together to form a protective casing for the small living animal. Later on—millions of years later—the animals made smooth shells for themselves, just as large shellfish do today. We can trace the development of the different kinds of forams throughout the millions of years of the past. Among them are some that apparently could not adapt to their environment, and therefore died out only a short time after they had evolved into their particular shape. Because the geologist knows to which geological age these forams belonged, we can date the age of the sediments containing them. We can also work out whether the sediments have been churned up, or whether they have been formed by a gradual settling of ocean rubbish from above.

We find other useful fossils in the cores as well—among them the *coccoliths*, minute disks of shell. Because the coccoliths help to tell the same story as the Foraminifera, we

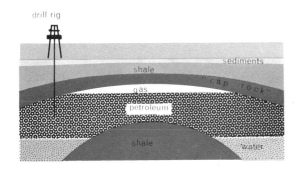

Oil deposits are often found at considerable depths and the techniques used in drilling for oil are helping oceanographers to develop improved methods of obtaining cores from deep in the Earth's crust and, possibly, in the mantle. The 1900-ton drilling platform shown above is located five miles off the Arabian coast in the Persian Gulf and can drill to a depth of 7000 feet. The diagram (left) shows one of the many kinds of "traps" in the structure of the crustal rock that are known to favor the development of oil and gas deposits.

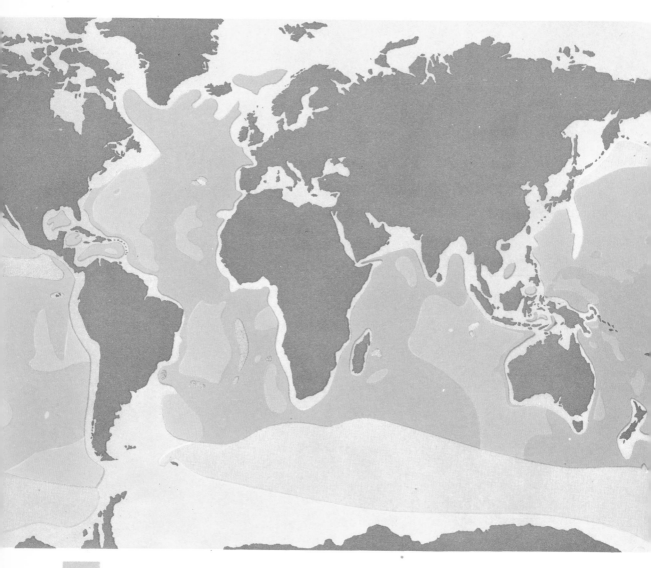

| | terrigenous deposits |
| --- | --- |
| | red clay |
| | *Globigerina* |
| | *Pteropod* ooze |
| | *Radiolarian* ooze |
| | *Diatom* ooze |

World map of the main ocean sediments. The muddy terrigenous deposits come from the land as do the red clay sediments. Globigerina, Pteropod (planktonic mollusks), Radiolaria, and Diatom oozes are formed from the skeletons of these floating animals and plants.

can use one as a check against the other. Although the study of coccoliths is comparatively new, they may extend the period of fossil dating. Coccoliths are so microscopic that it is necessary to coat them with gold and then take a photograph with an electron microscope before they show their special shapes, which are some of the most beautiful things in nature.

Professor Francis Shepard, of Scripps Institution of Oceanography, in California, has studied the sediments deposited in shallow water. He is trying to find out exactly how

Owing to its very small particle size, red clay (above), although derived from the land, may be carried far out onto the ocean floor by currents. The clay is stained red by minute traces of iron oxide, which it brings with it from the continents.

The widespread distribution of Globigerina (Foraminiferal) ooze (below) is evidence of the enormous population of these animals in the Atlantic, Indian, and Pacific oceans.

the various layers of silt and sand settle on the sea bed, and also what kind of animal remains are found in the shallow-water sediments. The oil companies are interested in this work because the same processes that are going on in the sea today produced the oil fields that are millions of years old. If they can discover exactly how the rocks are formed, they can bore down in the places that are most likely to contain oil.

A large amount of oil is now being found by drilling at sea on the continental shelves—in the Gulf of Mexico, for instance. This drill-

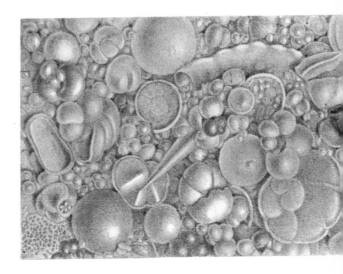

ing has had an important effect on oceano-graphy. It has shown how we can collect much longer cores than the 50- and 60-foot samples obtained with the Kullenberg corer. As soon as the oil companies learned how to drill from a floating ship, it became possible to drill through the deep ocean sediments— to any depth we like. To date, the record length of a core is more than 600 feet. This deep penetration of the ocean sediments was made as part of the Mohole Project, which is financed and sponsored by the National Science Foundation of the United States. As we saw in the first chapter, the aim of the Mohole Project is to drill right through the Earth's crust to the mantle. To achieve this a special drilling ship, the *Cuss I*, named after four oil companies, carried out trials in the Pacific Ocean near the island of Guadeloupe, where the water is nearly two miles deep.

Because it was not possible to hold the ship steady enough by anchoring it in such deep water, four giant outboard motors were fitted to the ship. Since their propellers faced in different directions, it was possible to make the ship move backward, forward, sideways, or to rotate. This meant that the ship could be kept in a fixed position above the drill.

A ring of "talking" buoys laid around the ship kept *Cuss I* informed about its exact position above the drill. On command from an underwater transmitter installed in the ship, each buoy sent out a signal. The time it took for the master signal to reach the buoys, and for the buoys to reply, gave an exact mea-sure of the distance of the ship from the buoys at any moment.

The drilling rig on board *Cuss I* consisted of a long string of rotating pipe that hung from a derrick on the ship and extended to the sea bed. At the end of the pipe a diamond cutting-bit chewed its way into the sediment floor. Because the bit had a hollow center, as it penetrated the sediment a core could be collected within the pipe. To bring the core to the surface a short hollow tube was lowered down the inside of the drill pipe then hauled back to the surface on a wire line. In this way sediment cores more than 600 feet long were collected. The only thing that stopped the

A special drilling ship, *Cuss I* (above), is being used for the Mohole Project. The hollow pipe extensions behind the derrick amidships are attached one by one to the drill as it bites deeper into the sea bed. Four giant outboard motors help to keep the derrick vertically over the drill, the exact position of which is fixed by signals from electronic buoys encircling the ship.

Right: The aim of the Mohole Project is to drill a hole through the sediments and crustal rocks and to bring up samples of the underlying mantle. If successful, it may tell us much not only about the nature and age of the sediment and rock layers but about the history of our planet as a whole.

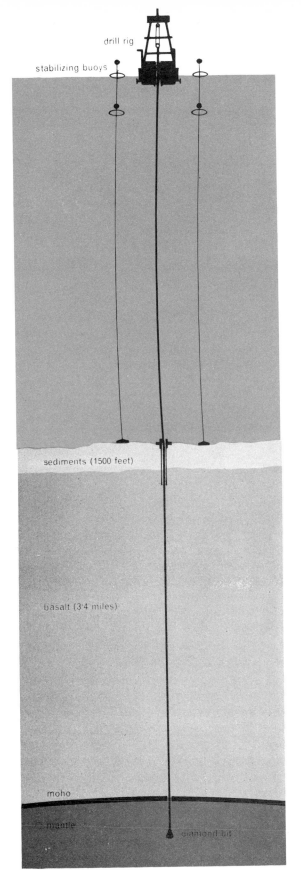

drill rig

stabilizing buoys

sediments (1500 feet)

basalt (3.4 miles)

moho

mantle

diamond bit

drillers from going even deeper was the fact that they struck rock. They were drilling on the side slopes of Guadeloupe Island and hit the volcanic rock that had poured out of the Earth when the island was first formed. When a larger ship capable of carrying more than three miles of drill pipe has been constructed, it will be possible to sample the entire thickness of sediment in any part of the oceans.

The sediment averages about 1500 feet thick in the deep parts of the ocean. There may, however, be greater thicknesses where turbidity currents have carried their debris out onto the abyssal plains. Although the rate at which red clay collects is slow, we might expect a thickness more like 10,000 feet —if we also assume that the oceans have always been pretty much the same as they are today. To find out where our assumptions have gone wrong we must go right through the sediment to the rock floor, examining the fossils and measuring their ages as we go.

This prospect is one of the most exciting things happening in oceanography today. When we finally manage to drill those holes in the sediment, we may find that the bed of the oceans is not the same all over. The surprising fact that the sediment carpet is much less thick than had been expected shows that the rate of accumulation is slower than had been forecast. Thus we are led to ask : What changes have taken place in the depths of the sea during the history of the Earth? Have the older sediments—older in geological time—been transformed into solid rock? Do these rocks form part of the crust of the Earth? What is the structure of these rocks? And at what period in geological time were they laid down? And by what agency were they brought about?

We need first of all a "geological calendar" obtained from the sediments themselves, which will give us the first dates in the history of the sea floor. These dates can be found out by analyzing the cores of sediment, thanks to the new techniques now available. The upper layers of the cores will give us the dates nearest to our own time; the deeper layers will give dates further back in our time table, till we find the date of the earliest deposits.

## 4   Under the Sea Floor

Before the sediments began to rain down and blanket the sea bed there was an exposed hard rock floor to the oceans. This rock floor is as much a part of our planet's crust as the continental rocks making up the land. When we consider the sea-floor crust there are two questions we should ask: (1) Is this crust made up of many different rock layers, as it is on the continents? (2) Is this crust underneath the seas thicker or thinner than the crust on the land?

During the past 20 years oceanographers have been discovering answers to these questions by probing into the rock layers with the help of sound waves. As we saw in the last chapter, the history of the sediments that have accumulated on the sea floor is different from the history of the sediments on the land. This is also true of the rock layers under the sea floor, which are not like those on the continents. Until the difference between the rocks making up the ocean floor and the rocks form-

ing the continents had been proved, many geologists believed that the same kinds of rocks stretched right around the world. They thought that when a continent sank beneath the sea a corresponding piece of the deep ocean floor would pop up somewhere nearby, like the push buttons of a radio.

### "Profiles" from Earthquake Waves

The seismologists—men who study earth-

Erupting volcanoes, such as the one at the far left, confuse the pattern of rocks beneath the oceans. Lava flowing down the side of the volcanic island forms a new rock layer on top of the sediments and crustal rocks, while the crustal and mantle rocks under the volcano become mixed with the magma by the force of the eruption. As they are much heavier than the water they replace, volcanoes tend to sink. In doing so the guyot (the flat-topped, no-longer-active volcano on the right of the picture) has forced the rock layers in the crust and mantle to dip beneath it.

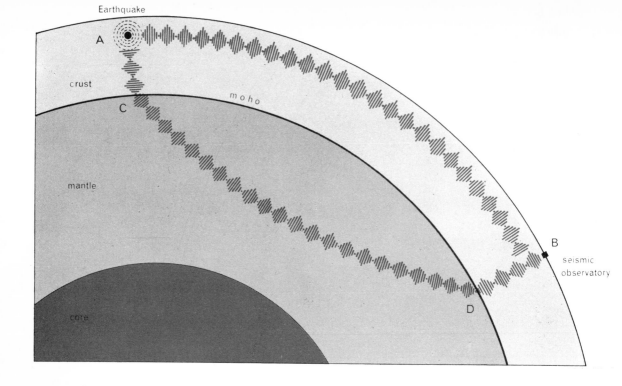

quake waves—were always suspicious of this theory, however. Could there be something out of the ordinary about the rock crust beneath the oceans? An earthquake, or any sudden shock in the crust, such as an underground explosion, produces two kinds of vibrations. There are the faster P or *push* waves, then follow the slower S or *shake* tremors. The P waves, for instance, travel through the Earth's crust at speeds of 5.0 to 5.6 miles a second. The slower S waves travel through the same rock at speeds of only 2.7 to 3.1 miles a second. Another difference between the two waves is that the S waves cannot travel through liquids. Experiments with S waves, by the way, have shown us that the Earth has an outer liquid core.

There is another group of waves that move even more slowly than S waves. These are the *surface waves*—and their name is very appropriate because they travel only in the top few miles of the crust. The surface waves are like the P and S waves in that their speed depends on the kind of rock through which they are passing. They travel more slowly, for example, in granite than they do in harder and more dense rocks such as basalt. Because some earthquakes occur under the oceans, the seismologist can record waves that have

traveled through the sea floor and distinguish them from others that have traveled through continental rocks. When the speeds of the waves along these two different paths are measured, we find that those taking the ocean route are faster than those that keep to the land. It is just as if the land contained a large proportion of granitic rock while the ocean floor was mostly of basaltic rock. And it turns out that this is so.

We can, then, get a rough idea of the type of rock that lies beneath the oceans and the continents by measuring the surface waves. But what about a more detailed picture? It is the P waves that really help us in this respect. Because we can record their travel time with great accuracy, they reveal the individual layers that comprise the great mass of rock beneath the oceans and under the continents. When we speak of *seismic waves* in the pages that follow we mean the P waves.

About 50 years ago the Croatian seismologist A. Mohorovičić used P waves to measure the thickness of the Earth's crust near his seismological observatory in Yugoslavia. Before Mohorovičić's time no one really knew that the Earth had a thin crust, although there had been some very peculiar ideas about the inside of the Earth—which led to the

Left: Sound waves travel faster through rock than through water; the denser the rock, the greater the speed. Thus the travel time of a signal along path ACDB may be less than along path AB. Below: The hammer striking the rod shows the different character of P waves and (lower diagram) the slower S waves.

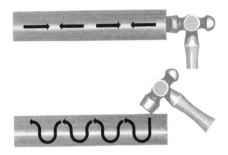

name "crust." Some people thought that the surface rocks were like pastry on an apple pie, and that inside the crust there was liquid. Some even believed that the Earth was hollow. Aristotle, for instance, thought that earthquakes were caused by winds that blew into caverns within the Earth. And earlier, Democritus supposed that the Earth's hollows collected water, and that when an excess of water sloshed from one place to another within the Earth, we experienced an earthquake.

Today we know that the Earth is not hollow; the fact is that solid mantle rock stretches down for nearly 2000 miles, and the surface rocks serve as an outer skin to the mantle. However, we still speak of the crust, and we call the place where the crust joins the mantle the *Mohorovičić discontinuity*. But since Mohorovičić had such a long and difficult name, we now call it the *Moho* for short.

How are we able to use earthquake waves to measure the thickness of the crust? Why is it that we do not use an echo-sounder, as we do to find the depths of the ocean? The echo-sounder sends out a pulse of sound, and we time how long it takes for the sound to echo back from the sea bed. In a similar way we might measure how long it takes earthquake waves to echo back from the Moho. But this is very difficult to do because our echo becomes confused by the many surface waves also sent out by the earthquake. Unfortunately we cannot make a short, sharp pulse in the ground as we can in the water. But we can find out about rock layers beneath us by using waves that have been *refracted*, instead of looking for waves that are reflected.

Refraction simply means bending. The paths of the sound waves from an earthquake are bent when they move from one rock layer to the next—provided that the velocities of the waves are different in the different rocks. The waves travel more quickly in the mantle than in the crust, so they are bent when they leave the crustal rocks and enter the mantle rocks. For example, the waves from the earthquake at A in the illustration can travel through the crust directly to the seismological observatory at B. But they can also go down to the mantle and follow the refracted wave path CD in the mantle before returning to B. Part of the path ACDB is down and up—AC and DB in the crust—but the longest stage is the path CD in the mantle. The wave traveling along AC is refracted at the Moho, in much the same way that light is refracted, or bent, when it goes from air to a glass prism. The seismic wave is bent again at D, so it travels upward to the seismograph in a symmetrical manner.

Although the path ACDB is greater in distance than that along AB, it does not necessarily take the wave longer to travel this greater distance. If the distance CD is great enough, the time saved in the fast mantle layer will more than make up for the down-and-up time in the slower crustal layer of rock.

To make these refracted waves work for us, we must either have the right equipment ready when an earthquake occurs—or we must create our own seismic waves by setting off underground explosions. And to find out about the different rock layers through which the waves pass, we must measure their travel times through at least two small and at least two large distances. If the distance is small, the first wave that is recorded at the station will be the one that has traveled directly through the crust to the recording station. However, when the distance is large, the wave that travels down through the mantle will arrive at the seismograph first. The time that a wave takes to get from A to B tells us many things about the crust and the mantle rocks. We can calculate the speed of the seismic wave in both the crust and the mantle, and from the speed we can work out the thickness of the crust.

This way is better than echo sounding, because it tells us something about the composition of the rocks as well as the depth to which they extend. And if there are two or more layers of rock, we can work out the depth of each layer and the kind of rock of which each is made. But labeling the particular rock is not always easy.

Unfortunately there are many different rocks that produce about the same seismic-wave velocity. Moreover, seismic waves may travel at different speeds through one particular kind of rock, depending on how the individual grains in the rock are packed together. In general, low velocities like 1.1 miles a second are characteristic of soft materials such as clay. Limestones may have values of anywhere from 1.9 to 3.8 miles a second. Granites vary from 3.1 to 3.8 miles a second. Basalts are higher at about 4.2 miles a second. There are some very hard rocks through which waves travel as fast as 5.1 miles a second. The earthquake results that Mohorovičić obtained showed a change from 4.2 miles a second for the crust to 5.1 miles a second for the mantle rock beneath.

The seismic-refraction method has been used to probe into the layers of rock that form the sea bed. Maurice Ewing of the Lamont Geological Observatory, Columbia University, has used it at sea for many years. He uses two ships, one for setting off the explosions, the other to carry the listening apparatus that detects the waves after their long journey to below the sea bed and back again. The ships steam away from each other, one firing charges every few minutes, until they are nearly 100 miles apart. This distance is great enough to allow the waves refracted from the mantle to be recorded.

It is expensive for oceanographers to run two ships at the same time, and for this reason Dr. Maurice Hill of Cambridge University puts his listening instruments into buoys. Four buoys are laid in a long line by the ship, which then steams away to fire its charges. The advantage of Hill's method is that only one ship is needed, yet the buoys serve as four additional ships. Accordingly, each charge that is fired is recorded by all four buoys. The seismic waves that reach the buoys are transmitted to the ship by radio and are recorded on photographic paper.

From these miniature earthquake records that Ewing, Hill, and others have been collecting, we can work out a picture of the different rock layers that make up the sea floor. The top 1000 feet or so of the sediments is soft clay, but below this is a layer that must be hard rock. We know this because the seismic-wave velocities are too high for clay. This hard part of the sediments is called simply *Layer 2*, because we are not certain whether it is solidified volcanic lava that has spread out from fissures in the sea bed, or whether it is limestone that has been formed from the shells of animals in the sea. One day we hope to find the answer to this question by drilling right through the sediments and bringing samples of Layer 2 to the surface.

Underneath the sediment is the *basaltic layer*. This is made of hard rock that formed the original ocean floor before any dust or turbidity-current mud started to rain down from above. It is this layer of ocean floor that is most different from the continental rock. The granitic rocks that form a large part of the continents are entirely missing under the deep oceans. Although we have not yet drilled into the basaltic layer, the travel times of seismic waves tell us that it cannot be granite. We also know that the basaltic layer is only three or four miles thick before we come to the Moho, which marks the top of the mantle rock.

As we saw earlier, if we attempted to reach the Moho by drilling down through the land, we would have to drill through 20 miles of limestones, sandstones, and granites before we reached the mantle. You can consider these great continental blocks of rock as "rafts" of lighter material floating on a basaltic layer. The weight of each continental raft pushes the basaltic layer down into the mantle, so that the continental rock combined with the basalt beneath it floats in the mantle rock in much the same way as an iceberg floats in the sea. How is it possible for "solid" mantle rock to allow anything to sink into it? One thing to remember is that the Earth is thousands of millions of years old, so there has been plenty of time for such a sinking to take place, even though very slowly.

While time is on the side of this slow sinking process, so is the nature of the mantle rock itself. Perhaps many of you have seen a special kind of plastic that has been on the market for some time now. It looks like a lump of putty, but it contains silicon. You can roll it into a ball or squeeze it flat. If you throw a lump of it onto the floor it bounces like a rubber ball, which means that it is elastic. But it also behaves as a brittle material; if you hit it with a hammer it breaks up into small fragments as if it were a piece of china.

The behavior of this putty depends, then, on how quickly you do things to it. Subject to a slow squeeze, it is soft and will flow, but under a sharp blow it is hard. Ice behaves in a similar way. If you hit it with a hammer it shatters like glass, but in glaciers ice flows like a very slow river. It is the same with the material making up the mantle. The rock is solid, so far as transmitting earthquake waves is concerned, but over many thousands of years it has given way like putty under its tremendous load of continental rock.

## Measuring Gravitational Attraction

In addition to seismic waves, there is another kind of measurement that tells us about the rock layers beneath us. Dr. J. Lamar Worzel of the Lamont Geological Observatory has for many years been measur-

Seismic-refraction methods teach us much about the nature and thickness of rocks under the sea floor. One method uses two ships steaming away from each other to a maximum distance of about 100 miles. One ship sets off underwater explosions every few minutes and the other receives the resulting sound waves after they are refracted in passing through the rock layers.

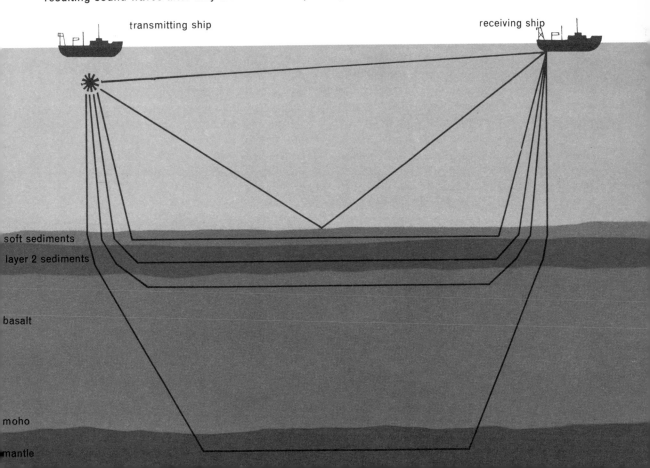

transmitting ship

receiving ship

soft sediments

layer 2 sediments

basalt

moho

mantle

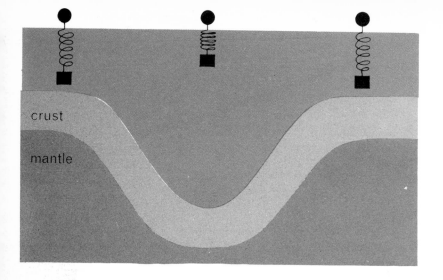

The gravitational attraction of rocks increases with their density, and gravity meters help to show both the density of rocks and their tendency to rise or fall. The deep ocean trench (left) has below-average attraction, perhaps because the light crustal rocks, in buckling downward, have pushed aside the denser mantle layers. Parts of Scotland are rising, like this raised beach (right), to seaward of the coast road on Great Cumbrae Island.

ing gravitational attraction at sea. The results provide a check on the picture that seismic-wave experiments have given us of the sea floor and continents. Isaac Newton showed that an apple falls to the ground because the apple and the Earth attract one another. In the same way, a plumb bob is pulled downward—but at the same time it is pulled sideways by the attraction of a range of mountains. In fact every irregularity on the Earth's surface causes some change in the force of gravitational attraction.

Since water is much lighter than rock, the force of attraction should be less at sea than it is on land. However, the measurements of Worzel and other oceanographers show that gravitational attraction is much the same over the oceans, where there are three miles of water, as it is on the continents, where there are 20 miles of crustal rock resting on the mantle. Why should this be so? There are two reasons: (1) The mantle rock is heavier—or more dense—than the crustal rocks; (2) as we have seen, the mantle comes much closer to the Earth's surface under the sea floor than it does on land. The extra attraction of the mantle under the oceans just makes up for the smaller attraction of the three miles of water. Even if we had not known that the crust is thinner under the sea than on land, we could work this out from the gravity readings that we make at sea.

Measuring gravity, even on land, is a very delicate operation. It is like weighing something correctly to one part in a million; and

you can imagine how difficult this must be when a ship is tossing about on the sea. This is why the first measurements of gravity at sea were made in a submarine. If you go below the sea surface for about 100 feet you are below the action of the waves and all is calm. But submarines are expensive to operate. Recently engineers have designed a specially stabilized platform that allows for the effect of all the motion due to the waves so that gravity can be measured from a ship.

When we find gravity values that are out of the ordinary, we try to imagine some picture of the rock layers to fit the results. In this way we learn a little more about the slow changes that are going on all the time within the Earth. The deep ocean trenches, for instance, are below average in their gravitational attraction. This is probably because they have been formed by a downward buckling of the light-weight crustal rocks. Because the light crust has pushed away some of the heavier mantle rock—in a way we do not yet understand— the gravitational attraction in the deeper trenches is below average. But just as a piece of wood tends to bob back to the surface when you press it underwater, so the crustal rocks of the trenches try to rise up. The gravity measurements tell us the extent of this upward push.

The combination of the seismic experiments (which tell us the thickness of the rock layers) and the gravity measurements (which show whether the rocks are trying to rise up or to sink) gives us some idea of what has

happened to the Earth's crust in the past. Take Scandinavia as one example: We now know that during the last great ice age Scandinavia was weighted down by millions of tons of ice. Under this great pressure the land was pushed down deeper into the mantle, but when the ice started to melt the land began to rise. Gravity measurements and sea-level observations show that Scandinavia is still rising today, at the rate of half an inch a year. And it will continue to rise until the raft of land reaches its natural level.

The upward and downward movements of the continents make it possible for rocks to be worn down and redeposited as sediments in shallow water to form new rock layers. But more important for the oceans is the possibility of the continents moving sideways as well as up and down. If the raftlike continents have drifted in the past, they will have done something to the rock under the sea floor.

## Wandering Continents

Any globe or world map shows that the east coast of South America projects where the west coast of Africa curves inward. Returning to our idea of wandering continents, we can ask if these two land masses were ever joined together as one large continent. A careful piecing together of all of the continents shows that this could have happened. It is important, of course, to fit the pieces of this jigsaw puzzle at the true edges of the continents, which, as we saw earlier, are the submerged edges of the continental slopes and not the coast lines. If we push North America over to Europe, and South America over to Africa, we get rid of the Atlantic Ocean and find that the two land masses fit fairly well together. We can go a stage further and slide India beside Madagascar on the east coast of Africa, and we can fit Antarctica and Australia in

SOUTH AMERICA

AFRICA

Wandering continents? The coast of northeast South America fits neatly into Africa's Gulf of Guinea. The continental-drift theory is also supported by the similarity between the older rock strata on both sides of this part of the Atlantic.

The idea of continental drift may also be supported by the rift valleys and associated regions of crustal weakness in eastern Africa (right) that may gradually be splitting away in a northeasterly direction.

the area now occupied by the Indian Ocean.

Some geologists believe that all the continents once formed one large land mass, and that certain forces inside the Earth caused this giant of a continent to break apart. We can see some signs of splitting now—in the geological faults that formed the rift valleys of Africa, and in the Red Sea. There is also the sequence of valleys (mentioned in the second chapter) that the echo-sounder picked out in the Mid-Atlantic Ridge. The ridge itself winds down the center of the Atlantic Ocean, rather as one would expect if the continents on either side had moved away from it.

It is not only the shapes of the continents on either side of the Atlantic that can be fitted together. The rock layers themselves can be related—at least those more than 100 million years old. The newer layers are different on the two sides of the ocean, but this is just what we should expect if Africa and South America parted company about 100 million years ago. Other signs that continents might have moved about in the past are provided by the coal de-

posits found in Antarctica. We know that coal is made from compressed plant matter and that Antarctica's coal could not have been formed in a cold climate like that found on the continent today. How can we account for this polar coal? Either the Antarctic must have been nearer the equator at one time, or else our planet must have gone through a period of warming up. It seems more likely that the continents have moved about, because other parts of the world show indications of cold climates at the time when Antarctica was warm. There are many other odd things about the climates of different parts of the world millions of years ago, yet they seem to make sense if we accept the idea of drifting continents.

If Africa and America were joined together as recently as 100 million years ago, then the Atlantic is only a new ocean compared with the age of the Earth. This at first sight seems to contradict our picture of oceans and continents being quite different from each other. We have said that continents have always ex-

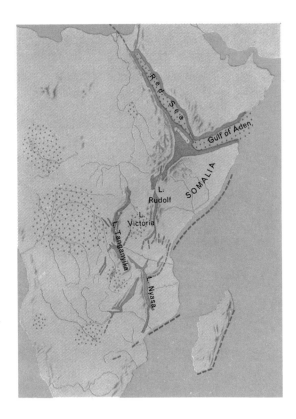

continents were still one great land mass.

Many oceanographers do not believe in the theory of drifting continents. They point out, for example, that the carpet of sediments on the Pacific floor is no thicker than that on the floor of the Atlantic. If the Pacific has always been an ocean it must be thousands of millions of years old. But the Atlantic, according to the drifting-continent theory, is less than one tenth of this age, so its floor should have fewer sediments. Against this, it can be argued that the Atlantic collected its sediments very quickly as the continents on either side drifted apart. We shall not know the answer to this until we drill many bore holes right through the sediments of both oceans.

## Atolls and Lava Flows

We saw in the second chapter that the deep ocean floor is spotted with rock formations called *seamounts*. Although in many cases the seamounts tower above the sea bed to the same height as the continental rafts, they are not made of the same kinds of rock. The seamounts are the result of volcanic eruptions, and their history tells us many interesting facts about our planet's crust.

Dr. Russell Raitt of Scripps Institution of Oceanography in California has made seismic studies of many of those beautiful islands called *atolls* in the Pacific Ocean. An atoll is a coral reef enclosing a lagoon. A typical atoll is one named Funafuti in the Ellice Islands of the Pacific. The lagoon here is about 120 feet deep and 10 miles across. Its ring of coral has gaps that enable ships to enter the lagoon, and part of the ring consists of submerged coral reefs. The reefs making up the ring have a constant width of about 100 yards and they are never more than a few feet above sea level. How are these peculiar reefs and lagoons formed?

To understand this it is first necessary to know something about the way coral rock is produced. The beautiful columns and fantastic shapes of coral are the outside clothing of small marine animals called *coral polyps*. They grow rapidly in clusters and the spaces between the thousands upon thousands of the

isted. We have also said that they may sink at times and become covered by shallow water, but they never sink to a depth of as much as three miles so that their surface would be on a level with the true ocean floor. If we keep the "raft" idea in mind, however, there is nothing to worry about.

The sequence of events could have been something like this. First there was the big continental raft, which was broken into separate pieces, each of which drifted slowly to the places where we find them today. During the process they must have been subjected to great forces as they were pushed about. This may account for the long line of the Rocky Mountains running down from Canada through the United States and on into South America as the Andes. These mountains could be the result of buckling that occurred when the rafts were pushed into the Pacific Ocean. If we accept this picture, the Atlantic Ocean has grown at the expense of the Pacific, which must have been much larger than it is now in the days when all the

individual polyps are filled in by small plants called *algae*. Polyps helped by algae produce large quantities of calcium carbonate—chalky, shell-like material—and they are sensitive to their living conditions. They need the water to be warm (over 70°F.) and they need light. They do not thrive in water more than 200 feet deep, because at greater depths the life-giving rays of light are filtered out. The coral polyps need sea water and they dislike mud, so they avoid the fresh water of river estuaries. Coral grows around most tropical islands, and forms fringes of shallow reefs in many parts of the world.

To return to the formation of our atoll, let us begin with an island that has been built up by volcanic activity. All around its edge, away from the shore, are coral reefs that have been forming over the years. Suppose now that, due to a weakness in the sea floor, the island begins to sink slowly into the sea bed. As it does so year after year, the coral polyps near the edge build up higher and higher, keeping pace with the sinking island. The growth next to the shore will be poor, because of fresh water and mud brought down by rain from the island. On the other hand, the outer edge of the reef will be strong and healthy. Gradually, as the island continues to sink, a stretch of lagoon appears between the island and the outer rim of the reef. The island of Bora-Bora near Tahiti looks like this. Eventually our original island disappears beneath the sea; all that is left is a circular coral reef enclosing a shallow lagoon. In the course of time the reef wears down into sand and soil with some vegetation, and ends up as a broken ring of reefs forming the atoll.

Seismic experiments carried out in the lagoon at Bikini Atoll by Raitt showed that there was a thickness of about 3000 feet of coral rock, beneath which was the original volcanic island that had started the whole process. Atolls, then, are rather like gravestones that mark the positions of ancient volcanic islands.

At this stage you might ask why the flat-topped guyots described in the second chapter did not become atolls. Possibly they sank too quickly to be populated by coral polyps; or

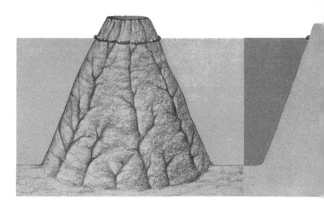

These three views and cross sections show the development of an atoll. Above, a coral reef grows around the shore of a volcanic island.

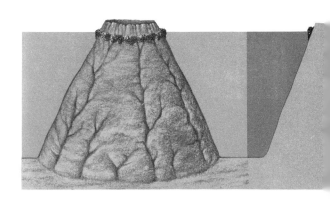

As the volcano sinks the coral grows upward, keeping pace with the rise in sea level and becoming a barrier reef.

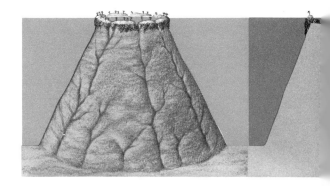

The island has sunk below the waves leaving an atoll—a ring of upward-growing coral encircling a calm lagoon.

Calcareous coral, the basic material of atolls, is formed by tiny marine animals.

possibly at the time they sank the ocean waters were too cold for coral polyp growth on a scale necessary to fringe the island. We still do not know for sure.

Seismic experiments show that the old volcanic islands and atolls are rather like teeth. Each one has a root—of volcanic material embedded in the rock beneath. This deep-seated rock seems to be a mixture of the mantle and basaltic layers—at least we think so because we believe that volcanoes start as molten rock somewhere beneath the Moho. Measurements of gravity show that new and active volcanic islands like the Hawaiian Islands are pressing down on the sea floor and are slowly sinking. This is to be expected since this great mountain of volcanic rock erupted out of the sea floor and is much heavier than the water that was there before the mountain building took place. The older islands and the atolls have normal gravity values because they have sunk down until they are floating in the mixture of the basaltic layer and mantle. Because the sinking needed to reach this final state of rest is about three miles, the Moho lies at a deeper level near islands (about 10 miles down) than it does beneath the flat parts of the oceans (about seven miles down).

When lava erupts from volcanic islands, some of it flows down the side of the island into the sea and out onto the sea floor, where it spreads for miles around. This forms a new rock layer on top of the basaltic layer—also on top of any sediments that happened to be there before the volcano erupted. It is not difficult to see how this complicates our simple picture of the various layers of material making the sea floor—sediments, basaltic layer, and the mantle beneath. Any lava-flow material will be counted as part of the rocks forming Layer 2. The reason for this is that the travel time of seismic waves through the lava material falls between those for soft sediment and basalt. Also, the travel time of the waves through lava-flow material is difficult to distinguish from the travel time through any hard limestone layer that might be present in the sediment. Although we are not sure what Layer 2 is made of all over the ocean, we do know that it is composed of lava flows near islands, seamounts, and atolls. Away from the islands on the open sea floor, Layer 2 may be made up of lava-flow material that has poured up through cracks in the crust, or it may be made up of limestone, or a combination of both.

What is the source of all the volcanic rock that pushes its way upward from the sea floor? The old atolls show that there were volcanic eruptions on the Earth many years ago; and the eruptions that take place every few years in Hawaii, for instance, show that volcanoes are very much a part of our lives today. For every island or atoll that we can see at the ocean surface, there are probably 20 seamounts that have either sunk out of sight or that never grew quite tall enough to reach above the waves.

The lava that issues out of volcanoes does not start from the liquid core of the Earth. The core is 2000 miles down, and there is a solid rock layer of mantle to stop the core material from leaking out. There is not even a liquid layer in the mantle that produces the lava. Each volcano begins as a small pocket of lava about 50 to 70 miles inside the Earth. We can tell that there are no enormous underground "lakes" of lava by the way the eruptions break out. A series of eruptions may last for a few days, or a few weeks, and then die down, having drained the small pocket of lava. New eruptions will not begin until more hard rock inside the Earth has been melted to produce another burst of activity. If there

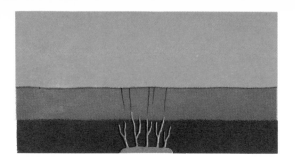 

These four pictures illustrate the "life cycle" of a typical undersea volcano. First, cracks form in the ocean floor and widen as molten rock forces its way upward through the crust. Then the volcano breaks through the sediments, its erupted lava cooling to form a rising cone.

were a vast layer of liquid rock that fed volcanoes, our studies of earthquake waves would reveal it; so far, we have not discovered such a layer.

## Measuring the Earth's Heat

How do the deep underground rocks become so heated that they melt? We now think that the heat that melts hard rock inside the Earth comes from small quantities of radioactive atoms that are sprinkled throughout the mantle and the crust. Even though the amounts of radioactive heating are very small, the heat produced is stored inside the Earth because the rocks are very poor conductors. In fact they act as a blanket. In addition to this general heating by radioactivity there is local heating as well. Volcanoes in the oceans form long lines, like the 1500-mile chain of islands and seamounts stretching northwest from Hawaii. Forming a heavy ridge in the sea floor, the rocks composing the mountain chain provide an extra blanketing effect and therefore prevent heat escaping from the mantle. In addition, the ridge itself produces heat. It does this as its rock bends and cracks.

We know these crackings of rock as earthquakes, but only a small part of the energy of the earthquakes travels out as earthquake waves. Much of the energy is turned into heat when the rock is ground and crushed against itself. This extra local heat, combined with the total radioactive heat and the blanketing effect, melts rock in certain areas and forms small pools of molten rock called *magma*. Gases are mixed with the magma, and as this rock-gas mixture begins to expand, pressure is built up. At this stage the magma is forced toward the surface. When it pours out over the surface of the land or over the sea bottom as lava, all the heat that has been collecting over the years is suddenly released. The slow process of accumulating heat then starts all over again.

You might think that we could tell from the kind of lava a volcano produces whether the melting pot is in the mantle or in the basaltic layer. We can make a guess, but we cannot be certain. Some separation of the various rock materials may take place when the rock melts, just as slag separates from steel in a blast furnace. Also, on the way up the magma may melt and mix with the rock through which it is pushing its way. If we knew what the mantle rock were made of, we could then work out what would happen when it melted. We would also like a sample of the mantle rock in order to make accurate calculations about the radioactive heat produced.

On the land we can measure the amount of heat being conducted out of the Earth. By going down a mine shaft, or by lowering a thermometer down a bore hole, we can measure the increase in temperature. We know that the deeper we go, the hotter it gets. At the bottom of a west Texas oil well the temperature is 380°F. We can also measure the conductivity of different kinds of rocks in the laboratory and then calculate the amount of heat that must be flowing upward to the land surface. But it is only in the last 10 years that we have tried to make such measurements at sea. Prof. E. Bullard and Dr. A. Maxwell

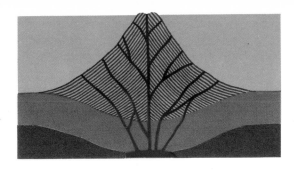

The volcano rises out of the ocean, forming an island. Eventually it becomes inactive, begins to sink—in this case too quickly for an atoll

to develop—and its summit is flattened by the action of waves. The resulting guyot is half a mile or so beneath the ocean's surface.

made a 10-foot temperature probe which they forced into the sea bed in much the same way as we force corers into the sediments. The experiment was difficult for many reasons. Because the thermometers were so sensitive and recorded the frictional heat set up when the probe was pushed into the sea bottom, the probe had to be left in position for half an hour so that it could cool. Another difficulty was caused by the three miles of cable attached to the probe. Dragged this way and that by currents, the cable tugged at the probe and bent it through a right angle! In most parts of the ocean it was found that the heat flow from within the Earth to the sea bottom was about the same as on the land, but in a few places he found abnormally high heat flows. Probably these are areas where rock is melting deep within the crust or in the mantle and is ready to burst out.

The heat-flow measurements in the oceans are very important because the area of sea is thrice that of land. It is the *total* amount of heat lost by the whole Earth that we need to know when we try to work out the past history of the Earth. It now appears that our planet is heating up, not cooling down. This probably sounds surprising in the light of what was said in the first chapter—that the young Earth cooled from a molten mass and separated out into a liquid core, solid mantle, and crust. But there is no reason why the planet could not have cooled that way, then followed its cooling period by a period of heating up.

Some very careful earthquake measurements have shown that there is a layer about 50 to 100 miles inside the Earth—well into the mantle—where the seismic waves travel more slowly than expected. This may mean that a band of rock in the mantle has been growing soft. As more and more radioactive heat is produced and stored inside the mantle, this soft band may move upward until it is near enough to the surface to break through in an enormous outburst of volcanic activity. This would release a great deal of heat very rapidly, so that the Earth would settle down to another period of slow warming up. In the meantime the more fluid mantle might allow the continents to drift about, while the sea-bed sediments were being baked to form a hard rock, such as our Layer 2.

We do not know whether these things have happened in the past, or whether they are probable in the future. All we can do is continue to make observations, and measure such things as the heat flow through the sea bed and so to gather more figures to use in our calculations. Then at some stage we shall have enough information to know which theory is right, or whether we shall have to develop a new theory. Much of the history of the oceans, then, and of the Earth itself is bound up in the rock layers hidden beneath the sediment carpet of the sea floor.

In the last chapter we saw how valuable it would be to have a sample of the sediments from top to bottom. It would be just as valuable to have such a sample of the rock layers underlying the sediments. Although seismic waves can tell us much about these rocks, drilling all the way through them—through to the mantle—will be of enormous value.

A gigantic tsunami hurls itself upon a coastal
village in Japan. Caused by earthquakes often
located thousands of miles from the eventual
point of damage, tsunami are by far the
most destructive and terrifying of all sea
waves and have been known to attain a height
of as much as 200 feet.

# 5 Waves of the Sea

$$T = \frac{L}{V}$$

In this chapter we shall talk about another group of waves—the waves that travel on the surface of the sea. Compared with the high-speed seismic waves and the sound waves sent out by the echo-sounder, the waves of the sea travel quite slowly. Ocean waves, of whatever kind, have a great deal in common with other wave forms, but they also have important differences. For one thing, because they travel comparatively slowly and are also visible to the naked eye, we can learn much about them just by watching them.

You will find waves on any pond, or even a puddle, because they need only a small water surface on which to travel. If you watch the rings of waves formed by a stone dropped in a pond, you will find that the outer rings are more widely spaced than those near the center. This is because the stone produces both long and short waves, and the long ones, which travel faster, move out ahead of the short ones.

Even in a quiet pond, then, waves can be quite different in character. Before looking at these and other differences in more detail, we should first know the meaning of some of the terms the oceanographer uses when he describes a wave.

## The Anatomy of Waves

The highest point of a wave is called the *crest*, and the lowest point between two crests is called the *trough*. The time it takes a wave (from crest to crest) to pass a fixed point is called the *period* of the wave. The period is the same whether we measure it as the time taken for the water surface to make one complete up-and-down movement at our fixed point or for two successive crests to pass that point. The period depends on two things: (1) the speed of the wave; (2) the distance between one crest and the next, which is called the *wave length*. To put it as a simple equation, where the period is $T$ (for time), the speed is $V$ (for velocity), and the wave length is $L$,

Unless you have special instruments, it is easier to measure the period of a wave than either its speed or length. You can, for instance, use a watch to count how many seconds elapse between each crest passing a rock on the shore. Suppose that you do this on a coast where rollers from the open ocean are tumbling onto the shore and you find that the wave period is 10 seconds. Now the problem is to find the wave length and speed. You can work out the speed if you remember that it equals 5.14 times the period (in seconds). The speed of our rollers, then, will equal $5.14 \times 10$, or 51.4 feet/second. Now that you know the period and speed, your next problem is to work out the wave length. You can do this by multiplying the square of the period by 5.1—in this case $100 \times 5.1$, which equals 510 feet for the wave length. So if you stand on the shore armed with a watch—and a gift for mental arithmetic—you can find out not only the period of waves, but also their speed and length. As you can see, the speed of waves varies considerably and depends on the wave length: As a wave lengthens, its speed increases, but not in direct proportion.

Travel time of ocean waves

| Period seconds | Speed feet/second | Wave Length feet |
|---|---|---|
| 1 | 5.14 | 5.1 |
| 3 | 15.4 | 46 |
| 5 | 25.7 | 127 |
| 6 | 30.8 | 184 |
| 7 | 36.8 | 250 |
| 8 | 41.0 | 326 |
| 10 | 51.4 | 510 |
| 15 | 77.0 | 1148 |
| 20 | 1030 | 2040 |

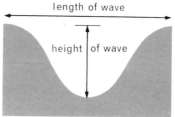

length of wave

height of wave

Above: A wave breaks upon the sea shore. As it runs into shallow water it is slowed by friction against the bottom. At this point the wave becomes higher and much steeper, so that even gentle swells often produce waves that are several feet high when they break.

Left: Diagram of wave height and wave length. The length of a wave can also be determined by measuring the distance between one trough and the next. The period of a wave is the amount of time it takes for two successive crests or troughs to pass a fixed point.

(You will find a table of periods, speeds, and wave lengths in the table on the previous page. The speeds are given in feet per second, but if you want to convert these into miles per hour you simply multiply them by 0.7. The exact figure is 0.675 but the factor 0.7 will give a good approximation.)

If you watch carefully a seagull floating on the sea, you will discover that it moves in a circular pattern as well as up and down. The crest of the wave carries the seagull upward and slightly forward, then the trough takes it down and back to its original position.

The reason for this is that the water itself has a circular movement. This disturbance, caused by the action of waves, is greatest in water near the surface. At a depth of one ninth the wave length, the movement of the water is halved. The longer the wave, then, the deeper the disturbance, and we shall see later in this chapter that there are waves so long that they move the water right down to the bottom of the deepest oceans. But these waves are exceptional. As a rule, waves in the deeper parts of the ocean affect the water close to the surface; it is only when they enter shallow water near the coasts that they make themselves felt along the bottom.

How are waves formed? If you have ever seen a storm at sea, you probably know that it is the wind blowing upon the water that builds up the waves. The interesting thing is that, no matter how strong it may be, the wind has to blow for a certain distance over the water in order to produce a wave of a certain length. You can see this if you watch a puddle on a windy day; however hard the wind blows, the wavelets skimming across the water remain close together because the puddle is not large enough to allow long waves to develop. If, on the other hand, you look at the waves formed by the wind in a large swimming pool, you will see that the waves are quite short at the end from which the wind blows, and that they grow longer the nearer they get to the other end. Even in this small area of water, the wind has had time to stretch the waves.

The largest ocean waves, those reaching heights of 50 to 60 feet, can be produced only by winds of gale force blowing over hundreds of miles. Wave experts will tell you that large waves breaking upon the coasts of Europe sometimes owe their size and speed to storms that formed them near the coast of the United States, two or three thousand miles away.

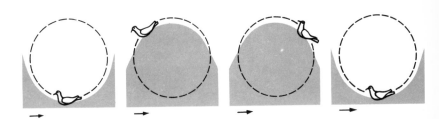

Right: The seagull shows the circular movement of water in a wave (reading from right to left). On the leading edge of the wave the gull's movement is up and slightly backward. Then, passing behind the crest, it dips down and forward to its original point in the trough.

The distance that waves run under the force of the wind is called the *fetch*. The greater the fetch, the longer and faster are the waves.

Although the length of waves depends upon the fetch, it is the speed of the wind that determines how slowly or quickly waves develop to their maximum size. As well as stretching the waves, the wind increases the height of their crests. The height is, however, limited by the length of the waves, for when a wave reaches a height one seventh of its length it becomes so steep that it breaks. If the wind is very strong the top of a wave may be blown off before it becomes this high, hence the whitecaps, or white horses—patches of white, foaming water seen on a stormy day.

When ships are in the middle of a storm they are buffeted by a complicated pattern of waves. While some new waves are being formed locally by the wind, others reach the ship from areas of the storm a few miles away and grow in size so rapidly that they break. Mixed among these waves are others still—long, high ones that have come across distances of hundreds of miles. We call the confused mixture of waves in a storm a *sea*.

Surprisingly, quite high waves are often found in areas where there is hardly any wind. The reason for this is that large waves retain their energy for a long time and are able to travel hundreds of miles away from the storm center where they were formed. Once away from the driving force of the wind, they "sort themselves out," becoming evenly spaced with smooth, rounded crests. Wave patterns of this kind are called *swells*, and from an airplane they resemble calm water.

Such swells soon become obvious when they meet any sort of obstacle. On reaching shallow water (shallow compared with the wave length) friction against the bottom slows the waves down. The waves that arrive first slow down first, so that those following behind begin to pile up. The result is that the wave length becomes shorter. The period, however, stays the same because the speed and the length become smaller together. This squeezing-up process makes the waves a little higher and much steeper, so that what looks like a flat calm out at sea often produces waves several feet high when they break upon the shore.

Waves break when their height from crest to trough is about one and a half times the total depth of water, and it is now that an important change takes place in the motion of the water. Many of us, as we watch waves approaching a beach from the open sea, imagine that the waves carry the water itself along at the same speed. This is not so. Water waves are, in this respect, very like sound waves. As you talk, the sound you make does not travel wrapped up in a small parcel of air. The atmosphere is a medium through which sound waves travel; and although they disturb the air locally, they do not push it about from point to point. Similarly, sea waves disturb the water locally (in a circular movement) but they do not push the water along in their path—except when they break upon the shore. At this moment the wave carries the water with it and hurls it up onto the beach. This is why surf riding can be done only on waves that are beginning to break. Farther out at sea the water, however mountainous the waves, would simply carry the surfboard in a circular motion.

The slowing-down of the waves in shallow water explains why the breakers come onto the beach parallel to it rather than at an angle. A wave may be moving at quite a steep angle to the shore when it is in deep water, but the part nearest the shore reaches shallow water first, and slows down, so that the wave as a whole turns in toward the shore. This is the same refraction effect that bends seismic waves in the layer of rock under the sea floor, and it can produce beautiful patterns on the ocean surface—as you may see if you ever fly over a rough sea near the coast.

The slowing-down of waves as they approach the shore was used during World War II to measure the depth of water off enemy coast lines. This was necessary because storm waves are constantly changing the profiles of beaches. To find the depth, aerial photographs were taken in pairs a few seconds apart when a good swell was running. By measuring the distance each wave had traveled between the time of the first and second photographs, the speed of each wave could be calculated. The depth of the water could then be determined from the extent to which the wave speed had slackened.

In the open ocean the currents act as

Surf riding can be done only when waves break and carry the water along with them. Surfing is best on gently sloping beaches, where waves begin to break well away from shore.

obstacles that break up wave patterns and create new patterns in turn. It is a combination of current and waves off the north of Scotland that is responsible for the wild tumbling seas so dangerous for small fishing boats. Currents, by deflecting and confusing wave patterns, sometimes act as breakwaters in much the same way that engineers make concrete breakwaters to reduce the energy of the waves and so provide a calm-water anchorage.

Even out in deep water the energy of the waves is being worn away all the time by friction. Rain helps to beat down waves; and in extreme northern and southern latitudes,

Although a wave may approach the coast at an angle, the leading part slows when it reaches shallow water. The whole wave then turns in toward shore, breaking parallel with the beach.

when the water starts to freeze, particles of ice help to calm the restless surface. For centuries men have poured oil onto the water to smooth the surface when rescuing shipwrecked sailors in a storm. The oil forms a thin skin on the water and helps to flatten the crests of the breaking waves that might capsize small boats. The film of oil cannot, of course, flatten large waves, but the large waves are not always a nuisance because they are fairly regular and their up-and-down motion can be anticipated. In order to build up a wave, the wind needs small ripples of water on which to bite; so if these small ripples can be held down by a skin of oil there will be fewer of the choppy three- or four-foot-high waves.

## Measuring and Forecasting Waves

The more we know about how the wind forms waves, the easier it is to forecast the size of waves that will reach certain coasts at certain times. Accurate wave forecasting greatly helped the Normandy invasion operation of World War II, when it was essential that landing craft should not try to approach the beaches at a time when the waves were too large. Today there are many oceanographers who keep detailed records of

meteorological reports and compare them with records of waves reaching the coasts. After hundreds of such observations have been made, it is possible to relate the period, height, and speed of waves to the wind fetch.

We can use this kind of knowledge in many parts of the world. In the Gulf of Mexico, where oil companies drill for oil many miles off shore, a permanent wave-forecasting service has been in use for many years. The forecaster warns the drillers if a storm is likely to interfere with the barges carrying equipment to the giant drilling platforms that stand on legs extending 50 to 100 feet to the bottom. Perhaps more important, the forecaster tells the oil company when it is safe to move these giant platforms, because they could easily be wrecked when being moved from one drilling position to the next.

Year by year research scientists are improving methods for calculating wave height from wind speed. The waves that are recorded are divided into groups of short, medium,

and long periods. The short-period waves are important in comparatively small sea areas such as the Persian Gulf, because here the fetch is too small for the wind to build up waves of longer than eight- to nine-second periods. On the other hand, the long, 15- to 20-second-period waves that reach certain coasts of Europe are often formed thousands of miles away by great storms in the Atlantic. Our knowledge of waves is now so good that the wave-observer can sometimes tell the meteorologist about a storm that had not been reported because it was in some area of the ocean seldom used by ships.

This gap in our knowledge of weather conditions in large areas of the oceans means, of course, that we are to some extent hampered in our ability to forecast wave conditions. The accuracy of wave forecasting can be only as good or as bad as the weather information that happens to be available. Like the meteorologist, the wave forecaster looks forward to the day when more weather satellites

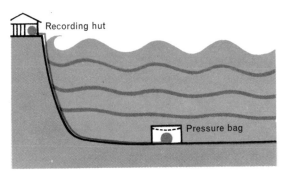

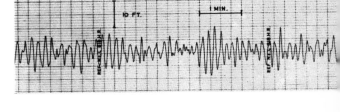

Left: Test tanks reproducing wave movements help in design of harbors. Top: A pressure-measuring instrument on the sea bed enables a shore station to plot the height, length, and period of waves on the graph shown above.

will be orbiting the earth and sending a continual record of meteorological conditions at sea. One satellite, for example, would be able to record the complete pattern of winds over most of the Pacific in a matter of minutes.

Although we know how to forecast the height and length of waves from weather charts showing wind speed and direction, we know much less about the way in which the wind builds waves. In an attempt to understand exactly what happens, it has been necessary to measure the waves themselves. On page 71 we have already described a simple method of estimating the period and calculating the length and speed of waves. But this is too slow in practice and it is much more useful to make continuous and more accurate records by using automatic instruments.

The rise and fall of the sea surface, as the waves travel over a given area, cause the pressure on the sea bed to increase and decrease. If we know what height of wave is represented by a particular pressure, we need only place a pressure-measuring instrument on the sea bed in order to see exactly what the waves are doing. The simplest type of pressure recorder is a rubber bag filled with air and connected by a hose to a recorder. The pressure of the wave on the sea surface above squeezes the bag and makes the air work a pen, which marks the height of the wave on the recorder paper. When the period and the height of the waves are read in the laboratory from the recorder paper, the waves can be

separated into groups of different periods by using an instrument called an analyzer.

Dr. M. J. Tucker of Britain's National Institute of Oceanography has invented a recorder than can be worked from a ship. As the ship moves up and down, the wave recorder automatically records the height of the waves passing by. It does this by measuring the pressure at the bottom of the ship in much the same way that the waves are measured by the pressure recorder on the sea bed in shallow water. But because the ship itself is heaving about and changing the pressure, the movement of the ship must also be measured. This is done with an "accelerometer," an instrument that accurately records the motion that produces the "sinking feeling" you have in a heavy sea. Once we know the motion of the ship, we can correct the pressure readings and calculate true wave height and period.

From time to time we hear stories of gigantic seas said to reach "above the horizon" even when viewed from the crow's nest. It is very difficult not to exaggerate the size of big waves if you attempt to measure them by eye from the deck of a pitching and rolling ship. Exactly what "above the horizon" means is difficult to say, since the ship might have been heeling over sharply at the time. Furthermore, it all depends on exactly what is meant by "gigantic." Tucker's wave recorder has, however, shown waves more than 60 feet high. In the North Atlantic 50-foot waves are quite common during winter. In 1954 a weather ship stationed about 250 miles

This aerial view of St. Margaret's Bay in the Strait of Dover shows a section of chalk cliff that has collapsed after constant pounding by the sea. Large waves unleash tremendous power as they break against cliff, beach, or breakwater. The "White Cliffs of Dover" are very easily broken down by wave action, because chalk is a soft, friable substance. Wherever the earth is friable, the sea erodes. At points along Britain's east coast the shore line once extended much farther out to sea.

south of Iceland reported that the average height of waves for the month of February was 23 feet, and waves of 50 feet were not uncommon during this period. The ship-borne recorder is particularly useful because it can measure waves far out at sea at the time they are building up in a storm. This type of research will gradually teach us more about how the wind builds waves.

The ship-borne wave recorder also teaches the marine engineer more about ship design. Although it is sometimes possible to plot a course through calm water by taking the wave forecaster's advice, bad weather is often unavoidable. But the effect of rough seas can be reduced by improving the design of the ship itself. One big improvement has been the fitting of stabilizing fins to ships. As the ship rolls, the fins rotate in such a way as to move it in a direction that anticipates the roll. Ships waste much of their power by building waves that stream away from their bow and fall behind in the wake. Research into these wave movements helps ship designers make more efficient shapes of hull. The marine engineer is also being helped by those who study the shapes of fish and small whales and the way they move through the water.

## The Force of Waves

The large waves that break against the shore carry a tremendous amount of energy with them. You have probably seen how they

The sea carves curious and beautiful shapes in some cliffs made of soft rock. The photographs on this page, taken at different places, show a typical progression in the eroding action of the waves. Top: A cave hollowed out of a soft cliff face near Flamborough, England. Center: This arch in Scotland is the last remaining section of a cave. Bottom: These stacks (also near Flamborough) are all that is left of an arch after its "keystone" has collapsed into the sea.

wear away a coast, sometimes reducing boulders and cliffs to small particles of shingle and sand.

Because of the softness of the rocks, one section of the coast of England is being washed into the sea by waves at the rate of 17 feet a year. Other sections of the English coast are made of tougher rocks and are able to stand up to the pounding of the waves. But the action of waves is not always of a destructive nature. Waves sometimes build up coasts in such a way that the land gains over the sea. This can happen when sand washed onto a beach by the waves is blown above the high-tide line and is formed into dunes. Certain plants take root in the dunes and before long an area that was once sandy coast becomes firm grassland.

Sometimes we must fight against nature by building sea walls and groins to protect our coasts and stop the waves from carrying away sand and pebbles from a beach. If you visit a stony or shingle beach after a big storm you will see that a wall of stones has been thrown high up onto the back of the beach. These stones have probably been moved by *longshore* movement of the sea from another part of the coast; so that while one part of a coast is being built up, another is being eroded; this is called littoral drift.

The force of waves beating against the shore has been known to move blocks of concrete weighing more than 1000 tons, so great skill is needed to design breakwaters that will stand up to the constant pounding of the waves. The art of building breakwaters is not to challenge the waves head on, but to wear them out gradually by making them flow in a zigzag path; for example, through the interlocking feet of concrete structures called Tetrapods. Whenever a new harbor is being planned, wave measurements throughout each month of the year must be collected so that adequate breakwaters and docks can be constructed. Only one bad storm is needed to wreck a weak sea wall completely.

There is a particular kind of wave—fortunately not very common—that can destroy any structure made by man. If an earthquake occurs within the sea bed, the movement of rock sometimes disturbs the water with enormous and sudden force. The result is what is popularly called a "tidal wave," but what is scientifically called a *tsunami*, a word from the Japanese, who suffer from more of these waves than anyone else. The tsunami consists of a group of three or four waves and has a wave length much greater than the three-mile depth of the ocean—the result being that it behaves as a shallow-water wave all the time. As we saw earlier, the longer the wave, the faster it travels, so the tsunami, with a length of more than three miles, is a very fast traveler.

In fact, some tsunami reach speeds of 500 miles an hour over deep water. Surprisingly they are only a foot or two high as they rush across the open Pacific, and out there they do no damage; in fact very few people on board ships overtaken by a tsunami ever notice the waves. But as one of these waves nears shore and begins to feel the drag of the shallow bottom, the front of the wave slows down to about 50 miles an hour, and the water behind quickly piles up and builds to heights of more than 100 feet. (There have been reports of a tsunami more than 200 feet high.) When the wave finally crashes against a low-lying, unprotected shore it creates sudden and devastating havoc. This is how one eyewitness described the tsunami disaster on the island of Hawaii in 1946:

"The waves of the tsunami swept toward shore with steep fronts and great turbulence. Between crests the water withdrew from shore, exposing reefs, coastal mud-flats and harbors' bottoms for distances up to 500 feet or more from the normal strand-line. The outflow of the water was rapid and turbulent, making a loud hissing, roaring, and rattling noise. At several places houses were carried out to sea, and in some areas even large rocks and blocks of concrete were carried out onto the reefs. People and their belongings were swept out to sea."

Because tsunami have wave lengths up to 100 miles, they create turbulence right down to the sea floor. This turbulence may cause a shifting-about of sediments on the sea bed and may trigger off turbidity currents from

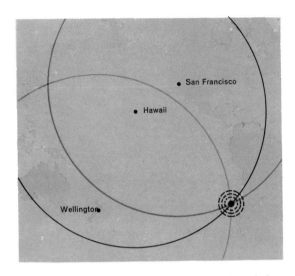

Tsunami early-warning system: An earthquake's position is plotted by seismic stations in San Francisco, Honolulu, and Wellington.

time to time. An interesting sidelight on tsunami is that if we did not know how deep the oceans were, we could calculate their depth from the speed of these long waves.

A confusing feature of a tsunami is that it seems to concentrate its strength in some places but avoids others. This is why Dr. Walter Munk, of the University of California Geophysical Institute at La Jolla, is measuring these long waves in the sea and trying to find out how they behave. In 1960 a wave that was set off by the great Chilean earthquake spread over the Pacific and was overtaken by the Early Warning System that helps us avoid disasters. The wave was only about one foot high by the time it reached some isolated islands in the South Pacific, but for some reason it concentrated in the channel leading to Hilo Harbor in Hawaii, and roared in as a 40-foot high wave smashing shops and houses and even bending parking meters to the ground. In other parts of the island the wave was not nearly so high, perhaps because much of its energy had been reflected and refracted by the steep coast line.

A warning system for tsunami can be made to work because earthquake waves travel

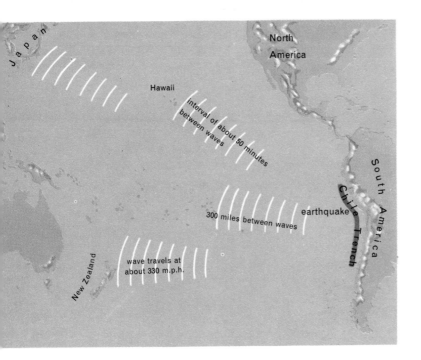

Japan

North America

Hawaii

interval of about 50 minutes between waves

300 miles between waves

earthquake

South America

Chile Trench

New Zealand

wave travels at about 330 m.p.h.

The tremendous speed and range of tsunami emphasize the need to trace as soon as possible their source and direction of travel. When these are known, meteorological stations can be alerted in areas like Hawaii and Japan.

across the oceans in only a matter of minutes and can be picked up by seismograph stations that can plot the center of the earthquake. As soon as the seismologists detect an underwater earthquake they first of all alert the meteorological departments in places like Japan and Hawaii. Then they keep in touch with stations that watch for any rise in water level in various parts of the ocean. With such information at hand it is usually possible to give several hours' warning to the authorities in any area that lies in the path of one of these destructive waves.

Winds of great strength—hurricanes, cyclones, and tornados—are also devastating. In sea areas such as large bays, hurricanes can pile up water at one end of the area and cause terrible flooding of low-lying beaches and coastal towns. This sometimes happens in Texas and elsewhere on the Gulf of Mexico. Along the Gulf coast there is a difference of only a few feet between normal low and high water, so the coastal towns and villages are built almost at sea level and are poorly defended against the sea. As a result a hurricane need raise the water only five feet or so to create extensive damage.

In Europe, the North Sea has caused many disasters on the coasts of England and the Low Countries. On the last night of January 1953 a gale-force wind from the Arctic drove the water of the North Sea in a southerly direction until it reached the Strait of Dover, where the "pincer" effect of the strait caused the water to pile up. At the same time, this *storm surge*, as it is called, was reinforced by high spring tides. Consequently, the water level rose rapidly in the southern part of the North Sea and the hitherto adequate sea defenses in many parts of eastern England and the Netherlands were swamped. Today, our knowledge of waves and water movements is much greater and our warning systems are becoming more efficient, yet they did not prevent a terrible loss of life and destruction in the Hamburg floods of February 1962.

As we have seen, apart from the comparatively rare freaks of nature, waves are a surface feature of the oceans. There are, however, other water movements of a completely different nature. These are the currents, so important to marine navigators and to people everywhere, for they strongly influence our climate and our way of life.

# 6 Currents of the Sea

The waves that are raised by the winds carry with them a great deal of energy. But, as we have seen, the water making up the waves does not move along with them, except when the waves reach very shallow water. The wind does, however, produce a continuous slow movement of the water—of a few miles an hour—on the surface of the sea, and these slow movements are called *currents*.

A man can walk as fast as the currents move, so they are much slower than waves, which may roll along at the speed of a car. Most of the currents move at a speed of half a knot or so, but some, such as those found at or near the equator, move faster, at speeds of one or two knots. It is as well to point out here that 1 knot = 1 sea-mile per hour = 6080 feet per hour = 1.15 miles per hour.

For the most part surface currents are not very deep, their effects fading away at a depth

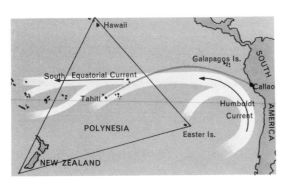

The courses of the slow-moving surface currents of the oceans are determined by the direction of the prevailing winds in each hemisphere. In his attempt to prove that the earliest settlers in Polynesia could have sailed from South America (see map at left) Thor Heyerdahl relied on the Humboldt and South Equatorial currents in the South Pacific to carry his raft *Kon-Tiki* from Peru to Polynesia. Above, *Kon-Tiki*, named after a Polynesian sun god, approaches a dangerous coral reef in the Tuamotu Archipelago, east of Tahiti, after a journey of 4300 miles from Callao.

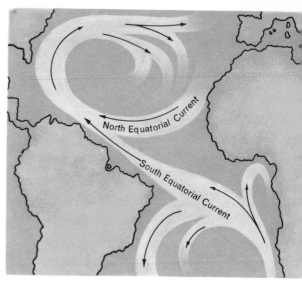

of only a few hundred feet below the surface. The warmer the water the less dense it is and many surface currents carry warm water. The Gulf Stream for instance brings tropical water from the Gulf of Mexico at a speed of four to five knots at the surface, but where at depths of about 1200 feet it meets and mixes with a colder, denser countercurrent most of its force is lost.

The chart on page 88 shows the great surface-current systems of the world. In the days of sailing ships these currents were of much service when the navigator's course could be set in the direction of the stream, but if the course lay counter to the current, care had to be taken to sail either to one or the other side of it. Captains of ocean liners and merchant ships today make use of the currents wherever possible, especially if trying to make a record run, but the force of the currents is strong enough to impede even these great ships. Although modern steamships can push their way against currents, their captains usually avoid them, so as to save time and fuel. The tankers and cargo ships that run up and down along the eastern seaboard of America use their knowledge of the Gulf Stream by keeping close inshore when going south.

Long before man appeared on the scene

If you blow across the top of a pan of water the current thus formed turns to the right and left at the opposite side and moves back toward you along the edge of the pan. Similarly, the South Equatorial Current, blown westward by trade winds across the Atlantic, is deflected on approaching the American coast.

currents were playing an important part—as they still do today—in distributing plants and animals around the world. For example, as soon as there is enough soil on the coral atolls described on page 66, floating coconuts washed ashore can take root and grow into palm trees. Larvae floating in the plankton may be carried far away from their original habitat, and if conditions are favorable can colonize the distant shores. Insects and seeds can also travel far on floating logs.

We have already seen that man has always made use of the currents for getting across the sea. Some centuries ago, it is thought by some anthropologists, Easter Island in the Pacific might have been reached by the Incas from Peru by using the prevailing winds and currents to carry rafts. The Norwegian, Heyerdahl, and his team of five set out to show that this was possible by floating on a raft of balsa wood from Peru to the Society Islands, a distance of 4300 nautical miles, which they covered in 101 days.

The steady trade winds that blow toward the equator—from the northeast in the Northern Hemisphere, and from the southeast in the Southern Hemisphere—make the surface layers of water move across the Atlantic and Pacific oceans from east to west. These great westward-moving currents (known as the North Equatorial and South Equatorial currents) give rise to the many complex movements of the water in other parts of the oceans. Sandwiched between these two large currents is a countercurrent flowing in the opposite direction. It is found in all three oceans—the Atlantic, Pacific, and Indian. When the North and South Equatorial currents approach the barriers of the continents they are deflected and swing toward the south in the Southern Hemisphere and toward the north in the Northern Hemisphere.

You can make a current of your own by blowing gently across the top of a pan of water. The current continues moving across the pan until it reaches the opposite edge. Here it is stopped, swirls around, and begins moving back along the edge toward you; it must get back to replace the water that you have blown away along the "equator" of the pan. Much the same thing happens in the oceans, only in a much more complex way. This is why the Gulf Stream moves from west to east across the Atlantic; it is bringing water back to replace that which has been blown from east to west by the trade winds along the equator. While this picture explains the general surface motion of the water in the Atlantic and Pacific oceans, the picture in the Indian Ocean is quite different.

In the Indian Ocean the monsoon winds are not steady like the trades, but change their direction with the seasons. From August to October there is a southwest monsoon that sets up an eastward-flowing current, then during the winter when the monsoon reverses, there is a westward-flowing current. Because of this complex wind pattern it is more difficult to work out a regular current system for the Indian Ocean than for the Atlantic and the Pacific. We know that changes in wind can bring about changes in the currents, but it is more difficult to measure how large the

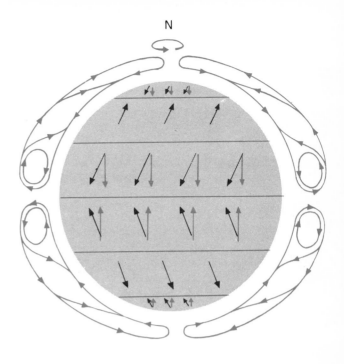

Diagram of world air circulation and winds. When heated by Sun at the equator, air (red) rises and moves toward the poles, where it cools, sinks, and turns back toward the equator. Blue arrows show prevailing wind directions if the Earth did not rotate. Black arrows show actual wind directions caused by rotation.

wind effect is and how long it takes to build up.

Winds are caused by a combination of heating by the Sun's rays and the rotation of the Earth. These same two forces—heat from the Sun, and spinning of the planet—also play a direct part in the circulation of the currents in the oceans. The rotation of the Earth affects the currents and winds by making them twist around slightly—to the right in the Northern Hemisphere and to the left in the Southern Hemisphere. It is not difficult to understand why this happens if you imagine an observer sitting on the North Pole. He would be turned around once every 24 hours as if he were on a turntable. At the equator there would be no turntable effect, only a rushing through space like a stone whirled on the end of a string. But if our observer attempted to rush from the equator to either the North or South Pole he would be unable to travel in a straight line; he would be pushed slightly sideways because the Earth is spinning. This

is just what happens to the ocean currents and to the winds, and the force causing it is known as the *Coriolis effect*. While it is felt most strongly at the poles, it becomes less and less as you travel toward the equator, until at the equator itself the force is zero.

William S. von Arx of the Woods Hole Oceanographic Institution, Massachusetts, has made large models of the Northern and Southern hemispheres in flat dishes of water, which he rotates. At the same time as the dishes are rotated, jets of air are blown onto the surface of the water to represent the winds. The dishes contain plaster models of the continents, and the currents that are caused by wind and rotation are bounced off the land in a most realistic way. The great advantage of models is that the various forces of nature can be switched on or off at will, or they can be made large or small. It is with such experiments that we find that the rotation of the Earth is needed to explain the troublesome current that creeps down the coast of Labrador from the icy north. The Labrador Current is well known to fishermen because when it meets the Gulf Stream dense fogs occur, and it also carries icebergs down into the shipping lanes, thus creating hazards to transatlantic liners.

Above: This revolving model simulates flow of wind-driven currents in the Northern Hemisphere. Different shades of dye in the water trace the courses of the currents.

Below: Deep-water currents in the Atlantic between Antarctic (left) and Arctic. Water-salinity figures at specific points are given in parts per thousand by weight.

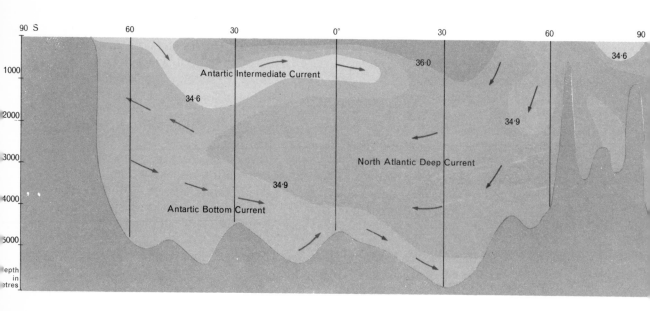

## Temperature and Salinity Cause Currents

All the time that the wind is setting up waves and surface currents, and the rotation of our planet is twisting the currents, the Sun is heating the water in the tropics while ice is cooling the sea in the polar regions. This difference in temperature in different regions of the oceans also helps to set up currents. When water is warmed it becomes lighter and tends to rise; when it is cooled it tends to sink. This means that deep movements of water throughout the oceans must be taking place. The two great sources of cold water are the Weddell Sea in the Antarctic and the water in the extreme north of the Atlantic. This cold water from the north and south polar regions sinks and flows along the bed of the oceans toward the equator. Along the entire length of the Atlantic the cold water flowing from the north retains its identity, but when it reaches the south Atlantic and meets the even-colder bottom water flowing north from the Weddell Sea something happens. Because the south polar water is colder, it wedges itself beneath the lighter north polar water. Temperature, then, in addition to wind, is another cause of the mixing of the ocean water. Both of these deep, slow-moving currents, by the way, help to replace surface water that has been blown westward along the equatorial currents. So one current that takes away water must have compensators, either at the surface or deep down, that return water from another source.

There is something else that makes ocean water heavy or light and which, like the temperature, makes it sink or rise. This is the amount of salt that the water contains. The more salty the water is, the heavier it becomes, so that salt water usually sinks under fresh water. In the Mediterranean, which is an enclosed sea except for the narrow entrance at the Strait of Gibraltar, the evaporation is large because of the hot sun that beats down for most of the year. This causes a greater evaporation of water than can be replaced by rainfall and river run-off. During the winter, the surface water of the Mediterranean is cooled and sinks, which

means that there is a constant supply of deep water made dense by its high salt content. This deep water, which is much denser than neighboring water on the Atlantic side of the strait, slides out of the strait into the Atlantic as a heavy current. Once in the Atlantic this dense, salty current sinks to a depth of half a mile or so and spreads out into the north Atlantic.

This constant loss of deep Mediterranean water must be replaced from the Atlantic. The result is a surface current flowing into the Mediterranean through the strait but on top of the heavy salty water flowing out. This is why ancient sailors of the Mediterranean found it so hard to battle their way out into the Atlantic—the surface current was against them. In World War II submarines trying to get into the Mediterranean had trouble, for at their operating depths they met the salty water flowing out. Italian submarines, however, could drift out in the deep current without giving themselves away by the noise of their propellers.

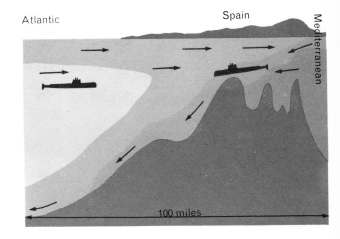

Water made salty and dense by evaporation in the Mediterranean Sea sinks and flows westward through the Strait of Gibraltar. It is replaced by lighter, less saline water flowing above it from the Atlantic. During World War II Italian submarines used both currents to drift with engines silenced past Britain's naval base at Gibraltar.

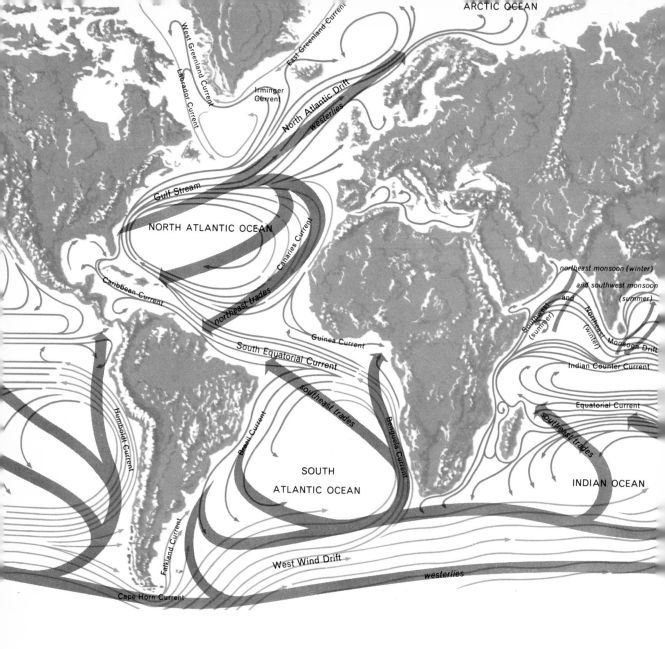

This map of the world illustrates how the warm (red) and cold (blue) surface currents of the oceans are driven by prevailing winds (grey) in the Northern and Southern hemispheres. The trade winds blowing from the northeast in the Northern Hemisphere and from the southeast in the Southern Hemisphere are responsible for the great North and South Equatorial currents flowing westward across both the Atlantic and Pacific oceans. On reaching the continents the currents are deflected northward in the Northern Hemisphere and southward in the Southern Hemisphere. Between the equatorial currents is a countercurrent that flows eastward and is found in the Indian Ocean as well as the Atlantic and Pacific. The pattern

of currents in the Indian Ocean is complicated by monsoon winds that are not steady like the trades. From August to October the southwest monsoon drives an eastward-flowing current, while in the winter the monsoon reverses, setting up a current that flows westward. Several of the warm equatorial currents are connected with cold currents flowing from the polar regions. In the Northern Hemisphere are the Oyashio Current from Bering Strait, the Labrador Current from Smith Sound and Baffin Bay, and the California and Canaries currents joining the North Equatorial currents. South of the equator are the Humboldt (Peru), Benguela, and West Australian currents, and the west-wind drift that encircles Antarctica.

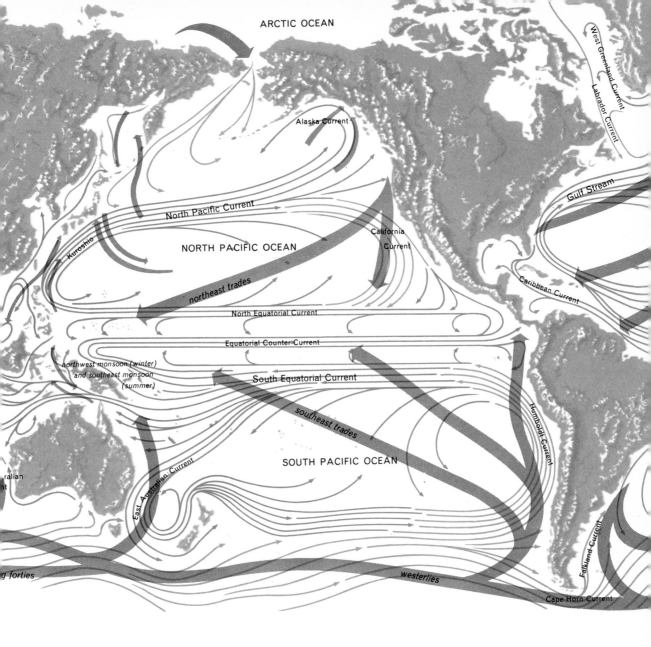

ARCTIC OCEAN

West Greenland Current

Labrador Current

Alaska Current

Gulf Stream

North Pacific Current

Kuroshio

NORTH PACIFIC OCEAN

California Current

Caribbean Current

northeast trades

North Equatorial Current

Equatorial Counter Current

northwest monsoon (winter) and southeast monsoon (summer)

South Equatorial Current

southeast trades

Humboldt Current

SOUTH PACIFIC OCEAN

ralian t

East Australian Current

g forties

westerlies

Falkland Current

Cape Horn Current

Circulation of deep cold-water currents (right) follows a different pattern from surface currents. The two main sources of cold water are the Weddell Sea, in the Antarctic, and the North Atlantic to the south of Greenland. Large well-defined currents flow from the poles toward the equator. These slow-moving currents help to replace warm surface water that has been blown westward along the equatorial currents in the Atlantic, Pacific, and Indian oceans.

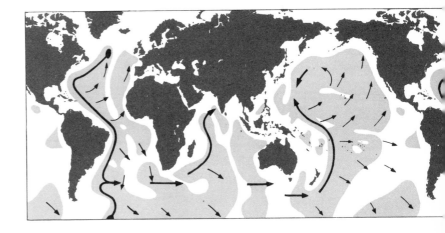

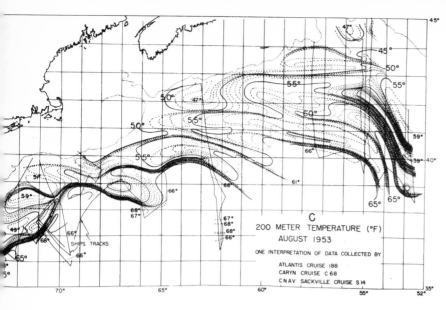

C
200 METER TEMPERATURE (°F)
AUGUST 1953
ONE INTERPRETATION OF DATA COLLECTED BY
ATLANTIS CRUISE 188
CARYN CRUISE C 68
C N A V SACKVILLE CRUISE S 14

Benjamin Franklin's map of the Gulf Stream (see page 21) showed a broad, smooth-sided current flowing across the North Atlantic. This chart, made in 1953 by the Woods Hole Oceanographic Institution, suggests that it is not one but a series of currents that overlap and at various points meander like a river.

You might suppose that water of different temperature and salinity would rapidly become mixed into one uniform mass of sea by the action of waves and currents. This does not happen because the waves affect only the top hundred or so feet, while the currents are sluggish and tend to flow as a uniform mass of water. Apart from regions near certain coasts, where deep water wells up from the bottom, most of the mixing occurs in a thin layer at the surface and along the borders of currents.

Yet both of these areas are small compared with the total volume of all the currents. The mixing is so slow, in fact, that measurement of the temperature and salinity of any current provides a label by which we can follow the current during its travels. For example, the water near the bottom in the tropical part of the Atlantic is only a few degrees above freezing; this is because it has flowed there from the polar regions. We know that this is so because we can trace, say, the North Atlantic deep water back to its source in the Arctic by taking temperature and salinity readings all along the way. We can also use temperature and salinity measurements to follow the salty deep Mediterranean water as it flows out into the Atlantic and joins forces with colder water. Temperature and salinity measurements are very useful because they show the positions of the top and bottom of the currents as well as marking their sides.

While some of the great "rivers" of the oceans are more sharply bounded than others, of course none of them is held in by "banks" as are the rivers on land. Even the Gulf Stream, one of the fastest and most sharply defined of all the large ocean currents, wanders about from month to month, and sends off little eddies of side currents as it goes along. This is one of the reasons why old charts show the Gulf Stream to be wider than we now know it to be. Those who plotted it included *all* observations on a single chart, and in this way confused the stream's several different paths as one large river. They probably also made mistakes in their position fixing, and in this way reported the stream outside its real course.

The Gulf Stream is strong and narrow where it funnels out of the northeast corner of the Gulf of Mexico. The feeding water for the Gulf Stream comes from both the North and South Equatorial currents—but mainly from the North—which are driven across the Atlantic by the trade winds. (The small Equatorial Countercurrent in the doldrum area of calm between the northeast and southeast trades reduces the general westerly flow by only a small amount.) The top part of South America deflects the equatorial current system up to the Gulf of Mexico. When you look at a map of surface currents in the Atlantic the bending of the equatorial cur-

rents when they meet the land seems perfectly natural. Fed by this water that has traveled westward across the Atlantic, the Gulf Stream passes out through the Strait of Florida as the Florida Current, then swings northeastward up the coast before turning toward Europe and being helped along by the westerly winds that blow in the northern latitudes. By this time the Stream has become four currents. The Labrador Current pushes down from the north, but the Gulf Stream continues on its way, the northern part reaching beyond Iceland and Scandinavia, and the southern part completing the circuit by rejoining the North Equatorial Current that flows back westward.

A similar circular pattern of currents occurs in the South Atlantic. The Brazil Current has to share the westward flow of the South Equatorial Current with the Gulf Stream, and so it is smaller than its famous opposite number in the north. The cold Benguela Current, which runs up the west coast of Africa, is the best known of the South Atlantic currents.

As you might expect, from the way in which the ocean currents are produced, the pattern in the Pacific is much like that in the Atlantic. The North Equatorial Current is the largest in the world and runs for 9000 miles from Panama to the Philippine Islands. Although it has no trap, like the Gulf of Mexico, to catch it, it is nevertheless de-

19th-century print of a hurricane in Japan. Storms of this kind usually develop on the western sides of oceans in regions of warm equatorial currents—in this case the Kuroshio Current off Japan's southeast coast.

flected. The general right-hand twist in the Northern Hemisphere, produced by the Earth's rotation, combined with the continental barrier of Asia, deflects the North Equatorial Current to form the Japanese current known as the Kuroshio. This current flows all the way across the North Pacific Ocean, spreading out and turning south in much the same way as the Gulf Stream. The Pacific also has a current similar to the cold Labrador Current—the cold Oyashio Current which sweeps down along the Kuril Islands from Alaska. In the South Pacific the system of currents is more muddled because the South Equatorial Current is continually being broken up into large eddies by the many islands in its path. The famous cold Humboldt (or Peru) Current brings water from the Antarctic up the western coast of South America to form the link with the equatorial current that carried the *Kon-Tiki* on its long voyage.

**Measuring the Currents**

The detailed maps of the surface currents

Chart (above) of mid-Atlantic floor (vertical scale much exaggerated) shows temperature variations at different depths measured by sampler bottles like the one below.

available to us today have been pieced together over the years from reports of drift by ships. By dead reckoning a ship's captain can tell where his ship *should* be, based on his compass reading and the ship's speed through the water. But a favorable current may have added to this speed. If the navigator takes star sights at the end of the day he can plot the ship's *actual* position. The difference between the actual position and the dead reckoning position will reveal the speed of the current carrying the ship along. A current of even one knot may put the reckoning out by 24 miles a day, so it is not difficult for merchant ships to make reliable estimates of the speeds of the currents.

The oceanographer adds to the knowledge obtained from ships' reports by sampling the sea water and so tracing the currents. We have seen how the temperature and the amount of salt in the sea can act as labels for a current, so that the paths of the currents, both on the surface and at great depths, can be followed. By tracking the currents a general plan of the ocean circulation can be drawn, and this helps to fill in those places on the current maps where there are few or no reports.

There is yet another way of using water samples. From temperature and salt measurements the density or "weight" of the water can be calculated. Because heavier water tends to sink and light water tends to rise, it is possible to work out where there *should* be deep currents and how fast and in what directions they are flowing. This is similar to the way in which meteorologists calculate the direction and speed of winds from charts that show differences in barometric pressure over large regions. Air moves from areas of high pressure to areas where the pressure is lower; and the greater the difference in pressure, the faster the wind blows. In a similar way areas of denser water in the sea flow toward areas of less dense water, an example being the dense salty water flowing out of the Mediterranean into the less dense water of the Atlantic. It is for this reason that oceanographers spend many tedious months at sea, hove-to in rough weather more often than not, lowering strings of carefully spaced water bottles, each fitted with 2 thermometers, until they have obtained a water sample from every few hundred feet from top to bottom of the sea.

By studying the density readings taken at all depths in the oceans, oceanographers like Professor Henry Stommel of Woods Hole can calculate not only the surface currents, but also those that are flowing unseen below the surface. Sometimes the observed flow at the surface is not matched properly by the density readings. In such cases it has been necessary to "invent" extra currents between the surface and the sea floor. In this way oceanographers forecast that there *should* be a current flowing beneath the Gulf Stream, but in the opposite direction. And again, there *should* be a thin ribbon of current flowing eastward—against the normal surface current —below the equator in the Pacific. True to prediction, both these currents were found in due course.

It is not easy to check these deep currents by direct measurement from a ship, because the ship itself drifts with the wind and with currents flowing at the surface. J. C. Swallow of Britain's National Institute of Oceanography has overcome this problem by inventing an instrument capable of following the water movement at any depth in the oceans. Special metal floats that can be made to sink to any desired depth are thrown overboard and track the movement of a deep current by

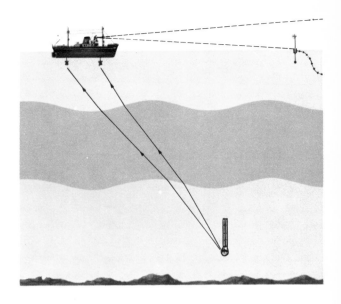

Movement of currents is measured by Swallow floats transmitting signals as they drift at various depths. The listening ship's position is fixed by radar from an anchored buoy.

transmitting "pings" to the listening ship. The floats are made of aluminum tubes sealed at the ends. They sink at first because they are weighted to be heavier than water. As the float sinks deeper it is compressed by the pressure of the water around it. But because the aluminum tube is less compressible than the surrounding sea water itself, there comes a depth at which the water becomes dense enough to support the tube. When the float reaches this depth it remains there drifting in whichever way the currents take it. The depth to which the float is intended to sink is determined by the way the float is weighted before it is thrown overboard.

The pings transmitted periodically by the float are received by hydrophones hung over the side of the listening ship. But in order to plot the changing position of the moving float the ship must in some way "anchor" itself in a fixed position. Since a depth of three miles of water is too great for conventional anchors, the listening ship fixes itself by radar from a buoy that is anchored to the sea bed by a line. Even this is not quite good enough, because a buoy on the end of a three-mile line can wander about. To overcome this difficulty the buoy is checked every few hours by careful soundings of the sea bed all around it. If the buoy drifts, then the whole pattern of the surrounding sea bed will appear to have changed. Once the amount of drift of the buoy is known, corrections can be made and the exact position of the ship determined.

Swallow's floats have shown that the currents are faster and more variable at all depths than had been expected. For example, the deep current that flows from the Arctic down the Atlantic gradually moves southward, but on any particular day it may stop or flow to the north for a while. For this reason, the floats, which at first operated only for a few days, have been made to last for a month or more. This means that good average values of the currents can now be obtained. The Swallow floats are being used all over the world now to check the calculations made by other oceanographers. No attempt is made to recover the floats, by the way. It is less expensive to make new ones than to try to recover the old ones.

There are many reasons why oceanographers want to know how the currents flow in deep water. For one, radioactive wastes from power stations may one day be dumped far out at sea, "safely" away from human habitation. But it would be useless doing this if deep currents were to bring any of the material that leaked out of the containers back to land again after a few months or years. Another reason for wanting to understand the movement of deep water is the effect the currents have on life in the sea. The amount of food available to fish depends largely on the movement of the water. There is very little plant or animal life in the open expanses of the Pacific, for example, even though the water there is pleasantly warm and light. This is a "desert" region, so to speak, because the top few hundred feet of water are stagnant. The warm water at the surface tends to stay at the surface and not mix with deeper water because it is lighter than the cold water underneath. Under these conditions all the phosphates, nitrates, and carbonates that the sea plants and animals need are very quickly used up.

The places where things grow extremely well in the seas are where the deep ocean

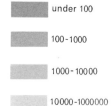

| | under 100 |
| | 100 - 1000 |
| | 1000 - 10000 |
| | 10000 - 1000000 |

Effect of upwelling currents. The key indicates number of diatoms (tiny plant organisms on which sea animals feed) in top 35 fathoms of an area of "fertile" water off the coast of California in June 1938.

currents meet each other and force water from the sea bottom up to the surface. If ever we are to grow large quantities of food in the sea we must know how best to make use of these "upwelling" currents as they are called; and to do this we must know how to measure and calculate the behavior of the deep currents.

The climate in many parts of the world is very much influenced by the oceans. Great masses of warm water may be carried into cold latitudes, while cold water is carried into warmer latitudes. The warm water of the Gulf Stream helps to keep the British Isles warm. Norwegian harbors enjoy ice-free conditions because warm water from the Atlantic flows into the Norwegian Sea. The monsoon gathers summer rain from the sea for India, while the cold Labrador Current makes the water so cold at American seaside resorts in the northeastern state of Maine that bathing is extremely uncomfortable.

As the Sun beats down on our planet, about two thirds of its heat is collected by the sea. The sea does not, however, get very hot in summer or very cold in winter. At the western

Icebergs carried by currents from the polar regions are often a menace to shipping in the North Atlantic and Antarctic. These icebergs in the west-wind drift were photographed between Cape Horn and Graham Land.

approaches to the English Channel, for instance, the surface temperature of the sea ranges from 47°F. in winter to 61°F. in summer; while off the Florida Keys it varies between 70°F. in winter and 82°F. in summer. The small temperature ranges of the oceans are accounted for by the fact that water stores heat more effectively than the land. The waves and the currents help out to a large extent by mixing cold water with warm over many parts of the oceans. Even the slow-moving deep currents that flow toward the equator from the poles act as giant stirrers for the oceans' heat.

If we are to understand the weather properly, and especially if we are to explain why changes in climate occur over periods of tens or hundreds of years, we must find out more about the currents, particularly about the deep ones. It used to be thought that the deep water of the oceans moved along at about one mile a day. But these results do not agree with the measurements made of Carbon-14. A study of Carbon-14 can provide an "age" for the water near the sea floor by revealing how long it has been removed from exposure to carbon dioxide contained in the atmosphere. In other words, great masses of surface water can sink to the bottom and retain their identity as they flow slowly over the sea floor. The Swallow float measurements in the eastern Atlantic give the rate as one to three miles a day. The more rapid transport of water masses at great depths may account for the fact that there are discrepancies in the values of Carbon-14. New apparatus, then, and the observations that are collected with it, introduce new problems to be solved—but each time we come to understand a little more about the oceans.

# 7 Movement of the Tides

The movement toward and away from the shore is the most obvious effect of the tides, so much so that we say that the tide is "coming in" or "going out." The early mariners of the Mediterranean, accustomed to the relatively small tides found in that sea, were greatly surprised when they ventured beyond the Mediterranean and for the first time encountered impressively large tides. Alexander the Great's men found themselves in serious trouble during their voyage down the Indus when they anchored their boats near the mouth of the river and were caught by the rising tide of the Indian Ocean. In 55 B.C. when Caesar beached his ships on the Kentish coast of England a 20-foot spring tide, reinforced by wind, left his fleet high and dry.

In addition to the regular to-and-fro motion of the tides there are local effects called *tidal currents.* Unlike the ponderous currents that slowly wind their way across the oceans, tidal currents are relatively fast streams that run for a few hours in one direction and then swing around the other way. Tidal currents can be very dangerous when they flow in narrow channels, like the Maelstrom by the Lofoten Islands in the Norwegian Sea, which swirls around so fast that whirlpools are formed. Boats have been known to be sucked down into these giant eddies. One of the most dangerous tidal current areas in the world is found in the Aleutians off Alaska in

the Akutan and Unalga passes. Many ships have been caught by these swift currents and dashed against the rocks. Other dangerous tidal currents are found in Discovery Passage and in Seymour Narrows, British Columbia. Compared with the speeds of the tides of the open oceans, which move at about a quarter of a knot, some tidal currents move at 10 knots or more.

## Tides, Moon, and Sun

The tides, like the great currents, occur all over the oceans, although their effects are much stronger in some parts of the world than in other parts. Tides are caused by the gravitational attraction between the Moon and the Earth, the effect of which is more strongly shown by the water than the land. As the Earth rotates, this attraction is more marked at that part of the Earth's surface nearest to the Moon. As a result the water "bulges out" toward the satellite. On the far side of the Earth, away from the Moon, this "pull" acts more strongly on the solid part of the sea bed, and so the water is "left behind" —the diagram below will make this clear— so that for one rotation of the Earth two high tides occur.

If you imagine the Earth chopped in half you can see why this happens. The half nearest to the Moon must be considered first. In this case the land part of the Earth (the

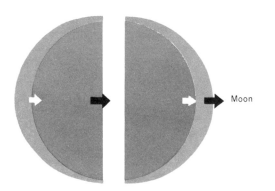

Tides are caused by the gravitational attraction between the Moon and the Earth. On the side of the Earth facing the Moon the sea is raised slightly toward the satellite. On the other side of the world the ocean floor, being nearer the Moon than the oceans themselves, is attracted more strongly. As a result the sea is "left behind" and bulges outward slightly from the Earth. The oceans most affected by the Moon's pull are those nearest the axis of the Moon and Earth.

In most rivers the flood tides raise the water level evenly or in a series of rolling waves (see diagrams at right). In certain estuaries, however, a large wave called a tidal bore rushes upriver for some miles. The Tsientang River (above) in China carries the world's largest, forming a destructive 11-foot vertical wave.

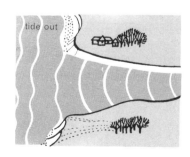

tide out

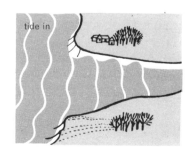

tide in

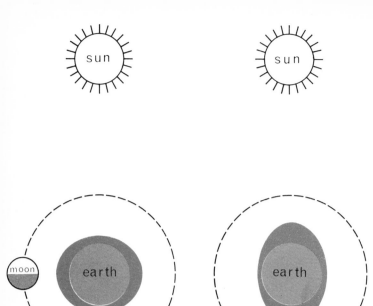

sea floor) is farther away from the Moon than the water on top; therefore, the water is attracted more than the land and so it "bulges" very slightly up toward the Moon. But the reverse happens on the other half—far side —of the Earth. Here it is the land part of the Earth that is closer to the Moon, so the land is pulled closer to the Moon than the more distant water at the surface. The sea is "left behind," so to speak, and therefore "bulges" outward slightly from the Earth.

In most places there are two high and two low tides in every *lunar* day of 24 hours and 50 minutes. (We call such tides semi-diurnal.) But since we reckon our *solar* day as 24 hours exactly, each high tide will be 25 minutes late by our time scale, and the total time lag per day will be 50 minutes. Why are the lunar and solar days different?

The reason for the 50 minutes' difference is this: The Moon takes 28 days to complete its orbit round the Earth, while the Earth makes one rotation in 24 hours. During the time that the Earth is completing one rotation, the Moon has moved along a little farther in its orbit. A point in the middle of the Atlantic Ocean does not reach its position directly under the Moon again until it has caught up with the distance that the Moon has traveled

in its orbit over a period of 24 hours—at a rate of about 50 minutes a day.

Why is it that the Sun, which is 27 million times heavier than the Moon, does not control the tides? The answer is that the Sun is hundreds of times farther away from the Earth than the Moon is, and the attraction due to gravitational forces becomes weaker and weaker as the distance between the attracting bodies increases.

When we take both mass and distance into consideration we find that the tide-raising force of the Sun is less than half that of the Moon. But this does not mean that the Sun does not affect the tides at all. It does. Sometimes the pull from the Sun reinforces the Moon's control of the sea; at other times the Sun and Moon work at cross purposes. This is why the tides vary in height from one week to the next. When the Sun, Moon, and Earth all lie along a line, which is at new moon and full moon, the combined attraction of the Sun and Moon produce high *spring tides*. When the Sun and Moon are pulling at right angles to one another we have the smaller *neap tides*. There are many other things, such as the Moon's varying distance from the Earth, and the Earth's varying distance from the Sun (because the orbits are not exact

circles), that produce smaller effects in the height of tides.

The shape of the coastlines and the varying depths of the sea floor together produce marked effects on the tides. In the large oceans, thousands of miles away from land, the pull of the Moon and the Sun on the water raises the level by only a foot or so, while in the middle of the English Channel the difference between high and low water may be as great as 30 feet.

A child on a swing goes higher and higher if someone pushes him each time the swing reaches its high point. It is much the same with the tides. Each lake, sea, and ocean is pulled very gently first one way and then the other twice a day; and 12 hours is just a good time interval to build up a large swing, or *oscillation*, for some parts of the sea. The semidiurnal tides occur with varying degrees of rise and fall. Local conditions affect the tidal height. Along the rocky coasts of the Bristol Channel, which is open to the Atlantic Ocean surging, there is a large rise and fall, and in some places no shore or beach is exposed at low tide, but the tide marks can be seen on the cliffs. Along gently sloping coastlines, on the other hand, long flat stretches of sand appear as the tide goes down. In the

Mediterranean the rise and fall of the tide is usually hardly noticeable, although a semidiurnal tide of up to six feet occurs in the Gulf of Tunis. On the other hand, in the Adriatic the tide only reaches to a height of about three feet. The Gulf of Mexico is another area of sea with a small tidal range of only a foot or two. And here the rhythm of the tides is different, as it is in the Adriatic. The tides are *diurnal*, producing only one high and one low water a day.

In parts of the open ocean the tides are hardly visible at all. You can demonstrate why this is so if you make the water oscillate in your bath. While the water rocks up and down at both ends, the rocking motion leaves a patch of water in the middle where there is hardly any up-and-down movement at all.

Some islands in the Pacific are in the middle region of oscillating water, so the people there experience only a very gentle daily change in sea level. They say that you can tell the time of day at Tahiti by examining the water line on the beach. This is because the island is situated in a place where the tide-raising effect of the Moon is canceled out. The Sun, then, is the controller of the water movement, and its gravitational force acts with an exact 12-hour period instead of the 12 hours and 25 minutes of the Moon. Sometimes the combined forces of Sun and Moon make one high tide very large and the next one very small. In places where this happens it appears as if there is only one tide a day, but careful measurements would show traces of a second smaller one.

On the whole the tides in the oceans are quite small. It is only when they swing back and forth in restricted places such as the Bristol Channel and the Bay of Fundy that they are able to build up to impressive rises and falls of 30 to 50 feet. The most spectacular effect of the tides is the *tidal bore*. When a strong tide enters a river it floods the river, causing its level to rise. Sometimes the flood builds up rapidly but evenly; at other times, depending on the river, a train of low, rolling waves travels upstream. But in other cases a large foaming wave forming a destructive wall of water several feet high rushes up the

Storm surges, caused by hurricanes during periods of very high tides, can raise water several feet higher than normal tide forecasts. This one occurred on the Florida coast in 1957.

river for several miles at considerable speed.

The world's largest tidal bore is found in the river Tsientang, which empties into the Bay of Hangchow in China. As the bore churns its way up the river it forms a wall of water reaching a height of 8 to 11 feet and extending almost in a straight line across the mile-wide river. Because this bore moves forward at about 14 miles an hour it is a most impressive sight and can be dangerous to boats on the river, yet some Chinese boatmen use the wave to carry their junks up river.

Danger from the tides can occur when gales or hurricanes strike during periods of particularly high tides, causing what is known as a *storm surge*. At such times the water may reach several feet higher than that predicted by normal tide forecasts. Early in 1953 wind and spring tides combined to cause a piling-up of water in the North Sea. Storm waves in addition to the abnormally high tidal water—which in some places was 10 feet higher than the predicted level—washed

houses into the sea and took the lives of more than 2000 people along the low-lying coast of Holland and the eastern coast of England. Learning from such disasters, oceanographers and meteorologists attempt to pool their knowledge of the tides and the weather. When they see that high tides are due at the same time as bad storms are expected they advise people living in low-lying areas nearby to leave.

In the Gulf of Mexico, where the normal rise and fall of tide amounts to only a foot or two, many houses are built on low ground and are in danger of being flooded during the hurricane season. To do extensive damage the water has to rise only a few feet. Many years ago a great flood catastrophe occurred at Galveston, Texas, but in 1961 a similar situation was forecast and people were evacuated before the hurricane with its floods arrived.

## Measuring the Tides

It would be very difficult to work out the tides for all the estuaries and other odd-shaped bodies of water found around the continents, yet sometimes oceanographers are asked to do just that. A ship's pilot, for instance, must know about tidal currents in relation to shallow channels, sandbanks, or other obstructions at the river mouth where a port may be situated. Large ships such as the *Île de France* and the *United States* have to wait for slack water before tying up at the pier in New York, so a precise knowledge is needed of how the tides will affect the flow of water at the Hudson River mouth at any particular moment.

From an economic point of view, a knowledge of the water movement that the tides cause in river estuaries, and in channels leading to ports, is as important as a knowledge of the overall rise and fall of the water level. British oceanographers are now measuring the flow of water in the estuary that is used by ships bound for the docks at Liverpool, one of the busiest ports in Europe. The complex flow of water is partly due to the river water running out and partly to the ebb and flow of tidal water that mixes with it. The different types of water are traced and labeled by measuring their density and salinity, in much the same way that the ocean currents are traced.

The study of river estuaries is important because it shows how channels can best be kept clear of mud brought down by the rivers. As we saw in the chapter about waves, it is essential that the harbor engineer has a knowledge of wave action along the coast where a harbor installation may be planned. In a similar way a knowledge of the effects of local tides is essential.

Ever since shipping began mariners have had to keep records of the tides as their ships moved from one port to another. Tide tables can be calculated with great accuracy, provided that at least one month's observations are available. It may seem surprising that a record extending over only one month is long enough, considering that the tides are exceptionally high at certain times of the year and low at others. However, it is the pattern of the rise and fall that we need to know, and this can be obtained by making observations for one complete revolution of the Moon around the Earth. And because astronomical calcula-

Harbor authorities at great ports like New York City on the Hudson River must keep exact records of tidal currents, river channels, and sandbanks to ensure safe docking for ships.

tions provide accurate information about the paths that the Moon and Earth will follow throughout the year, we can work out the way tide-raising forces will operate.

What we do not know, however, are the resonance effects that determine the exact way in which the water oscillates in the oceans. And we cannot always calculate in advance what to expect as the oceanic tides lap around the coastlines of the continents and meet shallow water. But there are tide-predicting machines that analyze one month's readings from a port. The machines separate the effect of the Moon and the Sun and also allow for local effects due to the resonance of the bays and estuaries. They produce *tidal constants* for the port, and these constants tell us whether the two tides a day are equal, or whether the Sun or the Moon is in control. They also tell us by how much resonance effects have increased the tide from normal, and how much build-up can be expected to take place due to funneling of the water. We can use these constants to calculate by how much the normal tide for any one month in the year may be altered by local conditions. With the help of tide-predicting machines tide tables can be prepared quickly and accurately for ports and other such places of interest the world over.

If you want to know the time and the height of high water for a place that is not shown in the tide table you merely have to look at the nearest place on either side for which values are given and make an estimate for your place in between. A tide table may be needed for an island well out to sea; for example, in the middle of the Persian Gulf where offshore drilling for oil is being carried out. In such cases it is advisable to take a month's readings because the tides may be quite different in the center of a basin of the sea from those at ports around the edge of the same basin.

Although the tide tables showing average conditions are perfectly satisfactory for the purposes of navigation, oceanographers take every opportunity to learn as much as they can about the detailed pattern of the water movement all over the world, especially from isolated islands. During the International Geophysical Year many new measurements of the tides were made. In some places automatic tide gauges that measure the continuous rise and fall of the tide were set up, yet in other places it was possible only to camp ashore for a month and make periodic readings of the changing height of water against a marked pole placed in the sea. Although the readings of the tide pole are good enough for working out average conditions calculated by machine, a continuous recorder shows additional things of interest to the oceanographer.

A continuous record can show small variations as well as the smooth curve of the main rise and fall of the tide. These small variations may be due to oscillations of the water contained in a bay or a harbor and set up by the wind or by changes in the barometric pressure. The time of oscillation for any body of water depends on its size. For the bathtub it can be a few seconds, for a large lake it may be several hours, and for a basin of the sea, longer still.

Considering the enormous amount of sea that is being moved back and forth every day by gravitational attraction of the Sun and Moon, why is it not possible to harness this water movement to provide electrical energy to light homes and operate factories? Tidal-power schemes, such as the Severn Barrage, have been discussed over the years and rejected because of cost, but the French have now begun constructing a new type of electricity generating plant at St. Malo on the English Channel.

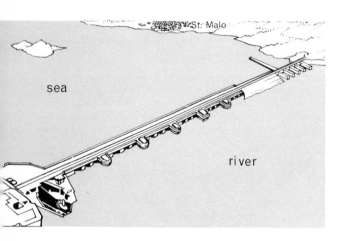

Above: In a scheme to harness tidal energy for hydroelectric power, a dam across the Rance estuary at St. Malo will have turbines that are operated by both incoming and outgoing tides. Below: The map indicates tidal ranges in feet at St. Malo and other English Channel areas.

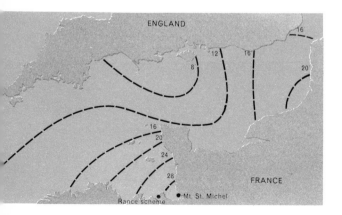

Above: At low tide the fortified abbey of Mont St. Michel, off the coast of Brittany, is left high and dry amid a vast expanse of sand flats. Below: At high tide the sea rapidly floods the surrounding plain and the abbey becomes an island connected only by a causeway to the mainland.

When it is completed a dam will extend across the estuary of the river Rance and will trap the large tides of 30 feet or more. Two-way turbines built within the dam will be operated first by the incoming tide, then by the outgoing tide. Because the turbines need a certain amount of water to operate them efficiently, they will also serve as pumps to adjust the water level in the estuary that may be required in order to resume power generation between tides. A reserve head of water can also be built up for times when peak power loads occur.

The Rance project will be very expensive, and there are only a few places in the world where the rise and fall of the tide is great enough to make this method of generating electricity worthwhile. Yet the French regard this project as the forerunner of a much more ambitious one. One day they hope to build a similar dam across the bay of Mont St. Michel and with this one dam supply enough power to satisfy half France's present needs for electricity.

| MAGNESIUM BROMIDE | ·22% | 365,000 tons |
| CALCIUM CARBONATE | ·34% | 564,000 tons |
| POTASSIUM SULFATE | 2·46% | 4 million tons |
| CALCIUM SULFATE | 3·6% | 6 million tons |
| MAGNESIUM SULFATE | 4·74% | 8 million tons |
| MAGNESIUM CHLORIDE | 10·88% | 27 million tons |
| SODIUM CHLORIDE | 77·76% | 120 million tons |

Above: The ocean's minerals come mainly from the land. The sea's surface water, evaporated by the Sun, forms clouds that bring fresh-water rain to the land. The water runs off the land into rivers that carry mineral-rich sediments down to the sea, where they accumulate. Below the ocean surface another cycle is in progress. Currents of upwelling water bring nutrients from the bottom, which support plant and animal life. Dead plant and animal matter sinks again and contributes further minerals to the sea floor. The table (left) shows relative quantities of the main minerals to be found in a cubic mile of sea water. The total quantity of salts in all the oceans is about 54,000 million million tons.

## 8 Minerals of the Sea

All the time that the oceans are being stirred by the currents, tides, and waves, they are being fed with minerals and other materials from the land. In an earlier chapter we saw how the sediments of the sea are formed by erosion of the continents, and how the shallow-water sediments of the continental shelves are sometimes thrust up to form new land. Meanwhile certain areas of the land gradually become shallow seas and in their turn accumulate sediments.

The forces that transport the sediments are wind and water—a seemingly endless supply of water that washes off the land century after century. As the Sun beats down on the oceans it evaporates the surface water. Water vapor rises into the atmosphere where it cools and forms clouds. Water droplets formed within the clouds fall as rain or snow, which feed the rivers, which return the water to the sea where the cycle begins all over again. During this endless process the land

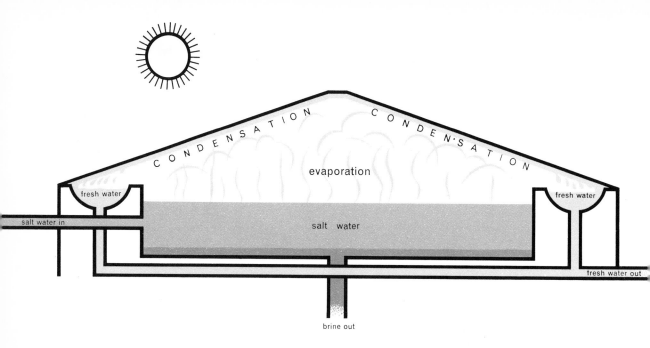

In this diagram the labels read: CONDENSATION, CONDENSATION, evaporation, fresh water, fresh water, salt water in, salt water, fresh water out, brine out

In this plant man exploits natural processes to help him obtain fresh water from the sea. The Sun's heat makes the sea water in the tank evaporate and rise, as fresh-water vapor, to the sloping glass roof of the plant. Here it condenses and runs into collecting troughs at both ends of the roof. Later, the brine is pumped from the bottom of the sea-water tank.

is constantly being worn away by the rivers, and minerals are deposited in all of the oceans.

There must have been a time in the Earth's history when there were no sediments—when the land consisted only of bare rock that had yet to be broken down by wind, frost, and rain. It was the weathering of the original rocks that provided the first sediments. The rains washed minerals into the great basins that were gradually filling up and becoming oceans—oceans quite different from those we know today. Millions of years before man had evolved into an intelligent maker and user of tools, the mineral content of the oceans had been accumulating.

As they have for millions of years, the rivers today continue to carry dissolved salts down to the sea as they scour the land. Even our drinking water contains small quantities of salts, some of which are deposited as "fur" on the inside of kettles. Yet the saltiness of the rivers is nothing like that of the sea. The

reason is that the cycle of events that moves water from sea to land and back again only involves the water and not the salts contained in it. Evaporation of water from the surface causes the salts to be deposited in the sea to enrich it. The rate at which the salt content of the sea has accumulated during the Earth's history is not known. Perhaps it was rapid at first; perhaps it was always a slow process. Now the less soluble residues are being carried down and today the total annual addition of salts to the ocean is less than a ten-millionth of those already present.

We make use of this one-way process when we "purify" water by distilling it—boiling the water to make steam and then condensing the steam back to water again. During the process the dissolved salts are left behind. A colored mixture of water and blue ink, for example, distills off clean, pure water. Nature's distillation plant works only at the surface of the oceans where the Sun supplies heat enough to form pure water by evaporation. As with the "fur" in the kettle, the concentrated salts are left behind in the sea water, and so the sea keeps its residue of salt. Small particles of sea salts, however, are sometimes carried onto the land by the wind, but the amount is infinitesimal when we consider the total quantity of salts in the sea.

It is quite surprising how much solid matter exists in a dissolved state in ocean

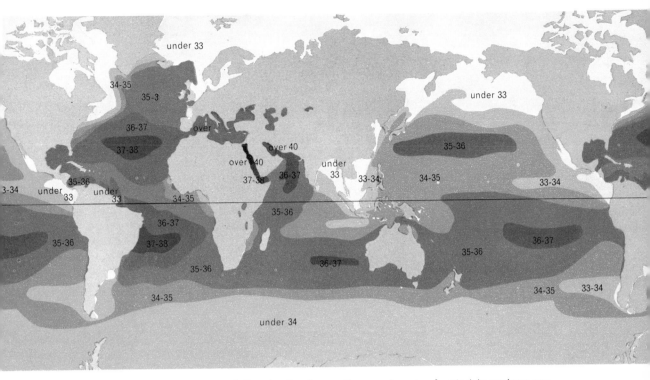

This map shows the salinity (parts of dissolved salts per thousand parts of water) in various sea areas. Land-locked waters in hot, dry regions are the saltiest; open seas in cold, rainy regions are least salty. Even so, the salinity variation between two such regions is usually quite small.

water. If you evaporate a pint of sea water you will have left about a teaspoonful of salt, but not clean, white table salt because the salt in the sea is a mixture of many different substances. A teaspoonful in a pint corresponds to about three and a half parts in a hundred. Considering that the oceans are enormous—covering about two thirds of the Earth's surface and averaging about three miles deep—when we work out the total weight of salts that the sea holds we find the staggering figure of 54,000 million million tons. If all of these salts could be extracted from the sea water they would form a layer more than 160 feet thick when spread over the entire surface of the Earth.

Perhaps it is easier to understand the wealth of the oceans if we consider the amount of various minerals that are dissolved in one cubic mile of sea water. A cube with a mile for each of its sides would hold 4000 million tons of water. But a cubic mile is only a small sample of the oceans, for there are 300 million cubic miles of sea. However, by thinking of one cubic mile at a time we avoid numbers which are so large that they are meaningless to most of us.

Our cubic mile of sea contains 166 million tons of salts altogether, ordinary table salt (sodium chloride) being the most abundant chemical compound making up this total. Because table salt accounts for about 120 million tons per cubic mile of water, this explains why sea water tastes as it does. The next most common compound is magnesium chloride. Most of the rest of the 166 million tons is made up of magnesium sulfate, calcium sulfate, and potassium sulfate. There are, for example, about eight million tons of magnesium sulfate (epsom salts) in one cubic mile of sea water. These five compounds account for all but a small fraction of the total salts of the sea. In addition there are small amounts of aluminum compounds together with compounds of iron, manganese, copper, lead, and even small traces of radium.

## Parts per million

| | |
|---|---|
| Chlorine | 18980 |
| Sodium | 10561 |
| Magnesium | 1272 |
| Sulfur | 884 |
| Calcium | 400 |
| Potassium | 380 |
| Bromine | 65 |
| Carbon | 28 |
| Strontium | 13 |
| Boron | 4·6 |
| Silicon | 4·0 |
| Fluorine | 1·4 |
| Nitrogen | ·7 |
| Aluminium | ·5 |
| Rubidium | ·2 |
| Lithium | ·1 |
| Phosphorus | ·1 |
| Barium | ·05 |
| Iodine | ·05 |

## Parts per thousand million

| | |
|---|---|
| Arsenic | 10·2 |
| Chromium | ·2 |
| Iron | ·2 |
| Manganese | ·10 |
| Copper | ·10 |
| Cadmium | ·05 |
| Cobalt | ·03 |

## Parts per million million

| | |
|---|---|
| Zinc | 5·0 |
| Lead | 4·0 |
| Selenium | 4·0 |
| Cesium | 2·0 |
| Uranium | 1·5 |
| Molybdenum | ·5 |
| Thorium | ·5 |
| Cerium | ·3 |
| Silver | ·3 |
| Vanadium | ·3 |
| Lanthanum | ·3 |
| Yttrium | ·3 |
| Nickel | ·1 |
| Scandium | ·04 |
| Mercury | ·03 |
| Gold | ·006 |
| Radium | 0·0000002 |

The main elements in proportion to given parts of sea water. The figure for gold represents several thousand tons throughout the oceans.

In fact almost all the metals are found in the sea, including gold and silver. This is just what we would expect, knowing that the oceans have long been gathering minerals from the weathering away of rocks on the land.

In some parts of the oceans there are more salts than in others. In general, those regions of the oceans receiving high rainfall, or located near the mouths of large rivers, are the least salty; and those that have a high rate of evaporation are the most salty. In spite of such local differences, the proportion of each metal or salt seems to vary by only small amounts in the oceans; these minerals have been well mixed by the stirring effect of the currents, tides, and waves. But these small differences are very important in the economy of the seas and must be carefully assessed. Today there are accurate methods of analyzing sea water and calculating the percentages of the dissolved salts.

Analyses are also made to determine the amount of silicate and phosphate present in sea water, for these compounds are very important to the population of the oceans. On this page is a list of the elements which occur in very small quantities—in parts per million, per thousand million, and per million million —but which nevertheless are becoming increasingly recognized as having a role to play in maintaining marine plant and animal life. We shall return to this point later in this chapter, on pages 114-15, where the concentration of rare elements is discussed.

### Measuring the Minerals

Salinity measurements through the entire width and depth of the oceans have been carried out systematically over the past 80 years or so. To carry out this task today oceanographers use strings of sampling bottles that can be lowered to any desired depth. When the bottles are in position a small "messenger" weight is allowed to slide down the wire supporting the bottles. As the weight hits the first bottle it trips a switch that makes the bottle reverse, close, and so trap a water sample. As the first bottle reverses it

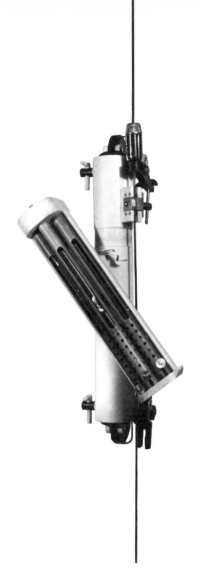

When an immediate chemical analysis is wanted, the total salt content of each sample can be measured in the oceanographic ship's laboratory. The disadvantage of chemical analysis, however, is that it is rather slow and one man can do only about 30 samples a day. For this reason another method is sometimes used—that of measuring the electrical conductivity of the water sample. In this way about 200 samples a day can be measured. It is also possible to tow instruments on a cable behind the ship and so obtain a continuous measurement of the electrical conductivity of the water at depths down to about 500 feet. This method of continuous measurement can be more useful to the oceanographer who is tracking currents than to the chemist, who usually prefers to bring a sample to the surface for more accurate measurement of its salt content.

Another method of finding out the concentration of salt and trace elements in sea water is by evaporation. If you pour a small amount of table salt in water it dissolves and forms a solution. If you then begin to evaporate the water there comes a time when there is too much salt for the remaining water to hold. When this stage is reached the salt collects itself together and forms solid crystals again. The crystals can then be separated from the water by filtering.

If you had started with a solution containing several different kinds of chemical compounds you would find that first one compound then another crystallized out as solids. This is because some compounds dissolve more easily than others. If they dissolve easily, then they only need a relatively small quantity of water to keep them in solution. When the water is removed by evaporation the compounds that need only a small quantity of water stay in solution longest. If you tried an evaporation experiment with sea water you would find that the first solids to crystallize out of solution would be calcium and magnesium carbonates; next would come calcium sulfate, and then sodium chloride mixed with chlorides of magnesium and potassium. Because the proportion of any one salt to another may differ from sea to

Suspended on a wire, Nansen reversing bottles collect water samples and temperature readings. Actuated by trip weights, each bottle closes while the thermometers reverse sharply to break the mercury thread.

breaks the mercury thread in the thermometers and releases a second small weight that slides down the wire and trips the second bottle, and so on down the line to the last bottle. In this way a salinity profile from the surface to the bottom, anywhere the oceanographer wishes to work, can be obtained. When the string of bottles is raised to the surface the water in each bottle is transferred to a special container so that it can be analyzed weeks or months later, or it can be immediately analyzed on board ship.

sea, the order of precipitation would be different for a sample of water taken from the Dead Sea, say, than for a sample taken from the Mediterranean.

There are salt flats in India where shallow ponds of sea water collect at certain times of the year. The hot Sun and the dry wind cause rapid evaporation of the water and salt is precipitated. This process has been going on in various parts of the world for millions of years, which is why we find layers of salt on the surface of the Earth and underground. Sometimes the salt is brought to the surface by mining, in much the same way that coal is obtained, but usually it is easier to bore a hole into the salt bed and pump in water to dissolve the salt. The salt can then be brought to the surface in a concentrated solution.

In some places, where very hot desert conditions exist today, there are hundreds of

At this French works at Batz, Brittany, sea water is guided along a system of channels and into rectangular salt pans. Here the water is lost by evaporation and the salt collected to be refined for commercial use.

feet of salt layers that over the centuries have crystallized out from ancient seas that have gradually dwindled to small lakes as they evaporated. All the time, as these old inland seas became more concentrated, different chemical compounds crystallized out so that now we find not only common salt but also valuable deposits of compounds containing bromine and iodine, which can be mined or dissolved and pumped to the surface for industrial use.

Britain, France, Germany, Poland, and the United States all have rich deposits of

Evaporating rapidly, shallow Kara Bougas Gulf receives salty water from the Caspian Sea. Its salinity is exceptionally high—160 parts per thousand against 13 for the Caspian and an average of 33-37 for the oceans as a whole.

This salt-mine tunnel in Louisiana, United States, is 500 feet below ground. It is possible that the Kara Bougas Gulf will eventually dry up completely, allowing similar industrial exploitation of its salts.

salts, some of them up to 2000 feet thick. Searles Lake in California and the exposed Salt Flats of Utah are outstanding examples. Today Searles Lake, which has a salt crust 50 to 70 feet thick, supplies about 50 per cent of all the lithium salts and borax mined throughout the world.

The Dead Sea is an example of this concentration process going on today. The water of the Dead Sea is so salty (26 per cent, most of which is magnesium chloride) that animals cannot live in it. Every year the river Jordan continues to bring down more salt, which is

rapidly accumulating as the water evaporates. One day the concentration of salts will become so large that crystals will form and sink to the bottom and make a new deposit of valuable materials. The Dead Sea is already being extensively worked for the rich supply of salts.

Even though the sea has with the passage of time returned some chemical compounds that we need in neatly packaged form as layers of solids on land, by far the greater proportion of the dissolved salts remains in solution in the oceans. Many of the layers of salt that are laid down by the evaporation of inland seas are eroded away and taken down to the ocean again by rivers. The wealth of the world, then, lies for the most part in the seas, but since nature's evaporation processes require millions of years to operate, chemists spend a great deal of research effort to discover their own means of concentrating and extracting the valuable chemical compounds directly from the oceans. After oceanographers have found out what quantity of each compound is available, methods are developed for extracting the minerals economically.

## Mining the Oceans

Mention has just been made of the Indian Salt Flats worked from the earliest historical times. The crystals of salt left by evaporation are simply raked into heaps and dried ready for market. In the future salt may be a valuable by-product of fresh-water distillation plants such as those at Kuwait, on the Persian Gulf, which until the last few years had to bring its water by ship from the mouth of the Euphrates River.

As the population of the world increases, many other areas of the world will find themselves short of fresh water, and water itself will become one of the most important products of the sea. Experimental distillation plants, operated by energy from the Sun, are being tested now in the United States, North Africa, Australia, Spain, and Italy. But what of those densely populated regions of the world that do not have a hot sunny climate to operate solar distillation plants and will need more fresh water than can be supplied at present? One day there may be a cheap supply of heat from atomic power stations, so that stills will not have to depend on heat from the Sun. This may make it worthwhile to run sea-water distillation plants on an enormous scale.

During the Second World War magnesium was in great demand for aircraft construction, as it still is today, because it makes a light and strong alloy. Scientists of the Dow Chemical Company in the United States

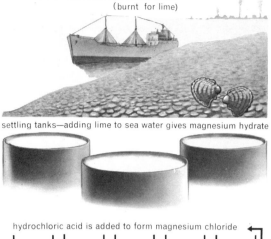
raw materials—sea water and oyster shells (burnt for lime)

settling tanks—adding lime to sea water gives magnesium hydrate

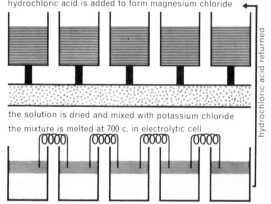
hydrochloric acid is added to form magnesium chloride

the solution is dried and mixed with potassium chloride
the mixture is melted at 700 c. in electrolytic cell

hydrochloric acid returned

floating metal scum is removed and formed into ingots

The diagram shows the various stages in the Dow Chemical Company's method of producing magnesium from sea water on the Texas coast. Oyster shells, providing calcium carbonate (lime), are collected from the Gulf of Mexico.

developed a process for tapping the four million tons of magnesium that are dissolved in every cubic mile of sea water. Now that the process has been perfected there is no need to rely entirely on geologists to discover concentrations of magnesium hidden in the rocks on land.

Another valuable mineral that is successfully extracted from sea water is bromine. There are about a quarter of a million tons of bromine in every cubic mile of sea water, so that there is a good supply of this raw material, even though the concentration is very small. The process of concentrating the

This plant at Hartlepool, England, using a different process, produces magnesium from sea water and calcium magnesium carbonate (dolomite). Each cubic mile of sea water contains about four million tons of dissolved magnesium, which is the third commonest element in the oceans.

bromine was invented in 1933, and apart from common salt this was the first mineral to be obtained from the sea by chemical methods. The bromine problem was solved first because there was a large demand for this material in the photographic industry and for producing anti-knock compounds for gasoline. In the future, when the need arises, chemical engineers will probably find ways of extracting other materials from the sea. In the meantime oceanographers are collecting all of the basic information that the engineers will require.

The prospect of mining gold from the seas has long fascinated men. Although estimates of the amount of gold contained in a cubic mile of sea water differ widely, there is more gold in the sea than there is in use now. This may sound like a way of getting rich quickly —until we work out the simple economics of concentrating and extracting the gold. An Australian set up a gold processing plant in 1935. Although he managed to extract a few ounces of the metal, the cost of processing the vast quantities of water and the chemicals needed to separate the gold were far greater than the value of the gold itself.

A much more ambitious investigation was

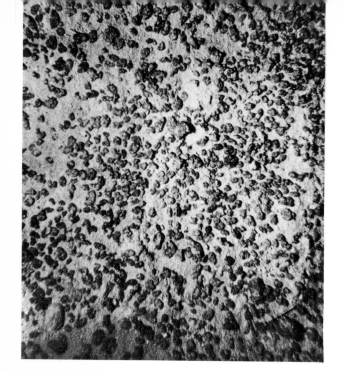

Above: Manganese nodules, first discovered by the *Challenger* expedition, are present in vast quantities on the floor of the Atlantic and Pacific oceans. Varying in diameter between about half an inch and ten inches, nodules build up layer by layer from a small hard core. Below: This section through a nodule is 37 times actual size. It shows several nuclei, including a piece of shark's tooth near the top and, below it, a fragment of volcanic rock.

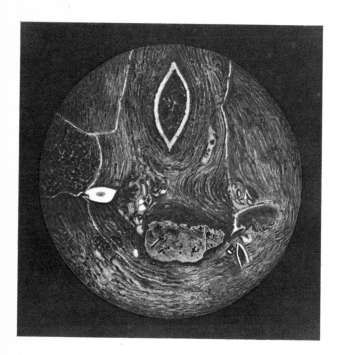

made by the Germans in 1924. They hoped to extract enough gold from the sea to pay their war debts, but before they went to the expense of building a concentrating plant they decided first to sample the oceans carefully. When they sent the oceanographic ship *Meteor* on a three-year research cruise in the Atlantic, they discovered that the concentration of gold was not great enough to pay for the cost of extracting it. Although this part of the *Meteor* expedition was disappointing, the expedition as a whole was to reveal an enormous amount of information about the oceans that has been of great value ever since.

There are some processes of concentration of minerals that are going on all the time in the sea, although we do not yet understand how they work. R. S. Dietz and the Coast Geodetic Survey, Washington, study manganese and phosphate rocks—or nodules, as they are called. Scattered over the sea bed, these nodules were discovered by the *Challenger* expedition of the last century. The peculiar lumps of valuable minerals are somehow built up layer by layer rather like an onion. Usually there is an inside hard core consisting of a piece of volcanic rock. It looks as if these hard cores have acted as seeds, rather like the seeds that are used to make oysters form pearls in the Japanese cultured-pearl industry. Millions of square miles of the Atlantic and Pacific sea floors are carpeted with nodules, waiting for someone to make his fortune by mining them from a depth of two to three miles or so in the sea.

Although iodine is one of the scarcest non-metals in the sea, men have been able to obtain it easily. For some reason, seaweed concentrates iodine from the water, so all we have to do is collect the weed and process it in order to extract the iodine. The following figures show the low concentration of iodine in water, and its high concentration in seaweed: While there is only about one gram of iodine in 80 tons of ocean water, we can extract one gram from only 200 grams of certain dried seaweed. One of the reasons why biologists and chemists study plants and animals in the sea is to find out how they concentrate materials needed by man. So far, iodine is the only mineral that has been obtained commercially from plants or animals of the sea, yet we know that other valuable substances are also concentrated by marine organisms. Lobsters manage to collect cobalt, even though its concentration in the sea is less than that of gold, and the oyster has an ingenious way of gathering stray atoms of copper from the water. Several animals that live on the sea bed concentrate the rare element vanadium, used in the steel industry, in their blood. How these animals manage the delicate separation process we have yet to find out.

The formation of petroleum is still another mystery associated with the sea. It may seem strange, but scientists have yet to explain just how petroleum is formed. We have a good idea that the formation process has something to do with the sea, because we always find petroleum near rocks that we know have been formed by the settling of sediments on the sea bed. One explanation is that petroleum is the remains of sea animals that died, drifted to the bottom, collected in the mud, and finally decayed, eventually producing a mixture of gas and crude oil. If this theory turns out to be correct, then the Black Sea may be a petroleum "breeding" ground today. While the upper levels of the Black Sea are rich with life, the bottom region cannot support life because of a concentration of hydrogen sulfide, which is poisonous. This means that as the surface creatures die and sink to the bottom there will be no scavengers to devour their remains. Their bodies then will become part of the bottom ooze, decay, and await the chemical process that converts them into gas and petroleum.

Other oil scientists think that it is the waxes and oils from the leaves of plants on land that are the main source of supply of the raw material for oil. If we knew the answer to the question of how oil is formed it would be easier to pinpoint the places where it lies waiting to be found. For this reason oil companies take a great interest in sampling the sea bed, and in trying to understand how the chemistry of life in the sea operates.

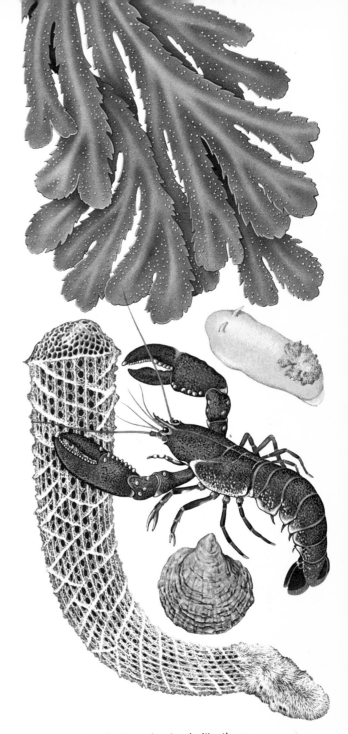

Many marine plants and animals like these concentrate minerals from sea water. From the top: seaweed (concentrates iodine), sea slug (vanadium), lobster (cobalt), oyster (copper), venus flower basket (silica).

# 9   The Great Chain of Life

Men have relied on the sea for food from time immemorial. Oysters and lobsters, considered a luxury today, in ancient times were probably a staple and very monotonous diet for many coastal people. Today the range of sea food available to us is much wider than it was in ancient times, simply because over the years improved techniques of fishing have been developed.

The problem today is not that of catching enough fish, but the possibility of catching too many and temporarily reducing the number of a particular species to a dangerously low level. In years ahead, as the world population continues to increase, this problem may become even greater. In 1950 the world population was 2550 million; at the present rate of increase it is expected to be more than double that—about 6300 million—by the year 2000.

How many fish of a given species are there in the sea? How do they feed? And what are their various life cycles? Finding the answers to such questions is the work of marine biologists. They study everything that lives in the

Plankton is caught in fine-mesh towing nets for study by marine biologists.

Left: The picture (with explanatory key, right) shows part of the great "food chain" in the sea. (*Neither fish nor water depth are to scale.*) Algae (1) near the surface are eaten by animal plankton (2), which include euphausians (3), main diet of whalebone whale (4). Herring (5) hunted by fishermen and sea birds are also eaten by cod (6), which are a common target of blue sharks (7) seen here with remora, or sucking fish (8), attached to them. In deeper water a giant squid (9) is fighting a sperm whale (10). On the (shallow) sea bed a lobster (11) and hermit crab (12) fall prey to a sand shark. Nutrients from decaying bones (13) near by may eventually be carried to the surface to provide food for algae, the first link in the chain.

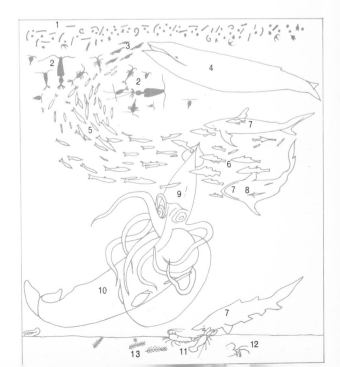

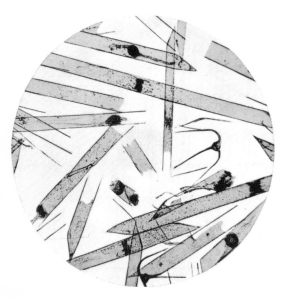

Above: Photomicrograph of phytoplankton. The long cylindrical shapes are diatoms, the smaller ones with whiplike tails are flagellates. These and other microscopic marine plants and animals form the main diet of much of the animal plankton.

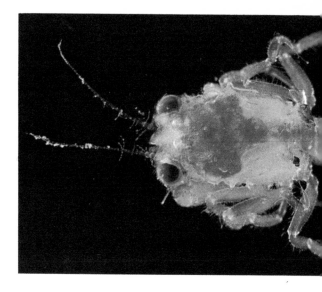

Above: Lobster larva about an inch in length. Animal plankton includes many adult creatures but also contains larvae of some forms that become bottom dwellers when fully grown, e.g., lobsters, crabs, and mollusks.

Left: Young angler fish, another sea animal that is planktonic during its early life.

sea and are helped in their work by oceanographers studying other aspects of the sea—currents, the temperature and salinity of the water, sediments, and the shape of the sea floor, all of which affect the lives of marine animals and plants.

The chain of life in the sea stretches from the smallest plant and animal forms, which are microscopic, up through forms of ever-increasing size to the great fish and whales. Each link in the chain is vital to the support of the whole, and the whole is ultimately dependent on the supply of nutrients kept in circulation by the current systems of the ocean. Before the nutrients can be utilized, some source of energy is needed and this is provided by the Sun, just as it is on land. These nutrients must be built up into the highly complex organic materials required by

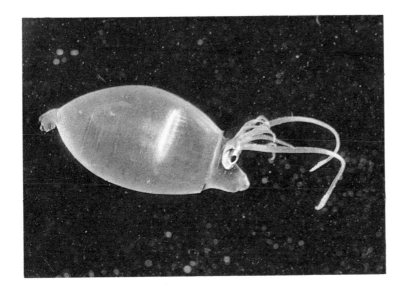

This young squid, about an inch long, is also a planktonic creature. On reaching adult size it becomes too agile to be caught in a plankton net.

living organisms. The first stage in this process, of which we have some understanding, is performed by the unicellular planktonic plants which compose the *phytoplankton.*

Plants require light, carbon, oxygen, water, and nutrients to build up carbohydrate which is the basis of living matter, and they contain green or brown pigments by means of which they absorb radiant energy from the sun to do this. They also give off oxygen and carbon dioxide (the latter in the absence of sunlight) during the process.

The sun's rays are both scattered and absorbed when they enter the sea, and submarine light is only effective for use by phytoplankton down to depths of about 650 feet. Nevertheless, there is a considerable population of microscopic plants in the depths of the ocean. We have evidence to show that different species of phytoplankton are found at almost all depths.

*Diatoms* and *flagellates* live in the upper surface layers and build up carbohydrates. *Blue-green algae* and *coccoliths* occur in large numbers from depths of about 1000 feet down to at least 12,000 feet. They probably live on decaying organic matter, because they are found below the limit to which sunlight penetrates. All these unicellular plants are of immense importance to the animal life of the

sea, the *zooplankton*, because animals cannot exist unless they can eat other living organisms. All animals are predators, even the unicellular groups, two of which have already been mentioned in connection with the deepsea oozes: Foraminifera and Radiolaria. A third group of importance is the Ciliata, small highly mobile forms. Eggs, larvae, and adults of crustaceans, worms, mollusks, and fish are all represented also in the zooplankton, some only at certain stages in their life-history, others throughout their lives. Some are herbivores, grazing down the phytoplankton; many are carnivores. All are important links in the food chain of the sea.

### Distribution of the Plankton

Zooplankton is the food of the larger fish and of the great whales, both of commercial importance today. Knowledge of the distribution and abundance of plant and animal plankton therefore helps toward locating good fishing grounds. Most commercial fish live on the continental shelf, that is at depths above 450 feet, well within the zone of greatest penetration of sunlight and of plankton production.

Within this zone daily movements of both plankton and fish take place, a sinking by day and a rise toward the surface at night. Seasonal movements are also made by the fish, in connection with their breeding habits, and by whales as well.

In those areas of the sea where currents of upwelling water occur, bringing essential nutrients from the depths, phytoplankton production is at a maximum with a consequent abundance of zooplankton as well. Temperature and salinity also play a part in influencing the growth and limiting the distribution of some forms. Plankton communities tend to group themselves in layers, particularly during the day, and spread upward and surfaceward at night; and where the communities occur their predators are also found.

There are no areas of the oceans and seas where planktonic plants and animals are entirely absent. Wherever they are abundant the fishing will most likely be good. Large shoals of herring, for instance, are likely to be found where there are large quantities of small zooplankton. The clues that help us find the dense plankton populations are the various nutrients necessary to their diet. The most important of these is phosphorus; others are the salts of such metals as cobalt, copper, and vanadium. All forms of life require at least some phosphorus for healthy growth; in vertebrates it is most important for the development of bones. Plants and animals in the sea obtain their supply of phosphorus from the decayed bones and flesh of dead fish and other materials forming the sediments.

The largest concentrations of phosphates are found on or near the sea floor where the dead matter finally comes to rest and is decomposed by bacterial action. A flourishing crop of phytoplankton rapidly exhausts the available phosphates and other minerals in the water at or near the surface unless the supply is constantly replenished by currents welling up from the sea floor to the surface.

It is only when sunlight and upwelling currents occur together that we can expect to find the huge supplies of plankton necessary to attract large quantities of fish. In the polar regions, and in more temperate latitudes around the North Sea and Canadian coasts, the seas are stormy in winter, so that the water

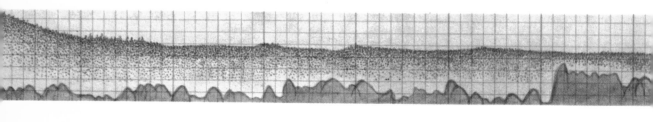

Vertical migrations of the plankton can be traced by echo sounding. The longer echogram shows this vertical movement over a 24-hour period beginning at night. The other diagram shows the distribution in greater detail.

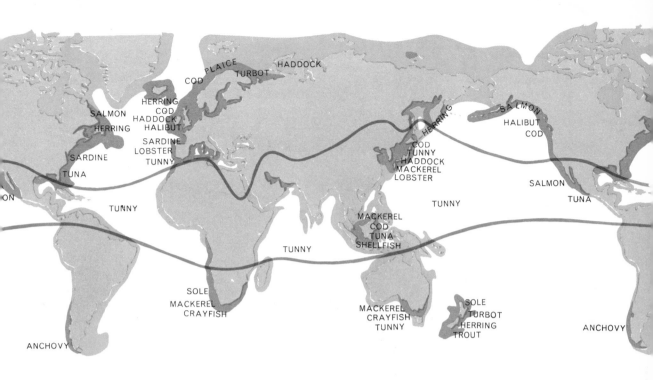

This map shows the world's main fishing grounds. Note their distribution in relation to the world's human population, half of which lives between the two red lines.

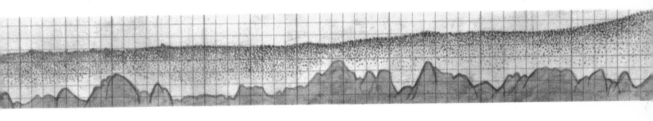

Canadian fishermen land a net of cod in the **North Atlantic**. Modern fishing craft often operate hundreds of miles away from base on trips that may last several weeks.

African fishermen on the Congo River use these trawl-shaped baskets to catch fish in the rapids.

is well mixed and rich in phosphates and other minerals. But the plankton cannot thrive at this time of the year because of the shorter days and the scarcity of sunlight. In the spring, with increased sunlight and warmer water, the phytoplankton begins to develop and the zooplankton communities to breed. The surface water that becomes warmed by the Sun stays on top of the colder water, and in areas where there are no upwelling currents to stir this water, the algae stop multiplying after they exhaust the phosphates.

It is for this reason that regions of the sea at the equator, where one might expect a rich growth of plant life, are in fact the "deserts" of the ocean. In these regions, which are heated by the Sun throughout the year, a layer of lightweight warm water rests permanently on the colder water below, so hardly any mixing occurs. As a result the waters have too few phosphates to support a thriving plankton population. In general, conditions are much better to the north and south of the equator.

We can now tell quite accurately which

parts of the oceans most encourage the large-scale growth of plankton, thanks to the work of physical oceanographers and marine biologists, and so we know where heavy catches of fish can be expected—knowledge of great value to the fishing industry.

Antarctic waters have long been known to produce phenomenal quantities of plankton, enough to support all the whale, seal, and bird populations found in these regions. Occasionally, these rich waters can have catastrophic effects. The cold Benguela Current sweeps out of the Antarctic seas up along the southwest coast of South Africa, past Walvis Bay. The sea bed in this area has practically no plant or animal life, but consists of diatom ooze, and the water over it contains no oxygen. Sulfate-reducing bacteria can therefore flourish. If a shift of the wind from north to west occurs, the cold Benguela Current is diverted or slowed down, allowing very saline, warm tropical water from the Atlantic to enter the coastal region. Unable to survive the higher temperatures, the fauna and flora of the colder water die, the dead matter sinking to the sea bed. This enrichment encourages the growth of the sulfate-reducing bacteria, which produce hydrogen sulfide in large quantities. This cuts down the amount of oxygen in the surface waters so that all the fish are killed as well. This formation of "poison water," as it is called, is common in Walvis Bay.

The shallow waters of Walvis Bay would be a fertile fishing ground were it not for this seasonal toxic rhythm, where we find sudden death striking the fish population in one of the potentially richest areas of the sea. Sometimes, also, an invasion of flagellates in enormous numbers occurs here. This "red tide" will be discussed in more detail on page 130.

## Life Cycles in the Sea

Life for most fishes consists mainly of eating and trying to avoid being eaten. It is an

The enormous growth in local fish populations caused by upwelling currents may lead to disaster if the oxygen in the water becomes exhausted (see light band in diagram above). This happens at Walvis Bay, South Africa, where the men (below) are gathering pilchards that have suffocated in their millions and have drifted ashore.

Above: A fish paralyzed by the stinging cells of a sea anemone is drawn into the body of the animal. Below: Leg movements of a copepod create currents that guide food to its mouth.

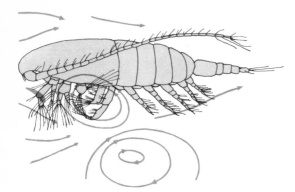

endless struggle for survival in an environment where many animals in the zooplankton are carnivorous. In many cases the food chain is a long one: Phosphates carried from the sea floor are absorbed by a flagellate, which is eaten by a copepod, which in turn is consumed by a herring, which disappears inside a squid, which succumbs to a shark. If the shark is lucky enough to escape being eaten by one of its own kind and die of old age, it decomposes and fertilizes the water for the flagellates to begin the cycle anew.

Other cycles end in the fishing nets. If a herring escapes the tentacles of a squid, it may end up on your breakfast plate or be used as bait to catch the equally tasty but much larger tuna. The sardines you may have for lunch are the natural prey of the mackerel you enjoy for dinner. Every animal in the sea has its enemies—sometimes others of its own species. Underwater photography has shown that squid (or cuttlefish) prey on one another. Cannibals may be common. It has been estimated that only one fish in 10 million dies a "natural" death. Most end up in the stomachs of larger fishes, and only a small proportion is caught by man for food.

One fishing industry that has been studied most systematically in order to improve the quality and limit the number of catches is that of whaling. Whales have been hunted for hundreds of years; there are accounts of whaling voyages by the Norsemen as early as the ninth century A.D. It is only in the last 150 years, however, that whaling on a large scale, mainly in Antarctica, has developed. Today a large whale may be worth as much as 10 new automobiles. Its oil is a most valuable product and is used in the production of margarine and other food fats. In some countries the meat is eaten; properly cooked it tastes rather like beefsteak. The bones and other parts of the whale are ground up and used to make fertilizers.

Until recently, many nations sent large whaling fleets yearly to the Antarctic, but now that the whale stocks have declined so much, fewer ships are employed. In the first 30 years of this century, this decline was already the concern of scientists, who feared that the antarctic whales would suffer the same fate, by over-fishing, as that which overtook the arctic population in the previous century. An International Commission of the nations engaged in whaling was formed and regulations were drawn up to limit the number, size, and species of whales to be caught and also the number of ships to be employed in the industry. It was hoped that in this way the whale population might be conserved; but the killing of whales has remained excessive.

This program of regulation necessitated more knowledge of breeding habits, growth rate, and movements of whales, for when the Commission was first discussed, information on these subjects was scanty. Today we know a great deal about the life cycle of these animals. Whales are *mammals*, not fish; they are warm-blooded and the females suckle their young. Whales are the largest animals alive today. Of the whalebone whales, the blue whale generally measures over 80 feet, but larger ones over 90 feet have been taken. The fin, sei, and humpback whales measure 75-80, 60, and 50 feet respectively. The sperm whale, the largest of the toothed whales reaches about 60 feet. Because of their size, these whales are of commercial importance, but there are other smaller kinds as well. Whales breed slowly, producing one "calf" every two years; twins are rare. The "cows" suckle their young for about eight months.

Whales appear to be at least as intelligent as horses. The smaller dolphins for example can be tamed to some extent and taught to recognize signals meaning food. All whales are able to "talk" to each other by transmitting and receiving sounds through water. It seems they have a built-in "echo-sounder" to

Small crustaceans called euphausians or krill (below) are the principal food of the whalebone whale. Krill are trapped in bristles fringing the whalebone and are then swallowed. Adult whales need several tons of this "shrimp soup" every day.

This painting captures the danger and skill of small-boat sperm whaling in the mid-19th century.

help them navigate around obstacles and to detect fish. One early Australian explorer reported that Aborigines made use of the dolphins' ability to hear and understand underwater sounds. The fishermen made noises under the water and so encouraged the dolphins to chase shoals of fish toward their nets. The fishermen even gave names to the various dolphins, just as farmers have names for their sheep dogs.

Although there are many different kinds and sizes of whales, they all fall into one of two main groups distinguishable by their very different habits of feeding. The first are the *whalebone* whales, which feed exclusively on shrimplike plankton called "krill" by the whalers. The whalebone takes the place of teeth by serving as a plankton filter. To gather these animals, the whale takes in a great mouthful of water then presses it out through the whalebone on the roof of its mouth. The plankton animals are trapped among bristles fringing the bone and are then swallowed. Whales need tons of this "shrimp soup" every day, which means that they must follow the plankton.

The whalebone whales leave the krill-rich antarctic waters when winter comes and the plankton population is reduced. Fattened by their summer-long banquet they move toward

blue whale

porpoise

killer whale

fin whale

bottle-nosed dolphin

man

beluga or white whale

narwhal

sperm whale

pilot whale

the equator so that their calves can be born in warm waters. The calves of the largest whales weigh about two tons at birth and are about 20 feet long. Each calf drinks nearly a ton of its mother's milk every day. At about seven or eight months a calf may weigh 40 or 50 tons and in two years it can begin to breed, although it is not yet fully grown. The life span of whales has not been determined with any degree of accuracy.

The whales forming the second group are not plankton feeders. Among them are the killer whale and the sperm whale, both fierce animals equipped with great teeth. The favorite food of the sperm whale is the squid

The order *Cetacea* includes over 70 species. Nine of them, together with a human figure, are shown to scale above. Cetacea are in two main groups—whalebone and toothed whales. The first group—the plankton eaters—includes the blue whale (which with a length at most of 100 feet and a weight of up to 150 tons is the largest animal that ever lived) and the rather smaller fin whale. The second group includes not only the sperm whale (up to 60 feet long and with large teeth in its lower jaw) but also the other whales shown here, and the dolphin, porpoise, and narwhal.

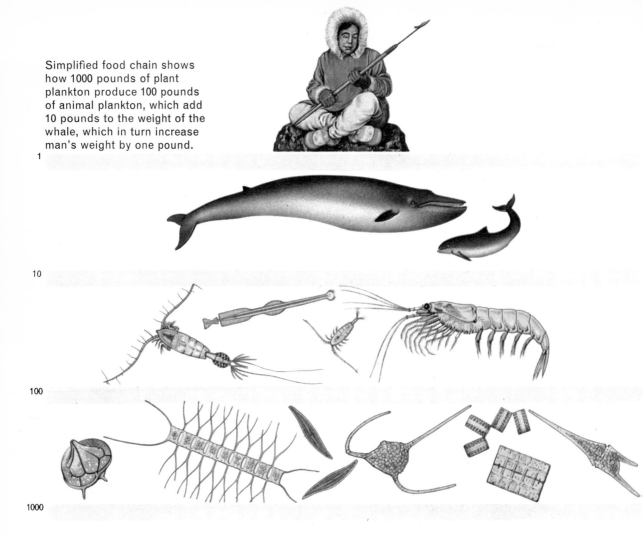

Simplified food chain shows how 1000 pounds of plant plankton produce 100 pounds of animal plankton, which add 10 pounds to the weight of the whale, which in turn increase man's weight by one pound.

1

10

100

1000

which abound in the mid-water layers of the sea, and which are usually 3 to 6 feet long. Occasionally much larger squid are taken. These have already been described on page 15; they do not form the staple food, though sperm whales caught after battles with a "giant" squid have had tentacles up to 30 feet long in their stomachs and have borne the scars of giant suckers on their bodies. Congregated in schools, the sperm whale is usually found in more temperate latitudes than the whalebone whales. Although never as important commercially as the whalebone whale, it is hunted both for *spermaceti oil*, found in a cavity in the head, and for the blubber oil. Sperm oils are used in various industrial processes, and in cosmetics.

The food chain in which the sperm whale takes part is more complex than the one in-volving its plankton-eating relative. Its favor-ite diet, the squid, lives on fish, which in turn live on smaller fish, so eventually the food chain leads back to the plankton.

The food cycles involving different fishes are often very complicated; so also is the life cycle of certain marine creatures. One of the most curious and best understood life cycles is that of the eel. Born in the Sargasso Sea to the east of Bermuda, these animals swim thousands of miles to take up life in the fresh water of rivers, then return to the Sargasso to spawn. Every autumn adult eels leave the rivers of Europe and cross the Atlantic to the Sargasso where they lay their eggs and die. At least this seems to be their pattern of move-ment. A high proportion of the young eels are eaten by other fish, but thousands of others are carried slowly northward in the

Gulf Stream. Many of them find their way up rivers that flow into the sea along the East Coast of the United States. Others continue with the Gulf Stream and, after three or four years, reach the rivers of Europe and Africa. After living for four or five years in fresh water, they go to sea again to lay their eggs. We are not sure whether these eels ever succeed in returning to their mid-Atlantic birthplace. It may be that the successful breeders are only those eels that have to make the shorter journeys from American rivers. Certainly the latter could produce enough young to stock both American and European waters.

## Fisheries Research

At the present time our methods of fishing are slowly advancing, but many problems remain to be solved. Little attempt has been made to breed larger or tastier fish in the way that we have developed the selective breeding of cattle, pigs, and sheep; nor have we attempted to "cultivate" the sea to increase its animal and plant growth.

Until we know a great deal more about the life cycles of all the fish—which will enable us to predict more accurately where shoals of a particular fish are to be found at a particular time—we will stand little chance of improving fundamentally our methods of fishing. The more oceanographers study life in the sea, the more they are impressed by the complicated processes that encourage a species to increase its numbers or that lead to its depletion. When man engages in fishing on a large scale—as in the case of herring, cod, and whales—he always runs the risk of interfering with these processes and upsetting the population as a whole.

Many of the regulations limiting the catch of fish, whales, and other marine animals have been brought about on the advice of oceanographers. This advice has not always been taken. Some of the work on antarctic whales has been noted already. It also led to knowledge of whale food and its distribution. Stomach contents of animals processed aboard factory ships showed that crustacean "krill" is the staple diet of whalebone whales.

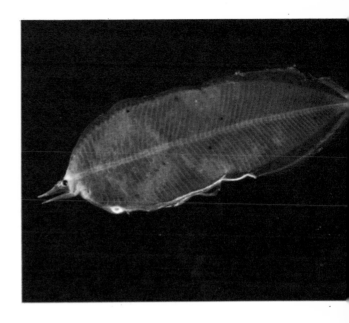

This eel larva (above) is about three inches long. Born in the Sargasso Sea to the east of Bermuda, eels swim thousands of miles to live in African and European fresh-water rivers (see map below), then attempt to return to the Sargasso to breed.

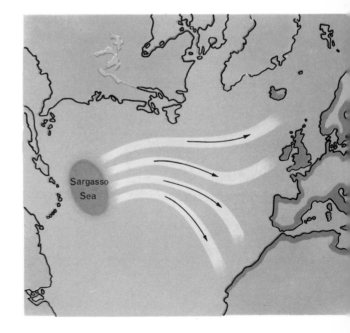

129

"Fish" (below), containing an echo-sounder, is towed behind a ship and is used to record depth and size of plankton colonies.

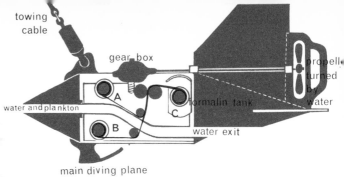

Hardy Plankton Recorder (above) is powered by a propeller that turns as the recorder is towed under water. Plankton is caught on two silk rolls, A and B, which are wound on to a drum C.

Fish such as these plaice are marked with tags that help marine biologists study their movements. The knowledge gained may aid the search for bigger and better catches.

Migrations are studied by shooting numbered marks into the blubber, the date and position of marking being noted. When marked whales are caught, the marks are found, and the new locality gives an idea of the area covered by the animal in the interval. But in spite of all this work, the International Commission has not been successful in preventing depletion of the stock of whales.

Because all other life in the sea ultimately depends upon the plankton, it is as important for marine biologists to study the behavior of these small plants and animals as it is to study the behavior of the food fishes we want to catch. With fine, cone-shaped nets samples of plankton are regularly collected at different depths of water all over the world.

As well as being beneficial to the animals that feed on them, planktonic plants can bring about the destruction of fish on an enormous scale. Abnormal amounts of sulfide in the water, attendant on high rainfall and lack of offshore winds, bring about the unusual swarming of flagellates in coastal waters with lethal effects. There may be as many as 50 million of them to a pint of water, which turns brownish red—hence the name "Red Tide." As well as in Walvis Bay, red tides occur in other parts of the world. In 1947, off the coast of Florida, millions of fish and crustacea were killed by a red tide. Such tides have been known since Biblical times.

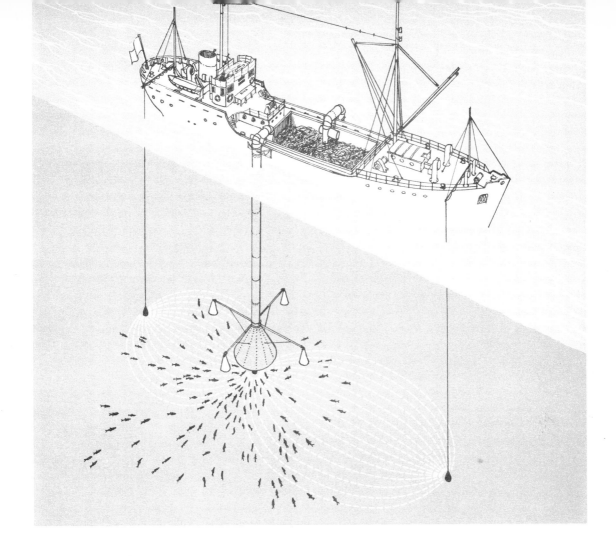

In this fishing system fish are attracted by underwater lights, paralyzed in an electric field, and then sucked into a tube and pumped into the ship's hold.

One of the best ways to detect the presence of large quantities of plankton or fish in the sea is by using an echo-sounder. Sometimes an echo-sounder recording shows "shallow water" in areas where the charts show deep water. The echo in this case is coming not from the sea floor but from a large concentration of plankton. This *deep scattering layer*, as the concentration is called, may be at a depth of 600 feet or so during the bright daylight hours; by bouncing echoes off it, it is possible to locate the position and depth of the plankton and the fish feeding on it.

Many trawlers now carry echo-sounders.

Modern technology is helping to increase our fish yield in yet other ways, some of which will become more and more important as larger harvests of fish are needed in the future. In the Caspian Sea Russian fishermen catch sardines through a large underwater pipe that sucks them on board by means of a pump. The sardines are attracted to the mouth of the pipe by bright lights suspended below the boats. An electric field set up in the sea paralyzes the fish so that they can be pumped aboard. Another method of fishing with artificial light is used by the Japanese, who mount their lights on a series of stakes planted in the sea at right angles to the shore. The lights are switched off one by one—starting with the most distant light—so that the fish are gradually led into the waiting

This Tonga Islander attracts fish to the
surface with the aid of a torch—a primitive
but effective version of the Japanese method
described in the text.

Right: This drawing (based on a photograph)
shows adult plaice trapped in a seine net
while younger ones escape through the meshes.
Indiscriminate catching of both marketable
and undersized fish is a needless waste of the
sea's wealth. Today international agreement
on the size of some net meshes is helping to
reduce this wastage.

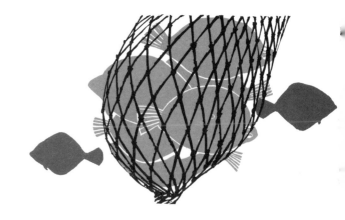

nets by the final inshore light. Meanwhile, the fishermen are sleeping comfortably in their beds as the lights are turned off automatically.

The fisheries industry presents the marine biologist with a double problem: How to catch more of a particular kind of fish that may be in demand—but without disrupting the food chain in which that fish forms an important link. However ruthless life in the sea may appear to us, nature maintains a balance among the fish population as a whole, and it is important to preserve this balance if we are to harvest the sea successfully and, what is more important, to do this continuously.

The fisheries of the world are concerned with control of the industry. Governments of various countries realize the importance of using nets of different mesh size, depending on the kind of fish being caught. A net with a very fine mesh catches young fish as well as fully grown ones. Because this is one of the quickest ways to reduce the population of a particular species, marine biologists measure the full-grown fish and recommend the particular mesh that will hold them, but at the same time allow the young to escape so that they may reproduce.

The annual growth of fish amounts to about 1000 million tons, the balance in the fish population being maintained from year to year by the consumption of smaller fish by larger ones. We catch roughly 30 million tons a year, so there is plenty of room to expand the fishing industries of the world, so long as we understand the effect our fishing activities have on other forms of life in the sea.

Nature maintains a balance in the population of the sea. Each species of animal life plays an essential part in it, from huge, commercially valuable whales to this small silvery fish, about two inches long. Limitless fishing is dangerous because it not only threatens the fish man likes to eat but affects the balance in the whole fish population.

# 10 The Future

What are likely to be the most important fields of oceanographic work in the future? We live in an age when science fiction seems only a few short strides ahead of science fact, and even our most imaginative predictions today may be left behind by the events of tomorrow. Technologically, man has come of age—he has the know-how and the resources to change the course of nature. This is why it is so vital that we have as complete a knowledge as possible of our planet's natural processes. The oceans, about which we still understand comparatively little, are one of the most important fields of scientific study, and they offer the few remaining "open spaces" on our crowded planet. We may not be able to live in the sea, but we can at least learn how to exploit its resources more systematically for the benefit of our land-based population.

## Power from the Sea

We saw in the chapter about tides how the French plan to obtain power from the rise and fall of the tide in the Bay of St. Malo. A similar idea but on a much larger scale is a

Underwater "houses" have been developed to allow men to stay several weeks beneath the waves. The painting (left) shows one such house, anchored by weights to the sea bed and with air lines to a surface vessel. Beside the house is a fish cage for studying live specimens, while in the foreground is a table fitted with lights. In the photograph (right) aqualung divers enter one of these houses through a hatch on the underside.

The Dutch have reclaimed
many thousands of square miles
of low-lying land. These fields of
rape seed are part of a large area
reclaimed from the southeast
corner of the Ijsselmeer.

Right: A Russian plan to dam
Bering Strait would alter the
arctic climate by pumping cold
water from the Arctic Ocean,
thus drawing warm Gulf Stream
water westward across the
polar basin. Wavy line shows
present southern limit of
permanent ice, dotted line the
limit of ice floes.

dams is that of the Russians to block the Bering Strait, which separates northeastern Siberia from Alaska. Permanent ice over the entire Arctic Basin affects the climate not only of Siberia and Canada but also that of much of northern Europe as well. At present there is a small flow of warm Pacific water into the Arctic Ocean through the Bering Strait. On the other side of America a much larger quantity of warm water flows northward between Greenland and Norway in a branch of the Gulf Stream. But before it reaches the Arctic Basin the Gulf Stream is chilled by the cold southward-flowing Labrador and East Greenland currents.

Soviet engineers have calculated that a Bering Strait dam, equipped with batteries of giant pumps, could draw vast quantities of cold water from the Arctic Ocean; that this would eventually reverse the cold currents flowing toward the North Atlantic; and that, as a result, the Gulf Stream could carry its warm water from the Atlantic across the Arctic Basin toward the Pacific. In time, the warm current would melt much of the polar sea-ice, and northern Siberia and Alaska would probably be cooled as a consequence of arctic water passing through the Bering Strait. To counteract this the Russian plan also includes pumping stations to draw warm Pacific water into the Chukchee and Beaufort seas.

The precise effects of this plan on the climate of countries on and around the Arctic Circle are still not known and would have to be studied very thoroughly before the dam could be built; however, if all the ice in the Arctic Ocean melted, there would not be any rise at all in sea level all over the world. The main value of the project, however, would be to convert much of the bitterly cold, ice-bound coasts of Siberia, Alaska, and northern Canada into more temperate areas easier of access to shipping and industries. There is also vast mineral wealth in these regions. A study of their rocks shows that the regions enjoyed a hot climate millions of years ago. This is how some of the islands in the Canadian Far North have come to contain rich deposits of oil. At present it would be

proposal, made by several engineers, to dam the English Channel at the Strait of Dover, which is about 21 miles wide. The amount of hydroelectric power that such a dam could produce would be about 100 times as much as that provided by the St. Malo scheme. The water in the North Sea flows almost entirely from the north. The dam would use this pent-up tidal water to drive turbines. There would be a higher average level of water in the North Sea in consequence with some loss of shore-line. It would of course be a very expensive project, but it would provide as good a link between Britain and France as either the proposed Channel Bridge or Tunnel.

The most ambitious of these plans for large

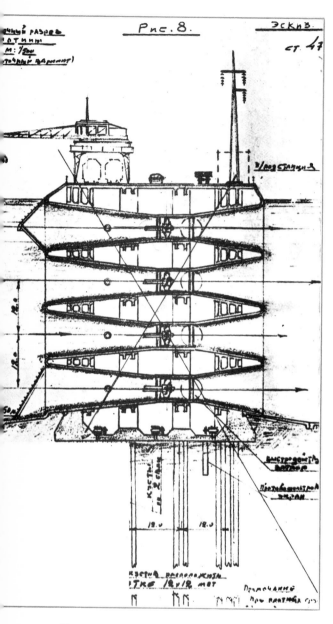

The Bering Strait dam is the visionary, but as yet untested, plan of a Russian engineer, Pyotr Borisov. His sketch above illustrates his proposal that the dam be made of ferroconcrete pontoons equipped with gigantic pumps. The dam would provide a road and rail link between the United States and Soviet Union.

expensive to ship the oil to the markets of Europe and elsewhere because these islands are icebound for all but two months of the year. If the Bering Strait plan worked, however, the ice fields around many of these islands would be turned into navigable waterways for oil tankers.

Obviously the dam would present tremendously difficult engineering problems. At its narrowest point, the strait is more than 45 miles wide, and the average depth of water is about 150 feet. Possibly the earth-moving capacity of nuclear bombs could be used in preparing the foundations of the dam. Soviet engineers have suggested that the dam itself should be prefabricated in huge reinforced-concrete sections mounted on pontoons that could be towed to the site.

There have been other visionary proposals of this kind. The one that would have the most profound economic and political effects is the Atlantropa project proposed in 1928 by the German architect Herman Sörgel. The Mediterranean Sea loses millions of cubic feet of water through evaporation every year. Neither the rainfall in this region nor the rivers feeding the sea are sufficient to make up this loss, two thirds of which is replaced by water flowing from the Atlantic Ocean through the Strait of Gibraltar at a rate of over three million cubic feet per second. If a dam were built across the strait, thus shutting off the supply from the Atlantic, the level of the water in the Mediterranean would fall, according to Sörgel, more than three feet each year. To ensure this rate of fall, another dam would have to be put across the Dardanelles to seal off the Black Sea.

The dam linking Europe to Africa would be about 18 miles long and would need to stand in about 1000 feet of water. At the end of 100 years the water of the Mediterranean would have fallen about 330 feet. It would then be possible to dam the Strait of Messina between Italy and Sicily and also the gap between Sicily and the coast of Tunisia, thus converting the whole of the eastern half of the Mediterranean into a gigantic lake. On Sörgel's reckoning a further drop of 330 feet in the water level of this area could then be

The Atlantropa Project, first proposed by the German Herman Sörgel in 1928, is a plan to convert the Mediterranean into a vast lake by damming the Strait of Gibraltar (left) and the Dardanelles. As a result of these dams there would be a drop of 33 feet every ten years in the sea's water level, providing a source of abundant hydroelectric power at the mouths of rivers flowing into the Mediterranean.

achieved in considerably less than a century.

The benefits gained from the Atlantropa project could be impressive. In the first place, all the countries bordering the Mediterranean would acquire huge areas of new land for cultivation and industrial development. Estimates vary, but it is likely that as much as 220,000 square miles of land would be reclaimed, notably in the Adriatic Sea and around the coasts of Tunisia, Sicily, Sardinia, and Corsica. Secondly, dams built across mouths of rivers flowing into the sea would provide a source of cheap hydroelectric power on a gigantic scale.

The scheme might also affect countries on the eastern Atlantic seaboard. As we saw in the chapter about currents, there is a deepwater current that passes through the Strait of Gibraltar—this one flowing from the Mediterranean into the Atlantic. If this current were blocked by a dam, there might be changes in the circulation of water in the eastern Atlantic, and this in turn might alter the climate in parts of Western Europe. We must therefore consider very carefully the cumulative effects that man-made alterations to the geography of one region might have on other regions before we begin large schemes of this kind.

Apart from its great cost, the Atlantropa project would present formidable problems. All the present seaports in the Mediterranean would be left high and dry—Trieste would be as much as 300 miles from the sea if the water level dropped more than 600 feet—so new harbors would have to be built along the new coastline. Also, the land two or three hundred feet below normal sea level would have an unpleasant climate like that of the Dead Sea area. And there would be the ever-

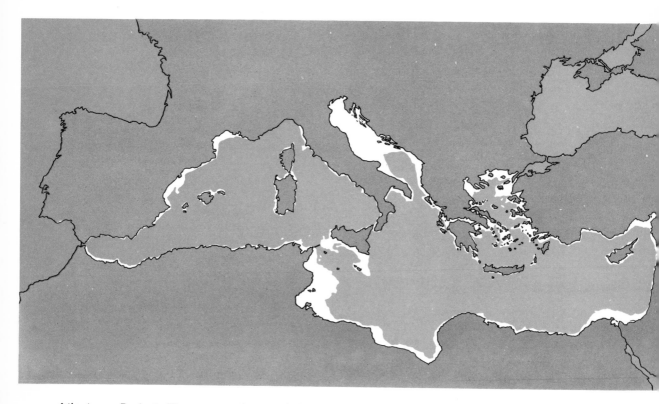

Atlantropa Project: The supposed area of the Mediterranean two centuries after damming the Strait of Gibraltar and the Dardanelles is shown in blue; the black outline shows the present coast.

present danger of catastrophe if the main dam collapsed.

The heat stored by the oceans provides us with another possible source of power based on a principle known as *heat exchange*. Suppose you wished to provide yourself with a cheap and simple electricity generator in the polar regions. The air temperature in the Arctic is often as low as −50° F.; yet the temperature of the sea water below the pack ice rarely falls much below 27° F.—a difference of 77° F. To provide yourself with a continuous supply of electric power all you would need would be a boiler, a condenser, a generator coupled to a small turbine, a simple water pump, and a liquid with a low boiling point. Butane would be suitable; it normally occurs as a gas, but it liquefies when cooled to 14° F.

The butane (in a liquid state owing to the low air temperature) is poured into the boiler, which is kept supplied with sea water pumped from below the ice. The water heats the butane sufficiently to make it boil, and the energy supplied by the butane vapor drives the turbine. From the turbine the vapor passes into the condenser where it cools, returns to a liquid state, and is pumped back into the boiler where the cycle begins again.

It might seem odd that we can extract heat (and therefore power) from the waters of the Arctic, which are far too cold to bathe in. This method of power generation, however, depends upon the *relative* temperatures of the working fluid (butane), the heating fluid (sea water), and the cooling medium (air). The Arctic is, in theory at least, an ideal location for a power plant based on this

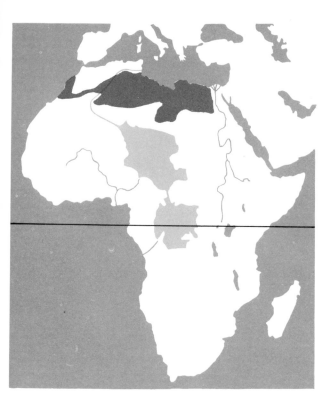

Sörgel also proposed to dam the Congo River near its mouth, the aim being to convert the Congo basin and Lake Chad into a great inland sea connected to the Mediterranean by a river.

method because it provides us with an abundance of heating and cooling materials whose temperatures differ widely. Very few practical experiments have been made to test this method on a large scale, but it might be developed as a supplementary source of power in polar regions.

## Our Changing Climate

The ability of the sea to store heat has a steadying influence on the temperature of our planet. On the Moon, which is without atmosphere and is without oceans or lakes, things are quite different. The land becomes extremely hot—near the boiling point of water—during the lunar day and at night the temperature falls to less than −150° F. On Earth heat is absorbed by the sea only

by the superficial water layers. These layers absorb a great deal of solar heat during the day, but give it up very slowly during the night. These important characteristics of water help to moderate the climate of the world by generally maintaining the land temperatures steady: Sea breezes during the day cool the land, while warm currents like the Gulf Stream carry tropical warmth to colder northern countries. Because the sea so greatly influences our climate on land, oceanographers are studying exactly how the exchange of heat occurs between sea and land.

We know that there have been times when thick sheets of ice were spread over parts of Europe and North America; the last of these "ice ages" reached its climax about 10,000 years ago. Studies of surface rocks and the marks on them suggest that two or three ice ages have occurred in the last 100,000 years. The ice sheets covering the affected regions were thousands of feet thick in places. A present-day ice age can be found in Antarctica, where the *ice cap* has a mean thickness of 6000 to 8000 feet. But ice ages are not confined to "recent" times. Evidence provided by the crustal rocks shows that ice ages have occurred regularly for at least 250 million years. From the fossils found in the rocks geologists have learned how to interpret the history of climatic changes.

One explanation for the development of ice ages might be that the warm and cold currents of the oceans were disturbed in some way. Another might have been a decrease of the heat from the Sun, which could have come about by some change in the Sun itself or in the Earth's atmosphere. The Earth's orbit might have been crossed by a gigantic cloud of cosmic dust that absorbed enough of the Sun's heat to allow extra-heavy snow storms to occur in the polar regions. It is not certain how quickly these events would cause changes of climate. But we do know that ice ages come and go at intervals measured in scores of thousands of years. Research at present being carried out in Antarctica is concerned with the rate of snowfall and of the *compaction* of snow into ice. The thick layers of snow pack tight to form great rivers

Above: Antarctic glacier on the coast of
Graham Land. Antarctica, covered by an ice cap
nearly two miles thick in some places, is
larger than the combined areas of the United
States and Mexico (see map below). If its ice
cap melted it would promote a catastrophic rise in
sea level all over the world.

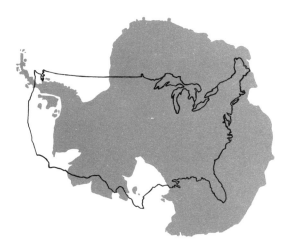

of ice. Much the same process must have occurred during the ice ages. The glaciers would
have shed great icebergs into the sea so that
in time the whole climate became colder.

There is another way in which the atmosphere can affect climatic change. Carbon dioxide in the air allows the Sun's short light
waves to penetrate to the Earth's surface.
As the ground is heated by this radiation it
sends back long-wave (heat) radiation into
the atmosphere. Unlike the short-wave radiation, the long waves cannot penetrate the
carbon-dioxide layer, with the result that a
heat trap, like that of a greenhouse, is
formed. If there were a large decrease in the

amount of carbon dioxide held in the atmosphere, this could lead to a period of cooling resulting in an ice age. As we saw in the last chapter, the sea absorbs carbon dioxide, which is used by plant life in the sea as the starting point in the great food chain. Suppose that the balance of life in the sea was upset so that much more carbon dioxide was absorbed by the sea. Could this set off another ice age? We do not know the answer yet, for much remains to be learned about the relationship between the seas and the atmosphere of our planet. The atmosphere is the connecting link in the exchange of water between land and sea. Water evaporating from the surface of seas and lakes is carried as water vapor into the atmosphere. Winds carry the water vapor across the Earth's surface and precipitate it onto the land as rain or snow. Some of the water is retained by the land for long periods. In the polar regions, for instance, it may become part of the permanent ice caps. In the Sahara Desert oil-drilling crews have discovered fresh water that has been held deep in the ground for thousands of years. But most of the water precipitated on land feeds rivers, which return the water to the seas and lakes where the cycle is renewed.

Another kind of catastrophe—the reverse of the ice-age process—would occur if all the land-ice in the polar regions melted. There is enough ice in Antarctica and Greenland to cause a 300-foot rise in sea level all over the world—enough to drown London, New York, Tokyo, and the other great coastal cities. If for some reason the carbon dioxide in the atmosphere increased, or if the Sun produced an extra burst of activity, the ice of the world would probably melt. Oceanographers and other scientists are watching carefully for all the signs that might indicate such a development.

It is only in recent years that studies have been made of the exchange of heat that occurs all the time between the land, the sea, and the atmosphere. The mechanics of this process are similar to that of the butane heat exchanger described earlier, but the Earth is so large that the exchange happens gradually

Engineers have found huge reserves of fresh water at depths of up to 6000 feet beneath the sands of the Sahara Desert. Under great pressure in its deep reservoirs, the water is hot when it gushes to the surface and is piped into these large cooling tanks.

and involves only small variations in temperature. We still do not know how great a change in sea temperature would be necessary to upset our climate—though some experts have calculated that an average rise of 4° F. throughout the world would be sufficient to bring about the melting of the polar ice caps. But to be certain, we would need to have temperature measurements covering a period of many thousands of years. As we saw in the third chapter, the sea sediments may provide us with the answer because the proportion of heavy and light oxygen atoms in the sedimentary layers helps us to determine the temperature of the sea at the time the layers were deposited.

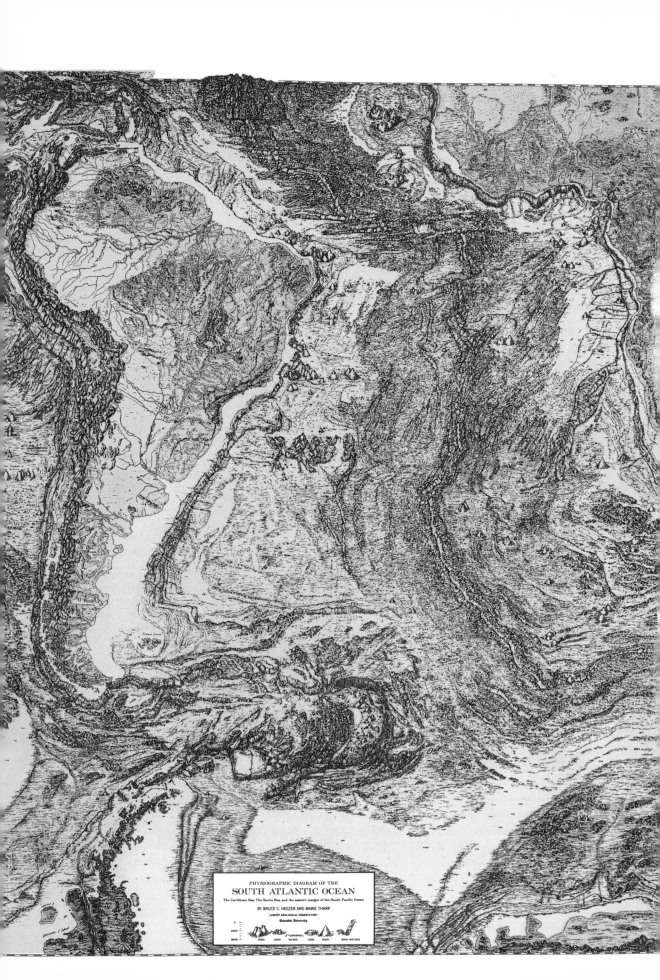

PHYSIOGRAPHIC DIAGRAM OF THE

SOUTH ATLANTIC OCEAN

The Caribbean Sea, The Scotia Sea, and the eastern margin of the South Pacific Ocean

BY BRUCE C. HEEZEN AND MARIE THARP

LAMONT GEOLOGICAL OBSERVATORY

Columbia University

## Research Beneath the Waves

The exciting thing about the Mohole project is that it will almost certainly tell us much about the history of our planet and about the forces that help to shape the ever-changing surface of the Earth. We shall be able to identify all the different layers of sediments and rocks that our seismic experiments have shown to be present beneath the sea. We shall also discover the age of the oceans, because we shall go right back into the geological past by probing through the very first sediments laid down on the original sea floor. To get as complete a picture as possible of the sediments, it will be necessary to drill holes all over the oceans of the world in order to see whether the oceans are of the same age, or whether some, like the Atlantic, may be comparatively new. Boreholes into the floors of the oceans will be routine operations one day, just as our seismic and gravity measurements are today.

But much of our knowledge of the sea's behavior depends on routine observations made year in and year out in all parts of the world. New methods of collecting data on currents and other water movements will one day include stations on the sea floor that will provide continuous records instead of the irregular reports from ships that oceanographers depend upon at present. These stations will collect valuable information about the slow-moving deep currents that are so important in maintaining the circulation of nutrients in the ocean. Attempts are now being made to attach instruments to the ends of old submarine telephone cables in the Pacific Ocean; later it should be possible for sea-bed seismograph stations to observe earthquake waves and to record the passage of tsunami across the ocean.

There will also be improvements in the seismic, magnetic, and gravity instruments that are being used in submarine investigation of the ocean floor. Already gravity meters are being carried on surface ships all over the world. This will enable many more measurements to be collected than was possible with the old gravity-measuring pendulum apparatus that was carried in submarines. The calculation of gravity and magnetic results is being speeded up by recording them on punched tapes. The tapes are then fed into a computer that makes all the necessary corrections for the ship's speed and position, and plots the final values. Many seismic soundings are now recorded on magnetic tapes that can be played back through electrical filters that help to eliminate external noises and interference that often blur these signals.

Gradually, many nations are showing a willingness to work co-operatively to discover the secrets of the oceans—the international program of oceanographic research taking place in the Indian Ocean being one example. Gravity and magnetic maps, together with detailed soundings of the sea bed, are being produced in much the same way as maps of ice thickness were made by joint international efforts in Antarctica during the International Geophysical Year in 1957-58. It is possible that artificial earthquakes will be produced by detonating powerful bombs underwater, in order to learn more about the structure of the Earth's interior and, in particular, to determine whether or not the "liquid" outer core contains an inner solid one, as we believe at present.

All the oceanographic data that is being collected will, one day, enable us to make maps of the sea bed as accurate as those we have on land. You may wonder why the shape of the sea floor is important to us when it is covered by two to three miles of water. There is no fear that ships will run aground —but submarines of the future might. Already submarines are being made so strong that they can cruise at depths of a thousand feet or more; one day there will be submarines capable of reaching the greatest depths. There will be entirely new problems

Oceanographers, using improved methods of sounding and recording, are gradually building up a more and more detailed picture of the ocean floor. This recent map of the South Atlantic was made by Dr. Bruce Heezen and Marie Tharp of the Lamont Geological Observatory.

of navigation. While gyrocompasses and inertial guidance systems will keep a submarine commander accurately informed about his position, he will still need detailed maps showing deep currents which may be faster than we think, and topographical maps showing the location of undersea mountain ranges, plains, and valleys. Detailed knowledge of the shape of the ocean floor is also needed for a proper understanding of the processes of sedimentation and mountain building that are constantly modifying the shape of the bottom. Reliable maps of small areas of the ocean floor made it possible to confirm the theory of turbidity currents and their part in transporting sediments from the land. What we need now are equally reliable maps of all the oceans.

Submarines and other special craft will be used more and more in the future to take oceanographers down to the bottom to see things at first hand. The bathyscaphe *Trieste* showed by its record dive that man can reach the deepest parts of the oceans. The problem now is to make more mobile bathyscaphes and fit them with equipment for surveying

While aqualungs have revolutionized exploration at depths of up to about 50 fathoms, man is still relatively slow and clumsy in water. Light and speedy "underwater scooters" like this help a diver to use more efficiently the limited time he can spend beneath the waves.

large areas of the sea floor.

Possibly the underwater equivalent of the airplane may eventually replace the bathyscaphe. The waterplane would drive itself down through the water by means of propellers and wings just as the airplane raises itself through the atmosphere. There is even talk of an underwater helicopter that would push itself down to the sea bed and, like the waterplane, if the engine failed, would "crash land" by floating up to the sea surface.

One task that underwater craft of the future would carry out is that of surveying the bottom. Although the murky water will make it difficult to see much down there, it will be necessary to discover in detail the real shape of the underwater hills, mountains, and valleys. This job is at present being done by echo-sounders operated from the surface of the sea. But when used from the sea surface the echo-sounder gives a "blurred" picture of the ocean floor—the main features are there but the details are missing. It is as if we tried to map the Earth's surface with a radio altimeter from an airplane flying at 20,000 feet. We should get the general shape of the land, but the sharpness of gorges and peaks would be smoothed out. If we carried the echo-sounder close to the sea bed in a special deep-sea craft we would be able to make very much more accurate and detailed maps.

## The Search for Fish

The most important problem facing marine biologists is to improve methods of catching fish. The Japanese catch six million tons of food fish every year. The Indians, with a population about five times as great, probably catch less than one million tons. National tastes in food vary, of course, but it is possible that the problems of feeding the rapidly expanding population of India could be partly solved by intensive fishing in the Indian Ocean. Little is known at present about the animal life of this ocean, but by the time the International Indian Ocean Expedition has completed its work it should be possible to advise Indian fishermen where to go for the best catches. The Indian Ocean may in fact turn out to be one of the richest fishing areas in the world, because its warm regions around the equator have no well-defined doldrum belts like those in the Atlantic and Pacific oceans. There is much greater mixing of the warm surface water with the cooler, phosphate-containing water below than there is in other tropical seas.

There are certainly some parts of the Indian Ocean where upwelling deep currents bring "fertilizers" to the surface. Russian scientists have reported seeing a mass of dead fish in the Indian Ocean—usually an indication that there has been an excessive outburst of life which temporarily upsets the living conditions of the population. A study of salinity and temperature together with some deep-current observations should indicate where this is likely to occur.

In addition to discovering new, rich fishing

The Japanese are one of the leading fishing nations, catching about six million tons of fish every year. In fish farms like this one they help to improve future catches by hatching and raising fish away from their natural enemies in the open sea.

grounds, man may one day be able to create his own on a small, local scale by fertilizing certain areas of the sea as he has of the land. If upwelling is so important to fish life because it brings minerals up from the bottom, why do we not produce some artificial currents to carry the fertilizer up from the sea bed in order to increase plankton growth at the surface? Phytoplankton, which is the "grass" of the sea, would then attract and support animals and fish. This idea has in fact been suggested. The heat from an underwater atomic pile might be used to produce a warm, rising current of water. Perhaps in the future, if atomic power is cheap and food is scarce, this will be a practical proposition, but the problems of such a scheme have not yet been studied in detail.

Another suggestion is to put the necessary phosphate and nutrients required by plants and fish into an enclosed basin of water, such as a coral atoll, and to use it as a giant breeding tank. It would certainly make catching the fish easier if they could be confined, and the fertilizing material would all be used instead of being spread wastefully all over the ocean floor. "Farming" the shallow seas is an alternative method of producing more fish. At present the fish we like to eat are seriously depleted in the early stages of their life history. Some of these fish come inshore to breed, others move out into rather deeper water, but they all spawn on the sea floor. The eggs and larval fish become the victims of their own parents and of bottom-living predators, such as starfish, which are known to feed voraciously on these young forms, as do many other fish. In a fish farm these predators could be carefully controlled, as long as the dangers of upsetting the balance of nature were always kept in mind. It might be worthwhile to breed special varieties of fish and introduce them from hatcheries into the fish farms.

Fish lay many thousands of eggs and although the mortality rate among eggs and young forms is high, normally a favorable crop of adults is maintained. Should circumstances arise which enable the predators to multiply to excess, the population recuperation is endangered. Experimental transplantation of young fish to areas with plenty of food and fewer enemies is being carried out at the present time. Some mollusks are also farmed. Oysters have been reared for many years in special beds where they can find suitable food and can be kept clean, and where they are protected against their natural enemies. The Dutch have developed better methods of growing and harvesting mussels, which grow around the Friesian Islands. The young mussels are transplanted in the rich waters of the deltas south of The Hague, where they fatten rapidly and are then col-

To improve their quality, young oysters are placed in these racks and laid in rich estuary waters near Conway, North Wales. The oysters, 400 to each rack, take about five years to reach maturity.

lected. Fish farming, then, is nothing new as far as shellfish are concerned; now it needs extending to include white fish.

Most fish are rich in proteins that are essential to our diet. At present, we do not take full advantage of the protein available to us —first because we are fussy about the kinds of fish we eat, and second because we throw away large quantities of fish "scraps" such as the heads, bones, and scales. These scraps are used in fertilizers to help farmers grow better crops. Recent research sponsored by the United States and by United Nations agencies has shown that scraps can be made into a highly nutritious "fish flour." The unpleasant fishy taste can be entirely removed by chemical processes so that the flour can be used in the preparation of soups, curries, and many other dishes. Fish flour could be used to help feed the undernourished peoples of the world, and it is quite possible in the future that it will become a regular part of everybody's diet.

The sea's minerals pose two main problems for the future. First is the relatively simple one of devising a method of collecting in bulk the rich deposits covering the sea floor. Second is to discover the processes by which plants and animals concentrate minerals from sea water. Once we have unraveled these complex processes we shall be able to develop man-made concentrating equipment to extract at least some of the minerals.

Man is out of his element in water, yet he has been able to conquer the sea and use it for his own purposes. He has explored the greater part of its surface and penetrated its greatest depths. He can predict its behavior with considerable accuracy and thus safeguard the lives of sailors and coastal dwellers. He is now beginning to learn something about the history of the Earth from the sediments and rocks beneath the sea. Yet the more we learn about the sea the greater, it seems, is the challenge to oceanographers. The sea is the last great frontier remaining on this planet. Its vast store of wealth in the form of power, food, and minerals has hardly begun to be tapped.

Two tons of cod, haddock, halibut, and skate from the Barents Sea pour onto a British trawler's deck—just a tiny portion of the rich harvest of food man takes daily from the oceans of the world.

# Index

# Illustration Credits

Key to picture position:
(T) top  (C) center  (B) bottom and combinations; for example (TL) top left, or (CR) center right.

Paul Popper Ltd.: title page
British Museum: 13(BL), 15, 18 (T)
Mansell Collection: 14(T and B), 84(TL)
Droemersche Verlagsanstalt Th Knaur Nachf: 17
Photo C. T. Goodworth: 20(B)
Based on illustration by Harry McNaught from *The Story of Geology* by Jerome Wyckoff © 1960 Golden Press Inc.: 26(B)
Reprinted with permission from Ph. H. Kuenen *Marine Geology* 1950, John Wiley & Sons Inc., N.Y.: 27(B), 42(R)
Official U.S. Navy photograph: 32
National Institute of Oceanography: 34, 36(B), 45, 77(R), 86(B), 93
Walt Disney's *20,000 Leagues under the Sea* by Jules Verne: 35 © 1934, 1951, by William Beebe: 37(R), 38(T)
Éditions de Paris: 38, 39(B)
Fairchild Aerial Survey, U.S.A.: 42(L)
Geological Survey photograph, Crown Copyright, reproduced by permission H.M. Stationery Office: 44(R), 79
British Petroleum Co. Ltd.: 46(B)
Secretary of Admiralty: 48, 130(L)
Shell photograph: 51(R), 54(R)
Shell Petroleum Co. Ltd.: 55
Photo J. K. Joseph, Crown Copyright reserved: 63
*Our Wandering Continents* by Alex L. du Toit, Oliver & Boyd Ltd., London: 65
Photo Percy Hennell: 72 and 73
Agent General for Queensland: 74
Aerofilms & Aero Pictorial Ltd.: 75, 95(T), 111(R), 142(T)

SOGREAH: 76
*Physics & Chemistry of the Earth*, Vol. 2, 1957, by L. H. Ahrens *et al.*, editors, London, Pergamon Press: 86(TL), 94, 120(B)
B. W. Robinson Collection: 91(T)
National Film Board of Canada: 99(TL)
USIS: 100(T)
The Port of New York Authority: 101
John Sykes: 102(T)
Based on Admiralty Chart No. 5058 with the permission of the Controller of H.M. Stationery Office and of the Hydrographer of the Navy: 102(B)
Photographie Aerienne Robert Durandaud: 103(T)
Ray-Delvert, Paris: 110
*Chemistry & You* © 1957 by Brady, McGill, Smith, Baker, published by Lyons & Carnahan: 112
The Steetley Co. Ltd.: 113
*Scientific American* December 1960, photo Dr. N. L. Zenkevitch, Moscow: 114(T)
Photo Peter David: 117(TR), 118(TR and BL), 119(T), 125
Photo N. Ingram Hendey: 118(TL)
*Scientific American* August 1962: 120, 121, 128
By courtesy of the British Trawlers Federation Ltd., photo Ray Dean: 121 (B)
Conzett & Huber, Zurich: 122
W. J. Copenhagen, Department of Agriculture, Pretoria, South Africa, from photo H. Offen, Swakopmund, Pretoria: 123(T and B)
Lennart Nilsson, Solna, Sweden: 124(T)
*The Story of Yankee Whaling*, American Heritage Junior Library: 126
Sir Alistair Hardy: 130(TR)
Ministry of Agriculture & Fisheries, Lowestoft: 130(BR)

Radio Times Hulton Picture Library: 132(T)
Scottish Home Office Fisheries Division based on photo by the late Commander Hodges, R.N.: 132(B)
© 1963 by France-Soir & Opéra Mundi, Paris, from drawing by Garel: 134
© 1962 by France-Soir & Opéra Mundi, Paris, photo by Jean Lattes: 135
KLM Amsterdam: 136, 137
*Russia Today* No. 119, 1960: 138, *Engineers' Dreams* by Willy Ley, Phoenix House Ltd., London: 139, 141(T)
Cliché Organisation Commune des Regions Sahariennes: 143
Geological Society of America: 144
Les Requins Associés: 146
Paul Almasy: 147
Keystone Press Agency Ltd.: 148
Photo Peter Waugh BTF: 149

## Artists' Credits

Barry Evans: 2/3, 11, 22, 24/25, 40/41, 70, 82/83, 97(T), 104/105, 116, 134.
Peter Sullivan: 12(T), 12(B), 20(T), 23, 26, 27(B), 30/31, 34, 42(R), 47, 49, 51(B), 52/53, 55, 58, 59, 61, 62, 72(B), 73(B), 77 (L), 80, 81, 83 (B), 84 (R), 86 (T), 86 (B), 87, 88/89, 98, 102, 120 (B), 120/121, 131, 136 (B), 139, 140.
Sidney W. Woods: 17, 28/29, 37, 38/39, 46 (T), 64, 65, 66, 68/69, 115, 127.
Joan Abbott: 85, 89 (B), 117 (B), 123 (T), 141.
Gordon Cramp: 94, 106, 107, 111 (L), 112.
Judy Hannington: 96, 97 (B), 103 (B), 104 (B), 121 (T), 123 (B), 124 (B), 130 (T), 132 (B), 142 (B).